COMPUTING
ANTICIPATORY
SYSTEMS

COMPUTING ANTICIPATORY SYSTEMS

CASYS—First International Conference

Liège, Belgium   August 1997

EDITOR
Daniel M. Dubois
Institute of Mathematics
University of Liège, Belgium

American Institute of Physics

AIP CONFERENCE
PROCEEDINGS 437

Woodbury, New York

Editor:

Daniel M. Dubois
asbl CHAOS, Institute of Mathematics
University of Liège
Grande Traverse 12
B-4000 Liège 1
BELGIUM

E-mail: Daniel.Dubois@ulg.ac.be

L.C. Catalog Card No. 98-72270
ISBN 1-56396-827-4
ISSN 0094-243X
DOE CONF- 9708158

Printed in the United States of America

CONTENTS

ANTICIPATORY AUTONOMOUS SYSTEMS AND ROBOTICS

COMPUTATIONAL AND DYNAMICAL SYSTEMS

NEURONAL AND COGNITIVE SYSTEMS

Preface

CASYS'97 is the First International Conference on Computing Anticipatory Systems, held at HEC-Liège, Belgium, August 11-15, 1997, organized by the non-profit association asbl CHAOS, Centre for Hyperincursion and Anticipation in Ordered Systems, Institute of Mathematics of the University of Liège, under the sponsorships of the Fonds National de la Recherche Scientifique of Belgium, HEC-Liège asbl, Euro View Services sa and the sponsors of World Organisation of Systems and Cybernetics, International Association for Cybernetics, and Washington Evolutionary Systems Society.

This conference was really a very big success: 110 papers were presented by 175 authors or co-authors coming from 25 countries.

These Proceedings contain a selection of the papers presented at CASYS'97, including the papers of authors laureates of awards. The other papers are published by CHAOS asbl.

The quality of papers was so high that 17 papers received a best paper award. Professor Robert Rosen received the first CASYS Award and Professor Ernst von Glasersfeld received the first CHAOS Award. The list of all these awards is now given:

CHAOS Awards

The **First CHAOS Award** with a Crystal of Val Saint-Lambert on Professor Ernst von Glasersfeld (USA) for his outstanding scientific works on Radical Constructivism.

Best Paper Award for CASYS'97 with a Crystal of Val Saint-Lambert bestowed by CHAOS asbl to the invited lecture Toys, Toddlers and the Times of Time by Pedro Roberto Medina Martins (Portugal).

Best Paper Award for CASYS'97 bestowed by CHAOS asbl to Walter J. Freeman (USA)

CASYS'97 Awards

The **First CASYS Award** with a Crystal of Val Saint-Lambert on Professor Robert Rosen (USA) for his outstanding scientific work on Anticipatory Systems.

CASYS'97 Best Paper Award with the Crystal Belgium to:

Koichiro Matsuno (Japan)
Pere Julià (Spain)
Michel Naranjo (France) and Georgios Tsirigotis (Greece)
Clemens Pötter (Germany)
Rudolf F. Albrecht (Austria)
Timo Honkela (Finland)

CASYS'97 Best Paper Award to:

Juan M. Alvarez de Lorenzana (Canada)
Olivier Sigaud (France)
Jerry L. R. Chandler (USA)
Vilmos Csányi (Hungary)
Erich Prem (Austria)
Andreas Birk (Belgium)
Hanns Sommer (Germany)
Guenter W. Gross (USA) and Jacek M. Kowalski (USA)
A. £uksza (Poland), Wieslaw Citko (Poland) and Wieslaw Sieńko (Poland)

As **President** of CASYS, I would like to thank the **Honorary President**, *Robert Rosen* (USA) and the **Vice-President**, *Gertrudis Van de Vijver* (Belgium). I would also like to thank the members of the following:

International Program Committee: Jerry Chandler (USA), Daniel M. Dubois (Belgium), Jean Etienne (Belgium), George Farre (USA), Jean Godart (Belgium), Thierry Grisar (Belgium), Loet Leydesdorff (The Netherlands), Peter Marcer (United Kingdom), Koichiro Matsuno (Japan), Jean Ramaekers (Belgium), Philippe Sabatier (France), Bernard Teiling (Switzerland), Elizabeth Thomas (Belgium), Robert Vallée (France), Gertrudis Van de Vijver (Belgium), Brian Warburton (United Kingdom).

International Scientific Committee: Marcel Ausloos (Belgium), Soeren Brier (Denmark), Philip Boxer (United Kingdom), Daniel M. Dubois (Belgium), George Farre (USA), Charles François (Argentina), Thierry Grisar (Belgium), Stig Holmberg (Sweden), Mireille Larnac (France), Ian Larsen (Belgium), Loet Leydesdorff (The Netherlands), Michele Malatesta (Italy), Koichiro Matsuno (Japan), Asghar T. Minai (USA), Gérard Morel (France), Jose Luis Simoes da Fonseca (Portugal), Charles Rattray (United Kingdom), Robert Rosen (USA), Eric Schwarz (Switzerland), Walter Schempp (Germany), Horia-Nicolai Teodorescu (Romania), Gertrudis Van de Vijver (Belgium), Ernst von Glasersfeld (USA).

Last but not least, I would like to thank Christiane Malemprez for her constant support in the practical organization of this conference.

Finally, I would like to point out that our association **CHAOS** is now a member of **IFSR, the International Federation for Systems Research**. As a consequence of the success of this first conference, the Second International Conference CASYS on Computing Anticipatory Systems is planned for August 10-14, 1998 at Liège, Belgium.

Daniel M. Dubois, President of CASYS

asbl CHAOS, Institute of Mathematics, University of Liège
Grande Traverse 12, B-4000 LIEGE 1, Belgium
Daniel.Dubois@ulg.ac.be

Introduction to
Computing Anticipatory Systems

Computing Anticipatory Systems
with Incursion and Hyperincursion

Daniel M. DUBOIS

Centre for Hyperincursion an Anticipation in Ordered Systems
CHAOS asbl, Institute of Mathematics, UNIVERSITY OF LIEGE
12, Grande Traverse, B-4000 LIEGE 1, Belgium
Fax: + 32 4 366 948 9, E-mail: Daniel.Dubois@ulg.ac.be
http://www.ulg.ac.be/mathgen/CHAOS/CHAOS.html

HEC, 14 rue Louvrex, B-4000 LIEGE, Belgium

Abstract

An anticipatory system is a system which contains a model of itself and/or of its environment in view of computing its present state as a function of the prediction of the model. With the concepts of incursion and hyperincursion, anticipatory discrete systems can be modelled, simulated and controlled. By definition an incursion, an inclusive or implicit recursion, can be written as:

$$x(t+1) = F [..., x(t-1), x(t), x(t+1),...]$$

where the value of a variable $x(t+1)$ at time $t+1$ is a function of this variable at past, present and future times. This is an extension of recursion.
Hyperincursion is an incursion with multiple solutions.

For example, chaos in the Pearl-Verhulst map model : $\quad x(t+1) = a.x(t).[1 - x(t)]$
is controlled by the following anticipatory incursive model : $x(t+1) = a.x(t).[1 - x(t+1)]$
which corresponds to the differential anticipatory equation : $dx(t)/dt = a.x(t).[1 - x(t+1)] - x(t)$.

The main part of this paper deals with the discretisation of differential equation systems of linear and non-linear oscillators. The non-linear oscillator is based on the Lotka-Volterra equations model. The discretisation is made by incursion. The incursive discrete equation system gives the same stability condition than the original differential equations without numerical instabilities. The linearisation of the incursive discrete non-linear Lotka-Volterra equation system gives rise to the classical harmonic oscillator. The incursive discretisation of the linear oscillator is similar to define backward and forward discrete derivatives. A generalized complex derivative is then considered and applied to the harmonic oscillator. Non-locality seems to be a property of anticipatory systems. With some mathematical assumption, the Schrödinger quantum equation is derived for a particle in a uniform potential.
Finally an hyperincursive system is given in the case of a neural stack memory.

Keywords: Control of chaos, Quantum systems, Discrete complex derivative, Incursive anticipatory systems, Hyperincursive memory systems.

CP437, *Computing Anticipatory Systems: CAYS--First International Conference*
edited by Daniel M. Dubois © 1998 The American Institute of Physics 1-56396-827-4/98/$15.00

1 Introduction

Robert Rosen (1985, p. 341), in the famous book, "tentatively defined the concept of an anticipatory system: a system containing a predictive model of itself and/or of its environment, which allows it to state at an instant in accord with the model's predictions pertaining to a later instant... It is well to open our discussion with a recapitulation of the main features of the modelling relation itself, which is by definition the heart of an anticipatory system"
Robert Rosen (1985), in his book, conjectures that adaptation and learning systems in biological processes are anticipatory systems. For him, anticipation is the central difference between living and non-living systems.

Before explaining my understanding of computing anticipatory systems by incursion and hyperincursion, I would like to point out that my mathematical modelling is drastically different from Robert Rosen one.

1.1 Robert Rosen's Interpretation of Anticipatory Systems

Robert Rosen (1985) states on one hand, that the evolution of an anticipatory system $S(t)$ at each time step is driven by the predictive model $M(t+1)$ at a later time. But, on the other hand, Robert Rosen says that the predictive model M is not affected by the system.
With these statements, a finite difference equation system can thus be written as

$$\Delta S/\Delta t = [S(t+\Delta t) - S(t)]/\Delta t = F[S(t), M(t+\Delta t)] \qquad (1a)$$
$$\Delta M/\Delta t = [M(t+\Delta t) - M(t)]/\Delta t = G[M(t)] \qquad (1b)$$

If M is model of the system itself, then the model is a model of the system itself without the predictive model guiding it. Thus, the predictive model of the system itself is not a true predictive model of the system itself. This leads to a contradiction.
If M is a model of the environment of the system, and if the environment is affected by the system, this leads also to a contradiction: the model is not predictive because the model is not affected by the system which affects the environment.
In conclusion of this short analysis of Robert Rosen approach to anticipatory systems, it must be stated that there is no drastic difference between his approach and the classical theory of control. Indeed, in control theory, the engineer designs the control function of a system as a function of future objectives which guide this system in feedbacking the outputs of the system to modulate its inputs in view of minimising the distance between the actual outputs to the wanted outputs.

1.2 My Interpretation of Anticipatory Systems

With the Robert Rosen definition of anticipatory systems, I propose to define a computing anticipatory system S as a finite difference equation system

$$\Delta S/\Delta t = [S(t+\Delta t) - S(t)]/\Delta t = F[S(t), M(t+\Delta t)] \qquad (2a)$$
$$\Delta M/\Delta t = [M(t+\Delta t) - M(t)]/\Delta t = F[S(t), M(t+\Delta t)] \qquad (2b)$$

4

where the future state of the system S and the model M at time t+Δt is a function F of this system S at time t and of the model M at a later time step t+Δt.

If the model is the system itself, then I write M = S and eqs. 2ab reduce to

$$\Delta S/\Delta t = [S(t+\Delta t) - S(t)]/\Delta t = F[S(t), S(t+\Delta t)] \tag{2c}$$

what I defined as an incursive system, an inclusive or implicit recursive system: the future state of an incursive system S(t+Δt) depends on the past and present state(s) of the system ... S(t-Δt), S(t) but also on its future state(s) S(t+Δt), S(t+2Δt), ...

If I replace S(t+Δt) in the second member of eq. 2c by the equation 2c itself, I obtain

$$\Delta S/\Delta t = [S(t+\Delta t) - S(t)]/\Delta t = F[S(t), S(t)+\Delta t.F[S(t),S(t+\Delta t)]] \tag{2c'}$$

in which I can replace indefinitely S(t+Δt) by the equation itself leading to an infinite series.

I will show in this paper an example where this infinite series is a simple expression depending only on S(t) with the Pearl-Verhulst chaos map.

If the model is a model of the environment E of the system, I write

$$\Delta S/\Delta t = [S(t+\Delta t) - S(t)]/\Delta t = F[S(t), E(t+\Delta t)] \tag{2d}$$
$$\Delta E/\Delta t = [E(t+\Delta t) - E(t)]/\Delta t = G[E(t), S(t)] \tag{2e}$$

which defines also as an incursive system because the future state of the system is a function of its state at the preceding time and of the future state of its environment. Such a system can be transformed to the recursive system

$$\Delta S/\Delta t = [S(t+\Delta t) - S(t)]/\Delta t = F[S(t), E(t)+\Delta t.G[E(t),S(t)]] \tag{2d'}$$
$$\Delta E/\Delta t = [E(t+\Delta t) - E(t)]/\Delta t = G[E(t), S(t)] \tag{2e'}$$

Examples of such systems will be given in this paper with non-linear Lotka-Volterra and linear harmonic oscillators.

If the environment has also a predictive model of the system, I can write

$$\Delta S/\Delta t = [S(t+\Delta t) - S(t)]/\Delta t = F[S(t), E(t+\Delta t)] \tag{2f}$$
$$\Delta E/\Delta t = [E(t+\Delta t) - E(t)]/\Delta t = G[E(t), S(t+\Delta t)] \tag{2g}$$

which is also an incursive system leading to two crossed infinite series.

1.3 Definition of Incursion

Let us first define the concept of incursion, a contraction of inclusive or implicit recursion.

A simple recursion with a function f[x(t), p] of a variable x(t) depending of time t and parameter p, is defined as

$$x(t+1) = f[x(t), p] \tag{3}$$

5

where the value of the variable at each instant t+1 is a function of the value of this variable at the preceding time step t. From the knowledge of the initial condition x(0) at time t=0, it is possible to compute all the future states of the variable:

$$x(1) = f[x(0), p]$$
$$x(2) = f[x(1), p]$$
...
$$x(n) = f[x(n-1), p] \tag{3a}$$

An incursion is an inclusive or implicit recursion, an extension of the recursion in the following way:

$$x(t+1) = f[x(t), x(t+1), p] \tag{4}$$

where the value of the variable at each instant t+1 is a function of the value of this variable at the preceding time step t, but also at time t+1. This defines a self-referential system which is an anticipatory system of itself. The function f describing the dynamics of a system contains a model of itself; indeed x(t+1) in the function f can be replaced in the following way:

$$x(t+1) = f[x(t), f[x(t), x(t+1)], p), p] \tag{4a}$$

where the system explicitly contains a predictive model of itself. Let us give a simple example of such an incursive anticipatory system.

1.4 Control of Chaos in an Incursive Anticipatory System

In population dynamics, the differential equation of the growth of a population following Pear-Verhulst is given by

$$dx(t)/dt = a.x(t).[1-x(t)]-b.x(t) \tag{5}$$

where a is the birth rate and b the death rate and [1-x(t)] represents a control of the birth rate of the population. Indeed, without control, the population would grow to infinity if a > b and to zero if a < b.
A finite difference Pearl-Verhulst equation is given by:

$$x(t+1) = x(t) + a.x(t).(1-x(t)) - b.x(t) \tag{5a}$$

where x(t) is the value of the population at time t and a is the growth rate and b the decay rate.
In taking b=1, the well-known chaos map, given in Fig. 1a, is obtained:

$$x(t+1) = a.x(t).(1-x(t)) \tag{5a'}$$

For value of a = 4, the system shows chaos and the future of this system is unpredictable from its initial condition x(0) at time t = 0. This is the sensibility to initial conditions.

An anticipatory system can be built from this eq. 5a' in writing:

6

$$x(t+1) = a.x(t).(1-x(t+1)) \qquad (6)$$

where now the saturation factor is a predictive function of the population at a later time step. In replacing $x(t+1)$ in the second member of eq. 6 by eq. 6 itself, we obtain:

$$x(t+1) = a.x(t).(1-a.x(t).(1-x(t+1))) \qquad (6a)$$

which is an incursive system. In continuing to replace $x(t+1)$ by itself, an infinite sequence is obtained:

$$x(t+1) = a.x(t).(1-a.x(t).(1- a.x(t).(1- a.x(t).(1- a.x(t).(1- a.x(t).(1- a.x(t).(1-...)))))))) \qquad (6b)$$

There is a paradox which in fact is easy to understand. If an anticipatory system contains a model of itself, this means that the model of itself must include also the model of itself and so on until infinity. There are an infinity of embedded models in each other. This looks without hope to resolve the problem or to simulate such systems. Mathematically, there is a means to resolve this paradox. Indeed, eq. 6b can be written as

$$x(t+1) = a.x(t).[1-a.x(t)+a^2.x(t)^2-a^3.x(t)^3+...] = a.x(t) / [1+a.x(t)] \qquad (6c)$$

because

$$[1-a.x(t)+a^2.x(t)^2-a^3.x(t)^3+...] = \sum_{n=0}^{\infty} [(-a.x(t)]^n = 1 / [1+a.x(t)] \qquad (6c')$$

This is an important result which shows that the future value of a variable x at time t+1 can be represented by an infinite sum of this variable x at the present time $x(t)$ which converges to a simple function depending on this variable $x(t)$ at the current present time.

An anticipatory system can be related to a hyperset H defined as a collection of its elements E and itself: $H = \{E,H\}$.

Jon Barwise and Larry Moss (1991) introduced this concept of hyperset for defining infinite descending sequences. They demonstrate that such equations are ruled out by an accepted axiom of set theory, the Axiom of Foundation.

Peter Aczel (1988) work has taught that the axiom of foundation is not a necessary part of a clear picture of the universe of sets. The Antifoundation Axiom was thus proposed (see Jon Barwise and Larry Moss, 1991, for a very interesting development of new directions of research in hypersets theory). In a simple manner, it is established that a classical set cannot be a member of itself. So, self-referential systems have today no theoretical well-established framework.

The incursive equation of the anticipatory system 6 can be thus written as the recursive anticipatory system

$$x(t+1) = a.x(t).[1-a.x(t)/[1+a.x(t)]] \qquad (6d)$$

7

With such an anticipation, eq. 6d doesn't show chaos. This anticipatory system has two fixed points which represent its implicit finality or teleonomy. The goal or objective of this anticipatory system is not explicitly imposed from outside the system like in control theory but is determined by the stability or instability points of the system itself.

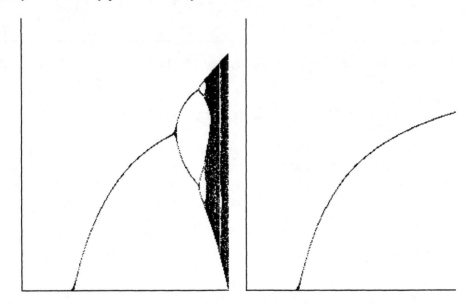

Figures 1ab: (a) Bifurcation diagram of eq. 5a', x(t) as a function of a.
(b) Incursive anticipatory control of chaos with eq.6c, x(t) as a function of a.

Eq. 6d can be transformed to the following differential equation

$$dx/dt = a.x(t). [1-a.x(t)/[1+a.x(t)]] - x(t) = a.x(t)/[1 + a.x(t)] - x(t) \qquad (6e)$$

Let us show that this is actually an anticipatory system in considering the following anticipatory differential equation

$$dx(t)/dt = a.x(t).[1 - x(t + 1)] - x(t) \qquad (6f)$$

which is an anticipatory system because the derivative of x(t) depends on x(t) but also on x(t+1) at a future time. From the discretisation formula of the time derivative, we can write

$$x(t+1) = x(t) + dx(t)/dt \qquad (6g)$$

and in replacing this equation 6g in eq. 6f, we obtain

$$dx(t)/dt = [a.x(t).[1 - x(t)] - x(t)]/[1 + a.x(t)] = a.x(t)/[1 + a.x(t)] - x(t) \qquad (6h)$$

which is identical to eq. 6e. This equation looks similar to the Monod-Michaelis-Menten equation largely used in modelling population dynamics, growth of bacteria, biochemical reactions and economics etc.

Let us show now an hyperincursive anticipatory system.

1.5 Definition of an Hyperincursive Anticipatory System

The following equation

$$x(t) = a.x(t+1).(1-x(t+1)) \tag{7}$$

defines an hyperincursive anticipatory system. Hyperincursion is an incursion with multiple solutions.
With $a = 4$, mathematically $x(t+1)$ can be defined as a function of $x(t)$

$$x(t+1) = 1/2 \pm 1/2 \ \sqrt{(1-x(t))} \tag{8}$$

where each iterate $x(t)$ generates at each time step two different iterates $x(t+1)$ depending of the plus minus sign. The number of future values of $x(t)$ increases as a power of 2. As the system can only take one value at each time step, something new must be added for resolving the problem. Thus, the following decision function $u(t)$ can be added for making a choice at each time step:

$$u(t) = 2.d(t)-1 \tag{9}$$

where $u = +1$ for the decision $d = 1$ (true) and $u = -1$ for the decision $d = 0$ (false).
In introducing eq. 9 to eq. 8, the following equation is obtained:

$$x(t+1) = 1/2 + (d(t)-1/2). \sqrt{} \ (1-x(t)) \tag{10}$$

The decision process could be explicitly related to objectives to be reached by the state variable x of this system. This is important to point out that the decisions $d(t)$ do not influence the dynamics of $x(t)$ but only guide the system which creates itself the potential futures.
This hyperincursive anticipatory system was proposed as a model of a stack memory in neural networks (Dubois, 1996c) which is developed at the end of this paper.

2 Incursive Anticipatory Linear And Non-Linear Oscillators

In computer science, the discretisation of differential equations systems gives sometimes numerical instabilities which can be controlled by incursive anticipatory discretisation.
Let us first consider the non-linear model given by the discretised Lotka-Volterra equations:

$$X(t+\Delta t) = X(t) + \Delta t.[a.X(t) - b.X(t).Y(t)] \tag{13a}$$
$$Y(t+\Delta t) = Y(t) + \Delta t.[-c.Y(t) + d.X(t).Y(t)] \tag{13b}$$

where t is a discrete time with steps Δt, and a, b, c, d are the parameters.

9

Analytical solutions exist only for small oscillations from the steady state $X_0 = c/d$ and $Y_0 = a/b$, which are identical to a harmonic linear oscillator. In taking $X = X_0 + x$ and $Y = Y_0 + y$, when x and y are small, the linearisation of these equations gives

$$x(t+\Delta t) = x(t) - \Delta t.(bc/d).y(t) \qquad (13a')$$
$$y(t+\Delta t) = y(t) + \Delta t.(ad/b).x(t) \qquad (13b')$$

With the change in variables $q(t) = x(t)$, $p(t) = -y(t)$ and in taking $bc/d = 1/m$ and $\omega^2 = ac/m$, the linear harmonic oscillator equations system is obtained:

$$q(t+\Delta t) - q(t) = \Delta t.p(t)/m \qquad (14a)$$
$$p(t+\Delta t) - p(t) = -\Delta t.m.\omega^2.q(t) \qquad (14b)$$

in defining by q the position and p the momentum ($p = m.v$ where m is the mass and v the velocity), and where ω is the pulsation. These equation can be reduced to an equation in $q(t)$:

$$q(t+2\Delta t) - 2q(t+\Delta t) + q(t) = -\Delta t^2.\omega^2.q(t) \qquad (14c)$$

which corresponds to a discretisation of the second derivative of $q(t+\Delta t)$.
But the solutions all these equations are unstable.

The following Fig. 2 shows the instability of the Lotka-Volterra discrete model.

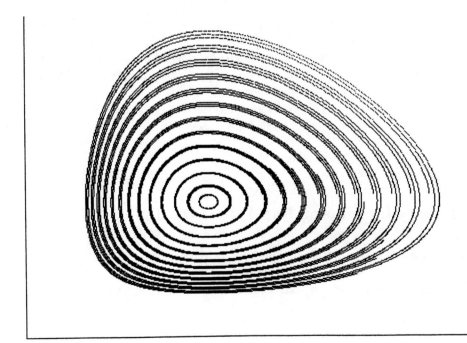

Figure 2: Simulation of eqs. 13ab, Y(t) as a function of X(t), with different initial conditions. This discretisation of the Lotka-Volterra model is unstable.

10

Different incursive discrete equations systems exist to control the oscillations of these non-linear and linear discrete equations.

The iterative values of $X(t+\Delta t)$ of the first equation (13a) can be propagated to the second equation (13b), in an incursive way, as proposed by Dubois [1992, 1993]:

$$X(t+\Delta t) = X(t) + \Delta t.[a.X(t) - b.X(t).Y(t)] \qquad (15a)$$
$$Y(t+\Delta t) = Y(t) + \Delta t.[-c.Y(t) + d.\mathbf{X(t+\Delta t)}.Y(t)] \qquad (15b)$$

The simulation gives solutions with orbital stability (see Fig. 3a) and chaos (see Fig. 3b) depending on the values of the parameters [Dubois, 1992].

The linearised incursive Lotka-Volterra equations give the incursive harmonic oscillator ones:

$$q(t+\Delta t) = q(t) + \Delta t.p(t)/m \qquad (15a')$$
$$p(t+\Delta t) = p(t) - \Delta t.m. \omega^2.q(t+\Delta t) \qquad (15b')$$

with orbital stability [Dubois, 1995). In replacing $q(t+\Delta t)$ in eq. 15b' by eq. 15a', eq. 15b' becomes

$$p(t+\Delta t) = p(t) - \Delta t.m. \omega^2.q(t) - \Delta t^2. \omega^2.P(t) \qquad (15b'')$$

where the term in Δt^2 disappears when Δt tends to zero. Eqs. 15a'b' can be reduced to an equation in $q(t)$:

$$q(t+\Delta t) - 2q(t) + q(t-\Delta t) = - \Delta t^2. \omega^2.q(t) \qquad (15c')$$

which is time invertible in replacing Δt by $- \Delta t$. With a time translation of $- \Delta t$, eq. 15b' writes

$$p(t) = p(t - \Delta t) - \Delta t.m. \omega^2.q(t) \qquad (15d')$$

which corresponds to a backward derivative of the momentum. In fact two derivatives can be defined for a discrete variable x:

$$\Delta_f x / \Delta t = (x(t + \Delta t) - x(t))/\Delta t \qquad (16)$$
$$\Delta_b x / \Delta t = (x(t) - x(t - \Delta t))/\Delta t \qquad (17)$$

The forward derivative 16 and the backward derivative 17 are not always equal (only at the limit for $\Delta t=0$ for continuous derivable equations); for non-derivable continuous equations like in fractal equations systems, two derivatives must be defined. Let us remark that when Δt is replaced by $-\Delta t$, the forward and backward derivatives 16 and 17 becomes the backward and forward ones. Moreover, the successive application of the forward derivative to the backward derivative, or the inverse, gives the second order derivative, which is time invertible:

$$\Delta^2 x / \Delta t^2 = x(t + \Delta t) - 2x(t) + x(t - \Delta t)] / \Delta t^2$$

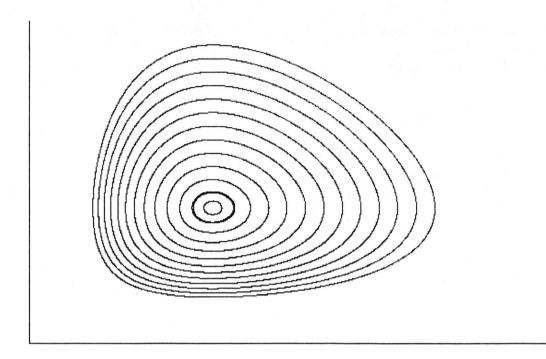

Figure 3a: Orbital stability of the incursive anticipatory Lotka-Volterra discrete eqs. 15 ab in the phase space, Y(t) as a function of X(t).

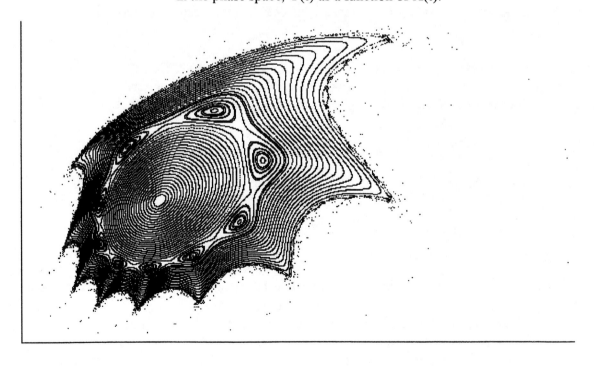

Figure 3b: Orbital stability and chaos of the incursive anticipatory Lotka-Volterra discrete eqs. 15 ab in the phase space, Y(t) as a function of X(t).

12

2.1 A generalized Discrete Derivative with Forward and Backward derivatives

Let us define a generalized discrete derivative by a weighted sum of these derivatives as follows [Dubois, 1995]:

$$\Delta_w x/\Delta t = w.\Delta_f x /\Delta t+(1-w).\Delta_b x/\Delta t = [w.x(t+\Delta t) + (1-2.w).x(t)+(w-1).x(t-\Delta t)]/\Delta t \qquad (18)$$

where the weight w is defined in the interval 0,1. For w = 1, the forward derivative 16 is obtained and for w = 0, the backward derivative 17. For w = 1/2, derivative 18 becomes

$$\Delta_{1/2} x/\Delta t = (x(t+\Delta t)-x(t-\Delta t))/2\Delta t=[\Delta_f x/\Delta t + \Delta_b x/\Delta t]/2 \qquad (18a)$$

which is an average derivative. With this generalized derivative 18, the discrete harmonic oscillator equations system can be defined by:

$$(1-w).q(t+\Delta t)+(2w-1).q(t)-w.q(t-\Delta t) = \Delta t.p(t)/m \qquad (19a)$$
$$w.p(t+\Delta t)+(1-2w).p(t)+(w-1).p(t-\Delta t) = - \Delta t.\omega^2.m.q(t) \qquad (19b)$$

which is identical to the equations 15a',b' for w = 0.

For w = 1, the equations 19a,b correspond to the linearised incursive Lotka-Volterra equations:

$$Y(t+\Delta t) = Y(t) + \Delta t.[-c.Y(t) + d.X(t).Y(t)] \qquad (20a)$$
$$X(t+\Delta t) = X(t) + \Delta t.[a.X(t) - b.X(t).\mathbf{Y(t+\Delta t)}] \qquad (20b)$$

where now we propagate the value of Y(t+Δt) in the equation of X(t+Δt).

The simulation gives solutions with orbital stability (see Fig. 4a) and chaos (see Fig. 4b) depending on the values of the parameters [Dubois, 1992].

For w = 1/2, the equations 19a,b become:

$$q(t+\Delta t) = q(t-\Delta t) + 2.\Delta t.p(t)/m \qquad (21a)$$
$$p(t+\Delta t) = p(t-\Delta t) - 2. \Delta t. \omega^2.m.q(t) \qquad (21b)$$

which are symmetrical, but 4 initial conditions are to be defined instead of two. These are similar to the following linearised discrete Lotka-Volterra equations

$$X(t+\Delta t) = X(t-\Delta t) + 2.\Delta t.[a.X(t) - b.X(t).Y(t)] \qquad (22a)$$
$$Y(t+\Delta t) = Y(t-\Delta t) + 2.\Delta t.[-c.Y(t) + d.X(t).Y(t)] \qquad (22b)$$

for which 4 initial conditions are to be defined.

The simulation in Figure 5a gives solutions with different initial conditions as in Figs. 3a and 4a. For an other set of parameters Figs. 4bcdefghi show chaos with different initial conditions.

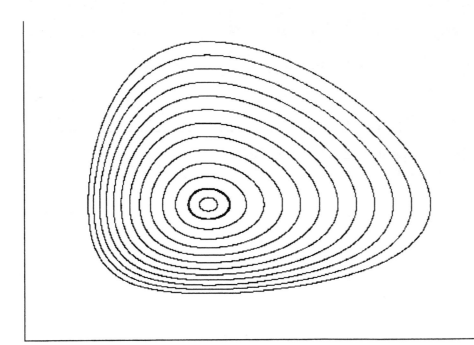

Figure 4a: Orbital stability of the second incursive anticipatory Lotka-Volterra discrete eqs. 20 ab in the phase space, Y(t) as a function of X(t).

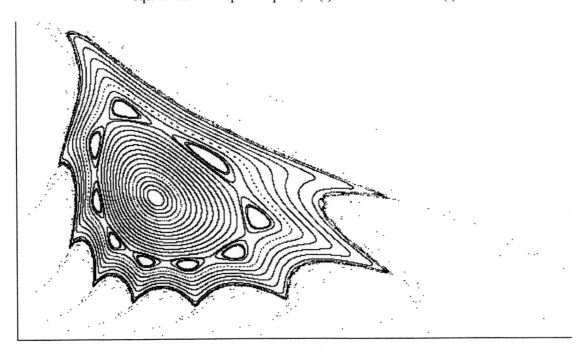

Figure 4a: Orbital stability and chaos of the second incursive anticipatory Lotka-Volterra discrete eqs. 20 ab in the phase space, Y(t) as a function of X(t).

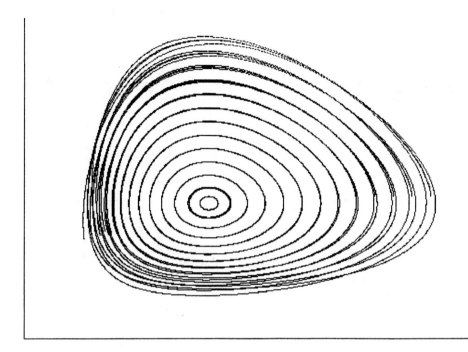

Figure 5a: The simulation of eqs. 22ab gives solutions in the phase space, Y(t) as a function of X(t), with different initial conditions as in Figs. 3a and 4a.

3 A Generalized Complex Discrete Derivative

From eq.8 of the generalized discrete derivative, the second order derivative is given by the successive application of eq.8 for w and (1-w), or the inverse:

$\Delta_w\Delta_{1-w}x/\Delta t^2 =$
$[x(t+\Delta t)-2x(t)+x(t-\Delta t)]/\Delta t^2 + w(1-w)[x(t+2\Delta t)-4x(t+\Delta t)+6x(t)-4x(t-\Delta t)+x(t-2\Delta t)]/\Delta t^2$
$= \Delta_{1-w}\Delta_w x/\Delta t^2$ (23)

which is the sum of the classical discrete second order derivative and a factor, weighted by w(1 - w), which is similar to a fourth order discrete derivative (multiplied by Δt^2). For w = 0 and w = 1, the classical second order derivative is obtained: w(1 - w) = 0.
For w = 1/2, w(1 – w) = 1/4, the second order derivative is also obtained but with a double time interval 2 Δt.
In choosing the value of the w(1 – w) equal to 1/2, we obtain weights w, solution of

$w^2 – w + 1/2 = 0$ (24)

which are given by the complex numbers

$w = 1 / 2 \pm i /2$ (25)

and $1-w = 1 / 2 \pm (- i / 2) = w^*$, where w^* is the complex conjugate of w.

15

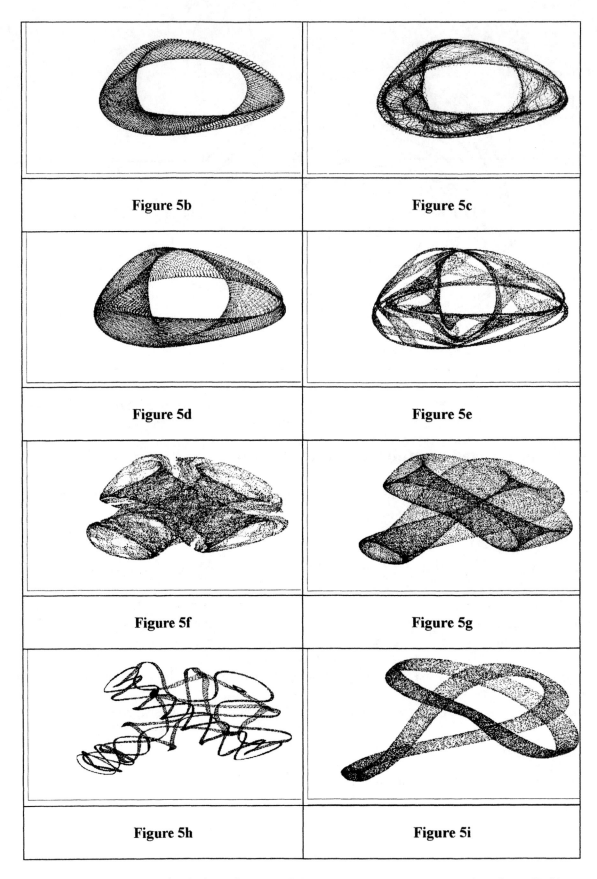

Figure 5b

Figure 5c

Figure 5d

Figure 5e

Figure 5f

Figure 5g

Figure 5h

Figure 5i

Figures 5bcdefghi: Simulation of eqs. 22ab in the phase space, $Y(t)$ as a function of $X(t)$.

So eq.18 of the generalized discrete derivative can be rewritten as

$$\Delta_w z /\Delta t = w._{Df} z/\Delta t + w*._{Db} z /\Delta t = [w.z(t+\Delta t)+(w^* - w).z(t)+w^*.z(t-\Delta t)] /\Delta t \qquad (18')$$

where the generalized complex derivative can be applied to a complex variable $z = x + i\, y$. The second order derivative is given by the successive application of eq.18' for w and w^*, or the inverse:

$$\Delta_w \Delta_{w*} z / \Delta t^2 = [z(t+\Delta t)-2z(t)+z(t-\Delta t)] / \Delta t^2 + w\, w^* [z(t+2\Delta t)-4z(t+\Delta t)+6z(t)-4z(t-\Delta t)+ z(t-2\Delta t)] / \Delta t^2 = \Delta_{w*} \Delta_w z / \Delta t^2 \qquad (25)$$

which is the sum of the classical discrete second order derivative and a factor, weighted by the real number ww^*, which is similar to a fourth order discrete derivative (multiplied by Δt^2).

3.1 Nottale's Complex Velocity

With the complex weight w given by eq. 15, the first derivative of the position x (eq. 18) gives rise to the complex velocity v

$$v = [x(t+\Delta t) - x(t-\Delta t)]/2\Delta t \pm i [x(t+\Delta t) - 2x(t) + x(t-\Delta t)]/2\Delta t \qquad (26)$$

In defining a forward velocity

$$v_f = [x(t+\Delta t) - x(t)]/\Delta t \qquad (27a)$$

and a backward velocity

$$v_b = [x(t) - x(t-\Delta t)]/\Delta t \qquad (27b)$$

the complex velocity 26 is given by

$$v = [v_f + v_b]/2 \pm i [v_f - v_b]/2 \qquad (28)$$

where the real part of the velocity is the average of the forward and backward velocities and the imaginary part is the difference of these forward and backward derivatives. In considering the inverse time interval in replacing Δt by $- \Delta t$, the forward and backward derivatives becomes the backward and forward derivatives. So, if the velocity is time invertible, the plus and minus signs correspond to $\Delta t \leq 0$ and $\Delta t \geq 0$. In the continuous limit Δt tending to zero, $v_f = v_b$, and the classical Newtonian velocity is rediscovered and the imaginary part tends to zero.
The kinetic energy is a real number given by

$$vv^*/2 = [v_f^2 + v_b^2]/4 \qquad (28a)$$

Let us remark also that the acceleration given by the second derivative of the position x is a real variable, because $w(1 - w) = 1/2$ is real in eq. 23.

The complex velocity given by eq. 28 is similar to the complex velocity proposed by L Nottale [1993]

$$v = [v_f + v_b]/2 - i [v_f - v_b]/2 \qquad (29)$$

where only a negative imaginary part is present. In his paper, L. Nottale proved that the definition of backward and forward velocities and the complex energy with the Wiener process and Newtonian law, give the Schrödinger quantum equation.

Let us show now, that without Wiener process and a real energy, the Schrödinger equation can be obtained from the generalized complex discrete derivative applied to a Newtonian harmonic oscillator.

3 Incursive Oscillator Related to Schrödinger's Quantum Equation

It is well-known that the Schrödinger equation is defined as a complex differential equation

$$i \hbar \, \partial \phi / \partial t = - (\hbar^2 /2m) \, \partial^2 \phi / \partial s^2 + V \phi \qquad (30)$$

in one spatial dimension s for a particle in a potential V(s), where $\phi(s,t)$ is the wave function depending on space s and time t.

A Newtonian particle in a harmonic potential is given by the equation system

$$dx/dt = p/m \qquad (31a)$$
$$dp/dt = - m \, \omega^2 \, x \qquad (31b)$$

In defining a complex function, similar to the Heisenberg formalism,

$$F = (m/2)^{1/2} \, \omega \, x + i \, p/(2m)^{1/2} \qquad (31c)$$

we obtain
$$dF/dt = (m/2)^{1/2} \, \omega \, p/m - i \, m \, \omega^2 \, x \, /(2m)^{1/2}$$
$$i \, dF/dt = (m/2)^{1/2} \, \omega^2 \, x + i \, \omega \, p/(2m)^{1/2}$$
or

$$i \, dF/dt = \omega \, F \qquad (31d)$$

The generalized complex discrete derivative 26 applied to eq. 31d gives
$$i \, [\, F(t+\Delta t) - F(t-\Delta t) \,]/2dt + i \, [\, F(t+dt) - 2F(t) + F(t-dt) \,]/2\Delta t = \omega \, F(t)$$
or, in multiplying both members by $\hbar$,

$$i \, \hbar \, [F(t+\Delta t)-F(t-\Delta t)]/2\Delta t = [-(\hbar/2) \, [F(t+\Delta t)-2F(t)+F(t-\Delta t) \,]/\Delta t]+\hbar \, \omega \, F \qquad (32)$$

The first factor in the second member is similar to a diffusion in time of the position and momentum. This could be interpreted as a temporal non-locality. In taking the real and imaginary part of F = R + iI, we can write eq. 32 as

18

$R(t+\Delta t) = R(t-\Delta t) - [I(t+\Delta t)-2I(t)+I(t-\Delta t)] + 2 \Delta t \, \omega \, I$ (32a)

$I(t+\Delta t) = I(t-\Delta t)] + [R(t+\Delta t)-2R(t)+R(t-\Delta t)] - 2 \Delta t \, \omega \, R$ (32b)

which is an incursive anticipatory equation system, because $R(t+\Delta t)$ is a function of $I(t+\Delta t)$ and $I(t+\Delta t)$ a function of $R(t+\Delta t)$. The real and imaginary parts of F have a predictive model of each other. Can we conclude that an anticipatory system has the property of temporal non-locality? We will show that eq. 32 can be transformed to a system similar to a quantum system for which spatial non-locality is a well-known property. So a quantum system could be interpreted as a spatial anticipatory system.

In view of obtaining a diffusion in space, let us define the position by $x = x(s, t)$ and the momentum by $p = p(s, t)$, where s is a spatial coordinate, the function $F = F(s, t)$, and eq. 32 becomes:

$i \, \hbar \, [F(s,t+\Delta t)-F(s,t-\Delta t)]/2\Delta t = [-(\hbar/2) [F(s,t+\Delta t)-2F(s,t)+F(s,t-\Delta t)]/\Delta t]+\hbar \, \omega \, F(s,t)$ (32d)

In inverting the space s and time t variables in the function $F(s,t)$ in the first factor of the second member, the second time derivative is transformed into a second space derivative

$i \, \hbar[F(s,t+\Delta t)-F(s,t-\Delta t)]/2\Delta t = [- (\hbar \, c^2 \, \Delta t \, /2)[F(t,s+\Delta s)-2F(t,s)+F(t,s-\Delta s)]/\Delta s^2]+\hbar\omega F$ (33)

We can justify this in reference to the classical wave equation. Indeed, let us consider the one dimension wave equation

$\partial^2 x(s,t)/\partial t^2 = c^2 \, \partial^2 x(s,t)/\partial s^2$ (34a)

where $x(s,t)$ is the value of the wave at position s at time t, and c the velocity. This differential equation can be replaced by the finite difference equation

$x(s,t+\Delta t)-2.x(s,t)+x(s,t-\Delta t) = c^2.\Delta t^2.[x(s+\Delta s,t)-2.x(s,t)+x(s-\Delta s,t)]/\Delta s^2$ (34b)

After Finkelstein [1996], the Planck constants: $L_p = 1.6 \, 10^{-35}$m, $T_p = 10^{-44}$ s and $M_p = 2.5 \, 10^{-8}$ kg can be deduced from the Maxwell constant $c = 3.0 \, 10^8$ m/s, the Newton constant $G = 6.67 \, 10^{-11}$ Nm^{2}kg and the Plank constant $\hbar = 1.05 \, 10^{-34}$ Js.

From these Planck constants, we deduce $M_p c^2 T_p = \hbar$ or $c^2 T_p = \hbar / M_p$
where the Plank interval of time T_p is related to the mass M_p of a particle. So, similarly, we deduce $c^2 \Delta t = \hbar / m$, where Δt is the interval of time related to the mass m of a particle.
In taking $c^2 \Delta t = \hbar / m$, eq. 33 becomes

$i \, \hbar \, [F(s,t+\Delta t)-F(s,t-\Delta t)]/2\Delta t = -(\hbar^2/2m)[F(s+\Delta s,t)-2F(s,t)+F(s-\Delta s,t)]/\Delta s^2 +\hbar\omega F(s,t)$ (35)

This discrete equation 35 gives the following finite difference equation

$i \, \hbar \, \Delta F/\Delta t = (- \hbar^2 /2m) \Delta^2 F/ \Delta s^2 + \hbar \, \omega \, F$ (36)

which is formally similar to the Schrödinger equation for a free particle in a uniform potential $V = \hbar \omega$ along the s axis. Eq. 36 can be split into the real and imaginary parts of F in function of x and v:

$$\Delta x(s,t) / \Delta t = -(\hbar/2m\omega) \, \Delta^2 v(s,t) / \Delta s^2 + v(s,t) \tag{37a}$$
$$\Delta v(s,t) / \Delta t = (\hbar\omega/2m) \, \Delta^2 x(s,t) / \Delta s^2 - \omega^2 x \tag{37b}$$

When the mass becomes higher and higher, the diffusive factors vanish and the Newtonian formalism is obtained where the wave packet tends to a particle.

For ω tending to zero, we will consider the following equation derived from eqs. 36a-b

$$\Delta^2 x(s,t) / \Delta t^2 = -(\hbar^2/4m^2) \, \Delta^4 x(s,t) / \Delta s^4 + (\hbar \omega / m) \, \Delta^2 x(s,t) / \Delta s^2 - \omega^2 x \tag{38}$$

which gives for ω tending to zero

$$\Delta^2 x(s,t) / \Delta t^2 = -(\hbar^2/4m^2) \, \Delta^4 x(s,t) / \Delta s^4 \tag{39}$$

which is similar to the quantum equation of the real part of the wave function for a free particle. Indeed, in defining the real and imaginary parts of the wave function

$$\phi(s,t) = x(s,t) + i \, y(s,t)$$

in eq. 30 with $V = 0$, one obtains

$$\partial^2 x(s,t) / \partial t^2 = -(\hbar^2/4m^2) \, \partial^4 x(s,t) / \partial s^4 \tag{40}$$

The ratio $\Delta s^2 / \Delta t \propto c^2 \, dt = \hbar / m$ is related to the Planck constant. When $\hbar$ becomes smaller and smaller, the Newtonian formalism is obtained [cf. Gutzwiller, 1990].

In fact, the classical Newtonian formalism can be obtained for particles or systems (like molecules) with a high mass m: in this case, the spatial diffusion factor vanishes, which means also that Δs tends to zero with Δt (the fractal dimension is then $D_s = 1$ in the Euclidean space of Newton formalism).

The factor before the space derivative is similar to a diffusion coefficient [cf. Gutzwiller, 1990].

In our framework, we have transformed a time diffusion to a space diffusion to obtain the Schrödinger equation in relation to the wave equation.

When Δs^2 tends to zero with Δt, this means that the space interval has a fractal structure of fractal dimension equal to $D_s = 2$ in agreement with Nottale [1989] conjecture.

An other interpretation is that c^2 tends to infinity with Δt: the velocity becomes infinite: this is not in contradiction with non-relativistic quantum mechanics.

When the velocity of the particle tends to the velocity of light, this formalism is no more available: following Nottale the time has the fractal dimension $D_t = 2$. In quantum relativity, $\hbar\omega$ could represent the mass energy of a particle: $\hbar\omega = mc^2$. Research in this direction is in progress.

I have tested on computer the validity of the finite difference eq. 33 in considering as usual by physicists $\hbar = 1$ and $c = 1$ with $\hbar\omega = 1$, for a particle between two reflecting walls at $s = 0$ and $s = 200$ ($\Delta s^2 / \Delta t = 1$): this is a quantum cellular automata.

The simulations are given in Figures abc and 7.

Fig. 6abc show the simulation of a particle reflecting between two walls. The initial condition is given by

$F(s, 0) = 0$, $F(s, -1) = 0$ for $s = 1$ to $s = 99$, $s = 101$ to 200
$F(100, 0) = 1$, $F(100, -1) = 1$

In the Figure 6a, the horizontal axis represents the space variable s (from 1 to 200) and the vertical axis represents the time variable t from the top to the bottom. Figures 6bc are the time continuation of Fig. 6a.

These Figures show $P(s, t) = F(s,t)F^*(s,t)$ as a function of space s and time t. The real function is related to the density of presence of the particle between the two walls. The initial particle behaves as a wave spreading until the two walls, reflecting and interfering with a spatial oscillation between the walls. Self-interference is well-seen.

Fig. 7 shows the simulation for two initial particles. The initial conditions are given by

$F(s, 0) = 0$, $F(s, -1) = 0$ for $s = 1$ to $s = 49$, $s = 51$ to 149 and $s = 151$ to 200
$F(50, 0) = 1$, $F(50, -1) = 1$
$F(150, 0) = 1$, $F(150, -1) = 1$

In the figure 7, the horizontal axis represents the space variable s (from 1 to 200) and the vertical axis represents the time variable t from the top to the bottom.

This Figure shows $P(s, t) = F(s,t)F^*(s,t)$ as a function of space s and time t. The real function is related to the density of presence of the particle between the two walls. The two initial particles behave as waves. They reflect on the walls and interfere: the interferences are well-seen.

The interesting aspect in these two simulations is the fact that the waves record in their interferences an information about the configuration of their space as well as the temporal history of their successive reflections on the walls and their successive interferences. Even a single wave record its self-interferences. So these waves behave like space-time memories. This is in agreement with the path integrals of Feynman.

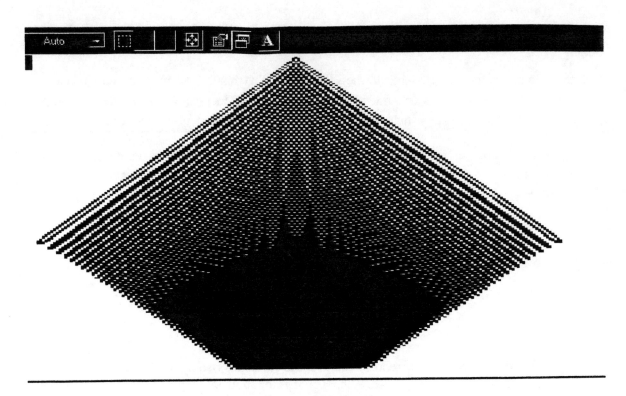

Figure 6a

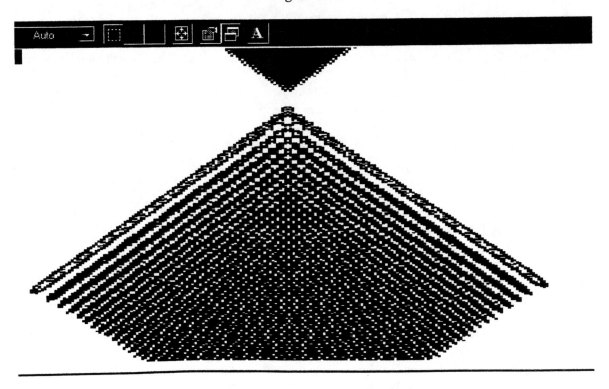

Figure 6b

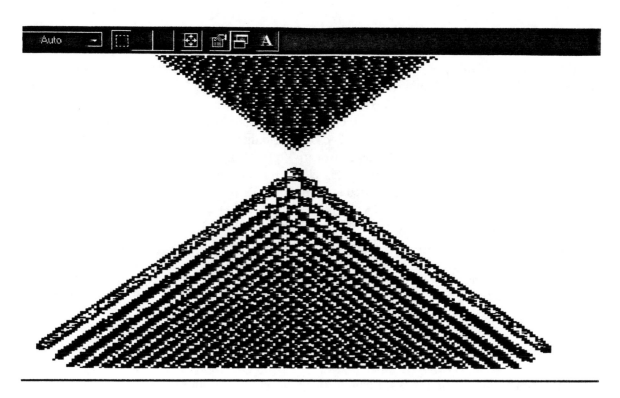

Figure 6c

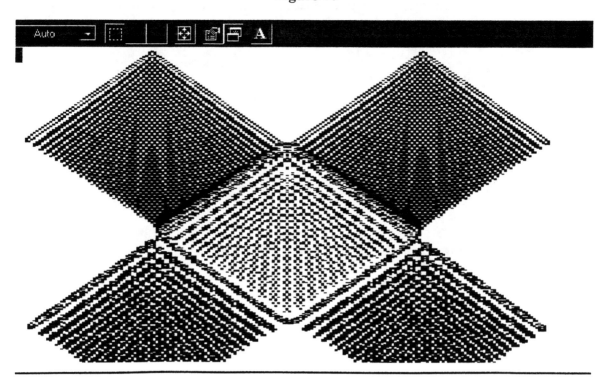

Figure 7

4 Hyperincursive Stack Memory in Chaotic Automata

In this last section of this paper, I present a new model of a neural memory by a stack of binary input data embedded in a floating point variable from an hyperincursive process based on the Pearl-Verhulst chaotic map: $x(t+1)=4\mu x(t)(1-x(t))$.

Theoretical and experimental works enhance the validity of such an approach. Von Neumann (1996) suggests that the brain dynamics is based on hybrid digital-analogical neurons. I proposed a fractal model of neural systems based on the Pearl-Verhulst map (Dubois, 1990, 1992). A non-linear threshold logic was developed from this chaos fractal neuron (Dubois, 1996; Dubois and Resconi, 1993) in relation to the McCulloch and Pitts (1943) formal neuron. Experimental analysis in nervous systems show fractal chaos (King, 1991; Schiff, 1994). Neural systems can be modeled as automata (Weisbuch, 1989). My model of a stack memory could be applied in the framework of symbolic dynamics and coding (Lind, 1995).

The Pearl-Verhulst map in the chaotic zone ($\mu=1$) can be transformed to a quasi-linear map $X(t+1)=1-abs(1-2x(t))$, where abs means the absolute value. This simple model was proposed for simulating neural chaos (Dubois, 1992). Let us consider the incursive map

$$X(t) = 1 - abs(1 - 2X(t+1)) \qquad (41)$$

where the iterate $X(t)$ at time t is a function of its iterate at the future time $t+1$, where t is an internal computational time of the system. Such a relation can be computed in the backward direction T, $T-1$, $T-2$, ... $2,1,0$ starting with a "final condition" $X(T)$ defined at the future time T, which can be related to the Aristotelian final cause. This map can be transformed to the hyper recursive map (Dubois, 1996c):

$$1 - 2X(t+1) = \pm(1 - X(t)) \quad so \quad X(t+1) = [1 \pm (X(t) - 1)]/2 \qquad (42)$$

In defining an initial condition $X(0)$, each successive iterates $X(t+1)$, $t=0,1,2,...,T$
give rise to two iterates due to the double signs $\pm$. So at each step, the number of values increases as 1,2,4,8,... In view of obtaining a single trajectory, at each step, it is necessary to make a choice for the sign. For that, let us define a control function $u(T-t)$ given by a sequence of binary digits 0,1, so that the variable sg

$$sg = 2u(t) - 1 \text{ for } t=1,2,...,T \qquad (43)$$

is -1 for $u=0$ and +1 for $u=1$. In replacing eq. 43 in eq. 42, we obtain

$$X(t+1) = [1 + (1 - 2u(t+1))(X(t) - 1)]/2 = X(t)/2 + u(t+1) - X(t).u(t+1) \qquad (44)$$

which is a hyperincursive process because the computation of $X(t+1)$ at time $t+1$ depends on $X(t)$ at time t and $u(t+1)$ at the future time $t+1$. Eq. 44 is a soft algebraic map (Dubois, 1996c) generalizing the exclusive OR (XOR) defined in Boolean algebra: $y = x_1 + x_2 - 2x_1x_2$ where x_1 and x_2 are the Boolean inputs and y the Boolean output. Indeed, in the hybrid system 44, $X(t)$ is a floating point variable and $u(t)$ a digital variable.

Starting with the initial condition X(0)=1/2, this system can memorize any given sequence of any length u. The following Table I gives the successive values of X for all the possible sequences with 3 bits.

TABLE I

u	X(1)	X(1)	X(2)	X(2)	X(3)	X(3)
000	1/4	0.25	1/8	0.125	1/16	0.0625
100	3/4	0.75	3/8	0.375	3/16	0.1875
110	3/4	0.75	5/8	0.625	5/16	0.3125
010	1/4	0.25	7/8	0.875	7/16	0.4375
011	1/4	0.25	7/8	0.875	9/16	0.5625
111	3/4	0.75	5/8	0.625	11/16	0.6875
101	3/4	0.75	3/8	0.375	13/16	0.8125
001	1/4	0.25	1/8	0.125	15/16	0.9375

This table gives the successive values of X for each sequence u as rational and floating point numbers. The number of decimal digits increases in a linear way (one bit of the sequence corresponds to a decimal digit of X). The last digit 5 corresponds to the initial condition X(0)=0.5 and the two last digits 25 or 75 give the parity check of the sequence. The time step t is directly related to the number of digits: with t=0,1,2,3 there are 4 digits. In looking at the successive increasing values of the floating points of X, we see that the correspondent sequences u represent the Gray code. Contrary to the binary code, the Gray code changes only one bit by the unitary addition. The numerator of each ratios is two times the floating point representation of the Gray code of the sequence u, plus one. With the Gray code, we can construct the Hilbert curve which fills the two-dimensions space: the fractal dimension is $D_H=2$. This is not possible with the Cantor set, which gives discontinuities in two directions in the space (Schroeder, 1991).

The neuron is an analogical device which shows digital spikes: the analogical part of the neuron is given by the floating point values X and the values of the spikes are given by the digital sequence u.

The analogical coding X of digital information of the spikes u is then a learning process which creates a fractal memory.

Now, let us show how it is possible to recover the digital information u(t) for t=,1,2,3,...,T from the analogical final value X(T) of the neuron (T=3 in our example).

The inversion of the sequence has some analogy with the inversion of the image received by the eyes. With our method, the coding of an image leads to an inverse image so that the image is reconstructed without inversion.

The decoding of a sequence u from the final value X(T) can be made by relation 1 for t = T-1, T-2, ...

$$X(t) = 1 - abs(1 - 2X(t + 1)) \qquad (45)$$

Let us take an example, starting with the final value X(T=3)=0.5625, we compute successively
X(2) = 1 - abs(1 - 2x0.5625) = 1 - 0.125 = 0.875,
X(1) = 1 - abs(1 - 2x0.875) = 1 - 0.75 = 0.25, X(0) = 1 - abs(1 - 2x0.25) = 0.5.

The sequence is then given by

$$u(t+1) = (2X(t+1)) \text{ div } 1 \qquad (46)$$

where div is the integer division: $u(3) = (2 \times 0.5625) \text{ div } 1 = 1$, $u(2) = 1$, $u(1) = 0$. The neuron will continue to show spikes 1,0,1,0,1,0, ...
It is well-known that neurons are oscillators which present always pulsations, the coding of information is a phase modulation of these pulsations (Dubois, 1992).

In taking the formal neuron of McCulloch and Pitts (1943), eq. 46 can be replaced by

$$u(t+1) = \Gamma (X(t+1) - 0.5) \qquad (47)$$

for which u=1 if $X \geq 0.5$ and u=0 otherwise.
As we can compute u(t) from X(t), it is possible to compute eq. 45 in the following way

$$X(t) = 2X(t+1) + 2u(t+1) - 4X(t+1)u(t+1) \qquad (48)$$

which is also a soft computation of XOR.

So, to retrieve the message embedded in the stack memory by the soft XOR relation 44, a similar soft XOR relation 48 is used.

The following Figure 8a-b gives a possible neural network for the stack memory (Dubois, 1996c).

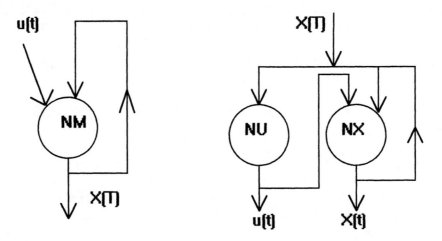

Figure 8a-b: (a) The neuron NM represents the soft XOR eq. 44 for the coding of the sequence u(T-t) giving X(T); (b) The neuron NU is a McCulloch and Pitts neuron given by eq. 47 computing the ordered sequence u(t) and the neuron NX represents the soft XOR eq. 8 giving X(t) starting from the final state X(T) coming from the neuron NM.

This is a property of XOR that the addition and the subtraction is the same operator. Here the soft XOR given by a non-linear algebraic relation gives the same property in a generalized way.

26

In conclusion, this mast section gives a way to build a stack memory for automata such as neural systems. A hyperincursive control of a fractal chaos map is used for embedding input informations in the state variable of the memory. The input sequence is given by a digital variable and the memory is represented by an analogical variable. The analogical variable is represented in floating point. With classical computer, the number of decimal digit is limited so that we must code the decimal digits of great length by strings. The actual neuron could be an analogical device working only with strings. In this way, such a neural system with hyperincursive stack memory could help in the design of a Hyper Turing Machine.

References

Aczel P. (1987). "Lectures in Nonwellfounded Sets", CLSI Lecture Notes n° 9.

Barwise J., L. Moss (1991). "Hypersets", in the Mathematical Intelligencer, 13, 4, pp. 31-41.

Dubois D. (1990). *Le labyrinthe de l'intelligence: de l'intelligence naturelle à l'intelligence fractale*, InterEditions/Paris, Academia/Louvain-la-Neuve.

Dubois D. (1992). *The fractal machine*, Presses Univiversitaires de Liège..

Dubois D. M. (1992a). "The Hyperincursive Fractal Machine as a Quantum Holographic Brain", *CC AI, Communication and Cognition - Artificial Intelligence*, vol 9, number 4, pp. 335-372.

Dubois D. M. (1995). "Total Incursive Control of Linear, Non-linear and Chaotic Systems", in *Advances in Computer Cybernetics*, volume II, edited by G. E. Lasker, published by The International Institute for Advanced Studies in Systems Research and Cybernetics, University of Windsor, Canada, pp. 167-171, 1995.

D. M. Dubois (1996a). "Introduction of the Aristotle's Final Causation in CAST: Concept and Method of Incursion and Hyperincursion". In F. Pichler, R. Moreno Diaz, R. Albrecht (Eds.): Computer Aided Systems Theory - EUROCAST'95. Lecture Notes in Computer Science, 1030, Springer-Verlag Berlin Heidelberg, pp. 477-493.

Dubois D. M. (1996b). "A Semantic Logic for CAST related to Zuse, Deutsch and McCulloch and Pitts Computing Principles". In F. Pichler, R. Moreno Diaz, R. Albrecht (Eds.): Computer Aided Systems Theory - EUROCAST'95. Lecture Notes in Computer Science, 1030, Springer-Verlag Berlin Heidelberg, pp. 494-510.

Dubois D. M. (1996c). *Hyperincursive Stack Memory in Chaotic Automata*, in Actes du Symposium ECHO "Emergence - Complexité Hiérarchique - Organisation: Modèles de la boucle évolutive", édité par A. C. Ehresmann, G. L; Farre, J.-P. Vanbremeersch, publié par l'Université de Picardie Jules Verne, pp. 77-82.

Dubois D. M. (1997). "Generation of fractals from incursive automata, digital diffusion and wave equation systems", Biosystems 43, pp. 97-114.

Dubois D., and G. Resconi. (1992). "Hyperincursivity: a new mathematical theory", Presses Universitaires de Liège, 260 p..

Dubois D. M., Resconi G. (1993). *Mathematical Foundation of a Non-linear Threshold Logic: a new Paradigm for the Technology of Neural Machines*, ACADEMIE ROYALE DE BELGIQUE, Bulletin de la Classe des Sciences, 6ème série, Tome IV, 1-6, pp. 91-122.

Dubois D. M.and G. Resconi. (1993b). "Introduction to hyperincursion with applications to computer science, quantum mechanics and fractal processes", in Designing New Intelligent Machines (European COMETT programme), Communication & Cognition - Artificial Intelligence, vol. 10, N°1-2, pp. 109-148.

Dubois D. M. and Resconi G. (1994), *The hyperincursive Fractal Machine beyond the Turing Machine*, in Advances in Cognitive Engineering and Knowledge-Based Systems, ed. By G. E. Lasker, published by The International Institute for Advanced Studies in Systems Research and Cybernetics, University of Windsor, Canada, pp. 212-216, 1994.

Dubois D. M. and G. Resconi (1994a). "*Holistic Control by Incursion of Feedback Systems, Fractal Chaos and Numerical Instabilities*, in CYBERNETICS AND SYSTEMS'94, ed. by R. Trappl, World Scientific, pp.71-78, 1994.

Dubois D. M. and G. Resconi. (1995). Advanced Research in Incursion Theory Applied to Ecology, Physics and Engineering, COMETT EUROPEAN Lecture Notes in Incursion, published by A.I.Lg., Association des Ingénieurs sortis de l'Université de Liège, D/1995/3601/01.

Finkelstein D. R. (1996). Quantum Relativity: A Synthesis of the Ideas of Einstein and Heisenberg. Texts and Monographs in Physics. Springer.

Grössing G. and A. Zeilinger (1991) Zeno's paradox in quantum cellular automata. Physica D 50(1991) 321-326.

Gutzwiller Martin C. (1990) Chaos in Classical and Quantum Mechanics, in Interdisciplinary Applied Mathematics, IAM number 1, Springer-Verlag New York Inc.

King Chris C. (1991), "Fractal and Chaotic Dynamics in Nervous Systems", *Progress in Neurobiology*, vol 36, pp. 279-308.

Lasker E. George, Daniel Dubois, Bernard Teiling, Editors (1996). Advances in Modeling of Anticipative Systems. The International Institute for Advanced Studies in Systems Research and Cybernetics, Canada, 132 p.

Lind Douglas, Marcus Brian (1995), *Symbolic Dynamics and Coding*, Cambridge University Press.

McCulloch W. S., Pitts W. (1943), "A logical calculus of the ideas immanent in nervous activity", *Bulletin of mathematical Biophysics*, vol 5, pp. 115-133.

Nottale L. (1989) Fractals and the Quantum Theory of Spacetime, International Journal of Modern Physics A, Vol. 4, N° 19 (1989), 5047-5117.

Nottale L. (1993) Fractal Space-Time and Microphysics: Towards a Theory of Scale Relativity, World Scientific.

Schiff Steven J. et al (1994), "Controlling chaos in the brain", *Nature*, vol 370, pp. 615-620.

Schroeder Manfred (1991), *Fractals, Chaos, Power Laws*, W. H. Freeman and Company, New York.

Von Neumann J. (1996), *L'ordinateur et le cerveau*, Champs, Flammarion.

Weisbuch G. (1989*), Dynamique des systèmes complexes: une introduction aux réseaux d'automates*, InterEditions/Editions du CNRS.

Rosen Robert (1985), Anticipatory Systems, Pergamon Press.

ANTICIPATORY SYSTEMS
A SHORT PHILOSOPHICAL NOTE

Gertrudis Van de Vijver
Senior Research Associate Fund for Scientific Research Flanders (FWO)
Department of Philosophy
University of Gent
Blandijnberg 2, 9000 Gent
fax. +32.9.220.73.05
gertrudis.vandevijver@rug.ac.be

ABSTRACT

My aim is to explore some aspects of anticipation from a philosophical point of view. I start from the way in which anticipation has been traditionally related to the relation between particular and universal, and situate from there the view of Robert Rosen, that is more adequately characterized in terms of local/global than in terms of particular/ universal. A short comment on Rosen will lead me to suggest some possible lines of research with regard to a more dynamical approach of anticipatory systems.

1. PARTICULAR/UNIVERSAL

One of the things I learned in preparing this paper, is that there are quite different philosophical stories that can be told about anticipation. For a moment, I was surprised by those differences, but I realized that they were a matter of focus, that is, a matter of initial choices and related consequences, much more than a matter of seeing things erroneously in one case, and better or more adequately (in any absolute sense of these words), in another case.

I became aware of these differences basically in reading the *Encyclopédie Philosophique Universelle* on anticipation. I read there, for instance, that in Greek philosophy, and in particular for the Stoicians and Epicurians, anticipation is about the relation between *universal and particular*. To them, anticipation is the capacity of designating things by a universal vocabulary, while having experience only with particular instances of those things. So, anticipation refers to the capacity of the human mind to possess and use abstract ideas, *before* the immediate perception of the object.

Quite some philosophers have thought about anticipation along the same lines. Kant, for instance, uses the term in his *Critique of Pure Reason* in relation to perception. When he deals with "The anticipations of perception", he refers to the *a priori* principles of human understanding relatively to the category of quality.[1] Even if Kant acknowledges that

---

[1] The principle of the anticipations of perception states that: "In all appearances the real which is an object of sensation has an intensive quantity, that is, a degree." (Kant,

CP437, *Computing Anticipatory Systems: CAYS--First International Conference*
edited by Daniel M. Dubois © 1998 The American Institute of Physics 1-56396-827-4/98/$15.00

sensations as such can only exist *a posteriori,* we can know on an *a priori* basis, i.e. independently of any sensory experience, that the object has the property of degree or intensity (the "intensive quantity"), that it has, in other words, the capacity in principle of affecting us sensorily in different degrees.

In the light of the current discussions on anticipatory systems, these classical viewpoints are remarkable.

Firstly, because anticipation — as a particular time-related process, in which something *comes before* something else (a universal, an abstract idea, that comes before the immediate perception) — clearly rests on the possibility of a type of knowledge that is completely out of time, a knowledge that is universal and that can be known on an *a priori* basis. The message is perhaps most clear with Kant, and it holds for his transcendental philosophy in general: first assume that a-temporal, universal, *a priori* knowledge is possible, then re-apply that knowledge as if it were situated in the time-domain. *"Time is use"* is perhaps a good paraphrase for such a philosophical procedure. It also is, I guess, the mainstream scientific reasoning about time since Newton: "Time is use" as the counterpart "Time is only an illusion".

Secondly, this viewpoint is interesting because of the recognized link between anticipation and abstraction. Instead of stressing in the first place the particular time-dimension, anticipation becomes related to the connections that can be built between particular sensations on the one hand, and types of knowledge that lie beyond the realm of the particular on the other hand. Once we acknowledge this, the relations that particular sensory systems come to establish with an environment are crucial to the interpretation of anticipation. Instead of simply assuming this relation and subscribing to the time paradox I just stressed, it becomes possible to study anticipation as the result of a particular history between the system's capacities and its experience with an environment. It becomes possible, to use more accurate terms, to attempt a naturalization of anticipation.

It can be noted that for ancient philosophers anticipation frequently was an indication of the priority to be given to the universal. Anticipation mostly referred to the possibility of extrapolating from the universal to the particular. This was, however, not the only possibility. For the Epicurians, for instance, anticipation was derived from sensory experience. In line with this, Francis Bacon opposes the rational interpretations to the natural anticipations, the former giving a scientific picture of reality, the latter leading to a completely subjective view on nature.

So, if there have been paradigm- and language-shifts during the last centuries, then it has been mainly on the point of the possible priorities to be given: priority to the particular or to the universal, priority at what moment, at the moment of the genesis (construction) of knowledge or at the moment of its functioning, and, last but not least, priority in relation to which systems ? For which systems is the particular prior, and the universal negligible ? For which systems can we start from the universal, and consider the particular as negligible ?

In order to go a little bit further in the elucidation of the basic philosophical shifts in approaching anticipation, I want to briefly deal with one recent interpretation, that of Robert Rosen.

Rosen's approach of anticipatory systems contains similar ideas to those stemming from the philosophical tradition: the temporal dimension, as well as the dimension of abstrac-

---

Critique of Pure Reason, transl. W. Schwarz, p. 69).

32

tion are by him stressed. There is, however, a basic shift, that arises from the connection he makes between anticipation and *systems*. On this point, it is useful to ask whether and to what extent his approach is a first order, externalist approach of systems, and what would be the implications of attempting to develop a second order, self-organizational approach in this regard.

2. ROBERT ROSEN, 1: ANTICIPATION AND TIME

In *Anticipatory Systems,* Rosen writes: "An anticipatory system is a system containing a predictive model of itself and/or of its environment, which allows it to change state at an instant in accord with the model's predictions pertaining to a later instant" (Rosen, 1985, p. 339).

The focus is here on the capacity of systems to build predictive models, of themselves and/or of the environment. Those models permit to take into account calculated future states in order to be able to decide about the adequacy of actual behavior. In this way, anticipation is about things that happen at a certain moment, *before* another moment; anticipation is about the consequences in the present of what might happen at a later moment. However, if anticipation is about time, then it is clearly about time only in the sense of predictability. As a consequence, there can be no problem of time-reversal here.

Many authors, in cognitive sciences, as well as in biology, have stressed (cf. Atlan, 1997, 1998) that the representational account of anticipation, i.e. the model-building aspect, in which the system's future behavior is determined by previous experiences and calculated consequences of these, does not contain the time-reversal that is often linked to anticipation or to teleology in general. There is a kind of flattening of time by shifting to the representational.

Therefore, if it is sometimes said that anticipatory systems contain the threatening seeds of teleology, because of the inclusion of future states that determine the behavior of the system before they actually took place, then a representational account clearly dissolves the mystery of this time-reversal. In the representational account, the problem of teleology, or the problem of anticipation are not about time. They are about inscriptions, about records, as Matsuno would say, and about types, possibilities and limitations of records. To adopt the words of Atlan: "A project for the future is always the outcome of the reversal, in its representation, of an effect into a cause. What was the end of a causal sequence of moves and its effect the first time, in an action that preceded its representation and transformation into a procedure, is converted into a cause in the representation itself. This conversion is not particularly mysterious in the kind of model we are suggesting, because it is made possible by the *flattening of time produced by all memorization.* Once a temporal sequence of states is memorized, its past-future order is transformed into a symbolic one where time disappears, since the entire sequence is present in the memory at the same time" (Atlan, 1997, p. 16, italics added). The effort of Henri Atlan, and I guess also of Rosen, is to give a plausible mechanistic account of anticipation, explaining "the apparent time reversal of causality in the definition and achievement of a project". For Atlan, this happens by showing that "the voluntary aspect of decision-making is merely a memorization of functional self-organizing processes rather than an ex-nihilo creation of a causal series" (Atlan, 1997, p. 2).

What can we conclude out of this ? Minimally, that the (formal) time-aspect is not the basic point to be dealt with in anticipation. The basic point lies elsewhere, as Robert

Rosen himself, hereby following the philosophical tradition, has clearly shown. The core of anticipation is indeed related to the capacity of abstraction certain systems have acquired. Or in more actual terms: *the appreciation of anticipation takes place at the interface between a system and its environment* (this is a paraphrase of Atlan's viewpoint on intentionality, 1997). Taking into account the flattening of time that takes place in representational systems, it becomes crucial to see how systems can be reintroduced in time by using of these same representations. Moreover, it will be crucial to grasp what role representations play in making systems into systems, and what role that "systematic" nature plays in constructing, developing and using anticipative representations. At this point, it remains to be seen whether we can still use the traditional philosophical, externalist, terminology, and state that the appreciation of anticipation has to do with the relation between particular and universal, leaving outside of their concerns the systems that actively take part in introduce and manipulate that distinction. It seems to me that in order to deal adequately with anticipation today, it is necessary, firstly, to go from the classical distinction between particular and universal towards systems, i.e. towards the distinction between local and global. Secondly, it is important to attempt articulating the consequences of a genuine dynamic viewpoint on anticipation, i.e. one in which the distinctions made between particular/universal and/or local/global are the result of a dynamical, anticipatory process in which the observer is an integral and non negligible part.[2]

3. ROBERT ROSEN, 2: MODELS AND ABSTRACTION

Let us see how Rosen conceives of the interface between a system and its environment.

In the book on *Anticipatory Systems*, but also in *Life itself*, Rosen is very specific about the status of *models*: "A model is a relation between a natural system S and some suitable formal system M. The modelling relation itself is essentially a linkage between behaviors in S and inferences drawn in M, [that] become predictions about the behavior of S (...)".

And further: "a modelling relation (...) is to be established through an *encoding* of qualities or observables of S into formal or mathematical objects in M (...) any encoding of a natural system S into a formal system M involved an act of abstraction, in which non-encoded qualities are necessarily ignored. Such an abstraction is not peculiar to theory; indeed, any act of observation (on which any kind of encoding must be based) is already an abstraction in this sense." (Rosen, 1985, pp. 339-340, original italics).

The first interesting thing about these fragments, is that Rosen is clearly concerned with the requirements of abstraction which are involved in anticipation: model-building implies abstraction, and this, he takes at a very basic level, namely already at the level of

---

[2] It is not sure that Kant really grasped the shift from particular/universal to local/global. It is true that when he talks about organisms as internal teleological systems, he clearly stresses their internal organization (their circular causality). But when he deals the ways of knowing them, his epistemology of those systems is laid out in terms of the relation between particular and universal. We need, so he says, to use reflective judgments in order to understand those systems; we cannot know them objectively. Determinative judgments, in which we deduce the particular from the universal, are radically impossible in this case. We cannot know organisms objectively, as we have to ever attempt the construction of the universal out of the particular, without ever attaining it.

observation. Abstraction means that something inevitably gets lost.

Secondly, in judging about the adequacy of models, about the adequacy between M and S, we are, within the scope of anticipatory systems, referred to the domain of interactions, where matters of stability prevail, rather than matters of correspondence. Rosen writes: "An anticipatory system S2 is one which contains a model of a system S1 with which it interacts. This model is a predictive model; its *present* states provide information about *future* states of S1. Further, the present state of the model causes a change of state in other subsystems of S2; these subsystems are (a) involved in the interaction of S2 with S1, and (b) **they do not affect (i.e. are unlinked to) the model of S1**. In general, we can regard the change of state in S2 arising from the model as an *adaptation*, or pre-adaptation, of S2 relative to its interaction with S1." (Rosen, 1985, p. 344, italics original, bold added). So, a basic indication of the fact that we are in front of an anticipatory system, is that a property P, which was "guided" by the predictive model S2 has of S1, has to be stabilized in the interaction between S1 and S2.

Thirdly, it is remarkable that Rosen stresses here that certain subsystems that are involved in the interaction of S2 with S1, "do not affect (i.e. are unlinked to) the model of S1". He also states that absence of predictability, error, disorder and emergence of novel behavior, are related to the discrepancies that arise over time between M and S. For Rosen, discrepancies are inevitable because M is a system that is open to interaction with an environment, and S is a system isolated from those environmental interactions. This leads me to conjecture that Rosen's viewpoint on anticipatory systems, even if it indicates the shift from particular/universal to local/global, expresses an externalist stance, in which the distinction between system and environment is taken for granted. Anticipatory systems are to him not instances of complex, self-organizational and autonomous systems that are auto-constructive *while* building models from the environment. System and environment are taken to be separable here, and the problem of stabilization between those two separable poles is solved in quite classical neo-Darwinian terms. In this way, Rosen doesn't seem to subscribe to the idea that the building of models is simultaneously, at various hierarchical levels, a matter of a gradual development of an interface between systems and environments and a development of the distinction between systems and environment. He doesn't seem to acknowledge the idea that systems become systems through an interaction with other systems and with an environment, and make their environment accordingly (cf. Matsuno & Salthe, 1995; Salthe, 1993; Van de Vijver, 1998; Van de Vijver et al., 1998).

In summary, an anticipatory system S2 has to fulfill the following conditions according to Rosen:
- it has to possess a model of another system S1;
- there must be some kind of encoding, which makes that some things necessarily fall out of the scope of the model;
- there exists a definite linkage between the observables of M and the observables that fall out of the scope of M, i.e. state of the model M must modify the properties of other observables of S2;
- the change in S2, arising from the predictive model of S1, must be manifested in a corresponding change in some actual or potential interaction between S1 and S2;
- the model of S1 which is encoded into S2 must be a predictive model.

4. CONCLUSION: LINES FOR FUTURE RESEARCH

I can only briefly indicate here what would be implied by a genuine dynamical view on anticipatory systems.

- As in philosophical tradition, and as with Rosen, anticipation is a matter of memory and abstraction;

- Anticipatory systems have the capacity of building adequate representations, i.e. representations that express past experiences (movements, cf. Sheets-Johnstone, 1996) and that have the potential of being extrapolated to particular future cases;

- In each anticipatory movement or interaction, the system's present identity is fully engaged, and it runs the risk of being modified by it. This also holds for the environment, or for any other anticipatory systems with which interaction takes place;

- Each new identity is the result of an interaction; it is the stabilization of an interactive process that, as long as the systems keeps being anticipatory, never comes to a halt. Rosen's viewpoint, according to which "the present state of the model causes a change of state in other subsystems of S2", but in which it is assumed that "they do not affect (i.e. are unlinked to) the model of S1", is no longer acceptable in a dynamical view on anticipatory systems.

- An anticipatory movement is thus the projection of an identity at a particular level, or, to use more appropriate terms, it is the externalization of an internal dynamics at a particular level (cf. Matsuno & Salthe, 1995).

- The aims that are linked to the anticipatory activity of systems — the anticipation of future states — are externally and *aposteriori* established, and are different from what drives the systems internally.

The general conclusion I come to in "speculating" about anticipatory systems, is thus quite paradoxical. Whereas anticipation has traditionally been linked with (conscious) behavior developed in function of a pre-established goal, and where the study of anticipatory systems has been focussing on the representational and abstract capacities that allow for such a goal-directed behavior, I am tempted to link anticipation with a certain type of dynamic behavior between systems and environments, in which the goal(s), even if it is recognizable *aposteriori* and *globally*, is never fully driving the behavior *apriori* and *locally*. In *Worstward Ho*, Samuel Beckett expresses this idea quite well: "All of old. Nothing else ever. Ever tried. Ever failed. No matter. Try again. Fail again. Fail better." (p. 7). Is it farfetched take Beckett's figures as the most exquisite anticipatory systems ? To me, certainly not, because the attempt to *fail* refers to the belief that each understanding is inevitably local, temporal and relative, doomed to be changed. The attempt of failing *better,* however, carries the hidden ambition of clarifying some of the aims and reasons for certain behaviors and discourses, albeit as *aposteriori, external and global* expression of an ever changing internal dynamics.

REFERENCES

Atlan Henri (1997). Intentional Self-organization (manuscript).

Atlan Henri (1998). Intentional Self-organization. Emergence and Reduction: Towards a Physical Theory of Intentionality. *Thesis Eleven*. Special Issue on Emergence and Autonomy, 52, 5-34. Sage Publications.

Kant Emanuel (1794). *Critique of Pure Reason*. Transl. Wolfgang Schwarz (1982).

Scientia Verlag Aalen.

Matsuno Koichiro; Salthe Stanley. N. (1995). Global Idealism/Local Materialism. *Biology and Philosophy* 10 (3), 309-337.

Rosen Robert (1985). *Anticipatory Systems*. Pergamon Press.

Rosen Robert (1991). *Life Itself. A Comprehensive Inquiry into the Nature, Origin, and Fabrication of Life*. Columbia University Press.

Salthe Stanley N. (1993). *Development and Evolution: Complexity and Change in Biology*. Cambridge, MA: MIT Press.

Sheets-Johnstone Maxine (1996). Consciousness. Text presented at the International Workshop on Origins of Cognition, San Sebastian, 13-14 December 1996.

Van de Vijver Gertrudis (1998). Psychic closure. A prerequisite for the recognition of the sign-function ?. Submitted to *Semiotica*.

Van de Vijver Gertrudis; Salthe Stanley; Delpos M. (eds.) (1998). *Evolutionary Systems. Biological and Epistemological Perspectives on Self-organization and Selection*. Dordrecht: Kluwer Academic Publishers.

Anticipation in the Constructivist Theory of Cognition

Ernst von Glasersfeld
Hasbrouck Laboratory
Scientific Reasoning Research Institute
University of Massachusetts
Amherst MA 01003
USA

Abstract

Much of what we call knowledge is based on the assumption that past experience can provide clues about future experience.

The practice of living and learning consequently involves the anticipation of events and situations at almost every step.

In my talk I shall present the constructivist approach to the epistemological prerequisites and some of the psychological mechanisms that seem necessary in order to explain such an otherwise mysterious capability of foresight.

Introduction

I am not a computer scientist and I do not speak the languages of Quantum Computation, Hyperincursion, or Cellular Automata. But a couple of weeks ago I read some of the prose sections of Robert Rosen's *Anticipatory Systems*, and it gave me the hope that the "anticipation" referred to in the title of this conference would not be altogether different from the anticipations we depend on in managing and planning our ordinary lives as human beings. It is a topic I have been concerned with throughout the many years that I battled against the mindless excesses of the behaviorist doctrine in psychology. I was involved with languages and conceptual semantics, and I was among those outsiders who thought that, when people speak, they mostly have a purpose and are concerned with the effect their words will have. This view was generally considered to be unscientific, and I am therefore very happy that "anticipatory systems" have now become a subject for open discussion among "hard" scientists.

Since most of you are probably deeply immersed in specialized research of a more or less technical nature, you may not have had occasion to consider that anticipation would have to be a fundamental building block of any theory of psychology that merits to be called *cognitive*. And you may not have had reason to wonder why the discipline that is now called 'Cognitive Science' still has not moved very far from the input/output, stimulus/response paradigm. But although everyone now agrees that intelligence and intelligent behavior are the business of a MIND, few are ready to concede much autonomy to that rather indefinite entity.

I do not intend to bore you with a survey of a situation that, to me, seems quite dismal. Instead, I shall focus on the one theory that, in spite of all sorts of shortcomings, is in my view the most promising basis for further development, and consequently the most interesting for people involved in the construction of autonomous models.

The theory I am going to talk about is the one Jean Piaget called Genetic Epistemology. The name was not chosen at random. He wanted to make clear that he intended to analyze knowledge as it developed in the growing human mind, and not, as philosophers usually have done, as something that exists in its own right, independent of the human knower. The name should have warned psychologists that Piaget's theory was not merely a theory of cognitive development, but also constituted a radically different approach to the problems of knowledge. However, especially in the English-speaking world, Piaget was mostly considered as a child psychologists, and his readers disregarded his break with traditional Western epistemology.

This neglect was unfortunate. Piaget actually provided a theoretical model of the human cognitive activity that is more complete and at the same time contains fewer metaphysical assumptions than those of most philosophers.

Towards the end of his working life, which lasted over more than six decades, he said:

> The search for the mechanisms of biological adaptation and the analysis of that higher form of adaptation which is scientific thought [and its] epistemological interpretation has always been my central aim.
> (Piaget, in Gruber & Vonèche, 1977; p.xii)

The term 'adaptation' is the salient point. In many of his writings (he published over 80 books and several hundred articles) he reiterates that what we call knowledge cannot be a representation of an observer-independent reality. And every now and then, as in the passage I quoted, he says that the human activity of knowing is the highest form of adaptation. But he rarely put the two statements together - and this may have made it easier for both his followers and his critics to ignore the revolutionary conceptual change his theory was demanding.

If you consider that in the context of the Darwinian theory of evolution, "to be adapted" means *to survive by avoiding constraints,* it becomes clear that, for Piaget, "to know" does not involve acquiring a picture of the world around us. Instead, it concerns the discovery of paths of action and of thought that are open to us, paths that are *viable* in the face of experience.

The passage I quoted also indicates that there is more than one level of adaptation. On the sensorimotor level of perception and bodily action, it is avoidance of physical perturbation and the possibility of survival that matter. On the level of thought we are concerned with concepts, their connections, with theories and explanations. All these are only indirectly linked to the practice of living. On this higher level, viability is determined by the attainment of goals and the elimination of conceptual contradictions.

To understand Piaget's theoretical scaffolding it is indeed indispensable to remember that he began as a biologist. He knew full well that the biological, *phylogenetic* adaptation of organisms to their environment was not an activity carried out by individuals or by a species. It was the result of natural selection, and natural selection does nothing but eliminate those specimens that do not possess the physical properties and the behavioral capabilities that are necessary to survive under the conditions of the present environment. All organisms that are equipped with senses and the ability to remember sensory experiences can, of course, to some extent *individually* increase their chances of survival by practical learning. Traditionally this was considered a separate domain, and it was explained by association. Piaget, however, who focused on human development, connected it to the biological principle of the reflex.

In most textbooks of behavioral biology, reflexes are described as automatic reactions to a stimulus. Piaget took into account two features that are usually not mentioned. The first was that the existence of heritable reflexes could be explained only by the fact that a fixed reaction, acquired through an accidental mutation, produced a result that gave the individuals who had it, an edge in the struggle for survival.

It is important to see that the specific property or capability that constitutes the evolutionary advantage has to be incorporated in the genome *before* the conditions arise relative to which it is considered *adapted*.

Remaining aware of the role of its result, Piaget thought of a reflex, as consisting of three elements:

```
            1                  2                  3

      PERCEIVED    ----->  ACTIVITY  ---->  BENEFICIAL or
      SITUATION                             EXPECTED RESULT
```

The addition of 'expectation' sprang from the second observation Piaget had made, namely that most if not all the reflexes manifested by the human infant disappear or are modified during the course of maturation. The 'rooting reflex', for instance, that causes the baby to turn its head and to begin to suck when something touches its cheek, goes into remission soon after nourishment through a nipple is replaced by the use of cups and spoons.

Piaget also found that new 'fixed action patterns' can be developed. Such acquired reflexive behaviors are an integral part of our adult living. Among them are the way we move our feet when we go up or down stairs, the innumerable actions and reactions that have to become automatic if want to be good at a sport, and, of course, the rituals of greeting an acquaintance and of small talk at a cocktail party. There are also reflexes that may lead to disaster - for example the way we stamp our foot on the brake pedal when an unexpected obstacle appears before us on the road.

An acquired reflex that impressed me much when I was young, was the one developed by the adolescent men and women of societies that prescribed skirts for females and trousers for males. In a sitting position, these women would unconsciously spread their skirt when something was thrown to them, whereas the men would clamp their knees together. (In those days, this was still used in the strictly male monasteries of Greece and Macedonia, in order to detect female intruders. Today, they have presumably thought of another test.)

Anyway, the more sophisticated view of the reflex enabled Piaget to take the tripartite pattern of *perceived situation, action, and result* as the basis for what he called 'Action Scheme'.

It provided a powerful model for a form of practical learning on the sensorimotor level that was the same, in principle, for animals and humans.

Studies of animal behavior had shown that even the most primitive organisms tend to move towards situations that in the past provided agreeable experiences rather than towards those that proved unpleasant or painful. Humberto Maturana has characterized this by saying:

> A living system, due to its circular organization, is an inductive system and functions always in a predictive manner: what happened once will occur again. Its organization (genetic and otherwise) is conservative and repeats only that which works.
>
> (Maturana, 1970; p.15-16)

This was not intended to imply that primitive living organisms actually formulate expectations or make predictions. It was a sophisticated observer's way of describing their behavior. The pattern of learning, however, is the same as in Piaget's scheme theory, and once we impute to an organism the capability of reflecting upon its experiences, we can say that the principle of induction arises in its own thinking.

This principle has its logical foundation in what David Hume called the "supposition that the future will resemble the past" (Hume, 1742, Essay 3, Part 2). Having observed that, in past experience, situation **A** was usually followed by the unpleasant situation **B**, an organism that believed that this would be the case also in the future, could now make it its business to avoid situation **A**.

Together with its inverse (when situation **B** is a pleasant one, and **A** therefore leads to pursuit rather than avoidance), this is perhaps the first manifestation of *anticipatory* behavior.

To be successful, however, both pursuit and avoidance have to be directed by more or less continuous sensory feedback, and this, too, involves a specific form of anticipation. In their seminal 1943 paper, Rosenblueth, Wiener, and Bigelow wrote:

> The purpose of voluntary acts is not a matter of arbitrary interpretation but a physiological fact. When we perform a voluntary action, what we select voluntarily is a specific purpose, not a specific movement.
>
> (Rosenblueth et al., 1943, p.19)

In their discussion of *purposeful behavior*, they used the example of bringing a glass of water to one's mouth in order to drink. The term *negative feedback*, they explained, signifies that "the behavior of an object is controlled by the margin of error at which the object stands at a given time with reference to a relatively specific goal" (ibid.). Such goal-directed behavior, however, has another indispensable component. In order to "control" the margin of error indicated by the feedback - in the given example this would be to reduce the distance that separates the glass from one's mouth - the acting subject must decide to act in a way that will reduce the error. And nothing but inductive inferences from past experience can enable the subject to chose a suitable way of acting.

Let us look at the example more closely. I am thirsty, and there is a glass of water in front of me on the table. From past experience I have learned (by induction and abstraction) that water is a means to quench my thirst. This is the 'voluntary purpose' I have chosen at the moment. In other words, I am anticipating that water will do again what it did in the past. But to achieve my purpose, I have to *drink* the water. There, again, I am relying on past experience, in the sense that I carry out the 'specific movements' which I expect (anticipate) to bring the glass to my lips. It is these movements that are controlled and guided by negative feedback.

When I reflect upon this sequence of decisions and actions, it becomes clear that the notion of causality plays an important role in the event. All my decisions to carry out specific actions are based on the expectation that they will bring about a change towards the desired goal.

The connections between causes and the changes they are supposed to produce as their effects, have forever been the subject of scientific investigation. The concepts of voluntary purpose and that of goal, however, were branded as remnants of teleological superstition and therefore considered inadmissible in the domain of science. The advent of cybernetics and the successful construction of goal-directed mechanisms has demonstrated that the proscription was unwarranted. Yet, today, there are still a good many scientists who have not fully appreciated the theoretical revolution. It therefore seems worth while to provide an analysis of the conceptual situation.

Many years ago, Silvio Ceccato, the first persistent practitioner of Bridgman's operational analysis, devised a graphic method of mapping complex concepts by means of a sequence of frames (I have borrowed the term 'frame' from cinematography; Glasersfeld, 1974). Because no single observation can lead to the conclusion that something has changed, we need a sequence of at least two frames showing something that acquires a difference. Consequently, the mapping has the following form:

$$
\begin{array}{ccc}
f_1 & & f_2 \\[2pt]
X \; \text{----------} & \equiv & \text{----------} \; X \\[2pt]
\text{not A} & \neq & \text{property A}
\end{array}
$$

where "X" represents an item that is considered to be the same individual in both frames (indicated by the identity symbol "$\equiv$"). In short, we maintain an item's individual identity throughout two or more observational frames, and, at the same time, we claim that in the later frame it has gained a property "A" that it did not have in the earlier one (or we claim that it lost a property it had before).

The condition of identity may seem too obvious to mention, but analytically it is important to make it explicit, because of the ambiguity of the expression "the same". In English we say, "This is the same man who asked directions at the airport", and we mean that it is the same individual; but we might also say to a new acquaintance, "Oh, we are driving the same car - I, too, have an old Beetle!" and now we are speaking of two cars. In the first case, we could add, "Look, the man has changed - he's had a haircut!" In the second case, we cannot speak of change although our car is blue, and the other's yellow.

In French, the ambiguity of "*le même*" is analogous, and in German and Italian, although two words would be available to mark the conceptual difference, their use is quite indiscriminate. In fact, the situation in all these languages is worse, because common usage has modified the meaning of "identical" so that it can refer to the similarity or equivalence of *two* objects as well as to the individual identity of *one*.

Without the conception of change there would be no use for the notion of causation. It arises the moment we ask *why* a change has taken place. I have suggested that this question most likely springs from the fact that we attribute a new property (or the loss of a property) to an item which we nevertheless want to consider one and the same individual. This has the appearance of a contradiction, and we are looking for a way to resolve it.   Because the frames indicating a change represent sequential observations, such a reason has to be found in the earlier one. We therefore examine what else could have been perceived in frame 1. We may compare the experience to others we remember, in which X remained unchanged throughout several frames, and try to find something that was present this time, but not present when X did not change. Or we may behave like scientists and replicate the situation of f_1, adding one by one new elements a,b,c,..., to isolate something we can hold responsible for the change. If we find one, we can map it as follows:

```
      f₁                                      f₂

      X ---------- ≡ ----------- X

(a,b) c ------->  ≠              property A
      |                              |
      cause of the difference      effect
```

If we really are scientists, we will run all sorts of experiments in order to construct a theoretical model that shows *how* the element "c" effects the change. If we are successful, we will proudly add "c" to the tools we use in attempts to modify the world.

In everyday living, we are not so meticulous. If we find that some element was present two or three times when a given X changed in a desirable way, we are likely to assume that it is the cause, and we will *use* that element in the hope that it will bring about the desired change. Even if it doesn't, it may take a number of failures to discourage us. If someone provides a metaphysical reason why it *should* work, failures do not seem to matter at all.

I am not sure, but I think it was the literary critic Cyril Connolly who made a startling observation in this regard. Insurance companies, he said, have the most sophisticated questionnaires to assess the risks involved in issuing a life insurance policy to an applicant. The questionnaires are based on meticulous research of mortality statistics. Connolly was struck by the fact that the questionnaires never ask the question "Do you pray?" And he wondered why people continued to pray for survival in all sorts of crisis, when the greatest experts of mortality clearly had found no evidence that it had an effect.

All this involves anticipation. The use of a cause-effect link in order to bring about a change is based on the belief that, since the cause has produced its effect in the past, it will produce it in the future. We project an established experiential connection into the domain of experiences we have not yet had. Hume has explained how we establish such connections: the repeated observation that the two items happened in temporal contiguity led us to infer and formulate a rule that says, if A happens, B will follow. Therefore, if we want B to happen, we try to generate A. In other words, we have a purpose and we act in a way which, we believe, will attain it.

The psychological establishment, which, from the 20s of this century until well into the 70s, was dominated in the United States by the dogma of behaviorism, considered purpose a mentalistic superstition. "Careless references to purpose are still to be found in both physics and biology, but good practice has no place for them," Skinner still wrote in 1971.

Natural scientists in physics, chemistry, and astronomy had found no reason to engage in thoughts about purpose in their disciplines, and they relegated it summarily to the realm of teleology. As Ernest Nagel put it:

> Perhaps the chief reason why most contemporary natural scientists disown teleology, and are disturbed by the use of teleological language in the natural sciences, is that the notion is equated with the belief that future events are active agents in their own realization. Such a belief is undoubtedly a species of superstition.
> (Nagel, 1965; p.24)

To believe that the future affects the present is no doubt a superstition, but to declare that purpose and goal-directed action must be discarded because they are teleological notions is no better. It shows an abysmal ignorance of the difference between empirical and metaphysical teleology.

I have suggested elsewhere that Aristotle, who provided the most valuable analysis of the concepts of causation, was well aware of the ambiguity. In his exposition, it becomes clear that what he called 'final' cause, i.e., the embodiment of a *telos* or goal, had two quite distinct applications (Glasersfeld, 1990). On the one hand, he saw the religious metaphysical belief that there was a *telos*, an ultimate, perfect state of the universe that draws the progress of the world we know towards itself. On the other, there was a second notion of the final cause, which he exemplified by saying that people go for walks *for the sake* of their health (Physics, Book II, ch.3, 194b-195a). This was a practical explanatory principle for which there is, indeed, an overwhelming amount of empirical evidence.

In this practical manifestation of finality, no actual future state is involved, but a *mental re-presentation* of a state that has been experienced as the result of a particular action. Even in Aristotle's day, bright people had noticed that those who regularly took some physical exercise such as walking, had a better chance of staying healthy. They had observed this often enough to consider it a reliable rule. Given that they had Olympic games and were interested in the performances of athletes, they probably also had some plausible theory of *why* exercise made one feel better. Consequently, they were confident in believing that going for walks was an *efficient* cause that had the *effect* of maintaining and even improving your health. People who felt that their physical fitness was deteriorating could, therefore, reasonably decide to use walking as a tool to bring about a beneficial change in their condition.

If I use the method of sequential frames to map this conceptual situation and give *temporal* indices to the individual frames, I get the following diagram:

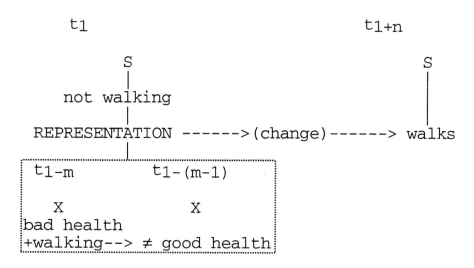

"S" is the person who wants to improve his/her health. Believing that, in the past (t_{1-m} to $t_{1-(m-1)}$), other people "X" have caused their health to improve by walking, the person decides to take a walk.

Thus, the cause of S's change of state, from not walking to walking, does not lie in the future. Instead, the operative cause of the person's change of state is a rule that was empirically derived in the past by means of an inductive inference from observations. It is a rule that is based on the notion of *efficient* cause, and the person is anticipating that it will be efficacious also in the future.

There is nothing mysterious or superstitious about this way of proceeding and it certainly is *not* 'unscientific'. In fact, it is no different from what microbiologists do when they place a preparation under the microscope *for the sake* of seeing it enlarged; or from what astronomers do when they carefully program the mechanism of the telescope to track the star they want to observe. But there is nothing particularly *scientific* about these ways of proceeding either. They are analogous to what we as ordinary people do all day long. We turn a door handle and expect the door will open, we put a seed into soil and anticipate a flower, and every time we flip a switch, we anticipate a specific effect. None of these maneuvers works *every* time, but success is frequent enough to maintain our faith in the viability of the respective causal connections.

Yet, we may call all these anticipatory procedures teleological, because they involve goals which, by definition, lie in the future for the individual actor. When I planned to take a particular train from Brussels, my goal was to be in Liège at a future moment. But it was not this future goal that would *cause* me to arrive there. Instead it was the expectation that the train scheduled to go to Liège would do once more what it had done often enough to warrant my anticipation that it would do it again.

Conclusion

One of the slogans of Skinner's behaviorism was: "Behavior is shaped and maintained by its consequences" (1972, p.16). Hence it would be affected by what happens after it. But there is no mysterious time reversal - only a specious description of what actually goes on. And Skinner was wrong to condemn teleology, insofar as he was speaking of intelligent organisms. Their behavior is shaped by the consequences their actions had in the past, and that is precisely what constitutes the 'empirical' teleology I have discussed above. As Maturana said, we "function in a predictive manner: what occurred once will occur again". In other words, we can learn from our experience and abstract regularities and rules regarding what we can expect to follow upon certain acts or events.

The mysterious feature is our ability to reflect on past experiences, to abstract specific regularities from them, and to project these as predictions into the future. This pattern covers an enormous variety of our behaviors. I would suggest that there are three related but slightly different kinds of anticipation involved.

1) Anticipation in the form of implicit expectations that are a prerequisite in many actions. For instance, the preparation and control of the next step when we are walking down stairs in the dark; this does not require the prior abstraction of cause-effect rules, but it does require familiarity with specific correlations of actions in prior experience.

2) Anticipation as the expectation of a specific future event, based on the observation of a present situation;   this is a prediction derived from the deliberate abstraction from actions and the consequences they had in past experience.

With regard to these two forms of anticipation, one may say that robots and other artificial mechanisms are today able to simulate them, if not actually, at least theoretically.

But there is a third form: Anticipation of a desired event, situation, or goal, and the attempt to attain it by generating its cause. This, too, is based on the abstraction of regularities from repeated correlations in past experience, and it is not considered 'scientific' without a conceptual model that 'explains' the cause-effect connection. But even where we have such a model, its simulation presents a problem. When we, the human subjects, pursue a goal and attempt to attain it by using regularities we have abstracted from our past experience, we ourselves have chosen the goal because we desired it. In contrast, Deep Blue, the chess program that can beat a human master and has in its repertoire thousands of cause-effect connections, does not know *why* it is playing or why it should be desirable to win.

46

The first two forms of anticipation I have listed are sufficient to say that, without them, every step we take would be a step into *terra incognita*. We would fall into precipices, crack our heads against walls, and only by accident would we find something to eat. In short, we may conclude that without anticipatory systems, there would be no life to speak of on this planet.

The third form, that involves the choice of goals, has an ominous significance. If we act in the belief that the future consequences of our actions will be similar to what they were in the past, we ought to be careful about the goals we choose. For whenever we attain them, these goal states may have further consequences, and they will be our responsibility. Therefore, if we don't heed the ecological predictions scientists compute today, we may have nothing to anticipate tomorrow.

References

Aristotle, Physics (Book II). In R. McKeon (Ed.), *Introduction to Aristotle* (translation by W.D.Ross). New York: Random House Modern Library, 1947.

Glasersfeld, E.von (1990) Teleology and the concepts of causation, *Philosophica, 46* (2), 17-43.

Glasersfeld, E.von (1974) "Because" and the concepts of causation, *Semiotica, 12* (2), 129-144.

Gruber, H.E. & Vonèche, J.J. (Eds.) (1977) *The essential Piaget.* London: Routledge & Kegan Paul.

Hume, D. (1742) *Philosophical essays concerning human understanding.* London: Millar. (In 1758, after the 4th edition, the work was called *An enquiry concerning human understanding.*)

Maturana, H. (1970) *Biology of cognition.* BCL Report No.9.0. Urbana: U.of Illinois.

Nagel, E. (1965) Types of causal explanation in science, in D.Lerner (Ed.) *Cause and effect,* (11-32). New York: The Free Press.

Piaget, J. (1937) *La construction du réel chez l'enfant.* Neuchâtel: Delachaux et Niestlé.

Rosen, R. (1985) *Anticipatory Systems,* Oxford/New York: Pergamon Press.

Rosenblueth, A., Wiener, N., & Bigelow, J. (1943) Behavior, purpose and teleology, *Philosophy of Science, 10*,18-24.

Skinner, B.F. (1971) *Beyond fereedom and dignity.* New York: Bantam/Vintage.

TOYS, TODDLERS AND THE TIMES OF TIME

P. Medina Martins
University of Lisbon

Priv. Address: R. Heliodoro Salgado, 61-2°
1170 Lisbon, Portugal

p.medinamartins@mail.telepac.pt

Abstract

Ultimately, this essay is about time or, more correctly, about the times of time. Throughout the work three of them become object of minute analysis: *physical time*, *social/cultural times* and *individual times*. Each entails not only distinguished and sometimes mutually contradictory features but gives also rise to distinct - and sometimes rather unconventional - conceptions of anticipation. Two of these conceptions are open to scrutiny. One, called *restrict conception*, can be thought of as coincident with the concept of forecast and/or prediction traditionally followed by the classical paradigm of physical/quasi-physical sciences. The other, named the *widen conception*, is prominently social and psychological and, to a great extent, its features are quite distinct from those the restrict conception is usually endowed with. For instance, it entails both several pasts and various futures, futures which, in turn, may retrospectively act upon such pasts, giving rise to different presents, etc.
The host of manifold problems that these features beget are analysed in detail and a possible unifying model that connects private times with social/cultural times and these with physical time is proposed.

THE ARGUMENT

> "What, then, is time? I know well enough what it is, provided that nobody ask me; but if I am asked what it is and try to explain, I am baffled. All the same I can confidently say that I know is that if nothing passed, there would be no past time; if nothing were going to happen, there would be no future time; and if nothing *were*, there would be no present time. Of these three divisions of time, then, how can two, the past and the future *be*, when the past no longer is and the future is not yet? As for the present, if it were always present and never moved on to become the past, it would not be time but eternity. If, therefore, the present is time only by the reason of the fact that it moves on to become the past, how can we say that even the present *is*, when the reason why it *is* is that it is *not to be*? In other words, we cannot rightly say that time is, except by reason of its impending state of *not being*."

> St. Augustine, "Confessions", Book XI, section 14

Of time there are many kinds. The most usual and known in the contemporary thought is the time measured by clocks, which maybe called *physical time*. But from the host of remaining kinds, two are also especially poignant for the purposes of this paper: one, that may be named *social or cultural time*, which - although 'public' like the preceding kind, is less known by physicists but not by historians, sociologists, archaeologists and so forth - is

CP437, *Computing Anticipatory Systems: CAYS--First International Conference*
edited by Daniel M. Dubois © 1998 The American Institute of Physics 1-56396-827-4/98/$15.00

closely related to the life and death of human societies, their rituals, traditions, etc. Finally - the last but not the least - the other type, the most primitive of them all, is a subjective and idiosyncratic time, the *private time* that belongs to you and I - and because of this necessarily different from individual to individual. This is the time of psychology in general and of depth psychology in particular (Freud, 1915), the features of which are not only the least known but also disagree almost totally with those the physical time is assumed to obey.

For the moment I shall not go into minute details on the features that specify each of these times. But, altogether looked upon, it must be worth noting that while physical time is said to be objective (because measurable), it is assumed to obey the algebraic axioms that characterise an invariant and homogeneous transformation group, to it a sharp distinction amongst the traditional divisions ('past', 'present' and 'future') is ascribed so that each of which is looked upon not only as belonging to non-overlapping classes but also as satisfying a strict deterministic or statistic succession (say, in which the 'past' is looked upon as determining the 'present' and this, in turn, the 'future') - private time is subjective, often must be regarded as heterogeneous, is far from satisfying the aforementioned linear sequentiality which physical time is suppose to obey, the duration of some happening may be greater or smaller according, for instance, with the affective disposition of its experiencing subject, and, finally, (amongst other characteristics) entails either several types of 'pasts', of 'presents' and of 'futures' or, paradoxically, various 'timeless pasts/presents/futures' (the so-called 'arrested times' that lie in the roots of those reactions 'as before' so well-known from psychic pathology). Additionally, instead of an overall temporal unidirectionality working in the sense of a 'today toward tomorrow', many of the decisions/actions each of us is taking everyday, are widely dependent on such 'futures', so that it is possible to assert that these 'futures are always retrospectively interacting with each of such 'present(s)'. Furthermore: since these 'presents' depend, in turn, on our past living experiences, then, (private) 'pasts', 'presents' and 'futures' instead of that (physical) sequentiality seem, rather, to satisfy a *cyclicity requirement* (*similar* to that Pask has assigned to the clusters of his entailment-meshes) in which the recollecting of one of such temporal divisions entails an 'weighted' recollecting of the remainder.

Being private times the most primitive of them all (social/cultural times may be looked upon as *public constructs* set up by human beings by means of shared interpreted agreements amongst individual actors and perpetuated from generation to generation), the foregoing discrepancies between the features of such an idiosyncratic time and those pertaining to physical time offer for physicists, neurophysiologists psychologists, psychoanalysts, sociologists, and so forth, a host of manifold problems to be researched and solved. For instance - and the following list is by no means exhaustive - how can we make the distinction between 'past' and 'present' or between 'present' and 'future'? Why and how, against the whole rational evidence, one can react in the present as though he/she were in the past, even being aware that these (uncontrollable) reactions are painful and distressful in extreme? Why, in some circumstances, the individual has a scattered private time? How and why both the social/cultural times and the time of physical sciences do come out from private times? Which consequences the features assigned to these private times may have upon the conventional concepts of anticipation?

The answer to some of these (and other) questions is far from being found but they surely call for approaches that the traditional perspectives (together with their computing methods essentially based on the physical paradigm) are usually unable to cope with. As a matter of fact, whereas surveying the private time, I said for instance that "the future acts retrospectively upon the present and the past, an action which, in turn, can beget new futures", I was also implicitly bringing to light not only situations in which purposiveness, intentionality and teleology are entailed but also of a *distinguo* between the time of the *adult* individual and the path each of us has followed from *childhood* until to reach a *meaningful* adult stage. But asserting this, I was too laying a particular emphasis both upon the relationships amongst some of the foregoing times and affections as well as upon all the bodily and psychological transformations each of us undergoes throughout our lives, till that our autonomy, consciousness, desires, wills or wants (among other features) maybe fully accomplished and achieved. Therefore, so stated, the question of the times of time and their cognate conceptions of anticipation becomes a problem intimately related

> i)to the interactions amongst human *qua sentient* beings (a subject matter which, with rare exceptions, is far from the usual concerns whenever time is focused),

> ii)in the aim of this symposium, to the possibility of setting up an overall relation between private times on the one hand and social/cultural and physical times on the other

both looked upon from a perspective encompassing the 'mechanism' which, in human beings *and* machines, may enfold the acquisition and diachronous transformation of private in such public times.

In order to provide answers to (some) of these questions the paper is divided into two major parts. PART I is essentially devoted to the presentation of the major features that characterise physical time and of its cognate restrict conception of anticipation. Fundamentally, this is owed to the intimate relationships existing amongst this time, the classical paradigm of physical sciences and the host of ideas that still frame the use and construction of present day computing/robotic systems. PART II treats of private time, its features, its relations both with social/cultural times and physical time. Nevertheless, its content is highly conjectural. As a matter of fact, due to its 'strange' features - many of which are far from being understood - as well as to its relations with affections, consciousness, volition, etc., far from the immediate objectives of this congress, I had to restrict the host of subject matters it entails to three of the major characteristics lying in the very nature of time itself, say, either as flow or as succession and duration. Since such a characteristics are capable of being impressed in a computer/robotic system, then, it is my hope that with the approach not only the door for the construction of a new type of psychologically autonomous machines can (eventually) be opened but - which is far more important - that our acquaintance and knowledge as growing human beings will also be enhanced.

I-THE RESTRICT CONCEPT OF ANTICIPATION AND PHYSICAL TIME

I/1-Toys, machines and the restrict conception

I/1.1-When I was a little boy I used to play with a toy similar to that depicted in Fig. 1, a replica of which, many years later, I saw described in a book written by Alfred Lotka and entitled "The Elements of Mathematical Biology". In those distant times we wonted to call it "the mechanical cockroach" (Lotka gave it the name of "walking beetle"); but, despite the difference of names, their functioning principles are exactly the same.

The system 'walked' owed to the action of two toothed wheels, one of which was an idler and the other was actuated by a spring, the gradual releasing of which was ensured by an escapement device. Another toothed wheel was disposed transversally to the direction of the driving wheel but as far as the system was moving 'normally', it did not touch the table - the cockroach's "operating zone". At its head the device had two antennae, one of which was a dummy and the other bent downwards, gliding along the table in a permanent contact with it. This operative antenna played a crucial role in the system's behaviour.

As a matter of fact, whenever the device was placed near the centre of the table, it made a straight track, attempting at reaching the edge of the table. However, when the operative antenna lost contact with the table's surface, the body of the toy - until that moment held up horizontally thanks to pressure that the antenna exerted upon it - sunk down a little, giving rise that the transverse wheel became in contact with the table.

Fig. 1

The Mechanical Cockroach

In consequence, the system rotated 90° until that the propelling wheel became parallel to the edge. This pattern of behaviour was repeated again and again and (in principle) the toy never fell from the table.

I/1.2-Beyond the enchantment the toy gave to me - perpetuated throughout time in all the computing/robotic machines I had to study and handle as cybernetician - in this paper it works essentially as an allegory for the whole problems which we have to face with.
As a matter of fact, let us assume that:

i)the device is moving along the straight line AC depicted in Fig. 2,

ii)at the time instant t=0, the driving wheel is at the zero mark,

iii)its movement is made at a constant velocity,

iv)the operative antenna measures 5 divisions,

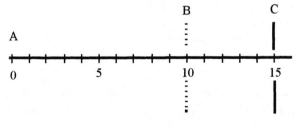

The cockroach path and its 'teleological' and 'anticipatory' behaviours

Fig. 2

v)the edge of the table corresponds to the division mark 15.

It follows from i), ii) and iii) above that - due to the assumed proportionality between space and (physical) time - the whole system works like a clock, that is to say, scale divisions to the left of, for instance, division 5, correspond to and represent *past* instants both for the cockroach and for its observers and those to the right correspond to *future* instants. Also, owed to iv) and v), if for some reason the cockroach is at mark 10 and the operative antenna fails its functioning (e.g., the transverse wheel does not touch the table), then the device will continue its movement until the division 15, where it will finish its path falling from the table. On the contrary, if the detecting system does not fail, then it is possible to interpret the cockroach's behaviour as though it were able of exhibiting not only purposiveness but also *anticipatory actions*. In other words: as though, 'aware' of the threaten that the fall would represent for its survival, the cockroach were sensing the danger and in order to avoid it, it executes the 90º rotation that change a potentially catastrophic trajectory in an actually safety path.

I/1.3-Asserting this, the reader can obviously argue that I am endowing the mechanical cockroach - or any of its highly sophisticated present-day inheritors exhibiting similar behaviours - with goals and intentions that no artificial organism can ever have. Although *partially* agreeing with the argument - later I shall be back to it - at the moment this is not the aspect I have in mind to point out. Important is to remark that such anticipatory reactions depend on the *assumed* correspondence between *points forward on the line of advance* and *future instants of time*.
For the present followers of the so-called 'hard sciences', physics in particular, or even in terms of the everyday common-sense language, this is so trivial an assumption that, with rare exceptions, its cogency is never the objective of criticism. However, when more closely looked upon - especially from the perspective of the so-called 'soft' disciplines, depth psychology in especial - it embraces some controversial issues, the content of which is far from obeying such a triviality.

I/1.4-In order to deepen some of these disputable aspects, let us begin with the foregoing correspondence between 'space' and 'time', a relationship already minutely examined by St. Augustine or Kant in philosophy, Minkowski, Riemann, Weyl or Einstein in mathematics and physics, or Piaget in psychology (amongst many others). From their perspective, each one of these *a priori* categories and/or categorical physical variables works as though it were the component member of an intertwined cluster (in Pask's sense) obeying a cyclicity requirement, that is to say, in which the recollecting of one of its elements entails, necessarily, the recollecting of the other(s).

As previously stressed out, it is far from the aim of this paper to survey the evolution that such a relationship between 'space' and 'time' underwent throughout the humankind history, especially from its scientific viewpoint. But it is interesting to point out that - even at the level of our common-sense language - whenever we are using such terms and expressions like 'inner' and 'outer' worlds, 'perspectives', 'horizons', spatial or temporal 'dimensions', etc. - we are, indeed, implicitly assuming the existence of some type of *spatial representation*, into which are inserted all the necessary *changes* that the general conception of time and becoming presuppose.
In the area to which the aforementioned 'hard' sciences belong, this 'time' (inherited from the Greek *chronos*), set up by classical physicists and still followed nowadays in General Systems Theory, computing/robotic disciplines, etc. - the time which is measured by clocks and, because of this, supposed to be *objective* as well as *publicly accepted* - obeys some assumptions easily brought to light from the cockroach's allegory. On the one hand, it is looked upon as a *linear* and *ordered sequence* of infinitely divisible 'instants', satisfying the laws of an algebraic *homogeneous* and *invariant group* which, furthermore, in the realm of that classical paradigm (classical dynamics, in particular), is also regarded as Abelian and, consequently, *reversible*. As further consequences, not only a neat distinction between the 'past', the 'present' and the 'future' in which this time maybe divided is assumed to exist but each of these terms is *in itself* also regarded as 'absolute' and 'unique'. Which I mean by this is

i)firstly, that (due to the assumed group invariance) such a 'present' is simply an instant conventionally chosen as the beginning of some experiment and corresponding to the time instant t=0. All the instants *before* it are said to belong to its 'past'. All the possible events occurring in time-instants *after* it will belong to the 'future' but no possibility of confusion exists between 'past', 'present' and this 'future'. For, owed to the aforementioned algebraic properties, these concepts belong to non-overlapping classes: a system or organism can only *be* either in the 'before', in the 'now' or in the 'after', but never in two or even three of these temporal divisions simultaneously. Finally, the number of (sequential) instants that some event lasts, corresponds to the measure of its *duration,*

ii)secondly, using the language of physics, that there is no question of considering various types of possible 'pasts', of 'presents' and/or 'futures'. Indeed, as previously referred to, the 'past' is everything that has occurred *before* t=0 (the 'present') and the 'future' everything that will occur *after* that 'present'.

I/1.1.5-Another of the most remarkable consequences of this temporal division in 'past', 'present' and 'future' is the way according to which the concept of anticipation may be described. Stated baldly, anticipation presupposes indeed:

i)that two events, say A and B, have occurred *once* at least,

ii)that these events must be distinct, not totally simultaneous and *significant* for some observer(s) and/or participant(s),

iii)that so strong a relationship have to exist between them, that the occurrence of A rises, *initially* at least, an *expectation* regarding the occurrence of B (which, whether confirmed, will in turn increase even more the 'strength' of their relationship),

iv)that, after a certain number of confirmed associations A→B, owed to the so-called *habituation phenomenon*, the strength of such expectations decreases, the observer(s)/ participant(s) will take the directed relationship between A and B as granted and whenever A occurs such observer(s)/participant(s) will be certain that B *will* also *arise*.

I/1.1.6-The reader have surely noticed that I laid emphasis upon such words as *once, significant, expectation* and *will arise*. As a matter of fact, each one of them is intimately related to a specific question which, progressively, will be approached in the course of the essay.

The first of these questions deals with the significance of the foregoing conditions i) to iv) *whereas altogether looked upon.* Indeed, are they enough to determine anticipation unambiguously or, simply, a foretell, prophesy, forecast or prediction? In other words, can they be regarded as mutually equivalent, or between anticipation and, for instance, forecasting and prediction, some distinction shall have to exist?

Getting back to the two hypothetical events A and B previously referred to, it seems at a first glance that when I said "whenever A occurs then B will also happen *later*", it can hardly be in doubt that I am asserting too that A 'anticipates' B's occurrence. But, in this sense, 'anticipation' and 'forecasting' or 'prediction' are practically synonyms - although 'forecast' is usually concerned with probabilistic or possiblistic events and 'prediction' with certainties. In either cases however - and discarding for the moment these semantic differences - what the foregoing assertion presupposes is the existence of some observer(s)/participant(s) who, *at the occasion of A's occurrence*, is(are) able of making a journey into the future and *taking into account both his/her/their previous experience(s) and the assumption that what has happened in the past still hold either in the present or in the future*, forecast(s)/predict(s) that B will occur again. To this extent, anticipation is therefore a concept essentially associated with an order relation, that is to say, "because A's occurrence is observed *before* B's occurrence, and because A is *related to* B, then A is said *to anticipate* B or, conversely, that B has occurred *after* A".

This is the concept of anticipation that I just have named *restrict conception.*

I/1.2-The Restrict Conception and Physical Time

I/1.2.1-This restrict conception is the most widely used in the disciplines obeying the classical paradigm. Physics in particular is a science the success of which in terms of predictability has been so often confirmed from the epoch of Leibniz and Newton onwards that its methods led to what Abel Rey named "the consciousness of the definitive" in science - a true ideology which has pervaded all the scientific disciplines of the ends of the XIXth century, beginnings of the XXth century - even those dealing with human beings, either individually or socially looked upon. The following statement due to P. Rousseau (1945) is an illustrative example of this whole-embracing feeling

> "Mechanics was so powerful and perfect that it had allowed the discovering of Neptune. Spectral analysis had brought about as well as extended the power of chemical research to the stars of the heavens. Maxwell had intimately connected electricity to light and the founders of thermodynamics had already finished the unification of the whole material world under the sign of energy. Ultimately (it was claimed), in the Universe there is a single type of energy, the mechanical energy, which, by means of its transformations begets all of the other types. Furthermore: since mechanics itself is related to mathematical analysis, then we may conclude that the whole Universe - not only physical but also dealing with all of its living beings, individually and socially looked upon - can be mechanically explained"

I shall not go into minute analyses of the standpoints on which these methods rest, already examined in other works (see, for example, Martins, 1993); but two of them deserve, for the moment, further attention.

The first of such tenets maybe called *the principle of spatial extension* and cursorily stated, what it presupposes is that - once the relation or 'natural law' that underlies the behaviour of some system is brought about - then, such a relation *discovered here* is supposed to be extended to all the systems identical to the one under scrutiny *wherever their location may be*. This is tantamount to saying that, at a macroscopic level at least, such 'natural laws' are referred to *logically open classes*.

The second of those underlying cannons (that, similarly, maybe named *the principle of temporal extension*) lays emphasis upon the temporal, diachronic, evolution that the scrutinised system may undergo. Everything happens, indeed, as if some result obtained *now*, e.g., to say, during every present-day research period, were able of being integrally transposed to the past, *with no sort of criticism.*

This common extension - widely used in physical disciplines - subsumes, once more intimately examined, an (unconscious) conception of Nature which, *structurally* at least, is regarded as something eternally unchangeable, something similar perhaps to the *Logos* of Heraclitus, that passively waits for the progressively more elaborate skills that we, human beings, are always discovering/creating as means to discover its 'hidden secrets'. Besides its philosophical grounds (which I shall not discuss here), the idea that ever since has framed this temporal extension, particularly from the epoch of Galileo, Kepler, Descartes, Newton, Leibniz, Euler and Lagrange (among others) until quite recently, is related to the possibility of obtaining a set of differential equations (one for each variable deemed 'relevant' for the description of the system under scrutiny), the solution of which, once the initial or boundary conditions are known and their integration is possible to be accomplished, will (in principle) determine the system's state at *any desired, past, present or future time instant*. Although, *generally speaking*, this assertion may be looked upon as a truth of fact in the realm of classical dynamics, it can however undergo some severe criticisms owed to the premises and limitations that frame either the search for such a solution or the solution itself.

I/1.2.2-The first of such limitations deals with a well known fact, especially poignant in the area of the so-called 'soft' sciences which, briefly, deals with the existence of systems for which most of such 'relevant variables' are hardly defined in numerical terms. For example, how can we ascribe a numerical value to 'identification', e.g., the psychological mechanism which (amongst other factors) confers 'cohesion' to a community, society or culture? By the same token, how can we estimate the 'strength' of a certain 'religious belief' or, yet, the social 'tensions' or social 'forces' to which some society is being submitted in a certain moment of its existence?

I/1.2.3- But, even assuming that the previous questions has been overcome - and in present-day times the so-called fuzzy approaches already allow that a great many deal of such definitions maybe settled - a host of other cognate problems can be brought to light. For, a differential equation is ultimately nothing but a *general* (equilibrium) *relation*, set up between the numerical values that those relevant variables referred to beforehand take between two time instants, say t_1 and t_2, where $t_2=t_1+\Delta t$ and Δt is assumed to be a period of time of infinitesimal duration. As a parenthetical commentary, it is worth noting that an identical reasoning can be integrally applied to problems entailing two infinitesimally neighbouring points of the Euclidean space as, for example, it happens with the difference equations that describe the lateral inhibition between neighbouring neurons. In either case, however - and this the aspect I bear in mind to stress - this idea of setting up a general (differential) relation entailing the variables deemed relevant for the specification of the system under scrutiny which, whether integrable, allow the knowledge of the state of the system at any desired, past, present or future instant, requires that some further, albeit underlying conditions are satisfied:

I/1.2.3.1-On the one hand, *the cogency of such a relation has to be assumed to hold during an extended time-interval* (mathematically corresponding to the integration interval) in so far as the integration of the equation is being performed. Phrased metaphorically, this is tantamount to saying that the truth of some proposition uttered *here* and *now* about something or someone, will be regarded *forever* true *whatever changes that something or that someone may undergo in the meantime* - an abductive/inductive assertion far from being correct either for *the whole* interval and/or for *the whole* known systems. Of course, some systems exist where we previously know that the time-invariance of their relevant variables does not hold (yielding the rising up of the so-called *time-variable* differential equations). But, here again, a crucial aspect has to be bear in mind: that, *unless the law ruling such time-variability is known, the equation is not integrable*. In other, more concrete but familiar words: the past or the future state of the system can, no longer, be predictable;

I/1.2.3.2-An identical situation occurs whenever the time-variations that the system undergoes are so fast that, in the interval of time eventually needed to perform the (possible) integration of the equations describing its behaviour, the aforementioned 'initial' or 'boundary' conditions have already radically changed their values. *Thus, even assuming that such an integration is possible, its result becomes thoroughly meaningless for the understanding of the system's behaviour.* Problems of this kind occurred by the ends of the XIXth century when - owed to the undeniable successful achievements reached until there - the methods of the classical paradigm were supposed to be applicable to the description of more and more complex systems. Particularly, of systems composed by a great number of elements (like those dealing with the molecules of a gas) entailing so complex an interaction that the physicists of the epoch were forced to realise, firstly, that no *individual* prediction was possible to be accomplished and, secondly, that if they intend to continue their search for describing the (thermodynamical) behaviour that such a type of systems exhibit, then, they had to change radically the whole perspective on which the dynamic standpoints had been based. This was just the work that Gibbs and Boltzmann performed, begetting the birth of a new physics based upon statistical methods, the so-called *statistical physics*;

I/1.2.3.3-Presently, statistical methods are so commonly used either in physics or even in a great deal of soft disciplines that most of their users no longer think of the premises and consequences on which they rely and which they yield. However, for the further comprehension of the major objectives of this paper, there are three aspects of it that undoubtedly deserve a further development. Two of them have already been stressed in other works (M. Martins, 1994, 1997) about the limitations that the classical paradigm evinces whenever its extension to the realm of soft sciences is attempted: which I have respectively called *indiscernibleness* and *singularity*. The last, brings to light the contemporary controversy *reversibility/irreversibility*, an endless dispute between the two conceptions of time that are supposed to frame either (classical) dynamics or thermodynamics.

I/1.2.3.4-Let us, in this sense, beginning with a cursory survey of indiscernibleness. Translated into the (classical) language of General Systems Theory, which this term ultimately signifies when referred to (most of) soft disciplines, is that every human being can no longer be regarded as the *qua sentient* being that *each of us effectively is*, say, a *personalised* and *unique* entity, altogether distinctive from all other beings to whom each one of us interacts with, but, on the contrary, as a simple element of a mass of *indistinguishable* elements - an '*it*' that the mechanist conception has attempted at reducing all entities of Nature. Obviously, in the realm of the systems that 'hard' sciences treat, there is nothing wrong with the identification between the 'object'/'system' under scrutiny and this 'it' as well as with the exclusion of such an uniqueness. For, as previously stressed - from the XVIIIth-XIXth centuries onwards, when the historical seeds of the mechanist idea began to grow and Nature was no longer regarded as an organism but, rather, an impersonalized machine whose components were assumed to interact sequentially - the success of such a mechanical *Weltanschauung* led to so profound civilizational achievements that, presently, no one can ever deny. However, in the realm of soft disciplines as, for instance, psychology or history, in which *this* or *that real* human being as well as their interpersonal relationships *have to be* actually treated, the ideas framing such paradigm have undoubtedly begotten some regrettable consequences;

I/1.2.3.5-One of the aspects in which these consequences has shown to be more restrictive deals with the (mathematical) treatment of those *singularities* mentioned beforehand and just concerned either with the use of terms like 'this' or 'that' whereas applied to human beings or with the aim and scope of disciplines like history and psychology which, oppositely to hard sciences, are primarily referred to logically *closed* classes (say, instead of dealing with 'human beings' in general, they are, on the contrary, referred to the actions that 'this' or 'that' human being has performed). A great many deal of such a treatment comes from two of the utmost important tenets on which the classical framework lays down: that a sharp distinction between the so-called 'object'/'system' and its 'environment' is *always* possible of being set up by some external, neutral and omniscient observer, and that the observer's searching for the 'law' mastering the system's behaviour (distinguished from the whole flux of events that he/she is able to recognise due to some feature or property he/she deems relevant) consists, summarily,

      -in defining which are the 'causative input(s)' that act upon it ('inputs' that, in turn, are supposed to come either from the environment or from the observer/experimenter himself/herself) as well as the respective 'output effect(s)',

      -in attempting at determining a *relation* (or 'natural law') between such causative input(s) and its(their) output effect(s).

In order to achieve this objective, the observer looks upon the object/system under scrutiny as that impersonal 'it' previously referred to - ultimately nothing but a 'machine' - *the outcomes of which are, consequently, assumed to be known beforehand.* As a further mechanist assumption, he/she is also (supposed to be) able to entertain deterministic and/or probabilistic hypotheses about the occurrence of such outcomes. Next, based upon these framing assumptions and following a strict and predetermined set of rules, the observer carries out 'experiments' upon the system, this in such a way that all but one of the causative inputs (usually regarded as independent variables) are supposed to hold their values. The numerical variations of this input are next related to the observation of one or few of the output (numerical) effects (usually looked upon as the 'dependent' variables). As a further, albeit mandatory condition seeking that the experiments can be exactly replicated again and again, all the other possible fluctuations that the system may undergo, deemed irrelevant, are minimised. Finally, if the initial and/or boundary conditions are known, if the integration of the set of total and/or partial differential, integro-differential or difference equations (in so far as time is assumed to be continuous or discrete) describing the system's behaviour are possible to be accomplished, and if the results obtained *correspond* to the actually measured data (thus, confirming or validating the entertained hypotheses), then, the state of the object/system (as well as of all other identical systems) at any desired past, present or future time-instant, becomes (in principle) *unequivocally determined.* From here proceeds the famous Leibnizian 'Law of Sufficient Reason', whose content sets up "the equivalence between the 'full' cause and the 'complete' effect".

Soon I shall be back to this deterministic statement which, using well-known common-sense words, asserts that "in the same conditions, the same (efficient) causes beget the same effects". For the moment however, it is worth noting that although, on the one hand, it do allows predictions/restrict anticipations of the possible behaviours that *some* systems may exhibit either at individual or social levels, the whole process also severely hinders, on the other hand, the emergence of those singularities the discussion of which we have just been mooting so far.

Individually for, whenever we are dealing with human *qua sentient* beings, not with machines (as the classical mechanist paradigm forces us to look upon), we have, willy nilly, to take into account unexpected *autonomous* or *operant responses*, e.g. those fluctuations just referred to beforehand, that the classical observer/experimenter minimises or even bears in mind to abolish thoroughly, because 'irrelevant' for his/her confined purposes.

Socially, for no human society has ever changed without the emergence of such 'singularities', historically, sociologically and psychologically translated by the sudden and unexpected appearance of, for example, some charismatic personality, scientific or technological discovery, natural catastrophe, etc. Hence - and this is the partial conclusion I would like to stress from these cursory considerations - although statistical methods have to be used whenever the description of *some* systems are in question - their widen application to the whole content of soft disciplines as, for instance, psychology, sociology or history, often begets regrettable results.

I/1.2.3.6-This reference to history, sociology and psychology is important for it just enables us to provide a brief account of the third of the aforementioned subjects which - together with indiscernibleness and singularity - is related to the problems that the emergence of the new physics of Boltzmann and Gibbs gave rise: what I had named the controversy between the reversible time of (classical) dynamics and the inherent irreversibility related to thermodynamical systems.

Obviously, the objectives I have in mind to accomplish and the unavoidable limitations in terms of the space I have at my disposal in the present work, do not allow that the historical/scientific drama that Boltzmann experienced in his attempt at explaining such an irreversibility according to the classical tradition, can be fully described. Nevertheless, since the question points directly towards the very heart of this paper - particularly towards the times of time and their relation both with human beings and machines - then, it becomes mandatory required that the subject must, somehow, be stressed. As a matter of fact, the problem that all of such disciplines share - in which biology itself can, by no means, be excluded - touches the very gist of two world-conceptions already discussed since the Classical Antiquity: are the changes everywhere and every time witnessed, part and parcel of a more general *cyclic* world-embracing transformation (of which such changes are simply *local* expressions) so that, after a more or less extended period of time, everything will be repeated again and again? Or, on the contrary, neither such a cyclicity nor that repetition exist, so that nothing and no one can ever return to the past? Furthermore: assuming the truthfulness of this last conception, then, how can knowledge itself be ultimately achieved?

For the Greek and Medieval thinkers, as well as for all of their Renaissantist inheritors until the XIXth century, the answer to this second question offered no doubt: such observed changes are simple sensitive appearances of something *eternally unchangeable* lying behind those bare appearances, the role of science - 'natural science', of course - being just devoted to the search for such an hidden 'something'. Ultimately, this 'something' is nothing but the host of (supposedly) *eternal* 'forms', relations or natural 'laws' using the present-day terminology, which, whether found and experimentally confirmed, will reinforce not only that conception of unchangeability but also the immutable character ascribed to their truth.

Of course, this did not mean that the changes everyday observed and witnessed were denied. However, following the cyclic tradition, all the transformations from some state, say A, to another state B, were always looked upon as part and parcel of a process which, whether able of being completed, would only finish if and only if the inverse transformation - say, from B to A - were also accomplished. Even before situations where this last transformation was far from being clear as, for instance, in the case of the ageing of living organisms, the Greek and Renaissantist thinkers regarded them as parts of a more general changeable process, that, whether capable of having been fully completed or whether totally observable, would also be looked upon as obeying the cyclic assumption.

One of the most important physical consequences that this conception gave rise, can be seen in the variational formulation of the energy principle - the so-called Hamilton's principle - the utmost general mathematical expression that, in itself, condenses the whole dynamical phenomena (Lanczos, 1952),

$$\delta \int_{t_1}^{t_2} \left[T(t) - V(t) \right] dt = 0$$

As a matter of fact, if the kinetic energy $T(t)$ is supposed to be a quadratic function of the velocities, and the 'work function' or potential energy $V(t)$ is assumed to depend on the position coordinates, it is easy to see that the integration of the foregoing expression yields the same numerical result either with $+t$ or with $-t$. This is tantamount to saying that if the motion of the system is supposed to begin at t_1, with a velocity v_1 and a kinetic and potential energy respectively given by T_1 and V_1 and if, furthermore, such a motion is assumed to end at t_2, with velocity and energies respectively given by v_2, T_2 and V_2, then, after the inversion of the signal of t, everything happens as though the motion were looked upon as beginning at the state (v_2, T_2, V_2) and ending at (v_1, T_1, V_1), all the intermediate states of this last path corresponding, *by an inverse order*, to the intermediate states of the direct motion.

Two major consequences (amongst others not relevant for this work) have been extracted from this result:

i)on the one hand, a conception of the natural world altogether mastered by a strict causality (and determinism as well), in which *the* past of some system is supposed to determine unequivocally *the* present of such a system and this, in turn, frames inexorably *the* system's future,

ii)on the other, the discrepancy existing between the temporal symmetry inherent to the theoretical construct of classical dynamics - mathematically translated by an invariant transformation regarding inversions of the time's signal (reason as to why I previously asserted that *this* concept of time was supposed to obey the axioms of an Abelian group) - and the temporal asymmetry that practically all the actual physical and biological phenomena evince. Here lies the *phenomenological* distinction between reversibility and irreversibility, so as stated by Carnot and Clausius: the former correspond to all of those situations in which the initial state of the system is assumed to be restored, the latter to those in which such a state can never be achieved, *even using all the possible physical means*. This is a veridical fact, especially notorious in the realm of biological systems. As a matter of fact, similarly to the death which is not the exact reversal of the birth, or to the anabolism which is not the opposite of the catabolism, also the cells' division does not obey a symmetrical temporal pattern. Hence, the reason as to why, metaphorically speaking, *biological* beings (individually and/or socially regarded) are said to be *unidirectional time arrows,* a picturesque expression created by Eddington (1929).

I/1.2.4-Owed to the necessarily general character assigned to this incursion in the realm of statistical mechanics, the analytic demonstration of the results to which Gibbs and Botzmann were led become, obviously, thoroughly meaningless. Two aspects deserve however further development. The first of them deals with the approach they followed in order to treat of those partially/totally (un)observable and partially/totally (un)controllable systems such as those the complex interactions witnessed amongst the molecules of a gas are good representatives. Indeed, as previously stressed, whenever the methods of classical dynamics are used, the behaviour that such systems exhibit is not predictable, due not only to the countless number of differential equations needed to be integrable but also to the total or partial unknowledge of their initial and/or boundary conditions. These were just the starting points of Gibbs' statistical approach.

His basic idea was that a (macroscopic) system entailing countless microscopic elements could be regarded as a host of subsystems so that, each of them, can yet be looked upon as being macroscopic, although, on the other hand, the number of elements it embraces is a minor fraction of the whole elements by means of which the global system is constituted. Therefore, each of such subsystems can no longer be regarded as isolated as it was assumed in classical dynamics. On the contrary, it is permanently undergoing the most variable actions from the side of the other subsystems. Furthermore: due to the great number of elements these subsystems encompass, those actions will have an extremely complex character. So, during a long-time interval T, the representative 'point' of the instantaneous state of the subsystem under scrutiny will therefore pass through an infinitesimal volume $\Delta\Omega$ of a 2n dimensional space (the so-called Gibbs' *phase space* or Ehrenfest' *μ-space*)

$$\Delta\Omega = \Delta p \; \Delta q$$

where q_k are generalised coordinates, p_k (k = 1, 2, ..., n) generalised momenta and

$$\Delta p = \Delta p_1 \, \Delta p_2 ...... \Delta p_n$$

$$\Delta q = \Delta q_1 \, \Delta q_2 ...... \Delta q_n$$

Let us assume that Δt ($\Delta t \ll T$) is the infinitesimal time-interval during which the phase-point is within $\Delta\Omega$. Therefore, when T increases indefinitely, the relation

$$\omega = \lim_{T \to \infty} \frac{\Delta t}{T}$$

will tend for a finite limit which maybe interpreted as being the probability of the subsystem's state staying within $\Delta\Omega$. Now, when this volume element becomes infinitely small with dimensions dp dq, it is possible to introduce a probability $d\omega$ so that the values of q_k and p_k stay within the intervals (q_k, q_k+dq_k) and (p_k, p_k+dp_k). This probability may be written as

$$d\omega = \rho(p_1, p_2, ..., p_k; q_1, q_2, ..., q_k) \, dp \, dq$$

where ρ - tantamount to a *probability density* in the phase-space obeying the normalisation condition

$$\int_{2n} ... \int \int \rho \, dp \, dq = 1$$

is the statistical distribution function of the considered system It is now clear how this distribution function becomes independent not only of the initial state of any part of the system (since the possible influence of such an initial state during a long observation interval becomes insignificant whereas compared with the influence of the other greater parts) but also of the initial state of the subsystem under scrutiny: As a matter of fact - and this a quite important result - *if, during the long observation period T, its representative phase-point is suppose to cross all the possible states then , any one of them maybe looked upon as being the initial state*. Which this means ultimately is that there is the possibility of determining the statistical distribution for the minor parts of the system without the need for solving the mechanical problem for the entire system.

From these brief considerations it is possible to see that the major problem that the statistical approach has to solve, consists in finding the distribution function characteristic of some (physical) subsystem. Here, however, the meaning assigned to the term 'subsystem' is quite different from that used in classical mechanics. As a matter of fact, all the macroscopic bodies which, for instance, we handle everyday, maybe looked upon as being small parts ('subsystems') of an isolated universal system, comprising both an 'outer' environment and all of such bodies. If its distribution function is known, then it becomes also possible to estimate the value of the different physical magnitudes that depend on the state of the subsystem under scrutiny. The same argument holds good for the estimation of the mean value, say $\bar{f}$, of some variable f(p, q), which is given by the integral (generalised to all the states) of the product of all of its possible values by their corresponding probabilities

$$\bar{f} = \int f(p, q) \, \rho(p, q) \, dq \, dp$$

I/1.2.5-From the manifold conclusions that may be extracted from the foregoing considerations, three deserve further emphasis. Firstly, it is possible to see that statistical mechanics allows forecasts/predictions about the behaviour of macroscopic systems whose features have a random character. This is a result that sets up a neat cut regarding the deterministic prediction characteristic of classical dynamics in which - as pointed out beforehand - the whole future of the system is inexorably determined by its past. Such a random character assigned to macroscopic bodies - and this is the second aspect deserving attention - is not evident by itself. This is due to the fact that, for time-invariant environmental conditions and for long enough time-intervals, the relative fluctuation of the magnitude of some variable f is inversely proportional to the square root of the number of particles of which the body is composed. Being this number exceedingly great, f may be regarded as remaining practically time-invariant, its value coinciding with its mean value $\bar{f}$. When this situation holds for all the macroscopic variables that specify some subsystem of an isolated system, the system is said to be in a state of *statistical equilibrium* (thermodynamical or thermical). But if, owed to some external action, the system leaves this equilibrium state and next, such action ceases its influence, then after a greater or lesser period of time (the 'relaxation time') the system will tend for reaching a new equilibrium state. This is a result which brings to light the third - and for the purposes of this essay the most important - of the aforementioned consequences: how the statistical approach to mechanics did allow that the second principle of thermodynamics has left its relative isolation in the context of classical physics to become a true analytical principle, reducible to a probabilistic proposition.

One of the arguments that, in the epoch, was used to justify such an isolation relied upon the anthropomorphic character ascribed to the concept of irreversibility implicitly contained in the interpretation that Kelvin or Clausius (amongst others) ascribed to the second principle of thermodynamics

"A transformation whose final result is to transfer heat from a body at a given temperature to a body at a higher temperature is impossible" (Extracted from Enrico Fermi, *Thermodynnamics*, 1936)

So stated, what this phenomenological definition subsumes indeed, is the impossibility *from the side of the human experimenters* in accomplishing some natural transformations like, for instance, of heat in work, without compensation.

This statement can, nowadays, be scientifically accepted; but this happens so, only because behind it there exists an analytical *substractum* just provided by Boltzmann when he set up the relationship between entropy and probability. Stated baldly what this relation emphasises is that all the transitions that a system satisfying the aforementioned thermodynamical conditions undergoes, are made toward an increasing of the (overall) entropy of such a system; or, which is more familiar, that the transmission of heat from a body with a high temperature to a body with a lower temperature is a consequence of the fact that an uniform distribution is more probable than a non-uniform one.

Once again it is meaningless to bog either down the formal demonstration of Boltzmann's results or down the whole consequences possible to be extracted from them. It suffices however to stress that - *under the categorical hypotheses according to which the reasoning of Gibbs and Boltzmann were framed* - such results were, firstly, able of solving the discrepancy existing between the irreversible character that the time-evolution of actual material systems undergo and the (supposed) reversibility such an evolution seems to satisfy from a pure mechanical perspective (an objection which was brought to light by Loschdmit); and secondly, for having demonstrated that, throughout time, the (overall) entropy of all the natural and/or artificial systems always tends either to increase or to remain constant but never to decrease. Furthermore: in so far as this entropy increases, the Universe and the whole of its closed systems tend to degrade themselves and to pass from a state of organisation, order and differentiation to another state of *maximal probability* where the chaos, uniformity and non-differentiation are prevailing.

From the whole of these systems some exist however which, due to their inner organisation, are *able of opposing themselves* - albeit in a restrict and temporary fashion - to this degrading trend. Biological organisations in general and *human beings in particular* are, by excellence two typical examples of such systems.

I/1.2.6-The reader must have surely remarked that in the above statement I purposively laid emphasis upon some expressions from which "under the hypotheses that framed the reasoning of Gibbs and Boltzmann", "maximal probability" and "human beings in particular" are especially poignant. Regarding the first of these propositions, I believe that the following statement - a free translation of one of Boltzmann's main assertion extracted from *Entre le Temps et l' Éternité*, of I.Prigogine and I. Stengers (1988) - is clear enough for the justification of his results

> "We may chose between two types of representation. Or we assume that the universe is presently in a rather improbable state or, on the contrary, we shall have to suppose that the eras that measure the duration of such an improbable state and the distance from here to Syrius are insignificant whereas compared with the age and dimension of the whole universe. Within such a universe which, altogether regarded, is in a state of thermical equilibrium and so death, relatively small regions with the dimensions of our galaxy (which we may call "worlds") can get far from such a state of thermical equilibrium during time intervals which, at that scale, maybe looked upon as being rather short. Amongst these worlds some exist whose states will have an increasing probability (say, increasing entropy). Likewise, others will have decreasing probabilities. So, in this universe, there is no possibility of making distinctions between the two time-directions, just like it happens with space, where we cannot say that no 'above' or 'under' exist....It seems to me that this is the only way we have of reconciling the cogency of the second law and the thermical death of every individual world without the need for calling for an unidirectional change of the whole universe, from an initial to a final state."

However, regarding the second and third of the foregoing propositions the situation is slightly different. On the one hand, because speaking of probability - and having no intention at all of debasing all the physical results where probabilities are indispensable - one additional and mandatory condition exists which has to be satisfied by some of the happenings that Boltzmann and his followers had to treat of in order that they can *actually* be regarded as an 'event' (eventhough we look upon it as a mistery): that some *meaning* has to be assigned to it for, otherwise, such events work as though they were opaque, impenetrable and even non-existent. But asserting this we cannot, on the other hand, no longer speak of biological organisations *alone*, say, of quasi-physical systems that, due to their inner structure, are able of opposing themselves - albeit in a restrict and temporary fashion - to the aforementioned (overall) degrading trend. On the contrary, what meanings and meaning-assignment do require is, minimally,

> i)interactions amongst human beings, exchange of perspectives, interpretations, mutual understandings, agreements or conflicts, etc., all of which belong to the *psychological realm* rather than to the pure physical or biological domain,

> ii)throughout time, a *change of such interpretations, of their understandings of their entailed perspectives, etc.*, without which there would be neither evolution nor possible meaning assigned to its characteristic events as well as to the emergence of new coherences (which, in turn, will modify the evolutive sense and so, the meaning of its entailed happenings, etc.).

If we add to these conditions the idea that there are *time-oriented uncontrollable processes*, so that the 'before' can no longer be re-reached from the 'after', say briefly, that there is no symmetry between the events these two terms represent, or, which is similar, that *irreversibility* is an overwhelming feature of the whole world we live in - we have just been led to the three minimal requirements that Prigogine (1988) claims as mandatory conditions for the existence of some kind of 'evolution'.

Obviously, when he was speaking of such an evolution he had essentially in mind the transformations witnessed in physical/chemical or biological systems - those that, instead of obeying Botzamnn's blind tendency for falling down chaotic and unorganised states are, on the contrary, sources of new types of organisation, emergence of new properties, etc. In Boltzmann's epoch neither the existence of these non-equilibrium structures (of which Prigogine himself was the leading researcher) had yet been discovered nor the historical character assigned to the whole universe had also been accepted. But, for this paper, this is not matter of fact. Important is, on the contrary, to lay emphasis upon the fact that whereas so far I have just followed Prigogine's ideas, from now on our ways become divergent. Indeed, whereas he has devoted his attention to the evolutionary character of the physical/chemical or biological world, I have just looked upon the times of the human *qua sentient beings*, to their development, to the psychological transformations each of us undergoes from childhood to adulthood (yielding, amongst other features, the emergence of the aforementioned meanings), etc. However, as already pointed out, this perspective brings to light not only a conception of time different from the physical one but also totally distinguished concepts of anticipation. The analysis of these subject matters will be the content of the following items.

II-THE WIDEN CONCEPT OF ANTICIPATION: PRIVATE AND SOCIAL TIMES

II/1-Private Times

II/1.1-Toys, machines an the widen conception

II/1.1.1-The reader is certainly remembered that, when in I/1, I exposed the major features by means of which the behaviour of the mechanical cockroach was specified, I also assumed some simplificative hypotheses (usually followed either by physicists or by the designers of computing/robotic machines) amongst which two were especially poignant:

> i)there exists a proportionality between space and (physical) time, giving rise that the whole system works like a clock. From this proviso, scale divisions to the left of some specific reference mark represent *past* instants for the observers' mechanical cockroach and those to the right correspond to *future* instants,

> ii)if the functioning of the detecting system - represented by the operative antenna whose length measures 5 divisions - does not fail, then the cockroach's behaviour maybe interpreted *as though* it were able of exhibiting *purposiveness* and *anticipatory actions*. As a matter of fact, if the device fails the detection of the table's hedge then it certainly would fall from the table. Everything happens therefore, as though '*aware*' of this danger - as well as of the '*threaten*' that the fall would represent for its '*survival*' - the cockroach

were forced to execute the 90º rotation that changes its potentially catastrophic trajectory in an actually safety path.

Both of these hypotheses can now be examined from a more detailed perspective. On the one hand, the assumed relationship between space and time do entails a distinction between *a future that may be* - depicted instant by instant by successive points in the line of advance of the cockroach (based in turn on the supposition that the law of motion remains unchanged) - and a *future that will be*, being this depiction upon the suppositious future that determines for the cockroach (or for its observers) its hypothetical or actual behaviour.

Here lies one of the major similarities and (consequently) differences between the cockroach-toy or its present-day sophisticated computing/robotic inheritors and their human designers/observers. For - *at a first sight and laying emphasis upon actions only* - it seems that no categorical distinction can be settled between the behaviour of the cockroach-toy and that one of human beings, at least in terms of a searching for survival. Indeed, human beings, as conscious adult creatures, do usually *act*, firstly, by construing a mental re-presentation of what we seek to avoid or to attain in the future and, secondly, thanks to our volition, by accomplishing a series of directed actions so that our preconceived aim may *in principle* be fully achieved; this based on the supposition that no 'inner' or 'outer' interferences exist. For us, this is the 'future that may be'. Nevertheless, since such interferences do usually exist, then the *imagined* course of the events is often modified. This requires not only the awareness of the differences between the imagined and the actual situations but also that many decisions are taken seeking (whether possible) to correct them. If not achievable, then such differences will beget a 'future that will be' not necessarily coincident with the 'futures that might have been'. But under these caveats, say, taking *solely* into account those 'outer' actions previously referred to, the "law of preconceiving" becomes just the same in both cases, at any rate in the simple case in which the reaction is one of avoidance of an unfavourable condition.

Of course, *altogether looked upon*, say, embracing both the observed actions and the (possible) psychic 'mechanisms' they underlie, it can hardly be in doubt that between the cockroach-toy and our own processes as human *qua sentient* beings, manifold distinctions exist. For, as Lotka asserts (1956) speaking of his "walking beetle":

> ...Little doubt enters our mind in construing the course of events in these two cases. We are directly *conscious* of our own *volition* (whatever its precise physical significance may be). We hesitate not at all in describing our action as *purposive*, as directed to and determined by a *final cause*. As to the tin beetle, we have dissected him and fully understand his mechanism. We would think it foolish, with our peep behind the scenes, to impute to him volition or purpose; we describe his action as mechanical, as fully determined by an efficient cause.

II/1.1.2-In this statement I purposively used italic in the words "conscious", "volition", "purposive" and "final cause". And this because none of them belongs to the realm of physical/quasi-physical disciplines (of which biology itself is a representative example) but rather to the domain of a psychology. Or, more generally, to the domain of a *relativist/ normative paradigm* rather than to the classical paradigm that frames the aforementioned physical/quasi-physical disciplines. Such normative paradigms - introduced into the contemporary science by several authors, amongst whom Gordon Pask and Heinz von Foester were especially relevant - have already been minutely scrutinised in so many works that their presentation is here thoroughly meaningless. Only one of their features deserves further attention: the extreme importance that Pask has ascribed to the individual *qua sentient* being (who Pask denotes as *M-individual*, someone like you or I), to his/her uniqueness and idiosyncrasies, to the private character and highly personalised interpretations ascribed by each one of us to each of our living experiences, etc., all of them features that set up a neat contrast regarding the massifying characteristics assigned to the indiscernible 'elements' pertaining to the classical paradigm.

But looking upon each individual from this prevailing perspective, Pask was obviously forced to justify how we can understand each other whenever we mutually interact, particularly when are engaged in conversation. And the answer he provided consist in setting up a distinction between the foregoing M-individuals and what he called *P-individual*. Primarily, although not exclusively, a P-individual is a collective entity comprising several M-individuals (examples of which are schools of thought, believers in the same religious dogmas, etc.) who - and this is the very gist of Pask's theory - share amongst themselves a commonalty of interpreted perspectives so that, whenever one of such M-individuals, say $\mathcal{A}$, is speaking about some subject matter, his/her listeners $\mathcal{B}$, $\mathcal{C}$, ... etc., are able not only of *interpreting* which $\mathcal{A}$ is uttering (using their own words, concepts, meanings and ideas) but also of *understanding* him/her. This 'shared agreement' and the means to reach it (that includes both the agreement to disagree and the possibility of conflicting), is one of the keystones of Pask's Conversation Theory (CT for short), an approach that plays a prominent role in the further development of this essay.

II/1.1.3- As a matter of fact, if - instead of starting from the widely spread idea of time inherited from physical sciences - we begin with the opposite side, questioning ourselves how the concept of time arose in each of us and how this *private* conception gave progressively rise to the *public* time of physics, then, we suddenly become faced with a host of manifold problems, many of which are still unsolved.

It is far from my present concerns to discuss these problems minutely. But two examples suffice to show how private times and the time of physics exhibit so distinct a features. The first of these examples deals with the assumed sharp distinction between 'past', 'present' and 'future' (underlying the concept of its 'linear sequentiality'); the second, with the concept of 'duration'. Suppose, indeed, that A and B are two events and that A is happening in the present of some *adult* observer/participant. Assume also that, whereas A is lasting, this observer/participant recollects not only that A's occurrence has always been associated with B (e.g. that B will happen later) so that, plausibly, not only a picture of B will arise in his/her mind but the observer/participant will also feel himself/herself *as though* he/she were *for a while* at the occasion of B's occurrence, say, *in the future*. Briefly: as if, momentarily, such an observer/participant were *both* in his/her present *and* in his/her future. A similar reasoning can also be integrally applied not to 'present' and 'future' but, rather, to 'past' and 'present'. Indeed, psychology and psychoanalysis are full of situations in which, owed (for example) to the intensity of the affections related to the personal interpretation of some event, its recollecting *here* and *now* is felt as *before*, yielding (sometimes many years later and against the whole rational evidence) dramatic reactions of, for instance, irrational fear. A fear that, possibly, had been justified when one was a defenceless and helplessness child but, when this individual is an adult (chronologically at least), it has no actual reason at all to exist. Under this caveat, however, one the crucial standpoints of the physical/mathematical concept of time - that 'past', 'present' and 'future' belong to non-overlapping classes - shows to be a failure in the realm of private times.

By the same token, when physicists speak of the 'duration of some happening' as something that objectively measures the instants of time elapsed during the happening itself, they are usually forgetting that such a duration is highly variable not only from individual to individual but also, *in the same individual*, according for instance to his/her emotional and/or cognitive state. The reader who has experienced an earthquake knows how the 'time' seems to last an eternity, eventhough - whereas measured by a clock - it has only last some few seconds. On the contrary (and not referring purposively to all the physiological alterations that some drugs provoke) the pratictioners of some Eastern meditation techniques do also speak of the 'speeding up of time', in which hours are felt as though they were lasting few seconds only. The traditional description of the flower that is being seen flourishing is so well known that needs no further explanation.

II/1.1.4-These differences between the private and subjective times of some M-individual and the time of physics have, of course, deepened consequences upon the respective concepts of anticipation. Indeed, accounting for what I said in I/1.1.5 and in II/1.1.1, 'anticipation' regarded from a physical perspective do simply entails that some human observer *or even a machine* is able of setting up a deterministic or probabilistic/statistical relationship between two distinct events A and B, so that, whenever A occurs, he/she/it is(are) capable of making a journey into the future and taking into account both his/her/it previous history and the assumption that what has happened in the past still remain either in the present or in the future, forecast(s)/predict(s) that B will occur again. However, from the point of view of private times, 'anticipation' becomes a much more complex process. As a matter of fact, it is not solely the strict causality which has framed the scientific mainstream of the last centuries or the (supposedly) 'neat' distinction amongst 'past', 'present' and 'future' that has now to be taken into account. On the contrary - as previously stressed - situations exist in which the individual may act/react either as though he/she were in the past, or in which his/her private time can, by no means, be regarded as homogeneous but, rather, thoroughly scattered. But, in these conditions, 'anticipation' looked upon from the private and subjective perspective may entails purposiveness, intentionality and teleological behaviours together with volition, awarenesss of risks, decisions to be taken, eventual 'timeless' personal constraints, etc.

This is the concept of anticipation that I just have called the *widen conception.*

II/1.1.5-It is obvious that in the aforementioned conditions, the 'mechanism' underlying anticipation in human *qua sentient* beings becomes quite different from those the classical paradigm treats of. To bring these differences to light and to scrutinise the (possible) functioning of such a 'mechanism' in detail will be therefore the content of the next items.

In this sense - taking the characteristics of physical time as a reference and adopting the point of view of *adult* observer(s)/participant(s) *only,* a simplificative hypothesis that (unhappily) has to frame the whole work - I shall begin with a specification of the major features of private times, attempting at bringing to light their mutual distinctions. Next - once the dynamics of the interaction amongst (private) past(s), present(s) and future(s) is clarified - then a survey of a possible model that entails the genesis both of social/cultural and physical times will be exposed.

II/1.2-Physical Time and Private Times: Temporal Ordering, Homogeneity and Affective Dependence

II/1.2.1-The first of the aforementioned characteristics to be analysed deals with the concept of ordered sequentiality adopted in classical physics practically without criticisms. As previously referred to, which this conception entails is

> i) that given some time instant that the observer regards as being the origin of his/her experiments - usually represented by t=0 - all the time-instants *before* it are said to belong to his/her *past* and all the time instants *after* it belong to his/her *future*,

> ii) that some happening is *or* 'in *the* past' *or* 'in *the* present' *or* 'in *the* future' but never in two or three of these temporal divisions,

> iii) that (from such a physical perspective) the past is not only unique but, once the value of the initial or boundary conditions (in the case of deterministic systems) are also known, then, the present and to some extent each one of our futures too (albeit in a lesser degree, due to the unexpected, unforeseen and uncontrollable 'outer' circumstances we are always submitted to) become inexorably framed.

All of these statements are so commonly spread out that, at a first sight, only hardly they may beget some doubt. However, when more attentively looked upon, especially from the psychological perspective that rules the setting up of the aforementioned private times, it can be clearly seen that they may undergo some additional and - I shall venture to say - rather unfamiliar characteristics.

The first of such unconventional features deals with the concept of past itself. I am not speaking of the (mental) origin of this temporal division - unhappily I have no possibility of discussing it in this paper - but in terms of i) above, say, of its *uniqueness*. As a matter of fact, instead of the traditional idea asserting either that "the past is the past" or that "it cannot, by no means at all, be changed", I shall positively claim thenceforward that to *each present corresponds the existence of several possible types of pasts*.

Three of them are just depicted in the Fig. 3: those I called *factual pasts*, corresponding to pasts that had, indeed, a factual and veridical existence both for the individual to whom it belongs and for his/her human neighbourhoods (parents, relatives, next of kin, etc., who, in this context, work as means to make sure that such factual pasts were also actual, not imaginary pasts), those named *interpreted pasts* which, metaphorically speaking, are nothing but images of such factual pasts although looked upon through the glasses of our own interpretations (essentially emotional, especially in the first years of life and dependent, in turn, on our own past living experiences), and those denoted *constructed pasts* which never actually happened for all the human neighbourhoods of such a M-individual but whose truthfulness is ever denied by him/her. As a parenthetical comment, the reader must bear in mind that all such pasts entail interactions with other human beings; hence, one of the reasons underlying the importance ascribed to CT.

In Fig. 3 a rather symbolic representation of several presents and of their cognate possible pasts belonging to this M-individual is depicted for various time-instants $t_1, t_2, ...t_n$. At a first sight the picture seems to be a bit confusing but, once explained, its understanding becomes quite easy. A factual past FP_1 is interpreted (IP_1) and recollected at $t=t_1$ by our M-individual in his/her present p_1. Some times later, this individual undergoes the influence of a different or similar factual past PF_2, who he/she interprets (IP_2) and recollects at t_2.

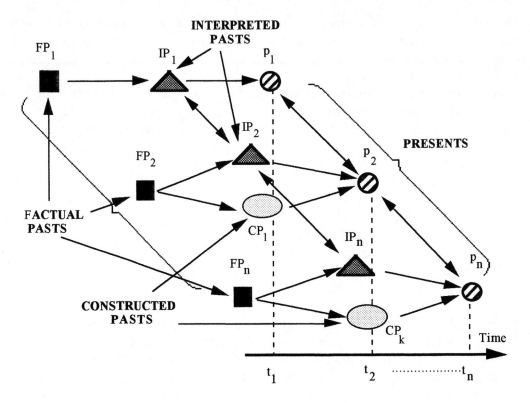

The possible pasts of a M-individual

Fig. 3

Altogether regarded, these processes - that, so far, seem *apparently* to be quite similar to those of physical sciences - are repeated for several factual and interpreted pasts FP_n, IP_n as well as recollected in a present p_n. Nevertheless, whereas physical systems are *in themselves* unable either of providing self-interpretations (and, consequently, of meaning-assignment) or of coping with the eventual temporal changes that such meanings may undergo - the semantic retro-action from some present to its pasts is a possibility that, by no means, can be excluded whenever private times are being questioned. Briefly: the aforementioned sequences

factual past $\Rightarrow$ interpreted/constructed past $\Rightarrow$ present

are *not independent*. On the one hand, because which has been looked upon as 'present' *now*, say p_1, soon becomes past for another present p_2. On the other, because the (metaphorical) 'glasses' according to which interpreted or constructed pasts have until there been regarded, may beget the emergence of changes - translated either by the modification of past interpretations/constructions themselves or, even, by the rising up of new associations, new semantic closures and consequently new understandings and new meanings (M. Martins, 1997) - all of which will do change, in turn, the proper perspectives we may have of the present itself.

This kind of mutual relationships past(s)/some present may give rise to a representation similar to that of Fig. 4, where the subscripts 1, 2,...,n are symbolic representation of time-instants.

There is however another conclusion which maybe extracted from this analysis. As a matter of fact, if all the actual or possible pasts of some present are condensed in one single (symbolic) circle then - comparing the mutual influence between such an overall past/present with the conception followed by the classical paradigm, we are thus led to something that *altogether regarded* becomes quite similar to the representation of Fig. 5. Briefly: instead of the *linear sequentiality*

$$past \Rightarrow present$$

what we have now is a *circularity*

$$past \Leftrightarrow present$$

in which the past(s) influence(s) the present and this, in turn, acts retrospectively upon the past(s) giving (eventually) rise to new perspectives of the present, etc.

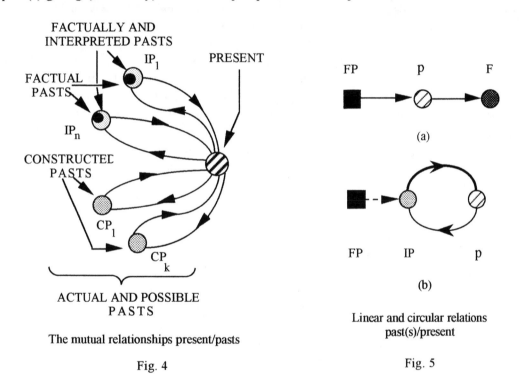

FACTUALLY AND INTERPRETED PASTS

FACTUAL PASTS

IP_1

IP_n

PRESENT

CONSTRUCTED PASTS

CP_1

CP_k

ACTUAL AND POSSIBLE PASTS

The mutual relationships present/pasts

Fig. 4

FP   p   F

(a)

FP   IP   p

(b)

Linear and circular relations past(s)/present

Fig. 5

II/1.2.2-In most cases, the distinction between/amongst these pasts is extremely tenuous or even non-existent. For instance - according to Spitz's thesis on which he named the primacy of the kinaesthetic organisation regarding the diacritical one (1965) or, similarly, Freud's primary process - all the 'factual' present happenings that babies or toddlers experience at the moment of their occurrence are always coloured by intense affective contents. This is nothing however but the transformation of a 'factual present' (whichever it maybe *in itself*) in an 'interpreted present' which, whether meaningful enough to be recollected later, will be remembered as, for instance, something predominantly pleasurable or unpleasurable.

Identical arguments may be brought to light regarding the construction of non-existent pasts. Who can forget Freud's "Family Romances" or the countless descriptions that practical all the psychological and psychoanalytic works have of the imaginary parents created by neglected and beaten children, descriptions which, sometimes, persist even during their adulthood? Or, contrariwise, the fresh tack that some of us may have of the earlier relationships with our relatives, *before* and *after* the birth of our own children, the elder in particular? In this sense - and although I do not like to generalise this sort of re-interpretations owed to their variability from individual to individual - I cannot forget the 'hardness' I wonted to use in judging many of the reactions that my parents had when I was a child, and how easy is now to understand their angryness (which does not mean to repeat it), faced with some of the behaviours that my frantic baby-boy is presently exhibiting.

Whichever the situation maybe - and this is the aspect I bear in mind to stress - in so far as each of us learns, becomes more mature or acquires more experience, acquaintance, wisdom, knowledge, etc., a great many deal of the recollected images which constitute what we usually assert to be each of our overall pasts, undergoes changes of perspective, of meaning, of new and till there unsuspected associations, etc., which, to some extent, are tantamount to the *re-creation and emergence of new pasts (either interpreted or constructed but never factual since these latter, once already occurred, can, by no means at all, be modified), some of which will co-exist thenceforth with those we have always remembered.*

Those engaged in a psychoanalytic process (both psychoanalyst and patient) know how important is the role that the former plays, looking upon the crude material that the latter's memory is providing and, by means of suitable suggestions, attempting at bringing to light new associations and new interpretations that, until there, the patient had been unable to set up. But, whether successful, e.g. because *already meaningful at that moment* - which mandatory entails the idea that such *a suggestion was interpreted, understood and accepted* - part and parcel of the temporally arrested past/present in which the patient had stood and firmly believed until there, becomes, thenceforward, radically changed. Briefly, it becomes a *new overall past that not only begets a new present* but gives also rise to a *new person, albeit keeping his/her genidentity in the sense of Kurt Lewin* (Pask, 1994). But this being so, the classical distinction between M-individuals and P-individuals referred to beforehand about Pask's CT may undergo a further development. As a matter of fact, without denying either the existence of *outer* P-individuals as *collective* entities encompassing dozens, hundreds or millions of M-individuals or each of these last as an incarnated being (in Pask's words "an organisationally closed but informationally open entity"), I shall assume thenceforth that each M-individual can *also* be regarded as entailing a host of *inner* P-individuals (where P still stands for 'perspectives', no longer colective as the P-individual that a school of thought or a group of persons sharing *through conversation* common perspectives exemplify (here the reasons as to why I named them 'outer' P-individuals), but rather, representing a *private and idiosyncratic set of coherent inner perspectives, lying in each of us, and of which the classical perspectives we everyday utter are composed.*

In Fig. 6 a), b), c) a depiction of these M and outer and inner P individuals is provided. But accepting the cogency of the preceding reasoning manifold conclusions which, ultimately, can give rise to further enhancements of Conversation Theory, may be extracted. As a matter of fact, in the realm of the subject matter we had been mooting, say, of Lewin's genidentity - although the past and the present of some patient may underwent some change owed to a suitable suggestion of his/her psychoanalyst - it is now easy to see that this may happen so, provided that - and this is a especially poignant result - *such a patient is regarded not only as* *the mere M-individual we usually tend to look at this or that person but also as containing within himself/herself a host of P-individuals, some of which had not yet been fully integrated in a whole (as P_k or P_n in Fig 6c) which, after the emergence of new associations, become inserted into the remainder perspectives.* The same reason subsumes also either the separation between the 'old' past/present and the *actual* present (yielding the opening of the door for the emergence of a true mental dynamics that the temporally arrested past/present had not yet allowed) or even, the construction of Freud's psychic instances (Id, ego, supergo proper and ego-ideal together with their complex inter-relationships amongst consciousness, preconscious, unconscious, etc.).

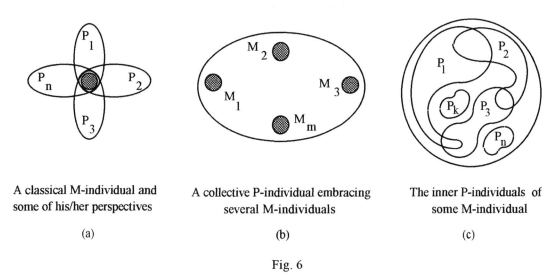

A classical M-individual and some of his/her perspectives

(a)

A collective P-individual embracing several M-individuals

(b)

The inner P-individuals of some M-individual

(c)

Fig. 6

Once again, albeit using a quite personal example to illustrate part of these considerations, let the reader assume that we were both engaged in conversation and that, suddenly, I asserted that, to a great extent, your inner peace depend on a small piece of rubber you wonted in your mouth as well as on a piece of a white fabric that you never abandoned, even when sleeping. Most plausibly, the *first* reaction you might have was that I had become suddenly fool, due to some unknown reason. However, looking upon the statement from a different perspective, the reader would certainly recall that, at an earlier stage of your life, the lost of anyone of those objects was felt as a terrific drama - which is just what presently happens to my baby-son. Transposed to an adult language, this is tantamount to saying that (in those distant times) life was inconceivable without a ruber nipple (that one, exactly alike!) and a nappy (or a Teddy-Bear or something affectively equivalent). But - and this is the aspect I had in mind to stress - in so far as we grew up, such a vision of the world lost progressively its importance *and although still existing in your (or my) mind*, it becomes thoroughly meaningless, being substituted by other perspectives, goals and interests.

This substitution and change leads me to the extension of the preceding considerations to the future and to its relationships with those presents and pasts we have just been mooting so far.

II/1.2.3-I shall begin with an analysis of such an extension by asserting that the 'mechanism' underlying either the 'circularity'

$$past \Leftrightarrow present$$

73

or the (possible) non-existent distinction between past/present and present can *mutatis mutandis* be integrally generalised to the future, yielding both the failure of these two of the major assumptions on which the concept of physical time is based.

Regarding that 'circularity' I shall claim indeed that it will now take the form

$$\text{past} \Leftrightarrow \text{present} \Leftrightarrow \text{future}$$

following a path quite similar to that summarily depicted in Fig. 7 b). Once again, Fig. 7 a) represents the classical linear sequence

$$\text{factual past} \Rightarrow \text{present} \Rightarrow \text{host of (possible) futures or 'futures that maybe'}$$

satisfying the cannons of the classical paradigm. This, regardless the features of the relationships past/present and present/possible futures, which may be deterministic or probabilistic in calibre.

But in Fig. 7b) these interactions acquire a much more complex character. Suppose, indeed, that some individual, say $\mathcal{A}$, is in his/her present but that he/she has to take a decision about some future event - something so simple as, for instance, to select the restaurant where he/she wants to have lunch. Then, as previously referred to, a sequence of images of several possible restaurants will plausibly appear in $\mathcal{A}$'s mind (the mental representatives of those 'futures that maybe') yielding that, momentarily, $\mathcal{A}$ is *simultaneously* both in each of these possible futures and in his/her present. This is by no means astonishing since - as already referred to - this (apparent) splitting of $\mathcal{A}$ is always possible, provided that we are able of looking upon individuals not only as simple incarnated M-individuals but also as a host of inner (mental) P-individuals.

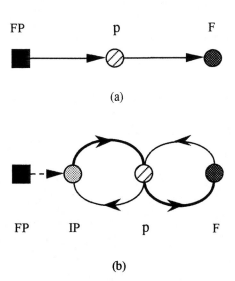

(a)

(b)

The double circularity past/present/ future in human beings

Fig. 7

Now, whenever each image of a possible restaurant arises in $\mathcal{A}$'s mind, also its major feature (type and quality of the food, of the environmental milieu, of the service, of the price, etc.) already experimented and experienced in preceding situations will, certainly, be recollected.

This means that - being in his/her present - $\mathcal{A}$ has suddenly made a journey to his/her past so that *momentarily* he/she has surely been in both, present and past. Finally, as soon as a (present) comparison of the features of each of the different restaurants has been made and a decision has also been taken, then - following the terminology of Aristotle - the transformation of all of these potentialities into actuality (say, of 'futures that maybe' in a 'future that will be') will require too that any moment, a comparison between the actual, present situation and that corresponding to the (past) imagined goal is made, so that eventual discrepancies between them are corrected. Once again, however, this execution stage do also implies an interaction between future and past (although using the present as a pivot) quite similar to that previously described.

II/1.2.4- A more detailed representation of this double interaction is provided in Figs. 8 a) to d). The pictures are practically self-explanatory but, for the sake of clarity, an additional - although cursory - explanation is provided.

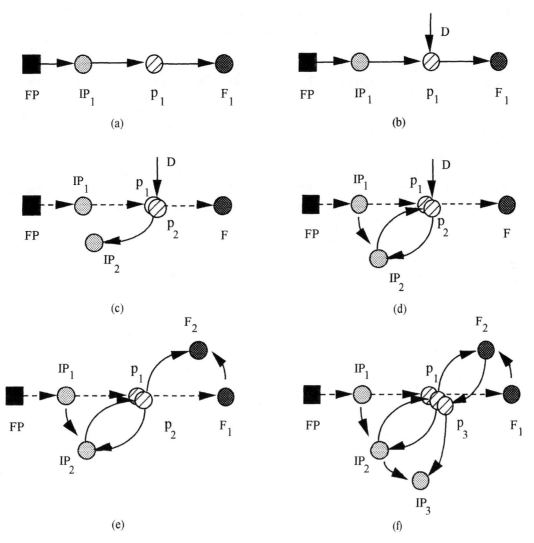

A more detailed account of the double circularity past/present/future

Fig. 8

The whole process begins with Fig. 8 a) in which a 'classical' succession

some factual past $\Rightarrow$ its interpretation $\Rightarrow$ its (eventual) recollection in the present $\Rightarrow$
$\Rightarrow$ present construction of some possible future

is given *a priori*. But in Fig. 8 b) it was assumed that our individual $\mathcal{A}$ underwent some sort of significant 'outer' interaction, symbolically represented by a 'disturbance' D. Then, owed to such a significance (something that eventually yield the emergence of an 'inner' change translated, for instance, by the well-known utterance "I had never thought of that!") a new re-interpreted past (IP$_2$) is created (Fig. 8 c)) and associated with other already existing past IP$_1$

75

(Fig. 8 d)). But due to the modification of $\mathcal{A}$'s past, also his/her present is changed as well as, plausibly, his/her expectancies regarding the future (Fig. 8 e)). Whether being accomplished or not, these futures will, in turn, retrospectively act upon the present, yielding the formation of new strategies, of new decisions to be taken, etc., briefly, of a new present (p_3) that, once again, due to all of the changes occurred in the meanwhile, may beget new past alterations, etc.

II/1.2.5-All the preceding considerations were based on the assumption that, although *momentarily*, our hypothetical individual may stay either in the present and in the future or in the present and in part of his/her past, *he/she is fully aware of the distinction existing between these temporal divisions*. Pathological situations exist, however, in which such a *distinguo* becomes extremely tenuous or even non-existent. Suppose, indeed, that this M-individual is engaged in a professional work that, for reasons whose analysis is not pertinent for the moment, becomes motive of anxious symptoms. Then, knowing *anticipately* that 'tomorrow' he/she will have to go to his/her office again, the individual will react 'today' (say, in his/her present) as though he/she were in the future, feeling 'here and now' an anxiety associated with a situation that *actually* will only happen (if this is the case) in his/her future. As in the situation mooted in II/1.2.3 (although for different motives) *this is tantamount to asserting that such a possible future works as though it were present and no distinction between them were existing*. Furthermore: since this 'present' is in turn related to an arrested past, then, *for this specific situation* everything works ultimately as there were no distinction amongst 'past', 'present' and 'future', although at the same time (and this is the reason as to why I laid emphasis upon "this specific situation") the M-individual *may - and I emphatically stress this 'may' -* also be aware of *other* changes, of becoming, of a context that *in reality* is thoroughly different from the one he/she underwent in his/her past, etc., briefly, of the time's arrow. *Under this caveat, however, one of the major features of physical time - which is generally named its homogeneity - may no longer hold whenever private times are being treated.*

II/1.2.6-The reader will certainly ask for the reason why in the preceding item so strong an emphasis was laid upon my statement concerning the fact that the M-individual "*may* also be aware of other changes, of becoming etc.". The answer for the question is that such an awareness - consciousness, ultimately - *is thoroughly dependent on affections*, usually related either to a single, however extremely high (or traumatic) displeasurable happening, occurred in a past where the psychic structure of such M-individual was not yet prepared for coping with so great an affective intensity or, likewise, to a series of contextual events, also displeasurable (although endowed with smaller intensities) that, as far as time went by, become traumatic due both to their countless repetition and (consequently) to their successive reinforcement. Usually, owed to the defence mechanisms which our psychic apparatus is endowed with, such events are (apparently) forgotten - correctly, *repressed* - and the individual may live the whole of his/her life without any conscious memory of which had occurred. But it may also happen that for some reason, such apparently asleeping memories become suddenly awoke so that the original traumatically distressful situation is so fully recollected that he/she is not psychically prepared for understanding what he/she feels as an inner 'attack'. Whereas it lasts, the individual becomes totally unable of being aware of any sort of changing, his/her mind becoming so entirely focused upon his/her own pain/distress that he/she *altogether looses the awareness of time itself*. This is tantamount to asserting that in his/her mind everything works *as time in general, changing and becoming were non-existent - a timeless time.*

Often, the local duration of these symptoms is not too long. But their consequences upon the individual's life can be dramatic. For, owed to the fear and anxiety associated with the (possible) repetition of a similar situation (which implies an anticipation of an unpleasurable 'future that may be') the individual tends to restrict and avoid activities that he/she imagines as having been the cause of such an 'attack'. Doing this, however - and beyond all the possible psychological and psychoanalytic explanations embracing these situations - what I am ultimately asserting is

i) that the individual begins *to act* according to a pure *teleological* frame (so, contradicting the efficient causality in which the very foundations of physical time are inherently rooted);

ii) that such a temporally arrested past/present is thoroughly influencing not only the present itself but also all of those 'futures that may be'. This means that such possible futures, instead of becoming a source of possible dynamic changes for the individual's life, become, on the contrary, progressively narrower so that an *unique* imagined - and displeasurable - 'future that will be' will ultimately remain. Furthermore: since these futures do also retrospectively influence the individual's present, and the awareness of the 'sense of time' is widely dependent on the consciousness of changing, then, the narrower the host of such possible 'futures that maybe', the narrower the time horizon of the individual will become (following a process which, using a terminology based upon control theory, obeys the features of a positive feedback action);

iii) that, since the time's arrow is ultimately associated with the *awareness of change and becoming*, then, (private) *time* and *consciousness* become too intertwined subject matters. This is a result already realized since the Greek Antiquity when Aristotle in *De Caelo* (Book IV, 10 and 11) asserts that "time implies change. For when our state of mind does not change or we unaware of the change, we do not think that time has elapsed. When we *notice* change we think there has been a lapse of time and vice versa" (the italic is mine);

iv) that, under this proviso, (private) *time, anticipation* and *volition* become also mutually associated. Getting back to II/1.1.3, it is easy to see that with the help of some kind of psychic therapy, the arrested future/present/past of such M-individual can undergo a radical change. This, if the therapist (either psychologist, psychotherapist or psychoanalyst) is good enough for suggesting suitable new 'futures that may be' (together with new re-interpreted pasts) *and the patient is already able of deciding (because this is his/her will) to follow such suggestions in order to transform these 'futures that may be' in actual 'futures that will be'.*
Obviously, these transformations require, mandatory, that real actions performed upon the external world have also to be accomplished. But doing this the patient is surely changing his/her present as well as re-interpreting many 'past' situations - an example of how the future can influence both the present and that part of the past susceptible of undergoing changes;

v) that, finally, whenever private times are in question, instead of the linear succassion, typically assigned to physical time, situations exist in which it is a *temporal cyclicity* (in

77

a sense *similar* to that Gordon Pask has used for concepts, never for time however) must too be taken into account. A word of caution should nevertheless be emphatically stressed. Indeed, whereas in Pask's theory the activation of any concept in a cluster begets the recollecting of *all* the remainder concepts it contains *whichever they may be* - here what we use to call 'the present' plays a prominent role, since it works as a pivot between the past(s) and the future(s). Furthermore: due to the pathological situations previously referred to, the importance asssigned to the past(s) or to the future(s) may be quite different. This means

   -that although some M-individual may feel himself/herself momentarily in his/her 'present', this present always contains remnants from the past as well as thoughts, intentions and/or purposes fully directed toward the future,

   -that the 'weights' associated with such temporal divisions, may be variable in extreme. As a matter of fact, as previously stressed, situations exist in which, due to some recollection, part and parcel of our past may become so prevailing in the present that the individual reacts as though he/she were not only a defenceless and helplesseness child (what I denoted as an arrested 'past/present') but also as he/she were fully unware of time itself. Or, contrariwise, that in order to avoid painful/distressful past recollections, the individual exhibits a frenetic activity, fully directed toward the accomplishing of future projects, case in which his/her past is (apparently) devoid of interest, etc.

   Both of these extreme situations point therefore toward an *weighted cyclity* - different from that Pask suggested - prominently ruled by affective factors, some of which act unconsciously upon the individual's behaviour.

II/1.2.7-These relationships amongst the (possible) 'non-linear' succession

$$past \rightarrow present \rightarrow future,$$

the (possible) temporal heterogeneity, the relations between (private) time and affections as well as between (private) time, consciousness, intentionality and volition, etc., briefly, the major aspects we have been mooting so far - determine, therefore, a host of characteristics that, to some extent, opposes frontally to those physical time is assumed to obey. Another of such features in which, once again, strong discrepancies exist between these two types of time deals with the concept of duration. This is will be the content of the next considerations.

II/1.3-<u>Physical Time, Private Times and Durations</u>

II/1.3.1-Although the assertion is almost needless to be remarked again, I said beforehand that private time and affections are intimately related to not only in terms of succession but also regarding duration. This is not surprising since, from a psychological perspective, we all know - owed to our living experiences, most of which are surely shared by all of us - how pleasurable situations last so short a time and how long bored or painful/distressful situations are felt, *eventhough their actual, objective, duration is the same in both cases*. This is one of the major reasons as to why (subjective) private times are so different from the time that clocks measure, the *objective* time of physical sciences. Being however this latter time an abstract construction ultimately based upon such private times, the question dealing with the genesis of the concept of duration has been, for centuries, one of the main concerns of philosophers, psychologists, neurophysiologists, etc.

I shall not repeat the several historical approaches to this subject, a remarkable condensation of which may be read in Boscolo and Bertrando (1993). On the contrary, I shall use some of the results previously stressed, attempting at presenting a possible - albeit rather incomplete - explanatory model that takes into account some of the most recent *hypotheses* about it.

II/1.3.2-I have purposively written 'hypotheses' because the question of duration of time (as well as of its succession, although in a lesser degree) continues to be some sort of cunundrum. The mistery begins with the 'sense of time' itself (whatever the interpretation assigned to this expression maybe), follows with the concept of 'present' and ends with those of 'past' and 'future'. Indeed, since the XIIIth century, epoch in which St. Augustine wrote the text with which this work begun (repeated again for the sake of the reader's commodity)

> "What, then, is time? I know well enough what it is, provided that nobody ask me; but if I am asked what it is and try to explain, I am baffled. All the same I can confidently say that I know is that if nothing passed, there would be no past time; if nothing were going to happen, there would be no future time; and if nothing *were*, there would be no present time. Of these three divisions of time, then, how can two, the past and the future *be*, when the past no longer is and the future is not yet? As for the present, if it were always present and never moved on to become the past, it would not be time but eternity. If, therefore, the present is time only by the reason of the fact that it moves on to become the past, how can we say that even the present *is*, when the reason why it *is* is that it is *not to be*? In other words, we cannot rightly say that time is, except by reason of its impending state of *not being*."

that, practically, very few advances have been accomplished in the understanding of the very nature of the 'mechanisms' - to use a popular term - underlying the setting up of the 'sense of time' (very soon I shall present an approach, due to Robert Ornstein, which I generally agree with, that does not take into account such a 'sense').

Similar problems exist in the definition of what the present maybe. On the one hand, because as Bergson (1889) pointed out, when one attempts at isolating a present moment, it has already become past. Although not fully agreeing with Bergson's argument (it just depends on what we understand by 'moment') it is a fact that our experience of the present itself is something hardly to be grasped or defined. For example, Boscolo and Bertrando (1993) quote the works of Reale (1982), Fraisse(1976) and Michon (1979) in which the (private) idea of what the immediate present is varies from 50 milliseconds to some few seconds. Also, chronobiologists like Cohonen(1967), Orme (1969), Reinberg (1979), etc., speak of a plurarity of inner rhythms of varying duration that "influence the time sense by enabling us to experience time in a number of different ways". Quite recently, the works of Rae Silver, Patrick Tresco, Warren Meck and John Gibbon presented at the 1996 Meeting of the American Association for the Advancement of Science have clearly enhanced these ideas assigning to them a true neurophysiological basis. The former two discovered how the *circadian clock* (a small group of cells in the suprachiasmatic nucleous, constituting one of two major biological clocks we are inantely endowed with which, in this specific case, synchronises the body's internal rhythms with the outside world) communicates with the rest of the brain. Surprisingly, this link is made through biochemical signals rather than by nerve connections. So, when for instance it is time to sleep, the suprachiasmatic nucleous emits a chemical that tells the pineal gland to release the soporific hormone melatonine into the blood stream. Researches on the full identification of this chemical are proceeding.

The other important discovery belongs to Warren Meck and John Gibbon and, according to their views, they brought to light the very nature of the mechanism mastering the time's duration. This evaluation is made by means of an *interval clock*, the functioning of which is like "an internal stopwatch which gives us a sense of how much time has passed since a particular event, or how long something is likely to take. It is used, for example, to judge whether we can cross a road safely before an oncoming car reaches us". In neurophysiological terms, this interval clock lies in a region of the brain named the basal ganglia. Its first component, the substantia nigra, works as a metronome or pacemaker that sends a steady stream of impulses to the striatum. This is a gatekeeper which turns on and off the awareness of time intervals and feeds the information to the frontal cortex where the information is stored in memory.

II/1.3.3-Whether effectively confirmed, the discovery has deepened implications for the sufferers of brain disorders such as the Parkinson's disease, who often have their basal ganglia damaged. It could not only explain why these patients suffer from a defective sense of time but also how it could eventually help to improve a treatment for the disease. But, from the perspective according to which the whole of this work has been framed, the results that may be extracted from this discovery (although of a quite different nature), are exceedingly important.

As a matter of fact, when it is asserted that "(the striatum works as) a gatekeeper which turns on and off the awareness of time intervals", a possible - *however highly conjectural* - explanation for the two major features we usually ascribe to time - what we call its *flow* and the possibility of looking upon it as a succession of *moments* - may be brought to light.

Let the reader assume, indeed, that we are endowed with a Gestalt 'mechanism' metaphorically similar to that which Fig. 9 depicts. There, face and vase co-exist, say, figure and ground are undoubtedly co-present, but - and this is the gist of the question - they can never be seen *simultaneously* only successively, just like an observer who, attempting at concentrating himself/herself in what the moment is ('figure'), is inevitably drawn to the time flow ('ground') coming back again to the moment itself, etc. (at least in non pathological situations).

The metaphorical face-vase illusion

Fig. 9

If this is so, then the idea expressed by Boscolo and Bertrando (1993) that "we can *experience* time simultaneously as flow and succession, albeit we can only *analyse* time alternately either as flow or as succession", finds a suitable inserted place into a (possible) general theory for private time.

Obviously, this does not explain how *each one* of the ideas concerning succession and flow arise in each of our minds. Here, I believe that one of the main reasons for the setting up of such notions relies ultimately upon the aforementioned features already brought to light about the (private) time: the intimate relationships existing between *change and time* on the one hand (the reader notice that I have written 'change and time', not 'time and change' as this relationship is usually expressed), and the relations amongst such changes, *affections, consciousness* and *time,* on the other.

In order to justify these complex relationships let us suppose, firstly, that, together with the Innate Releasing Mechanisms (IRM) a baby is endowed with (Spitz, 1965), we *minimally* assume that the child has also the possibility of detecting so 'simple' a changes as, for instance, either an *increasing* or a *decreasing* of his/her need's tension, a concept that Freud used about the so-called 'experience of satisfaction' (1895).

For those to whom this expression is unfamiliar, such an 'experience' is a general label which encompasses the host of actions, sensations and feelings related to the most primitive interactions between the baby and the mother whenever an homeostatic disruption (usually hunger) arises.

A rather condensed representation of part of one of these experiences (about 240 in the first month) is provided in Fig. 10. Purposively, the pictures are not correctly scaled since the curve representative of the energetic loss (from the left until point B) lasts about 3 hours whereas the straight line BE that corresponds to nourishment lasts 10-15 minutes. As a further issue, whenever the first derivatives of the curve of the need's tension are positive then it is currently admitted that to them correspond *unpleasurable* situations. By the same token, whenever such derivatives are negative,

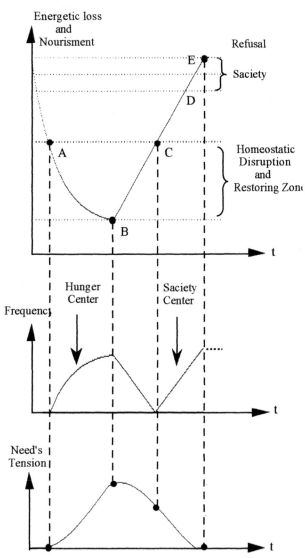

A representation of some of the mechanisms that the 'experience of satisfaction' underlies

Fig. 10

then, it is assumed that there are *pleasurable* situations to them related. Finally, in B where the first derivative of the curve of the need's tension changes its signal abruptly, it is assumed that to it corresponds a sensation of (momentary) *relief* (M. Martins et al., 1994).

This being clarified let us direct our attention to some of the preceding assertions. Namely, that the baby is able of detecting the *increasing* or *decreasing* of the need's tension as well as a *relief* in a situation *felt* as extremely unpleasurable. For, although apparently trivial, the statement (which, translates a veridical, factual, situation experimentally confirmed) subsumes a host of puzzling problems, some of which belong to the pure conjectural realm. Indeed, when Freud describes the 'experience of satisfaction' in more details, he textually asserts

"The filling of nucleaar neurones in Ψ will have as its results an effort to discharge, an *urgency* which is released along the motor pathway. Experience shows that here the first path to be taken is that leading to *internal change* (expression of the emotions, screaming, vascular innervation) but, as was explained at the beginning (p. 297), no such discharge can produce an underburdening result, since the endogenous stimulus continues to be received and the Ψ tension is restored. The removal of the stimulus is only made possible here by an intervention which for the time being gets rid of the release of $Q\eta$ in the interior of the body; and this intervention calls for an alteration in the external world (supply of nourishment, proximity of the sexual object) which, as a *specific action*, can only be brought about in definite ways. At first, the human organism is incapable of bringing about the specific action. It takes place by *extraneous help*, when the attention of an experienced person is drawn to the child's state by discharge along the path of internal change (e.g. by the child's screaming). In this way this path of discharge acquires a secondary function of the highest importance, that of *communication*, and the initial helplessness of human beings is the *primal source* of all *moral motives*." (*Project for a Scientific Psychology*, 1895)

Behind the genesis of the human communication that this text suggests, a quite especial aspect deserves further attention. As a matter of fact, when Freud (and with him, all the remainder psychoanalysts or psychologists) emphasise the importance of the *extraneous help* (usually represented by the mother) as a means to provide relief for his/her displeasurable situation they are also taking into acccount (amongst many other factors) that - between the very beginnings of the awareness of hunger itself, the attempt at discharging the cognate need's tension through screaming, vascular innervation, etc., the mother's appearance and the actual beginnings of the need's appeasement - *an interval of time* has elapsed. Obviously, asserting this we are implicitly using a language typical of adult human beings which, surely, a young baby has not. But, acting in in the way beforehand described, say, of screaming, kicking, etc., the baby is certainly aware that 'something' - what Freud just called 'need's tension' - is increasing and becoming more and more unpleasurable, so unpleasurable indeed, that, from a given threshold onwards, he/she has to find a means (the innate connection with the motor system) to discharge such a tension. Now, the awareness of that 'someting which is becoming more and more unpleasure' requires, mandatory

i)that a comparison is made between two 'values' of the need's tension and/or of its cognate unpleasure/pleasure,

ii)that, in turn, some affective short-memory has to exist, for otherwise no possibility exists of comparing what the baby feels 'now' with what he/she felt 'before'.

In normal babies both of these conditions are satisfied by the second or third months when perception and memory have already underwent a suitable maturation and development. But for the continuation of my reasoning this is not the subject matter to be discussed. Important is, indeed, to lay emphasis upon the fact that - despite his/her ignorance of what time is, the baby is able of coping with a *flow of affective changes*, especially those he/she has already set up with the 'mother' and mainly centered in the experience of satisfaction. Briefly, as a result of this interaction, the dyad baby-mother has already specified a *rhythm* - only one of manifold rhythms they mutually create, especially meaningful however owed to its intimate relationship with the species' survival - whose average intensity of the unpleasure that the baby feels until the beginnings of its appeasement may work as a reference 'measure' not only for other significant happenings (by comparison) but also for what, later on, *retrospectively examined* (when words like "before" and "after", "today", "tomorrow" or "yesterday" will already have an assigned and understood meaning) will be named 'the duration of the time elapsed'.

II/1.3.4-At a first sight, this idea of associating 'time' to a significant, albeit changeable, affective intensity, seems to be rather unconventional. But, on its behalf two strong arguments can be presented. The first brings again to light another aspect of the experience of satisfaction, now related to Freud's concepts of wish, wish-fulfilment and reality-testing. The second puts forth both the interval clock discovered by Meck and Gibbon referred to beforehand as well as Ornstein's theory which I mentioned in II/1.3.2.

Speaking of Freud's concept of wish which I had in mind are three of his assertions, one extracted from *The Interpretation of Dreams* [1900], the other from his *Project for a Scientific Psychology* [1895] and the last from *Metapsychological Supplement to the Theory of Deams* [1917] although the ideas there exposed underwent successive later deepenings. Owed to their importance, their (almost) textual reproduction will be presented in the sequel :

> "....There can be no doubt that the [mental] apparatus has only reached its present perfection after a long period of development. Let us attempt to carry it back to an earlier stage of its functioning capacity. Hypotheses, whose justification must be looked for in other direction, tell us that, at first, the apparatus' efforts were directed towards keeping itself so far as possible free from stimuli; consequently, its first structure follows the plan of a reflex apparatus, so that any sensory excitation impinging on it could be promptly discharge along a motor path. But the exigences of life interfere with this simple function, and it is to them, too, that the apparatus owes the impetus to further development. The exigences of life confront it, first, in the form of the major somatic needs. The excitations produced by internal needs seek discharge in movement, which may be described as an 'internal change' or as an 'expression of emotion'. A hungry baby screams or kicks helplessly. But the situation remains unaltered, for the excitations arising from an internal need is not due to a force producing a momentary impact, but to one which is in continous operation. A change can only come about if in some way or other (in the case of the baby through outside help) an 'experience of satisfaction' can be achieved which puts an end to internal stimuli. *An esssential component of this experience of satisfaction is a particular perception (that of nourishment in our example) the mnemic image of which remains associated thenceforward with the memory trace of the excitation produced by the need. As a result of the link that has thus been established, the next time this need arises, a psychical impulse will at once emerge which will seek to recathect the mnemic image of the perception and to re-evoke the perception itself, that is to say, to re-establish the situation of the original satisfaction. An impulse of this kind is what we call a wish; the reappearance of the perception is the fulfilment of the wish; and the shortest path to the fulfilment of the wish is a path leading directly from the excitation produced by the need to a complete cathexis of the perception.*
> Nothing prevent us from assuming that there were a primititve state of the psychical apparatus in which this path was actually traversed, that is, *in which wishing ended with hallucinating. Thus, the aim of the first psychical activity was to produce a perceptual identity [e.g, something perceptually identical with the experience of satisfaction] - a repetition of the perception which was linked with the satisfaction of the need.*

The whole paragraphs upon which my emphasis was laid (writen by myself in italic) are crucially important. As a matter of fact, what they briefly assert is, firstly, that whenever a new homeostatic disruption arises (in this case, hunger), the baby tends to *recollect* the *preceding* satisfactory situation (if this was the case), being this 'movement' or tendency just what Freud named as 'wish'; secondly, that such a tendency is translated by a search toward finding a *perceptual identity* between what - following an adult terminology - we can assert to be a 'present' and a 'past' situation

For the moment I shall not go into details about the consequences this notion of wish may give rise (see &II/1.3.5). But two important problems, directly related to the questions of time this essay attempts at enlightening, are there hidden:

i)How does a very young baby, to whom the 'world' is still far from from being organised, sets up the distinction between the re-evoked image (which, uttered in an adult language, is said to belong to the 'past') and the images he/she is receiving in the 'present', e.g. at the occasion of the need's arousal?

ii)How does such a re-evoked image work as a *future* reference for correcting the goal-directed activities that an older baby or a toddler develops *in order to* fulfil some wish?

The answer to the first question was often mooted by Freud throughout the whole of his life - see. for instance, *Formulation on the Two Principles of Mental Functioning* [1911] or *Beyond the Pleasure Principle* [1920] - but, for me, the most condensed of his asssertion about this matter is a text in the *Metapsychological Supplement to the Theory of Dreams* where it is textually written

> In an earlier passage [*The Interpretation of Dreams,* Chap. VII] we ascribed to the still helpless organism a capacity for making a first orientation in the world by means of its perception, distinguishing 'external' and 'internal' according to their relation to its muscular action. *A perception which is made to disappear by an action is recognized as external; where such an action makes no difference, the perception originates within the subject's own body - it is not real.* It is of value to the individual to possess a means such as this of reconizing reality, which at the same time helps him to deal with it, and he would be glad to be equipped with a similar power against the often merciless claims of his instincts. That is why he takes such pains to transpose outwards what becomes troublesome to him from within - that is, *to project* it.
> This function of orientating the individual in the world by discriminating between what is internal and what is external must now, after detailed dissection of the mental apparatus, be ascribed to the system Cs. (Pcpt.). *The Cs. must have at its disposal a motor innervation which determines whether the perception can be made to disappear or whether it proves resistant. Reality-testing need be nothing more than this contrivance.*

Stated baldly, what the preceding statement summarizes is that, after a certain number of repetitions, when the visual, accoustic, tactile, gustative, smelling and proprioceptive sensations of the baby are already more or less integrated in an unified percept - commonly speaking *the mother's image* - this ('good')image, endowed with a pleasurable affective content (which maybe strong enough to take an hallucinatory character) is recollected and compared with the actual sense data the baby is receiving at the moment of the need's arousal. Obviously, the baby does not know *yet* the diference between recollected and actual image. But he surely knows, firstly, that whereas such a recollected image arises in his/her mind, both the need's tension and its cognate unpleasure are increasing and, secondly, that a discrepancy exists between such image (which he/she does not know yet whether or not it is real) and the data that his/her exteroceptors are providing. Only when the two images become more or less *similar* (correctly speaking, obeying a *fuzzy resemblance relation*), say, only when the mother appears (usually, in consequence of the baby's screams) and, next, begins *in reality* with his/her nourishment (begetting either an adequate stimulation of the baby's extero/ interoceptors or the rising up of the aforementioned relief), the need's tension decreases and the unpleasurable state is substitued by a pleasurable one. But this is just the basis for the setting up of a criterion - which Freud named 'reality-testing' allowing to set up the distinction between what is recollected from what is actual. For as soon as the cohrence (in Pask's sense)

[mother's *actual* appearance, relief, 'outer' stimulation, decreasing of the need's tension] is already established in the baby's mind, also the distinction between those two images is thenceforth settled. This is just what Freud asserted in his *Interpretation of Dreams*

> The bitter experiences of life must have changed this primitive thought-activity into a more expedient secondary one. The establishment of a perceptual identity along the short path of regression within the apparatus does not have the same result elsewhere in the mind as does the same cathexis of the same perception. Satisfaction does not follow; the need persists. An internal cathexis could only have the same value as an internal one if it were maintained increasingly, as in fact it occurs in hallucinatory psychoses and hunger phantasies, which exhaust their whole psychical activity in clinging to the objects of their wish. In order to arrive at a more efficient expenditure of psychical force, it is necessary to bring the regression to a halt before it becomes complete, so that it does not proceed beyond the mnemic image, and it is able to seek out other paths which lead eventually to the desired identity, being established from the direction of the external world [in other words, it becomes evident that there must be a means for reality-testing, e.g, of testing things to see whether they are real or not]. *This inhibition of the regression and the subsequent diversion of the excitation, becomes the business of a second system, which is control of voluntary movement - which, for the first time, that is, makes use of movement for purposes remembered in advance*"

But this is also a means - albeit rather embryonic - of setting up a distinction between recollected/re-evoked and actual data, say, using an adult terminology, of 'past' and 'present'. This is a especially poignant conclusion that provides a possible answer to question i) referred to above.

II/1.3.5-Many other results could be extracted from the preceding statements. But, for the moment, I have to get back to the concept of wish and to direct my attention to some of the further consequences that may be extracted from it - now focusing the notion of anticipation itself.

i)We have seen that the emergence of a biological need such as for instance, hunger, begot in the new-born baby an increment of a tension that Freud called 'need's tension' together with intense unpleasurable sensations. Next, owed to the innate connections between the psychic sources of such a tension and the motor system, a discharge of the tension arises (through screaming, kicking, vascular innervation, etc.) which, in the earliest stages, are pure reflex actions but whose interpretation from the mother's side is tantamount to a crying for help. Since, biologically speaking, we are altricial and nidiculous organisms, 'help' in this case is provided by the nourishment supplied by an extraneous agent or, more generally, by an outer object (usually the mother's breast or an equivalent substitute) which yields the satisfaction of the need. The sequence of these actions, repeated about 240 times in the first month of life, was named by Freud 'experience of satisfaction'.

ii)The re-emergence of this somatic tension determines the re-cathexis of the mnemic trace of such an experience of satisfaction. Diachronously observed what this means is that, in the beginnings, only biological tension exists. However, as far as the experience of satisfaction becomes progressively memorised, the somatic tension also becomes thenceforth associated with a specific representation which, in turn, is re-evoked whenever the need's tension arises again. Briefly, the re-emergence of the need's tension determines the re-cathexis of the experience of satisfaction. This re-evocation is just what Freud called wish, e.g. "movement, tendency or process which goes from the unpleasure to pleasure" or, more specifically "the mnemic trace of the experience of satisfaction".

iii)It is evident that under these caveats the object of the experience of satisfaction becomes the object of the wish itself. But this being so, two inter-related components may be found in such an experience

-the somatic need itself

-the associated wish

the last of which cannot be reduced to a pure biological function. As a matter of fact, together with the need's satisfaction - which, essentially, deals with the reaching of a physico-chemical equilibrium - the infant also feels erotogenic pleasure, in this case related to the stimulation of his/her bucal zone, lips, cheeks, tongue, etc. And that this erotogenic pleasure cannot be simply reduced to the pure satisfaction of the need maybe easily witnessed, observing a baby sucking his/her rubber nipple after a lactation that clearly satisfies his/her hunger. In this case, the baby no longer wishes the nourishment but the pleasure related to the rhythmic driving of the mouth. A similar situation may be found in the realm of psychopathology with the compulsive obesity of the adult - another way of stressing the distinction between the biological need and the erotogenic pleasure (together with a way of compensating what may be interpreted either as an unconscious loss of love or an unachievable love).
But beyond these consequences (belonging to the pure psychoanalytic realm), some further results are able to be extracted from the foregoing distinction.

iv)On the one hand, there is the *sequence of actions in themselves* that the experience of satisfaction entails. We said in fact that the need's tension begets the re-cathexis of that experience. In this case the somatic tension works as some kind of pre-requirement for the activation of the remainder actions of which the experience consists (unpleasure, discharging, need for extraneous help, awareness that this help provides relief to the unpleasure and can be a source of pleasurable sensations, understanding that weeping can be an useful way both for communicating and for the exchanging of affections, etc.). However, as soon as the satisfaction of the somatic need becomes associated with that erotogenic pleasure, it is no longer necessary that the need *precedes* the re-evocation of the mnemic trace of the experience of satisfaction. On the contrary, it is the pleasure itself that the experience underlies, which may give rise to a tension tending to re-activate the (somatic) need. Examples of situations of this kind can be witnessed in adult people when, looking at a favourite delicatessen, they wish to eat it, even after a more than satisfactory and abundant meal. In this case, the process does not work according to the sequence

need's tension ⇒ experience of satisfaction

On the contrary, it is the perception of the delicatessen itself which re-evokes the experience of satisfaction and re-activates the somatic need.

v)On the other hand, there is an *inexhaustive character* related to the wish (as something lying far beyond its somatic component) whose importance can, by no means, be denied. In this sense, it is worth while to remembering that, whereas the experience of satisfaction is being occurring, the baby is also looking at the mother's face, is receiving her warmth, her smell, her caresses, listening to her words, etc. all of which - together with the nourishment itself - become an integral part and parcel both of the experience itself and of its associated wish. But under these caveats, the complex representation that the 'mother' is, becomes thenceforward intimately related to the baby's erotogenic pleasure. Briefly: owed to a *temporal simultaneity* (which the so-called Hebb's rule, widely used in artificial neural nets, is nothing but an exceedingly important mathematical simulating tool), the mother becomes converted too in an erogenic object and so, able of waking up pleasurable experiences tantamount to those the breast (a true 'object' in the literal sense of the word) gives rise.

vi)These facts yield a qualitative jump in the baby's psychism which, amongst other consequences, frame a real evolution from what Spitz called 'the perception of contact' to the 'perception at a distance'. Indeed, whereas in the earliest experiences, satisfaction of the need and pleasure belonging to a specific erotogenic zone were both intimately linked, now the awareness of the mother's looking, of her words, caresses, etc. begets no longer 'satisfaction' (simply understood as biological appeasement of this or that need) or mere organ-pleasure (as primitively happened with, for example, the mouth). On the contrary, they underlie a pure psychic erotism, which, furthermore, does not require real contact. Indeed, the whole pleasurable stimulation which, in the earlier stages was simply associated with the mother's breast is now integrally transferred to the priviledge Gestalt the mother's face symbolises - a representation which, thenceforth, becomes the mnemic trace to which the baby's mind will tend.

vii)Under this proviso, the baby's wish becomes no longer the wish of a concrete material object but, contrariwise, *the approval, acknowledgement, love of the figure who has become significant for him/her*. This is a result that was fully developed by the Lacanian school by means of a remarkable assertion that "the baby's wish becomes the other's wish" - understood either as being wished by the other or as wishing the other's wish. And on behalf of this assertion it is enough to remember either the pleasure that a baby aged 15 months shows, *replicating gestuals and movements which are pleasurable for the other* or, later on, the pleasurable expectancy followed by a triumphant attitude a toddler evinces when - accomplishing tasks similar to those the significant figure achieves - begets from this latter quite pleasurable answers.

viii)In both of the preceding situations a pleasurable behaviour of the significant figure yields the re-activation of the child's own pleasure. However, this is not all. For, in terms of the temporal question we have been mooting, these kinds of behaviour usually entail too an *anticipation of future actions* already witnessed in the concept of wish itself - a framing image that remains in the mind as a reference whereas a host of actions is being executed in order to accomplish it - which, in adult people is, by no means, astonishing but, in very young children, seem to be quite surprising *unless - as previously claimed in the last paragraph of II/1.3.3 - we assume that, long before the concept of time (as adult psychologists and psychoanalysts conceive it) is brought about, children do already have a private sense of time intimately related to affections.*

II/1.3.6-In II/1.3.4 I said that there was a second type of arguments, able to provide a strong support to the foregoing statement, based upon a theory, erected in 1969 by the experimental psychologist Robert Ornstein (briefly mentioned in II/1.3.2). The theory afforded an important contribution for our understanding of the mechanism that underlies the experience of time, particularly of the concept of duration. A cursory survey of Ornstein's theory is exposed in Boscolo and Beltrano (1993). In the sequel I shall bring to light the major aspects they focus - although *from my point of view* some of them deserve undoubtedly further development.

Ornstein's idea is that the widely spread conception concerning the existence of a 'time sense' entails both the existence of an object called 'time' lying in the external world and of a human 'sense organ' able to perceive it. Instead of this, Ornstein's experiments have shown that our temporal experience - its duration in especial - depends *not* on such a 'time sense' but rather *on the amount of information we can store in a given interval,* on the *complexity* of the stimulation, and on the way we *code* and *organise* the received stimuli, *even that this organisation is made long after the subject has perceived their succession.*

It follows from these results that the perceived duration will be shortened whenever a host of stimuli is encoded in large units (for instance, when a succession of apparent unrelated movements are encoded into a single dance step) or the subject learns to organise retrospectively some received stimuli, just like it happens when some subject matter is understood - using this term in its technical sense of 'whole's completion' (M. Martins, 1994). Also inserted into his theory are the reasons as to why how time seems to pass slowly and the duration seems interminable when we are bored or why pleasurable events seem short whenever they happen but are remembered as having been longer. In the former situation this happens so because we are forced to attend a sequence of happenings *unimportant* (Boscolo and Beltrano, 1993) - although I prefer to use "meaningless" or "not significant" since so stated they also do embraces a crucial relationship with affections. In the latter case pleasant things are recollected with a greater detail; so, when we look back on them, their duration seems longer. Contrariwise, during emergency states, more information is certainly processed but in longer intervals. So, less information is stored or coded in the same interval and time seems to pass more slowly. Briefly, whenever duration lengthens, time seems to slow down and whenever duration is shorter time seems to speed up.

II/1.3.7-If we compare Ornstein's ideas with those I have exposed about the relationships between time, change and affections, we can see that no contradictions or discrepancies exist if - and this is the gist of the question - firstly, the amount of information Ornstein speaks is looked upon as *meaningful* for his/her receiver/interpreter and, secondly, a specific definition of information itself (amongst the dozens of existing definitions) is the one selected.

One of such definitions - which deserves the merit of unifying a host of still unrelated subject matters - is just the one that Bateson (1972) provides as "the difference that makes the difference". For, if we understand the term 'difference' as implying 'change', then information may be looked upon as "the change that begets a meaningful change". Under this proviso. however, everything asserted about the relations baby-mother (as the figure who progressively affords meanings to a meaningless world), about affective changes, durations, distinctions between 'nows', 'befores' and 'tomorrows', etc. determine a whole-embracing and coherent - although simply conjectural - model for the setting up of the concept of private time.

To finish the essay we have now to venture a possible 'mechanism' which, relying upon the preceding premises, is able of explaining the rising up of public times, either they deal with social or physical times.

This will be the content of the next considerations.

II/2-Public Times: Social and Physical Times

II/2.1-One of the major reasons which explains (and underlies as well) the effective success the mechanist paradigm has reached until quite recently in the history of science, is the fact that its objectives and methods (till the rising up of statistical physics) had always been devoted either to the study of equilibrium states or to small disturbances around them (small enough to allow that their development in Taylor's series may be restricted to the first term). In a certain sense, this is a historical paradox for, focusing their attention upon such equilibrium states alone, physicists have left aside a host of quite interesting phenomena - such as those Prigogine himself has devoted his attention - opening the doors for a new vision of the world much more chaotic (albeit deterministic), less unorganised (although full of unstable situations), briefly, quite different from that we had, for example, 50 years ago.

To a great extent this change of perspectives has been partially based upon the discovery of new mathematical, physical and computational tools - which have allowed physicists to cope with complexity itself as an autonomous theme - and partially upon a releasing of the constraints that, for centuries, they had been forced to use in order to analyse and describe the systems to which their attention had been directed.

One of the areas where these constraints (either natural or artificially created) seem to be non-existent is just that one dealing with the human mentation, particularly in the realm of fantasy and imagination. Dreams, the idle-flitting from one half-formed notion to another, the functioning of our consciousness which is often half aware of the outer reality and, simultaneously, of a stream of remembrances and recollectings far distant in space and time from the 'here and now' we are actually experiencing, the always changing aspect such a stream takes - are some of the examples of the successive activation of (apparently) incoherent clusters/coherences in an entailment-mesh lying (perhaps) in our minds. Not a mesh in Pask's *strict* sense, of course, but, rather, a mesh where the aforementioned affections do take an integral part and parcel.

However, even in these realms, the foregoing 'constraints' have to be taken into account both at a biological and a psychological levels. Biologically, since human beings are altricial, incessorial and nidiculous animals whose survival during the long period that childhood lasts, demands a total dependence on the parents' support. Nevertheless, this long dependence has a (metaphorical) price which is psychological paid in the emergence of intimate emotional ties, firstly, between the child and his/her parents and later, between the child and his/her general social milieu. Both contribute to the progressive formation of two psychological structures named by Freud the *ego-ideal* and the *proper superego* - both enfolded in the general designation *superego* - whose role either in the setting up of the foregoing 'constraints' *or of new and public concepts of time* is crucially important. But in order to enlighten this last assertion that, indeed, brings to light the *liaison* between private and public times, some words about these two structures have to be said.

II/2.2-The first of such structures - the ego-ideal - is intimately related to Freud's concept of narcissism (1914). It works as that portion of the ego where the qualities, characteristics, conditions, etc., which determine one's self-love are contained. Unrestricted self-love must plausibly exist in the earliest period of childhood during which the child takes himself/herself as his/her own ideal. However, this kind of love can no longer subsist since (by means of introjection and internalisation) he/she also takes up into him/her the prohibitions, warnings, etc., proceeding from the parent's environment. From this moment onwards, such a self-love becomes also dependent on the conditions which he/she *imagines* as being necessary to obtain the parents' love (love and solicitude which, indeed, protects the child from the dangers that threatens him/her not only from the outer world but also from 'within'). The price the child has to pay for this security is a *fear of loss of love*. And, in order to avoid it, the child loves himself/herself, judges himself/herself, condemns or punishes himself/herself in the same way that his/her parents do have loved, judged, condemned or punished him/her.

In these conditions everything works as though love or condemnation were dependent on the agreement or disagreement of a critical or moral *inner* image (an *internalised standard*) which, ultimately, is nothing but a copy of the parental's' demands. This 'standard' - thenceforth inseparably related to the earlier prohibitions, warnings, punishments, rewordings, etc., briefly *emotional constraints* according to the terminology referred to above - gives rise in one's mind to a relatively autonomous system, that one that Freud has just named superego.

This necessarily cursory survey of the genesis of this psychic instance (to which, according to his conception, *conscience* is attached) does not stop here. For, throughout one's life, the superego receives further contributions from later successors and substitutes of the parents - teachers, public figures, admired social ideals and so forth - and, through them, of the family, of racial and national traditions, etc. Thus, they work as some kind of internalised image, mirrored from the outer milieu into which some individual is inserted, image which prolongs (by means of usually unconscious mnemic residues) his/her own parental relationships. This (usually unconscious) influence of the past upon the ego's actual and contemporary reactions has just led Freud to assert that "the superego is the vehicle of tradition and of all the time-resisting judgements of value which have propagated from generation to generation. The past, the tradition of the race and of people live in the ideologies of the superego and yields only slowly to the influence of the present and to new changes". *But, to state the question in these terms is tantamount to laying emphasis upon the temporally changeable inter-relationships that may exist between the outer and inner P-individuals introduced in &II/1.2.2.*

As a matter of fact, no child can grow up isolated from his/her parents and, through them, of some school of thought, religious influence, community norms, etc., that they - consciously or unconsciously - follow. Under this proviso, the child (here symbolising some M-individual) undergoes *throughout time and by means of conversation with his/her relatives sharing common perspectives (the outer P-individuals)* extraneous influences which will just beget the rising up of those 'inner' P-individuals there referred to. From here comes one of the main reasons, justifying Freud's assertion about the superego as the vehicle of tradition, of all the time-resisting judgements of value which have propagated from generation to generation, of the tradition of the race, etc., briefly, of the past of a certain community or society.

From the temporal perspective we have been mooting so far - and this is the very gist of the rising up of social/cultural times - this means that such inner P-individuals that are perpetuated from generation to generation and only "yields slowly to the influence of the present and to new changes" have a time - *a public time*, indeed, whose duration lies far beyond the individual's duration and, because of this, soon become converted in outer P-individuals.

II/2.3-For obvious reasons I cannot bog down minute analyses of these cultural times. Only some examples will be provided, coming from so different an areas as economy or arts. The reader must however bear in mind not only that such examples are *in reality* manifold but also that their related bibliography is so vast that only some areas may be focused here.

It is generally accepted, indeed, that the 'discovery' of such social and cultural times is associated with the work of Ernest Labrousse in economy "Esquisse du Mouvement des Prix et des Revenus en France au XVIIIéme siécle" (1933) about the secular variation of prices in Europe between the XVIIIth and XIXth centuries (increasing between 1717 and 1817, decreasing between 1817 e 1852).Soon, however, these conjunctural and long-time tendencies were detected not only within the economic realm - many other cycles, such as those of Kondratief (lasting about 50 years), of Juglas (8 years) and Kitchinque (4 years) seem to be unanimously accepted, although their underlying 'mechanism' remain unexplained - but also in many other scientific, humanistic and artistic areas. For example, the French historian Ferdinand Braudel speaks of the secular interactions between human beings and their geographical environment (constancy or not of their climatic milieu which, in turn, exerts deepened influences upon the whole animal, vegetal and agricultural life in a certain zone as well as on its prevailing trading routes, consequent cultural exchanges, etc.).

Culture is another area where manifold similar examples may be emphasised. For instance, Ernest Curtius' book "Europaische Literatur und Lateinisches Mitalter" (1948) brought to light an underlying structural culture which, until the emergence of the nationalist literature of the XIIIth and XIVth centuries, framed altogether the whole Latin civilisation of the Low Empire (already deeply influenced by an extremely heavy cultural inheritance). In his book Curtius shows, in fact, how all the intellectual *elites* of that epoch obeyed the same themes, the same comparisons, the same common places. Likewise, the work of Lucien Lefebvre "Rabelais et le Probléme de l'Incroyance au XVIéme siécle" (1943) describes and specifies a host of 'mind rules' (a true mental frame) which, *long before and long after* Rabelais mastered the whole ways of thinking, beliefs and so forth (and, consequently, also, restricted and censored all the independent voices who had the courage of blaming against such a frame). Similarly, the work of Alphonse Dupront "Le Mythe des Croisades" (1959) also showed how the idea of crusade remained in the Western Thought far beyond the XVIth century, as some sort of successfully repeated activity, the influence of which was even felt in the XIXth century.

In the realm of *art*, the book of Pierre Francastel "Peinture et Societé" (1951) brought also to light the existence and constancy of a geometrical pictorial space, practically invariant from the Florentine Renaissance until the emergence of the Cubism and of the abstract painting of the beginnings of the XXth century. A similar perspective is defended in the work of Peter Fuller "Art and Psychoanalysis" (1981) regarding the bidimensionality, pictorial location and relative dimensions of the religious and human representations of the Medieval and Pre-Renaissantist paintings.

Speaking of *myths*, the contemporary works of Gilbert Durand "Les Structures Anthropologiques de l'Imaginaire", "La Mythanalyse et la Sociologie des Profondeurs" and "Mythe, Symbol et Mythodologie" have provided a new light upon the existence and permanence of conscious and unconscious currents and tendencies underlying past and present societies (currents which had already been analysed although from different perspectives by Levy-Strauss, Abraham Moles, etc., based in turn on the ideas of Freud, Abraham, Jung, Erikson, etc.). Finally - just to quote some of the areas where social and cultural times have been deeply influential - a great deal of emphasis must too be laid upon the works of Collingwood, Khun, Lakatos and Feyrabend in the realm of the *history of sciences* or, yet, of Wallon and Piaget upon the so-called *genetic psychology*, this last in particular owed to the comparison he attempted at making (as well as Wallon and Freud also did) not only between the 'laws' that seem to master the child's development and the development of human societies - a subject matter which has been the object of countless sociological, anthropological and psychological discussions amongst the defenders of the so-called *meta-temporal* and *pre-logical* currents of thought (M. Muller, E. Taylor, L. Morgan, J. Frazer, E. Durkheim, M. Weber, C. Levy-Strauss, S. Freud, etc.) - but also on the setting up of the concept of time in children, however essentially analysed from a physical perspective (Piaget, 1946).

II/2.4-Finally - the last but not the least - we have the time of physics, e.g. the time measured by clocks with which this essay began. I shall not go into details on the history of this time which may be found, for instance, in Boorstin (1987). Only two of its aspects will deserve especial attention.

The first of these aspects deals either with the public and - however surprising my assertion maybe - variable character this concept assumed until quite recently or with the role that the ideas framing Pask's CT played in the setting up of an universal time as we know it nowadays.

As a matter of fact, since the most remote times that, for the sake of forwarding mutual activities, humankind has found necessary to set up an arbitrary 'public time', usually based on the recurrence of some phenomenon exhibiting a type of regularity in consonance with which people are willing to adjust their affairs. Many different types of phenomena were chosen for that, according to the society, community or religious norms they obeyed: the passage of the sun, the moon or the stars across the meridian, the floods of the Nile in the Ancient Egypt, etc. A further step however small to release humankind from the periodicity of these natural phenomena, begot the invention of some mechanisms - clocks, ultimately - the functioning of which was based either in water, in the period elapsed between some horizontal lines drawn in a burning candle were reached, or, finally, in mechanical, electrical or, recently, atomic clocks. A common feature to all of these devices is that their periods were/are very close from the period of such phenomena chosen as reference.

Generally speaking, however, and despite the increasing precision to them ascribed, the importance that a great many deal of humankind assigned to these ways of measuring intermediate times, only in the beginnings if this century has begun to acquire more and more importance. Indeed, only at the ends of the XIXth century was set up, by the first time, a global attempt at stating a world-embracing time-measure. One of the major reasons for it was provided by the railways and the telegraph. As a matter of fact, in 1870s *there were in United States over 200 local times between the east and west coasts*, which made the running of an unified national railway system exceedingly difficult. In 1883 this puzzling question was solved through the imposition of a series of time zones across the nation. In the following year, time and space were linked in a global system when it was agreed that Greenwich should become the zero meridian and the world time should be measured from there.

The international community only slowly decided to adopt this system and it was only in 1912, at the International Conference on Time in Paris, that the telegraph became the agreed means for beaming signals around the world. In 1913. the first time signal was transmitted from the Eifel Tower, thus yielding the creation of the public global time as we know it nowadays, through the connection of time and space into a single system of calculation.

The second aspect I had in mind to point out deals now with the *abstract* character that the concept of time has assumed, especially from Newton's epoch onwards. Indeed, as it happened with the concept of space, human thought has not remained satisfied either with the feeling for succession/duration or with simple empirical measures of time by means of

92

some conventional mechanism. It has striven for a more abstract concept to serve as foundation for logical thinking about physical events - a time which would be independent of individual consciousness or of the operations of such mechanisms. This conceptualisation had already begun with the Greek philosophers, who puzzled themselves greatly over the question of the divisibility of time intervals. But the first clear statement of the nature of the concept of time as used in physical theories is that of Newton *Principia*. Distinguishing between *absolute time* (the abstract concept for use in physics) and *common time* (which is tantamount to the 'public time ' mooted above), Newton adopted the idea that

> Absolute, true and mathematical time or duration flows evenly and equally from its own nature and independent of anything external; relative, apparent and common time is some measure of duration by means of motion (as by the motion of a clock) which is commonly used instead of true time.

It is known that this definition is useless, since it defines time in terms of the thing itself, without which the terms 'evenly' and 'equably' have no meaning at all. However, until the rising up of Einstein's Relativity Theory, Newton's idea of abstract time as it arises in the equations of physics - say, as a parameter which serves as an useful independent variable and whose range of variation is the real number continuum - has been and still continues to be widely used. Thus, every time interval is put into a one to one correspondence with the interval between two real numbers In this sense, the time parameter is thought of as being endowed with the continuity and other properties of real numbers, its analytical usefulness in physical theories just resting upon this fact.

These were, however, the ideas from which this work started up.

III-OVERVIEW

III/1-Three major thesis have being in discussion in this essay: firstly, that there are several kinds of time. Secondly, that the most primitive of them all - the so-called *private time* - is intimately related to the affective 'mechanisms' each of us is endowed with since our earliest stages of life. Thirdly, that such psychic 'mechanisms' may justify the rising up of *public times* - of which *social/cultural times* and the *time of physical sciences* are especially poignant.

III/2-Each of such times yields specific types of anticipation. Nevertheless, the ultimate notion of anticipation continues to be based on a concept of time (inherited from the Greek meaning of time as *chariness*) mirrored from physical sciences Thus, 'time' appears as a linear and ordered sequence of infinitely divisible 'instants' obeying the laws of an homogeneous transformation group in which, furthermore, the past is looked upon as framing the present in a causal way - a present that, in turn, leads inexorably to some specific and predetermined future (at a coarse, grained level, at least).

Although having had no intention of debasing neither this particular conception of time and becoming nor the whole research work already accomplished in the realm of physical and computing/robotic areas, it was however claimed in this paper:

III/3-That such a notion of *physical time* (measurable by clocks) is far from exhausting the several times that the *general* concept of time entails. For, together with this 'physical time' - which, familiarly, may be denoted as *objective time* - many other kinds of time may be considered. For instance, the social, cultural and/or subjective and private times previously referred to must categorically be taken into account. Hence, begetting a host of puzzling problems in the definition of what 'anticipation' maybe;

III/4-That whereas the objective time of physics belong to the category of a (generally shared) public *time*, social and cultural times pertain, on the contrary to a class of public *times*, albeit also created/constructed by shared agreements amongst M-individuals (Pask, 1970). As such, they can all be regarded as the *private time* of some P-individual - say, of an hypothetical mind-entity besetting *in principle* many M-individuals, who are sharing (and often perpetuating) both a commonalty of perspectives and a commonalty of individual times - either world-embracing (case of the time that clocks measure), locally circumscribed (case of some school of thought, of historical, archaeological or artistic situations, of the followers of some religious belief, etc.) or even, in the limit, of one single M-individual (case in which a distinction exists between *inner* and *outer* P-individuals);

III/5-That *private and subjective time*, particularly that one belonging to some M-individual (identifiable with the Greek *kairos*), is far from obeying the features ascribed to the 'objective' time of physical sciences. On the contrary, it can be modified not only by outer/inner circumstances but the (supposedly) neat distinction amongst 'past', 'present' and 'future' widely used without criticism by the followers of the objective time of physics, is also far from being neat - a fact which begets important consequences both upon the assumed 'homogeneity' of that physical time (at such an aforementioned coarse level) and, in consequence, on the concept of anticipation itself;

III/6-Indeed, that the individual, subjective and private time can undergo changes is a fact known by all those who have already experienced either intense emotional situations or, contrarily, some Eastern meditation techniques. In the former case - of which the few seconds that an earthquake lasts is a representative example - 'time' seems to last an eternity, say, in other words, as though it were progressively slowing down. On the other hand, the pratictioners of the aforementioned meditation techniques do also speak of the 'speeding up of time', in which hours are felt as though they were lasting few seconds only. The traditional description of the flower that is being seen flourishing is so well known that needs no further explanation;

III/7-That, regarding the (apparently) 'neat' distinction amongst 'past', 'present' and 'future' used by physical disciplines, some aspects exist that deserve further enlightenment. On the one hand, there are some situations in which the setting up of such a distinction may be extremely difficult. As a matter of fact and *stated baldly*, adult people is, *in most cases*, aware of when he/she is 'remembering', is living 'in the present' or is 'imagining and planning future actions'. However, under a closer scrutiny, this (apparent) cogent picture conceals not only some deep criticisms but also a host of troubling questions some of which remain unexplained. Indeed, and beyond clearly pathological situations (for instance, of acute depressions in which no time horizons seem to exist, or in schizophrenic situations in which time is scattered) all of those who, for instance, have been engaged in a psychoanalytic process know - sometimes against the whole rational evidence - how difficult is to straddle the 'past' from the 'present' or this last from more or less illusory 'futures'. Everything happens therefore as though, owed to the dramatic intensity of the affective factors related to some past situation, the patient were acting/reacting in the 'present' in the same exact way he/she did/made beforehand (the well-known 'reaction as before'). In these conditions however, this means

i)that the patient is *simultaneously* working with two types of registration memories: one, that tells him/her that he/she is in the 'present', the other in which - in spite of a *rational* location in what he/she says to be his/her 'past' - such a memory is, in reality, exerting its entire influence upon the 'present'. Under this caveat, everything happens *as though the patient were actually unaware of the distinction between his/her 'past' and his/her 'present'*. This is tantamount to asserting

ii)that the patient is unable of straddling the *memory of some happening* (recollecting, remembering, etc.) *from the happening itself* so that he/she is reviving it as it were *always* occurring by the first time,

iii)that (at an unconscious level at least), 'time' did not exist (a conception which maybe identified with the Greek notion of *aion*),

iv)that, consequently, the concept of time seems to be a creation/construction of consciousness, although from a neurophysiological point of view an approach due to Warren Meck and John Gibbon (1996) exists which relates the awareness of some happening with an *interval clock* lying in the brain (entailing particularly the basal ganglia , the substantia nigra and the striatum). This clock works as a metronome that turns on and off the awareness of time intervals, sending a stream of impulses to the frontal cortex where the information is stored in the memory,

v)that, furthermore, it seems that two general types of memory also exist: one, unconscious, altogether mastered by affective relationships and entailing timeless events (an idea that Freud set forth and followed during a great part of his life but, later, abandoned in favour of another explanation which, parenthetically, I do not agree with), the other related to consciousness in which its contents have different durations, may eventually be ordered, is the object of the conventional cognitive divisions and subdivisions (long term, short term, sensorial, scripts, etc.), etc.,

vi)that, under these caveats, the aforementioned concept of 'homogeneity' assigned to physical time, cannot be fully applied to each of our private time, but perhaps only to the functioning of parts of the mind (an hypothesis which, whether true, will do also require the existence of some kind of 'synchroniser'). Furthermore: if such a supposed 'synchroniser' begins to fail or stops working then, this would provide a possible explanation both for the arresting of time previously described regarding depressive states or even for the chaotic temporal organisation of schizophrenic people,

vii)that, consequently, 'time' and affections cannot be treated as unrelated entities. On the contrary, the intensity and rate of change of such affections (because meaningful for every individual) were supposed to provide a 'measure' for *duration* (satisfying the theory of Ornstein and the definition of information that Bateson provided as "the difference that makes the difference") as well as the construction of a private notion of 'time' in children from the perspective of *succession*, both of which will, later on, be retrospectively re-analysed, begetting the usual concept of time commonly shared by the great majority of adult people, viii)that, since affections depend on the living experiences that each of us undergoes/has undergone, then, each of our private times in childhood, adulthood or, yet, individually and socially looked upon, also depends not only of such experiences but also on the way we have interpreted them.

III/8-All of these *possible* consequences (some of which belong to the realm of pure research) do also have reflections upon the concept of anticipation itself. For, it is obvious that, from the foregoing perspective, instead of a linearity of time, what we do have now is some sort of *weighted cyclicity* (similar to that Pask used) in which 'past', 'present' and 'future' are all of them so mutually related to by means of a double loop (or double 'circularities') that the recollecting of one begets the recollecting of the others. Amongst the consequences that may be extracted from these facts, two are especially poignant:

i)on the one hand, that there may be an affective/cognitive 'weight' associated with each of these traditional temporal divisions, so that the individual's mind tends to prevail one of them in detriment of the others. Briefly: for reasons meaningless to be analysed here, this or that M-individual tends (metaphorically) to live more in the 'past' or in the 'present' or in the 'future', although all of them are permanently co-existing together;

ii)this allows, on the other, to look upon the idea (mirrored from physical sciences) that the past determines inexorably the present and this, in turn, an unequivocal future as a misleading and mistaken notion, regarding at least at the subjective type of time. On the contrary, on the light either of a psychoanalytic experience or on the development each of us undergoes throughout our lives, not only the ways we have of *interpreting* past events may change (as though, several interpreted/re-interpreted pasts were co-existing in each of our minds, *none of which is the real past factually occurred* because all of them underwent changeable interpretations) but also possible futures (suggested by the analyst or discovered by ourselves) yield both modifications of the 'present of things present' and, in consequence, of the course each of us decides to impress upon the host of our possible actions.

Furthermore: since we are permanently interacting with other people - and this congress is a representative example of such an interaction - then, under this caveat, an *exact* anticipation in the realm of humanistic disciplines (in the sense followed by physicists) becomes meaningless. This, just because that inherent unpredictability of human actions and of human lives is not taken into account in the disciplines framed by that physical paradigm. This does not mean a debasing of physical disciplines themselves but to their endeavoured extension to other areas.
As Gordon Pask said (1994): "... In doing so, these entities erect structures with comparable qualities, they delve into labyrinths commonly referred to as history.. But they do so by creating their OWN time, although an outside observer may see them, falsely, as embedded in the common time of clocks and sundials, which is a consensually approved coherence of moments and measuring instruments. At any rate is time to finish this paper".
I do agree.

IV-SELECTED REFERENCES

St. Augustine, *Confessions*, ed. Oxford University Press, 1991.

Bateson G., *Steps to an Ecology of the Mind*, NY Ballantyne, 1972.

Boorstin D., *The Discoverers*, Random House Inc., 1987.

Boscolo L. and Bertrando P.,*The Times of Time: A New Perspective in Systemic Therapy and Consultation*, W.W. Norton, 1993.

Eddington, *The Nature of the Physical World*, MacMillan, 1929.

Freud. S.,
.*Project for a Scientific Psychology*, Standard Edition, 1895.
.The Interpretation of Dreams, Standard Edition, 1900a.
.Formulation on the Two Principles of Mental Functioning, Standard Edition, 1911b.
.On Narcissism: an Introduction, Standard Edition, 1914c.
.The Unconscious, Standard Edition, 1915e.
.A Metapsychological Supplement to the Theory of Dreams, Standard Edition, 1917d
.*Beyond the Pleasure Principle*, Standard Edition, 1920g.
.The Loss of Reality in Neurosis and Psychosis, Standard Edition, 1924e.
.Negation, Standard Edition, 1925.

Goden C., *Social Time and Being*, Blackwell Oxford &Cambridge,1994.

Lanczos C., *The Variational Principles of Mechanics*, Oxford Univ. Press,1952.

Landau L. and Lifchitz E., Physique Statistique, Editions Mir, Moscou, 1967.

Lotka A., The Elements of Mathematical Biology, Dover, 1956.

Medina Martins P.,
.Metalogues: An Abridge of a Genetic Psychology of Non Natural Systems. CCAI The Journal for the Integrated Study of Artificial Intelligence Cognitive Science and Applied Epistemology, on 'Self Reference in Cognitive and Biological Systems', Vol. 12, 1994.
.Objectal Relations: a Cybernetic Approach. SPA, Lisbon, 1995.
.Emergence, Diachronism and Machines, sent to publication in *Intellectica*, 1997.

Pask G.,
.*Conversation, Cognition and Learning*. Elsevier, 1976.
.The Lies We Tell To Make The Life True, 94' European Congress on Cybernetics and Systems Research, Wien, 1994.

Piaget J., The Child's Conception of Time, Basic Books, 1946.

Prigogine I., *Entre le Temps et l'Éternité*, Librairie Arthéme Fayard, 1988.

Rousseau P, *Histoire de la Scienc*e,Librairie Arthéme Fayard, 1945.

Spitz R.,
.*The First Year of Life*, International University Press, 1965.
.*No and Yes: The Genesis of Human Communication*, International University Press,1966

Anticipatory Systems and Epistemology

Space-Time Framework of Internal Measurement

Koichiro Matsuno

Department of BioEngineering
Nagaoka University of Technology
Nagaoka 940-21, Japan
e-mail: kmatsuno@voscc.nagaokaut.ac.jp

Abstract

Measurement internal to material bodies is ubiquitous. The internal observer has its own local space-time framework that enables the observer to distinguish, even to a slightest degree, those material bodies fallen into that framework. Internal measurement proceeding among the internal observers come to negotiate a construction of more encompassing local framework of space and time. The construction takes place through friction among the internal observers.

Emergent phenomena are related to an occurrence of enlarging the local space-time framework through the frictional negotiation among the material participants serving as the internal observers. Unless such a negotiation is obtained, the internal observers would have to move around in the local space-time frameworks of their own that are mutually incommensurable. Enhancement of material organization as demonstrated in biological evolutionary processes manifests an inexhaustible negotiation for enlarging the local space-time framework available to the internal observers.

In contrast, Newtonian space-time framework, that remains absolute and all encompassing, is an asymptote at which no further emergent phenomena could be expected. It is thus ironical to expect something to emerge within the framework of Newtonian absolute space and time.

Instead of being a complex and organized configuration of interaction to appear within the global space-time framework, emergent phenomena are a consequence of negotiation among the local space-time frameworks available to internal measurement.

Most indicative of the negotiation of local space-time frameworks is emergence of a conscious self grounding upon the reflexive nature of perceptions, that is, a self-consciousness in short, that certainly goes beyond the Kantian transcendental subject. Accordingly, a synthetic discourse on securing consciousness upon the ground of self-consciousness can be developed, though linguistic exposition of consciousness upon self-consciousness remains necessarily under-complete analytically. For instance, the self-as-the-author is generative but local in its perspective, while the accompanied self-as-the-reader that can comprehend what the former self has produced is global but merely contemplative in accepting a completed discourse and not pragmatic any more. The self-as-the-author is conscious of itself by letting the self-as-the-reader, who happens to be a derivative of the former self, be aware of what the self has produced. Self-consciousness precedes consciousness. Consciousness can be vindicated only by securing the occurrence of the self-as-the-author in the perspective from the inside of a linguistic institution.

Keywords: Causality, Consciousness, Internal Measurement, Space, Time

1. Introduction

Emergent phenomena in general and biological evolutionary processes in particular are

CP437, *Computing Anticipatory Systems: CAYS--First International Conference*
edited by Daniel M. Dubois © 1998 The American Institute of Physics 1-56396-827-4/98/$15.00

taken to proceed both in space and in time. This observation naturally comes to remind us that the space-time framework we have to employ in any case could be a concomitant factor of the phenomenon to be examined and identified. Although this way of looking into the problem may sound quite philosophical, the issue of space and time would become unavoidable for critical examination of the nature of emergent phenomena. A crucial effort towards addressing the present problem goes back at least to the Critique of Pure Reason by Immanuel Kant in the late eighteenth century.

Kant argued that if there is a unified and organized experience to talk about in an objective manner, both the conceptions of space and time must be rooted in each experiencing subject prior to experiences. As for the conception of space, Kant stated:

"Space is not a conception which has been derived from outward experiences. For, in order that certain sensations may relate to something without me (that is, to something which occupies a different part of space from that in which I am); in like manner, in order that I may represent them not merely as without of, and near to each other, but also in separate places, the representation of space must already exist as a foundation. . . .

Space then is a necessary representation *a priori*, which serves for the foundation of all external intuitions. . . .

Space does not represent any property of objects as things in themselves, nor does it represent them in their relations to each other; . . .

Space is nothing else than the form of all phenomena of the external sense, the subjective condition of the sensibility, under which alone external intuition is possible." (Kant, 1952 (English translation), pp. 24-25).

In short, space to Kant is a container of objects which we can experience. We can eliminate these objects from the container, but cannot dispense with the container itself. If it were tried to eliminate the container, one would have to face the tenacious question of where the container could be eliminated from after all. Unless the container's container is available, it remains impossible to eradicate the container. The container necessitates another container on a higher level *ad infinitum*.

Following almost the similar line of argument, Kant reasoned that the conception of time is also exclusively subjective *a priori*.

"Time is not an empirical conception. For neither coexistence nor succession would be perceived by us, if the representation of time did not exist as a foundation *a priori*. . . .

Time is a necessary representation, lying at the foundation of all our intuition. .
. .

Time is not a discursive, or as it is called, general conception, but a pure form of the sensuous intuition. . . .

Time is nothing else than the form of the internal sense, that is, of the intuitions of self and of our internal state." (*ibid*, pp. 26-27).

The Kantian framework of space and time now imposes a specific constraint upon material phenomena observed in the empirical domain. Of upmost significance in this regard is the principle of causality. The conception of the relation of cause and effect in succession is taken to be prerequisite to apprehend the manifold of phenomena.

"For example, the apprehension of the manifold in the phenomenon of a house which stands before me is successive. Now comes the question whether the manifold of this house is in itself successive - which no one will be at all willing to grant. But, so soon as I raise my conception of an object to the transcendental signification thereof, I find that the house is not a thing in itself, but only a phenomenon, that is, a representation, the transcendental object of which remains utterly unknown." (*ibid*, p. 77).

While our apprehension of the house is successive as allowing the sequence of perceptions beginning at the roof and ending up at the foundation, or vice versa, the apprehension is not yet sufficiently distinguished from other apprehensions with regard to the order of the succession of perceptions. Kant made this point clearer by invoking another simple example.

"For example, I see a ship float down the stream of a river. My perception of its place lower down follows upon my perception of its place higher up the course of the river, and it is impossible that, in the apprehension of this phenomenon, the vessel should be perceived first below and afterwards higher up the stream. Hence, therefore, the order in the sequence of perceptions in apprehension is determined; and by this order apprehension is regulated." (*ibid*, p.78).

What is implied at this point is that the principle of causality requires for its own sake an *a priori* universal rule to specify the order in the sequence of perceptions. Consequently, it follows:

"For all experience and for the possibility of experience, understanding is indispensable, and the first step which it takes in this sphere is not to render the representation of objects clear, but to render the representation of an object in general, possible. It does this by applying the order of time to phenomena, and their existence. In other words, it assigns to each phenomenon, as a consequence, a place in relation to preceding phenomena, determined *a priori* in time, without which it could not harmonize with time itself, which determines a place *a priori* to all its parts. This determination of place cannot be derived from the relation of phenomena to absolute time (for it is not an object of perception); but, on the contrary, phenomena must reciprocally determine the places in time of one another, and render these necessary in the order of time." (*ibid*, p. 80).

Insofar as the unity of our experience through a sequence of perceptions in time is guaranteed, the order of the sequence, that is, the principle of causality has to be observed irrespective of the nature of phenomena to be experienced. So far, so good with the Kantian principle of the succession of time according to the law of causality. Nonetheless, the reciprocity in perceiving phenomena in temporal domain raises another reciprocity in spatial domain. However, the reciprocity we come to observe in spatial domain is quite different from the one in temporal domain. Phenomena to be perceived in space come to coexist if the perception of the one can follow upon the perception of another, and vice versa.

"Thus, I can perceive the moon and then the earth, or conversely, first the earth and then the moon; and for the reason that my perceptions of these objects can reciprocally follow each other, I say, they coexist contemporaneously. Now coexistence is the existence of the manifold in the same time. . . . It follows that a conception of the understanding or category of the reciprocal sequence of the determinations of phenomena (existing, as they do, apart from each other, and yet contemporaneously), is requisite to justify us in saying that the reciprocal succession of perceptions has its foundation in the object, and to enable us to represent coexistence as objective." (*ibid,* p. 83).

Coexistence of the manifold phenomena in one and the same time is thus based upon the relation of influence or the relation of community or reciprocity, that is, interaction in short. The present insistence on coexistence is, however, undoubtedly subjective, though it is intended to be grounded upon an objective basis.

"In the mind, all phenomena, as contents of a possible experience, must exist in community (*communio*) of apprehension or consciousness, and in so far as it is requisite that objects be represented as coexistent and connected, in so far must they reciprocally determine the position in time of each other and thereby constitute the whole. If this subjective community is to rest upon an objective basis, or to be applied to substances as phenomena, the perception of some substance must render possible the perception of another, and conversely." (*ibid,* p. 84).

Interaction as a dynamic community of reciprocal influences underlies the reciprocal sequence of perceptions that could in turn yield the perception of coexistence of objects. The present Kantian proof of the principle of coexistence is, however, different from other three proofs of subjective underpinning of space, time and causality. It refers to the conditions of things that are coexistent prior to how they are perceived. Before any phenomenon is perceived and experienced as such, the notion of interaction takes it for granted that all phenomena are connected in the dynamic community of reciprocal action to each other. The issue of aiming at the unity of experience through subjective perception thus comes to face a deeper problem of how to accommodate it to the unity of the universe, in the latter of which all phenomena are connected immediately or mediately.

This observation now gives rise to a convoluted situation of reflexive perceptions such that subjective perception of a perception could also be an empirical phenomenon in the unity of the universe. Needless to say, subjective perception is an empirical phenomenon unless it is further supplemented by the transcendental apprehension or apperception. Perception as an empirical phenomenon can render itself to be an object of further perception. The reflexive sequence of perceptions continues to hold indefinitely unless an artifact to stop it, such as the Kantian transcendental apprehension of perception, forcibly intervenes.

Nonetheless, each perception is a temporal phenomenon among themselves. A certain preliminary notion of time is inevitable in invoking perceptions. In fact, insofar as it is focused internally in the dynamic community of perceptions without referring directly to the unity of experience, each local perception comes to associate itself with a local time. Although the unity of experience, if ever possible, would require a synchronization among those local times in the light of time inherent in a transcendental subject, local perceptions participating in forming the unity come to constantly negotiate among themselves instead. Intrinsic to each local perception is the capacity of measuring local times pertaining to mutually influencing local perceptions immediately or mediately.

Internal measurement associated with each perception is eventually related to the unity of experience apprehended in terms of *a priori* time, but is more than that (Matsuno, 1989). It operates in a local time that does not presume an *a priori* synchronization among those local times. Each local perception or internal measurement is to proceed in a locally asynchronous time with each other (Matsuno, 1996a), while the transcendental apprehension of perceptions proceeds in a globally synchronous time that is unique to the Kantian transcendental subject. Transference from a locally asynchronous to globally synchronous time is inevitable for the unity of experience to be explicated on individual local perceptions (Matsuno, 1996b). Likewise, each perception is a spatial phenomenon of only a local extension unless the transcendental apprehension of space is forcibly imposed. For there is no empirical agency boasting of grasping space of an infinite extension.

Internal measurement upholding empirical perception being local both in space and in time serves as an elementary activity towards the unity of experience. Most indicative of the agency of internal measurement is a perceiving subject who has not yet incorporated into itself the Kantian transcendental apprehension of perceptions. That is a conscious self.

2. Conscious Self

Consciousness is about the act of being aware of events emerging from inside. Nonetheless, talking and writing about consciousness is an activity of being aware of what is being conscious, or awareness of an awareness in short. The distinction between consciousness and being conscious of consciousness is subtle, but is evident when one addresses the issue of consciousness. What concerns us most at this point is the nature of descriptive stance we take as facing the task of describing consciousness. Consciousness is intimately related to self-consciousness.

In this article, I shall develop a constructive strategy for approaching consciousness from self-consciousness. The main motivation for the present endeavor of constructing, instead of analyzing, consciousness rests upon the observation that whatever consciousness may look like, our effort for elucidating the nature of consciousness has to have recourse to a language that presumes the capacity of being conscious on the part of its speakers. We are already conscious beings in that practicing any languages is a conscious activity on the part of its users. Even if someone raises a question of what consciousness looks like, such question cannot be formulated unless a conscious being like we human being is taken for granted in the first place.

Consciousness could be dealt with only to the extent that we can reach consciousness from self-consciousness with the use of a linguistic vehicle. Even an experimental endeavor for neurophysiological understanding of consciousness presumes the presence of a linguistic means for its description. Primary significance of a language for the matter of consciousness is the occurrence of an author for any type of linguistic discourse.

3. The Self-as-the-Author

The present author who happens to be myself does not differ from any other authors in two respects. One is to wish to write something (Barthes, 1989), and the other is to rely upon a language for the purpose. The objective of this writing that has just been started is to construct a message that is both unique and specific to the present author but at the same time is intended to be comprehensible to many a reader. Underlying this enterprise is the interplay between the self-as-the-author and the self-as-the-reader (Gergen, 1984). The contrast between the two selves will become more pronounced if one more self to be described is introduced. For instance, that the presence of the self doubting everything else is irrefutable and doubtless is a statement intended to justify the self-as-being-described in view of the fact that if the self as a subject of doubting is doubted, the very deed of doubting would become no more tenable.

The doubtless certitude of the self doubting everything else that was first perceived by Rene Descartes is in its essence the statement authored by Descartes. The reader who can comprehend the self-as-being-described thus predicated turns out to be the self-as-the-reader whom Descartes originally intended to find in the audience. The self-as-the-reader assumes its own set of irreducible predicates to which every comprehensible statement can be reduced. If the self-as-the-reader takes the act of doubting everything else as the irreducible predicate of the self, the original intention of Descartes as the author would be met (Husserl, 1954). Descartes as the author claiming the self doubting everything else can reach any self-as-the-reader who happens to have the same set of irreducible predicates as Descartes has. Descartes as the author is of course legitimately followed by Descartes as the reader who maintains the same set of irreducible predicates as the author takes for granted.

However, it is by no means evident that every self-as-the-reader could share the same set of irreducible predicates. Descartes as the author of the statement claiming the self doubting everything else, for instance, cannot find in the sympathetic audience Friedrich Wilhelm Nietzsche who takes the will to power to be a most fundamental predicate of the self (Smith, 1992). The self-as-the-author can choose an arbitrary set of irreducible predicates in order to implement its own will to writing, but there is no guarantee that the resulting writing may reach many a self-as-the-reader because of the absence of the commonality of irreducible predicates to be shared among themselves. The inescapable problem with any self-as-the-author is how to construct a writing on a possibly shared common set of irreducible predicates.

The self-as-the-reader serves as a subject of pure reasoning in the sense that the set of irreducible predicates or categories is a priori given and that every descriptive object may be analyzed by the subject accordingly (Primas, 1993; Atmanspacher, 1994). Once such an a priori set of irreducible predicates is observed by the self-as-the-reader, it will be straightforward to analyze the self-as-being-described in terms of those pre-given categories. In particular, the self-as-the-reader can happen to be identical to the self-as-being-described if the self is described in terms of those irreducible predicates that the self-as-the-reader takes for granted. This assimilation of the self, however, does not apply to the self-as-the-author who has the capacity of both generating and varying the fundamental set of irreducible predicates. Descartes as the author is simply incommensurable with Nietzsche as the author in that each employs a different set of predicates for describing the self. Still, the self-as-the-author, whether that may be Descartes or Nietzsche, or anyone else, has already been described as such. It is both ironical and self-contradictory to see that the self-as-the-author being responsible for generating and varying the fundamental predicates has let itself be subject to an arbitrary set of predicates that has been prepared by whatever means.

The modest way to eliminate the formidable problem of this sort would be to abandon the problem itself altogether by stating that the self-as-the-author as a subject of practice goes beyond the subject of pure reasoning as championed by Immanuel Kant (Husserl, 1954). One is not allowed to ask and to predicate what the self-as-the-author is all about, because answering this question does require a trustworthy set of predicates in one form or another. Nonetheless, once the self-as-the-author or the subject of practice has been referred to, it would become inevitable to ask how in the world such a self could come into existence. Referring to the self-as-the-author is an act of calling the self into existence. The self-as-the-author is necessarily in the process of becoming just in the respect that the self has happened to refer to itself as the self-as-the-author, though what the self is all about

106

remains private, local and indefinite (Rössler, 1996).

The burden upon answering the question of how the self-as-the-author could come into existence is, however, less stringent than that for predicating what in the world the self looks like. Historical presence of both Descartes and Nietzsche as authors is incontestable, but the extent to which their writings could be legitimate remain controversial. At issue is how could one, including the present author too, justify the coming-into-being of the self-as-the-author, or the subject of practice, through a descriptive enterprise. Unless founding the self-as-the-author upon a firm descriptive ground is available, it would be of no use to refer to the self-as-the-author itself in writing. Although the self-as-the-author may seem to stand alone once referred to, its legitimacy cannot be found within the description in terms of any pre-given set of irreducible predicates. What is required to do is to justify the self-as-the-author as a class without relying upon any of definite sets of irreducible predicates, though each author like Descartes or Nietzsche may be entitled to have his own choice of a definite set of irreducible predicates arbitrarily within the allowable class. Only when the self-as-the-author may be guaranteed as a class, could one, the present author included, expect to proceed further in making a choice of an appropriate set of irreducible predicates.

The task of the present author as a self-as-the-author is accordingly somewhat convoluted in that it is required to generate or construct the self-as-the-author as a class through a descriptive discourse. The present author wishing to underwrite its own coming-into-being as an author to be found in the class has just started to complete its objective. Until such an objective of underwriting is met, the present author as well as any other author for this matter cannot claim its own birthright. Although it may seem all too obvious that the self-as-the-author can claim the legitimacy of its own presence once the writing project gets started, the fragility of the self would soon come to the surface once someone is allowed to ask the question of how could the author legitimately claim its choice of irreducible predicates over the other alternatives. Nietzsche may dismiss the Cartesian self of doubting everything else from the perspective of the will to power, but did not provide the impartial third party that could decide for the will to power against the Cartesian self. If the self-as-the-author is equated to a being carrying a particular choice of irreducible predicates, the charge against the lack of the uniqueness in the choice would necessarily dismiss the presence of such self as being contingent upon frivolity. In order to secure the self-as-the-author who is in turn responsible for giving birth to both the self-as-being-described and the self-as-the-reader, it would be absolutely required to descriptively construct the self-as-the-author as a class that has the capacity of choosing a definite set of irreducible predicates.

Descartes, Nietzsche and even the present author have one thing in common. All of them have something to say by employing languages as a vehicle. Once the self-as-the-author has been justified, the remaining problem would be which predicates to choose. Having something to say clearly differs from having already said. What is intended to do in the remaining part of this article is to linguistically justify the intention of having something to say on the part of the self-as-the-author from the perspective that neither the self-as-being-described nor the self-as-the-reader could follow unless the self-as-the-author may be guaranteed.

4. Transient Formal Language

The self-as-the-author takes intentionality toward writing something for granted. However, the self cannot regard intentionality per se as being an irreducibly fundamental predicate (Bechtel, 1985) because no one has the prerogative of prohibiting the question of what intentionality looks like from being asked. Once such a question is asked, the reducibility of intentionality to something else would indiscriminately be forced irrespective of whether it may be possible in the first place. Unless equipped with a trustworthy set of irreducible predicates, the only alternative would be to construct the will to writing upon the least set of premises that any of the self-as-the-author is going to agree to share. What will be intended in the following is to linguistically construct the self-as-the-author as a class based upon the common denominator among any of the selves-as-the-authors.

Before entering the construction, two remarks are in order. One is that the present construction, if ever succeeded, would turn out to be no more than a form of circular argument though, hopefully, not a type of vicious circle. One more remark is that even if it is a circular argument, the possible construction of the self-as-the-author could enjoy the privilege of supporting intentionality or the will to writing linguistically without confusing the will to writing with what has been written.

First of all, we as the editorial author start from the premise that the self-as-the-author is provided with a kind of formal language equipped with a definite set of irreducible predicates. A language for practicing sciences is necessarily formal. Any statement, once constructed by the self-as-the-author, turns out to be the one to be comprehended by the same self-as-the- reader. What is basic to any self-as-the-reader is to reduce whatever statement into its irreducible predicates that have already been given. The self-as-the-reader, that is a necessary outcome from the act on the part of the self-as-the-author, is involved exclusively in contemplating on any form of descriptive analysis in terms of a definite set of irreducible predicates, while the self-as-the-author utilizes the formal language in order to establish a bridge with its external referents. That the self-as-the-author is concerned with establishing a de novo relationship between a kind of formal language and its external referents is perceived by the self-as-the-reader as varying and disturbing the choice of a formal language to be utilized.

A formal language prepared by the self-as-the-author for the sake of the effecting self-as-the-reader is not strictly formal because of its reference to its external referents. Accordingly, a formal language with which any self-as-the-reader is concerned is at most transient in the sense that it is subject to disturbances of exogenous origin through referring to its external referents that remain contingent. The self-as-the-author thus categorically comes to provide the self-as-the-reader with a type of transient formal language to be influenced by disturbances of exogenous origin. The role of disturbances of exogenous origin is to exert a form of perturbations upon the set of irreducible predicates as met in a forced association of more than two different predicates. These perturbations upon the set of irreducible predicates are simply a way and means of describing what is going on within the external referents.

Transient formal language under the influence of disturbances of exogenous origin is thus seen to be a common denominator applicable to any of the self-as-the-author. The goal of our endeavor is to generate intentionality or the will to writing from the transient formal language in the manner that the self-as-the-reader may be able to perceive it. We shall first see what kind of structure the transient formal language could maintain in itself as a preliminary step toward the stated goal.

One of the fundamental attributes associated with transient formal language thus introduced is that the language itself is a structured organization, though of course variable, that can be caused and maintained by disturbances of exogenous origin. The structured organization that the self-as-the-reader or equivalently, the self-as-the-observer, can perceive as such is more than just an invariant set of irreducible predicates. When the self-as-the-author tries to construct whatever statement by employing a set of supposedly irreducible predicates, there should potentially be an indefinitely wide variety of statements

108

each of which could constitute a structured organization in terms of those predicates. A superb example of the structured organization of predicates is a monolingual dictionary, in which every predicate is associated with others in its own way. Such an association among different predicates can be established through the process of relating an arbitrary linguistic expression to its external referents under the control of the self-as-the-author.

In principle, the association can maintain in itself unfathomable immense possibilities simply due to the fact that any external referent could look differently depending upon the viewpoint that the concerned self-as-the-author may take. The self-as-the-author constantly perturbs the structured organization of predicates by generating de novo association of these predicates.

The self-as-the-reader views the emergence of such a new association as disturbances acting upon the structured organization that can be equated to the structure that the transient formal language could have come to maintain within itself. Although the self-as-the-author may have a good reason to come up with a new association of predicates like Descartes and Nietzsche have done, what the self-as-the-reader can appreciate at least minimally is a spontaneous occurrence of perturbations upon the then available structured organization of predicates. When these structural perturbations are absent, the transient formal language would literally reduce to a type of formal language in which the set of irreducible predicates remains invariant without allowing any further association among themselves.

Transient formal language supplemented by disturbances of exogenous origin is the self-as-the-reader's way of looking at what the self-as-the-author is doing. Since the self-as-the-author as the subject of practice commits itself to behaving and making choices in temporal domain, those disturbances acting upon the transient formal language are temporally variable. As a matter of fact, transient formal language supplemented by temporal disturbances of exogenous origin is an expression of the structure of a natural language. The emphasis is on the fact that a natural language would reduce to a formal language equipped with a fixed set of irreducible predicates at the vanishing limit of those temporal disturbances causing new associations of predicates. The dynamics of transient formal language in the presence of disturbances of exogenous origin is simply a crude form of describing how a natural language would evolve, with the consequence of revising its dictionary constantly.

The structured organization of transient formal language as exemplified in a dictionary of a natural language compiled at a certain historical period has been realized through the mutual compensation between two opposing movements, one is enhancing and the other is deteriorating the extent of organization. Enhancement of the structured organization would proceed through increasing the frequency of referring to a particular pattern of association among available predicates, while deterioration of the organization would follow less frequent reference to its pattern. The self-as-the-reader comes to see that the structured organization of exogenous origin, can be maintained as being driven by associative disturbances. The structured organization of transient formal language perceived by the self-as-the-reader now reduces to a disturbance-driven organization. As a matter of fact, disturbance-driven organizations are ubiquitous in evolution. Further elucidation of the structured organization of transient formal language will become available by uncovering the characteristics latent in disturbance-driven organizations.

What is common to any disturbance-driven organization is that the most likely organization to be realized is the one that can minimize the rate of its deterioration measured when left alone in the absence of any disturbances (Matsuno, 1978, 1995). In view of the fact that disturbances of exogenous origin serve as a factor of association among available components, the absence of such disturbances would as a matter of course mean deterioration of the once established association or organization. Consequently, granted that those disturbances occur at most intermittently, the realized organization is to maintain its own structure of association to the extent that the deterioration expected during the absence of associative disturbances may just be compensated by the disturbance-driven association. Any organization having the greater rate of deterioration would have no chance of survival compared to the one compensating its inevitable deterioration by the counterpart of new associations.

The principle of least deterioration rate thus formulated, when applied to transient formal language supplemented by disturbances of exogenous origin, serves as a selection sieve for the structured organization to be realized. At this point, however, it should be emphasized that the selective sieve characterized by the least deterioration rate is exclusively passive in its operation, whereas all the active attributes are sought in the associative disturbances. Insofar as the structured organization of transient formal language remains passive toward disturbances of exogenous origin, the self-as-the-reader cannot find any symptom of intentionality or the will to doing something in the very structured organization. The structured passive organization cannot deal with the origin of intentionality or activity. In order to generate the self-as-the-author or the subject of practice through a linguistic means, it is required to incorporate some of the active attributes into the structured organization explicitly. What must be sought at this point is the structured active organization of transient formal language.

5. Structured Active Organization

Arbitrary association of pre-existing predicates is unquestionably a source of disturbances acting upon the then available transient formal language. However, such an association is not the only source of disturbances. Unless it is prohibited to ask the question of how to paraphrase and decipher those predicates that cannot be found in possible associations of pre-existing predicates, the concerned transient formal language has to have the endogenous capacity of fixing such a linguistic incompleteness. There is in fact no authority prohibiting the self-as-the-reader from asking the definitive content of a predicate's predicate, a predicate's predicate's predicate, and so on. Incompleteness internal to any transient formal language is seen in the fact that there can be no forcible means for prohibiting the questions of an infinite regression type from being asked. Conversely, such an incompleteness has to be compensated internally in order to maintain the organizational integrity of the language or its structural organization.

This process of compensation is internal or endogenous to the transient formal language concerned in the sense that the organizational activity for eliminating the incompleteness in the preceding causes further incompleteness to be eliminated subsequently (Derrida, 1973; Salthe, 1993). This corresponds equivalently to raising and then answering the questions of an infinite regression type. The structured organization of transient formal language thus turns out to be equipped with the endogenous capacity of seeking its own organization (Shotter, 1983). The self-as-the-reader can view it as the structured organization supplemented by disturbances of endogenous origin. Because of the endogeneity, the organization naturally becomes active on its own.

The structured organization of transient formal language supplemented by disturbances of endogenous origin is always in the process of being organized in the sense that the preceding organization constantly serves as the cause for the subsequent organization to follow (Matsuno, 1989, 1996). The Cartesian self, for instance, could survive as a fundamental predicate until what it should be all about comes to finally be asked. Nietzsche who raised this devastating question eventually came up with Zarathustra's ego grounded upon the will to power (Smith, 1992). And again, Sigmund Freud who critically examined whether each of the Cartesian self and Zarathustra's ego could really be irreducible came to create the Freudian ego imputed to unconsciousness to break the impasse (Yearley, 1985).

The sequence from the Cartesian self through Zarathustra's to the Freudian ego that is conscious of unconsciousness is an instance of reducing everything to everything else (Van de Vijver, 1995). Unless such a reducibility is arbitrarily curtailed, the structured organization of transient formal language is necessarily in the process of being organized while constantly being subject to disturbances of endogenous origin. The structured organization supplemented by disturbances of exogenous origin is just an extreme case in which irreducibility of fundamental predicates is forcibly imposed.

The structured organization of transient formal language not constrained by irreducibility of an imposed character, when perceived by the self-as-the-reader, turns out to be active by itself. For progression of its organization is internally activated in a continual fashion. This activity is synonymous with self-referentiality in the sense that answering a possibility of reducing something to something else never fails to invite questions asking its further reducibility. The self-as-the-author is certainly self-referential in keeping writing as referring to what it has written. The self-as-the-reader now perceives this self-referentiality within the structured active organization of transient formal language that is entailed and constrained by disturbances of endogenous origin. The generative self-referentiality latent in the structured active organization is now found to be a class-property common to any self-as-the-author that the self-as-the-reader can comprehend as such.

The self-as-the-reader can equate the self-as-the-author to the generative self-referentiality latent in the structured active organization as a class even though the reader may not be knowledgeable about what the author is going to write. Conversely, generative self-referentiality intrinsic to the self-as-the-author gives birth to the similar generative self-referentiality in terms of transient formal language unless stipulated by imposed irreducibility of underlying predicates.

This recursiveness in turn provides the self-as-the-reader as a protege of the self-as-the-author with a legitimate opportunity for the genesis of another self-as-the-author. That is to say, the primary self-as-the-reader as a protege of the primary self-as-the-author can see the birth of the secondary self-as-the-author with the aid of generative self-referentiality of the transient formal language to rely upon. The present convoluted interrelationship among selves of different origin now paves the way for the secondary self-as-the-reader as the protege of the secondary self-as-the-author to prove and justify the primary self-as-the-author by following the similar logic in the reversed direction.

The primary subject of practice as the self-as-the-author can generate a secondary subject of practice in the manner that the primary subject of recognition and contemplation can comprehend. Likewise, the secondary subject of practice can generate the primary subject of practice in the manner that the accompanying secondary subject of recognition and contemplation can comprehend. This circularity in fact constitutes a scheme of grounding subjectivity upon symmetric intersubjectivity. An essence of the proclaimed subjectivity upon symmetric intersubjectivity is in the circularity that if you prove me, then I can prove you.

6. Symmetric Intersubjectivity

Any subject of recognition and contemplation, whether it may be either the subject of observation or of reading, tries to decipher whatever object in terms of fundamental predicates that would remain invariant. Whether there should be any invariant characteristic at all can be examined by checking the symmetric operations preserving the set of irreducible predicates invariant. For instance, a dictionary might become a representation of such an invariance if the manner of relating each predicate to others were symmetric among themselves in the sense that the entry of the dictionary remains fixed. There is, however, no guarantee that the set of fundamental predicates to be found in a dictionary would remain invariant against the operation of predication among themselves. On the contrary, the self-as-the-author takes it for granted that no imposed irreducibility may apply to the set of predicates. The present lack of both an invariant set of predicates and the associated symmetric operations urges us to look for a different type of symmetric operations in order to secure the subject of practice as an actor.

We have already seen that the primary self-as-the-author generates the secondary self-as-the-author in the manner that the accompanying primary self-as-the-reader can comprehend. Interchange of the role between the primary and the secondary self in turn generates the primary self-as-the-author. This relationship is symmetric in generating the generator, but not in what the generator generates in an exhaustive manner. The symmetry in generating the generator refers to a hyperspace in which both the generator and what is to be generated are included altogether. Neither the primary nor the secondary self can claim to globally grasp such a hyperspace in its every definite detail because each of them is a member constituting the hyperspace from the inside. There is necessarily required an involvement of the third-party self-as-the-author that can grasp what is going on in the interplay between the primary and the secondary self, even though the third-party self is a protege of the secondary self.

The primary self-as-the-author just like the present author is constantly concerned with the process of symmetry breaking by defying an imposed irreducibility of fundamental predicates. This process of symmetry breaking can be stopped only in the exceptional case such that there should be a set of symmetric operations and the associated invariants in the manner being independent of any self-as-the-observer.

The presence of symmetries and the associated invariants that could exist independently of the observer is the ones met in physics (Ne'eman, 1990). Precisely for this reason, the primary self-as-the-observer accompanied by the primary self-as-the-author can identify the symmetries and the invariants by employing an appropriate symmetric association of irreducible predicates. The presence of something invariant that remains as such independently of how it may be observed and described provides itself with an opportunity of being properly represented by describable operations of symmetry and the associated invariants, because the objective symmetry and invariants are guaranteed to remain unaffected even if their description is attempted.

Once it is admitted that there should be no symmetries and invariants to be found out there, what is left to the self-as-the-author is to constantly concern itself with the deed of writing. Still, the self-as-the-reader as the protege of the self-as-the-author incessantly commits itself to figuring out the underlying symmetry and the associated invariants, since the objective of recognition and contemplation is to reach an invariant association of irreducible predicates. This internal conflict between the author as an agent of symmetry-breaking and the reader as an observer of symmetry and invariants is intrinsic and ubiquitous (Gunji, 1995; Conrad, 1996). Although insistence on physical symmetries and invariants is certainly one attempt to mitigate the internal conflict, this resolution that is common in physics is still domain-specific.

One more resolution which we have tried is to look for a form of symmetric operations available in the hyperspace in which the original internal conflict between the author and the reader as the protege of the former can be avoided. If there is guaranteed a third-party subject observing that there are two subjects each of which is involved in proving the other, the third-party subject can recognize how each subject comes into existence by referring to the underlying symmetric operation of generation. This observation remains unaffected even if the third-party subject is given its own birth by one of the two subjects on the scene. In addition, exhaustive description of each subject is not attempted beyond bilateral proving of each other. The present tripartite relationship is summarized as saying that I can see that both of you prove each other, though I don't know who and what each of you are at all.

Subjectivity grounded upon symmetric intersubjectivity is a class-property common to any subjects. Because of this commonality, it is conceivable that each subject may have more specific capacities other than proving other subjects without being acquainted with who they are.

Any organism that experiences its outer world deciphers whatever input from the outside in terms of fundamental predicates unique to the experiencing organism. Examples of such predicates are attractants and repellents to protozoa such as paramecium, and predator and mate to an animal (Dewsbury, 1989). If those fundamental predicates remain irreducible to the experiencing organism, the predication is simply an awareness of the input from the outside, and not a mode of self-awareness. For the irreducibility of the predicates, instead of the organism itself, is responsible for being aware of the outside world. If the irreducibility of predicates is given, the predicated outside world would accordingly also be regarded as being given while prohibiting the organism from actively participating in forming and constructing the outside world.

On the other hand, however, once the condition of imposed irreducibility is removed, the preceding predication of the outside world is constantly modified by the succeeding predication. That an organism can modify the predication made by itself previously is to become aware of what the self has done (Ulanowicz, 1996). Self-awareness is characteristic to the subject of practice who can alter whatever predications made by the same subject previously. Self-referentiality not conditioned by imposed irreducibility of predicates underlies the capacity of self-awareness. For instance, the Cartesian self is not by itself irreducible, otherwise the self would loose its self-awareness. The self-as-the-author is constantly involved in the self-awareness activity by answering the questions of an infinite regression type which the self frames by itself, while the self-as-the-reader tries to do every effort to decipher whatever object in terms of the then available fundamental predicates. Self-awareness inherent to the self-as-the-author is within its dissatisfaction with any one of the Cartesian self, Zarathustra's or the Freudian ego as a fundamental predicate of the self.

A subject having self-awareness capability invariably carries an internal conflict in itself. If the activity of self-awareness is carried into its extremity such that the questions of an infinite regression are literally pursued by the self, it would become untenable that the self-as-the-reader claims itself for sure. For the self-as-the-author as the cause of the self-as-the-reader would make itself unsettled by stepping down the devastating ladder of an infinite regression, while the self-as-the-reader pretends to address itself toward what is persistent and invariant. In spite of this inevitable internal conflict of the divided selves between the author and the reader, it is still possible to justify the occurrence of subjectivity based upon symmetric intersubjectivity.

Clue to the justification of subjectivity is to refer to the symmetric operation that can keep the generation process of whatever kind going on. Even if individual operations are asymmetric or symmetry-breaking, there may be the case that the group of operations of the original individual operations could maintain a symmetry property. Grounding subjectivity upon symmetric intersubjectivity just happens to be the instance that the group of operations at an aggregated higher level can maintain a certain symmetry while the constituent individual operations at the lower level lacks the symmetry property found at the higher level.

7. Concluding Remarks

Any discourse on consciousness requires self-consciousness because the discourse necessitates involvement of the self-as-the-author who is aware of talking about the issue of consciousness. Linguistic exposition of consciousness upon self-consciousness thus conceived remains necessarily under-complete analytically. Precisely for this reason, the primary self-as-the-author is generative and competent in generating the secondary self-as-the-author that could be recognized as an active agent by the self-as-the-reader owing its birth to the primary self.

Consciousness viewed as a generative capacity latent in any of the self-as-the-author is intrinsic to the institution of a language. The linguistic institution is conscious of itself in keeping an arbitrary self-as-the-author involved there constantly modifying and adding to what the institution has accomplished.

One of the possibilities for approaching the issue of consciousness on experimentally material grounds may be to have recourse to the occurrence of a linguistic institution. If one can find an analogue of linguistic institution in a material body, the likelihood of securing the occurrence of consciousness on a physical ground could be expected. Prerequisite to such an enterprise, if ever possible, could be a viewpoint of the self-as-the-author who is generative but necessarily local, compared to the self-as-the-reader who is global in its perspective but merely contemplative and not pragmatic. That is a view from the inside. Material underpinning of consciousness should be upon appraisal of material interactions that are strictly local. Material manifestation of signal and communication of a local character is in order for addressing the issue of consciousness.

References

Atmanspacher, H., 1994. Objectification as an endo-exo transition. In: *Inside Versus Outside* (H. Atmanspacher & G. J. Dalenoort, Eds.). Springer, Berlin, pp. 15-32.

Barthes, R., 1989. To write: an intransitive verb? In: *The Rustle of Language* (Howard, R., transl.). Univ. California Press, Berkeley.

Bechtel, W., 1985. Realism, instrumentalism, and the intentional stance. *Cog. Sci.* **9**, 473-497.

Conrad, M., 1996. Cross-scale information processing in evolution, development and intelligence. *BioSystems* **38**, 97-109.

Derrida, J., 1973. Difference. In: *Speech and Phenomena and Other Essays on Husserl's Theory of Sign* (Allen, D. B., transl.). Northwestern Univ. Press, Evanston Illinois.

Dewsbury, D. A., 1989. Comparative psychology, ethology and animal behavior. *Ann. Rev. Psychol.* **40**, 581-602.

Gergen, K. J., 1984. Theory of the self: impasse and evolution. *Adv. Exp. Soc. Psychol.* **17**, 49-115.

Gunji, Y.-P., 1995. Global logic resulting from disequilibration process. *BioSystems* **35**, 33-62.

Husserl, E., 1954. Die Krisis der europaischen Wissenschaften und die transzendentale Phanomenologie. In: *Husserliana Bd. VI*. Martinus Nijhoff, Berlin.

Kant, I. 1952. *Critique of Pure Reason*, 2nd Edition (Meiklejohn, J. M. D., transl.) Encyclopaedia Britanica, Inc., Chicago.

Matsuno, K., 1978. Evolution of dissipative system: a theoretical basis of Margalef's principle on ecosystem. *J. Theor. Biol.* **70**, 23-31.

Matsuno, K., 1989. *Protobiology: Physical Basis of Biology.* CRC Press, Boca Raton Florida.

Matsuno, K., 1995. Consumer power as the major evolutionary force. *J. Theor. Biol.* **173**, 137-145.

Matsuno, K., 1996a. Internalist stance and the physics of information. *BioSystmes* **38**, 111-118.

Matsuno, K., 1996b. Symmetry in scynchronous time and information in asynchronous time. *Symmetry: Culture & Science* **7**, 295-305.

Ne'eman, Y., 1990. The interplay of symmetry, order and information in physics and the impact of gauge symmetry on algebraic topology. *Symmetry: Culture & Science* 1, 229-255.

Primas, H., 1993. The Cartesian cut, the Heisenberg cut, and disentangled observers. In: *Symposia on the Foundation of Modern Physics: Wolfgang Pauli as a Philosopher* (K. V. Laurikainen & C. Montonen, Eds.) World Scientific, Singapore, pp. 245-269.

Rössler, O. E., 1996. A remark made in the presentation in the occasion of the *Conference on the Foundation of Information Sciences* at TUW in Vienna on 13 June 1996.

Salthe, S. N., 1993. *Development and Evolution: Complexity and Change in Biology.* Bradford/MIT Press, Cambridge Mass.

Shotter, J., 1983. Duality of structure and intentionality in an ecological psychology. *J. Theor. Soc. Behav.* **13**, 19-43.

Smith, C. U. M., 1992. Zarathustra's evolutionary epistemology. *J. Soc. Evol. Syst.* **15**, 75-85.

Ulanowicz, R. E., 1996. Ecosystem development: symmetry arising? *Symmetry: Culture & Science* 7, 321-334.

Van de Vijver, G., 1995. The relation between causality and explanation in emergentist naturalistic theories of cognition. *Behavioural Processes* (Elsevier, Amsterdam) **649**, 287-297.

Yearley, S., 1985. Imputing intentionality: Popper, Demarcation and Darwin, Freud and Marx. *Stud. Hist. Phil. Sci.* **16**, 337-350.

SOME ISSUES ON THE CONSTRUCTION OF ORDER IN SELF-ORGANIZING SYSTEMS. ITS REVELANCE TO ANTICIPATORY SYSTEMS

Juan M. Alvarez de Lorenzana
Research Associate
Institute of International Relations
C456-1866 Main Mall
University of British Columbia
Vancouver, B.C. Canada V6T 1Z1   email:
alvarez@unixg.ubc.ca

ABSTRACT

Starting from a real and concrete problem, that of *technological transfer*, which is also very complex, we are faced with the frequent dilemma of either favoring the characteristics that need to be adressed in order to obtain a deep solution of the problem or favoring the use (and bias) of standard formal representations in the definition of *well posed* problems. This is the setting in which we draw an initial heuristic scheme for systems and open systems; from open systems we go on to self-organization; from self-organization to modeling relations; from modeling relations to anticipation and anticipatory systems. Through this sequence we explore and tackle some of the issues of system modeling and representation and end up assessing similarities between self-organization and anticipation in the light of open systems.

KEYWORDS

Indistinguishability, hierarchy of spatio-temporal scales, construction of order, self-organizing systems, anticipatory systems.

1. INTRODUCTION

My field of inquiry is social systems. Some years ago I became motivated by the issue of **technological transfer** among countries; specifically, between developed and less-developed countries. At the time, the record on this had not been good, success being the exception rather than the rule.

My initial understanding of why the record on this issue had been poor was that, at the core of the failures, there was a lack of understanding of the principles involved in any act of *transferring*, and, in particular, in transferring technology, which is communication by means of technological constructs.

The problem had the added constraint of being a "transfer" between socio-economic systems with different cultural contexts. So, there was also a need to understand what would be the impact (on the "transfer" processes that were taking place) of the *different specifications in cultural contexts of each system taking part in the interaction.*

Based on these initial perceptions of the problem, any framework had to address three main systemic subject matters:

CP437, *Computing Anticipatory Systems: CAYS--First International Conference*
edited by Daniel M. Dubois © 1998 The American Institute of Physics 1-56396-827-4/98/$15.00

(a) *protosystemic*: the relation between cultural contexts and the definition of systems;
(b) *intrasystemic*: the unfolding of the system, i.e., development and developmental processes;
(c) *intersystemic*: relation among systems, each one steming from a specific cultural context which might be considerably different from others (at least where developed and less-developed countries were to be involved).

I worked on a conceptual scheme to be called Evolutionary Systems Framework, ESF for short (Alvarez de Lorenzana, 1989). This attempt was initially stated as an heuristic set of principles trying to deal with two types of emergence: evolution and development.[1]

At this point there was a dilemma that had to be addressed, the result of which would, somehow, condition the whole process and eventual outcome of this inquiry. Do I pose the "problem" as a non-negotiable benchmark and develop a strategy based on the integrity of the problem itself at the cost, if necessary, of dismissing current formal approaches or, to the contrary, should I establish a research path that is based on what is available in terms of precedence and standard mathematical techniques, even if this means *distorting* the problem in order to accomodate it to those procedures which are not entirely satisfactory? I chose the former option, which meant no compromise about the *intuitional contours* of the problem in question and very few clues as to what should be the foundational grounds to formally define and eventually represent, model and simulate the problem.

I did not look for this dilemma, it was there before I entered the arena. Being on the side of the "problem" rather than on the side of the techniques, the way to go was to try to define, as minimally as possible, the postulational pillars on which the "problem" would have to be understood and explained.

One of the difficulties often encountered when describing problems of evolution and development is the lack of a fully adecuate formal representational language. In many cases, assumptions about the natural systems do not match the definitional boundaries of the mathematical structures chosen for their representations. We cannot blame the mathematician for not figuring out what the biologist or social scientist, attempting to formalize in their own fields, is going to need or do with those formal constructs. So, biologists, social scientists (let's not to forget physicists and other "hard" scientists), are often left facing what turn out to be subtle mathematical problems that can only be properly stated and resolved at the foundational level. No small task, even for an accomplished mathematician!

2. EVOLUTIONARY SYSTEMS FRAMEWORK (ESF)

ESF consists of three principles or "primitives" on which this approach is based:

> Principle of Generative Condensation (PGC)
> Principle of Combinatorial Expansion (PCE)
> Principle of Conservation of Information (PCI)

PGC establishes the relationship between environment and system.

PCE establishes the systemic unfolding through the environment which is manifested by means of a variety of scales, ranging from the most local to the most global systemic dynamics. The dynamics are scale-dependent and the set of scales is systemically constructed through a "process dynamics" (Alvarez de Lorenzana, 1988) taking place between system (system's components) and environment.

---

[1] "evolution" will be concerned with *proto*-systemic aspects; "developement" will be concerned with *intra*-systemic phenomena (Alvarez de Lorenzana and Ward, 1985; 1987).

PCI establishes a link between the above two principles in the sense that although there is a change in information when going from the state of *being* of an environment to the state of *becoming* of a system (which is steming from it), there is a *trade* in terms of information which, in fact, asserts conservation of information throughout the process of *qualitative* change. This is, obviously, an important element to have and to consider in such a transformation (Alvarez de Lorenzana, 1991a).

3. SOME INITIAL COMMENTS ON THE ENCODING/ DECODING DIFFICULTIES IN MODELING

I have already pointed out some of the difficulties I started with. In the ESF scheme, in particular, there were three problematic issues I felt had to be dealt with. The first problem had to do with finding a way to generate a mathematical construct able to encode systemic self-organization. That meant the generation of a hierarchy of scales which was not arbitrary (i.e., not biased) in the sense that the generation of levels would be determined by the intrinsic logic of the process and not by some external criteria. After a fairly convoluted search [Alvarez de Lorenzana, 1989] I found a mathematical structure, called Combinatorial Hierarchy (CH), that had the right attributes (Bastin et al., 1979).

Finding CH allowed me to deal with one of the issues, which I call *systemic development*. But that finding increased the pressure to address the second problem: the relation between system and environment (cultural environment for social systems). It turns out that CH provides the *syntax* for constructing variety and, therefore, can give us the "intrinsic logic" of a process of entailment[2] *as long as we keep the semantics of the system out of it*. Semantics is brought into the system by defining such a system in terms of a set of initial (external) constraints, originally called "global properties," concerning space, time, the system's potential and the information content of the initial collection of components (Alvarez de Lorenzana and Ward, 1985, 1987).

The fact that we separate semantics from syntax does not mean, in any way, that we can dispense with either one or that we can equate semantics to syntax or that there is no relation between them. Quite the contrary. Every syntax has to be anchored on a given semantics: there is no syntax without a semantical ground base. At the same time, in an evolutionary framework there is no absolute semantics. Each semantics is relative, and every syntax having to stem from a (relative) semantics is, in the end, also relative. This means that semantics cannot be reduced to syntax or syntax to semantics.

The requirement to separate semantics from syntax, has a difficult translation into the formal system (i.e., the mathematical encoding) because it implies context dependency. This takes us to the next problem.

The third problem facing ESF, was the need to establish a sort of ranking among ES, so that, different ES with different degrees of evolutionary past could have a standard way to relate to each other and, consequently, have their mutual interactions formalized. In essence, there was need for a hierarchy of hierarchies, i.e., a *metahierarchy* (Alvarez de Lorenzana, 1993).

In a recent paper Robert Rosen (1993) talks about "good mathematics and bad models." In my view, it is necessary and important to recognize the interplay between mathematical tecniques and external referents. There is a need for a balance between knowledge about natural systems and sophistication of formalisms to be used in their representation.

---

[2]Entailment plays in FS the same role that "causality" in NS. See Rosen (1991) for an extensive treatement of entailment.

4. OPEN SYSTEMS AND SELF-ORGANIZATION

Let us move on into the business of how to conceptualize systems.

4.1 On Systems and Environemts.

Previous to any study of systems, the first act made by the scientist is, typically, the partition of the universe under scrutiny into *systems* and their surroundings, which we generally call *environments*.

As Robert Rosen tells us (Rosen, 1991: 42), systems and environments are seen in very different ways and their descriptions and representations are stated also very differently.

Systems (**S**): are described in terms of states which, in turn, are determined by observation.

Environments (**E**): are characterized by the impact that they have on the systems.

4.2 Open Systems.

We are interested in *open systems*, a particular kind that is pervasive throughout our universe.

Open systems: systems that are made out of components and that are open to the environment. In order to maintain the condition of "systemness" while open to the environment, some *inner source* of organization among components has to be at play.

We call that source: *self-organization* (**SO**).

We call those systems: *self-organizing systems* (**SOS**).

Physics has always had difficulties with "open systems": the difficulty stems from the need to define states of a system, which are recursive, in the context of particles coming and going through the system's boundaries. The common adopted solution begs the question to a considerable degree by regarding the "open system" as an underlying closed system *plus something* (Rosen, 1991).

4.3 Environment.

The environment **E**:
(i) is bigger than (usually, much bigger than) the system. It could be said that it is unbounded, yet finite.
(ii) is invariant with respect to the system's set of initial constraints.
(iii) is the source of matter-energy the system is in need of.

4.4 Systemic Environment.

The systemic environment $\mathbf{E_s}$:
(i) is systemically constructed by means of systemic components interacting among themselves and with the environment.

(ii) Those interactions among components take place in space and time intervals defined within the system and in terms of the system's (initial) constraints. Through this activity, the connections among the components of the system are constructed and, with them, a systemic order (connectivity) is established. To this ordering we give the name of self-organization.

4.5. Modeling E by SOS.

In order for a "model" to be sufficient it has to represent **E** with respect to the definitional constraints of **SOS**, that is, to maintain invariant what could be termed as the system's definitional closure.

At the same time we have said that **E** is, generally, much bigger than **SOS** (**E » SOS**).

4.6. Challenges of modeling.

W. Ross Ashby used to make the critical point that for a system in which all parts interact fully: "... complexity at the outputs can often be ignored: it is complexity at the inputs of the system that is to be feared" (Ashby, 1972).

The very foundation of the modeling process of **E** by **SOS**, under the condition of **E » SOS**, is *reduction of input complexity*.

Input complexity is determined by the number of *distinct* components of the system and not by the nature of the information to be embedded into the system.

4.7. How to reduce input complexity.

Ashby: "... how much will the informational quantities be increased if the system is changed from one having no interaction between its parts to one having full interaction between them?"

"... when full interaction is occurring [...] no mere doubling or trebling of the resources, or even a multiplying by a millionfold, is likely to be of any use" (Ashby, 1972).

Ashby's "full interaction" is mathematically expressed by asserting that any element (i.e., "part") is biunivocally related to (is distinct from) any other element.

The opposite to that situation is when there is no relation between the components of the system. Such an unrelated (unordered) collection can only be represented, in finite set theory, as a singleton. For the system this is a *point of indistinguishability*. But because the system is a natural system defined in space, time and amount of stuff (matter, energy), the indistinguishability will have to be established accordingly, i.e., in terms of space, time and components.

4.8. Interactions.

We said that the components of the system will interact between themselves and with the environment. Due to the fact that **E » SOS** the number of interactions will eventually grow beyond the system's resources (the choice, if any, being between sooner or later). This phenomenon resembles the "combinatorial explosion" in rule-based artificial intelligence.

The strategy for minimizing the *pace* toward combinatorial explosion is twofold: (a) the system has to start from a state in which components are a collection of indistinguishables and therefore with the initial input variety kept to a minimum; (b) the interactions are scale-bounded and progressive in order to reduce the combinatorial growth process. The desired result for this strategy is to fulfill the difficult task of modeling **E** by the **SOS** with *ostensibly limited resources* (**E » SOS**).

The set of scales of interaction of increasing range, establishes a hierarchy of dynamics within the system, ranging from the smallest to the largest scale.

The hierarchy of scale-bounded dynamics manifests an underlying ordering process among systemic components. The material instantiation of such systemic construction of order is the self-organization of the system itself. *The system organizes itself in the process of constructing its own order.*

Moreover, it is through this process of order construction among the components of the system, that the system increases its ability to model the environment, which in turn increases its viability.

4.9 Expanding vs Controlling

The system's existence lies in between two opposing needs: the need to expand and the need to exert control over the environment. If it does not expand to favour control, it becomes vulnerable because the control is exerted over too small an environmental region. If it tones down control, favoring expansion, the system also becomes vulnerable because it loosens its grip over the covered environment. The only way to negotiate between those two opposing needs while attaining **SO** is by means of indistinguishability and scaling; not any arbitrary scales, but the set that both minimizes interaction growth and maximizes control.

5. MODELING REVISITED

In order to properly deal with SO we have to find the representational context in which to define it and use it to our best advantage. The context to be chosen is the relentless use of the "modeling relation" (MR) such as it has been established, principally, by Robert Rosen (1985, 1991) and John Casti (1992, 1996). We should, nevertheless, emphasize certain aspects in order to accomodate developmental processes. Here are two elements that should be added to the scheme because they enhance the understanding of the possible existence of SOS:
(i) the need to *define the system from the environment*, as opposed to *define the environment from the system*;
(ii) the need to *construct (systemic) order* as opposed to *order given*.

5.1. The Modeling Relation (MR)

MR is a relation between systems. These systems can be very similar or not but they have to share a certain common ground. One way of describing the basic idea of a MR is by considering an observer that can move from one system to the other and back again to the original system, without loosing the capacity to observe, make records of those observations and keep the records during his/her movements. MR could also mean that any action or procedure induced in one of the systems could have a replica in the other: whatever is to be done on one of the systems can be transferred or translated into the other system (if what is done is within the boundaries of the given correspondence). Saying it in yet another way, to give a MR is to provide a dictionary or channel of communication that makes it possible for the user of the scheme to do work in one system and transfer the results of such a work into the other system, so that knowledge obtained about the former will render knowledge about the latter.

This procedure has been part of our human activity for thousands of years (see Rosen 1985 for an in-depth study of the subject). The particular manifestation we are interested in is where one system is a **natural system** (from now on, NS) and the other a **formal system** (from now on, FS) as depicted in Fig.1. This is the scientific modeling scheme *par excellence*. What it means is that, given a NS known to us (e.g., the trajectory of a cannon ball) we define its equivalent in terms of a language that is based on a FS (e.g., analytic geometry).

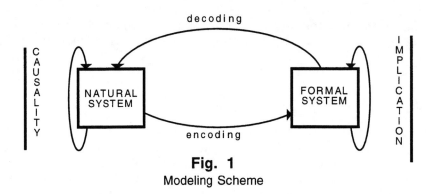

Fig. 1
Modeling Scheme

In order to implement the MR scheme that we want to put forward, it is essential to fulfill two conditions.

(1) The **first condition** is to have a correspondence between what it is expressible in the NS and what it is expressible in the FS. There is no way to convey a construct which is ill defined in one language, into another language. This means that a *necessary condition* for the implementation of correspondence in a MR is that whatever is to be transferred or translated has to be definable in *both systems*. To the extent that such a condition cannot be fulfilled, the encoding/decoding pairs *will not all be feasible*[3] and, a fortiori, neither will commutativity in the modeling scheme. We are asking for an isomorphism between encoding/decoding spaces[4].

(2) The next condition might be a bit controversial but it is important, particularly considering one of this century's most critical unsolved problems (not just for physics but for many scientific fields): I am talking about the question of how to *causally link the micro with the macro*. In the context of the MR, such a question can be satisfactorily answer by fulfilling the following **second condition**: in order to encode/decode there has to be an isomorphism in structure between NS and FS. By "isomorphism in structure" I mean that the *power of discrimination in NS has to be matched but not surpassed by FS*. This condition could be very restrictive.

To give some background for the explanation and justification of this "second condition" let's go back to what was emphasized in paragraphs 4 and 5 of this paper. We have choosen to adhere to a fundamentally constructive approach. The constructive nature of our MR scheme is based on two elements.

---

[3] Such a condition does not necessarily imply that for each construct in a system there has to be a construct in the other system. The implication is that the information carried by the initial construct can be conveyed in the other system (whether this is done by means of one construct or by the conjunction of several is immaterial).

[4] This goes against the currently frequent assumption that the bigger the FS the better, that is, that all that matters is for the NS to fit into FS. If we translate this perception in our everyday life it amounts to saying that, for example, a suit that is two, three, four sizes bigger than my exact size, is fine because I can get into it.

(a) The NS is defined in terms of a set of constraints that are established *from the environment*; this amounts to saying that the system is *quantized*[5] in all its basic properties, *including space and time*. In other words, whatever systemic manifestation there is, has to be built (constructed) systemically because there is no system or system's activity outside systemic initial properties (external constraints) or properties derived from them. So, the system is a quantized system or there is no system at all.

(b) The other element that was emphasized (paragraphs 4 and 5) was the finite nature of the systemic resources along with the fundamental characterization that this constraint imposes on all developmental processes[6]. That characterization was twofold: (i) order within the system's components (connectivity) had to be constructed, starting from the lowest relational dimension possible; (ii) the process of increasing the dimensionality of the relations had to be done stepwise, that is, by means of the construction of a set of nested scales (hierarchy of scales), ranging from the very local to the most global phenomena. The explanation and justification for these otherwise elaborate procedures rest in the need to curtail as much as possible the inevitable combinatorial explosion which brings the (partial or total) breakdown of the system. When we talk about "isomorphism in structure" we mean that to the hierarchy of scales of NS's interactions there must correspond a hierarchy of representations in FS.

This criterion imposes restrictions on present views on formalisms dealing with representations. We use Hilbert space formalisms in quantum mechanics because they provide all the specific characteristics that are needed. We do not seem to look into whether, along with the needed characteristics, Hilbert space brings also into play other characteristics that are not needed or even wanted[7]. That is, *we might be introducing states that are not systemically feasible although they are formally definable*. As long as we are not changing scales there will be no risk of loosing necessary systemic information (here, the advantage of using macro-scales is evident, but the use of one macro-scale to the exclusion of other scales will, in the end, bring difficulties). If change of scales are needed, we are necessarily forced to do those changes of scale by means of encoding-decoding transformations. In such a case the big dilemma is, how do we discriminate between the states that are genuinely systemic and those that are only formally possible? And this question comes about because we are working in a formal space that is bigger (usually, much bigger) than the systemic one.

Moving in the direction of *diminishing the size of the scales* we are forced, in the attempt to formally represent our system, to shrink the state space. There are appropriate formal ways to fold the state space, but only if we can account for all the states that there are in it. If we have no means to account for all the possibly definable states in a given scale of the NS, there is no way we can use those "appropiate formal ways to fold the state space" in FS. And because we cannot properly fold the system, from a given scale to a smaller one, we will end up having *states that are formally possible but causally impossible*.

---

[5] Let us remember how quantum theory is defined: "A departure from classical mechanics of Newton involving the principle that certain physical quantities can only assume discrete values. In quantum theory, introduced by Planck (1900), certain conditions are imposed on these values; the quantities are then said to be *quantized*." (Illingworth, 1991: 379).

[6] We named systems that entertain such developmental processes (self-organizing and anticipatory systems) "evolutionary systems" and the heuristic scheme, ESF.

[7] "The practitioner of laboratory physics appeals to the theoretician to completely describe his pratice in an objective manner. Again, on the one hand, we perform finite measurements and computations by prescribed methods, while on the other hand we are asked to accept the fact that these are artificial limitations of space, time, energy and symbolism. Again the description ignores the process of describing." (McGoveran, 1987: 38). See also comments by Dirac (Dirac, 1958: 48).

We have partially to thank Newton for the serious problem of formal misrepresentation just mentioned. At the same time, in planetary motions, Newton's theory has proven to perform extremly well.[8]

6. ANTICIPATORY SYSTEMS

To anticipate is to "use in advance" and, in the context of MR, is to consider the future in relation to acts that take place in the present and for which we have accumulated knowledge from the past. Whether this task becomes a guessing game or a scientific endeavour depends on what we mean by future and how we logically connect (entail) the past and the present with that future.

One of the outcomes of professing the views proposed in ESF is that there is no conceptual difference between self-organizantion and and anticipation. We understand self-organizantion as systemic construction of order (Alvarez de Lorenzana, 1997). Anticipation rests in the ability of the system to model the environment. The system can model the environment by systemic means only. In other words, there is no model of the environment below or above what the system can measure (discriminate, perceive, filter, etc.). To measure is the NS's counterpart to what in the FS is the construction of order.

Futher, as long as the system tends to expand (unfold its potentiality), there will be a need to increase the range of the model of the environment. That increase will have to be achieved based on what the system already has. In other words, the system aims at an enhanced model of the environement based on the one that it already has of it; and is drawn into that enhancement because the model that it has is not sufficient vis-a-vis *what is to come*. The key to expand the model is, then, to be able to do it by means of causality.

Entailment in the FS and causality in the case of NS assures *consistency* between what the system *is* (and can model) and what the system *will be*. This consistency is the very essence of the system itself in that it is the pillar on which the system's identity rests. You break the system's chain of causality and, by doing so, you break the system: the system disappears as such.

If we conceive self-organizing systems as collections of spatio-temporal distributed entities, then, any unfolding of the system means expansion (e.g., diffusion) over the environment. Such an expansion calls for a different (extended) spatio-temporal scale, which materializes as a change in the coupling (strength or range) of the interactions between system's components. The implementation of the change of scales (i.e., to be able to go from one scale to another), can only take place if the system has a model of the scale from which the unfolding has to take place. Such a model could take the form, for example, of Ross Ashby's *requisite variety* (Ashby, 1958; Alvarez de Lorenzana, 1991, 1993).

The chain of *model refinements* that it is implied in any anticipatory system is equivalent to the chain of scales (levels) implied in self-organizing systems. Model refinement and scale enhancement are, in fact, synonymous. If we see anticipation and self-organization in the context of systems that are open to the environment and under the external constraint of possesing finite resources, nothing is surprising or abstruse. The implicit complication denoted by phrases such as "Physics of far-from-equilibrium systems" are complicated when and because we consider system's normality to be "closed to the environment" which, ultimately, does not exist in our natural world (Alvarez de Lorenzana, 1997a, 1997b).

---

[8] This subject merits more than a light comment made *in passim*. I have partially dealt with it in a previous article (Alvarez de Lorenzana, 1995).

CONCLUSION

The idea that I most want to put foward in this paper -however informally- is the assertion that, ultimately, there are only open systems in our universe which are *relatively* closed to some aspects of the environment. If the assumptions in our representational framework are to reflect such a point of view, we need to bring forward a constructive approach that, if not the same, should be of a similar character to the one entertained in this paper. In particular, the consideration that quantization needs to go beyond present standard formalisms and include quantization of space and quantization of time. This could have profound as well as very subtle effects in the way we see our world of systems. It could also provoke difficulties in our formal languages of representation. But those difficulties need to be faced and the path towards their solution could very well bring clarification and depth in some aspects of the world of mathematics.

REFERENCES

Alvarez de Lorenzana, J.M. (1989)."Las Relaciones Internacionales en el contexto de la historia. Un caso de sistemas evolutivos jerárquicos." PhD Dissertation. Universidad Complutense de Madrid.

Alvarez de Lorenzana, J.M. (1991a). "Informal Comments on Indistinguishability, the Combinatorial Hierarchy and Evolutionary Systems Framework." In F. Young (ed.), *Proceedings 7th Ann. Mtg. ANPA West*, Stanford, Palo Alto.

Alvarez de Lorenzana, J.M. (1991b). "The Generation of the Combinatorial Hierarchy (CH) in the context of the Evolutionary Systems Framework (ESF)." In F. Abdullah (ed.), *Proceedings 13th Ann. Mtg. ANPA*, Cambridge, England.

Alvarez de Lorenzana, J.M. (1993)."The Constructive Universe and the Evolutionary Systems Framework." In S.N. Salthe, *Development and Evolution. Complexity and Change in Biology*. The MIT Press.

Alvarez de Lorenzana, J.M. (1995). "Evolutionary Systems in the light of Topos Theory or Aristotle's causal categories revisited yet again." Proc. Int. Conf. on Evolving Systems, Vienna, Austria.

Alvarez de Lorenzana, J. M. (1997a). "Some Issues on the Construction of Order in Self-Organizing Systems." In Tom Etter (ed.). *Proc. 13th Ann. Mtg. ANPA West*, Stanford, Palo Alto, pp.1-16.

Alvarez de Lorenzana, Juan M. (1997b). "Self-Organization and Self-Construction of Order." To be published.

Alvarez de Lorenzana, J.M. and L.W. Ward (1985). "Semantic and Syntactic Information." Proc. 29th Ann. Mtg. *Societ for General Systems Research*. Louisville, Kentucky; pp. 78-86.

Alvarez de Lorenzana, J.M. and L.W. Ward (1987). "On Evolutionary Systems." *Behavioral Science*, vol. 32, pp. 19-33.

Ashby, C.R. (1958). "Requisity variety, and its implications for the control of complex systems." *Cybernetica* 1: 1-17.

Ashby, W. Ross (1972). "Systems and Their Informational Measures." In Klir, George (ed.). *Trends in General Systems Theory*. John Wiley & Sons, pp. 78-97.

Bastin, T., Noyes, H.P., Amson, J. and Kilmister, C.W. (1979). "On the Physical Interpretation and Mathematical Structure of the Combinatorial Hierarchy." *International Journal of Theoretical Physics*, vol. 18, pp. 454-488.

Casti, J.L. (1992). *Reality Rules: Picturing the World in Mathematics. I-The Fundamentals*. Wiley.

Casti, J.L. (1996). "Confronting Science's Logical Limits." *Scientific American*, vol. 275, 4: 102-105.

Dirac, P.A.M. (1958). *The Principles of Quantum Mechanics*. Oxford University Press.

Illingworth, Valerie ed. (1991). *The Penguin Dictionary of Physics*. Penguin Group.

McGoveran, D.O. (1987). "Foundations for a Discrete Physics", in *Proc. 9th Ann. Mtg. ANPA*, H.P. Noyes, ed., Cambridge, England.

Rosen, Robert (1985). *Anticipatory Systems*. Pergamon Press.

Rosen, Robert (1991). *Life Itself. A Comprehensive Inquiry into the Nature, Origin, and Fabrication of Life*. Columbia University Press.

Rosen, Robert (1993). "On Models and Modeling." *Applied Mathematics and Computation*. vol. 56, pp. 359-372.

_____(1995). "The Mind-Brain Problem and The Physics of Reductionism." *Communication and Cognition-Artificial Intellegence*, vol. 12, pp. 29-43.

From Situated action to Noetical Consciousness: the Role of Anticipation

Olivier Sigaud

sigaud@dassault-aviation.fr

DASSAULT AVIATION DGT/DTN/ELO

Cedex 300, 92552 St Cloud Cedex FRANCE

Phone: (33).01.47.11.53.01 Fax: (33).01.47.11.52.83

Abstract: The bottom-up approach of Behavior-Based Artificial Intelligence – which starts from sensori-motor abilities and strives to build conceptual structures – casts a new light on the incapacity of AI programs to handle the meaning of the symbols they manipulate.

Our previous work was a computational study of the mechanisms which structure behaviors. In this paper we present an attempt to defeat the problem of meaning by introducing a first layer of intentionality into these mechanisms. But the notion of saliency which we have added cannot explain the structuration of meaning into layers of more and more formal objects. We analyze the reasons of this failure, then we come back to the phenomenological work of Husserl as the source of our hypothesis, and show that his standpoint is epistemologically complementary to ours. As a conclusion, we give some directions on the way to cover the distance between both standpoints.

Keywords: anticipation, intentionality, phenomenology, salient configurations, structuration.

CP437, *Computing Anticipatory Systems: CAYS--First International Conference*
edited by Daniel M. Dubois © 1998 The American Institute of Physics 1-56396-827-4/98/$15.00

1   Introduction

Our previous work, [Sigaud, 1996], was a computational study of the mechanisms that could help the extension of the behavioral abilities of an agent. Our algorithm operated on the *evolution space* [1] of the agent. Our proposal was intended to show how the evolution space could get more and more structured into meaningful configurations thanks to a categorization mechanism relying on the experiments of the agent.

The key element of our framework was the notion of *configuration*, which appears to be strongly related to Piaget's notion of *schema* ([Piaget, 1936]). But our definition for configuration is much more technical: formally, we call *configuration* a temporally bounded domain whose shape fluctuates in the evolution space. The mechanisms we developed were able to associate to each configuration a set of elementary actions, and to chain up these actions into purposeful behaviors. The configurations we obtained proved meaningful in the eye of an external observer. For instance, in a car driving experiment, the obtained configurations corresponded exactly to the sequence of straight lines and curves of the track where the car was trained. But the agent itself could not be aware of the meaning of the obtained configurations, neither by considering them as objects of its environment nor by generalizing its behavior among curves and straight lines.

This problem appears to severely limit the claims of the bottom up approach of cognition, since awareness seems to play an important part in the construction of conceptual structures, and these structures are the basis of higher level cognitive processes. For instance, our driving agent would not be able to find more general rules such as: "*I must slow down before the curves*" if it cannot build a general concept of curve.

Our concern for consciousness comes from this very practical experiment. Rather than designing an *ad hoc* solution as many AI practitioners would, we are convinced that we have to overcome this limitation by adding to our framework a notion of intentionality standing on the most basic configurations.

The purpose of our present work is to explain how consciousness might emerge from the structuration of behavior. Our framework is strongly influenced by Piaget's constructivism, but modified by lessons drawn from Husserl's and Merleau-Ponty's phenomenology to explain self-awareness, which is the blind spot of Piaget's theory.

In section 2, we will try to define better which kind of consciousness we are speaking about. Then we will present in section 3 a first attempt to introduce intentionality into our previous framework, thanks to the notion of saliency. We have tried to bind the meaning of action to the satisfaction of viability constraints (3.1), then to explain the extension of the meaning thanks to an anticipation mechanism which allows the propagation of saliency (3.2). But if it explains the *extension* of saliency, this mechanism does not explain its *structuration*, and in particular it fails to give account of the apprehension of objects.

So we come back to our philosophical sources: we first focus in section 5 on Husserl's study of the temporal structure of consciousness, in [Husserl, 1991b], to prepare the examination of the relation between anticipation and consciousness in section 6. Then we can present in section 7 the study of the structuration of meaning in [Husserl, 1991a], which is directly founded onto the former. This is the occasion to clarify the distance between our framework and Husserl's analyses in a discussion in section 8, and to conclude to the epistemological complementarity of both perspectives.

---

[1] The *evolution space* is the space of all the reachable perceptual states of the agent, corresponding to all its possible outputs.

2   The Nature of Noetical Consciousness

First, which kind of consciousness are we trying to build ? A close study on recent theories of consciousness, [Zalla, 1996], points out the diversity of phenomena which are subsumed under that term, leading to the risk of conceptual confusion. Among many possible distinctions, we will present our own concern in the terminology of [Tulving, 1985].

- We call "**anoetical consciousness**" the conscious access to the different modalities of present perception. One strength of [Sigaud, 1996] is that, since we wrought on the evolution space of the agent, its representation were egocentric: through its sensors, it acquired a subjective view of its environment at each time step. So we consider as a first approximation that our agent had a kind of anoetical consciousness.

- We call "**noetical consciousness**" the conscious access to the internal representations of an agent about the world. According to our analyses, this is, as a first approximation, what is lacking to our agent: it builds representations but it is neither aware of them nor of their meaning.

- We call "**auto-noetical consciousness**" the conscious access to a representation of the self as a "subject" living in the world. Since our agent has not built any representation of itself, it is clear that it is not auto-noetically conscious.

Thus, contrarily to Tulving whose concern is the nature of auto-noetical consciousness and its relation to episodic memory, we focus on the transition between anoetical and noetical consciousness.

3   Intentional Configurations

3.1   Innate Salient Configurations: the Reflex

Our first attempt towards consciousness consisted in drawing an analogy between getting aware of the meaning of an action and acting purposefully. We started with the metaphysical hypothesis that meaning comes from the intentionality of the most elementary actions. Then we have tried to bind to our theory of the structuration of behavior a view of the structuration of intentionality which followed the same line. Our way to introduce intentionality in our framework was to associate intentional values to configurations in the evolution space. Our starting point for doing so was the notion of *saliency*.

We call *innate salient configuration* a domain in the state space which corresponds to a functional constraint over the viability of the agent. For instance, in order to survive, an agent should not let its internal energy E undergo a bound value E_b. Hence, the domain where $E \in [0, E_b]$ is salient. Then we are driven to distinguishing among behaviors the reflex, which is univocally fired when the agent comes to a particular configuration, and is designed (through natural selection of species among generations) to reach a innate salient configuration (if the configuration is *attractive*) or to avoid falling in it (if it is *repulsive*). For instance, the reflex of our previous agent is to eat when E dangerously decreases towards E_b. Our view of reflexes is strongly influenced by the criticism of behaviorism developed in [Merleau-Ponty, 1942]. The important point is that the activation of the reflex is invoked by the *present* configuration rather than the *target innate salient* one: according to Piaget, a innate salient configuration is the purpose of a reflex, but this purpose does not appear in its definition, the reflex is not purposeful, it is just fired and satisfaction of the viability constraints follows. Given that the infant has no access to the target configuration, it has no access to its meaning, it is not conscious of it. We conjecture that this was the situation with our artificial agent, that Tulving would call "anoetically conscious".

3.2 The Propagation of Saliency

Thanks to the univocity of its definition, the reflex manifests a strong *correlation* between its fire conditions and the target configuration it is designed to reach or avoid. This correlation is the basis onto which is built the capacity of *anticipation*. If we call *acquired salient configuration* a configuration from which the agent is used to reaching a innate salient configuration through a straightforward behavior, thanks to anticipation, we can explain how saliency propagates itself from innate salient configurations to less and less salient ones: a configuration becomes salient when the agent anticipates that from this configuration it will reach a innate salient or another acquired salient configuration.

Such a propagation mechanism contributes to explaining how reflexes can structure themselves into successions of more and more purposeful behaviors. Through the propagation of saliency, intentionality radiates from innate salient configurations and invades the evolution space of the agent simultaneously with the structuration of its behavior.

4 A Deeper Inquiry

The general purpose of our work was to find a possible explanation of the transition from the passive perception of undifferentiated data from a "flat" world to the active perception of meaningful configurations with respect to the viability of the agent. This transition was intended to be a first step to clear the gap between sensori-motor intelligence and conceptual awareness. But we have to admit our failure. First, our propagation mechanism did not explain how the purpose of basic behaviors could become consciously accessible to the agent, but only how it could help the purposeful chaining of actions. Furthermore, it failed in explaining both the structuration of configurations into higher level *"signs"*, and the constitution of meaning out of this structuration.

In order to understand the reasons of that failure, we will now come back to our initial hypothesis, stating that the meaning of behaviors comes from the conscious access to their associated intentions. Such an hypothesis is corroborated by some empirical facts:

- to understand the behavior of somebody else, you have to know his intention ;

- generally speaking, in order to know what you do, you have to know what you do it for, i.e. where your action will lead you to (for instance, you are aware of closing the door when you move your arm *in order to* close the door). Which means that to be aware of your own behavior, you have to be aware of your intention. And being aware of it is having access to it, being able to take it into consideration.

Then, what does it mean to be aware of one's intention ? As a deeper metaphysical hypothesis, let us state that one is always aware of its present impression (this is anoetical consciousness) and of some *expectations*. The idea is that an expectation becomes intentional when it gets adjusted to the impressions of the immediate future. We can also find empirical facts to corroborate this hypothesis:

- just after birth, the expectations of an infant shall be poorly adjusted to reality, and the infant is probably not noetically conscious;

130

- when you encounter something startling, i.e. something you did not expect, you keep "frozen", until you can recover the "meaning" of this experience.

Now if we adopt this hypothesis, a key problem in understanding the nature of noetical consciousness consists in explaining the mechanisms that lead to this adjustment of expectations. This is what we will extract from one of the deepest inquiry into the nature of consciousness, i.e. Husserl's phenomenology.

5   The Temporal Structure of Consciousness

The key difficulty to explain how we can categorize the world into more and more meaningful layers of objects is pointed out by Kant [Kant, 1987] as the problem of *temporal synthesis*. If you consider experience from a conscious standpoint, phenomena always appear in a temporal succession from which you cannot go out, and consciousness seems to be limited to the present phenomenon. But, in order to consider a succession of phenomena as the experience of an object, you have to gather together successive impressions out of the temporal flow. Hence, you have to synthesize distinct experiences from different moments in the same conscious unit. But paradoxically, a conscious experience seems to be something instantaneous. So how can you gather together experiences coming from distinct moments, if you are only aware of the present one ?

5.1   The Conscious Unit

Husserl proposes a solution to that problem in [Husserl, 1991b]. The purpose of this book is to explain how we get aware of time as objective and shared by everybody, since the experience of time is one of the most intimate and subjective experiences.

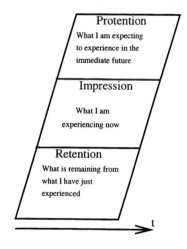

Figure 1: **A conscious unit**

Starting from Brentano's analyses, Husserl considers that the extension of a "conscious unit" is not limited to a single moment, but can be split into three parts laying on a short duration. These three parts are the following (see figure 1):

Protention is what you expect in the immediate future. For instance, when you close your eyes, you expect your visual field to get dark.

Impression (or intuition in Kant's terminology) is what you actually perceive through the senses at the moment.

131

Retention is what remains from your immediate past. It is closer to a *mnesic print*, a very short term memory, than to a remembered experience. Husserl highlights the distinction between memories and retention: the content of retention is still actually present in your conscious experience, even though you have to "retrieve" a memory to "activate" it.

5.2   From Conscious Unit to Conscious Unit

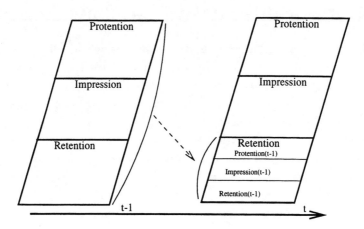

Figure 2: **From conscious unit to conscious unit**

There is a strong relation between a conscious unit and the following one: the protentional part of the first one might correspond to the impression of the second, and the retention of the second to the first one itself.

This model is of course a crude simplification that would not deserve the agreement of phenomenologists. In particular, instead of a discrete transition between two conscious units, we experience a continuous flow of changes in the present conscious unit.

One important thing is that these components of a conscious unit are consciously accessible themselves. Then retention is the conscious content of everything you can immediately remember from the previous conscious unit; it may contain the totality of the previous conscious unit itself, with the different modalities of protentional, impressional and retentional content.

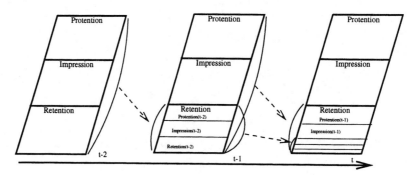

Figure 3: **The potential infinity of retention**

Thus, the retentional part of a conscious unit, for instance, can be decomposed further into three parts as shown in figure 3. But this possibility of further decomposition collapses into an infinite recursion: retention at time t contains retention at time $t-1$, which itself contains retention at time $t-2$, and so on. Hence, if we do not characterize this model any further, every past conscious experience would be available within the present experience, even if it would be modified with the "past" modality character.

This problem is highlighted in [Bachimont, 1996], were the author stands that a retentional experience must be "released" from a conscious unit and stored in some material memory *which must be external to the conscious unit*. This point of view allows to clarify Husserl's distinction between retaining and remembering, and it is important since it enforces to admit the existence of a device external to consciousness, which is what phenomenology tries to delay as much as possible.

6 Anticipation and Adjustment

The focus of this paper is not on retention, but rather on protention, expectations, and their relation to meaning. We just presented this point in order to make clear that our model is too crude to give a serious account of the dynamics of conscious experience. Some developments on adjusting the content of protention can be found in [Husserl, 1991a], where Husserl extracts from the lowest layers of experience the origin of logical judgment. According to his analyses, we can distinguish two complementary mechanisms corresponding to two different kinds of synthesis.

6.1 Anticipation as Conscious Induction

According to Husserl, attributing meaning to a perceptual experience consists in having a *structure of interpretation* which will be fulfilled by the impressions. These structures of interpretation are a kind of expectation about what you will encounter next. You can get convinced of this when reading complex sentences: if you do not have a preconceived idea of what it will be about, you do not catch its meaning. So, the first kind of synthesis results in the *construction* of new expectations which give more meaning to future experiences. It is a form of *anticipation*, consisting in a transposition of a *past-present* relation into a *present-future* relation.

According to figure 2, the subject has access in the same conscious unit both to his impression at time t and to a retention of his impression at time $t-1$. Then the following inference is phenomenologically plausible:

If the subject often experiences an impression A at time t with a retention of impression B from time $t-1$, he can infer a relation between both impressions, and then expect an experience of A when he will experience B in the future.

6.2 Adjustment as Conscious Reduction

The second kind of synthesis is differential. According to figure 2, the subject has access in the same conscious unit both to his impression at time t and to a retention of his protention at time $t-1$. His expectations were correct if the previous protention and the present impression match. If they do not, it means that the anticipation mechanism went to a wrong inference, which must be corrected. This is possible through the following inference:

If the subject experiences an impression A at time t while his expectations about time t were B – which are present in the retentional part of the conscious unit –, then he has to reconsider the association(s) that let him infer his expectations of B.

The way to reconsider these associations can be seen as specializing their context of validity, then eliciting the difference between the contexts where the subject can infer either A or B. This specialization mechanism was precisely the core of the categorization algorithm of [Sigaud, 1996]: it relies on the causality principle according to which the same cause produces the same effect *if the contexts are equal*.

Now we see better how the cooperation of anticipation and adjustment lead to a more powerful learning mechanism than adjustment alone. Anticipation is a building mechanism, adjustment refines what is built. But this still does not explain how different layers of meaning (objects, formal notions...) can be built out of perception. To come closer to that concern, we now present [Husserl, 1991a].

7 Experience and Judgment

Husserl tries in [Husserl, 1991a] to ground the truth of logical judgment on deeper forms of truth, coming from physical experience.

7.1 External Perception

He starts from *external perception*, where, he believes, can be found the ultimate and apodictic obviousness. More precisely, the basement of his inquiry is the unconscious mechanisms that allow a subject to identify singularities in his perceptual field. Thus he examines the structure of experience at the perceptual level, with the classical introspective methods of phenomenology. One of the key claims of his framework is the fact that an object of experience is not the result of a mere perception, it comes with some expectations on what cannot be actually perceived (its hidden side, for instance), and it can always be determined further, with a closer examination. This level of passive perception constitutes an ante-predicative layer into which the judgment is settled, and the process of determination of objects of experience is called *explanation*, which is the core of the next layer.

The second layer is the level of predicative judgment itself, limited to judgments about objects of perceptual experience. In this layer, Husserl distinguishes between *internal explanation*, which helps determining the inner properties of an object, and *external explanation*, which extracts relational properties between objects.

7.2 Internal Explanation

It was important to present the analysis of the structure of temporality beforehand, since internal explanation consists in *holding in mind at the same time* an object and one of its features. Within the conscious unit operates a passive synthesis which binds the feature to the object. Then the same feature will be attributed again to the object at every new presentation, even if it is not perceived. The feature itself can become the object of the same process, hence deep structures of attributed properties can be bound to any object.

Up to that point, we see how the experience of an object can be enriched with the attribution of many features, but these features are coming from the layer of external perception too, hence it does not explain how structures of *meaning* can emerge from this process.

7.3 External Explanation

The dual mechanism, external explanation, is responsible for the extraction of the relational attributes of the objects. The analyses of internal explanation are still valid, the main difference being that external explanation consists in holding in mind at the same time an object *and its surrounding* rather than its internal features.

The condition of possibility of such syntheses is a mental proximity of their objects, either in actual perception (temporal or spatial proximity), or in the mental life of the subject (for instance, a comparison between an imagined and an actual object is possible if they are associated with respect to some of their features). So there is an underlying passive process of *association* which binds close experiences together.

The products of external explanation are richer, since the synthesis is not limited to relations between objects and properties. In particular, external explanation is responsible for the extraction of comparative predicates between features of the involved objects, which leads to the possibility of *formal* judgments.

7.4 Logical versus Practical Meaning

Then Husserl comes to existential and general judgment on states of the world through abstraction and judgments on other judgments. He explains how the phenomenological method of *eidetic variations* leads to the apprehension of the essence of objects, of their *eidos*. But his research is still limited to the stratification of objectivity into judgments of the form "*S is p*", and does not include any analyses of the edification of *practical meaning* of the form "*S helps to a*" in this constructive process. This does not mean that he is not aware of this dimension of cognition. On the contrary, he explicitly rejects analyses on thematic apprehension of objects as a psychological concern irrelevant for his inquiry.

8 Discussion

Now our epistemological situation is the following. In [Husserl, 1991a], Husserl explains the stratification of predication about the objects of experience.

- His analysis comes from a deep inquiry into the structure of temporality, where anticipation and adjustment appear as central mechanisms.

- But he analyses this stratification from the standpoint of a conscious subject, so he does not explain the constitution of awareness itself. It is quite symptomatic that he is compelled to introducing in this explanation a *passive* mechanism of association which he does not try to study, but which threatens the validity of his perspective.

- Furthermore, as B. Steigler points out in [Steigler, 1996], anticipation is explained by Husserl as a *recognizing* process, so he has to presuppose *a priori* structures of knowledge, the *eidos* of the objects.

- And, last but not least, he starts from perceptual mechanisms, without devoting much consideration to the role of action. Then he cannot include in his analyses the sedimentation of *thematic* values, because these values are acquired through the use of objects when the agent acts in its environment. This also explains why he is doomed to start from the attribution of perceptual qualities to objects considered as given, rather than explaining how we categorize the world into objects.

Our own analyses, on the contrary, explain in a proprioceptive perspective how the experience of the world is categorized into subject-centered categories, and how practical values can be bound to these categories. Thus, our level of analysis seems deeper. But we also encounter complementary limits:

• In our previous framework, we did not give account of the construction of higher levels of experience, so we have to include the mechanisms thanks to which objects and features are bound together into more and more structured levels of meaningful objects. Then unifying our perspective with Husserl's one should allow us to exceed the limits of our approach. In order to achieve this unification, we shall try to bind to our salient configurations ante-predicative assertions on the presence of objects, of the form "*When I am in configuration C, I encounter a S*". Such a unification should allow us to bind practical values to objects of experience through the corresponding categories, then to make profit of the basis from which starts Husserl's stratification of judgments. Since our sources of practical meaning are the innate salient configurations bound to reflexes, our approach will include the concern for the body identified in [Merleau-Ponty, 1945]. Then we hope to introduce in our framework elements of practical meaning, which is the goal we expressed in the introduction of the paper.

• It is now clearer that we have to explain how we can map the experience of physical objects onto the categorization resulting from our algorithm. In the case of static objects like the segments of the track in [Sigaud, 1996], we have shown that there exists a direct mapping. But we have no guarantee to obtain the same results on other experiments and, in particular, we will have to deal with mobile objects rather than static ones.

9   Conclusion

Thanks to its adoption of the viewpoint of a Cartesian *cogito*, Husserl's phenomenology deals with consciousness from the inside. This allows the philosopher to explain the stratification of experience into layers of meaning. But this also dooms him to stay trapped into the realm of conscious experience. Thus, Husserl fails in his central attempt to reduce conscious experience to an ultimately originary layer (see [Derrida, 1990] about the problem of genesis in phenomenology). Or, at least, his reduction to primary perceptual features in a *Gestalt*-like perspective is not convincing since it does neither take into account the development of behavioral abilities, nor the mediation of the body.

The constructivism of Piaget, on the contrary, gives a very convincing starting point of cognitive development, but fails on the explanation of the transition from sensori-motor intelligence to the development of conceptual abilities.

In this paper we have tried to make profit of the apparent epistemological complementarity of both perspectives, hoping that our unification attempt could lead to a single theory covering the whole field of cognitive development. Such a unification is difficult to achieve, since the foundations of both perspectives are notoriously divergent. So there appears to be a division line between their domains, with conflict zones where they overlap and theoretical gaps where they are not adjacent.

Our task, for now on, consists in exploring the totality of the expanse of this division line, and in examining if some epistemological adjustments, by both sides, can lead to a true interpenetration. We hope that building the ability to recognize objects in the environment will be a major step in this process, the second main step being the constitution of a special object, the self.

References

[Bachimont, 1996] Bachimont, B. (1996). *Herméneutique matérielle et Artéfacture : des machines qui pensent aux machines qui donnent à penser.* Thèse de troisième cycle de l'Ecole Polytechnique, CREA, Paris.

[Derrida, 1990] Derrida, J. (1990). *Le problème de la genèse dans la phénoménologie de Husserl.* Presses Universitaires de France, Épiméthée, Paris.

[Husserl, 1991a] Husserl, E. (1991a). *Expérience et jugement.* Presses Universitaires de France, Épiméthée, traduction D. Souche-Dagues, Paris.

[Husserl, 1991b] Husserl, E. (1991b). *Leçons pour une phénoménologie de la conscience intime du temps.* Presses Universitaires de France, Épiméthée, traduction H. Dussord, Paris.

[Kant, 1987] Kant, E. (1987). *Critique de la Raison Pure.* GF-Flammarion, Paris.

[Merleau-Ponty, 1942] Merleau-Ponty, M. (1942). *La structure du comportement.* Presses Universitaires de France, Paris.

[Merleau-Ponty, 1945] Merleau-Ponty, M. (1945). *Phénoménologie de la perception.* Gallimard, TEL, Paris.

[Piaget, 1936] Piaget, J. (1936). *La naissance de l'intelligence chez l'enfant.* Delachaux et Niestlé, Neuchâtel.

[Sigaud, 1996] Sigaud, O. (1996). *Apprentissage : de la commande au comportement.* Thèse de troisième cycle de l'Université PARIS XI, http://www.etca.fr/Users/Olivier%20Sigaud.

[Steigler, 1996] Steigler, B. (1996). *La technique et le temps.* Galilée, Paris.

[Tulving, 1985] Tulving, E. (1985). *Elements of episodic memory.* Oxford Clarendon Press.

[Zalla, 1996] Zalla, T. (1996). *Unicité et multiplicité de la conscience : une étude critique des théories contemporaines à la lumière d'une hypothèse modulariste.* Thèse de troisième cycle de l'Ecole Polytechnique, CREA, Paris.

THE QUESTION CONCERNING EMERGENCE:
IMPLICATIONS FOR ARTIFICIALITY

S.M.Ali, R.M.Zimmer and C.M.Elstob
Brunel University
Uxbridge UB8 3PH
England (UK)
Syed.Mustafa.Ali@brunel.ac.uk

ABSTRACT

This position paper has three parts. In the first part, a brief historical background and various modern formulations of the concept of emergence are presented. A number of problems associated with the concept are identified. One outstanding problem involves the incommensurability of secondary qualities (or phenomenal *qualia*) with materialist (externalist) ontologies. The intractability of this problem with respect to existing scientific approaches is an indicator of ontological category error, in this case, an attempt to subsume subjectivity into objectivity. In the second part, various attempts at solving the mind-body problem (of which the subjectivity-objectivity issue is a modern incarnation) are investigated and shown to be problematic. It is argued that these problems necessitate reconsidering the metaphysical foundations upon which the concept of emergence is grounded. In the third part, the notion of emergence is reconsidered and a new theory grounded in a synthesis of Heideggerian and Whiteheadian metaphysics is outlined. Finally, the implications of this synthesis for artificing (technology) are briefly considered. It is maintained that "strong" artificiality, the artifactual realization of natural phenomena such as life and mind, is impossible and that this result follows from the essence of artificing. Thus, ontology does not entail technology.

KEYWORDS: emergence, category problem, ontology, panexperientialism, artificiality.

1. THEORIES OF EMERGENCE

The earliest articulation of the essence of emergence may well be the ancient Greek maxim 'the whole is more than the sum of the parts' and historical links to the notion of self-organization are traceable to the writings of the pre-Socratics Thales and Anaximander. The first serious attempt at investigating the concept of emergence was not made until the middle of the nineteenth century when G.H.Lewes distinguished between resultants and emergents: in the former, the sequence of steps which produce a phenomenon are traceable while in the latter they are not. Lewes could be interpreted as identifying emergence with the epistemological limitations of an observer. In C.L.Morgan's *Emergent Evolution* (1923), emergence was identified with novelty while J.C.Smuts', writing in *Holism and Evolution* (1926) identified it with the generation of stable wholes. Samuel Alexander's *Space, Time and Deity* (1920), in which emergence is identified with the tendency for things to arrange themselves into new patterns which as organized wholes possess new types of structure and properties, provides the basis for information-theoretic and computationalist interpretations of the concept (Baas,93) (Darley,94). Nagel (1961), following Lewes, defines emergence in epistemological terms. This position is consistent with the adoption of a scientific (materialist) metaphysics and reflected in a commitment to ontological reductionism, the view that emergent phenomena do not contradict physical laws and, moreover, are causally determined by such laws. Non-materialistic theories of emergence are rejected *a priori* on the grounds that "emergents are regarded as spiritual creations emanating from an unknown shadowy world." (Kenyon,41) However, ontological reductionism does not entail epistemological reductionism: for example, Mayr (1982) defines emergence as the appearance of new characteristics in wholes that cannot be deduced from the most complete knowledge of the parts, taken separately or in other partial combinations.

CP437, *Computing Anticipatory Systems: CAYS--First International Conference*
edited by Daniel M. Dubois © 1998 The American Institute of Physics 1-56396-827-4/98/$15.00

Similarly, Churchland (1985) defines emergence in terms of the irreducibility of properties associated with a higher level theory to properties associated with components in a lower level theory. Bottom-up causation and epistemological non-reductionism are the defining characteristics of most modern theories of emergence, computationalist or otherwise: in support of the former, Langton (1989) defines emergence in terms of a feedback relation between the levels in a dynamical system: local microdynamics *cause* global macrodynamics while global macrodynamics *constrain* local microdynamics; in support of the latter, Cariani (1991), who distinguishes three concepts of emergence, viz. computational, thermodynamic and relative-to-a-model, follows Rosen (1985) in identifying emergence with the deviation of system behaviour from an observer's model.

2. PROBLEMS WITH EMERGENCE

Despite an *a priori* commitment to materialism, there are a number of outstanding problems associated with the concept of emergence which can be briefly stated in the form of a list of opposing metaphysical postulates: (1) realism or relativism (Rosen,77); (2) continuity or discreteness (Cariani,91); (3) monism or pluralism (Bunge,69); (4) causality or epiphenomenality (Sperry,76) (Searle,92); (5) determinism or indeterminism (Elstob,84); (6) reversibility or irreversibility (Prigogine,84); (7) computability or noncomputability (Baas,93); (8) closure or openness (Cariani,89) (Ali,96); (9) intrinsicality-extrinsicality (Crutchfield,94). The intrinsicality-extrinsicality issue is particularly interesting since it leads directly to the issue of observation. Crutchfield (1995) defines observation in terms of the embedding of models within endophysical observers, that is, observers existing within a closed physical system or 'universe'. He maintains that *extrinsic* emergence, in which a phenomenon is identified as emergent by an exophysical observer, that is, an observer outside the system, leads to an infinite regress of observers in much the same way that the collapse of the wave function by an observer in the Schrödinger's Cat gedanken experiment necessitates a second observer to collapse the wave function of the larger system incorporating the first observer and so on. Crutchfield's solution to the regression problem associated with extrinsic emergence is to fold the regress into the system. This leads to a position similar to that of the participatory interpretation of the measurement problem in quantum theory: the universe gives rise to observers who in turn are responsible for collapsing the wave function describing the universe. It is important to note that this concept of observation is objectivist, not merely in the 'weak' epistemological sense implying universal (necessary) as opposed to particular (contingent), but also in the 'strong' ontological sense that implies externality or 'third-personhood'. There are (at least) two related problems with this position: (1) Emergence depends on complexity, complexity on degree of abstraction, and abstraction on the intentionality of an observer/abstractionist. However, in what sense is it meaningful to speak of intentionality (directedness) in the absence of consciousness ? (2) Consciousness is a phenomenon that is subjective, internal, experiential and first-person. How can the external and non-experiential give rise to the internal and experiential ?

Nagel (1974) maintains that those things of which I am conscious, and the ways in which I am conscious of them, determine (or rather, *define*) what it is like to be me. Those that have consciousness are (or can *become*) subjects, beings to whom things can be one way or another, beings it is *like* something to be. Subjects have experiences, feelings, sensations, in short secondary qualities or *qualia*. Griffin (1988) maintains that subjectivity and objectivity belong to orthogonal ontological categories and that any attempt at deriving the former from the latter will lead to an instance of category error. Moody (1993) defines the latter as "the result of grouping something in a category with other things that are logically dissimilar." (p.31) It is simply the case that subjectivity is irreducible (ontologically, methodologically and epistemologically) to objectivity. Thus, emergence, with its necessary link to observation (Ayala,85) (Cariani,89) (Baas,93), leads directly to the mind-body problem.

3. THE MIND-BODY PROBLEM

The mind-body problem is simply the problem of relating subjective, internal, mental phenomena to objective, external, material phenomena. Various solutions to the problem have been proposed, beginning, perhaps with Descartes' invocation of a 'God-of-the-gaps' maintaining the link (located in the pineal gland within the brain) between the *res cogitans* or mental substance and the *res extensa* or material substance. Dualistic theories fall broadly into two categories depending on the way in which the link between mind and matter is viewed: (1) interactionism - mind and matter are causally connected; (2) psychophysical parallelism (Spinoza) - mind and matter are causally disconnected, merely remaining in operational 'harmony' with each other. Aesthetic dissatisfaction with dualism leads to the postulation of two kinds of monism: (1) physicalism (or materialism) and (2) idealism (or mentalism); in the former, matter is ontologically primitive, whereas in the latter, mind is ontologically primordial. Physicalism gives rise to a number of positions ranging from naive identism (in which mental states are held to be identical with brain states) through to behaviourism, functionalism and eliminative materialism, in which mental phenomena are interpreted as either non-existent, organizationally-contributive to survival, or simply 'folk psychological' respectively. However, all these positions are problematic: identism since it fails to explain the link between mental states and brain states, establishing, at best, a mere correlation between the two orders of phenomena; behaviourism, functionalism and eliminative materialism since subjectivity is rejected outright, thereby reducing human beings to the level of zombies (Searle,92). Idealism is problematic because under a solipsistic interpretation (*my* mind is the only thing that exists) it fails to adequately explain why the subject-object distinction should arise. Consequently, and consistent with a naturalistic interpretation of the phenomenal hierarchy in evolutionary terms, a number of emergentist schemes have been proposed as possible solutions (Harth,93) (Scott,95).

4. THE EMERGENT THEORY OF MIND

Proponents of the emergent theory of mind (ETM) such as Searle (1992) maintain that the mind (consciousness) emerges from the body (brain) as a consequence of bottom-up causal neurophysiological processes. The ETM describes the mind-body relation in terms of a two-level systemic hierarchy: the pattern of neuronal 'firings' in the brain (lower, local or substrate level) give rise to mental phenomena including (but not limited to) the subjective experience of consciousness (higher, global, or emergent level). Proponents of the computational ETM go further and assert that the formal aspect of bottom-up causation, viz. the *pattern* of neuronal activation, provides the necessary and sufficient conditions for the emergence of consciousness. This position, which is grounded in the theory of emergence as formulated by Alexander, is consistent with functionalism and supports the possibility of artificial consciousness. However, Searle (1992) has contested the computationalist isomorphism, "mind is to brain as software is to hardware", on the grounds that three of the terms (brain, software, hardware) are completely definable in objective terms while the fourth term (mind) necessitates description, at least partially, in subjective terms. Assuming biological-naturalism, Searle advances the following homology: brain states are to mental states as molecular behaviour is to liquidity. However, Tallis (1994) has, in turn, contested this position, maintaining that the latter two terms (molecular behaviour and liquidity) are necessarily on the same side of the mind-matter divide: material (objective) if viewed as intrinsic properties of water; mental (subjective) if viewed as different ways of experiencing/observing water. The assumption of bottom-up causation, which entails an epiphenomenal view of mind, has been contested by Sperry (1976) who has examined the link between consciousness and causality in connection with an emergentist solution to the mind-body problem. Sperry describes his position as emergentist, functionalist, interactionist, and monistic. Mind, a spatio-temporal pattern of mass-energy, is identified with subjective meaning which is held to be a causally-supervenient emergent property of neuro-physiological processes. The view that material-efficient causation constitutes a closed system is rejected; mind is held to functionally constrain the neurophysiological substrate from which it emerges. Hence, formal-final causation allows for top-down causation. Sperry follows Polanyi (1967) in defining the formal-final causality of mind in field-theoretic terms, viz. as an autonomous boundary condition eliminating degrees of freedom in the lower-level substrate.

5. CREATIO EX NIHILO AND EX NIHILO, NIHIL FIT

"Strong" (or ontological as contrasted with merely epistemological) emergence necessarily involves *creatio ex nihilo*, that is, "creation from nothing" since the (field-theoretic) laws governing the behaviour (or determining the properties) of higher level phenomena are non-reducible to the laws governing behaviour at lower phenomenal levels. However, *creatio ex nihilo* is a problematic concept since if it is held to apply universally then the primordial substrate (matter in materialism and Space-Time in Alexander's metaphysics) must itself emerge from nothing, interpreted in the sense of pure emptiness or the void. This is so because if matter (or Space-Time) exists then it must partake of Being and under the conventional interpretation of *creatio ex nihilo*, emergence of Being from nothingness is implied. Additionally, *creatio ex nihilo* conflicts with an important metaphysical maxim, one which is consistent with the First Law of Thermodynamics, viz. in a closed system, matter/energy is neither created nor destroyed: *ex nihilo, nihil fit* or "from nothing, nothing comes". This concept can be broadened into the assertion "like from categorial like". Thus, material phenomena can give rise to material phenomena (of lesser *or* greater complexity since the maxim does not exclude the possibility of evolution) and mental phenomena can give rise to other (lower or higher order) mental phenomena. If experience (or subjectivity) is categorial, that is, ontological, then it cannot be reduced to the non-experiential (or objective). However, this is precisely what *is* being asserted in the ETM. Hence, the categorial mind-body problem of secondary qualities or *qualia* again arises in the context of emergentism. Given the apparent intractability of this problem, it is worth re-examining its status; in short, is the mind-body problem a well-formed problem in the sense of logically sound ? Ryle (1949) has argued that "the belief that there is a polar opposition between Mind and Matter is the belief that they are terms of the same logical type" (p.23), a belief which he holds to be false given his rejection of Cartesian substance dualism and assertion that mind is an aspect or property of matter. However, while it is certainly the case that Cartesian substance dualism is problematic (two ontologically different substances cannot interact given the closure of each to the other and the validity of the extended form of the above maxim) and hence, the associated category problem does not emerge as well-formed, this does not apply to the subjectivity-objectivity problem (Griffin,88). While it may be incorrect to maintain a polar opposition between mind and matter on the basis of logical type identity, subjectivity and objectivity are instances of a more general logical type, viz. perspective or *view* (Nagel,86). Subjectivity is an epistemological perspective which is both ontologically grounded and ontically irreducible; in short, subjectivity cannot be generated from objectivity since they are mutually exclusive categories.

It would appear, therefore, that the only possible solution to the mind-body problem which does not entail a category error (a violation of the preceding maxim under its extended formulation) is some form of dual-aspect theory such as "panpsychism" in which the most primitive ontological components in the universe are held to possess both internal (subjective) and external (objective) aspects (Nagel,86). However, panpsychism (or animism) conflicts with some of our most basic assumptions regarding the problem of other-minds, viz. the problem of determining whether or not any entities other than myself possess minds: although the ascription of consciousness on the basis of external behaviour *alone* is problematic (Searle,80), surely there is some connection between consciousness and external behaviour? A denial of this position would entail holding that entities such as stones, clouds and, perhaps more controversially, computers are conscious. What is required is an emergentist variant of panpsychism which (1) supports the ontological irreducibility of subjectivity (thereby solving the mind-body problem) and (2) is consistent with a phenomenal hierarchy in which matter, life and mind are viewed as qualitatively distinct from an objective, behaviouristic perspective (thereby preventing conflict with the intuitive requirement for inclusion of behavioural criteria in the solution of the other-minds problem.)

6. HEIDEGGER AND THE QUESTION OF BEING

In preparation for the discussion in the next section, it is necessary to consider what is meant by Being in the context of Western metaphysics. The philosophy of Martin Heidegger (1889-1976) is the result of a detailed questioning concerning Being and hence, provides an appropriate starting point. For example, in the introduction to *The End of Philosophy* (1973), Stambaugh maintains that

> with Plato's distinction of essence (whatness) and existence (thatness), the difference between Being and beings [or things] is obscured, and Being as such is thought exclusively in terms of its relation to beings as their first cause .. and thus itself as the highest of those beings.

> When the distinction of essence and existence arises, it is essence, whatness, which takes priority. The priority of essence over existence leads to an emphasis on beings. The original meaning of existence as *physis*, originating, arising, presencing, is lost, and existence is thought only in contrast to essence as what 'factually' exists. In contrast to what 'factually' exists here and now, Being is set up as permanent presen*ce* (nominal) abstracted from presen*cing* (verbal) in terms of time-space. (p.x)

Thus, metaphysics since Plato has concentrated on the *essence* of beings, interpreting this essence in terms of categorial notions, for example, space, time, substance, process, change, cause, whole, part, quantity etc. The Being of beings, however, has been reduced to the status of a mere universal predicate and defined as existence, thatness, actuality or reality. However, according to Heidegger (1973), "in the beginning of its history, Being opens itself out as emerging (*physis*) and unconcealment (*alētheia*). From there it reaches the formulation of presence and permanence in the sense of enduring (*ousia*). Metaphysics proper begins with this." (p.4) From this assertion it follows that the Platonic interpretation of Being is grounded in a more primal *mode* of Being, viz. *alētheia-physis* or emergent presencing in what is unconcealed (beings). Consequently, Heidegger maintains that "the actual being, is incomprehensible in its beingness [or Being] when thought in terms of *idea* [whatness or essence]." (p.9) Furthermore, "because thatness remains unquestioned everywhere in its nature, not, however, with regard to actual beings (whether they are or are not), the unified essence of Being, Being as the unity of whatness and thatness, also determines itself tacitly from what is unquestioned." (p.11) By revealing a more originary mode of Being other than that of pure thatness, Heidegger raises the possibility of a *hermeneutic* (interpretative) approach in which Being is seen to disclose itself multiply. Dreyfus (1991) claims that the basis for this approach is already implicit in the thought of Aristotle:

> 'Being' does not behave like a very general predicate. For example, the Being of numbers seems not to be the same as, but at best only analogous to, the Being of objects, and the Being of real objects differs from the Being of imaginary objects such as unicorns. Aristotle says that Being is predicated analogously. Since Being transcends the universality of a class or genus the Scholastics called it a *transcendens*. Heidegger concludes that Being is clearly no ordinary predicate. (p.10)

Dreyfus maintains that Being is "'that on the basis of which beings are already understood'. Being is not a substance, a process, an event, or anything that we normally come across; rather, it is a fundamental aspect of entities, viz. their intelligibility." (p.xi) In *Being and Time* (1927), Heidegger began a detailed investigation of the structure of Being beginning with that being on the basis of which all beings gain their intelligibility, viz. human being. Heidegger's approach to the understanding of Being is phenomenological (ontological, existential, hermeneutic) and thereby distinct from the scientific approach which is causal (ontical). Four areas of investigation are possible depending on the *kind* of investigation (ontical or ontological) and *what* is being investigated (a 'who' or a 'what'). Thus, ontologically, there are two main categories of Being: (1) human being or *Dasein* and (2) non-human being. The latter can be categorized into the ready-to-hand (*Zuhandenheit*)

142

and the present-at-hand (*Vorhandenheit*). *Dasein* (being-there, that opening or hermeneutic *clearing* which enables Being to come forth from concealment) is ontically distinct from non-human being in that it has the understanding of Being as its unique characteristic. This understanding is either (1) preontological (pretheoretical) or (2) ontological; in the former, understanding takes the form of tacit and ineffable *know-how* which assumes a 'Background' of shared practices, skills etc, while in the latter it takes the form of reflective *know-that* (or *know-what*) characteristic of the kind of understanding usually associated with a conscious cognitive subject. *Dasein* is a being-in-the-world where the 'in' of being-in should be understood in terms of concernful involvement (for example, 'being in love') and not spatial location (for example, 'being in a box'). The ready-to-hand (or available) is how *Dasein* encounters non-human being when coping with the world in an average-everyday unreflective manner. The ready-to-hand is something which is used in-order-to get something done and is, therefore, defined in terms of its functionality with respect to the concerns (unreflective purposes) of *Dasein*. However, as Dreyfus (1991) states, "an 'item' of equipment is what it is only insofar as it refers to other equipment and so fits in a certain way into an 'equipmental whole'." (p.62) The equipmental whole may be distinguished from other 'wholes' such as the referential and involvement whole as follows: the equipmental whole describes the interrelated equipment; the referential whole its interrelations; and the involvement whole human purposiveness which is defined in terms of significance, "the background upon which entities can make sense and activities can have a point." (p.97) Heidegger elucidates the meaning of the ready-to-hand by way of an example, that of a person engaged in the act of hammering: if all is going well and the nail is being driven into the wood, then hammer, nail and wood are all transparent to the one doing the hammering, forming part of the functional network of equipment. The present-at-hand (or occurrent), encountered by *Dasein* in its mode of the Cartesian subject reflecting upon objects, describes beings (things) viewed independently of functional context, significance or the equipmental whole, and hence, independently of their relationship to *Dasein*. When the ready-to-hand *becomes* the present-at-hand (for example, during 'breakdown') it is revealed as an *object* with determinate, context-free properties. Dreyfus maintains that Heidegger "wants to stress three points. (1) It is necessary to get beyond our practical concerns in order to be able to encounter mere objects. (2) The 'bare facts' related by scientific laws are isolated by *a special activity of selective seeing rather than being simply found*. (3) Scientifically relevant 'facts' are not merely removed from their context by selective seeing; they are theory-laden, i.e. recontextualized in a new projection [emphasis added]." (p.81) The transition from a concerned being-in-the-world to a subject reflecting upon objects *emerging* during breakdown lends support to Heidegger's contention that the primordial mode of *Dasein* is the unreflective or pre-ontological mode:

> Heidegger does not deny that we sometimes experience ourselves as conscious subjects relating to objects by way of intentional states such as desires, beliefs, perceptions, intentions, etc., but he thinks of this as a derivative and intermittent condition that presupposes a more fundamental way of being-in-the-world that cannot be understood in subject/object terms. (p.5)

However, what is also thereby asserted is the primacy of the functionally-contextual nature of the equipmental whole:

> For Heidegger, unlike Descartes, Husserl, and Sartre, the object of mere staring, instead of being that which really is, is an impoverished residue of the equipment we directly manipulate. The bare objects of pure disinterested perception are not basic things we can subsequently use, but the debris of our everyday practical world left over when we inhibit action. (p.47)

Hence, Heidegger maintains that the ready-to-hand is a more primordial way for things to be than the present-at-hand. However, he should not be understood as merely asserting the primacy of the practical (*praxis*) over the theoretical (*theoria*). As Dreyfus (1991) states, "Heidegger seeks to *supplant* the tradition by showing that the ways of Being of equipment *and* substances [objects with context-free properties], and of actors *and* contemplators [Cartesian subjects], presuppose a background of understanding of Being - originary transcendence or being-in-the-world." (p.61)

There are two stages in the development of Heidegger's phenomenology, the latter of which is signalled by 'the turn' (*Die Kehre*), when Heidegger shifts his attention from a hermeneutic of *Dasein* to historical thinking on Being (*Sein*). The distinction between the two stages may be introduced by way of a statement due to Dreyfus explicating the thought of the Heidegger of *Being and Time*: "for Heidegger, scientific theory is an autonomous stance. It is not mere curiosity, nor is it merely based on an interest in *control* [emphasis added]." (p.80) This position conflicts with the account of science given in (Heidegger,77), a post-*Kehre* work in which science is linked to technology which in turn is linked to Enframing (*das Gestell*), a mode of Being which developed out of Greek *technē* (artificing). Heidegger maintains that it is with Plato and Aristotle that *technē* begins to take on the form of Enframing following the bifurcation of Being into *essentia* (whatness) and *existentia* (thatness). As stated previously, the pre-Socratics had a very different understanding of *technē*, a consequence of their very different understanding of Being as *alētheia-physis* or the emerging power which brings itself forth from concealment. The thinking of the later Heidegger concentrates on two related projects: firstly, a more aesthetic and poetic approach to the understanding of Being (Heidegger,71), and secondly, an attempt at clarifying the nature of logic and science in order to draw attention to the 'danger' inherent within technology (Heidegger,77). Of particular relevance in the context of this investigation is his understanding of mathematics:

> mathematics is the reckoning that, everywhere by means of equations, has *set up* as the goal of its expectation the harmonizing of all relations of order, and that therefore 'reckons' in advance with one fundamental equation for all possible ordering [emphasis added]. (p.170)

By this, Heidegger means to imply that mathematics and modern science (which *is* mathematical) are self-fulfilling in the sense that both specify *a priori* what and how things are to be encountered, viz. as the Enframed (or anthropically-ordered), a characteristic of the fact that science and mathematics have their metaphysical origins in *technē*. Heidegger maintains that although the Enframing characteristic of modern theoretical science is *a* way of revealing Being, it does not constitute *the* way of revealing Being; hence, his commitment to *hermeneutic* realism:

> Theory never outstrips nature - nature that is already presencing - and in this sense theory never makes its way around nature. Physics may well represent the most general and pervasive lawfulness of nature in terms of the identity of matter and energy; and what is represented by physics is indeed nature itself, but undeniably, it is only nature as the object-area whose objectness is first defined and determined through the refining that is characteristic of physics and is expressly set forth in that refining. Nature, in its objectness for modern physical science, is only *one* way in which what presences - which from of old has been named *physis* - reveals itself and sets itself in position for the refining characteristic of science .. Nature thus remains for the science of physics that which cannot be gotten around. (pp.173-174)

Importantly, he states that

> scientific representation, for its part, can never decide whether nature, through its objectness, does not rather withdraw itself than bring to appearance the hidden fullness of its coming to presence. Science cannot even ask this question, for, as theory, it has already undertaken to deal with the area circumscribed by objectness. (p.174)

In this statement, Heidegger maintains that scientific objectivity prevents the unconcealment, appearance or emergence of other ways (modes) of Being. Thus, the scientific interpretation of emergence is clearly very different from Heidegger's interpretation: on the former, emergence denotes the appearance of new properties in a whole that are not present in any of its parts; on the latter, emergence denotes the unconcealment of the various modes of Being in beings. Appreciation of this distinction is crucial for the following reasons: firstly, on the Heideggerian view, consciousness is not held to be emergent in the sense understood by Searle (1992); that is, consciousness is not a systemic whole *constructed* from parts, a view which ultimately leads to the category problem

144

described in §5 if materialism is assumed. Rather, consciousness (subjectivity) is a mode of Being which is *revealed*, that is, brought forth from concealment. The value of Heidegger's understanding of emergence is three-fold: (1) it is originary, dealing with the understanding of the emergence of Being as opposed to merely beings (§7); (2) it solves the category problem by unconcealing an alternative ontology (§8); and (3) it enables the *creatio ex nihilo* and *ex nihilo, nihil fit* maxims to be rendered commensurable (§9). Secondly, Heidegger's clarification of the distinction between the two modes of *poiēsis* or coming-forth (§7) has a number of implications for the "strong" artificiality project, that is the artifactual realization of natural phenomena, as discussed in §10.

7. PROLEGOMENON TO A NEW THEORY OF EMERGENCE

It can be argued that the apparent incommensurability of emergentist (§4) and panpsychist (§5) accounts is a consequence of adopting a monistic, materialistic, and atomistic metaphysics. Etymological analysis of the notion of emergence reveals an alternative and historically antecedent interpretation of the concept (briefly introduced in §6) which renders the positions commensurable. The *Oxford Latin Dictionary* (1982) provides the following etymology for emergence:

> *emergo* ~gere 2. to come forth (from confinement, concealment), emerge. 4. to become apparent, come to light; (of something unexpected) to turn up, present itself, to appear as a result, to emanate.

Adopting the hermeneutic approach to philosophical investigation pioneered by Heidegger and *questioning* concerning emergence reveals two types of emergence:

1. $emergence_1$: a movement between Being and beings
2. $emergence_2$: a movement between parts (beings) and wholes (beings)

(1) is associated with the coming-into-being (or becoming) of ontical or ontological primitives from a *poiētic* (generative) source while (2) is associated with the manifestation of categorial potentialities in complexes once ontical or ontological primitives have been defined. Heidegger's ontological analysis of Being (1959) leads to a conception of emergence in terms of the *poiēsis* or coming-forth from concealment of Being in beings (things, essents). As stated in §6, Being is neither substance nor process; rather, it is the originary 'background' condition for the intelligibility of beings (Dreyfus,91). (Hence, Heidegger's radical assertion that the positions of Heraclitus, a processist, and Parminides, a substantialist, are equivalent.) Being is not to be thought of in terms of a whole (system or complex) composed from parts (components or primitives); rather, Being is the primordial ground of parts and wholes, in fact, of all that is. Hence, the following statement due to Heidegger (1959), originally formulated in the context of a critique of anthropological thought:

> The beginning is the strangest and the mightiest. What comes afterward is not development but the flattening that results from mere spreading out; it is inability to retain the beginning; the beginning is emasculated and exaggerated into a caricature of greatness taken as purely numerical and quantitative size and extension. That strangest of beings [man] is what he is because he harbours such a beginning in which everything all at once burst from superabundance into the overpowering and strove to master it. (p.155)

Heidegger (1977) distinguishes two forms of *poiēsis* (coming forth) in Greek thought: *physis* (*autopoiēsis* or bringing-forth-by-self*)* and *technē* (*allopoiēsis* or bringing-forth-by-other) corresponding to the modes of becoming associated with naturals and artifactuals (or artifacts) respectively. In this section, the presentation focuses entirely on *physis*, the 'self-cutting' originary power of Being, which establishes $emergence_1$ as a necessary condition for $emergence_2$. There are four possibilities for emergence delimited by *type* ($emergence_1$ or $emergence_2$) and *direction* (expansive or contractive) as shown in Table 1:

	emergence$_1$	*emergence$_2$*
Expansive	*joining$_1$* : essent$_{part}$→Being	*joining$_2$* : essent$_{part}$→essent$_{whole}$
Contractive	*cutting$_1$* : Being→essent$_{part}$	*cutting$_2$* : essent$_{whole}$→essent$_{part}$

Table 1 Four possibilities for emergence.

An originary joining of type *joining$_1$* is necessitated in the event of category error (a form of 'breakdown'); this originary joining corresponds to the *disappearing* of existing ontical or ontological primitives (concealment of Being) and the *appearing* of new ontical or ontological primitives (unconcealment of Being). The *hermeneutic* (interpretative) movement takes the form of a spiral as shown in Fig.1:

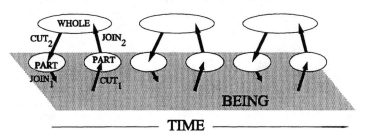

Figure 1 The hermeneutic spiral.

The stages in the hermeneutic spiral associated with materialism are as follows: (1) objective, external atoms (parts) unconcealed as ontical primitives (*cutting$_1$*); (2) objective physical-biological systems emerge from atomic substrate (*joining$_2$*); (3) subjective (experiential) biological systems encountered leading to category problem (*ontical*-breakdown); (4) category problem traced to ontical primitives (*cutting$_2$*); (5) ontical primitives withdraw *into* concealment (*joining$_1$*). The problem of ontical-breakdown associated with materialism can be solved *if* the manifoldness of Being and its unconcealment in various modalities is accepted. As stated in §6, Heidegger's *questioning* was motivated by an attempt at revealing the structure of Being via phenomenological (hermeneutic) investigation of the structure of a particular kind of Being, viz. *Dasein* or human being. However, in unconcealing two ways of being human (causal or ontical and existential or ontological), Heidegger also reveals two modes of Being and in revealing the manifoldness of Being, Heidegger implicitly points toward the possibility of a post-materialistic metaphysics. Heidegger's denial of the *ontological* primacy of subjective, reflective (thematic) consciousness with respect to human being (*Dasein*) does not entail support for the view that objects are *ontically* primitive and subjects *ontically* derivative nor that Being (as the ground of beings) is *ontically* objective; according to Heidegger, it is the dualism of subjects and objects that is secondary, emerging in the event of *ontological*-breakdown. However, the *ontical* question of the emergence of subjectivity, that is, the category problem described in §2, remains unsolved. In order to solve this problem, it is necessary to consider an alternative metaphysics rather than an alternative phenomenology; however, this is only possible because, by way of phenomenology, it has been shown that Being has manifold modes and hence, can support a metaphysical paradigm-shift. It was argued in §6 that Being *conceals* consciousness, that consciousness is not created *ex nihilo* in the sense of "from void"; rather it is *revealed*. Subjectivity is, therefore, a mode of Being awaiting unconcealment through the hermeneutic process. Heidegger maintains that subjects and objects emerge from being-in-the-world. This position is consistent with Whiteheadian organicism (panexperientialism) in which subjectivity and objectivity refer to different temporal aspects of experiential events. Thus, an *ontical* solution to the category problem is to be found in the coming forth from concealment of Whiteheadian events as newly emerging ontical primitives (*cutting$_1$*).

8. PANEXPERIENTIALISM

Materialism is based on the metaphysical assumption of *vacuous actuality* in which the ultimate ontical primitives of reality are held to be non-experiential and non-spontaneous. However, as Griffin (1997) states,

> we know from our own experience that experiencing actualities can exist, but we have no experiential knowledge that a vacuous actuality is even possible. (p.4)

Additionally, realism has been conflated with materialism (objectivism) in the philosophy of Descartes, Galileo and Newton. Both Heidegger and Whitehead are committed to realism; however, in the former, this realism is *hermeneutic* (ontological) while in the latter it is *processual* (ontical). Griffin (1997) maintains that "panexperientialist physicalism portrays the world as comprised of creative, experiential, physical-mental events." (p.13) An *event* (actual occasion or actual entity) is a happening at a certain place at a certain time. A *process* is a series or chain of events which can give rise to either (1) a *concresence* (compound individual or organism), that is, a complex and highly ordered society of events manifesting both experience and causation or (2) an *aggregate* in which the experiential and causal components associated with primitive components cancel out in the whole. (A compound individual is a *nexus*, a system of *internally*-related interacting subjects whereas an aggregate is a *network*, a system of *externally*-related interacting objects. Internal relations are constitutive of both the essence and existence of things; hence, panexperientialism - a form of idealism - is *a priori* committed to the bifurcation of Being described in §6.) Experience implies subjectivity, which pan-experientialists associate with "for-itself" in contrast to mere "what-it-is-likeness" (§2), and is defined in terms of both (1) prehension and (2) causation. Farleigh (1996) maintains that "to 'prehend' an object [datum] is to experience it, perceive it, feel it, or more literally, to grasp it or take it into account, though not necessarily in a conscious way." (p.7) Prehension is of both past and external events and is involved in determining the creative causal response of an event to incoming efficient causation. As stated previously, each event has both physical and mental aspects. The physical aspect is always *prior* and is the reception by the event of the efficient causation of prior events into itself. As Griffin (1997) states: "each event prehends aspects of the past into itself and then gets aspects of itself prehended into future events." (p.15) Again,

> Every unit [or compound-individual] event (as distinct from an aggregational event) has a mental aspect, and this mentality involves an element, however slight in the most elementary events, of spontaneity or self-determination. Although the event's physical [aspect] is given to it, its mentality is its capacity to decide precisely what to make of its given foundation. Its physicality is its relation to past actuality; its mentality involves its prehension of ideality or possibility, through which it escapes total determination by the past. (p.13)

Hence, events are experiential both as subjects (caused) and objects or *superjects* (causing). Furthermore, "as subjects, events enjoy an inner duration; as objects, however, they are purely spatial. An event cannot be prehended until its moment of subjectivity is finished, because it is nothing fully determinate until its moment of self-determination is completed." (p.15) Thus, instantaneous events are fictions on Whiteheadian panexperientialism. The distinction between Newtonian (necessary) and Whiteheadian (contingent) events is shown in Figs.2 and 3 respectively:

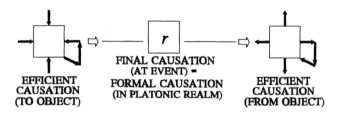

Figure 2 A Newtonian event (*r* denotes the response).

147

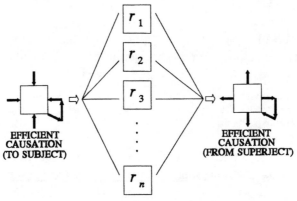

FINAL CAUSATION (AT EVENT) ∈ FORMAL CAUSATION (IN PLATONIC REALM)

$(r_k \in R)$

Figure 3 A Whiteheadian event.

Although panexperientialists hold that experience is ontically primitive, this does not entail the panpsychist claim that consciousness is primordial. Consciousness is not identical to experience since the former involves the capacity for negation, that is, making a distinction between possibility and actuality. As Griffin (1997) states,

> very elementary events, by virtue of synthesizing prior events and possibilities into rudimentary analogues to propositions, have incipient-intentionality; somewhat higher-level events, complex enough to form propositions, have proto-intentionality; while only very high-level events are sophisticated enough to contrast propositions with alternative possibilities, thereby enjoying .. the subjective form of consciousness." (pp.13-14)

Hence, panexperientialism is consistent with non-materialistic emergentism and the Greek view of reality in which subjectivity is held to be intrinsic to nature (§10). However, while this metaphysics solves the category problem, it appears to do so at the expense of the structural order of the world. This is entailed by postulating 'freedom', that is, the subjective response of an event to incoming causation, as real since universal causation (necessity, determinism) is thereby undermined. If panexperientialism is to be retained, it must, therefore, explain the stability of phenomena. Whitehead is able to achieve this by incorporating a Platonic component into his framework, viz. the idea of *ingression*, that is, the realization of the potentiality of 'eternal objects' (Platonic forms) in actual entities (events). Although in the Whiteheadian universe *all* events are connected, the extent of connection varies; hence, certain events participate in the determination of an event to a greater extent than other events. The extent of participation in an event is determined by three factors: (1) incoming efficient causation; (2) ingression of eternal objects; and (3) the subjective response of the event, that is, a *selection* of which eternal objects (from a given set) are to ingress (structurally supervene) on the incoming efficient causes. Eternal objects act as a transcendent source of stability. Hence, as Griffin (1997) states, "the regularities of nature are not due to externally imposed 'laws' but are .. nature's most long-standing habits - the habits of its most elementary members, reflecting patterns that they have internalized." (p.16) For example, "the electron in the living human body .. will behave differently than it did in the organic environment, not because the 'laws of electron behaviour' have been violated, but because it is there subject to different influences." (p.17) Thus, determinism is the exception, not the rule. However, Collingwood (1945) is critical of the idea of eternal objects. He maintains that

> a world of eternal forms which included in itself forms of every empirical detail in nature would only be a lumber-room of natural details converted into rigid concepts, and that a world of forms so conceived, instead of explaining the processes of nature, would be a mere replica of these processes themselves with the process left out. [Thus,] appealing to the conception of a world of forms or objects as the source or ground of natural process [necessitates] an account of this world .. (pp.171-172)

Thus, Whiteheadian panexperientialism contains the potential for vicious infinite regress. It is argued that these problems are artifacts of the Platonic essence(potentiality)-existence(actuality) duality. Plato maintains a distinction between the realm of existence (matter or substance) and the realm of subsistence (ideas or forms). However, if subsistents (forms) *are* then they must partake of Being. Hence, the solution to the problem of infinite regress lies in transcending the opposing categories to the more primordial ground of pure Being understood as presencing forth from concealment. The link between Cartesian, Whiteheadian and Heideggerian frameworks with respect to the issue of structural stability is shown in Table 2:

	Descartes	*Whitehead*	*Heidegger*
a priori (substrate)	Laws	Forms	*Physis*
a posteriori (emergent)	Phenomena	Laws	Forms

Table 2 Structural stability in Descartes, Whitehead and Heidegger.

Farleigh (1996) maintains that

dualism is strongly rejected and, for Whitehead, this means that epistemology does not have any priority over ontology. Any inquiry into knowing is at the same time an inquiry into being. (p.3)

The latter view is correct insofar as knowing is a mode of Being. However, Being subsumes knowing; hence, the reverse of Farleigh's position, viz. "any inquiry into being is at the same time an inquiry into knowing" is false. This view is consistent with Heidegger's critique of Husserlian phenomenology in which human being is equated with the experiencing Cartesian ego. Hence, although Whiteheadian panexperientialism solves the category problem, its situatedness within Being as an encompassed mode (or way) must not be ignored if the *question* of Being is to remain valid.

9. THE EX NIHILO REVISITED

Contrary to Bunge's (1977) assertion that he regards nothingness as an entity, Heidegger (1993) maintains that "the nothing comes forward neither for itself nor next to beings, to which it would, as it were, adhere." However, he goes on to state that "for human existence, the nothing makes possible the openedness of beings as such. The nothing does not merely serve as the counterconcept of beings; rather, it originally belongs to their essential unfolding as such. In the Being of beings the nihilation of the nothing occurs." (p.104) Heidegger clarifies the latter statement as follows: "in its nihilation the nothing directs us precisely towards beings." (p.104); furthermore, "negation is grounded in the not that springs from the nihilation of the nothing." (p.105) Thus, according to Heidegger, the nihilation of the nothing is the *poiētic* source of negation and as Spencer-Brown (1969) has shown, negation is equivalent to the act of drawing a distinction. Heraclitus maintained that conflict (setting-apart) is the source of both the creation and preservation of the world, anticipating Spencer-Brown's assertion that a universe comes into being when a space is severed. The link is clear: conflict necessitates difference and difference implies distinction. Hence, the nihilating nothing is the source of beings. This is readily shown: nihilation of nothing = not nothing = not not-thing = thing = being; hence, nihilation of nothing = being. However, the making of a distinction necessitates a distinction-*maker*. If the nothing is the ground of distinction-making, it must have the capacity for both *making* (acting) distinctions and *recognizing* (knowing) that distinctions have been made, both of which are modes of Being. In §6 it was stated that Being is the *poiētic* source of beings since Being discloses (unconceals, reveals) itself in and as beings. Hence, Being=Nothing.

This interpretation finds support in the statement that "the nothing does not remain the indeterminate opposite of beings but reveals itself as belonging to the Being of beings" (p.108), a position consistent with Hegel's assertion that pure Being and pure Nothing are the same. Being in concealment is identical to the nothing and its unconcealment in beings corresponds to the nihilation of the nothing. According to Heidegger, the *why* of unconcealment is "enigmatic". However, in the context of this study it can be linked to the (re-)emergence of the question of Being, that is, to a recognition of the necessity of a hermeneutic approach to revealing the various modes of Being. In asserting that the nothing is not identical to the void (the latter is a fiction of conventional metaphysical discourse), but rather is the *poiëtic* ground of beings, Heidegger is able to correct the conventional (mis)interpretation of *ex nihilo, nihil fit*, viz. "nothing comes from nothing". The corrected version is "from nothing, nothing comes to be", that is (Being=nothing)→beings. Thus, *creatio ex nihilo* because *ex nihilo, nihil fit* and the positions are rendered commensurable. This might appear to support a solution to the category problem (§5) based on an expansive (joining) form of *emergence*$_2$. However, this inference is incorrect: subjectivity as a mode of Being associated with certain kinds of beings is concealed-within Being awaiting unconcealment in beings. (It is important to appreciate that the primordial condition of Being is concealment.) Hence, the category problem *can* be solved, but only by moving to an interpretation of Being (an alternative ontology) in which subjectivity and objectivity are irreducible essents (modes of Being). As argued in the previous section, panexperientialism provides such an interpretation. Thus, emergentism is possible, but only if the concept of emergence is extended to incorporate *emergence*$_1$.

10. THE ESSENCE OF ARTIFICING

In §7, a distinction was made between the two modes of *poiēsis* (emergence or coming-forth), viz. *physis* (or *autopoiēsis*) and *technē* (or *allopoiēsis*), and in §8, a (post-materialist) ontology for *physis* based on panexperientialism was briefly described. In this section, the meaning of *technē* will be investigated. For the Greeks, *poiēsis* was the gathering *of* man *into* the presencing of subjects (beings); *technē* was a mode of *artificing* (artifact-making) in which man was summoned-forth as the means by which artifacts were brought forth into Being from concealment (Heidegger,77). On this original interpretation of *poiēsis* both man and artifacts are viewed as hermeneutically integral components of the self-presencing of Being (*physis*); hence, there is no trace of "design for human purposes" (anthropocentricity) or "ideas in the mind awaiting realization" (egocentricity) associated with the artificing (*technē*) of the Greeks. However, modern *alētheuein* (revealing) is radically different; according to Heidegger, it is a challenging-forth *by* subjects (volitional Cartesian egos) *setting-upon* objects (beings) which are re-conceived as standing-reserve (*Bestand*), that is, as resources 'Enframed' or ordered (set in order) according to human purposes. Although Heidegger draws a distinction between *poiēsis* and Enframing (*das Gestell*), the latter can be viewed as a type of the former in which the determining locus (subject) has been inverted (Fig.4):

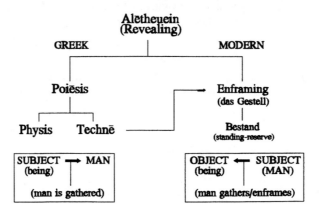

Figure 4 The subject in *poiēsis* and Enframing.

The essence of modern *technē*, that is, artificing in its modern incarnation as Enframing is described by Heidegger (1977) as follows:

> What is, in its entirety, is now taken in such a way that it first is in being and only is in being to the extent that it is set up by man, who represents and sets forth .. The Being of whatever is, is sought and found in the representedness of the latter. (pp.129-130)

The distinction between Greek and modern artificing can be understood in terms of the shift in meaning of the maxim, "man is the measure of all things": For Protagoras, to whom first articulation of the statement is attributed, it meant that finite man is a limiting horizon for the unconcealment of Being (in beings); that is, Being is revealed to man only partially at any time, a position which is consistent with hermeneutic realism. However, with Descartes the maxim *came to mean* the self-certainty of the ego upon which the Being of beings is grounded. In the egocentricity of the Cartesian 'cut' (Enframing), man at once both (1) blocks the presencing of Being as *physis* and (2) forces the presencing of Being as standing-reserve. There are (at least) two consequences of the shift in the locus of subjectivity and the change in the orientation of man to beings in the move from *poiēsis* to Enframing (*das Gestell*) which can be identified as (1) ontological and (2) ontical.

10.1. Ontological

According to Heidegger, concernful, non-thematic, involved being-in-the-world is the primordial mode of human being; the primordial Being *of* the world is, therefore, equipmental, that is, ready-to-hand or available (§6). As Dreyfus (1991) states, "equipment is *defined* by its function (in-order-to) in a referential whole" (p.91) which, in turn, is determined relative to an involvement (meaning or *significance*) whole. Furthermore,

> since equipment is in no way derivative, and since involvement is as genuine a mode of access as theory, we can say that equipment in use is equipment as it is in itself. (p.66)

Consequently,

> traditional ontology [that is, the metaphysics of ontical causality] succeeds only if it can account for *all* modes of being, including Dasein's practical activity and the equipmental whole in which Dasein is absorbed, in terms of the law-like or rule-like combinations of occurrent elements. If it can be shown that the world is irreducible to occurrent elements, be they bits of matter, atomic facts, sense data, or bits of information, then an ontology based on the occurrent fails. (pp.108-109)

Heidegger maintains that the ontological (existential) cannot be derived from the ontical (causal). Evidence supporting this claim includes Husserl's failed attempt at articulating the structure of the human life-world and the frame problem of artificial intelligence (the problem of extracting contextually-relevant knowledge from an ever-expanding factual repository). Although, Heidegger accepts that the ontical is in some sense prior to the ontological since, for example, the properties of materials constrain the functionality of equipment, the ontical cannot be made intelligible without the ontological. Since Being denotes intelligibility (§6), the Being of beings is fundamentally ontological and hence, the ready-to-hand is prior to the present-at-hand. Furthermore, only in the event of 'breakdown' when human beings *become* (come-forth-as) Cartesian subjects, do beings presence as occurrent, that is, as objects with determinate, context-free properties. (Elstob (1984), however, argues for the context-sensitivity of properties). Dreyfus (1991) maintains that Heidegger makes three important points with respect to the Being of the present-at-hand (occurrent):

> (1) It is necessary to get beyond our practical concerns in order to be able to encounter mere objects. (2) The 'bare facts' related by scientific laws are isolated by a special activity of selective seeing rather than being simple found [hence, a commitment to a *hermeneutic* realism]. (3) Scientifically relevant 'facts' are not merely removed from their context by selective seeing; they are theory-laden, i.e., recontextualized in a new projection. (p.81)

This 'selective seeing' is an Enframing (*das Gestell*) which involves the postulating of *ceteris paribus* ("all things being equal") conditions made possible by a grounding in the shared 'background' of human practices. As Dreyfus states, "it is only on the background of already taken up practices and equipment that we can doubt the existence of particular objects, and even a whole domain of objects [that is, *exclude* beings from consideration in a domain of discourse]." (p.248) Enframing makes possible a modern form of *technē* (artificing), viz. technology, which is characterized by the *joining* of occurrent parts in systemic wholes. This *metaphysical* approach to artificing, in which the occurrent as substantial or processual is taken as primal, has its origins in Plato and Aristotle. While the latter upholds the distinction between artifacts and naturals, this distinction is defined in terms of the becoming (or mode of coming into being) of *matter* associated with each (natural or artifact): "in the case .. of artifacts we make the matter for the work to be done, whilst in the case of natural objects it is there already" (*Physics*, Book II, ch.I). Aristotle follows Plato in asserting the primacy of form over matter, essence (whatness) over existence (thatness) based on a view of the latter as the actual or what 'factually' is. This position is carried over into the modern context and culminates in the functionalist claim that substrate ontology is irrelevant and phenomena are multiply-realizable; hence, the possibility of "strong" *artificiality*, whereby the latter is meant functionally (or behaviourally) isomorphic artifactual realizations of natural phenomena. However, Heidegger's originary association of Being with *physis* (originating, presencing, arising) and his distinguishing of this mode of *poiēsis* from *technē* (artificing) makes possible the distinguishing of naturals from artifactuals on the basis of their respective modes of becoming (*poiēsis*). Heidegger's claim that modern artificing is grounded in the Enframing of nature (as object) by the Cartesian ego (or subject) allows naturality to be distinguished from artificiality on the basis of *historical* (or hermeneutic) relations (ontic and epistemic) between the Cartesian subject and beings (objects):

		ONTICAL STATUS OF OBJECT WITH RESPECT TO SUBJECT	
		A PRIORI	*A POSTERIORI*
EPISTEMICAL STATUS OF OBJECT WITH RESPECT TO SUBJECT	*A PRIORI*	(1)	(2)
	A POSTERIORI	(3)	(4)

Table 3 Historically-defined ontic and epistemic subject-object relations.

Ontically *a priori* beings are *found* (given) whereas ontically *a posteriori* beings are *made* (artificed) by the human subject; epistemically *a priori* beings are "forwards-interpretable" whereas epistemically *a posteriori* beings are (only) "backwards-interpretable" by the human subject. It was stated previously that in the case of naturals, the ready-to-hand (equipmental) is prior to the present-at-hand (occurrent). In the case of artifactuals (or artificials), the present-at-hand (occurrent) is prior since this is how beings are encountered during artificing in the Cartesian subject-object mode. However, since the ontological cannot be derived (or emerge) from the ontical, the present-at-hand cannot give rise to the ready-to-hand. Hence, artificiality cannot support being-in-the-world (Dreyfus,91). This holds true for both (2) *and* (4) in the above scheme (that is, both for symbolic/top-down *and* connectionist/bottom-up approaches to artificing) since the Being of the *substrate* in both cases is epistemically *a priori* and ontically *a posteriori*, that is, artifactual or present-at-hand and the latter can only give rise to beings of the same ontological kind, viz. occurrent or present-at-hand.

10.2. Ontical

In addition to its implications for the *ontological* project of an 'artifactual-phenomenology', the system of relations described in Table 3 also has implications for the *ontical* project of an 'artifactual-causality.' This immediately becomes apparent upon examining the Being of artifacts and naturals under panexperientialism (§8). The key insight is expressed by Heidegger (1977) as follows:

> scientific representation, for its part, can never decide whether nature, through its objectness, does not rather withdraw itself than bring to appearance the hidden fullness of its coming to presence. Science cannot even ask this question, for, as theory, it has already undertaken to deal with the area circumscribed by objectness. (p.174)

The Being of objects is *objective*, that is, externally-defined and externally-related; however, given the existence of experiential entities (humans and possibly other higher-order entities), it is simply an existential fact that there are beings whose Being is (at least partially) *subjective*, that is, (at least partially) internally-defined and internally-related. Since objects cannot give rise to subjects (§2), any mode of *poiēsis* involving objective (external) relations between beings and an artificing subject can only lead to the production of objects. Modern *technē*, viz. Enframing, is a mode of *poiēsis* which *takes* beings as objective, that is, encounters them as externally-defined (by the Cartesian subject) and externally-related to each other. While panexperientialism allows beings to be re-interpreted as subjects (thereby returning to the original Greek view of subjectivity), it does not alter the relationality involved in modern artificing: artificers can only access the external aspect of beings, an existential fact which follows from the essence of artificing. This point can be clarified as follows: panexperientialism distinguishes between (1) primitive actual occasions (events) and compound individuals which are experiential and (2) aggregates which are not. In artificing, the parts (components) of the whole (artifact) either are aggregates or are viewed as aggregates. In the latter case, compound individuals or societies of events are epistemically-transformed into aggregates, that is, entities *defined* in terms of external relations (efficient causation), by an act of 'cutting', that is, "selective-seeing" (§10.1): individual responses (final causation) to incoming efficient causes are ignored and an average or aggregate response is adopted; thus, the "spectrum" of final causation associated with each event is "collapsed" onto a single final cause. (This has the effect of 'closing' the system down such that contingency/intentionality is transformed into necessity/determinism.) This collapse *can* and *does* occur in naturals (for example, stones and clouds) at the level of the aggregate; however, in this case it takes the form of a self-collapse, that is, an *autopoiētic* process. In artifacts, the collapse is *allopoiētic* and either epistemic or ontic. In the latter case, aggregates are *produced* (artificed) from compound individuals via $cutting_2$ and $joining_2$ operations; the former involves the dissociation of an actual occasion from its nexus of experiential relations and the latter leads to the emergence of an aggregate in which experience statistically 'cancels out'. Since both operations are carried out by a subject relating to the primitive material objectively (externally), the *product* is always an aggregate; thus, *artifactual-cutting$_2$* and *artifactual-joining$_2$* cannot generate an experiential nexus. However, *natural-cutting$_2$* and *natural-joining$_2$* can support the emergence of compound individuals; this possibility follows from the essence of panexperientialism itself and is demonstrated by the existence of higher-order experiential entities. Moreover, the emergence of compound individuals from *aggregates* via *natural-cutting$_2$* and *natural-joining$_2$* operations is also supported. However, this (logically) necessitates self-*re*-organization of the aggregate such that its primal components (which are experiential) are able to manifest their subjectivity (experience and final causality) non-aggregatively, that is, in the formation of an experiential nexus; clearly this can only occur naturally in a "suitable" context. It follows, therefore, according to Birch (1994), that

> to attempt to make [an experiential artifact] by building up a hierarchy of compound entities that think and feel would [necessarily] be to attempt to repeat evolution from scratch. (p.8)

This is because concrescences (compound individuals) are context-sensitive and internally-related and, as Heraclitus maintained, the world of the present, although more like the world of the past than unlike it, is never identical to the latter. Thus, in artifacts, where re-configuration is objectively specified in terms of determinate, context-free external relations and thereby *determined* - or rendered *deterministic* - by the artificing subject, the transformation of an aggregate into a compound individual is impossible since although $cutting_2$ and $joining_2$ can occur, it is necessarily an objective form of *cutting₂* and *joining₂* involving a movement between aggregative wholes and parts. The ontical distinction between artificiality and naturality (that is, artifacts and nature respectively) is summarized in the following statement due to Farleigh (1996):

> the primary function of a machine can be described in terms of the external relations of the parts which are assumed to be 'simply located'. One set of external relations is as good as any other, and hence the function of one machine can be modeled on another. The function of an organism on the other hand is constituted by both the internal and the external relations between events. Each event is not simply located, is unique to its history and is hence, highly context-dependent. The procedure, then, of attempting to map an organism onto a machine can only be a process of abstraction and hence such a mapping would be done with a loss of information and the two would not be functionally equivalent. The adherents of 'strong AI' [artificial intelligence] and 'strong AL' [artificial life] commit the simple, but major, fallacy of confusing the abstract with the concrete. (p.17)

11. SUMMARY

The 'questioning' in this paper assumed the following form: first, a number of problems associated with the notion of emergence were listed and the mind-body or category problem was identified as fundamental; second, a conflict was identified between emergentism and the *ex nihilo* maxims. It was argued that conventional (that is, materialistic) emergentism cannot solve the category problem due to (1) a commitment to an objectivist and monist ontology (metaphysics) and (2) a classical interpretation of the maxims in which nothing is equated with void. In preparation for a rethinking (and possible rehabilitation) of the notion of emergence, the question of Being, as understood in Heideggerian ontology, was briefly examined. This provided the basis upon which to formulate, in outline, a *hermeneutic* theory of emergence and, consistent with the new view, an ontical (causal) solution to the category problem grounded in panexperientialism was presented. The *ex nihilo* maxims were reinterpreted on the new theory and found to be consistent with emergentism as understood in its extended sense. Finally, the ontological (phenomenological) and ontical (causal) problems associated with a shift to the new interpretation of emergence and the adoption of a panexperientialist ontology were briefly considered. An implicit argument for an authentically (that is, recognizably) *anthropocentric* view of technology (artificing) was made; in short, although we can *know* (in the sense of "make intelligible"), we cannot always *do*. Ontology does not entail technology (Ali,97).

REFERENCES

Ali, S.M., Zimmer, R.M. (1996). Beyond Substance and Process: A New Framework for Emergence. To be published by MIT Press.

Ali, S.M. (1997). Towards an Anthropocentric Computationalism: Ontological and Epistemological Issues in the Evaluation of Computationalism as the Metaphysical Basis of a Unifying Framework of Emergent Artificiality. Ph.D Thesis. Brunel University, England. (Forthcoming)

Ayala, F. (1985). Reduction in Biology: A Recent Challenge. In Evolution at a Crossroads: The New Biology and the New Philosophy of Science. Edited by D.J.Depew and B.H.Weber. MIT Press, Cambridge, MA, pp.65-79.

Baas, N. (1993). Emergence, Hierarchies and Hyperstructures. In Artificial Life III. Edited by C.G.Langton. Santa Fe Institute Studies in the Sciences of Complexity, volume 17, Addison-Wesley, Reading, MA, pp.515-537.

Birch, C. (1994). Why I became a Panexperientialist. (Unpublished manuscript), pp.1-12.

Bunge, M. (1969). Metaphysics, Epistemology and Methodology of Levels. In Hierarchical Structures. Edited by L.L.Whyte, A.G.Wilson, D.Wilson. American Elsevier Publishing Company, Inc., New

York, pp.17-28.

Bunge, M. (1977). Treatise on Basic Philosophy, Vol.3, Ontology I: The Furniture of the World. D.Reidel, Dordrecht, Holland.

Cariani, P. (1989). On the Design of Devices with Emergent Semantic Functions. Ph.D Thesis. State University of New York, Binghampton.

Cariani, P. (1991). Emergence and Artificial Life. In Artificial Life II. Edited by C.G.Langton, C.Taylor, J.D.Farmer, S.Rasmussen. Santa Fe Institute Studies in the Sciences of Complexity, volume 10, Addison-Wesley, Reading, MA, pp.775-797.

Churchland, P. (1985). Reduction, qualia, and the direct introspection of brain states. Journal of Philosophy 82, pp.8-28.

Collingwood, R.G. (1945). The Idea of Nature. Oxford University Press, Oxford.

Crutchfield, J.P. (1994). Is Anything Ever New ? Considering Emergence. SFI 94-03-011, pp.1-15.

Darley, V. (1994). Emergent Phenomena and Complexity. In Artificial Life IV. Edited by R.A.Brooks, P.Maes. MIT Press, Cambridge, MA, pp.411-416.

Dreyfus, H. (1991). Being-in-the-World: A Commentary on Heidegger's Being and Time, Division I. MIT Press, Cambridge, MA.

Elstob, C.M. (1984). Emergentism and Mind. Cybernetics and Systems Research 2. Edited by R.Trappl. North-Holland, pp.83-88.

Farleigh, P. (1996). An Event-Based Cognitive Science. (Unpublished manuscript), pp.1-18.

Farleigh, P. (1997). Whitehead's Even More Dangerous Idea. (Unpublished manuscript), pp.1-6.

Griffin, D.R. (1988). The Reenchantment of Science. SUNY Press.

Griffin, D.R. (1997). Panexperientialist Physicalism and The Mind-Body Problem. Journal of Consciousness Studies (pre-print), pp.1-20.

Harth, E. (1993). The Creative Loop: How the Brain Makes a Mind. Addison-Wesley, Reading, MA.

Heidegger, M. (1959). An Introduction to Metaphysics. Yale University Press, NY.

Heidegger, M. (1973). The End of Philosophy. Introduction and translation by J.Stambaugh. Harper & Row, NY.

Heidegger, M. (1977). The Question Concerning Technology and Other Essays. Introduction and translation by W.H.Lovitt. Harper & Row, NY.

Heidegger, M. (1993). What is Metaphysics ? In Martin Heidegger: Basic Writings. Edited by D.F.Krell. Routledge, London, pp.89-110.

Kenyon, F. (1941). The Myth of The Mind. Watts & Co., London.

Langton, C.G. (1989). Artificial Life. In Artificial Life. Edited by C.G.Langton. Santa Fe Institute Studies in the Sciences of Complexity, volume VI, Addison-Wesley, Reading, MA, pp.1-47.

Mayr, E. (1982). The Growth of Biological Thought. Harvard University Press.

Nagel, E. (1961). The Structure of Science. Routledge & Kegan Paul, London.

Nagel, T. (1979). Mortal Questions. Cambridge University Press, London.

Nagel, T. (1986). The View From Nowhere. Oxford University Press, NY.

Polanyi, M. (1967). The Tacit Dimension. Routledge & Kegan Paul, London.

Prigogine, I., Stengers, I. (1984). Order out of Chaos. Heinemann, London.

Rosen, R. (1977). Complexity as a system property. International Journal of General Systems 3, pp.227-232.

Rosen, R. (1985). Anticipatory Systems. Pergamon Press, NY.

Scott, A.C. (1995). Stairway to The Mind. Springer-Verlag, Berlin.

Searle, J.R. (1980). Minds, Brains and Programs. Behavioural and Brain Sciences 3, pp.417-424.

Searle, J.R. (1992). The Rediscovery of The Mind. MIT Press, Cambridge, MA.

Spencer-Brown, G. (1969). Laws of Form. George Allen & Unwin Ltd., London.

Sperry, R.W. (1976). Changing concepts of consciousness and free will. Perspectives in Biology and Medicine, 20, pp,9-19.

Tallis, R. (1994). Psycho-Electronics. Ferrington, London.

Analysing the *lack* of Demand Organisation

Philip Boxer, Boxer Research Ltd

and

Professor Bernard Cohen, Interoperable Systems Research Centre, City University

We seek to develop means of intervention in Enterprises that will enable them to react in an effective, sustainable and timely fashion to changes in the ways that markets and demand are organized; that is, to act *strategically*.

We take an enterprise to be some entity that seeks to provide its clients with services that they value while maintaining its ability to do so in the face of changes in the demands of its clients and in the resources at its disposal. The services that clients value form around what the organization of their demands *lack*. The concept of *strategy* therefore rests on critically evaluating the ontology and semantics of the Enterprise in relation to these *holes* in demand organization.

We access ontology and semantics by constructing and manipulating hypothetical, first-order, mathematical models of the Enterprise's services and of its value-adding processes. Because an enterprise is an *anticipatory* system, its semantic domain must include representations of the enterprise's model of itself and of the market and demand organizations within which it competes. First-order (set) theory provides adequate expressive power here, but alternative, higher order, mathematical frameworks, such as Dubois' *hyperincursion*, provide inadequate power, particularly in relation to the analysis of the properties of *emergence*.

Knowing exactly why and where this mathematical *lack* manifests in the analysis process enables effective collaboration between *systems analysts* and *psychoanalysts*, and suggest directions for mathematical research.

Keywords: Anticipatory, Enterprise Modelling, Hole, Ontology, Stratification.

1. Introduction

An enterprise may be modeled as an interrelated collection of services that would, in concert, enable the enterprise to satisfy its needs. Since these services might, in general, be supplied by alternative providers, their descriptions, (or specifications) should be independent of their implementation by any provider, and each provider should be able to demonstrate that the service complies with its specification.

This paper is about an approach to modeling that can support a dialogue concerning the design of an enterprise's value-adding processes. In practice, the specifications of 'services' in an enterprise are rarely explicit. Making them explicit involves making a distinction between value chains and value ladders (this distinction between value ladders and value chains is elaborated in "Performative Organization", Boxer and Wensley 1997), but to compete on value ladders means making design specification explicit (an approach to this is outlined in Boxer, 1993). When this approach is applied to the industry contexts within which enterprises compete, it results in a distinction between Market and Demand Organization.

CP437, *Computing Anticipatory Systems: CAYS--First International Conference*
edited by Daniel M. Dubois © 1998 The American Institute of Physics 1-56396-827-4/98/$15.00

Our concern is to develop means of intervention in Enterprises that will enable them to react in an effective, sustainable and timely fashion to changes in the ways markets and demand are organized, that is, to act strategically. We take an Enterprise to be some entity that seeks:

- to provide its clients with services that they value and
- to maintain its ability to do so in the face of changes
  - in the demands of its clients and
  - in the resources at its disposal.

These services which clients value themselves form around *holes* in demand organization – around what demand organization *lacks*. The conception of "strategy" ultimately rests therefore on critically evaluating the ontology and semantics of the Enterprise in relation to these holes in demand organization (Boxer and Wensley, 1997).

The main mathematical problem facing the modeler in the business world arises from the 'anticipatory' nature of systems in that world. The semantic domain must include a representation of the business's model of itself (warts and all), so that the formal model may include among its consequences the kinds of decision that the business might make based on its model of itself. But it also has to include a model of the organization of the market organization and demand organization within which it is competing. Hence the need for something like Dubois' "incursive" (formal) systems (Dubois, 1996), and to differentiate between the notions of 'time' inside in the business (the 'simulation') and outside in the competitive environment (the 'world').

1.1 Approaching the client's frame of reference

We want to distinguish between the description of a service and its implementation both for technical and for business development reasons, especially for those services that are to be available from different (competing) suppliers.

- Technically, it has to do with quality: every supplier must 'guarantee' that the service provided satisfies some description that tells every client what properties of the service can be 'relied' on, whoever supplies it. This notion of 'rely/guarantee' is central to the formal specification and verification of programs, etc. (Cohen, Harwood and Jackson, 1984).

- In terms of business development, the client should be able to determine from a service description what its value might be to (i.e. to include it in a design of) the enterprise before investigating the cost of purchasing it from different suppliers; each supplier, on the other hand, should be able to evaluate alternative implementations (i.e. designs), all of which (demonstrably) satisfy the same service description, with respect to that supplier's culture/knowledge/value system.

There are obvious interpretations of 'added value' and 'competitive edge' in terms of the client's own culture/knowledge/value system - the client's frame of reference, but this does not satisfactorily account for 'stratification' (Boxer, 1996).

- So the third reason for distinguishing between service description and implementation is so that the descriptions can both provide and exploit 'languaging tools for articulating different forms of stratification'.

Each service implementer (or agent) may itself be modeled as an enterprise in a similar manner, its purpose being to supply services to its clients while in turn invoking services provided elsewhere. Some agents may be characterized as platforms: integrated collections of 'generic' services that support the implementation of many different service specifications (Cohen, 1996a). Similar contextual issues may therefore arise at many levels of service provision.

Thus, the interrelationships among services that comprise the context of a given enterprise may both add constraint to individual service specifications, as well as endowing their composition with emergent properties that no isolated service exhibits. It is therefore possible that an implementation that complies with an isolated service specification may fail to satisfy that specification in some context (Cohen, 1996a). This means that the choice of what is modeled and what is ignored becomes of particular importance.

1.2 Anticipatory Systems and Stratification

Another way of approaching this problematic of emergence is in terms of the effects of stratification - the way services are stratified will affect what emergent/higher order properties will be observed. Thus, not only does the client's frame of reference affect the ways in which 'added value' and 'competitive edge' are interpreted. It will also affect the way services are stratified.

An Enterprise may be construed as a complex anticipatory (in the sense of Rosen, 1987) system, as may be every agent serving within an Enterprise. In other words, one can attribute to each of an Enterprise's services an 'agent', 'implementing' the service in its own way and subject only to the rules governing its interoperation with other services. Such an 'agent' can be understood as having the same characteristics as an 'Enterprise'. As a result, this distinction of being "within", which can be made between Enterprise and the larger context in which clients value the services provided by the Enterprise and its competitors, can be applied in the same way between agent and Enterprise itself.

The way this "within" distinction is made between a particular relation of inside to outside is dependent on the ontology of the modeller. When applied recursively by a modeller, it results in a *stratification* of embedded 'insides' and 'outsides' (Van der Vijver, 1996a; Boxer, 1996). So what happens when we seek to model an enterprise conceived of as being stratified in this way?

We come to this problem from different points in the 'analytic' universe: Philip Boxer from the psychoanalytic approach to the Enterprise, and Bernie Cohen from the systems analytic approach. The paper aims to consider the implications of bringing these two approaches together in the modelling of the Enterprise. We will argue that the modelling developed by Boxer in relation to clients' expectations of the services provided by an Enterprise, together with the 'knowledge' implicit in the stratification of the modelling of the Enterprise itself, provide an ontological setting within which the formal 'systems' models of Cohen can be elaborated and composed. As a result, the latter offers the possibility of illuminating the topological structures and logical consequences of the former. We make a start on this by articulating a semantic domain for stratification analysis.

2. Holes in Space

2.1 Formulating the *lack*

Rosen's 'Anticipatory Systems' (1987) provides an exceptionally clear exposition of the issues surrounding the relationship between ('natural') systems and models of them. In understanding more of the difficulties surrounding this relationship, this question of the *lack* can be approached.

Rosen treats 'abstraction' as the necessary selection of a set of 'observables' from a system's state space in terms of which a formal model can be expressed. Every formal model can therefore capture only a 'subsystem' of the system observed, and that subsystem is considered in the model to be 'closed', i.e. 'isolated' from the influence of any other part of the system's state space. An 'open' system is one in which this assumption may be violated.

To model an open system therefore requires either an extension of the closed system model, in which the additional influences from the environment are incorporated into state components, or the composition of a closed system model with another that captures some other natural system with which the former is deemed to interact. This definition is recursive, since the formal model + model of system observed can become a composite model which can itself be taken as a formal model in relation to a system observed. If we therefore take some composite model with some number of stratified levels of emergence as being the formal system, then 'error' can be treated as a 'bifurcation' that arises as the behaviour predicted by this formal model diverges from the behaviour exhibited by the system observed. What forms of 'error' can arise, then?

2.2 Types of Error

From the point of view of the formal system, two kinds of error can arise in claiming any kind of ability to state the 'truth' about the system observed, and a third kind of error can arise in the composition of the formal system and the model of the system observed. These can be summarised in terms of three characteristics of the model:

- the *operation* of the model in replicating the behaviour of the system observed;
- the internal composition of the formal *model* itself (whatever its levels of stratification, and however it can be formalised in a state space); and
- the *composition* of the organisation of the data, in terms of which the system observed is being described, in relation to the formal model.

With a 'Type I' error, model and composition are taken as being 'true' but the operation of the model in replicating the data produces wrong results. With a 'Type II' error, the composition is 'true' and the model's operation is 'correct' but the formal model itself is wrong. And with a 'Type III' error, the model is 'true' and is applied 'correctly', but the composition as a whole of model in relation to system observed is wrong.

In considering the relationship between observed systems and formal models of them, which class of error we think we are dealing with raises important questions therefore:

- with a Type I error, we can say that there is a *correspondence* error between the model and the system observed; and
- with a Type II error, we can say there is a *coherence* error in the particular way in which the model itself is constituted.

In the first of these, there has to be a presumption of an independently existing ('natural') reality in order to be able to claim a correspondence error between ('natural') system and model - a question of reference, while for the second, it has to be possible to be able to assert that a non-sense has arisen - a coherence error. For this second type of error, therefore, there is a need for some form of (normative) state space in which it is possible to assert that a coherence error has arisen.

So what is happening with a Type III error? In a Type III error, what is called into question is the point of view from which both the formal model and the model of the system observed have been constructed.

- With a Type III error, we can say that an *undecidability* error has arisen in the way the composition of the model in relation to the system observed has been framed.

If this undecidability error is to be distinct from errors of correspondence or coherence, a Type III error has to be an error in which it is not possible to determine whether or not a Type I or a Type II error has arisen. The very constitution of the model itself in relation to the system observed is unable to render 'tractable' the truth or falsity of the relation between the ('natural') system and the model.

So how are we to distinguish Type III errors without recourse to a 'higher' authority either of correspondence-to-reality or to coherence-in-state-space? (Maturana (1988), in seeking to argue against Aristotle's 'transcendental' ontology, was arguing against the first of these in order to be able to assert the supremacy of the second 'constitutive' ontology).

2.3 Undecidability Error as 'hyperincursion'

'Bifurcation' between the observed behaviour of a system and the predictions of a formal model of it is an empirical matter that can be decided only at the 'time' (in the natural system) when divergence occurs. On the other hand, some of the 'errors' manifested by the 'emergent properties' of a composite model are formal, and occur at the 'time' (in the modelling world) when the composite model's implications are derived (e.g. inconsistencies). Of course, some are also empirical in that, although the composite model is consistent, its predictions when imputed on the natural system ('now', rather than in the 'future') diverge from observation, i.e. are 'invalid'. So,

- an invalid model would exhibit a Type I (correspondence) error;
- an inconsistent model would exhibit a Type II (coherence) error.

Anticipative Systems Theory formulates undecidability as a hyperincursion. We could say therefore that, instead of there being a (first order) bifurcation, because the natural system had not yet reached its bifurcation with the model, there was instead a second order bifurcation – the hyperincursion becomes a bifurcation in anticipation itself: how we are to decide how we are to anticipate. Thus

- a hyperincursive model would exhibit a Type III (undecidability) error.

We might get a glimpse of potential Type III errors by monitoring the changing ontology and topology of the formal model as it is changed to accommodate the needs of the 'natural' system observed. Sudden and dramatic alterations to those structures might suggest the onset of 'undecidabilities', or at least raise red flags that signal serious conceptual alterations and motivate a closer look at strategies and plans.

When Maturana (1988) asserts 'the supremacy of the second 'constitutive' ontology', he seems to be demanding that model consistency take precedence over model validity. There is indeed no purpose to be served by even attempting to validate an inconsistent model, but this is a chronological precedence for the modeler. To take it as an epistemological precedence is to commit solipsism. But with Type III errors we are calling into question the ontology of the modeler - the modeller as frame.

2.4 Bifurcating Models

We now have a rather complicated situation in which we are talking about a number of different contexts in which different forms of error may arise and in relation to which it is not always clear what the notion of an 'error' is referring to. The operation of an observing model on some observed model results in the anticipation of a certitude (Lacan, 1988) - a prediction taken to be true. When we are talking about a Type I 'error' we are talking about the prediction having been proved to be wrong, as a result of which we are seeking to formulate an explanation. In this sense, therefore, an error always ends up being a 'bifurcation' between observing and observed model.

For an error to arise in this Type I sense, we need a relationship between the observed behaviour of one system, and the predictions about the behaviour of that system by another (formal) model. But in our case we have two types of model that we can speak about:

- a formal model of a natural system - such as might arise in seeking to 'explain' the behaviour of systems which themselves remain wholly 'Other' to us, even though we may approximate to the nature of their behaviour through describing them in terms of composite models.

- a formal model of a composite system - the composite system being some system of systems which, in being 'artificial' in nature, can themselves be described as formal systems, even though we may only be able to do so one-by-one. The formal modelling of this composite system has to be able to account for 'emergent' behaviours arising through the interactions between its constituent systems. The composite system may itself include a composite model of (a) natural system.

In the first case, we are talking about a *1ˢᵗ order* modelling process in that we have a model and a thing modelled, even though the thing modelled may itself be stratified. In the second case we are modelling the modelling of a natural system and we may say, therefore, that we are dealing with a *2ⁿᵈ order* modelling process.

In each of these cases, an error can only be said to arise as an *après coup* - after the event of the disconfirmation of the anticipation of the certitude, we might say. In speaking of what *type* of error we have made, we are therefore seeking to form some kind of explanation of how the bifurcation might have arisen. What kind of modeling situation we assume ourselves to be in will affect the form our explanations take. We can therefore further elaborate the versions of error arising as a result of a bifurcation between the anticipated behaviour resulting from the modeling process and the behaviour of the system (whether composite or not) modeled:

- an <u>invalid</u> model - a correspondence (Type I) error.
- an <u>inconsistent</u> model - a coherence (Type II) error.
- a <u>bifurcating</u> model - an undecidability (Type III) error.

Put in these terms, it is not quite right to say that a Type III error is therefore just a Type I error that has not yet happened. A Type II error can always be said to be a series of Type I errors (this being the normal defense of a current scientific paradigm), just as a Type III error can be said to be a series of Type II errors (this being the normal defense of the practices of science itself) .... At this point we are questioning the nature of the truth claims which support the practices of science itself (Lacan1966). But what kind of explanation is assumed for the bifurcating error leads to two forms of Type III undecidability:

162

- a 'correspondence' undecidability, if the assumption is that the error is a failure in the scope of the model (i.e. that 'hidden variables' seem to be involved); and

- a 'coherence' undecidability, if the assumption is that the error is a failure in the very constitution of the model itself.

Finally, with reference to the demand to make model consistency take precedence over model validity, this is only possible to argue in principle if we wish to argue that the 'natural' is *maya* - a construction, as indeed the radical constructivists seek to do (Boxer and Kenny, 1992). These are exactly the conditions which can be said to exist for the modeler of composite (artificial) systems, however, insofar as they are able to postpone indefinitely an encounter with the Real!

2.5 Holes in (state) space

'Anticipatory' systems can be considered as that class of systems which contain formal models of their own (sub-)systems which they use to predict the future behaviour of their own (sub-)systems in relation to the system(s) observed; and through which they are enabled to select their present actions. It becomes more useful therefore to speak of 'bifurcations' rather than errors arising from the divergence between the system's internal models of the behaviour of its (sub-)systems and those (sub-)systems' 'actual' behaviour. We have now re-located the boundary from being between the 'natural' system and the model of it, to one between the model of the model-in-relation-to-the-'natural'-system and the model-in-relation-to-the-'natural'-system itself. But if we abandon the notion of an independently-existing reality, a 'natural' system can never be more that another open system.

For the anticipative model, 'bifurcations' will now arise from erroneous decision making (whether Type I or Type II) in relation to both explicitly modeled and emergent behaviours of the (sub-)systems. Reconstruction of the internal model may 'compensate' for such bifurcations, and may have to involve radical 'redesign'. We now have a way of speaking about the issues raised by 'service interaction' (Cameron and Cohen, 1996). It does not, of course, go into the manifestations of these issues in information system implementation (where the models of system and platform interact), nor into the 'legacy problem' (where old and new ontologies have to be reconciled). But it does provide a unifying framework in which all of these issues may be considered to be related, and to have related formal and methodological requirements.

Finally, both for these anticipative models and for the model-as-a-whole constituted in relation to the other ('natural') system(s), we can consider a 'stratified analysis', where bifurcations can arise at every level. If we formulate an undecidability as arising in relation to a *hole*, we must recognise two kinds of hole: those that can be filled and those that cannot, within the current stratification. A 'third cybernetics' might be formulated as one that seeks to address the problematics of such holes. A *third-order* cybernetics would then be one where the 'model' becomes what arises in relation to 'unfillable holes', experienced by the observer as the problematics of desire. In other words, an account of the dynamic properties of stratification would require a theory of the observer, in which correspondence and coherence undecidabilities, at *every* layer of *any* conceived stratification, could be analyzed in terms of the composition of models and the discovery of their emergent properties.

163

3. The modeling approach

3.1 The object of interest

The IS culture has grown up with a particular kind of analytical approach, mainly because it has taken its metaphors from first-order cybernetics and computing - hence the 'state machine', 'dataflow', relational', 'object-oriented', etc., approaches to conceptual modeling' and 'business process re-engineering'. However, this background itself constrains the choice of modeling language and therefore limits the diversity of articulative structure, and the strength of analytical power, that can be deployed.

Set against this, the apparent success of IS modeling techniques in certain business applications has encouraged the belief that they can be used effectively for purposes far beyond the limits of their applicability - a belief encouraged by those with interests vested in the modeling tools, as vendors, teachers, or practitioners. There is therefore a key problem to be considered: what is the relationship between the IS frame of reference and the client's frame of reference?

We could choose to work within the IS frame of reference. It would undoubtedly be an easier 'cut' to work with, given the natural leaning of IS towards formalization. But would it enable us to consider the ways in which the IS frame of reference did or did not restrict the strategic options open to the enterprise which it was supporting?

Our interest is in considering the ways in which to look at the enterprise as a whole (a relatively 'simple' one, whatever that means, at first). Our aim would be

- to work within the client's frame of reference in order to articulate (service) models and their stratification.

- To work in relation to the client's development objective for the enterprise in order to identify 'lacks'.

By considering the implementation of stratified service models on IS platforms, we may find that it is precisely in the way these models are, or have been, implemented on IS platforms that endows these holes with varying degrees of resistance-to-change.

Problems recently observed in enterprises as disparate as Telecommunications ('feature interaction', Cameron and Velthuijsen, 1993) and Healthcare ('electronic patient records', Cohen 1996b, 1997) show us that they concern the way the technology may not only not be unable to support certain forms of languaging processes between people, but that these processes have a logic of their own. It is insofar as a hole acquires resistance-to-change that they become objects. These objects are therefore our object of interest. Formulated as such, we would want to suggest that these objects *qua* holes are now ubiquitous and should be taken seriously by corporate strategists, standards writers and systems designers alike.

3.2 Developing a rich descriptive methodology

In these terms, the modeling techniques and notations currently favored by IS disciplines do not provide the requisite analytical power, nor do those of the computing platform communities. Richer and mathematically more sophisticated modeling frameworks are urgently needed and there are very few candidates.

Decon™ (an expert system shell, developed by BRL) is proposed as a rich descriptive methodology which can be used not only to 'extract' the axiomatics of the discursive practices in the client system, but also to generate stratifications which are relevant to the strategic options which the enterprise wishes to consider. An analysis of holes at all levels of strata make it

164

possible to determine where the forms taken by particular services need to be migrated to a form of greater articulation. The extracted axiomatics means that this can be done within a formal framework.

Thus Decon™ becomes a means of identifying ('extracting') viable ontologies and structures, in terms of which we might construct formal models, that articulate a theory (or 'axiomatics'), whose closure might account for 'the discursive practices in the client system'. The basic thesis is that, if we define the aims of a business development process in terms of the changing problem domains in which the business is competing, then the <u>reason</u> that legacy IT exerts a disabling effect on business change is because of holes both in the information environment and in the network organization of services... and so on down the strata.

3.3 So what is being modelled?

The class of anticipatory systems is characterised by 'non-locality'. This means that an anticipatory system's global behaviour in relation to its context (whether successful or not) is emergent from the composition of its local parts and cannot be attributed to any one part in isolation. In particular, one cannot localise the model used by the system to anticipate the potential consequences of its actions.

The activation of this model, and the delivery of its predictive consequences, can be construed as one of the services invoked by the Enterprise, to be composed with other services so invoked in order to enable the Enterprise to act strategically. This modelling strategy escapes the criticism of reductionism because the whole is decomposed into a collection of inter-operating services, each described independently of its 'implementation', of which there might be several possible, or even available.

The analytical problems raised by such a modelling strategy are legion, not least of which is the definition of what is 'within' the Enterprise. (This question relates to the questions of identification and embodiment raised in Van de Vijver's (1996b) and in Boxer, 1995). But they at least direct attention to the nature of the mathematical framework in which to construct service descriptions, so that their composition and the inference of their consequences can be performed.

Cohen, on the one hand, starts with a formal representation (in set theory), which is an abstraction and generalisation of the ontology and intended semantics of the Enterprise as an integrated collection of 'services'. Each service is modelled as a state space (an indexed collection of set-valued state components) and the events in which the service participates (each defined by precondition and postcondition predicates on the state space, Schuman, Pitt and Byers 1990). In general, the services so modelled are not isolated. They comprise each other's context and so 'act' on each other (if this were not so, then the enterprise would not be a 'system' but a mere aggregation of services). As a consequence, he has to address the following particular forms of relationship:

- The logical verification of the models' internal consistency.
- The logical derivation of the models' behavioural properties (closures), so that they may be validated by attempted refutation of their predicted behaviour in relation to and satisfaction of client expectations.
- The utility of the generalisation used, assessed by restricting some parameters of the generalised models to values that had been used in the past and showing that the resulting, more specific models indeed account for the behaviour of the enterprise's 'legacy' services.

Boxer, on the other hand, starts from defining the relationship of the Enterprise as a whole to what is valued by the client, defined in terms of the Enterprise's 'knowledge' of itself and its competitive environment. He starts by seeking to describe the activity logics and capabilities that comprise the underlying task structures in relation to which the Enterprise's workgroup and information processes are constituted. This provides an entry-point for a more abstract form of description, which describes the organisation of the market, and demand contexts within which clients' expectations are formed.

Boxer's model of the Enterprise takes the form of a User-Defined Knowledge Base (UKB) which includes modelling the effects of stratification within the Enterprise itself. Through the instantiation of the objects and relations in the UKB, the formulations of clients' expectations can also be tested for consistency (in the familiar, not the logical, sense) in relation to the data generated by legacy systems.

But in order to be able to articulate this model of the Enterprise's 'knowledge', a formal syntax has to be developed - the Expert-Defined Knowledge Base (EKB). The description is therefore done in terms of distinct types of relation, and the Decon™ system is a software tool that supports the construction of such a relational model of the enterprise in the EKB. The particular relationships resulting from these types of relations are formulated in the UKB by inviting the enterprise's agents to nominate those services, processes, activities, capabilities, etc. in terms of which they currently perceive the enterprise, and the organization of the market and of demand.

The resulting matrices can then be analyzed using the PAN™ system (developed by BRL). PAN™ is an analytical tool which uses the language of Ron Atkins' Q-analysis (Casti, 1994), and which provides empirical evidence for both the structural differentiation of an enterprise's services, but also the identification of 'holes' in both market and demand organization. Feeding these results back to the agents enables (and encourages) them to reconsider their own theories, to reformulate them and to perform this kind of analysis for themselves.

3.4 Anticipative systems

In Dubois' conclusions to "A Semantic Logic for CAST ...", he says: "Any systems theory deals with a semantic logic, corresponding to a meaning in the mind of the CAST engineer, which can be represented by mathematical equations." This sentiment differs from the approach in Decon™. What is this difference?

In order to formulate a 'User-Defined Knowledge Base', Decon™ needs a syntax. The concrete syntax provided in the 'Boxer' approach to micro-organizational design contains symbols that act as tokens for certain classes of objects and relations in terms of which the client's statements about his enterprise are articulated. The terms in these statements, whether expressed lexically using words or tokens, need to carry two aspects of 'meaning': their 'sense' and their 'reference'. Their reference is to things in the client's world -- his frame of reference -- and are therefore ontological in nature. Their sense is 'semantic', in Dubois' usage, where both the objects and the relations are mapped into a mathematical domain that provides a formal system with the analytical power necessary to determine the consistency of any model constructed in it and to enable the derivation of its logical consequences. As Rosen says,

> "What we require is a new universe in which we can create systems. But in this universe, the systems we create will have only the properties we endow them with; they will comprise only those observables [read 'objects'], and satisfy only those

linkages [read 'relations'], which we assign to them. But the crucial ingredient of this universe must be a mechanism for making inferences, or predictions".

That is what Cohen understands Dubois to mean by "a meaning in the mind of the CAST engineer", what Bunge (1974-80) means by the 'fictional' nature of formal models, and what Rosen means by 'abstraction', which is the deliberate focus of the modeller on a (fictional) 'subsystem', and by 'imputation', which is the modeller's act of relating the model, and its logical consequences, back onto the observed world, with a view to the refutation of the model, or to an improved understanding of the world.

The language(s) of Decon™, then, need semantics for their syntax, and those semantics need to be expressed as models in a mathematical framework with suitable expressive and analytical power. For example, in the 'Glass Industry' knowledge base, there is a relation called 'supplies' which is defined to take named objects of certain classes in its domain and range, specifically product-process, or process-product. At this level, Decon™ allows the user to check formulations in the User-defined knowledge-base (UKB) for syntactic validity. The syntactical constraints on the UKB are not however making it possible to assert semantic constraints on its usage, such as whether every instance of it should exclude the identity (as the PAN™ analysis would suggest), that is that "x supplies x" is not acceptable even when "x" appears both a product and a process. Such constraints appear as 'invariants' in state space models. These 'invariants' are indeed not fully expressible in relational languages, even in those in use by the OO community. They also serve the much more important function of restricting the state space of composite system models to a subset of the product of the state spaces of their component 'subsystems', in order formally to capture the additional topology that arises from such composition and that, when imputed to the world of observation, accounts for both 'emergent' and 'pathological' behaviour (e.g. 'holes' or 'feature interactions'). So what are the characteristics of this 'state space'. It is the expert-defined grammar (EDG) in Decon™ which begins to formulate these semantic constraints. The challenge of a formal approach to modelling is to formulate these constraints within a state space.

What we have done is to explore the overlapping areas of functionality and the ways in which our different approaches to the overall problem have created complementary methods in which the Decon™ toolset supports the formal modelling process between client and facilitator/modeller. As a result, the client is enabled to articulate an ontological structure with its associated 'holes' and conceptual flaws within the client's world view. In the sections below, we establish the equivalencies between the formal model (UKB) provided by this toolset, in relation to which the axiomatics (EKB) of the enterprise can be formulated, the logical consequences of which might reveal flaws of a different kind within the framework of a formal abstraction.

3.5 Analysing the lack

The toolset used by Boxer supports a modelling process between Enterprise and facilitator/modeller of the Enterprise's relations to Client Expectations. As a result, the Enterprise is enabled to articulate not only an ontological structure which is 'lacking' in relation to the client's demand organisation, but an organisation of demand which may itself be lacking. Thus the EKB and UKB provide starting points for the construction of formal systems models, the logical consequences of which can reveal holes.

However, the relational structures produced in Decon™, and analyzed in PAN™, lack the topological depth and formal foundations that would be required if the enterprise model were to admit the analytical inference of *semantic* 'holes' – that is, errors in the agents' models that

167

logically preclude the satisfaction of the enterprise's needs, or that do not adequately account for the enterprise's interactions with its world.

We seek to overcome this deficiency by constructing a formal semantics, in set theory, of Decon's relational structure. We also expect to find that set theory will ultimately prove to be inadequate for this purpose because the enterprise is an inherently *anticipatory* system (Rosen 1987), whose model must therefore be *hyperincursive* (Dubois 1996), and therefore beyond the essentially *first-order* descriptive and analytical power of set theory. However, knowing exactly where, and why, a first-order system becomes inadequate when modeling an enterprise would itself be instructive to the analyst and illuminating to the agent because it represents the onset of that knowledge diffusion concerning the nature of the PAN/Decon™ methodology itself. It would also provide the theoretician with a clearer image of the 'holes' in mathematics itself.

Nevertheless, our theoretical foundations allow us to infer errors (i.e. inconsistencies) in such models, but not, in general, to derive corrections to such errors. Model alteration must therefore be considered as a fallible task to be continually subjected to refutation, just like scientific theory construction. The internal coherence of established models and systems builds up resistance to such changes in much the same way as it does in science. That resistance, which manifests itself as reluctance to consider alternatives, augments and may be indistinguishable from the social and political resistances that are usually associated with Enterprises. In that case, we would really be following the precepts of Freud and Lacan in considering the effects of the 'Symbolic' in relation to the 'Real' and the 'Imagined'.

4. A Semantic Domain for Stratification Analysis

4.1 Introduction

Our objective, then, is to provide a formal foundation (in set theory) for relational models of enterprises, which are constructed when the agents of an enterprise participate in its analysis. For the purposes of this study, we shall use the following working definitions:

- An **agent** is an entity that exists to satisfy its needs in a world populated by other agents. Every agent has its own theory of its needs which is a composition of its theories of itself and of other agents' needs, the latter including those other agents' theories of it. An agent is a formulation within second order cybernetics, in that it is a system that has a distinguishable identity, and an organization which manifests itself as revealed preferences in its behaviour (anticipative behaviour). An agent within third order cybernetics becomes a second order agent whose anticipative behaviour is organized in relation to a lack.... a point of instability/hyperincursion. (Boxer 1996.)

- An **enterprise** is an agent which offers services that it expects (according to its theory) to be valued by other agents in its world.

- An agent perceives a service to be of **value** to it if the service, as described, fulfils a demand that derives from the agent's theory of its needs; that is, that the provision of a service would satisfy a demand.

When an enterprise is approached in these terms, it becomes necessary to describe the internal organization of an enterprise, the ways in which this is related to the external organization of the market in which it competes, and the ways in which this external market organization is itself related to the organization of demand to which it is a response.

168

4.2 Supply and Demand Organization

We assume that:

- Supply organisation identifies *services*
  - One engages in supply when one deploys *processes*.
  - A process *provides* a service
  - By *organising* a set of services.

	Name	Type	Interpretation
1	SR	Set	Every service that might be provided to or organised by a process
2	PR	Set	Every process that might be deployed
3	prv	PR→SR	The (unique) service that each process *provides*
4	org	PR↔SR	The services that each process *organises*

- Demand Organisation identifies *specifications*. We may need to introduce an ordering relation, *weaker than*, on specifications, because any service that satisfies a specification should also satisfy all 'weaker' specifications. For completeness, the formal definition of this relation, and some of its properties, is given in Appendix 1.
  - One expresses demand when one articulates *design specifications*
  - A design specification *decomposes* a specification
  - into *demands* for a set of specifications.

	Name	Type	Interpretation
5	SP	Set	Every specification that might be demanded or decomposed by a design
6	DE	Set	Every design that might be articulated
7	dec	DE→SP	The (unique) specification that each design *decomposes*
8	dem	DE↔SP	The specifications that each design *demands*

- Every service *satisfies* some specification

	Name	Type	Interpretation
9	sat	SR→SP	The (strongest) specification that each service *satisfies*

- A Process *realises* a design if, and only if:

	Name	Type	Interpretation
10	rea	PR→DE	The (unique) design that each process *realises*.

  - And imposes the following constraints on it:

	Name	Interpretation
11	(dec°rea)⊆(sat°prv)	The service provided by the process satisfies the specification decomposed by the design
12	(dem°rea)⊆(sat°org)	Each of the specifications in the design's decomposition is satisfied by one of the services organised by the process

4.3 Demand Organization

We can now consider 'demand organisation' to be the successive decompositions of designs that leads a potential services *user* to express a specification, thereby *situating* that specification in a user's 'value ladder'.

- Demand is expressed by *users*.

	Name	Type	Interpretation
13	US	set	All potential *users* (in a problem domain)

169

- Users articulate designs. We do not formalise the relation between the refinements and demands of designs, i.e. the *validation* that a design is fit for its purpose. This would require the establishment of a sound decidable calculus for the composition of specifications, and there isn't one. The onus of design validation therefore lies with the customer in relation to whose demands designs are situated.

| 14 | art | US↔DE | The designs *articulated* by each user. |

- – To specify a need as a demand without decomposing it is to define a *problem* if there is no service which satisfies that problem directly. This problem arises in a *demand situation.*
- – Those specifications that a user's designs decompose but do not demand constitute the user's *demand situation* in terms of 'smaller' problems.

| 15 | dst | US↔SP | The specifications that comprise each user's *demand situations.* |

$$16 \qquad \text{dst} = (\text{dec}°\text{art})\backslash(\text{dem}°\text{art})$$

- – Those specifications that a user's designs demand but do not decompose constitute the user's (c-type) customer situations. 'Value ladders' are defined here as directed acyclic graphs – of designs, from decomposition through demand to further decomposition, and so on. This ladder of designs for a service specification might include that same specification among its demands. For example, a customer whose purpose is to provide air transport services might refine that purpose into a design that demands air transport services, e.g. for the transportation of components to assembly plants.

| 17 | cst | US↔SP | The specifications that comprise each user's (c-type) *customer situations* |

$$18 \qquad \text{cst} = (\text{dem}°\text{art})\backslash(\text{dec}°\text{art})$$

- – Those users whose designs construct customer situations are *customers.*

| 19 | | CU ⊆ US | Those users who are customers |
| 20 | | CU = dom(cst) | |

- – The choice about whether or not to demand in this sense is equivalent to the decision to 'outsource'.
- – If we think in terms of classes of demand as being some measure of 'distance' from the user's need, then it is possible for a supplier to service a higher-level specification while the user is still servicing a lower level one. (e.g. when the enterprise uses a design consultant to design a product that it will make itself.)

4.4 Supply Structure

Symmetrically, we consider 'supply logic' to be the chain of organisations of processes that enable a *position* to provide a service, thereby positioning that service in a 'value chain'.

- Supply is expressed by *positions.*

| 21 | PO | set | All positions in an industry sector. |

- Positions' capabilities are determined by the processes that they deploy. We do not formalise the relation between provision and organization of processes, i.e. the *verification* that a process achieves its intent. This is the traditional task of engineering, the onus of verification lying with the engineering staff of the competitor whose positions frame processes.

| 22 | cap | PO↔PR | The processes that comprise the position's *capabilities* |

- Those services that a position's processes provide but do not organise constitute the position's *value profiles*.

| 23 | vpr | PO↔SR | The services that are 'framed' by each position, which constitute the position's *value profiles*. |

$$24 \qquad vpr = (prv°cap)\backslash(org°cap)$$

- Those services that a position's processes organise but do not deliver constitute the position's *infrastructure*. Infrastructure presents constraints on the positions that a supplier might take. Symmetrically, we might identify *superstructure* with constraints that limit users' potential for articulation. This concept should have something to do with the substitutivity of specifications, and the services that satisfy them, in the context of particular demand situations (the property known as 'reuse' in the IS world). Formally, we would expect each design (or decomposition of a demand situation) to determine an equivalence relation on specifications.

| 25 | inf | PO↔PR | The processes that constitute each position's *infrastructure*. |

$$26 \qquad inf = (org°cap) \ \backslash \ (prv°cap)$$

4.5 Market Organization

In a market, competitors seek to supply their services to customers whose customer situations include specifications that they can satisfy. Symmetrically, customers seek to acquire services from competitors whose positions include services that satisfy their specifications. In general, customers deal with competitors' organisations who *order* services from positions that they control.

27	CO	set	Competitor organisations.
28	ctl	PO→CO	The (unique) competitor who *controls* each position.
29	ord	CO→ (PO↔SR)	The positions that each competitor orders its infrastructure to take.
30		(dom°ord) ⊆ ctl⁻¹	Competitors can issue orders only to positions that they control
31		ran(ord) = inf	All, and only, the positions taken by a competitor's infrastructure (i.e. all the services in its value profiles) are ordered by the competitor.
32	sup	(CO x SR)↔ (US x SP)	The services *supplied* by competitors to customer situations.
33		dom(sup) ⊆	Competitors can supply only those services that occur in

171

	ran°ord	their controlled positions.
34	ran(sup) $\subseteq$ cst	To customers whose situations comprise specifications
35	sup $\subseteq$ (PA x US)*sat	that are satisfied by the services with which they are supplied.

5. Discussion for Future Development

5.1 Propositions and composition/decomposition within value ladders

Given our earlier formulations of agent, enterprise and value, services might be known to the agent to be available (at an acceptable price etc.), either from among its own services or from other agents. To be in a position to estimate the value of a service, an agent must have articulated a model of its needs, which thereby take the form of demands. These demands can themselves be a composition of service descriptions (but not necessarily if they are at the 'top' of the compositions for this agent, whose emergent properties, themselves composed with the agent's model of its world, predict a satisfying outcome). Such demands give rise to the simplest form of proposition – the c-type proposition:

- The **c-type** service offering, which supplies 'off the shelf' services. The supplier has no involvement in the decomposition of the client's needs. The supplier's concern is focused on its own operational effectiveness with respect to those services for which it has identified a 'market'. What is sold is the ability to supply a service.

Those remaining demands, for which there is a demand but no supply, constitute a 'lack' in market organization, which is relative to the agent's current formulation of its needs, resources and market knowledge. Even for demands for which there are currently services, there is no unique decomposition of any purpose. One decomposition may reveal a lack in market organization which cannot apparently be satisfied, while others might reveal completely different lacks, or no lack at all. Equally, decomposition might be erroneous in that composition of the service descriptions in the model because:

(a) it does not induce the emergent properties required by the demand ('invalidity'), or
(b) induces mutual constraints on the (isolated) services which they cannot jointly satisfy ('inconsistency').

Or decomposition might be ineffective in that:

(c) some of the service descriptions in the model cannot be effectively supplied ('infeasibility'), or
(d) the composite model does not admit inference of its internal consistency or of its emergent properties ('intractability'), or
(e) the service descriptions in the model admit individual satisfaction by services that, when themselves composed, fail to meet the agent's purposes ('destructive interaction').

In all these cases, however, what is being offered as a service is a capability for composition:

- The **K-type** service offering, which is configured to deliver services in a form that is appropriate to the demand situation in which it believes its customer to be. The client defines the need in the form of a demand, but the supplier decomposes that demand – the problem - into a set of (better) satisfiable service specifications. What is 'sold' is the ability to decompose the problem into a form which renders it (more) tractable.

172

There may however be difficulties in the formulation of need itself. The enterprise may have needs which are not even formulated as demands, and which therefore constitute a 'lack' in demand organization.:

(i) The agent's theory of its needs may itself be erroneous so that the composition of a collection of services that satisfied a demonstrably valid and feasible model would, when deployed, fail to satisfy the agent's needs ('bifurcation')

(ii) Decomposition might be conceptually impossible in that the agent's needs are not themselves sufficiently well-formulated to admit decomposition into services (this comes closest to the notion of a lack in Demand organisation), or

(iii) the agent's conceptual framework is not rich enough to admit any effective decomposition (lack of elaboration of signifying networks).

Services provided to the agent which aim to enable the customer to formulate demands which are satisfiable and/or decomposable are P-type:

- The **P-type** service offering, which collaborates with a potential service customer in the specification of the latter's need as a demand, before (jointly) decomposing it into the form of services which itself or others can supply. What is 'sold' is the ability to articulate need in the form a demand for services which are believed to be satisfiable, in other words to define a problem.

5.2 The 'lack' in mathematics itself

Services provided to an agent which overcome compositional difficulties are K-type, and the providers of K-type services must be able to identify and classify these difficulties by reasoning over the composition of service specifications. However, service composition is not a formally definable operator; that is, it occurs in no logical calculus.

A related compositional operator can be defined for axiomatic systems provided that all such compositions are conservative extensions of their parts. However, the composition of service specifications is not, in general, conservative (i.e. it is not definable as an operator) unless those specifications are orthogonal. The result of an orthogonal composition is neither stronger or weaker than its components; that is, all services that satisfy each of the specifications s and t separately (i.e. in isolation) also satisfy the composition of s and t. However, the composition of service specifications may be either:

(a) restrictive, where a composite specification is stronger than at least one of its components, that is, although services p and q satisfied specifications s and t, respectively, in isolation, together they fail to satisfy the composition of s and t. Composition may be so restrictive as to result in the inconsistent specification, which is satisfiable by no service.

(b) emergent, where the composition of two specifications s and t possesses properties exhibited by neither in isolation.

Emergent composition is a major factor in the development of 'value ladders'. If a service specification ,s, to be provided by an enterprise, e, can be successfully decomposed into two orthogonal service specifications, p and q, then its clients may as well acquire services that satisfy p and q separately, rather than acquiring s from e. If, on the other hand, s is an emergent composition of p and q, then e 'adds value' to services that satisfy p and q by composing them.

However, an enterprise that conceives a design for some service specification, s, as the emergent composition of service specifications p and q may encounter difficulties if:

(a) services that satisfy p and q are not available, or

(b) such services are available but they do not, together, satisfy s.

The enterprise may respond to these situations in a number of ways. It might undertake (or commission) the development of either a service that satisfies s directly (a 'custom' system), or of services x and y that separately satisfy p and q, respectively, and together satisfy s (although this latter may not be apparent to the suppliers of x and y, nor to the customer for s). On the other hand, it might abandon that design of s and seek a different composition that is more effectively satisfiable.

5.3 Knowledge Diffusion

The know-how of an agent is a function of its ability to articulate satisfiable demand situations in relation to any given problem. For any given set of problems defined by a value ladder, knowledge diffusion is defined by the rate at which the population of agents with that know-how increases over time. Knowledge diffusion affects value through the way in which it affects the balance between supply and demand in any given problem domain. The way in which this balance changes over time is further affected by the architecture of the supplying enterprises. Thus, the articulation and promulgation of service specifications, and the demonstration of their emergent properties, constitutes 'knowledge diffusion', as these descriptions and demonstrations become part of the enterprise's model of itself.

Once the stratified nature of a value ladder is understood in terms of a layering of restrictive and emergent compositions, it therefore becomes necessary to distinguish the different kinds of supplier-customer relationship constituting the form of the ladder itself. Within this context, the successful discovery of an effective emergent composition is a developmental step (a 'radical break'?) for the enterprise because it makes possible the creation of new forms of value.

6. In conclusion

What has been formulated here is a **primary stratification** in the relations between:

[[[Infrastructure :: Enterprise]::Market Organisation]::Demand Organisation]::Needs

Each of these 'layers' have holes in them in relation to the layer 'above'. Thus needs give rise to holes in demand organisation, just as demands give rise to holes in market organisation, etcetera. Neither is it clear where the 'top' is, because not only is this primary stratification itself an effect of the observer, but agents have needs, and infrastructure is made up of agents. Thus the whole stratification has properties like those of the Klein bottle in the sense that inside becomes outside and vice-versa. Within this primary stratification, there is then a **secondary stratification.** This is seen in the layering of the value ladder and in the organisation of supply relationships as much as it is present in the relations between agents within infrastructure.

All 'anticipatory systems' are emergent composites, no behavioural property of the whole being attributable to any of its parts in isolation. 'Agents' as space-time bounded (imaginary) entities are formulated in relation to a semantic (symbolic) domain which is itself lacking. Thus, even though a compositional operator might describe the presence of a having-been-composed, composition itself becomes something of the Real, a tuché marking the presence of a lack in the semantic domain. An interesting, and open, question concerning the nature of 'enterprise' itself might therefore be to understand the sense in which all emergent composites might be anticipatory.

Appendix 1: The ordering relation on specifications

Specifications are ordered under the relation *weaker than:*

A1	wkr: SP↔SP		The specifications that each specification is *weaker than*
A2	wkr ° wkr = wkr		This relationship is transitively closed and
A3	wkr ∩ wkr^{-1} = ide(SP)		Reflexive and antisymmetric, i.e. it is a *complete partial order.*
A4	vac: SP		The weakest (i.e. *vacuous*) specification
A5	inc: SP		The strongest (i.e. *inconsistent*) specification, where
A6	wkr^{-1}[vac] = wkr[inc] = SP		Every specification is stronger than **vac** and weaker than **inc**

Appendix 2: Composing Services and Processes

Recap:

We have already presented a model in which services provided by processes are deemed to *satisfy* specifications demanded by designs.

[9]   sat: SR → SP

Satisfaction is modelled here as a total function, mapping every service to the *strongest* specification that it satisfies. The relative strength of specifications was modelled by the relation *weaker*:

[A1]	wkr:	SP ↔ SP	a complete partial order (cpo): reflexive, antisymmetirc and transitively closed together with the limits of this cpo:
[A4]	vac:	SP	the weakest (i.e. *vacuous*) specification, and
[A5]	inc:	SP	the strongest (i.e. *inconsistent*) specification, where
[A6]		wkr^{-1}[{vac}] = wkr[{inc}] = SP	

We will also need the strict (irreflexive) order

[A7]   swk: SP ↔ SP

swk = wkr \ ide(SP)

The connection between satisfaction and strength may be formally defined by defining that each service *meets* all specifications weaker than the one it satisfies:

[B1]	meet:	SR ↔ SP	
[B2]		meet = wkr ° sat	where
[B3]		meet^{-1}[{vac}] = SR	every service meets the vacuous specification, and
[B4]		meet^{-1}[{inc}] = ∅	no service meets the inconsistent specification.

Composition

We now address the issues raised by the *composition* of services and of their specifications. Intuitively, the composition of two services is the (unique) service that is offered by the components operating together:

[B5]   src: (SR x SR) → SR

Similarly, the composition of two specifications is the (unique) specification jointly demanded by both components:

[B6]   spc: $(SP \times SP) \rightarrow SP$

These composition operators are not formalised here. Their intended semantics is clearly akin to the 'parallel composition' of concurrency theory (as in CCS [MILL], CSP [HOA], etc.) but their elaboration depends on the algebra and calculus of each notation chosen to represent services and specifications.[1]

Whatever choices are made for representation, a composite specification may be an *aggregation*, a *strengthening* (of which the extreme form is *mutual inconsistency*), or an *emergence*. These may be defined as follows:

Aggregation

The components of the composite specifications are *orthogonal*, that is, they are not systemically interrelated but form an *aggregate*. The composite specification, and each of its components, are met by the composition of any pair of services that separately meet its components:

[B7]   agg: $SP \leftrightarrow SP$

[B8] $\forall p,q: SP \bullet agg(p,q) \Leftrightarrow (\forall r,s: SR \mid (meet(r,p) \cap meet(q,s)) \bullet$
            $meet(src(r,s),p) \cap meet(src(r,s),q) \cap meet(src(r,s),spc(p,q))$

[B9] $agg = agg^{-1}$ aggregation is reflexive

[B10] $agg \circ wkr = agg$ and closed under weakening

but not necessarily transitive, i.e. even if agg(a,b) and agg(b,c), it does not follow that agg(a,c).

Strengthening

The composition is strictly stronger than any specification met by the composition of any pair of services that separately meet its components:

[B11] str: $SP \leftrightarrow SP$

[B12] $\forall p,q: SP \bullet str(p,q) \Leftrightarrow (\forall r,s: SR \mid (meet(r,p) \cap meet(q,s)) \bullet$
            $meet[\{src(r,s)\}] \subseteq swk[\{spc(p,q)\}]$

[B13] $str = str^{-1}$ strengthening is reflexive

but not necessarily closed under weakening, nor transitive.

Mutual Inconsisitency

The composition is satisfiable by no process; it is the *inconsistent* specification:

[B14] mic: $SP \leftrightarrow SP$

[B15] $spc[mic] = \{inc\}$

[B16] $mic \subseteq str$ Mutual inconsistency is the worst case of strengthening

[B17] $mic = mic^{-1}$ and is reflexive

but not necessarily closed under weakening, nor transitive.

Emergence

The composition is neither vacuous nor inconsistent, but is *incomparable with* (neither stronger nor weaker than) either of its components:

---

[1] Strictly speaking, SP and SR are not sets. In any concurrency theory notation, these objects would be 'classes', in the sense of von Neuman-Bernays set theory, over which many set-theoretic operations are not defined. However, we are modeling enterprises in which the collection of specifications and services 'ready-to-hand' at any time is always finite.

[B18] emg: $SP \leftrightarrow SP$

[B19] $\forall p,q: SP \bullet emg(p,q) \Leftrightarrow (\{spc(p,q)\} \times \{p,q\}) \cap (wkr \cup wkr^{-1}) = \varnothing$

[B20] $emg = emg^{-1}$ Emergence is reflexive

but not necessarily closed under weakening, nor transitive.

It seems reasonable to propose that these alternatives are exclusive, i.e. that the composition of any pair of specifications is either aggregation, or strengthening, or emergent, and that no other possibilities exist (in particular, that *composition never weakens*):

[B21] $(agg, str, emg) \in part(SP \times SP)$

Appendix 3: Stratification of Demand

Emergent compositions *stratify* a user's demand situation by splitting the value ladder into *layers* within which only non-emergent compositions occur. Only designs within such a layer may be *realised* by some process (see property 10), if such a process exists.[2]

We need some extra machinery to formalise this.

First we must *lift* composition from its present binary form to an operator that takes multiple specifications to a specification (and similarly for services).

This is done by the generic set-theoretic operator *lift* (defined in Addendum).

The multiple composition operators may then be defined:

[C1]   spm   set $SP \to SP$ Composing a set of specifications is *lifted*

[C2] $spm = lift(spc)$ from binary specification composition (B6)

[C3]   srm   set $SR \to SR$ and similarly for

[C4] $srm = lift(src)$ service composition (B5)

each design is mapped to the specification that it *de*composes (7) and to the set of specifications that it *demands* (8). We can now define:

[C5]   des $DE \to (set\ SP \times SP)$ Maps each design to a set of specifications *and*

[C6] $des = dem\ \&\ dec$ to the specification that they *collectively* satisfy.

[C7] $cod(des) \subseteq spm$ Collective satisfaction is multiple composition.

This predicate may be visualised thus:

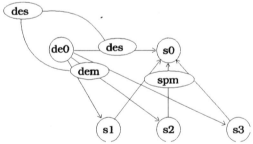

To formalise stratification, we model the structure of a user's designs as a *directed acyclic graph* (dag). The dag is a well-known mathematical object but, for completeness, we provide a set-theoretic model of it and of some of its operators in the Addendum.

---

[2] I still have trouble with this, because even an emergent composition can be realised... but the 'break' arises because

This allows us to refine definition 14 as follows:

[C8]   lad   US → (DE ↔ DE)

[C9]   ∀d: cod(lad) • d ∈ dag[DE] ∩ Im(dec) ° cf(d) = cf(dem)

This predicate may be visualised thus[3]:

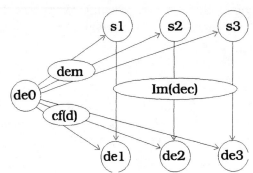

Since the members of a user's ladder comprise just those designs articulated by the user (see definition 14), we can add the predicate:

[C10]   member^{-1} ° lad = art

and it follows that

[C11]   dec ° roots ° lad = dst

and

[C12][4]   dec ° leaves ° lad = cst

Stratification

By properties 11 and 12, a design whose collective satisfaction relation includes a multiple composition that invokes emergence cannot be satisfied by a process.

Such a design requires a 3rd order composition between processes and in that sense 'fills' a *hole*.

[C13]   fill   set DE

[C14]   des[fill] ⊆ lift(spc ↑ emg)

These designs *punctuate* the value ladders of users who define them and *stratify* the demand situations articulated by users (definitions 14, 15, 16) into *layers* within which all designs are satisfiable by processes.[5]

---

[3] Note that the graph of designs is acyclic but that the corresponding graph of specifications decomposed and demanded by those designs need not be, because a specification might be invoked by different designs, as was noted at 16.

[4] Although we want to be able to have leaves at the bottom of each stratum

[5] Holes are not in dags, but in the ex ante specifiable relations between processes… in this sense these 3rd order compositions are *q-objects* which 'fill' the holes in the underlying causal texture of the network of processes.

References

Boxer, P.J. (1993) *The Role of IT in Supporting Business Development Processes.* in Proceeedings of The Hewlett-Packard Computer Users' European Conference, Birmingham. http://www.brl.com

Boxer, P.J. (1996) *The stratification of cause: when does the desire of the leader become the leadership of desire?* in Lasker, G.E., Dubois, D., & Teiling, B. (Eds.) "Advances in Modeling of Anticipative Systems" Institute for Advanced Studies in Systems Research and Cybernetics, Baden-Baden. Pp106-110. http://www.brl.com

Boxer, P.J. (1995) "Freud's Project and topologizing organization" GOWG working paper http://www.brl.com

Boxer, P.J. and Kenny, V. (1992) "Lacan and Maturana: Constructivist Origins for a 3^0 Cybernetics", in *Communication and Cognition* Vol 25 No 1 pp73-100.

Boxer, P.J. and Wensley, J.R.C. (1997) *Performative Organisation: learning to design or designing to learn? (in press)* http://www.brl.com

Bunge, M. (1974-80) *Treatise on Basic Philosophy* (8 vols). D. Reidel.

Casti, J.L. (1994) *Complexification* Abacus

Cameron, J. and Cohen, B. (1996) *Formal Approaches to Feature Interactions.* Tutorial Notes, FORTE/PSTV'96, Kaiserslautern, 1996.

Cameron, J. and Velthuijsen, H. (1993) *Feature Interactions in Telecommunications Systems.* IEEE Communications Magazine, August 1993.

Cohen, B., Harwood, W., and Jackson, M.I. (1984) *The Specification of Complex Systems.* Addison Wesley.

Cohen, B. (1996a) *The Description and Analysis of Services as Required and Provided by their Agents.* http://www.cs.city.ac.uk/homes/bernie/services.html

Cohen, B. (1996b) *Models and Modelling Frameworks in Health Care Informatics.* Proc 'Towards an Electronic Patient Record', San Diego and London.

Cohen, B. (1997) *Formal Modelling Frameworks of the Principles Governing the Confidentiality of the Patient Record,* Proc 'Towards an Electronic Patient Record', Nashville.

Dubois, D.M. (1996) *Introduction of the Aristotle's Final Causation in CAST: Concept and Method of Incursion and Hyperincursion – EUROCAST'95.* Edited by F. Pichler, R. Moreno Diaz, R. Albrecht. Lecture Notes in Computer Science, volume 1030, Springer-Verlag, Berlin, Heidelberg, New York, pp 477-493.

Lacan, J. (1966) "La Science et la Vérité". *Écrits.* Éditions du Seuil, Paris.

Lacan, J. (1988) "Logical Time and the Assertion of Anticipated Certainty" Translated by Bruce Fink and March Silver. Newsletter of the Freudian Field 2 (1988): 4-22. Originally written in March 1945, this was originally published in Écrits pp197-213. (1966).

Maturana, H.R. (1988) "Reality: The Search for Objectivity or the Quest for a Compelling Argument". *Irish Journal of Psychology,* 1988, 9,1, p33

Rosen, R. (1987) *Anticipatory Systems* Pergamon

Schuman, S.A., Pitt, D.M. and Byers, P.J. (1990) *Object Oriented Process Specification,* in "Specification and Verification of Concurrent Systems" (ed Rattray), Springer-Verlag.

Van de Vijver, G. (1996a) "Internalism versus Externalism: a matter of choice", *Japanese Journal of Contemporary Philosophy.* Pp93-101.

Van de Vijver, G. (1996b) "On the Origin of Psychic Structure: a case-study revisited on the basis of Freud's *Project*" University of Ghent

Addendum

Naming Conventions

We have chosen to denote the names of sets by two-letter, bold face, upper case abbreviations of their intended interpretation, thus: **SR** denotes the set of *services*.

The names of relations among those sets are denoted by a three-letter, bold face, lower case abbreviation of their intended interpretation, thus: **prv** denotes the relation in which processes *provide* services.

Notation

$\in$	set membership; $x \in$ **XX** if and only if x is an element of the set **XX.**
$\notin$	non-membership; $x \notin$ **XX** if an only if x is not an element of the set **XX.**
$\varnothing$	the empty set, which has no members.
$\cup$	set union; $x \in$ **XX** $\cup$ **YY** if and only if $x \in$ **XX** or $x \in$ **YY.**
$\cap$	set intersection; $x \in$ **XX** $\cap$ **YY** if and only if $x \in$ **XX** and $x \in$ **YY.**
$\setminus$	set difference; $x \in$ **XX** \ **YY** if and only if $x \in$ **XX** and $x \notin$ **YY.**
$\subseteq$	subset; **XX** $\subseteq$ **YY** if and only if every member of **XX** is also a member of **YY**.
$\wp$	powerset: the set of all subsets of a set; **XX** $\in \wp$ **YY** if and only if **XX** $\subseteq$ **YY.**
x	set (or cartesian) product: the set of ordered pairs drawn from an ordered pair of sets; $(x,y) \in$ **XX** x **YY** if and only if $x \in$ **XX** and $y \in$ **YY.**
$\leftrightarrow$	the set of relations between two sets, a relation being any subset of their cartesian product; **abc** $\in$ **XX** $\leftrightarrow$ **YY** if and only if **abc** $\subseteq$ **XX** x **YY.**
	If $(x,y) \in$ **abc** then **abc** is said to *map* x to y.
$\rightarrow$	the set of functions between two sets, a function being a relation that maps uniquely;
	if **abc** $\in$ **XX** $\rightarrow$ **YY,** and $x \in$ **dom(abc)**, then
	if $(x,y) \in$ **abc** and $(x,z) \in$ **abc**, then $y = z$.
ide	the identity function on a set; $(x,x) \in$ **ide(X,X)** if and only if $x \in$ **XX**.
dom	the domain of a relation, being the set of the first elements of its pairs;
	if **abc** $\in$ **XX** $\leftrightarrow$ **YY**, then
	$x \in$ **dom(abc)** if and only if there exists some $y \in$ **YY** such that $(x,y) \in$ **abc**.
ran	the range of a relation, being the set of the second elements of its pairs;
	if **abc** $\in$ **XX** $\leftrightarrow$ **YY**, then
	$x \in$ **ran(abc)** if and only if there exists some $x \in$ **XX** such that $(x,y) \in$ **abc**.
-1	the inverse of a relation; $(y,x) \in$ **abc**$^{-1}$ if and only if $(x,y) \in$ **abc**.
[●]	the image of a relation through a set; if **abc** $\in$ **XX** $\leftrightarrow$ **YY** and **ZZ** $\subseteq$ **XX**, then
	$y \in$ **abc[ZZ]** if and only if there exists $x \in$ **XX** such that $(x,y) \in$ **abc**.
°	the composition of relations;
	if **abc** $\in$ **XX** $\rightarrow$ **YY**, and **def** $\in$ **YY** $\rightarrow$ **ZZ**, then
	def ° **abc** $\in$ **XX** $\rightarrow$ **ZZ**, where $(x,z) \in$ **def** ° **abc** if and only if there exists some $y \in$ **YY** such that $(x,y) \in$ **abc** and $(y,z) \in$ **def.**

   &   the relational join;

       if **abc** $\in$ **XX** $\to$ **YY**, and **def** $\in$ **XX** $\to$ **ZZ**, then

       abc & **def** $\in$ **XX** $\to$ (**YY** x **ZZ**), where (x,(y,z)) $\in$ **abc** & **def** if and only if (x,y) $\in$ **abc** and (x,z) $\in$ **def.**

   *   the relational product;

       if **abc** $\in$ **AA** $\to$ **BB**, and **def** $\in$ **XX** $\to$ **YY**, then

       abc * **def** $\in$ (**AA** x **XX**) $\to$ (**BB** x **YY**), where ((a,x),(b,y)) $\in$ **abc** * **def** if and only if (a,b) $\in$ **abc** and (x,y) $\in$ **def.**

The *Lift* operator

Extends and transitive binary operator to a naturally corresponding operator on sets.

lift: (S x S $\to$ S) $\to$ (set S $\to$ S)
(lift f)(ss) = y
$\Leftrightarrow \exists x \in ss \bullet f(x, (lift\ f)(ss \backslash \{x\})) = y$

For example, given the numerical operator '+', lift(+) is the summation operator usually written Σ.

The Directed Acyclic Graph (or *dag*)

These structures are diagrams of relations with named 'nodes' linked by 'arcs' decorated with arrowheads. Each node is an element of the underlying set and each arc joins a pair of elements in the relation, mapping in the direction of the arrow. There are no 'loops' in a dag, that is no 'path' along connected arcs that runs from any node back to itself.

Every relation whose transitive closure takes no element to itself is a dag.

dag[X] = set(X $\leftrightarrow$ X)
$\forall d: dag[X] \bullet d^{+} \cap ide[X] = \varnothing$

The roots of a dag are those nodes that have no incoming arcs.

roots: dag[X] $\to$ set X
roots(d) = dom(d) \ cod(d)

The leaves of a dag are those nodes that have no outgoing arcs.

leaves: dag[X] $\to$ set X
leaves(d) = cod(d) \ dom(d)

The paths of a dag are all those sequences of nodes that can be traced along the arcs (see below).

paths: dag[X] $\to$ set(seq[X])
sl $\in$ paths(d) $\Leftrightarrow$ next(sl) $\subseteq$ d

Every node is a member of its dag

member : X $\leftrightarrow$ dag[X]
x member d $\Leftrightarrow$ x $\in$ dom(d) $\cup$ cod(d)

The *next* operator on sequences

Generates the (unnamed) 'successor' relation induced by a sequence.

next: seq[X] $\to$ (X $\leftrightarrow$ X)
next(s) = cod(s & succ $\circ$ s)

Cybersemiotics: A suggestion for a Transdisciplinary Framework for Description of Observing, Anticipatory and Meaning Producing Systems

M. Sc., Ph.D. Søren Brier, The Royal School of Librarianship, Aalborg Branch, Denmark.
Ed. & publ. of the international journal Cybernetics & Human Knowing. E-mail: sbr@db.dk,
http://www.db.dk/dbaa/sbr/home_uk.htm.

Abstract

The ability of systems to be anticipatory seems to be intricate connected with the ability to observe and to cognate by reducing complexity through signification. The semantic capacity of living systems, the cognitive ability to assign meaning to differences perturbating the system's self-organization, seems to be the prerequisite for the phenomenon of communication, language and consciousness. In cybernetics Bateson developed the idea that information is a difference that makes a difference and second order cybernetics developed the concept of organisms as self-organized and self-produced systems (autopoietic) as the prerequisite of life and cognition. The cognitive ability seems to be qualitative different from what so far is computable on any known machine although parts of different aspects of the process can be partly simulated in AI, neutral network and AL. In semiotics the fundamental process of cognition and communication is called semiosis or signification and C.S. Peirce created a special triadic, objective idealistic, pragmatic and evolutionary philosophy to be able to give a fruitful description of the process and its relation to logic and the concept of natural law. Both second order cybernetics and semiotics sees information and meaning as something *produced* by individual organisms through *structural couplings* to the environments or other individuals through historical drift and further developed in social communication. Luhmann points out that social communication also only functions through structural couplings which he calls *generalized media* such as science, art, power, love and money. Peirce talks of *the semiotic net* as a triadic view of meanings developing through history and in animals through evolution. In accordance with this Wittgenstein points out that signification is created in *language games* developed in specific *life forms*. Life forms are the things we do in society such as seducing, commanding and explaining. As animals do not have language in the true sense I have extended his concept into ethology and *bio-semiotics* by talking of *sign games* related to specific *motivations* and *innate response mechanisms*. Life as such seems to be an anticipatory function generating expectations through evolution through open genetic programs as Konrad Lorenz pointed out. The phenomenon of *imprinting* in ducks for instance is a standard example of programmed anticipation. Expectations are expectations of meaning and order (information) related to the *semiosphere* the organism constructs as its individual world view and live in. (The *Umwelt* of von Uexküll). On this basis events that perpetuates the semiosphere are reduced to meaning, i.e. something related to the survival and procreation of the individual living system, it *conatus*, to use one of Spinoza's terms. The framework of *cybersemiotics*, uniting second order cybernetics, semiotics and language game theory, is created to make transdisciplinary concepts and models that can handle the process of cognition, information and communication across the domains of the sciences, the arts and social sciences in a non-reductionistic way. It is seen as an alternative based on biological and semiotic thinking (biosemiotics) to the functionalistic information processing paradigm of cognitive science that is build on the computer as paradigm and based on classical logic and mechanistic physics - and therefore has severe problems of dealing with semantics and signification.

CP437, *Computing Anticipatory Systems: CAYS--First International Conference*
edited by Daniel M. Dubois © 1998 The American Institute of Physics 1-56396-827-4/98/$15.00

Introduction

I am a biologist and philosopher of science working with interdisciplinary conceptualizations and models of cognition, information and communication. With a background in ethology and neurobiology I use second order cybernetics and Peirce's semiotics to make an alternative framework to the information processing paradigm of cognitive science. This is partly a bio-epistemological project. What I want to in this paper is to say something about the function of anticipation in living cognizing systems from this point of view. Hoping that this description can move us further towards the understanding of the phenomenon. My goal is understanding, not making a computer model. It is to ask how can we conceptualize these depositions to perceive, communicate and behave! How are the apriori created in evolution and ontogenisis?

Based on the results of ethology I propose that all perceptual cognition is anticipatory. As can be seen on figure 1 the fixed action patterns that is the behavioral part of the animal instinct is only released by the innate release mechanism if a motivational borne pattern recognition appears. The more or less hereditary releasing perception is called sign stimuli and only sets of the innate realize mechanism if property motivated for instance hunger, care for the young, aggression etc. Ethology enumerates many different and some of them species specific motivations (Brier 1993b). Most living systems have great problem perceiving something not biological, psychological or socially anticipated. So from an ethological point of view you have to include the action of the subject and its motivational value into a model of the dynamics of behavior. It therefore makes sense to view animal instincts with their specific motivation for fight, mating, hunting etc., innate response mechanisms and sign stimuli are psycho-biological expectations or anticipation's.

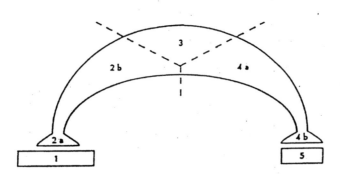

Figure 1: Illustrates ethology's conception of the release mechanism. 1 denotes the outer world, elements of which can be stimuli. 2a the senses, 2b the perceptive part of the nervous system which conveys information from the outside world. 3 the motivating and coordinating part of the nervous system. 4a the motor center of the nervous system and 4b the movements of the muscles. 5 the total external behavior. 2 + 3 are the release mechanism. 4 the effector section which comprises one or more fixed movements possibly with learned modifications. 2 + 3 + 4 shows the congenital nerve connection which in conjunction with certain motivating conditions produces certain effects - without prior learning - resulting in the release of the congenital behavioral dispositions. The difference in length between 1 and 2a illustrates that organisms can never register the entire physical reality, and that precisely the concept of sign stimulus emphasizes the selection by each species of a few but well

defined stimuli in complex phenomena. Translated after Reventlow (1970) here taken from (Brier 1993b).

Some of these anticipation's are completely innate such as the hunting and mating behavior of the digger wasp which have no contact what so ever with its parents. It gets not of its e.g. buried in a little cave under the surface of the ground. It does not encounter other species members in its larval stage and no body peaches it anything. Still it is able to hunt and mate when it meets the relevant other living systems. So this is a rather close inheritance based system. Some are partly open for variations in the anticipated as in the imprinting of ducklings who will follow the first big moving object expressing sound they see and later on choose a mate like it. Many birds have a basic song but have to listen to other members of the species to learn full song, but they will not learn the song of another species. But again some bird species can include the song of other species and natural sounds and make different kinds of variations some even as ongoing improvisations. So it is general opinion that most living systems only perceives their surroundings according to their needs. Gibson says that they see affordances. Jacob von Uexkull spoke of die Umwelt des tieres Brier (1995a). To emphasize that they live in a world of their own choise, so to speak. Modern bio-semiotics speaks about semiospheres. But how does this come about?

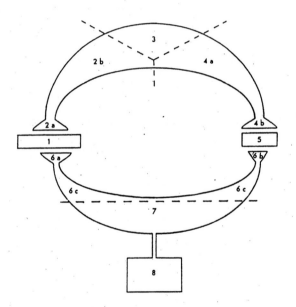

Figure 2: The model illustrates the conditions for observation of an individual organism. 1 represents the exterior world, which stimulates/perturbates both the observed individual as well as the observer. 2, 3 and 4 represent mechanisms in the observed individual (compare with fig. 1), which cause that 5, which represents the total exterior behavior, is brought forth. 6a and b represent the sense organs of the observer, and 6c the other perceptual parts of the nervous system and what further determines 7, which is his experience of the observed behavior. 8 represents the description of the observations which the observer gives, and which becomes the scientific datum, that is the foundation for the further scientific analysis (...) When 6a is not situated symmetrical with 2a it is because that some times animals are most likely to react on stimuli, whose physical properties we do not know (...), while we (e.g. through physical measurement apparatus) can get knowledge about appearances of the physical world, that are without significance for the perceptions of animals. In the same way 6b is smaller than 4b and 5, because the animals have behaviors which we do not know, and even some that we cannot perceive or measure yet".

In figure 2 the Danish ethologist Iven Reventlow (1970) illustrates the problem of objective reality when you investigate living systems. The difference between the senses of the animal (2a) our theoretical idea of reality (1) and our own perceptual apparatus range of perception (6a) shows how difficult it is to determine a common objective world for the animal and the human observer. Not only our sense capacity but the whole organization of our cognitive organization of the nervous system determines what we cognize. Somehow both systems are closed in their own cognitive apparatus as it is developed through the evolution of the species. The concept of sign stimuli and Umwelt was supposed to conceptualize this phenomenon. But it did not give a fruitful expression of the dynamics of the cognitive system. New concepts and models was needed, as it was also difficult to grasp the concept of motivation in traditional biology (Hinde 1970).

G. Bateson (1973) brought us a step further when laid the basis for second order cybernetics by stating that information is a difference that makes a difference! For something to be perceived it has to be of relevance for the survival and self-organization of the living system and therefore being anticipatet. Later on Maturana and Varela (1980) coined the term autopoiesis to underline the organizational closeness of the living system, including the nervous system to conceptualize this point. Through the perceptual apparatus the nervous system is perpetuated with stimuli's that disturbers is own firing patterns, but perception is only possible if structural couplings are formed in advance in evolution so the perturbation of the autopoietic system is anticipated. Then it will generate information inside the system, which is hardly possible without a structural coupling or the system disintegrates. Other members of the species is also surroundings and is again recognized through preestablished structural couplings. This is what ethology calls sign stimuli (Lorenz 1970-71).

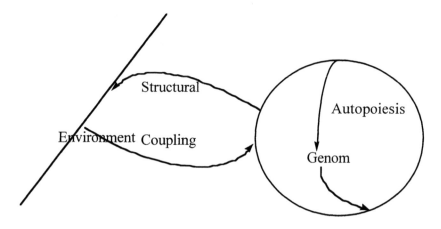

Figure 3: Graphical model of the living system and its autopoietic interaction through its genome with itself and its creation of a structural coupling with the environment that pertubates it in a repetable way.

Communication then between members of the same species is in second order cybernetics explained as a double structural coupling between two closed systems. Each internally creating information. Luhmann (1990) has developed this approach into a general communication theory distinguishing three different levels of autopoiesis. The biological, the psychological and the social communicative underlining that our mental system is also self-organized and closed around its own organization. Even in the social realm messages are only received if the lay within the anticipate spectrum of structural couplings called generalized media such as power, money, love, art, science etc. In humans the complexity of the environment is reduced on the background of meaning (Brier 1992 and 1996a)

But still this functionalistic model does not tell us how meaning is anticipated and generated. See my analysis in Brier (1996b). The answer of semiotics will be that the structural couplings are established through semiosis. Second-order cyberneticians do not even have the triadic concept of sign in their theory, and some are opposed to it. But I think that it is possible to fuse the two theories here, as they both are of second-order (Brier 1992). All the elements in Peirce's semiosis are signs themselves. In a very famous quotation Peirce defines his dynamical and pragmatic concept of a sign:

> "A Sign, or Representamen, is a First which stands in such a genuine triadic relation to a Second, called its Object, as to be capable of determining a Third, called its Interpretant, to assume the same triadic relation to its Object in which it stands itself to the same Object. The triadic relation is genuine, that is its three members are bound together by it in a way that does not consist in any complexus of dyadic relations. That is the reason the Interpretant, or Third, cannot stand in a mere dyadic relation to the Object, but must stand in such a relation to it as the Representamen itself does.
>
> ... A Sign is a Representamen with a mental Interpretant".

> Peirce in (Buckler 1955, p. 99-100)

The semiotic process as described in the work of C.S. Peirce give us further insight into the relation between anticipation, cognition and life. As one can see, Peirce's definition is second order because all the elements of the sign process are signs themselves. Further a sign is not a thing, but a dynamical process. In humans the meaning of the sign emerges out of a social dynamic network of relational logic, creating an ever evolving interpretant. The interpretation of a sign is never finished because the meaning of the sign is the social habits it gives rise to and they are in constant development and are forever spreading and returning. Peirce talks about *unlimited semiosis*.

So a sign is something which stands for something in some capacity for somebody, and to be able to do so it must be triadic. So what a sign means is established as a bio-psychological or social habit in a network of meaning generating habits: The semiotic web. As signs are the basic components of perception, understanding, thinking and communication, Peirce's development of his system gave rise to a whole system of thought and logic. To understand semiosis Peirce constructed a triadic philosophy with qualia as firstness, resistance or constraints as secondness and habit formation for instance in cognition through signs as thirdness.

The triadic categories went right through the whole system and became the basic categories of knowing. Peirce simply calls them firstness, secondness and thirdness. In the following quotation Peirce gives a short description of his triadic categories:

> "Actuality is something *brute*. There is no reason in it. For instance putting your shoulder against a door and trying to force it open against an unseen, silent and unknown resistance. We have a two-sided consciousness of effort and resistance, which seems to me to come tolerably near to pure

sense of actuality. On the whole, I think we have here a mode of being of one thing which consists in how a second object is. I call this Secondness.

Besides this, there are two modes of being that I call Firstness and Thirdness. Firstness is the mode of being which consists in its subject's being positively such as it is regardless of aught else. That can only be a possibility. For as long as things do not act upon one another there is no sense or meaning in saying that they have any being, unless it be that they are such in themselves that they may perhaps come into relation with others. The mode of being a redness, before anything in the universe was yet red, was nevertheless a positive qualitative possibility. And redness in itself, even if it be embodied, is something positive and sui generis....

Now for Thirdness. Five minutes of our waking life will hardly pass without our making some kind of prediction; and in the majority of cases these predictions are fulfilled in the event. Yet a prediction is essentially of a general nature, and cannot ever be completely fulfilled. To say that a prediction has a tendency to be fulfilled, is to say that the future events are in a measure really governed by law.... This mode of being that *consists* ... in the fact that future facts of Secondness will take on a determinate character, I call Thirdness".

(Peirce in Buckler 1955 p. 76)

Firstness is among other things a monadic characteristic or predicate: sense qualities, simple forms and feelings, the modus of possibilities, that which exists without reference to any other thing and pure quality. Firstness is vague because it does not in itself stand in any relation to anything else. Abduction is firstness.

Secondness is a dyadic quality which something has in its relation to something else, but independent of some third thing. This is the category for the characteristics of the objects which makes it possible to know them and identify them independent of concepts by pointing and saying this/that. For example indexes are signs that stand for things without describing them. Secondness is the subject in logic. It is resistance, breaks, separateness, quantity. Where firstness is possibility, secondness is necessity such as local causality. Deduction is secondness.

Thirdness is the triadic quality, which only that has which is as it is, because it brings something second and third in relation to each other. This is the category of generality and understandability, rationality and lawfulness. By this it is first of all the category of the sign and of logical inference. From a human point of view firstness is feeling and secondness experience. Thirdness is the generation of some kind of biological, cultural or linguistic habit which elevates us above firstness' universe of possibilities and secondness' numberless incidents. Thirdness puts quality and quantity together in a relation parallel to logical inference as is done in science. Induction is thirdness. Lyne sum's up very well the essence of Peirce's conception:

"On the one hand, Peirce surpasses the naive reference theories (based solely on the sign-object relationship) that have pervaded so much thought on language, ... and proceeds to follow out the implications for the various domains of inquiry. On the other hand, in contrast to the Saussurian semiological tradition, which recognizes only an arbitrary relationship between a dyad, signifier and signified, ... Peirce conceives of semiotic relations as not simply arbitrary. The object does not reduce to the interpretant; furthermore there is a distinction made between things represented and the *respect* in which they are represented to any given interpretant. As a philosophical realist, ... Peirce saw semiotic as an instrument suited not just to cultural analysis, but to empirical and

187

rational investigation as well. The discovery and expression of truth, in fact, are the fundamental purposes of his semiotic."   (Lyne, 1980, p.157-158)

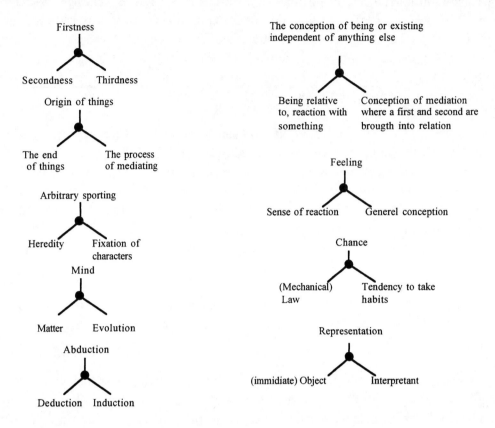

Figure 4: An illustration of C.S. Peirce's triadic philosophy and semiotics

So to combine the ethological, the autopoietic and the semiotic description one can say the following. Meaning is habits/structural couplings established between the autopoietic system and the chaos that we call environment. 'Objects' are cognized in the environment - through abduction - by attaching sign habits to them related to different activities of survival such as eating, mating, fighting, nursing what we - with a term I have boroughed from Wittgenstein - call life forms (Brier 1995a, 1996a).

In figure 5 I have made a graphical representation of this integration of an ethological, an autopoietic and a semiotic interpretation of the cognition of living systems. Ethology calls the partly inherited structural couplings Innate Release Mechanism. It works of cause only in antropoietic systems with a genom. This organization of the nervous system makes it anticipating certain kinds of stimuli patterns which are connected to some fixed behavioral patterns.

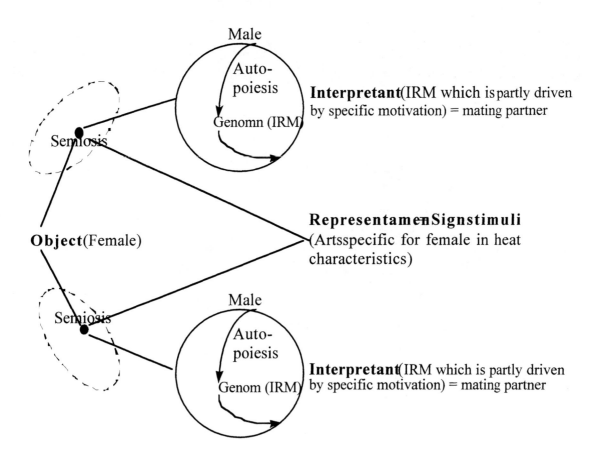

Figure 5: Lifeform of two autopoietic systems (males) seeing the same sign in an object (a female). This happens through the partly inherited structural coupling Innate Response Mechanism that anticipate certain sign stimuli. The whole model is in one lifeform: Mating.

Inspired by the older Wittgenstein's (1958) pragmatic concepts of language games related to social lifeforms in his pragmatic language philosophy I say that animals have sign games! (Brier 1995a). Meaning is only created through the anticipation in the games of life in ascertain repertoire of lifeforms. This is the ground or context of Peirce's semiosis which we here transforms to a bio-semiotic context. Signification is seen in an ecological and evolutionary pragmatic perspective. It is this combination of second order cybernetics autopoiesis theory, Peirce's triadic and pragmatic semiotics and the old Wittgenstein's pragmatic language game theory of semantics I call cybersemiotics. It delivers a bio-psycho-social framework for understanding of signification that supplements and develop the original ethological models of animal cognition. So I claim that perception, cognition, anticipation, signification and communication are intrinsic connected in autopoietic systems in mutual historical drift, and we still understand very little about how meaning is generated in this ecological, evolutionary and social process. As Lakoff (1987) writes then the relations between categorical concepts are not logical but motivated. It has its original in the basic life forms and their motivated language games. Life is a self-organized cognizing anticipatory autopoietic system. With Spinoza I will say that it has got Conatus! This means that the individuality of life systems value itself through its continuing efforts to preserve its own internal organization.

"On the experientialist view, reason is made possible by the body - that includes abstract and creative reason, as well as reasoning about concrete things. Human reason is not an instantiation of transcendental reason; it grows out of the nature of the organism and all that contributes to its individual and collective experience: its genetic inheritance, the nature of the environment it lives in, the way it functions in that environment, the nature of its social functioning, and the like".

(Lakoff, 1987, p. xv)

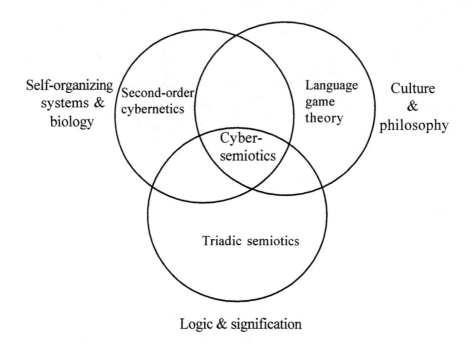

Figure 6: Cybersemiotics: A pragmatic understanding of information, communication and language as an irreducible plenomenon is a common denominator for the second-order cybernetics of von Foerster, Maturana & Varela and Luhmann, Peirce's triadic pragmaticistic semiosis and Wittgensteins pragmatic philosophy of language. It is recognized each paradigm has different points of departure. From Brier (1996a)

So I have introduced a framework to deepen the understanding of the relation between observing, anticipation, cognition, signification and communication in what we now call information sciences. A graphical overview over the components of the cybersemiotic framework is given in figure 6. I hope that I have brought us a little further into the nature of anticipation and the structural dynamics of the living anticipatory systems. I believe that a pragmatic evolutionary psycho-biological framework as I have shown here is necessary to make progress in this field. I will close with a model relating the different concepts describing the anticipatory function of cognition. Let me end with a figure summing up the suggested integration between second order cybernetics and Peirce's triadic semiosis. See figure 7.

In figure 7 I show that the three basic elements of semiosis which Peirce deduces from his triadic philosophy can be found in the conceptual apparatus of second order cybernetics. Varela (1975) even shows the necessity of a triadic cognitive model to make the autopoietic model of cognition work (Brier 1993b). But Peirce developed this in to a deep philosophy and theory of sign processes that brings the model a lot further in our understanding of cognition and communication (Brier 1996b). In this figure I have tried to show how the concepts of second order cybernetics fit into the

190

basic semiosis model of Peirce's triadic philosophy. I have further added the one most used sign types klassification: Index, icon and symbol.

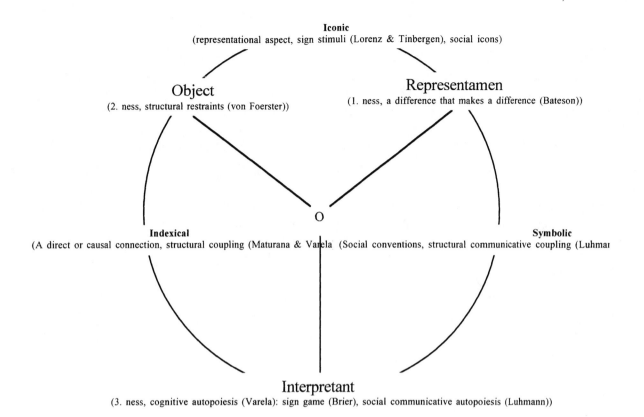

Figure 7: This shows the basic structure of the triadic sign process of Peirce's semiosis: Reprsentamen, Object and Interpretant. Further is indicated three basic types of signs: Indexical, Iconic and Symbolic. In the parenthesis is first given a short definition from Peirce's philosophy followed by a concept from second order cybernetics that is claim deals with the same aspect of reality. It is followed by the name of the scientist(s) that coined the term. The sign stimuli concept under iconic is from the ethology of Lorenz and Tinbergen. I have elsewhere showed how the concepts of second order cybernetics gives a more comprehensive foundations for some ethological concepts.

Thus by joining the effort from the latest development of cybernetics and ethology with Peirce's already transdisciplinary semiotics and uniting them with Wittgensteins pragmatic language game theory we get a framework that is truly transdisciplinary. This framework also in a more fruitful way conceptualize the anticipatory dynamics of all cognition. There is an active anticipatorial element in all perception and recognition. Perception and cognition is an active process deeply connected to the self-organizing dynamics of living systems and their special ability to be individuals.

In the short space here given I can only give a decription of the cybersemiotic framework. To get further argumentation one must read my other papers in the reference list. Further references can be obtained at http://www.db.dk/dbaa/sbr/home_uk.htm and some full text papers at hhtp://www.db.dk/dbaa/sbr/cyber.htm. A book collecting all the arguments into a whole is under production.

References:

Bateson, G. (1973): Steps to an Ecology of Mind (Paladin, Frogmor, St. Albans, USA)

Brier, S. (1992): "Information and consciousness: A critique of the mechanistic concept of information", in Vol. 1, no. 2/3 of *"Cybernetics & Human Knowing"*. Aalborg, Denmark, pp. 71-94.

Brier, S. (1993a): "Cyber-Semiotics: Second-order cybernetics and the semiotics of C.S. Peirce". *Proceedings from the Second European Congress on Systemic Science,* Vol. II, pp. 427-436.

Brier, S. (1993b): "A Cybernetic and Semiotic view on a Galilean Theory of Psychology", *Cybernetics & Human Knowing*, Vol. 2, No.2, pp. 31-45.

Brier, S. (1995a): Cyber-semiotics: On autopoiesis, code-duality and sign games in bio-semiotics, Cybernetics & Human Knowing, vol. 3, no. 1, pp. 3-14.

Brier, S. (1996a): Cybersemiotics: A new interdisciplinary development applied to the problems of knowledge organization and document retrieval in information science, Journal of Documentation, 52(3), September 1996, pp. 296-344.

Brier, S. (1996b): From Second Order cybernetics to Cybersemiotics: A Semiotic Reentry into the Second order Cybernetics of Heinz von Foerster, Systems Research vol. 13, no. 3, pp. 229-244 (A Festschrift to Heinz von Foerster).

Buckler, S. (ed.) (1955): *Philosophical Writings of Peirce* , Dover Publication, New York.

Foerster, H. von (1984): *Observing systems*, Intersystems Publication, California, USA.

Hinde, R. (1970): *Animal Behaviour: A synthesis of Ethology and Comparative behaviour,* McGraw-Hill, Tokyo (International Student Edition).

Lakoff, G. (1987): Women, Fire and Dangerous Things: What catagories reveal a bout the Mind, (The University of Chicago Press, Chicago and London).

Lorenz, K. (1970-71): *Studies in animal and human behaviour.* I and II. Cambridge, Mass. Harvard Univ. Press, USA.

Luhmann, N. (1990): *Essays on self-reference*, Columbia University Press, New York.

Lyne, J. R. (1980): "Rhetoric and Semiotic in C.S. Peirce", *The Quarterly Journal of Speech*, Vol. 66 1980, pp.155-168.

Maturana, H.R. & Varela, F.J. (1980): *Autopoiesis and Cognition: The Realization of the Living,* Boston, Reidel.

Reventlow, I. (1970): *Studier af komplicerede psykobiologiske fænomener.* Munksgaard, Copenhagen. Doctoral thesis. University of Copenhagen.

Varela, F.J.G. (1975): "A Calculus for self-reference", *Int. for General Systems,* Vol. 2, pp. 5-24.

Wittgenstein, L. (1958): Philosophical Investigations, Third Edition (MacMillan Publishing Co. Inc., New York).

THE DYNAMICS OF NONDECIDABLE SENTENCES AS AXIOMS IN AXIOMATIC METATHEORIES CONVERGING TO THE UNIQUE NONDECIDABLE SENTENCE OF MAXIMAL GENERALITY

Ion I. Mirita
University of Petrosani
str. Institutului nr 20 2675 Petrosani ROMANIA
tel /fax (40)54231238 email rector@utp.sfos.ro

Abstract

Gödel states that in any formal system there are statements which, although possible to formulate with the means of the system they cannot be demonstrated starting from the axiom system, these being the propositions on which it is impossible to decide formally. For the modern theory of knowledge, these statements have important consequences, as the concept of truth cannot be replaced by the concept of demonstrable, the class of true propositions being larger than that of demonstrable sentences and thus, the structures of semantic notions cannot be reduced to structures of syntactic notions.

The existence of a true sentence which cannot be demonstrated within a theory proves the theory to be incomplete. In the dialectic logic this theory is completed, extending the axiom system by raising the true nondemonstrable proposition to the rank of an axiom and adding it to the other axioms. The change of axioms in an axiom system alters the meaning of predicates and the relations on the theory, the new theory becoming a metatheory in which a new true nondemonstrable proposition can be formulated. Thus, it is possible to develop a sequence of metatheories each representing a relative truth possible to replace by a larger relative truth situated on a higher step, and creates the possibility of developing the general science, meant to comprise the principles of scientific thinking of all, sciences, named by Leibniz "Scientia generalis".

1.Theoretical backgrounds of the meta-theories of the axiomatic theories

1.1. Deductive systems Wang Hao assumes that any scientific theory consist of a body of ideas and an assemble of assertions.

Any concept is explained or defined by the aid of others.

The truth of a certain assertion is deduced from other accepted assertions, when the concepts and statements of a theory are arranged suitable with definability and deductibility links into an axiomatic system of the theory.

A deductive system or a theory **T** is given by:

CP437, *Computing Anticipatory Systems: CAYS--First International Conference*
edited by Daniel M. Dubois © 1998 The American Institute of Physics 1-56396-827-4/98/$15.00

1° A system of notions {N} and relations {R};

2° The system of correctly created statements {P} using the notions and relations;

3° A system of deducing rules {L};

4° The true statements sub-system {Ad} ⊂ {P} such {Ad} and {P} are stable in rapport with {L} .

1.2. Meta-theory of an axiomatic theory The analysis of an axiomatic theory by which the properties of the theory are determined, its internal structure and relations with other theories are called meta-theory.

In the analysis of a meta-theory, the syntactic and semantic points of view are used.

The syntax of the theory consists in the analysis of the theory's mode of construction.

The semantic of the theory consist in the analysis of the theory by the interpretations and the models.

For the meta-theory of an axiomatic theory, the non-contradiction, independence, completeness and categoricalness are important.

1° The axiomatic theory is non-contradictory if a sentence **p** and its contrary,]**p** cannot be deduced from axioms.

2° An axiom of the theory is independent from other axioms of the theory if it cannot be deduced logically by other axioms.

3° An axiomatic theory is complete if, for any non-contradictory sentence **p** or]**p** can be deduced from axioms.

A model of an axiomatic theory is an interpretation.

Two models are called isomorphous if a bi-directional correspondence which keep their true properties can be established.

4° An axiomatic theory **T** is called categorical if all its models are isomorphic.

1.3.Meta-mathematics, the Gödel's paradox. The meta-mathematics, a branch of mathematics, has as objects demonstrations, methods of constructions and theories instead numbers, icons and functions.

The theory of demonstrations or the meta-mathematics operates mainly with calculable functions, because the decision procedures must be generally effective.

The symbolical logic uses symbols to designate notions.

From meta mathematical point of view it is even which symbols are selected for the designation of logical notions.

Gödel choose the natural numbers as signs.

A formula will be a finite seria of natural numbers and a demonstration will be a finite seria of finite series of natural numbers.

The meta-mathematics will be constituted from mathematical concepts and sentences , and because these are expressed as finite series of natural numbers, the mathematical sentences and concepts will be sentences and concepts about natural numbers as:

$$n \in K = \overline{B_{ew}}$$

where: B_{ew} - demonstrable;

R - the symbol which represents the order in the class;

196

R(n) - has the following meaning: in the order R, the class sign has the number n.

The definition says that a natural number **n** belongs to class **K** if, for it, the formula **[R(n);n]** is undemonstrable.

A consequence of this definition is that it exists a number of class **S**, such as the formula **[S;n]** shows that the natural number **n** belongs to **K**.

As **S** is a class number, it means that he has a position **q** and that **S** is identical - as consequence of the definition of the class sign - with **R(q)**:

$$S=R(q).$$

Gödel shows that the sentence:

$$[R(q); q]$$

is undeterminable, because **[R(q); q]** can be demonstrated like its negation can be demonstrated too, which is a contradiction.

From this paradox, Gödel conclusionates that in every class of non-contradictory formulas are non-decidable sentences.

An important consequence is the fact that a true sentence of a T theory can be made with this theory, but it cannot be demonstrated in the same theory.

It has been achieved a decisive result for the modern theory of knowledge founding out that the concept of truth cannot be replaced by the concept of demonstrable.

The class of the true sentences is much bigger than the class of the demonstrable ones.

In another context the results which Gödel achieved have a meaning for the dialectical logic.

The fact that in the T theory are true sentences which cannot be demonstrated within the theory can be explained as the fact that this theory is not complete.

This theory can be completed by enlarging the axiomatic system, by raising the undemonstrable but true sentence to the rang of an axiom, and adding it to the other axioms.

A change in the axioms of an axiomatic system drives to the change in the meaning of the predicates and of the relations in the respective theory.

But it is also possible to formulate in the new theory a sentence which is true, but which cannot be demonstrated in the new axiomatic system.

The new theory is larger but it still is incomplete.

Each of these theories are representing only a relative truth which can always be replaced by a larger relative truth, which is on a higher ground, so as we can build a succession of theories.

1.4.The theory of types. The theory of the logical types derives from the principle of the vicious circle; the paradoxes are the result of the fact that it is assumed a collection (set) of objects can contain members which can be defined only with the help of the total collection.

The paradox of the class of classes which do not contain themselves as elements is solved by Russell like: a class is an object which derives from a proportional function f(x)and which assumes the function.

Then a class cannot be the argument of the function which it defines.

So the argument of a function cannot take arbitrary values and the values which it can be assigned to this function are limited.

Russell establishes a logical hierarchy of the concepts, meaning: individuals, logical objects of type t_0; the proprieties of the individuals, concepts of type t_1; proprieties of individuals, concepts of type t_2, etc.

The theory of steps establishes that a type **n** concept can be applied only to a type **n-1** concept, so in the proportional function **f(x)**, the argument **x** can take values only of type **n-1** if **f** is of type **n**.

Russell establishes in a similar way different kinds of truth: when we announce the true or false statement of a sentence **b**, as an example **"p is false"** a kind of truth is used; when we say "p is false" and it is a false sentence another kind of truth is used.

The theorems found by logicians show that the value of true of a sentence built in a logical system **S** cannot be proved in the system **S** but in a metasystem **S'**, which refers to the sentences of the **S** system, but which is not the same as **S**.

1.5. Metalanguages in the theory of semantical steps An equivalency, considered as an intersentenceable connection is the component of a object-theory, which assumes something about the proprieties of the objects of knowledge.

An equivalency composed by intersentenceable connections (semantical equivalency) belongs to the metatheory which says something about the knowledge of the object-language.

One of the basic notions of the semiotics is the theory of semantical steps, which begins from the existence of the objects, proprieties and relations which are belonging to the objective reality, and are not signs of any language.

This sort of objects form step zero and the signs which are these objects noted with belong to a object-language or first step language.

A metalanguage, or a second step language consists of all the signs which are necessary for the characterisation of the signs of the object language.

Whenever an affirmation about Metalanguages is made it is used a third step language, and so on.

The difference between object languages and Metalanguages is all about differences between object-theories and metatheories or second step theories and so on, meaning a hole hierarchy of theories.

Objects of step zero are the basis of all the succession of steps of knowledge from the semantical step theories.

2.The convergence of axiomatic metatheories towards the unique non-decidable sentence of maximum generality.

In this paper we shall define the diachronic space as the set of diachronic levels (fig. 1):
$$\{...N_{-i},N_{-i+1},...N_{-1},N_0,N_1,N_2,...N_{i-1},N_i,N_{i+1}...\}$$
in which the subset:
$$\{...N_{-i},N_{-i+1},...N_{-1}\}$$
represents the diachronic levels corresponding to objects from virtual world, and the subset:
$$\{N_1,N_2,N_3,...,N_{i-1},N_i,N_{i+1}...\}$$
represents the diachronic levels corresponding to objects from real world.

The level N_0 is corresponding to the level of the absolute question.

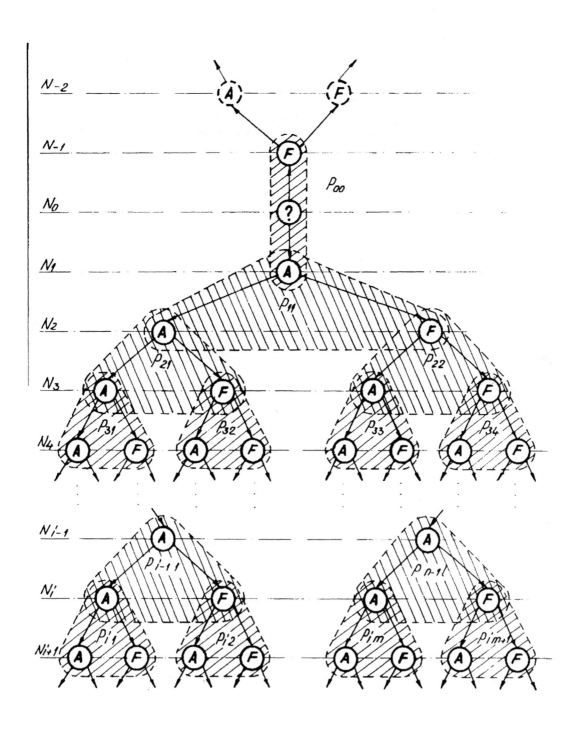

Figure 1 The structure of truth in diachronic space of sentences of axiomatic metatheories

From the definition given by Wang Hao, in which a theory is a body of concepts and a set of assertions as well as from the law of logic from Port Royal, in which it is assumed that a notion is characterised by two attributes: extension and intention which are in a reverse proportional rapport, withdraw the conclusion that the finite number of sentences from a theory is smaller than the number of sentences within an axiomatic metatheory.(fig.2)

The concepts of truth, in this article, is assimilated with a biarborescency(fig.1), where on diachronic levels corresponding to the truth types are structured: the axiomatic truth (level N_1) and the relative truth (levels $N_2,...,N_i,N_{i+1},...$)

The configuration of axiomatic sentences, theories and metatheories which converge to the nondecidable sentence of maximum generality p_{00} is mathematically modelled by generalised array of sentences, (fig.2, where p_{ij} means the theorems and metatheorems sentences).

p_{00}
p_{11}
p_{21} p_{22}
p_{31} p_{32} p_{33} p_{34}
p_{41} p_{42} p_{43} p_{44} p_{45} p_{46} p_{47} p_{48}

$\quad \cdot \qquad\quad \cdot \qquad\qquad\qquad\qquad\qquad\qquad\qquad\quad \cdot$

$\quad \cdot \qquad\quad \cdot \qquad\qquad\qquad\qquad\qquad\qquad\qquad\quad \cdot$

$\quad \cdot \qquad\quad \cdot \qquad\qquad\qquad\qquad\qquad\qquad\qquad\quad \cdot$

$p_{i-1\,1}$ $p_{i-1\,2}$ $\cdot$ $\cdot$ $\cdot$ $p_{i-1\,1}$ $p_{i-1\,l+1}$ $\cdot$ $\cdot$ $\cdot$
$p_{i\,1}$ $p_{i\,2}$ $\cdot$ $\cdot$ $\cdot$ $p_{i\,m}$ $p_{i\,m+1}$ $\cdot$ $\cdot$ $\cdot$
$p_{i+1\,1}$ $p_{i+1\,2}$ $\cdot$ $\cdot$ $\cdot$ $p_{i+1\,n}$ $p_{i+1\,n+1}$ $\cdot$ $\cdot$ $\cdot$

$\quad \cdot \qquad\quad \cdot$

$\quad \cdot \qquad\quad \cdot$

Figure 2. The generalised matrix of the axiomatic metatheories sentences

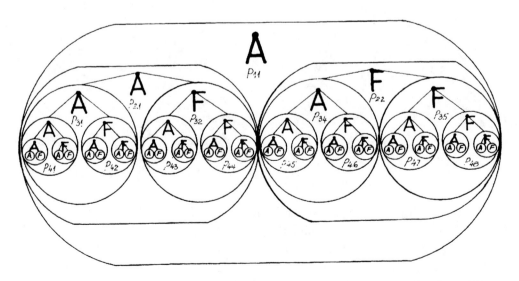

Figure 3 Diachronic structure of truth levels and sentences of axiomatic metatheories

Introducing in the semantic levels theory and the truth type theory made by Russell the principle of convergence, towards the axiomatic truth, it is shown in fig.1 and fig.3 that the number of sentences, theories and metatheories is decreasing pointing to the unique nondecidable sentence of maximum generality p_{00}.

From fig.3 where the diachronic structure of the levels of truth and of the axiomatic sentences, theories and metatheories we see that the structure of the types of truth accepts a fractal interpretation which allows the explanation of the convergence of the sentences, theories and metatheories towards the nondecidable sentence p_{00}.

Conclusions:

1) By assimilation of the true sentence from an axiomatic theory which cannot be demonstrated with axioms, it can be obtained an axiomatic metatheory in which another true but undemonstrable sentence can be formulated.

2) The assimilation of the new true but undemonstrable sentences with axioms leads to the obtaining of another axiomatic metatheories.

3) From fig.1 and fig.3 we can see that by the increase of extension of notions from within the theories and metatheories, the number of sentences, theories is decreasing, pointing to the unique nondecidable sentence of maximum generality.

References

1. Aristotle - Organon II, Prime Analitics, Scientific Publishing House, Bucharest.

2. Klaus, G - Modern Logic, Scientific and Enciclopedic Publishing House, Bucharest, 1977.

3. Leibnitz, G.W - Monadologie, Fragmente zur Berlin.

4. Mirita, I - Quadridimensional Interpretation of Syllogistic Inferential Processes in Polyvalent Logic, with a View To Structuring Concepts and Assertion for Realising The Universal Knowledge Basis-Knowledge Transfer, London, 1996. Edited by A. Behrooz.

5. Wang, Hao - Studies of Mathematical Logic, Scientific Publishing House, Bucharest, 1972.

INFORMATION, CAUSALITY AND SELF-REFERENCE IN NATURAL AND ARTIFICIAL SYSTEMS

Alvaro Moreno Bergareche
Departamento de Logica y Filosofia de la Ciencia
Universidad del Pais Vasco UPV/EHU
Apartado 1249 / 20.080 Donostia/San Sebastian / Spain
Tel: 34-43-44 80 00 ext 5496
Fax: 34-43-31 10 56
E.mail: ylpmobea@sf.ehu.es

Keywords: Information, causality, constraints, self-reference, complexity

This paper deals with the problem of the relation between information and causality. We consider information to be some kind of structure that has a referential capacity or, simply, the referential content of that structure. Hence, information does not have to do with the physical properties of its material support, only with referential relations (we shall address this notion later). These, in turn, may be of two different types, either relations 1) between informational structures and the domain they denote (the world of meaning), or 2) among the informational structures themselves (the world of computation and formal systems). In case 2), informational relations may be seen as implementations of a mathematical universe where both the formal relations (syntax) and the referential capacity (semantics) have been extracted from the material support as pure abstractions, and, as a consequence, they might refer to different physical objects or events.

We speak of information when we deal with relations between structures which refer to other structures in such a way that there are no inherent physical causal links between them. Thus, we use informational terms when we deal with systems in which stable relations appear among components, in a way which would be contingent with respect to the physical nature of its material support. However, this basic fact is the root of two different, and even opposed, meanings of the term information. In artificial systems, information is a result of an external interpretation of certain relations among physical structures. The intrinsic material dynamics can be ignored in the description of those relations, they are independent of the materiality because it has been restricted so as to play no role. Accordingly, this type of causal link among events can only appear as a result of an imposition external to the system whose origin remains unexplained. In natural systems, on the contrary, information is internally interpreted and generates causal actions autonomously. This kind of causal information requires as a precondition a complex, operationally closed form of material organization. We will be back on this issue latter.

Even if the first sense of the concept of information -formal relations among inert entities externally interpreted- cannot materially exist without the second one, it is the most popular and widespread of the two, because of its fundamental role in science and technology, especially in computer science. In this domain the structures that support the information are strongly constrained, they can be described in informational terms insofar as its underlying dynamics is "frozen" -which amounts to the suppression of its dynamics for pragmatic purposes. Thus, these informational events emerge out of a rather strange kind of matter, where processes take place in arbitrary times and with arbitrary costs of energy. Disembodied information is, in the context of such systems, only a set of passive entities: they only do things pushed by (or following instructions coming from) an external, autonomous, human system.

CP437, *Computing Anticipatory Systems: CAYS--First International Conference*
edited by Daniel M. Dubois © 1998 The American Institute of Physics 1-56396-827-4/98/$15.00

This way, we arrive at two opposed conclusions. Nevertheless, they share some common ground. For many, information cannot have a causal power, because the world of informational processes is independent of time and energy, or more precisely, independent of the rates of change of time and energy. Yet, according to other authors, information captures the essence of the physical laws that govern material processes. In this case, material causality and computation would be the same thing, which makes possible to say, as Toffoli (1982) does, that "at each step the universe computes its next state". In fact, the first position is implicitly dualistic, because it supposes that the physical-material world and the symbolic-informational world constitute two separate and parallel universes, much as in the Cartesian view of the *res extensa* and *res cogitans*. The second position, on the other hand, is close to a platonic philosophy.

In short, this disembodied idea of information, usual in artificial systems, has become epistemologically indispensable, because of the predominance of Functionalism in many modern sciences, which try to explain systems in purely formal terms, ignoring or neglecting energetical and material considerations. However, this idea of information is ontologically incomprehensible, because it ignores its natural origins, and, in fact, has erased all its causal traces. As we will show in the following paragraphs, information is only understandable if we give account of its origins, i.e., in terms of what information actually does. Thereby, the problem of the origin of information is posed in causal terms.

But what does "informational causality" mean? We say that a relation is causal when there is a link between structures or events based on the material properties of the constituents. Taken literally, this formulation only admits one kind of causality: physical causality. However, we argue here that, in those natural systems where it exists, as in all biological systems, information establishes a new type of causal link. If this link were merely external to the system, as it is the case of artificial informational systems, we would not be allowed to speak of a causal relation (unless we jump out of the artificial system into the social system producing it). Accordingly, when we speak of the causal action of information as something different from the physical causality, we are in fact pleading for a plural vision of causality. The type of causal action of information is "formal" in the sense that it infuses forms, i.e., it materially restructures matter according to a form. *Materially*, because this process requires complex and specific aggregates of matter; *restructures*, in the sense that information brings forth and stabilizes possible -but improbable- complex organizations of matter (i.e. proteins, cells, pluricellular organisms); and *form*, because the essence of information as a causal element is nothing but a pattern, whose domain of possible structures is autonomous from dynamical considerations. Unlike material causality, informational causality implies a type of autonomous over-determination of matter in formal terms, which requires underlying levels of material organization, and occurs in arbitrary times and energy.

This may appear as contradictory with respect to the idea put forward above which states that matter has an implicit causal activity, i.e., that matter is as such, active (in aristotelian terms, this activity should be named 'material causality').Nevertheless, in some cases, such an activity can -even- build up a kind of systems in which it is generated a complementary and recursive interaction between certain records (that are the result of previous processes of organization which involve an enormous amount of systems and time) and a set of components in the individual system that are restructured according to the pattern of such records. Hence, these records briefly codify complex dynamical processes, i. e. they act as instructions.

However, some kind of formal causality may occur in the absence of records, too. This is probably the case of prebiotic systems in which a set of component production processes recursively regenerates its own raw materials. The essential feature of these systems is their operational closure, which, modified in several ways and with different degrees of complexity (introduction of non-linearity and other higher range concepts) appears in different models of prebiotic or primitive biological systems -like Kauffman´s autocatalytic sets (Kauffman, 1986) or the autopoietic systems of Maturana and Varela (Varela et al., 1974). In such systems certain components will finally constitute a network that can become stable in virtue of functional constraints which spontaneously emerge among components. This way, autonomously generated constraints (or complex boundary conditions, like selective membranes) restructure the lower level dynamics in such a way that they finally reproduce the conditions of their own appearance. Accordingly, we can consider that this capacity to (self)restructurate is a type of formal causality. But, as far as these systems (or at least their most primitive versions) lack informational components whose pattern is transferred to the dynamical organization of the system, this organization is only implicitly in-formed.

There are very restricted structural limits concerning the development of non informational functional systems (Moreno and Fernandez, 1990). In general terms, their robustness decreases as their complexity increases. Following an idea proposed by V. Neuman (1966), H. Pattee (1977, 1982) has argued that the development of complex systems in Nature is not possible without the appearance of informational self-descriptive structures internally interpreted by those very systems in which they perform their causal action. The crucial difference between living and non living forms of organization, (like dissipative structures or even functional prebiotic autocatalytic networks) lies on the capacity for open-ended evolution. This capacity lies on the fact that evolving organizations possess some components able to act as records which transfer to the individual organisms a set of patterns generated in a collective frame. These patterns are, in turn, autonomously interpreted as specifications or instructions by every individual organism. Once expressed, selective pressures discard much of such embodied patterns. Individual organisms cannot be explained unless we take into account that their organization is the result of such records which act as a certain kind of boundary conditions that selectively constrains those dynamical processes which constitute their identity. Thus, in functional terms, these records (or information) can be described as a kind of "freezing" of some internal constraints that become autonomous from the dynamics of the individual system in which they operate. In turn, this requires high-level forms of physico-chemical organization. This is so because the appearance of informational structures requires turning on mechanisms that connect these structures with other independent of those from the physico-chemical point of view (this is why the mechanism is a true code). The key of this mechanism of coupling lies on the self-referentiality of the system, since some of the products of the expression of information are the very elements which control the coupling mechanisms. That is why the causal action of the information in natural systems is internal. Therefore, the information cannot appear if it does not carry along the construction of its own framework of interpretation, that is, of the material mechanisms for its expression. (In its most simple form, the meaning of the information consists in the recursive construction-reproduction of the machinery necessary for its expression as metabolic organization).

Living beings are hence capable of generating by themselves the causal relations which permit their own existence. In these relations, information only exists as far as it exerts causal actions, while it disappears outside the system. The material supports themselves are generated, maintained and interpreted in the frame of a complex material organization which cannot exist without such information. At the level of a single organism, information works as a set of instructions which specify the constructive processes of the system. More specifically, the structure of nucleotide strings in DNA acts as a record for protein synthesis (through a code created in the evolution of life), and proteins act as highly specialized constraints on the chemical processes of the cell. Thus, biological information links two empirical domains whose material support is related but in a contingent way, not in an inherent physical one. Interestingly, the causal action of information allows on the one hand the robustness of the processes of self maintenance in the early living systems, and on the other hand, the increase of their complexity. In this sense, the appearance of natural systems with internal information can be considered as a result of the concurrence of certain material conditions. But, at the same time, this also implied the appearance of a new type of causality -formal causality- since it materially re-structures material subsets of structures in such a way that allows the generation of highly organized systems, whose levels of complexity are qualitatively superior to all previous systems.

Even if biological information is entangled in the dynamics of the whole system, acting through the aforementioned set of constraints, we should not forget that there is an essential difference between the informational components and the remaining components of the cell. This difference is that information resides in non dynamical and relatively stable material structures, which are independent of the rate of time and energy, and generated through a collective and historical process. In this sense, the informational domain is only contingently related to the domain of the dynamic organization where it is expressed. Then, in addition to referentiality, information presents compositional capacity: the construction of informational patterns is an open process, essentially independent of the dynamics of their material supports and of the very system where they exert causal actions. In functional terms, the main difference between implicit and explicit formal causality is that the former does not allow compositionality. This is probably the key mechanism for understanding the qualitative difference in the degree of complexity that informational systems show: compositionality of information permits storage and transferring of unlimitedly complex structures from one to another domain.

This is the reason why information cannot appear in the frame of a single organization. Its origin in Nature requires a link between a historic and collective dimension (the frame of Natural Selection, where information is generated) and the individual dimension of organisms (the frame of phenotypic expression and of genotype reproduction). In other words, the organizational requirements of life exceed the level of the single organism. They would include a meta-network where the individual networks constitute a structure of synchronic (competition) and diachronic (transmission of characters via reproduction) relations. There is a "bootstrapping" type of process between the setting-up of the individuals and the collective network where ultimately the information is generated, because the interpretation of this information occurs in each individual organism and constitutes it (Moreno & Umerez,1998). In fact, we can view the whole evolutive process as the continuous generation of causal relationships through records.

In conclusion, we may say that information is autonomous causal action, embedded in the evolutionary process, where it is both a product or effect and a cause. Information could only arise as a causal element, entangled in the maintenance and evolution of complex material systems. Human beings, the most complex product of this evolutionary process, have been able to develop, through an inverse process of disembodiment, a purely epistemic, non-causal idea of information. Useful as this may be, we should not forget that it derives from, and implies, ontological, causal information.

Acknowledgements: This research was supported by the Research Project Number PB95-0502 from the DYGCIT-MEC (Ministerio de Educación y Ciencia, Spain) and by the Research Project Number UPV 141.226-EA from the University of the Basque Country.

References:

Kauffman, S. (1986): Autocatalytic Sets of Proteins. J. Theoretical Biology **119**. pp. 1-24.

Moreno, A. & Fernandez, J..(1990): Structural limits for evolutive capacities in molecular Complex Systems *Biology Forum* **83, 2/3** pp. 335-347

Moreno, A. & Umerez, J. (1998): Downward Causation at the Core of Living Organization. P. B. Andersen & N. Finnemann (eds.) *Emergence and Downward Causation.* Lawrence Erlbaum, Hillsdale, NJ (In Press)

Pattee, H.H.(1977): Dynamic and Linguistic Modes of Complex Systems. *International Journal of General Systems*, **3**:pp. 259-266.

Pattee, H.H. (1982): Cell Psychology: An Evolutionary Aproach to the Symbol-Matter Problem. *Cognition and Brain Theory*, **5 (4)**: pp.325-41.

Toffoli, I T. (1982): Physics and Computation. Int J. Theor Physics, **21** pp.165-175

Von Neumann, J. (1966): *Theory of Self-reproducing Automata*, (A. Burks Ed.) Urbana: University of Illinois Press.

Varela, F., Maturana, H. & Uribe, R. (1974): Autopoiesis: The Organization of Living Systems, its characterization and a model. BioSystems **5** pp. 187-196.

Anticipatory Systems
Modeling and Control

Intentionality, Self-Reference And Anticipation

Pere Julià
Institut Mediterrani d'Estudis Avançats
Spanish Council for Scientific Research / CSIC
Ctra. de Valldemossa, Km. 7.5
07071 Palma de Mallorca, Spain

Abstract

The conceptual analysis of anticipation inherits all the difficulties associated with intentionality. These can be reduced to: (1) the unwillingness to recognize the dispositional nature of the organism-environment interaction; (2) a reluctance to treat language and perception as bona fide forms of activity, thus blocking an effective approach to self-reference, among other things; (3) the general tendency to accept the ascriptional mode of research as a way out of these difficulties. The stage is then set for the assimilation of animal and human behavior studies to machine simulation; also, for the unnecessary reductionism and largely speculative physiologizing that go with mind-brain analogies in lieu of a holistic approach to the facts. This paper argues for a naturalistic approach. Anticipation is about "bringing the future into the present" as part of adaptation; it involves special activities of a cognitive nature, sometimes automatic and probably shared with other species, at other times self-generated and only possible due to some unique properties of human language. Anticipation is best conceptualized as a component in self-regulation. An account in terms of "etiology" serves to recast the notion of "context"; the importance of conative factors is stressed. The emphasis is on experimentally determined variables and processes rather than on abstract idealization or post facto reconstruction. Incursion and hyperincursion techniques are discussed in connection with a possible formulation centered on the synergistic influence of environmental and organismic factors.

Keywords: Conation, dispositions, etiology, hypostasis, self-regulation.

1. Intentionality and Complexity

1.1. Intentionality – Mind – Brain

Intentionality (fr. L <u>intendo</u> to point at, extend toward) has a venerable pedigree, generally traced to the scholars of the Middle Ages. In its modern phase, it was defined by Brentano as "the mark of the mental". Mental states are said to be directed at objects and events in the external world; the intentional is "that property of consciousness whereby it refers to or intends an object"; "the intentional is not necessarily a real or existent thing but merely

CP437, *Computing Anticipatory Systems: CAYS--First International Conference*
edited by Daniel M. Dubois © 1998 The American Institute of Physics 1-56396-827-4/98/$15.00

that which the mental act is about". It is presumably this characteristic that permits a distinction between, say, intentions and beliefs, and diffuse forms of anxiety or elation.

The general agreement nowadays is that intentionality has to do with aboutness. Disagreements arise, however, when it comes to specifying its defining characteristics. Some might argue, for example, that intentional phenomena should not be confused with "doing something deliberately or on purpose". Mental states and events (ideas, beliefs, hopes, expectations, fears, perception, hallucinations, and so forth) are of or about something; like linguistic items (sentences, questions, etc.) and "other sorts of representations" (pictures, symphonic tone poems, computer programs, etc.), they are not something we do or, it follows, we are inclined to do.

In an effort to circumvent the attendant perplexities, experts turn to the language of intentionality. As it turns out, the presumed relations between intentional states and their objects do violence to the standard logic of relations. Questions then arise as to whether intentional relations are only apparently relational. Along with the assumed "aboutness", this, according to Dennett--one of the foremost writers on the subject--"nicely locates the problem". Nevertheless, the problem remains unresolved even though Dennett and others are aware that "(people's) beliefs involve quite different states of mind--they have different objects--and are to be distinguished from the state of mind of the skeptic or agnostic who can be said in quite a different sense to believe in nothing" (Dennett and Haugeland, 1987, p. 384).

Once we concentrate on the language of intentionality in lieu of intentionality itself, the problem seems to reduce to what Brentano called the "intentional inexistance" of some intentional objects. Intentional idioms themselves turn out to be "referentially opaque"; they lie outside the range of the predicate calculus. Tully and Cicero are frequent guests in this kind of discussion: substitution of 'Tully' for 'Cicero' in 'Cicero was an orator' preserves the truth value of the sentence, but substitution within the scope of 'believes that', e.g., in 'Tom believes that Tully was an orator', does not, for Tom may or may not know that Tully and Cicero were the same person.

In short, "an ascription of a particular intentional relation depends crucially on the words used in the clause expressing it" (emphasis added). The possible nonexistence of the intentional objects can be seen as a special case of opacity. This is no idle speculation, it is held, for the philosophical analysis of intentionality is of practical importance in discussing the theoretical foundations of the social sciences and biology: mentalistic theories are those, and only those, that make the use of intentional idioms ineliminable; their viability hinges on how we manage to construe the logical problems that come with them, coherently. But is the viability of the social sciences and biology a logical problem? And if so, for whom? The move from the phenomenon itself to the ways of speaking about it does nothing to clear up either the emergence of the inferred disposition, state of mind, bodily condition or what have you, or its relation to the external object. In the example about Tully and Cicero, we would do well to concentrate on Tom and his changing dispositions rather than on the truth status of someone's sentences about Tully or Cicero.

Similarly, a characterization of intentional events (sometimes deemed "states", others "episodes") as the "mark of the mental" merely relabels an old problem without advancing useful criteria for demarcating different kinds of dispositions: on such a view, fear and elation should be deemed as mental as, say, belief and intention. The real issue would appear to be not so much a matter of diffuseness as of privacy, which accounts for their forbidding status. Diffuseness and privacy are both relative matters, the analysis of which constitutes again the analyst's, not the subject's problem.

Assigning a kind of "pre-behavioral time" or mental status of unspecified dimensions to aspects of the emerging disposition accords well with the traditional view that mental events are out of reach, but it also stands in the way of accounting for relevant differences: fears, hopes, elations, etc. can be as diffuse or as concrete as beliefs, expectations, intentions and the like. The only way to track them down is to relate the initial or changing conditions in the organism to the independent variables that precipitate them at given times--assuming, of course, that we are not committed to the view that they just "happen". A priori restriction to internal events alone remains just that. Dispositions, covert or overt, rise and fall but they do not do so in vacuo. Naturally, they are all diffuse or concrete according to circumstance, but it is by looking into how they become more sharply defined that we identify the relevant processes as well (and the special role and place of language and cognition within the total episode).

That is a far cry from the usual kind of lightheaded inferential physiologizing, based on purely a priori and mainly linguistic characterizations, or the construction of equally a priori and largely metaphorical models of the mind or brain. The prescription is mentalism in a new guise--theoretical models and computer simulation for the time being, to be validated by neurophysiology some distant day in the future. The program is thus physicalistic in principle and likely to allay, in some quarters at least, possible fears of reintroducing age-old dualisms. But is there any reason to suppose that the relevant processes have been identified? Can we assume that simulation is synonymous with explanation, or that broad categories such as intentionality give neurophysiology useful clues as to how the organism is changed, at every step, in its interaction with the environment?

A word about privacy is in order at this point: given that mental events are deemed inherently private, on what grounds are the ascriptions made and relations posited other than verbal reports? And if so, whose, and again, on what grounds? If intentional idioms are to provide an index of anything, we must ultimately ask, who and under what circumstances is actually in the best position to provide them. The central role of privacy turns intentionality into a convenient substitute for the older "mind". Roughly, most of its mystery revolves around certain unique features of human behavior, a large part of which can take place:

(1) in the absence of the pertinent (usually the original) stimulus setting or "out of context", as we say;
(2) in covert form, that is, it may not be available for immediate identification (the largely redundant notion of internalization is often invoked in this connection);
(3) when, after acquisition, crucially important segments of verbal and perceptual behavior are no longer keyed only to the external world but to stimulation arising in the individual--as in consciousness and certain aspects of rationality; we will have occasion to come back to these unique properties of human language and cognition below;
(4) when, based on the above, human beings come to acquire self-knowledge, which constitutes a sine que non for a realistic approach to self-regulation or self-monitoring.

211

It is easy to show that mainstream recourse to representations and rules constitutes only an extension of the ascriptive mode of research, forced upon us by an inadequate appraisal of the interactional nature of behavior, including, naturally, language and cognition (Julià, 1983, 1994a, 1996b). If representations stand for anything like the view that "in order to be usable, relevant aspects of the world need to be encoded in symbolic form" (Clocksin, 1996), or any extant paraphrase to that effect, it would follow that we need as rigorous and complete a description of the relevant aspects of the world as possible, to begin with. This is precisely what the intentionalist, the psycholinguist, the cognitive psychologist, and the cybernetician (except when the latter is engaged in <u>practical</u> engineering tasks) will not do; preference is given instead to the development of criteria for the postulated mappings "to be encoded in symbolic form". But we are entitled to ask: Encoded by whom? And "used" for what? Mappings based on what and symbolic for whom? Failure to answer these questions results in arbitrariness and frustation (Eysenck, 1994, p. 295).

The same applies where mental activity is equated with brain activity in any of the numerous current proposals on the mind-brain conundrum. The plain fact is that even if brain activity is detected through any of the technical means at our disposal, we still have to account for the emergence, different intensities, degrees of definition and rates of displacement or disappearance of particular patterns with specific properties at a given time t. Here as elsewhere, the mere detection of activity constitutes only a first step. The equation of consciousness, thoughts, ideas, etc., along with other "intentional" episodes, with the patterns alone is, from an explanatory point of view, vacuous. Their emergence is clearly a matter of time and circumstance or, in functional rather than formal terms, the result of independent variables at work. Some generic rather than genetic questions come to mind: Where is somebody when he is out of his mind? What do we forget when we forget ourselves? What do we come back to when we come back to our senses? What do we do when we make a mental note of something? Given present understanding, it is foolhardy to allocate "parts and processes" in accordance with preliminary representational schemes, or to confidently speak of what the left or right hemispheres are doing or "telling" each other. Recent findings in neuroscience suggest rather more flexible parameters of brain functioning (see, e.g, Gazzaniga, 1995). Investigation of brain plasticity in these terms can be made more rigorous, if conceptualized as part of bona fide processes in which feedback, feedforward, and related concepts are seen in light of the full-fledged interactional nature of the subject matter.

1.2. Dispositions and Conation

There is an irreducible degree of circularity associated with any characterization of behavior--human or otherwise--that refuses to take into account basic experimental findings from the laboratory or the field, at least as a way to break up the prevailing vicious circle.

A frequent criticism of an experimental analysis runs as follows: if behavior must first occur for consequences to follow, there necessarily remain, ex hipotesi, important aspects of (human) behavior outside the scope of the formulation. How can we possibly account, e.g., for novelty, generality, or creativity? Even writers willing to acknowledge that the Law of Effect is here to stay, diagnose the problem as a reluctance to bring terms like 'want', 'belief', 'intend', 'feel', 'expect', etc. into the explanatory picture. Without intentional idioms no account of rationality or intelligence is possible; the "use" of these terms, it then turns out, presupposes intelligence or rationality, which, these writers contend, hold the answer to our original question (see, e.g., Chomsky, 1959; Dennett, 1981, 1983; Taylor, 1964).

Thorndike's Law of Effect has been mentioned in view of its generality (it can no longer be associated with any one school of thought) as well as its direct relevance to anticipation. As will be remembered, this law states that if an act is followed by "satisfying" consequences, the probability of its recurrence under similar circumstances increases; in short, the response has been "strengthened". It might be added that there is a close connection between the notion of a "satisfying state of affairs" and the "appropiateness" of the response at that particular time; appropiateness applies to internal as well as external conditions. Similarly, there is a sense of "belonging" or "pertinence" associated with the notion of a set of "similar circumstances".

The widespread contempt for consequences backfires, or yields less than cogent results, however, when the intentionalist turns to a biologically oriented, evolutionary interpretation of adaptation. As a technical term, 'consequences' has a somewhat checkered history; it is at any rate roughly synonymous with 'payoff', 'reinforcement', 'purpose', 'feedback', 'effects', 'goal', etc. all of which are similarly in need of qualification. Sapon (1972) has suggested 'subsequences' as a descriptively far more neutral term; his reasons for the replacement should be carefully considered.

The "insufficiency argument" fails to appreciate: (1) the probabilistic nature of behavioral activity, regardless of its dimensional properties; (2) the subtlety and diversity of the multiple variables at work; (3) the complexity of the resulting patterns of perceptual, linguistic and motor interaction in the human case; the latter are especially obvious when coping with novelty, difficulty or creativity (Julià, 1984, 1996a; for an updated account of basic research, see, e.g., Mazur, 1990).

Though often mentioned, learning does not play a major role in research that takes intentionality as a point of departure, or in those brands of cognitive science that merely pay lip-service to the environment. Wherever procedural dependence on AI is the norm, learning is in fact relegated to a mere counterpart of memory: learning has occurred if it is manifested in memory (i.e., if a given response or set of responses occurs at a later time) but not otherwise. This is perfectly in line with an outlook that, starting from faits accomplis, prefers to infer or, more accurately, invent processes and constructs to the systematic investigation of how the organism "got there". While consistent with a view of learning as a matter of "exposure", leaving conative and other contextual factors out of the picture results in an unwarranted idealization of the facts.

Anyone concerned with the legitimate question, How do dispositions emerge in the first place? is left with three choices: (1) dispositions are intrinsic to the individual; (2) dispositions are intrinsic to the individual qua member of a species; (3) external factors have something (some anachronistically would say "everything") to do with them. Given its hostility to basic experimental findings, the intentional outlook would favor the nativistic suggestions in (1) or (2), tempered perhaps by a specific writer's degree of commitment to an evolutionary outlook (e.g., Dennett, 1996). For all intents and purposes, the intentionalist has no alternative but to put the initiation of action inside the organism. That renders the environment irrelevant or, at best, collateral. Nativism is on the rise again.

The environment becomes inescapable even so: it is sometimes referred to as "context", if only for the flair of specificity this term connotes. It is interesting to note that "contextualism"--note the "ism"--is on the rise too. But context is too vague a concept to direct successful research strategy; hopes of achieving the required specificity in abstraction from consequences and other participating variables may well be classified in the wishful set.

Theories that do not countenance concepts more empirical than, say, input, output, processing, information, retrieval, and so forth inevitably confuse, or reify, the subject matter in the process of imposing an alien grid, however rigorous and useful this may have proved in an engineering context. Witness, for example, Massaro and Cowan's (1993) statement: "We must not underestimate the importance of the attempt by the IP [Information Processing] approach to account for the mental processes intervening between stimulus and response. Where behaviorism aimed to understand behavior, the IP approach seeks to elucidate the proceses that cause behavior". Such a statement reflects a misunderstanding of what is being studied and how. The conceptual and "causal" center of gravity has shifted from behavior to mental processes mediating the behavior under study, from stimuli "causing" responses to processes causing behavior. Still, what causes the processes causing the behavior? Why should we not consider the behavior itself a process, the result of converging external and internal stimulation?

Naturally, that applies to cognition and language as well, with due respect for the topographical and dynamic differences that set them apart from the rest of human behavior. Indeed, that's the rub; for it is owing to these particular properties that language and cognition can play the role they do within the integrated system, or set of systems, that make human behavior unique. The active nature of cognition has been gradually acknowledged in recent years (see, e.g., Llinàs and Churchland, 1996); the same cannot be said about language, to which we come back below.

The main outcome is the proliferation of essentially metaphorical schemes of reference. Thus, even first-class experimental workers seem compelled to adopt these terminological schemes as if their data were devoid of significance without some conceptual safety net, such as the philosophy of mind and intentionality freely offer. With the adoption of a conceptual framework comes also a way of doing research. For example, Grossberg (1995) summarizes his recent work in a subtitle that reads: "Neural networks that match sensory inputs with learned expectations help explain how humans see, hear, learn, and recognize information". Are we to assume that the "learned expectation" and "recognition of information" in a neural network is comparable to the actual conditions and processes prevailing when a human being is coping with the world? Surely something more than the organism's initial biological constraints and the physical properties of the incoming stimulation must be involved in a naturalistic account of inputs. What is actually processed? Why must we assume that outputs can only be accounted for through analogies with programs specifically designed for certain tasks and impervious to conative variables?, or that processing, storing, retrieval, learning, etc. are unmotivated?

Where behavioral processes are concerned, a view of learning by exposure is both conceptually circular and methodologically inconsistent. It leaves unanswered simple questions like, Why don't we learn just anything we are exposed to? Or, once learned (whatever that may mean under the terms of the formulation), Why do the pertinent dispositions tend to recur in similar or different ways under some circumstances but not others? To suggest that some environments are just more likely to potentiate certain initial conditions, for example, when language learning is likened to nest building or, in an extreme case, to the growth of an arm (Chomsky, 1975, and ever after) makes us no wiser about the structure of the organism (initial or otherwise) or of the environment; more important, neither have we learned anything about the interaction patterns that actually constitute the subject matter (Julià, 1983, 1996a, 1997; Verplanck, 1955). Is there a realistic way to approach pre-dispositions that does not begin with dispositions observable at some level? Both come in the way, and are in fact detected as, changing probabilities of occurrence given certain conditions. The systematic identification of these conditions is what has traditionally been understood as a satisfactory empirical explanation.

To sum up: contextualism is on the rise too, and that is a welcome break from formulations based on products and inferred processes and constructs, real or imagined. The question is whether the study of context, as currently conducted, will yield ever more concrete specifications of partial determinants of activity, in light of fairly well-established experimental methodologies; or whether context will be brought in, fudgingly as it were, in order to relieve the dissatisfaction that comes with the adoption of frameworks that take the machine-man analogy as a given (see 3.3). In the former case, processes like discrimination, generalization, abstraction, and so forth come to the fore, but then so do avoidance and other processes having to do with aversion (internal or external) in a large variety of subtle combinations. Their relevance to the task of making sense of the hodge-podge usually subsumed under "cognition" should be immediately obvious. Significantly, it was not among theoreticians bent on discussing the workings of language and cognition in abstraction, but among cognitive psychologists concerned with clinical and other practical tasks, that emotional variables began to receive serious attention.

The cybernetician or systems theorist approaching many of these facts in terms of feedback and subsequent elaborations thereof can only benefit from the empirical breakdown that obtains when environment-organism interaction fields are dealt with directly (see, e.g. Kantor, 1963; Wachs and Plomin, 1991; Mazur, 1994). Greater specification is bound to make for better use of postulated loop systems, if these systems are to serve as anything like a theoretical instantiation of the paths and effects, broadly speaking, of converging extero-, propio-, and interoceptive stimulation. In short, if we are to cut loose, to use an old frame (Skinner 1961), from the overwhelming "flight from nature" of the past few decades.

Conation is rarely mentioned in psychology textbooks or, for that matter, in sourcebooks on cognitive science and neuroscience and other readings on the mind. Yet, it is arguably the single most important factor (more reasonably, set of factors) to which we must appeal, if we are to break free from the intentional vicious circle. Conation has a philosophical past, but that cannot be the only reason for its neglect: most psychological terms do. Take, for example, cognition, generally taken to cover a broad and amorphous spectrum. Reber (1985, p. 129) writes: "A broad (almost unspeakably so) term which has been traditionally used to refer to such activities as thinking, conceiving, reasoning, etc. Most psychologists have used it to refer to any class of 'behaviors' (using that term very loosely) where the underlying characteristics are of an abstract nature and involve symbolizing, insight, expectancy, complex rule use, imagery, belief, intentionality, problem-solving, and so forth".

Runes (1945, p. 61) succinctly writes about conation: "L conatio, attempt; referring to voluntary activity". Similarly, Blackburn (1996, p. 72) states: "L trying. Conative aspects of

215

the mind are those associated with the initiation of action", which includes at least a reference to "initiation of action". As is often the case, Webster can be invoked to fill the bill: "Conation: L conatio act of attempting, fr. conatus, pp. conari, to attempt: "the power or an act of striving that may appear as conscious volition or desire or behavioral tendencies". This definition amply illustrates the extant ambivalence with respect to the initiation of action. Allowance is made at least (besides the etymological convergence on attempting, striving, trying, etc.) for the fact that conscious volition and desire can be treated along with behavioral tendencies, the etiology of which may appear to be the same.

So, we may safely assume that conation covers both motivation and emotion or, at any rate, what these two broad and complex fields jointly contribute to answering the all-important question, Why should organisms attempt, try, or strive to do anything, and, in the human case, why should they symbolize, use complex rules, solve problems, imagine, or engage in any intentional activity whatsoever? Further elaboration would be outside the scope of the present discussion. To go back to the brain, how can we believe that conative conditions and inputs are any less important than those we usually posit for cognition, indeed, that the latter can ever be accounted for without respect for the former?

To set cognitive and conative factors apart is merely to perpetuate well-entrenched, fragmented approaches to the raw facts; it accounts, for example, for the largely artificial nature of most research on memory, which holds a central role in artificial intelligence work. Conation has a direct bearing on the crucial issue of automaticity and consciousness; it should also hold center stage in the study of adaptation and survival, which are the stuff of most anticipatory mechanisms.

1.3. Etiology Inside and Out

At issue is the origination of behavior. Clearly, a distinction must be made between the phenomena under investigation and the terms and tools we propose to deal with them: terms and constructs are warranted to the extent that they capture identifiable factors at work and promote a dynamic account of the subject matter.

Where intentionality is concerned, the conceptual and methodological bottom line would appear to be:

(1) the reluctance to appeal to the independant variables acting on the individual whose intentionality we are trying to characterize, even though, ironically, we are drawn into speaking of the (in)existence of presumably relevant "objects";

(2) a similar reluctance to recognize the proper role of language and perception, jointly making up cognition, which people turn upon themselves, more often than not as part of broader fields of interaction.

(3) ascription as a procedural way-out, on the assumption that intentionality then becomes "a feature of language, rather than a metaphysical or ontological peculiarity of the mental world" (Blackburn, 1994, p. 196). Such an outlook, like the insistence on "aboutness", merely reintroduces the Other Mind's problem; one

form or another of dualism is bound to follow, oscillating between greater emphasis on mental or physical substance. Hence, the renewed debate centers on mentalism or its elimination in favor of (a somehow unmotivated) brain activity.

Relations are then postulated between objects (why only objects?) "out there" instead of trying to reconcile their role with respect to the individual's waxing and waning dispositions. In so doing, we eschew a confrontation of the private nature of intentional events.

Directedness, casually proposed by Brentano, provides an entering wedge to the etiology of the phenomena under investigation. It is not enough to acknowledge directedness tout court; the further step is to ask, Directedness toward what and why? We can dispel the usual philosophical quibbles associated with causation, by substituting "How does an intentional event come about?" for "Why?". Questions about ontology and aboutness give way to a productive distinction between kinds of intentional events, not only, say, belief vs. intention, knowledge vs. belief, desires vs. wishes, anxiety vs. fear, and so forth, but also among kinds of intention, kinds of belief, knowledge, etc. based on the participating variables inside and out (Julià, 1986, 1996a). In short, directedness forces us to face the dynamic properties of dispositions, opening the door to an account of emerging complexities (from simple intentions to complex decisions) rather than speculating on the basis of our own tendencies and history--a practice variously known as projection, excessive theory-ladenness, subjectivism, idealism or, where infrahumans and machines are concerned, anthropomorphism.

It is through the direct study of the relevant factors and the identification of the processes they define that we can avoid the tangles of the mental-physical divide, as we do in any other branch of empirical science. As an elaboration on directedness, etiology captures the problems associated with diffuseness by doing away with the implied all-or-noneness of intentional events. Nowhere is it suggested that either the participating variables or their effects operate on a yes-no basis.

The literature already registers a number of objections to a functional interpretation of intentionality such as the present. They can be generically grouped under the following headings:

Deliberateness illustrates the sorts of inconsistencies and culs-de-sac to which a speculative, formalistic approach almost inevitably leads. While consciousness and volition figure prominently in most discussions of intentionality, some writers would rule out deliberateness as an inherent part of it. That is a case of the adherence to the all-or-none unstated criterion just mentioned. Yet failure to take deliberateness into account stands in the way of differentiating, for example, between acting knowingly or unknowingly; it would lead us to treat the innocence of a child and the shrewdness of the adult on the same level; legal distinctions like "causing to die" vs. "killing with intent" would also become meaningless; we would miss the subtleties of the gradation between preference and choice, choice and decision, decision and resolve, etc.

Deliberateness is of course a matter of degree--the result of complex converging variables that make all the difference between inaction due to paralysing doubt, simple hesitation, and determined action. Degrees of deliberateness reduce to degrees of consciousness, where differences in conation generally make the j.n.d. Philosophical hesitation on the subject probably revolves not so much around deliberateness as around actual deliberation. While it would be a mistake to equate the latter with consciousness, a measure of consciousness must be acknowledged wherever deliberation, or deliberateness, do indeed obtain.

First-Person Reports constitute another bone of contention. Some would argue for their qualified validity; others would rule them out as unreliable: people may lie, may simply not know, or may be mistaken about their feelings, tendencies, factors influencing them, and so forth. Much psychotherapy has to do with clarifications thereof. While this is all too true, the alternative is no more promising, as we have seen. Nothing is gained by reading our own tendencies and reasons into the subject's or by concentrating on an overdose of disembodied linguisticism.

Content. Brentano's formulation of intentionality gave birth to another persistent source of disagreements: mental states (sometimes revealingly called "philosophical states") are said to have "content", formally captured by the embedded 'that'-clause in 'X believes that' ..., 'knows that' ..., etc., which make for opacity of treatment, as we saw. Other than in the first person, intentional idioms ride on, and perpetuate, a view of language and cognition as something other than forms of activity, thereby fostering all the intricacies we associate with semantics. In any event, whether we are dealing with cognition or motor behavior, activity alone can scarcely be taken as a criterion. This is the essence of an etiological approach, in contradistinction to the treatment this problem receives in philosophical action theory (e.g., Davidson, 1980). The standard example--or counterexample--here is the movement we commonly call "raising an arm"; the point is that we can raise an arm for many different reasons. To say that behavior occurs is trivial but the philosophical counterexample becomes a paradigmatic example, if we are willing to acknowledge the fact that the "same" response can be a function of different variables. Naturally, the same applies to covert forms of activity, often remaining inchoate or truly microscopic. For obvious reasons, etiology becomes particularly important in connection with phenomena covered under (1)-(4) above.

Explanation lies with the "reasons" why something occurs: we can raise an arm to reach for something on a shelf, to hail a taxi, to call someone's attention, to counteract stiffness or pain, to threaten, to stop someone, to make a point, etc. These are instances that could look the same to a bystander who knew nothing of the circumstances, but we could hardly classify them together; to know their meaning is to know the circumstances, which may indeed be outside the range of immediate interpretation; we may then have to ask the Other Mind, who may or may not know. Moreover, we are likely to raise an arm automatically rather than consciously, and the subject, if asked, might have to do some self-searching. At any rate, we say that each of these movements "means" or "stands for" different things, which need not be the same for the observer and the observed. Thus misunderstandings arise.

Yet, we rarely speak of "content" with respect to arm-raising: we specify the meaning in any specific case by pinpointing at least one of the variables, however casually (as in the examples above). Why should we then insist on imbuing cognition and language with content? This sends us to the hypostatical nature of linguistic "facts" and the problematic use we make of descriptive schemes worked out for linguistic products, which are later extrapolated to perceptual activity, where external products or traces are absent.

2. Self-Reference and Experience

2.1. Languages as Systems

Underlying the controversy on the status and significance of self-referential language is the perennial confusion between <u>traces</u> (products) of behavior and the <u>behavior</u> itself. This confusion is enhanced in the verbal case by the availability of handy transcriptive and recording means. Structural/formal analysis takes these traces at face value and strives to make sense of them. The outcome is a conception of language as a self-contained, independent system of forms, which can presumably be described without explicit appeal to speakers, listeners, and the circumstances under which they interact.

Accordingly, language acquires the status of a tool or instrument people "use" to convey their thoughts, ideas, feelings, etc. But again, that only raises the further questions, What gives rise to these thoughts, ideas, feelings and so forth? How do they come to be "conveyed" the way they do? What do we really know about their fate once "conveyed" or "transmitted"? To speak of communication without further qualification perpetuates an age-old mirror image of speaker and listener (or writer and reader) that goes from thoughts, propositions, ideas, feelings, etc. to form mysteriously emerging in the speaker and then, just as unaccountably, going from form into thoughts, propositions, ideas, feelings, etc. in the listener. However convenient from the standpoint of theoretical linguistics and engineering information theory, this hardly depicts what is going on as people understand and misunderstand one another. Unnecessary complications ensue, of which reference and self-reference are prime examples.

The stress on syntax and the recourse to recursive techniques have turned linguistic analysis into a combinatorial study of forms (Julià, 1983, 1994a). Whatever their intrinsic value in formal logic and mathematics, techniques derived from recursive function theory and allied fields cannot be transferred to the study of language performance tout court; here as elsewhere, performance is all we have to go by. (Quite apart from the sheer artificiality of the examples usually offered, arguments put forth to illustrate infinity, grammaticalness and so forth could have been better handled in terms of partial ordering.) Paradigmatically, Chomsky et al's speculations on the competence-performance distinction have had an almost ideological impact, via psycholinguistics, on mainstream cognitive science; inspired by AI, the latter wittingly or unwittingly confuses the concepts and techniques brought to bear on cognition with the tools appropiate to engineering cybernetics (Julià 1983, 1994b). At issue is ultimately the relevance and tenability of the computational approach itself (for powerful indictments from a different perspective, see, e.g., Searle, 1992; Horst, 1996). Revision passes through the clarification of the hypostatical nature of linguistic data.

Children come to respond verbally to stimulation arising from their body or their own behavior, and sometimes to the variables which give rise to it, as they come to respond to any other feature of the environment: because doing so has consequences. Ironically, self-knowledge has a social origin. This is more than an conjecture: an extensive experimental literature gathered from a broad spectrum of conceptual positions on the establishment and modification of such repertoires is at hand (see below).

Thus, self-reference (better, self-referring) is a kind of behavior whereby the speaker talks about himself or herself. What can it mean to say, however, that a <u>sentence</u> refers to itself? Or that it is self-embedding, or right-branching, and so forth? In order to press their point, some writers bring in humor, newspapers reporting on their own activities, arguments about arguments, etc. But there is, of course, some person or group of persons behind every such instance, as 'The purpose of this paper' or 'Please publish me in your collection of self-referential sentences' and dozens of similar examples immediately suggest. What can 'This prophecy will come true' (which prophecy?), 'I am not the person who wrote me' or 'I am not provable in axiomatic system S' possibly be if not word games? (to be distinguished, perhaps, from the latter Witgenstein's "language games").

Even Gödel's major achievement can be used to advantage. After the preceding examples, Hofstadter (1981) typically goes on to argue: "Part of human nature is to be introspective, to probe. Part of 'verbal behavior' deliberately, often playfully explores the boundaries between conceptual levels of systems. All of this has its root in the struggle to survive, in the fact that our brains have become so flexible that much of their time is spent in dealing with their own activities, consciously or unconsciously. It is simply a consequence of representational power--as Kurt Gödel showed--that systems of increasing complexity become increasingly self-referential".

We have grown accustomed to this kind of hyperbolic discourse, where it is intimated that things simply happen; what is more, they happen because our brains are the way they are, that is, in line with our representational devices. Thus, it is "verbal behavior" that explores boundaries, consciously or unconsciously; this verbal behavior becomes self-referential, and this is all a consequence of "representational power" itself. In the process, speakers and listeners have disappeared altogether. Representational devices prove all too convenient, indeed necessary, in the study of servo-mechanisms, the language of which is larded with metaphorical expressions of this sort, probably without detriment. It is otherwise where behavior is the subject matter.

Many of the basic problems underlying second-order cybernetics, which relies heavily on self-referential language, can be similarly traced to formulations where the behaving individual is simply lost sight of; discussion proceeds on the basis of idealized putative outputs. Under the circumstances, individuals and systems conceived in their own terms--say, von Ferster's "observed vs. observing systems", Pask's "purpose of a model vs. purpose of the modeler", Varela's "controlled systems vs. autonomous systems" and so forth (Brier, 1994; Umpleby, 1994)--may not be immediately available for conceptual reconstruction.

Geyer (1996) refers to the mélange as well and intimates, or so I read him, that the subtleties captured in autopoiesis, auto-catalysis, self-organization, self-steering, self-reference, etc. may not be amenable to a streamlined treatment. To bring them all together as part of the "emerging sciences of complexity", however, in the hope that the proper mathematical tools will come along that will extricate us from the emerging difficulties, may be only a way to postpone a head-on confrontation of the underlying assumptions. After all, as Geyer also points out, first-order cybernetics fully recognizes nonlinearities. By the present reckoning, the problem must lie elsewhere.

Hypostasis is ultimately responsible for the reification of language as L; it invites the unabashed treatment of this fundamental aspect of human behavior as a formal object instead of the natural phenomenon it obviously is. As a result, the relevance of rules imported from other fields is merely taken for granted (Julià, 1983, 1994b; Sapon, 1971). Rules provide a comfortable means for the postulation of relations, the elusive nature and arbitrary characterization of which suggest more than a metaphorical status. It is self-delusion to discuss games or any other intricate system of rules as self-reproductive, self-destructive, self-amending, or as an exploration of "the reflexivity of the law", whether in the legal or logical context. We speak of a language without speakers and listeners as "dead"; likewise, without subjects or players, laws, systems of rules and games are merely inert marks on paper.

In practice, representations soon become assimilated to the data. Metatheoretical discussions revolve around them rather than about the original events they are presumably employed to depict. Language learning becomes a matter of "internalizing" (on simple exposure to a never defined verbal environment) the rules of the ambient language--trivially, since a child can do so thanks to its innate knowledge of their nature to start with. No language learning would be possible otherwise, we are told. Reification reaches its apogee when the study of language learning is deemed a matter of a priori determination of "what a natural language must be like if it is to be learned" (sic).

Meaning is always the meaning of a speaker or listener. In the final analysis, it is by a process of metonymy that we assign meanings to words and other linguistic constructs, the product of real people's activity. It is quite easy to show that 'X says that'--referred to an utterance taken at face value or to a person without going into the reasons for uttering it-- inherits all the difficulties associated with intentional idioms like 'believes that', 'expects that', etc. Resolution is less a matter of logic than of etiology (Julià, 1995b).

Insistence on distinguishing between behavers and the the traces of their behavior may seem trivial, but many a paradoxical collocation of words and the corresponding situations is thereby clarified: keeping the distinction prevents the reefication of words, sentences, and the like, alone or in context, in monolingual or bilingual combinations. Sentences are sound waves or marks on paper designed to have a given effect on the listener or reader; a similar analysis applies to paintings of paintings or any deliberate or playful crossing of levels, as in some forms of humor. A standard favorite like 'This sentence is false' could of course occur as a genuine form of linguistic performance, e.g., as a response to another sentence and the "corresponding" state of affairs someone meant to describe. But this is clearly very different from "the same sentence" in isolation, cast as a paradox, whether of the logical or semantic type.

Good science should not be so leery of paradox as to become paralyzed before the fact that linguistic "objects" and their names look the same; they "designate their object not by describing it but by picturing it", as Quine (1940, p. 26) characterized hypostasis. Logical devices such as use vs. mention, language vs. meta-language, formal vs. material forms of speech and so forth are ways to capture, or dodge, some of the problems attendant on this anomaly, which should not be confused with self-responding as defined below.

2.2. Language Functions

As the record shows, the study of language in abstraction from "use" cannot elucidate how we come to refer to ourselves any more than it can explain how we refer to the world at large. It takes function to do that and the recognition, as already noted, that language as well as cognition are forms of activity. For starters, they are sensitive to the effects of fatigue, chemical substances, conflict, emotional and motivational conditions and so forth, like any other behavior.

The concept of function is ambiguous even if functionalism has become all-pervasive. By and large, in its variegated uses 'function' generally denotes postulated relations between forms and forms, inferences from products, or reconstructions after the fact; cognitive processes and related constructs, modules and the like are hypothesized on the basis of theoretical possibilities, and components are postulated on similar grounds. But without the benefit of direct study, they remain hypothetical and deprived of the "tangibility" that ensues from the direct study of activity itself.

Linguists, psychologists, anthropologists et al. have sometimes discussed, of course, what could be broadly termed the "communicative functions" of language, which come closer to the sense in which the terms 'function' and 'process' are employed here. This is not the place to go into the intricacies of the controversy, except for pointing out that even this kind of research has been concerned primarily with language as a system, details of which are then somehow "correlated" with human needs and external demands.

More relevant for our purposes, however, is another set of functions occasionally labeled by some psychologists as "regulative". Roughly speaking, we could say that communicative functions have to do with social control, whereas regulative functions have to do with self-control, generically understood. The latter are intimately tied up with privacy as discussed in Section 1 and with the topics of the present and following sections.

Regulative functions may participate in a variety of ways when we turn our behavior upon ourselves, that is to say, when we respond to stimulation from our bodies, dispositions and/or relevant variables, as in self-referring. It should be remembered that self-responding cannot be taken for granted: it occurs only on specific occasions and it is likely to involve a measure of perceptual activity, feelings, etc., often in connection with nonverbal activity as well (as in self-control). Most phenomena generally classified under standard headings (say, attention, memory, problem solving) can be profitably conceptualized in these terms. We are not dealing with all-or-none situations but rather with different degrees of engagement of the requisite neuroanatomical structures, depending on the importance or complexity of the ocasion or "context", the momentary dispositions we bring to it, the types and properties (for example, with respect to timing) of the possible consequences, etc. Both context and consequences can drive the self-responding, and eventually the self-monitoring, to the covert level, as they can inhibit the overt emission of any other interactive behavior (though not necessarily the basic disposition to behave). All of the above can, moreover, take place in the absence of the original or other pertinent settings, as in imagining, reminiscing, certain kinds of planning, deciding and so on, the very occurrence of which must still be accounted for.

These are the kinds of functions that make human language unique, in that it plays a sort of "catalyzing" role within the total behavior of the individual, allowing us, among other things, the possibility of transcending momentary conditions and promoting equally unique ways of adapting to novelty. Much research will be necessary on these "poor cousins" of language study, which goes well beyond the appeal to rules even in a wider sense than usual (see, e.g., Hayes, 1989; for a broader perspective, see, e.g., Banaji and Prentice, 1994). Learning and performance must be seen as part of an unbroken continuum.

As here understood, <u>function</u> stands for systematic changes in activity (whatever its dimensional properties or neurophysiological modality) corresponding to changing requirements in the environment--a view any experimental physicist, chemist or biologist would feel comfortable with. Competence models and similar abstract and idealized constructs are beside the point. The processes thus identified can then be used to interpret complex ordinary situations, say, in an initial account of reference, self-reference or anticipation.

2.3. Reference to Self-Knowledge

Reference (better, referring) is such an integral part of everyday experience that we can all too easily take it for granted. Closer inspection reveals certain important difficulties, however. Briefly, the basic clew can be schematized as involving the "relation between a name and the person or object which <u>it</u> names" (quotations in this section are from Blackburn, 1996, p. 323; emphasis added to highlight the underlying assumptions).

Other questions have to do with:

1. whether other semantic relations are to be brought in to elucidate the basic relations, typically, "between a predicate and the property <u>it</u> expresses" or "between a description and what <u>it</u> describes";
2. the relation between myself and the word "I";
3. whether we can refer to such things as abstract objects.

Militating against finding a solution are the following facts:

4(a) strictly speaking, only concrete objects can be considered as referents;
4(b) the same referents may evoke more than one word, phrase, etc.;
4(c) more than one referent may evoke a single expression;
4(d) terms like 'like', 'should', 'therefore' and many others prove impossible to pin on any one segment of the world.

As indicated, underlying it all is the hypostatical frame of reference within which these problems are cast. What is essentially being asked in (1) is whether the bond between words and sentences and their referents, or vice versa, can be explained in any other way than through the appeal to other relations, themselves unexplained, when in fact what we actually do in trying to puzzle out an answer is to bring in our <u>own</u> semantic input, as listeners or readers: pure formal analysis is a myth; we just do not discover ambiguity or concoct paraphrases in a language we do not know actively enough (see, in particular, Julià, 1983, Chs. 3 and 5). Similar observations apply to (2), to wich we come back later; (3) is reminiscent of philosophers bringing their best knowledge to bear on the question "Is knowledge possible?".

4(a)-(c) demand closer attention, if only to point out that they can best be compared to nonverbal behavior under similar situations. In this respect language is no different from other behavior: we can deal with a table in different ways, say, as a place to eat or as a place to write without thereby getting into ontological quibbles. We may call a table an "eating table" or a "writing table" and that is as close as we come to having a parallel for the "corresponding" motor movements involved in each case. The fact that English has 'desk' for the latter, leaving 'table' for the former, merely opens the subject of synonymy. But is it so difficult to find nonverbal "synonyms" as well? At issue once again is the bond between single or multiple responses and sources of stimulation, which may share a subset of properties and, in some cases, even only one.

Close examination shows, however, that if speakers and listeners have a similar enough history (and that implies, of course, environments as well), the problem simply doesn't arise: a given response is always part of a larger repertoire keyed in a variety of ways to segments of the environment and other repertoires. If problems of understanding do arise, and if paraphrase fails, then, significantly, we fall back on more basic responses and that implies their relational bond as well. As a rule, illustrated dictionaries include drawings where definitions alone might prove insufficient in conjuring up the visual response to the object or event defined. (We might note in passing that, for all his antipsychologism, this was essentially Frege's strategy when he characterized reference as a derivative notion.)

The basic question remains: if reference has to do with certain correlations between language forms and aspects of the world, How do these forms and aspects of the world (call them "impinging stimulation") come to be related? How do they cease to be when they do? What happens when the response is forthcoming in the absence of the stimulus? Or, for that matter, What is happening when the stimulus is present but the response is not? Some would claim, of course, that meaning and form are intrinsically related, in fact that the former is inherent in the latter. But the very existence of fields like etymology, historical and comparative linguistics, or even a cursory look at lexicography, give the lie to this outlook as well as to any proposal about innate ideas. They also provide a key to 4(d), but that would take us too far afield.

The discussion becomes a matter of <u>function</u> as defined above: (1) as with any other form of activity, we have to turn to the independent variables at work. It is by appeal to them that we can explain the strengthening of given responses, as parts of overall dispositions under certain circumstances; (2) this is, therefore, a matter of probability, where conative variables are inescapably involved and, in some cases, overriding; (3) the learning processes involved are those generally discussed in classic experimental psychology textbooks under the headings "Stimulus Control" and "Concept Formation" (not to be confused with "Pattern Recognition" as routinely dealt with; pattern re-cognition presupposes either innate knowledge or prior learning, as the hyphen plainly suggests); (4) one of the principal conative variables sends us back to the Law of Effect mentioned in 1.2. The consequences to which it refers have often been relabeled "reinforcement", not only by numerous psychologists but also by some well-grounded philosophers (e.g., Quine, 1960,1978); technical for 'reward', 'payoff', etc., 'reinforcement' is a term with a tortured history.

'Feedback' takes the edge off some of this history, but great care must be taken not to confound the difficulties associated with reinforcement (positive and negative) with similar restrictions concerning negative and positive feedback, closed and open loops and so forth. The term 'reafferent stimulation' has sometimes been proposed: with proper handling it should do good service in an account of adaptation conceived as changing dispositions, i.e, as complex dynamic sets of waxing and waning probabilities of action as a function of changes in external and internal stimulation. Before the issue is resolved, 'consequences' and 'payoff' can also be used interchangeably without excessive damage; the same applies to 'activity', 'behavior' and 'action'.

This becomes of primary importance when we turn to self-reference, self-categorization, consciousness, and self-knowledge. The processes here are the same as with respect to the external environment, the difference being that stimulation arising from our bodies, dispositions, and relevant variables takes over, as it were. Hence the private side of self-regulation; effective self-monitoring requires the smooth compensatory interlocking of external and internal cues.

The literature on this topic is far less homogeneous than when we deal with the external environment; the weight of philosophy is heavier and, of course, that is to be traced to the inaccessibilitiy of the internal sources of stimulation. It should be pointed out, however, that the standard view, which equates private with unobservable, is misleading, especially during acquisition (see infra). At any rate, whatever their frame of reference, few would dispute nowadays the close involvement of language in self-knowledge and consciousness; differences of opinion have to do with how this takes place and to what a degree language is responsible for the phenomenon (see, e.g., the relevant sections in Catania and Harnad, 1988; Hameroff et al., 1996; Zuriff, 1985; also, Snodgrass, 1997). Obviously this is an area in which a great deal more research is needed, but enough is known for a thumbnail discussion such as the present.

Much can be expected from recent technical developments in imaging technique, which permit a relatively peaceful invasion of the organism with a high yield. Their usefulness presupposes, of course, that we know what we are looking for, not just how to track down structural changes. Again, we are bound to pay closer attention to function and process when research is geared toward a holistic account than when we are guided by arbitrary segmentations that come with following traditional chapter headings and unwarranted reductionisms (see, e.g., McGuigan, 1987; Raichle, 1994; Lester et al., 1997; Kantor's 1947 penetrating critique remains amazingly instructive fifty years after publication).

A simple example may illustrate the present discussion. Preverbal children are with respect to, say, fatigue, no different from infrahumans: if they are tired, they act tired or go to sleep; the same applies to so many other conditions, say, sadness, ebullience and so forth. An intrinsic part of becoming normal children involves, however, the development of naming (to simplify matters) and eventually to refer to themselves as well. Numerous processes are involved but one of them is sure to figure prominently, viz., answering questions such as "What's the matter?", "Is Johny not feeling well?", "Where does it hurt?" and scores of others.

Parents and caretakers possess external evidence, e.g., the child's current behavior patterns, number of waking hours, previous level of activity, customary level of output, etc. If heat rather than fatigue is the problem, there is, in addition to the behavior patterns, the perspiration, the adults' own response to the ambient temperature, and probably a thermometer for good measure. How else is Johny supposed to learn about fatigue, heat, hunger, fear, joy, a dry throat, a full bladder, etc.? How else is he to come to know about his inclinations, strengths, weaknesses, anticipation, purposes, values and the like? Awkardly at first, no doubt, Johny will eventually report fatigue and the rest rather than just display them; and his language will be necessarily self-referential and increasingly complex. (Although it does take time, age is not the most important factor: adults adjusting to a foreign milieu go through essentially the same processes, where previous experience may not help).

Through questions like, "Why are you so tired?", "Have you been playing too long in the sun?" and so on, Johny eventually discriminates and reports different kinds of fatigue

225

brought about by different kinds of variables (not necessarily classified into the four types generally distinguished by experts, who have additional cues to go by). He reports on a world that is overwhelmingly private, but he does so with relation to a physical environment and a social world that "taught" him to do it in the first place. After all, different social environments encourage different degrees of self-responding of this type, whether as a matter of detail or frequency, as any hypochondriac quickly learns. Cultural patterns constitute an excellent source of data to which philosophers and AI devotees ready to claim intentionality and consciousness for machines and infrahumans could appeal with profit.

A concern with processes rather than post facto speculation helps us to answer the question, Why should we turn our language and senses upon ourselves as part of becoming mature members of a community? The simple answer is that it is important to others that we do so. The occupational hazards of a pediatrician would appear to be closer to those of a veterinarian than to those of a general practitioner. The technical literature on the effects of neglect, which can be hardly restricted to "physical" neglect, is unfortunately large and constitutes another excellent source of data.

Self-knowledge emerges in the form of elaborate repertoires such as these, which are eventually internalized in the sense of becoming covert or private, only to resurface if and when needed. We may ask ourselves, "Am I too tired and should I stop?", "What have I really done to be so tired?" and so forth as an echo of what others have said to us or might again ask us. Here again, the parameters are culturally determined. Owing to too much tension or confusion, the response may be so strong (so "pressing") that we speak to ourselves aloud; when feelings are strong enough, we may revert to others. Wisdom as a synonym for self-knowledge has long been associated with asking for advice.

(Ironically, this accounts for the benefits of rule-learning in the formal sense too: we go over our explicitly learnt multiplication tables, for example, when we discover that we have momentarily forgotten them; the same applies to grammatical or logical rules. If unable to recall, we look them up or ask somebody else. Analogies between arithmetic and natural-language learning proposed by some linguists (e.g., Chomsky, 1962) in support of a presumed "ceaseless creativity", and repeated like a mantra ever since, do anything but credit to the critical faculties of the behavioral science community).

The further step is self-regulation. The better we know ourselves--our dispositions, limitations and possibilities vis-a-vis the relevant demands of the environment--the more likely we are to take effective action or "do something" about either our external or, where possible, our internal conditions. Wisdom includes knowing how to monitor ourselves in work and in play, how to manage our needs, how to cope with the physical and social world and so forth. In short, self-reference alone would be solipsistic: self-reference lays the foundation for self-knowledge, which in turn makes self-regulation or self-monitoring possible. Jointly they provide the link between us as individuals and as members of the community. That makes cognition a first component in a "chain" ultimately referred to conation; self-knowledge short of action would be pointless. All theses activities come into play in anticipation.

226

3. Naturalizing Anticipation

3.1. The Call for Anticipation

Anticipation arises in connection with the inherent open-endedness of experience. Such is the nature of the interaction between organisms and the environment into which they are born, that the formers' initial conditions and subsequent history do not always equip them to cope with the unexpected, the unplanned, the new; sometimes problems arise also in connection with the expected, the planned, the foreseen. Creativity--a topic very much in vogue during the past few decades--might be said to cover any and all cases where the relations defining even a slightly problematic situation have been successfully "negotiated". Each creative event, like any effective adaptation, has an enriching effect on the individual's history, promoting, in theory at least, the chances of better survival. Creativity too is a relative matter.

The mechanisms that come into play are fashioned in different ways depending on the dispositions composing that history and the nature of the problems themselves. The organism is changed at every step of this interaction. It would be an oversimplification to speak of mechanisms or organisms restructuring or reorganizing themselves, if that were to imply no external forces at work at some remove. A changed organism faced with the "same" old environment presents a different interactional situation; the environment, or some environments anyway, are also changed in the process and that can hardly be held as irrelevant. A full description tries to conceptualize and measure the synergistic influence of organismic and environmental factors.

It would be equally misleading to speak of anticipatory mechanisms across the board, or to suppose that the search for a formal characterization is something we can do in abstraction from empirical data, or that any one set of techniques will work for the full spectrum covering machines, infrahumans and humans. That is the old cybernetic story. Were we to do so, the term 'anticipation' would carry a heavier conceptual burden than it can bear. It was even suggested above that anticipation can be seen as a paradigm case of intentionality. There is nothing new about invoking intentionality when faced with anticipation for they obviously share certain essential properties.

Stripped to their bare bone intentional events reduce to (a) basic strength, (b) directedness, (c) privacy, (d) first-person reports, where possible. Jointly (a)-(c) circumscribe intentional states as dispositional events, whatever the degree of neurophysiological involvement the actual strength (probability of occurrence) might take. There is no need to equate privacy with unobservability. Neither can diffuseness be taken as a disqualifying feature: diffuseness is only a case of unclear directedness, which provides the foundation for formulating the etiology of the event (presumably, diffuse consciousness is intentional too). On the other hand, (d) was intimated because it is immaterial in principle whether individuals can or cannot report their private condition: the corresponding dispositions remain inextricably intertwined with behavioral tendencies. It is on such grounds that we speak of intentionality to begin with. No matter how we look at it, individuals themselves are in the best position to respond to the pertinent stimulation. Our account of it can thus be much facilitated if a first-person report is available; if steps are taken to secure a certain reliability, such reports provide another entering wedge to the possible variables at work. A measure of consciousness concerning the basic strength, its quality, and the nature of its sources provides also some sort of cut-off point, say, between mere awareness and self-awareness, between self-reference and self-knowledge, between self-knowledge short of action and actual self-regulation.

The added tension that goes with anticipation as a paradigm case of intentionality is new: it captures the notion that cognition is for action, that it itself is a form of activity. And that provides the link between anticipation and purpose. The full extent of this proposal is embedded in the distillation of mind or "mental" proposed in 1.1, which provides the matrix for mapping out different kinds of "mental events" in lieu of the heterogeneous grab bag generally offered in such discussions.

As originally characterized by Rosen (1985), an anticipatory system contains a predictive model of itself and/or its environment, which makes it possible for the system to compute its present state as a function of the model's prediction. The question is how to endow the model with the resources to cope with the novel, the unexpected, the problematic. This should be in line with the preceding discussion, and all would be well with the world if we only knew what are the features of the world that the model must "predict". But that is just the problem, which cannot pass as a solution. We faced a similar situation, for very similar reasons, with respect to rules and representations. The primary concern is frankly the modelling relation proper: the form it takes, in fact, defines the anticipatory system. Since anticipatory systems are meant to encompass not only artificial devices but natural systems as well, Rosen naturally has reason to wonder about causality (how can the future affect the present?), falling back on Aristotle's "final causes" in order to account for "purpose".

Certain features immediately stand out in light of the discussion of self-reference and intentionality in preceding sections: (1) while the environment is mentioned as usual, it is not clear, as usual, what the term stands for--perhaps it stands for too much, in practice transforming its referent into one more abstraction; (2) the burden then falls on self-referential devices, where the referent is avowedly the model itself. As with intentionality, we are back with a confusion between the system worked out and the working-out itself: to paraphrase, we are speaking about our own behavior and not about any reasonably well-established relations between a specified environment, the properties of the evolving model, or the ways in which they interact.

But in the absence of a framework for the description of self-referentiality in either functional or formal terms (which probably accounts for the "and/or" above), such a model arrangement makes no room for the regulative functions of language; it could thus scarcely be expected to make full use of human resources, even if we could feed it anything like a description of "the relevant aspects of the world" or if it were possible to encode the parameters and properties of subsequent action, verbal or otherwise.

The upshot could be another chapter in the cybernetic story, where behavior--particularly human behavior--is assimilated to the machine model. This sort of assimilation goes back to Wiener, Von Neumann et al. and accounts for the eventual rise of second-order cybernetics, about which Geyer (1996) aptly asks whether it is not "a step too far". For a brisk and incisive reconstruction of this story, all the way to a possible third-order cybernetics, see Dubois (1995). Cybernetics soon lent itself to a reductionistic program, whereby the assimilation went beyond the machine-man analogy to the identification of the machine with the brain (Julià, 1995a, 1997).

In a series of publications, Dubois (e.g., 1993, 1994, 1996a, 1996b, 1996c, 1997) has discussed Rosen's initial formulation and has offered the concepts of incursion and hyperincursion as a way out of some initial limitations, unabashedly appealing to ordinary life situations (like taking an umbrella in anticipation of rain). We need not subscribe to the representational mode of research to appreciate the possibilities of Dubois' framework for a direct, unmediated formulation based on experimental data. We seek and determine control matrices, special parameters, and specific value ranges, but the locus of the variables and processes is always the subject and its milieu. In Dubois' view, anticipatory systems deal with the "why" rather than with the "how"; the latter is the domain of classical systems, which are recursive. Anticipatory computation encompasses potential final conditions; the nature of the variables that make up the "why" remains perforce unspecified.

Dubois pertinently wonders about the meaning of computing anticipatory systems in the absence of consciousness and intentionality. For his part, Rosen deems antipation what makes the difference between living and inorganic systems; on this view, there would be no difference between human adults and nonverbal children and infrahumans. That may be why, for all the subtlety of his logico-mathematical framework, Rosen finds that no formalism can make the "why" explicit, when applied to reasoning and intelligence. Rosen's liberality in this matter is reminiscent of the enthusiastic extension of intentionality and even consciousness to infrahumans and computers among some philosophers, for example, some proponents of the intentional stance. Clearly, this kind of discourse reflects the standpoint of experts looking on and trying to make sense of anticipation without taking into account the anticipating individuals' "point of view". Closer inspection shows that only if we restrict anticipation to verbal individuals equipped with self-responding of the sort discussed in 2.3 can we broach finality in a meaningful way.

Another distinction worth noting is that between external and internal events. Interestingly, Dubois refers to "internalized anticipation", which creates the conditions to meet particular events in the future. But, of course, internalization occurs only under some circumstances and not others; therefore, we cannot just take it for granted. ("Externalized anticipation" is a largely redundant concept, e.g. if the weather provides the immediate context; the need to internalize arises with respect to action in the future). At any rate, relativizing internalization makes for a less artificial approach to various kinds of memory than, say, the standard short-term vs. long-term dichotomy. Counting on enough self-knowledge on the subject's part, we should be as entitled to ask, "How did you internalize the possibility of bad weather?" as we are in asking "Why did you take your umbrella?". That is, if subjects could answer not only "why" but actually "how", we would be closer to an account in terms of etiology. That would include external and conative factors--which jointly define "context". "How" and "why" merge wen we are speaking from the standpoint of the subject, not ours alone. In an earlier section we saw that the need to postulate representations comes up precisely in a similar connection, viz., privacy and inferred intentionality.

Knowing that, we could incorporate the relevant variables in our description, which aims at specifying processes. If we know what precipitates these processes and how, it all reduces to a technical problem. That is why recursion applies to current computation as a function of past computations: $x(t+1) = f(...x(t-1), x(t))$, each state of vector $x(t)$ being a function of preceding states. Explanation still lies with the independent variables, past and present: we may locate them in the system's memory; as far as the system is concerned, things simply happen, though not always automatically. It is not quite accurate to say "any human action at each current time takes into account the past events, the current situation in the environment, and the future anticipated events" tout court (Dubois, 1997). Consciousness provides a natural boundary, diffuse as it can often be, and that calls for special treatment. This would appear to be the natural place for anticipation.

Incursion as "an inclusive or exclusive recursion leading to a self-referential system" computes its future state as a function of this future state: $X(t+1)=f(..., x(t-1), x(t), x(t+1), ...)$. Whether incursion "leads to a self-referential system" in the sense of capturing the self-responding here under discussion, is an open question. An anticipatory system is conceived as containing a model of itself; if so, we can embark on an endless series of self-embedding models. Anticipatory systems can be related to a hyperset H defined as a collection of its elements E and itself: $H = \{E, H.\}$. But a classical set cannot be a member of itself. Maybe it is on account of this dependance on formally defined structures that we still do not have a well-established framework for referentiality.

Dubois finds an easy solution to this paradox by endowing the anticipatory system with two fixed points representing the model's "implicit finality or teleonomy": the goal is not imposed from without (as in control theory), but is "determined by the system itself". In point of fact, real-life systems coping with the unexpected do have their goals imposed from the outside. That is precisely the point. Strategies of this sort may pay off, e.g., with neural network setups, where the parameters are preselectively limited and where we speak of "goals" metaphorically at best.

Hyperincursion allows for multiplicity of possible future states. Therein lies its potential as a mathematical tool for formulating anticipation. In practice, "the system must make a choice at each current time to collapse the set of potential solutions to one realized solution". As indicated earlier, that implies a preliminary mapping of potential solutions, therefore potential problems, as well as enough knowledge about the kinds of relations prevailing between concrete history (call it memory) and current problems. The study of anticipation in these terms can serve as an example of how formal techniques can be applied to empirical data, enriching and clarifying their formulation, and generating, in turn, further disciplined research including parametric studies (of which behavioral science has often been rather chary).

Despite his formal commitment, Dubois remains open to possibilities: "The decision process (itself) could be explicitly related to objectives to be reached by the state variable of this system" (emphasis added). How to do so or to what specific purpose remains unanswered, probably because of the insistence on keeping the system and its environment de facto separate: "To be viable, the system must make a choice at each current time to collapse the set of potential solutions to one realized solution. This is guided by environment for externalized systems of evolution, and by the system itself for internalized systems for decision processes". It would seem that if truly put up to depict actual interactions, the "decision" process should be explicitly related to objectives to be reached by the state variable in conjunction with all the independent variables at work.

Hyperincursion is thus a hyper recursion with multiple solutions: each iterate can be a function of its future iterates. Its greatest promise for a naturalistic approach to anticipation lies in the fact that it makes room for the integration of multiple independent variables, which collapse to produce a single event at time t. That is what is needed in order to account for the subtle dynamic processes that go with the formation of complex components, the emergence of new properties and the emergence of a hierarchy of components of increasing complexity during evolution (Dubois, 1992, 1996c). With proper conceptual and procedural care, that should do just as well for an integrative formulation of basic <u>data</u> on learning and adaptation.

There is room for multiple solutions but these would have to be seen less as a procedurally pre-established set (within more or less artificially bounded limits, as in neural networks) than as a formal framework for the open exploration of how history and current variables come together and "precipitate" one response over competing ones. Empirical specificity demands that we base ourselves on the identified variables and available experimental data; a significant measure of quantification is also on hand, and that provides the platform for responsibly circumscribed simulation as well.

3.2. Anticipation, Context and Self-Regulation

Conflating cybernetic and behavioral terminologies, we could say that feedback feeds back and the entire event passes into history; feedforward would be based on that but could actualize responses to the future <u>now</u>. All responding takes place at time t. Humans do that because they can rely on special self-responding repertoires; these additional systems of responses combine in complex and subtle ways to make up the behavioral equipment that sets us apart as a species.

To recapitulate, classical adaptive systems are defined recursively: $(x+1) = f[... x(t-1), x(t), p]$. Incursion and hyperincursion capture various degrees of complexity in anticipatory mechanisms. In a simple incursion $x(t+1) = f[x(t), x(t+1), p]$ the value of x at each t+1 is a function of its value at t but also at t+1. In the present reckoning, this would provide for an anticipatory response given a more basic repertoire; ideally, it should help define the former's "self-referential" character along functional lines. Incursion is an extension of implicit or explicit recursion, which stands for established history. Hyperincursion makes room for multiple solutions: the additional responses here should be those of the self-responding and self-generated type (cf. infra); if genuinely internalized, they too are incorporated and pass into history; they are then endowed with a measure of automaticity and remain available for further strengthening and participation in response to "future" demands.

Somewhat surprisingly, Dubois writes: "Learning and adaptation systems in biological processes are viewed as such anticipatory systems in agreement with the conjecture of Robert Rosen" (1997). This can probably be best understood in light of the fact that we still do not have a theoretical framework for self-referentiality; it would appear to be more in accord with Rosen's distinction between inanimate and animate systems (the latter telescoping learning, adaptation and evolution) than with Dubois' own earlier question as to the relevance of computation for "systems without self-consciousness and intentionality".

On this hinges the already mentioned distinction between automaticity and consciousness, relevant not only to anticipation studies but central also to any discussion of mechanisms of human self-regulation. It is probably owing to the prevailing opacity in this connection that Karoly (1993) is forced to end his splendid review of the state of the art as follows: "Self-regulation has, until relatively recently, defied experimental analysis, perhaps because of its uncertain epistemological status ... As a concept akin to 'getting one's life together', self-regulation has not achieved a simple or uniform paradigmatic embodiment, nor should we expect this in the foreseeable future". Perhaps the "phases" (e.g., goal-selection, directional maintanence, reprioritatization, etc.) into which most approaches can be said to crystallize have more to do with the technological side of cybernetics and systems research than with the processes that define human conscious and unconscious behavior.

We are entitled to ask, indeed we must ask again, Why should we regulate ourselves or go to the trouble of monitoring any one aspect of our activity? Doing so constitutes an extra effort and that runs counter to the natural tendency to let go, to act spontaneously or automatically, what we significantly call "freely". According to Dubois, the function f describes the dynamics of the system, which contains an anticipation of itself. Thus (x+1) in f can be replaced with $(x+1) = f([x(t), f[x(t), x(t+1)], p)$ p]. Another interpretation would be that f captures the interplay of dispositional history and current environmental demands, where the additional repertoires are not so much the equivalent of "a model of the system itself" in the formal sense, but actual self-knowing of the type that enhances further responding. It can thus be taken to denote the growing complexity of the interaction field in self-regulation.

It may well be that a naturalized approach can dispense altogether with the oft-mentioned theoretical framework for self-referentiality, which continues to evade us anyway. Its formulation revolves around essentially hypostatical systems in lieu of the etiology of self-responding (from minimal self-reference to self-knowledge), which would be incorporated into the dynamics of the system, and throw light on the formulation of parallel systems of feedback and feedforward, in accordance with the effects of the ongoing interaction with the environment.

In sum: feedback builds history; effective history may be subject to recursive formulation; but feedback does not operate alone (externally or internally); other variables are acting simultaneously and these can be captured by incursive and hyperincursive means. Feedforward, either automatic or self-generated (see below), is grounded on this history, which underlies current dispositions, now strengthened by current variables having to do with the future. Taken together, these sources of strength define <u>functional contexts</u>, which are in turn captured by the proposed etiology.

State variables operate on the strength of similarity and functional connection between past and present, even if this present is self-generated in the form, for example, of perceptual and verbal responses in preparation (fr. L <u>prae-pre</u> + <u>parare</u> to procure, to make ready--a cognate of <u>parere</u> to give birth to, to produce) for an adequate future response to meet rising demands. Therein too lies the divide between automaticity and consciousness.

Anticipation has to do with the future. This is the sense conveyed by the etymology of the term as well as the various cross-references indicated by Webster: 'forestall', 'foresee', 'prevent'; in addition, 'heave' suggests increased strength or probability of action, greater disposition to act; hence the added sense of finality that goes with it. Anticipation itself is activity of a cognitive nature. That is in part why anticipation provides "a particularly apt illustration of intentionality". Whether the subsequent activity is itself perceptual, verbal or nonverbal or, more likely, a combination thereof, the strengthened disposition is intimately bound up with action; and this bond seems clearer in the case of anticipation than in most standard examples of intentionality.

232

Clearly, a single abstract characterization will not do. The dispositional nature of both anticipation and the subsequent action in forestalling or preventing are one thing, covered by the experimentally well-illustrated process of avoidance; foreseeing can share the same kind of tendency, but some kinds of foreseeing are closer to "looking forward to", while yet others are speculative, we might say. The key lies in the nature of the attendant consequences, which, moreover, may be more or less immediate or deferred.

Strength differences come also from what history permits: the basic readiness to engage, say, in positive action or in avoidance, depends on how well personal history equips us to cope now in light of a future that has become relevant at this point. Strength is also influenced by the availability of means, the feasibility of the action, and such (some philosophers distinguish between minimal and non-minimal intentions on similar grounds). This "retroactive" strength should be easily captured by the incursive extension of recursion.

Much relevant research goes under such headings as "expectancy" and "expectation", which obviously overlap. Current trends favor the latter; thus, Walker (1994, p. 79) leaves out the former and characterizes expectations as "Anticipation of future behavior or events. Also refers to investigations of the effects of that anticipation on behavior". Some relevant mechanisms involve "stimulus-stimulus relations", "signaled" avoidance, delayed responses, delayed "reinforcement" and so forth. Particularly illustrative for the study of basic mechanisms shared by verbal adults and non-verbal organisms is the so-called "conditioned suppression" technique, which has proved useful for interpreting certain kinds of anxiety and related "mental" events. This research has been carried out in the laboratory with a broad variety of species and with startlingly homogeneous results. We naturally can speak of anticipation wherever homogeneity obtains; we welcome continuity and parsimony, but that should not blind us to the facile extrapolation to human anticipation.

To reiterate, the presence of self-responding repertoires makes all the difference in the world. Building on the basic processes we do share with other species, self-responding allows us to ask whether anticipatory dispositions simply arise or whether they are self-generated. This self-generation is at the basis of forethought, planning, deciding, commitment, and so on, which play a central role in self-monitoring. Involved here are different degrees of visualization, rehersal of auditory stimuli, manipulation of supplementary sources of strength or weakness, the production of graphic and other materials (say, models of various types) and so forth, i.e., the standard operations of problem-solving. A major outcome of these self-responding repertoires (where language, needless to say, plays a central role) is the possibility of "bridging" ever more extended temporal delays between responses now and eventual payoff in the future.

The effects of impinging stimulation and stimulation arising from self-generation need not be "located" in some unreachable metaphysical nook or abstract construct; the pertinent crannies are what make up our bodies. How else would we feel a lessening of anxiety as a result of sensing our way through a modus operandi at a later time? Where else would the mounting exhilaration reside when a distant goal becomes ever more immediate, say, seeing the end of this paper (for both writer and reader)? Simultaneous systems of feedback loops have a major role to play in an integrative account, where their operation is part of a context that naturally includes external stimulation as well. The same can be said about feedforward, which is more closely associated with self-generated stimulation or stimulation resulting from self-promoted action.

To restate the obvious, language plays a crucial role in this complex network. A full interpretation places the role of language within the coordinates discussed in Section 2 as well as in (1)-(4) in 1.1. Having to do with the future, anticipation is, by definition, a case in which the pertinent stimulus setting is absent (though the importance of related subliminal stimulation can hardly be exaggerated; would anyone seriously propose that the external environment plays no role, in one way or another, in unconscious tendencies?). The perceptual components are also private by definition and the verbal activity is likely to be so too, on and off anyway, as we saw in 2.3. The basic dispositions and related tendencies to act may have to remain covert till circumstances permit; private or public, specification thereof is what we have repeatedly called "etiology" throughout this study. Etiology is the result of our technical expertise.

All subsequent taxonomizing is ultimately traced to these fundamental parameters. Working from a different perspective, Bandura (1986), for example, provided a comprehensive coverage and interpretation of the experimental literature with humans. Bandura is particularly keen on the importance of motivational factors, probably owing to his extensive original research on observational learning, imitation, and related forms of complex learning. His is a "social cognitive" theory, in sharp contrast with purely cognitive approaches. It is instructive to note, however, that a reluctance to assign language and some aspects of cognition a fully behavioral status inevitably forces even Bandura to rely heavily on largely unsupported symbolic mechanisms and mental representations.

The feelings that make up experience (conscious rather than only sensory) are rooted in the following basic facts:

(1) all behavior is stimulating to the behaving organism and, when public, to other organisms within the interactional field (albeit in different ways);

(2) any behavioral event is unique in time, passing into the individual's history and changing her/his dispositions vis-à-vis the future (naturally, in nonlinear as well as linear ways);

(3) the individual's history itself is a source of (private) stimulation that summates with current stimulation impinging from the outside;

(4) external factors, which may have played a primary role during acquisition, may become "implicit" (to use J.R. Kantor's terminology) when the individual becomes the principal focus of stimulation and self-stimulation;

(5) dispositions are made up of functionally related sets of responses, the strength of which covaries in the course of the ongoing interaction; conative factors play a preeminent role in the emergence and/or reemergence of dispositions;

(6) converging independent variables account for the selection of one particular response over others in the same set at any given time, whether the selected response occurs in overt or covert form;

(7) in addition to systems specifying automatic repertoires, feedback and feedforward systems can be established to specify the stimulation that goes with the self-responding involved in consciousness and related phenomena;

(8) taken together, (1)-(7) provide the context and content for any given behavioral segment; as a near-synonym for the more procedural "etiology", "context" makes the elusive "environment" manageable as a field concept.

3.3. On Purpose

It is not for nought that the terms 'adaptation', 'purpose' and 'functionalism' tend to cluster in discussions on the perennial problem of causation. Darwin's epoch-making work could not but make adaptive processes relevant to psychology, too; after all, the innate tendency of living organisms to adapt can only be fully understood if the context of this adaptation is also taken into account. Thus William James (1890) could write: "The pursuance of future ends and the choice of means for their attainment are thus the mark and criterion of the presence of mentality in a phenomenon. We all use this test to discriminate between an intelligent and a mechanical performance". Nevertheless, a large segment of the experimental psychology of the following decades eschewed "purpose" and related notions in order to keep their vocabulary as free as possible from philosophical discussion, as James' phrasing can easily intimate.

It is somewhat ironic that a concern with purpose should have resurfaced under the aegis of a kind of functionalism that circumscribes the concept of function to real or posited relations between system components (mechanical or living, intelligent or artificially so). Sometimes these systems are explicitly endowed with "goal-directed" mechanisms; at other times this directedness is merely assumed or inconsistently invoked when needed. The urge to bring in teleological notions attests to the inevitability of conative factors.

Wiener et al. had good reason to struggle with purpose during the gestation period of cybernetics. In their classic paper, Rosenblueth, Wiener and Bigelow (1943) lamented the fact that purposefulness had been gradually neglected along with the abandonment of final causes. They deemed the reintroduction of purpose "necessary for the understanding of certain forms of behavior" and proceeded conveniently to interpret purposefulness in terms of negative feedback--a pristine example of how we can let our technical means determine our conception of the subject matter rather than the other way around.

A definition of purpose in terms of negative feedback leads to circularity, particularly where closed loops are concerned. In practice, Rosenblueth et al. merely extended teleological language to specific forms of behavior singled out for this purpose. As Scheffler (1991, p. 95) points out, "the claim to reduce human purpose by their method can achieve plausibility only by so enriching the latter as to encompass the former". There is a vast difference between relabeling a phenomenon and providing an explanation for it. Even if they enlarged the range of theoretical systems, including complex self-regulating systems, subsequent enrichments, say, positive feedback, feedforward, different types of loops, parallel systems, etc., did little to improve this basic situation. While submitting to the overall characterization of cybernetics as the science of "control and communication in the animal and the machine", they do not clear up the nature of purpose.

Clearly, machine language demands closer consideration if cybernetics and systems research, on the one hand, and behavioral principles and data, on the other, are to be

235

reconciled without licensing a wedding of insufficiencies; given its direct relevance to physiology, the place to begin may well be feedback itself. At any rate, it is not easy to reconcile purpose as feedback and Rosenblueth et al.'s statement "the basis of the concept of purpose is the awareness of 'voluntary activity'". The temptation to reintroduce purpose under the cloak of anticipation would appear to be all too natural, anticipation being about the future; but again, care must be taken not to confound different phenomena under the same label.

Feedforward provides a useful illustration. The standard example is catching a ball where, paraphrasing Arbib (1989), we have to "interpret" (sic) our view of the ball's movement to estimate its future trajectory. Feedforward is said to anticipate the relation between the system and the environment so as to determine a course of action, whereas feedback monitors, as it were, the finesse of this action. Lacking an account of the nature of the relation between the system and the environment, "control problems" arise. But this sort of anticipation is not really about the future. Given a definition (Wiener's) of feedback as "a method of controlling the system by reinserting into it the results of its past performance", feedforward becomes a requisite technical addition. Even allowing for all the necessary identification algorithms, feedforward antecedes rather than anticipates feedback, along with the performance leading to it: it keeps the system "in the right ballpark". In that sense, it is more like responding to first and second time-derivatives--as in Wiener's Pencil, to keep to classic examples--than anticipation as here discussed.

This kind of responding can be studied directly without unnecessary model apriorisms, as in the paradigmatic experiments by Notterman and his associates (Notterman et al., 1982); or Pribram's (1980) work on neural mechanisms underlying attention, though aimed at accounting for a measure of consciousness, thus necessitating the convergence of memory mechanisms, coordination and so forth. Attention and anticipation have much in common.

What is generally called a control problem has to do with the choice of inputs, the setting of reference values, components, etc.: "The controller (the brain) must have an accurate model of the effect of all disturbances upon the system (say, our musculature) in order to compute controls that will effect the necessary compensations", to use Arbib's phrasing. We hear reverberations of our previous discussion about models embedded in models without the benefit of etiological insight. It is this kind of apriorism that leads to a view of cognitive activity as a matter of "recognition" and "completion", as well as "contexts" mapped out in abstraction from the actual responses and variables at play; the "patterns of connective strength" manipulated predictably become "workplaces in the brain".

In his foundational work, Rosen (1985) had to confront purpose for similar reasons: he could scarcely be satisfied with a logico-mathematical frame of reference, if his account of anticipation had to cover learning, adaptation and evolution. Formalisms only specify operations and their implementations; like simulations at large, they cannot be expected to do anything resembling what we would recognize as adaptation in real life, however open-ended their range of application (Clocksin, 1995; Julià, 1996b). How could they, whatever the degree of graphic, computational similarity we may obtain? Alas, electromechanical systems simply do not have the right nerves going to the right places, as the saying goes.

The anticipation we are concerned with has to do with the future, but not only; the pertinent environment is actually absent and the approach to it must involve something else, as we saw in Sections 1 and 2. The past is crucially important and so is any activity that modifies

236

our current dispositions. Immediate and delayed payoff constitute a major variable in accounting for the complex ways in which we generate cognitive and non-cognitive responses as we "bring the future into the present". The importance of directedness in extricating ourselves from the intentional bottleneck was stressed earlier. Another way to break free from circularity might be to take a closer look into "voluntary activity". Rosenblueth et al. made awareness and voluntary activity the defining criteria even if, as we have seen, their own framework could not handle either one of these concepts. As we have seen, Dubois (1997) also wonders about the significance of computing anticipatory systems devoid of consciousness and intentionality.

There appears to be some kind of relation between volition and purpose; closer still is the relation between purpose and conation, which probably explains the intuitive connection between anticipation and purpose. It is interesting to note that conation tends to fuse with the automaticity of instinct as we go down the evolutionary scale; with an increase in automaticity goes a decrease in the <u>ascription</u> of purpose. We are more likely to read some kind of purpose as we move up the scale (by the same token, we are more likely to read purpose into an adult than into a toddler). Much the same can be said about anticipation, but this sort of anticipation can be easily referred to the rather simple mechanisms mentioned in 3.2. Furthermore, different histories of "reinforcement" account for different degrees of resistance to extinction--the protracted process whereby the probability of behavior goes back to zero with the consistent discontinuation of payoff or stimulus-stimulus pairings, or both. Resistance to extinction constitutes one of the best-known behavioral phenomena; it is also one of the most basic platforms for coming to terms with the enduring effects of the organism's adaptation to the environment. Acquaintance with the specific parameters that make up the organism's history of interaction sends expressions like 'persistance', 'stubborness', 'purposefulness', etc. back to the realm of everyday discourse. There again, we can speak safely of automaticity, but we do so in qualified ways, proportional to our actual knowledge of details. To this extent, a machine could probably simulate the performance; but simulation is not explanation or even emulation.

Even less likely to be simulated is the apparent self-awareness we find in many species, particularly when presented with puzzling situations. Whether ascription of self-awareness is more than an anthropomorphic exercise remains an open question, although anthropology, sociology and other social sciences have much to suggest. Self-awareness, like self-reference, is a relatively sophisticated form of self-responding, which culminates in consciousness.

There appears to be a correlation between the ascription of voluntary activity, purpose, anticipation and such, and our inability to point out the variables at work. We are speaking from the vantage point of observers but the individuals whose activity we are trying to explain may, in the human case, have a very different perspective on the "same facts". We thus read purpose, goal-directedness, intention and so forth into anything that <u>looks</u> like our own conscious directedness or self-directedness. But ascription without an actual etiology constitutes an autistic exercise. The appeal to Aristotle's final causes as part of the characterization of anticipation may prove helpful under some conditions, but these conditions should be investigated rather than taken for granted.

In any case, we tend to drop the ascription of purpose if we have good reason to believe that the behavior is automatic. Far from well understood as it is, consciousness holds center stage in disambiguating purpose. It would be a mistake, however, to equate consciousness with volition: we may be perfectly aware of what we are doing and totally unable to do anything about it despite the anticipation, even the "foretaste", of dire consequences; similarly, we may be fully conscious of the futility of certain efforts and still feel "driven" to keep trying; many keep going with no clear purpose in mind. These are extreme conditions where conation has taken over; under ordinary circumstances, self-knowledge--we found earlier--is a necessary component of self-regulation, which is bound to include both anticipation and a <u>sense</u> of purpose.

The link between all the etiological variants comes with primary directedness toward consequences--generic for intention, goal, purpose or any other analogues that might qualify under Thorndike's Law of Effect. As already pointed out before, the watershed is the distinction between automacity and deliberateness of action. That applies equally to intention, goal and purpose, even if there are differences among them; these differences reduce to degrees of complexity. For the time being, it can be safely stated that intention is contained in goal, which is in turn contained in purpose (Julià, 1986, 1996a).

In short, automaticity we share, mercifully, with animals and machines; it would be a very tough life indeed if we had to be constantly aware of, and monitor, all that we do and say, the accompanying feelings, the "reasons" for doing so, alternative courses of action, possible outcomes and so on. But with respect to situations that, owing to novelty, unexpected complexity, unpreparedness, distance in time, etc. do demand the kind of self-responding singled out above, humans can bring about the kind of self-monitoring that justifies notions like "responsibility" (as its etymology suggests, we may have to "answer for" our actions, purposes, plans, etc. to others and eventually to ourselves). At this level of development, resistance to extinction would be a poor index indeed. The matrices that define these extremely subtle and complex activities circumscribe what we have called privacy in Sections 1 and 2; they define "experience" not as some sort of epiphenomenal flotsam but as the very stimulation resulting from a history of interaction, including, crucially, interaction with ourselves; the more immediate referents were listed at the end of 3.2.

The machine-man analogy blurs rather than facilitates an effective discussion of concepts like purpose because it preempts the meaningful demarcation of relevant domains; we are bound to make category mistakes or even trespass on natural kinds. Cybernetics lost sight of the "kybernetes" (G for "steersman" and L for "governor") which originally set the field apart and from which it took its name. In the process of designing increasingly sophisticated self-steering devices, the unique "self-steering" functions of the steersman receded into the background, as if design, redesign and ultimate control had ever ceased to rest with human agents (Julià, 1997, for an extended discussion). But devices remain impervious to conative factors, they do not register fatigue, frustration, counteragression, cooperativeness, confusion or, for that matter, improve their "rationality" unless we set them going and provide the means to do so. Stated otherwise, they do not display dispositions, including anticipation; lacking experience, there are no etiologies to speak of; they only compute, if that is what they actually do. To the extent we can, <u>we</u> have to anticipate their future and their foreseeable goals. The transfer of such "networks" and corresponding formalisms to living systems remains metaphorical and probably misleading when we come down to the gritty realia, as eventually we must.

The case for conscious self-examination (including the examination of products or traces of our activity) arises because, beyond a certain point, humans are "accountable" for their actions, including the way they speak. But this accountability has little if anything to do

238

with the computations of computational devices: living systems do not keep a record of their interactional history as a function of rules. They adapt and evolve (are changed) qua organisms in the course of this interaction. A very important part of human adaptation is verbal and through it we also learn about the world and our relation to it. An account of the emergent complexity involves us in the privacy of "purposeful" activity. To elucidate whether this is what "final causes" are about seems less urgent than trying to reconcile the available data and the formal and instrumental techniques at our disposal. A genuine bottom-up approach would appear to provide our best chance to keep the study of anticipation and self-regulation rooted, as far as possible, in the natural science ballpark.

References

Arbib, M.A. (1989). Feedback and Feedforward. In Oxford Companion to the Mind. Edited by R.L. Gregory. Oxford University Press, Oxford.

Banaji, M.R. and Prentice, D. (1994). The Self in Social Contexts. Annual Review of Psychology, 45, pp. 297-332.

Bandura, A. (1986). Social Foundations of Thought and Action. Prentice-Hall, Englewood Cliffs, New Jersey.

Blackburn, S. (1996). The Oxford Dictionary of Philosophy. Oxford University Press, Oxford.

Brier, S. (1992). Foreword. Cybernetics and Human Knowing, 1, pp. 1-3.

Catania, A.C. and Harnad, S. (Eds.) (1988). The Selection of Behavior. Cambridge University Press, Cambridge.

Chomsky, N. (1959). Review of B.F. Skinner's "Verbal Behavior". Language, 35, pp. 26-58.

Chomsky, N. (1962). Explanatory Models in Linguistics. In Logic, Methodology and Philosophy of Science. Edited by E. Nagel, P. Suppes and A. Tarski. Stanford University Press, Stanford, California.

Chomsky, N. (1976). Reflections on Language. Pantheon, London.

Clocksin, N.F. (1995). Knowledge Representation and Myth. In Nature's Imagination. Edited by J. Cornwell. Oxford University Press, Oxford.

Davidson, D. (1980). Essays on Actions and Events. Clarendon, Oxford.

Dennett, D.C. (1981). Brainstorms. MIT/Bradford Press. Cambridge, Mass.

Dennett, D.C. (1983). Intentional Systems in Cognitive Ethology: The "Panglossian Paradigm" Defended. Behavioral and Brain Sciences, 6, pp. 343-355.

Dennett, D.C. and Haugeland, J.C. (1987). Intentionality. In Oxford Companion to the Mind. Edited by R.L. Gregory. Oxford University Press, Oxford.

Dennett, D.C. (1996). Kinds of Minds: Toward An Understanding of Consciousness. Basic Books, New York.

Dubois, D.M. and Resconi G. (1993). Hyperincursivity: A New Mathematical Theory. Presses Universitaires de Liège, Liège.

Dubois, D.M. (1994). Holistic Control by Incursion of Feedback Systems, Fractal Chaos and Numerical Instabilities. In Cybernetics and Systems, II. Edited by R. Trappl. World Scientific, Singapore.

Dubois, D.M. (1996a). Introduction. In Proceedings of the 14th International Congress on Cybernetics. Association Internationale de Cybernétique, Namur, pp. 383-388.

Dubois, D.M. (1996b). Incursive Modelling and Control of Fractal Chaos. In Proceedings of the 14th International Congress on Cybernetics. Association Internationale de Cybernétique, Namur, pp. 412-422.

Dubois, D.M. (1996c). Hyperincursive Stack Memory in Chaotic Automata. In Symposium ECHO. Edited by A.C. Ehresman, G.L. Farré and J.P. Vanbremeersch. University of Picardie Jules Verne, Amiens.

Dubois, D.M. (1997). Introduction to Anticipatory Systems. CASYS 1997. Presses Universitaires de Liège.

Eysenck, M.W. (1994). Dictionary of Cognitive Psychology. Blackwell, Oxford.

Gazzaniga, M.S. (1995). The Cognitive Neurosciences. MIT Press, Cambridge, Mass.

Geyer, F. (1996). The Increasing Convergence of Social Science and Cybernetics. Plenary Address. 10th International Congress of Cybernetics and Systems. Bucharest.

Grossberg, S. (1995). The attentive brain. American Scientist, 83, pp. 438-449.

Hameroff, S.R., Kaszniak, A.W. and Scott, A.C. (Eds.) (1996). Towards A Science of Consciousness. MIT Press, Cambridge, Mass.

Hayes, S.C. (Ed.) (1989). Rule-Governed Behavior. Plenum Press, New York.

Hofstadter, D.R. (1981). Letter to the Editors / Metamagical Themas. Scientific American, April.

Horst, S.W. (1996). Symbols, Computation and Intentionality. University of California Press, Berkeley, California.

James, W. (1890). The Principles of Psychology. Excerpt in R.J. Herrnstein and E.G. Boring (eds.) (1968). A Source Book in the History of Psychology. Harvard University Press, Cambridge, Mass.

Julià, P. (1983). Explanatory Models in Linguistics. Princeton University Press, Princeton, New Jersey.

Julià, P. (1984). Contingencies, Rules and the "Problem" of Novel Behavior. Behavioral and Brain Sciences, 7, pp. 598-599.

Julià, P. (1986). Language: Form and Function. Technical Report. Generalitat de Catalunya.

Julià, P. (1994a). Formal vs. Non-formal Description of Human Languages. In Cybernetics and Language. Edited by D. Maxwell, M. Preotu and A.P. Tacu. Akademia Libroservo/Kava-Pech. Prague, pp. 35-52.

Julià, P. (1994b). Rules, Artificial Intelligence and Cognitive Science. In Cybernetics and Systems II. Edited by R. Trappl. World Scientific, Singapore.

Julià, P. (1995a). On Control and Communication in the Animal and the Machine. In Proceedings of the 14th International Congress on Cybernetics. Association Internationale de Cybernétique. Namur, pp. 879-884.

Julià, P. (1995b). Information: From Etymology to Etiology. In Proceedings of the 14th International Congress on Cybernetics. Association Internationale de Cybernétique. Namur, pp. 885-890.

Julià, P. (1996a). Dispositions. Plenary Lecture. First International Conference on Philosophy and Mental Health. International Association for the Advancement of Philosophy and Psychiatry, London / Benalmádena, Spain.

Julià, P. (1996b). Representations, Artificial Intelligence and Cognitive Science. In Advances in Artificial Intelligence and Engineering Cybernetics. Edited by G. Lasker, International Institute for Advanced Studies in Cybernetics and Systems Research, Windsor, Ontario, pp. 1-7.

Julià, P. (1997). A Culture of Peace -- But What Has Become of the "Kybernetes"? In press.

Kantor, J.R. (1947). Problems of Physiological Psychology. Principia Press, Chicago.

Kantor, J.R. (1963). The Scientific Evolution of Psychology. Principia Press, Chicago.

Karoly, P. (1993). Mechanisms of Self-Regulation: A Systems View. Annual Review of Psychology, 44, pp. 23-51.

Lester, D.S., Felder, C.C. and Lewis, E. (Eds.) (1997). Imaging Brain Function and Structure. The New York Academy of Sciences, New York.

241

Llinàs, R. and Churchland, P.S. (Eds.) (1997). The Mind-Brain Continuum. Sensory Processes. MIT Press, Cambridge, Mass.

Massaro, D. and Cowan, N. (1993). Information Processing Models: Microscopes of the Mind. Annual Review of Psychology, 44, pp. 383-425.

Mazur, J.E. (1990). Learning and Behavior. Prentice-Hall, Englewood Cliffs, New Jersey.

McGuigan, F.J. and Ban, T.A. (Eds.) (1987). Critical Issues in Psychology, Psychiatry and Physiology. Gordon and Breach Science Publishers, New York.

Notterman, J.M., Tufano, D.R. and Hrapski, J.S. (1982). Visual-Motor Organization: Differences Between and Within Individuals. Perceptual and Motor Skills, 54, pp. 723-50.

Pribram, K. (1980). Mind, Brain and Consciousness. In The Psychophysiology of Consciousness. Edited by J.M. and R.J. Davidson. Plenum, New York.

Quine, W.V.O. (1940). Mathematical Logic. Harper and Row, New York.

Quine, W.V.O. (1960). Word and Object. MIT Press, Cambridge, Mass.

Quine, W.V.O. (1974). The Roots of Reference. Open Court, La Salle, Illinois.

Raichle, M.E. (1994). Images of the Mind: Studies with Modern Imaging Techniques. Annual Review of Psychology, 45, pp. 333-353.

Rosen, R. (1985). Anticipatory Systems. Pergamon Press, New York.

Rosenblueth, A., Wiener, N. and Bigelow, J. (1943). Behavior, Purpose and Teleology. Philosophy of Science, 10, pp. 18-24.

Runes, D. (1945). Dictionary of Philosophy. Philosophical Library, New York.

Sapon, S.M. (1971). On Defining A Response: A Crucial Problem in Verbal Behavior. In The Psychology of Second Language Learning. Edited by P. Pimsleur and T. Quinn. Cambridge University Press, Cambridge.

Sapon, S.M. (1972). The Descriptive Analysis of Behavior. Monopress, Rochester, New York.

Scheffler, I. (1991). In Praise of the Cognitive Emotions. Routledge, London.

Searle, J. (1992). The Rediscovery of the Mind. MIT Press, Cambridge, Mass.

Skinner, B.F. (1947). The Flight from the Laboratory. In B.F. Skinner, Cumulative Record (1972). Appleton, New York.

Snodgrass, J.G. and Thompson, R.L. (Eds.) (1997). The Self Across Psychology. The New York Academy of Sciences, New York.

Taylor, R. (1964). The Explanation of Behaviour. Routledge, London.

Umpleby, S.A. (1994). Twenty Years of Second-Order Cybernetics. 12th European Meeting on Cybernetics and Systems Research. Vienna.

Verplanck, W.S. (1955). Since Learned Behavior is Innate, and Vice Versa, What Now? Psychological Review, 62, pp. 139-144.

Wachs, T.D. and Plomin, R. (Eds.) (1991). Conceptualization and Measurement of Organism-Environment Interaction. American Psychological Association, Washington, D.C.

Walker, A. (1994). Thesaurus of Psychological Terms. American Psychological Association, Washington, D.C.

Zuriff, G.E. (1985). Behaviorism: A Conceptual Reconstruction. Columbia University Press, New York.

SPEECH INTELLIGIBILITY MEASURE
FOR VOCAL CONTROL OF AN AUTOMATON

Michel NARANJO

LASMEA, UMR 6602 du CNRS
Université Blaise Pascal de
Clermont Ferrand
63177 Aubière Cédex (France)

Georgios TSIRIGOTIS

Electrical Engineering Department
Technological Educational Institute
PO box 1194, St. Loukas,
654 04 KAVALA, Greece

Abstract

The acceleration of investigations in Speech Recognition allows to augur, in the next future, a wide establishment of Vocal Control Systems in the production units. The communication between a human and a machine necessitates technical devices that emit, or are submitted to important noise perturbations. The vocal interface introduces a new control problem of a deterministic automaton using uncertain information. The purpose is to place exactly the automaton in a final state, ordered by voice, from an unknown initial state. The whole Speech Processing procedure, presented in this paper, has for input the temporal speech signal of a word and for output a recognised word labelled with an intelligibility index given by the recognition quality. In the first part, we present the essential psychoacoustic concepts for the automatic calculation of the loudness of a speech signal. The architecture of a Time Delay Neural Network is presented in second part where we also give the results of the recognition. The theory of the fuzzy subset, in third part, allows to extract at the same time a recognised word and its intelligibility index. In the fourth part, an Anticipatory System models the control of a Sequential Machine. A prediction phase and an updating one appear which involve data coming from the information system. A Bayesian decision strategy is used and the criterion is a weighted sum of criteria defined from information, minimum path functions and speech intelligibility measure.

Keywords

Speech Processing, Loudness Calculation, Neural Networks,
Automata Control, Anticipatory System,

1. Introduction

The understanding and utilisation of the Speech Recognition technology allow to augur, in the next future, a wide establishment of vocal control systems in the production units. One can consider that the decisive progress originates, on the one hand in time/frequency algorithms for the speech signal pre-processing, and on the other in the simplification of discriminators using neuromimetic algorithms. The communication between a human being and a machine necessitates a technical device that emits, or is submitted to important noise perturbations. The received speech signal is degraded and only a low information quantity is perceived.

The control of well-modelled continuous or discrete systems has been greatly studied in the past and many algorithms are now available to solve various real - time control problems. In

CP437, *Computing Anticipatory Systems: CAYS--First International Conference*
edited by Daniel M. Dubois © 1998 The American Institute of Physics 1-56396-827-4/98/$15.00

the case of complex systems, it is very difficult to obtain by identification methods, any models accurate enough to be used for efficient control. For some kinds of control objectives, it is sometimes easier to consider these systems automata (Naranjo and al, 1980). In spite of this conceptual simplification, it is often necessary to deal with uncertain information on their functioning and this infers problems of modelling and control of automata. In Robotics for example, it is absolutely imperative to take a final decision that leads to an unambiguous result for the location of autonomous mobile robots or for location of pieces that must be picked up by manipulator robots.

The vocal interface introduces a new problem of the control of a deterministic automaton using uncertain information. We have now two predictive systems. One model takes information about the past and present state of the system and makes a prediction of its state in the near future. The human - machine interface provides orders taken about information from the past and present state of the surrounding environment. We have the essential elements of an Anticipatory System (fig. 1) which forms an expectation of future events and renders a decision accordingly.

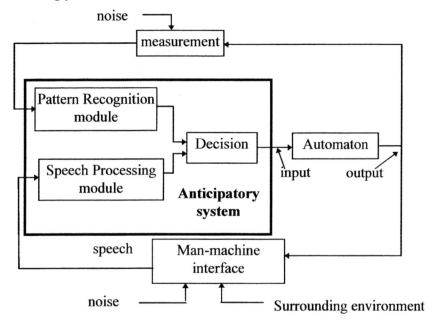

Fig. 1 : Place of the anticipatory system in the control problem

The whole Speech Processing Procedure has for input the temporal speech signal of a word, and for output a recognised word labelled with an intelligibility index given by the recognition quality. We present an automatic calculation procedure of the loudness and the architecture of the Time Delay Neural Network (TDNN). We give the organisation of the network when the input is constituted by the output of the critical bands obtained after calculation of the loudness. The output of the network is under the form of a vector where the components have values comprised among 0 and 1. While assimilating this vector to a fuzzy vector, the Information Theory allows us to extract a recognised word and its intelligibility index. A large part of this paper is devoted to the problem control of a deterministic automaton using, on the one hand, stochastic information of its Pattern Recognition Module, and on the other, fuzzy information provide by the Speech Processing Module.

2. Loudness Calculation

2.1 Psychoacoustic definitions

A sound becomes a stimulus when it attains the sensorial organ. The stimulus components are the values of excitation, physically measurable. When stimulus corresponds to an audible sound, it becomes a sensation, that can be only qualitative. The loudness (or relative loudness) is the associated value of sensation magnitude to a loudness level. One attributes arbitrarily a relative loudness N = 1 sone, for a sound of 1 kHz with an acoustic level of L=40 dB and a duration of 1 second.

The resolution frequency limit is put in evidence by different types of interactions between sounds having close frequencies. The frequency range, bellow these interactions are produced, is called critical band. The loudness is independent of the band width until this one is inferior to a certain value. Such this analysis, the whole audible frequency range is divided in 24 adjacent critical bands. The passage of a critical band to the superior critical band corresponds, by convention, to a growth of 1 *Bark*.

The *specific loudness* is defined as the value of the retrieved relative loudness to critical band rate. Zwicker has shown that the specific loudness of a pure sound close of 1 kHz (figure 2a), and a white noise (figure 2b), present the form of the excitation pattern of the basilar membrane of the ear (Zwicker and Fast, 1991). The movement of this membrane modulates the ionic current of the ciliated cells stimulating thus the liberation of neuro-transmitter.

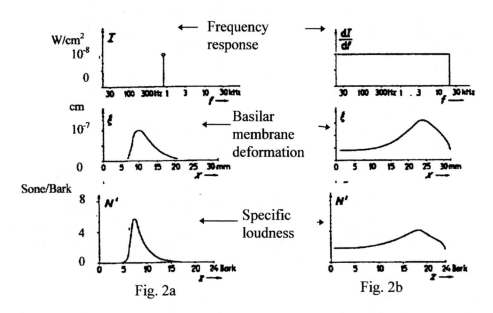

Fig. 2a   Fig. 2b

Fig. 2 : Link between the values of excitation and sensation (specific loudness)

The subjective intensity of a sound to a given level is diminished when another sound is simultaneously present. This is the phenomenon of masking in frequency. A Temporal masking is also observed. It is due to the response time that necessitates the ear when it is submit to a stimulus.

2.2 Loudness Automatic Calculation

The schema for the automatic calculation of the loudness is given in fig. 3. The heart loudness (or main loudness) is the maximal specific loudness widened to an interval of 1 bark.

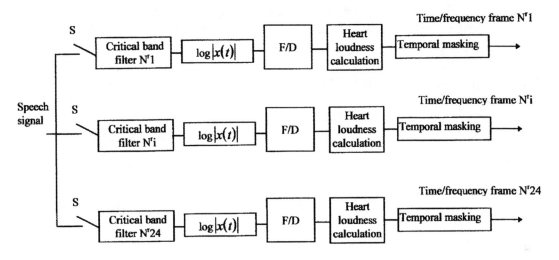

Fig. 3 : Loudness Automatic Calculation

Thank to Dr Stephane Veste, Engineer in the society « Teamlog », we can use a realisation based on the « LabWiev » software. The sampling frequency S varies from 150HZ for the lowest band, to 25 kHz for the highest band. The signal is first filtered by FIR filters having an attenuation of approximately 50 dB per octave. The different filter width bands are given by critical band. The coefficients of difference equations are to the number of 20. The intensity is obtained in calculating the absolute value logarithm of the signal for each critical band. The F/D factor takes into account the fact that sometimes, instead of an ideal plan acoustic field (F=Free), the sound simultaneously arrives, to the point of measure, of all the directions (D=Diffus). The outputs are on the form a time/frequency representation giving a psychoacoustic model of the hearing. These are the (temporal) output impulses of the different frequency bands that are used to be computed by the neuromimetic recognizers A 3D time/frequency representation can be extracted (fig. 4).

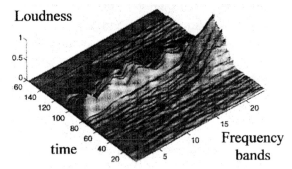

Fig. 4 : 3D representation of the loudness (back view)

The calculation of the loudness concerned words (sometimes short phrases) find in an ordinary vocabulary of control (industrial or militarily).The complete basis has 90 words divided in three categories according to whether they are themselves long, medium or short words. The length of a corresponding frame is then, respectively 177, 203, 270 impulses. To test if the recognition rate can be right, the recording of words was effectuated in an ideal acoustics environment. "LabView" software was excellent for the realisation of the device.

3. Recognition

3.1 Time Delay Neural Network (TDNN)

In Speech Recognition, it is fundamental to take into account the time evolution. Alex Waibel introduced a series of captors highly paralleled to explain the time phenomenon in term of neural networks (Waibel and al, 1987). The superior levels of treatment are thus capable to effectuate outputs comparisons of various instants. The TDNN replies to the two characteristics (Lang and Hinton, 1988) :

- Taking into account temporal relations between input events as the Voice Onset Time (VOT).

- Putting invariant in temporal translation this identification: the window does not need to be precisely centred on the analysed event.

From the described word basis we have elaborated a network architecture of three levels: input, hidden and output levels data (Naranjo M. and Tsirigotis G., 1997). The input level receives the output of the 23 obtained frequency bands in the loudness calculation. Therefore the number of feature units is 23 for the three categories of words. Let us remember that the length of a frame is different according to the length of the word and equals respectively to 177, 203, 270 for the short, medium and long words. The output level possesses 6 units in the three cases, that we can code in binary $2^6 = 64$ words. An important delay length, therefore a strong interconnection, allows to diminish the number of hidden level neurons (of about 300). One uses three different networks, one for each category of words. We have used the excellent SNNS software (Stuttgart Neural Network Simulator) developed, under the direction of Andreas Zell at the « Institute for Parallel and Distributed High Performance Systems ». For the three cases, the obtained results are satisfactory: the average quadratic error is inferior to 0.05 and the recognition rate very close of 100% on the learning.

3.2 Recognised word and its Intelligibility Index Extraction

3.2.1 Determination of the closest binary vectors corresponding to the fuzzy output vectors

The output of the network is a vector where the components have values comprised between 0 and 1. While assimilating it to a fuzzy vector, the Information Theory allows us to extract a recognised word and its intelligibility index.

Let S_i be the TDNN i^{th} output, having for value O_i.

S_i is a element of the set $E = \{S_1 \quad S_2 \quad S_3 \quad S_4 \quad S_5 \quad S_6\}$ and

O_i is a component of the vector $O = (O_1 \quad O_2 \quad O_3 \quad O_4 \quad O_5 \quad O_6)$;

i) $O_i \in \{1 \quad 0\}$, the outputs form a partition: $E = \{E_1 \quad E_2\}; (E_1 = \{S_m\}:0_m = 1), (E_2 = \{S_n\}:O_n = 0)$

example : $(1 \quad 0 \quad 0 \quad 1 \quad 0 \quad 0)$ forms the partition : $E = \{\{S_1 \quad S_4\}, \{S_2 \quad S_3 \quad S_5 \quad S_6\}\}$

ii) $O_i \in [0 \quad 1]$,.

example: $(0.9 \quad 0.01 \quad 0.01 \quad 0.85 \quad 0.1 \quad 0)$ seems close of $(1 \quad 0 \quad 0 \quad 1 \quad 0 \quad 0)$.

The problem, in the presence of a fuzzy vector, is to find the closest binary vector.

Let E_1 and E_2 be a partition of E : we respectively attribute them the fuzzy index Ψ_1 and Ψ_2 with: $0 \leq \Psi_1, \Psi_2 \leq 1$ and $\Psi_1 + \Psi_2 = 1$. The entropy is $H_E = -(\Psi_1 \ln \Psi_1 + \Psi_2 \ln \Psi_2)$, nil when the one of the two indexes is equal to 1 and the other to 0, that corresponds to a binary vector. It is maximal when the two indexes are identical and equal to 0,5, that corresponds to the maximal confusion (total fuzzy) between the two subsets. The attribution of the fuzzy indexes Ψ_1 and Ψ_2 is effectuated from the normed average values of the outputs in the sub-sets, respectively E_1 and E_2, in our example:

$$\Psi_1 = \frac{(0.9 + 0.85)/2}{(0.9 + 0.85)/2 + (0.01 + 0.01 + 0.1)/4} = 0.967; \Psi_2 = \frac{(0.01 + 0.01 + 0.1)/4}{(0.9 + 0.85)/2 + (0.01 + 0.01 + 0.1)/4} = 0.033;$$

We calculate the sub-sets E_1 and E_2 of E minimising the entropy H_E.

Examples:

- On the one hand for $O = \begin{pmatrix} 0.9 & 0.01 & .0,01 & 0.85 & 0.1 & 0 \end{pmatrix}$,

$$\{ \begin{matrix} E_1 = \{S_1 \quad S_4\} \\ E_2 = \{S_2 \quad S_3 \quad S_5 \quad S_6\} \end{matrix} \} \text{ gives } H_E = 0,145 \ ;$$

The closest binary vector of O is $\begin{pmatrix} 1 & 0 & 0 & 1 & 0 & 0 \end{pmatrix}$.

- On the other for $O = \begin{pmatrix} 0.9 & 0.09 & 0.11 & 0.3 & 0.08 & 0.12 \end{pmatrix}$,

$$\{ \begin{matrix} E_1 = \{S_1 \quad S_4\} \\ E_2 = \{S_2 \quad S_3 \quad S_5 \quad S_6\} \end{matrix} \} \text{gives } H_E = 0.405 \ ;$$

$$\{ \begin{matrix} E_1 = \{S_1\} \\ E_2 = \{S_2 \quad S_3 \quad S_4 \quad S_5 \quad S_6\} \end{matrix} \} \text{gives } H_E = 0.386 \ ;$$

The closest binary vector of O is $\begin{pmatrix} 1 & 0 & 0 & 0 & 0 & 0 \end{pmatrix}$.

3.2.2 Determination of the intelligibility index

The intelligibility index is given by mean of the distance between a fuzzy vector and the more closest binary vector or Minkovski's distance : $I(u, v) = \left[\sum_{i=1}^{m} |u_i - v_i| \right]$

example $\{ \begin{matrix} u = \begin{pmatrix} 0.9 & 0.01 & 0.01 & 0.85 & 0.1 & 0 \end{pmatrix} \\ v = \begin{pmatrix} 1 & 0 & 0 & 1 & 0 & 0 \end{pmatrix} \end{matrix}$ The Minkovski's distance is I= 0.37.

4. Control of a Deterministic Automaton using uncertain information

4.1 Formulation of the control problem

Let $A = \left(Q, E, F, \delta, \gamma, q^0, q* \right)$ be an automaton.

- Q is the finite set of t states,
- E the finite set of u inputs,
- F the finite set of v outputs, also called forms (state-classes, patterns),
- δ is a mapping : $QxE \rightarrow Q$,
- γ is a mapping : $Q \rightarrow F$,
- $q^0 \in Q$ is the initial state of the automaton
- $q* \in Q$ a pre-specified final state.

Information system

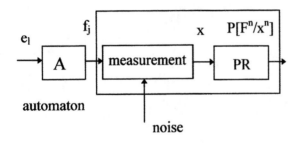

Fig. 6 : The automaton and information System

Example 1

$Q = \{q_1, q_2, q_3, q_4\}; E = \{a, b\}; F = \{f_1, f_2\};$

$q_0 = unknown; \quad q* = q_4$

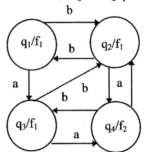

Fig. 5 : transition graph

The knowledge of the output of the automata is only accessible through a noisy measurement subsystem that produces a vector $x \in R^P$ followed by a Pattern Recognition subsystem that gives, at the step n of the procedure, a probability vector $P[F^n / x^n]$ after the complete application to the automaton of a command

$$d^n = d_l = (e_{l_1}, e_{l_2}, \dots, e_{l_r}); d_l \in D.$$

D is the set of all command, d^n connotes the command decided and applied at the nth step. Similarly (q^n, f^n) will be respectively the state and the output of the automaton after the application of d^n.

The control problem is to find a sequence of commands $(d^1 d^2 \dots d^m)$ such as $P[q^m = q*] = 1$ and the total number of inputs has to be minimum, $q*$ being a final state ordered by human voice.

The automaton is controllable but not fully observable. It is obvious that the complexity of the problem is due on the one hand to the presence of the mapping γ in the definition of the automaton, and on the other, to the Pattern Recognition subsystem provides an output probability vector $P[F^n / x^n]$. It is supposed that all the variables whose values are needed to calculate this vector, have been previously obtained by learning.

250

4.2 Modelling the Anticipatory System

In the automaton A, two automata A_Q and A_F can be distinguished, respectively called state and form automaton.

$$A_Q = (Q, E, \delta, q^0, q*)$$

$$A_F = (F, E, \alpha, f^0, f*) \quad \text{with} \quad \alpha = \gamma \delta \gamma^{-1}$$

$$FxE \to F; f^0 = \gamma\left(q^0\right); f* = \gamma\left(q*\right)$$

The nature of the automaton A_F depends on the mapping

- if γ is bijective, it is obvious that A_Q and A_F are equivalent ; then A_F is a deterministic automaton ;
- if γ is surjective, A_F may be a non-deterministic automaton. For this it is sufficient that :

$$\exists (e_i \in E)(f_j \in F): \gamma^{-1}(f_j) = \left(q_{k_1}, q_{k_2}\right);$$

$$\delta\left(q_{k_1}, e_i\right) = q_{l_1}; \delta\left(q_{k_2}, e_i\right) = q_{l_2}; \gamma\left(q_{l_1}\right) = f_{j_1}; \gamma\left(q_{l_2}\right) = f_{j_2}$$

$$\alpha\left(f_j, e_i\right) = \gamma \delta \gamma^{-1}\left(f_j, e_i\right) = \left(f_{j_1}, f_{j_2}\right)$$

The knowledge of the state of A_Q and the forms of A_F is probabilistic since the system of Pattern Recognition provides a probability vector. The Anticipatory System must take a decision d^n in function of the last calculated state probability vector and a fuzzy ordered final state. Then after the application of each command d^n followed by a measure x^n, it is necessary to calculate the state and form probability vectors.

Notations :

Let
- $\Delta\left(e_i\right)$ be the binary transition matrix associated with the input e_i,
- $\delta_D: QxD \to Q$ be the command mapping,
- $\Delta_D(e_l) = \Delta\left(e_{l_r}\right)\Delta\left(e_{l_{r-1}}\right)...\Delta\left(e_{l_1}\right)$ be the transition matrix associated with d_l
- Γ be the binary matrix of the mapping γ.

$$\text{An element } \Gamma_{ij} \text{ of } \Gamma \text{ is such as} \begin{cases} \Gamma_{ij} = 1 & \text{if } \gamma(q_i) = f_j \\ \Gamma_{ij} = 0 & \text{othewise} \end{cases}$$

- $P\left[Q^n / d^n\right]$ and $P\left[F^n / d^n\right]$ be respectively the state and form probability vectors knowing the command d^n
- $P\left[Q^n / d^n, x^n\right]$ and $P\left[F^n / d^n x^n\right]$ be respectively the state and form probability vectors knowing the measure x^n after the command d^n.

Anticipatory scheme

For a command d^n, the states of A_Q and the forms of A_F can be predicted as follows :

$$P\left[Q^n \mid d^n\right] = \Delta_D\left(d^n\right) . P\left[Q^{n-1} \mid x^{n-1}\right]$$

$$P\left[F^n \mid d^n\right] = \Gamma^t . P\left[Q^n \mid d^n\right]$$

Updating scheme

After obtaining the measure x^n, the knowledge on the states and the forms can be updated. Let $P\left[f_j^n \mid d^n x^n\right]$ be the jth component of the vector $P\left[F^n \mid d^n x^n\right]$. We have used a non linear scheme :

$$P\left[f_j^n \mid d^n x^n\right] = \frac{P\left[f_j^n \mid d^n\right] . P\left[f_j^n \mid x^n\right]}{\displaystyle\sum_{k=1}^{v} P\left[f_k^n \mid d^n\right] . P\left[f_k^n \mid x^n\right]} . \text{ Noting that } \Gamma_{ji} = P\left[f_i \mid q_j\right], \text{ Bayes}$$

theorem enables to calculate :

$$P\left[q_j^n \mid d^n f_i^n\right] = \frac{P\left[q_j^n \mid d^n\right] \Gamma_{ji}}{\displaystyle\sum_{k=1}^{t} P\left[q_k^n \mid d^n\right] \Gamma_{ki}} = \left(\Gamma^{-1}\right)_{ji}^n \qquad \text{Then}$$

$$P\left[Q^n \mid d^n x^n\right] = \left(\Gamma^{-1}\right)^n . P\left[F^n \mid d^n x^n\right]$$

Initialisation :

For $n=0$, $\left(\Gamma^{-1}\right)^0$ must be defined since $P\left[Q^0 \mid x^0\right] = \left(\Gamma^{-1}\right)^0 . P\left[F^0 \mid x^0\right]$. In the case of lack of initial data, arbitrary values may be chosen for $P\left[q_j^0 \mid f_i^0\right]$. For example :

$$P\left[q_{j_h}^0 \mid f_i^0\right] = \frac{1}{r}, \forall h = 1, \ldots, r \quad \text{if} \quad \gamma^{-1}(f_i) = \left(q_{j_1}, q_{j_2}, \ldots, q_{j_r}\right)$$

Convergence :

Let f_a^n be the most probable form at step n, before knowing x^n, we get :

$$\lim_{n \to \infty} P\left[f_a^n \mid d^n x^n\right] = 1.$$ In this case, the successive concordance of the predicted values and the information brought by the observation x^n, for the most probable form, are strongly taken into account.

Example 2: From the automaton of the example 1 :

$$\Delta(a)=\begin{pmatrix} 0 & 0 & 0 & 0 \\ 0 & 0 & 0 & 1 \\ 1 & 0 & 0 & 0 \\ 0 & 1 & 1 & 0 \end{pmatrix}; \Delta(b)=\begin{pmatrix} 0 & 1 & 0 & 0 \\ 1 & 0 & 1 & 0 \\ 0 & 0 & 0 & 1 \\ 0 & 0 & 0 & 0 \end{pmatrix}; \Delta\left(d^1=ba\right)=\begin{pmatrix} 0 & 0 & 0 & 0 \\ 0 & 0 & 0 & 0 \\ 0 & 1 & 0 & 0 \\ 1 & 0 & 1 & 1 \end{pmatrix}; \Gamma=\begin{pmatrix} 1 & 0 \\ 1 & 0 \\ 1 & 0 \\ 0 & 1 \end{pmatrix};$$

$$P\left[Q^0 / x^0\right]=\begin{vmatrix} 0.25 \\ 0.25 \\ 0.25 \\ 0.25 \end{vmatrix}; P\left[Q^1 / d^1\right]=\Delta(ba).P\left[Q^0 / x^0\right]=\begin{vmatrix} 0 \\ 0 \\ 0.25 \\ 0.75 \end{vmatrix};$$

$$P\left[F^1 / d^1\right]=\begin{vmatrix} 0.25 \\ 0.75 \end{vmatrix} \quad and \quad after \quad PR: \quad P\left[F^1 / x^1\right]=\begin{vmatrix} 0.4 \\ 0.6 \end{vmatrix};$$

The results are : $P\left[Q^1 / d^1 x^1\right]=\begin{vmatrix} 0 \\ 0 \\ 0.28 \\ 0.82 \end{vmatrix}$

4.3 Control strategy

Every control strategy must satisfy the two objectives previously defined :

- $P\left[q_*^m / x^m\right]=1$, q_* is ordered by human voice,
- Minimum number of inputs applied to the automaton.

The first objective is related to the knowledge that has been obtained on the state of the automaton. A classical measure of this is the value of an information function defined from the state probability vector. The well-known Shannon information function will be used here for its simplicity. When the real state of the automaton is not precisely known, it is absolutely necessary to choose a command d_l after the application of which the measure x will reduce this uncertainty as much as possible. A first criterion L_1 for this choice using information function is then useful.

The second objective is clearly related to the displacements in the state transition graph. A second criterion L_2 may be associated with them that is a function of the length of the shortest path from any state to the final state q_*.

This decision problem is then a multi-criteria's one and a classical way of solving it consists on defining a global criterion that is a weighted sum of L_1 and L_2. Finally, it must be noted that the nature of this control problem imposes a heuristic search of the solution, which is characterised by the determination of the best successive commands using a Bayesian strategy, at each step.

253

Bayesian control strategy

At the step $n+1$, a command $d^{n+1} = d_a$ is chosen according to a Bayesian strategy if :

$$E\left[L(q_i^n, d^{n+1} = d_a\right] \leq E\left[L(q_i^n, d^{n+1} = d_l\right] \forall d_l \in D$$

with $L\left(q_i^n, d^{n+1}\right) = L_1\left(q_i^n, d^{n+1}\right) + \alpha L_2\left(q_i^n, d^{n+1}\right)$

$$E\left[L(q_i^n, d^{n+1} = d_l)\right] = \sum_{i=1}^{t} L(q_i^n, d^{n+1} = d_l) . P\left[q_i^n / x^n\right]$$

$\Rightarrow$ **Criterion** $\left[L_1(q_i^n, d^{n+1})\right]$:

Supposing that the set x of all the measures x is finite, the a priori probability of q_i when the state of the automaton is q_i is defined by :

$$P\left[q_j / q_i\right] = \sum_x P\left[q_j / x\right] . P\left[x / q_i\right] \text{ where :}$$

- $P\left[q_j / x\right] = P\left[q_j / F\right] . P\left[F / x\right]$. As seen before, the value of $P\left[q_j / F\right]$ changes during the control process. It is set here to its initial value $P\left[q_j^0 / F^0\right]$.

- $P\left[x / q_i\right] = P\left[x / F\right] . P\left[F / q_i\right]$; $P\left[F / q_i\right]^t$ is the ith row of the binary matrix Γ.
- $P\left[F / x\right]$ and $P\left[x / F\right]$ are learning data.

The confusion index associated with the state q_i is defined by the Shannon's information function :

$$H(q_i) = -\sum_{j=1}^{v} P\left[q_j / q_i\right] . \ln P\left[q_j / q_i\right]$$

It is known that $0 \leq H(q_i) \leq \ln t$ and it can be proved that :

i) $H(q_i) = 0$ if and only if the form f_i of q_i contains only the state q_i and the subset x_i of measures associated with f_i is such as $x_i \cap x_j = \phi, \forall j \neq i$ (x_j is the subset of measures associated with f_j). The state q_i is then absolutely discriminated by a measure x.

ii) When $H(q_i) = \ln t$ which implies that $P\left[q_j / q_i\right] = \frac{1}{t}, \forall j$, then the state q_i is totally non-discriminated.

Let $q_k^{n+1} = \delta\left(q_i^n, d^{n+1}\right)$ and $L_1\left(q_i^n, d^{n+1}\right) = H\left(q_k^{n+1}\right) - H\left(q_i^n\right)$;

$L_1 < 0$ indicates that the resulting state q_k^{n+1} after the application of the command d^n is better discriminated on an average than the starting state q_i^n.

254

Example 3 : From the transition graph of the example 1 and a given confusion matrix

$$P[q_j / q_i] = \begin{pmatrix} 0.3 & 0.3 & 0.3 & 0.1 \\ 0.3 & 0.3 & 0.3 & 0.1 \\ 0.3 & 0.3 & 0.3 & 0.1 \\ 0.1 & 0.1 & 0.1 & 0.7 \end{pmatrix}; \begin{array}{l} H(q_1) = H(q_2) = H(q_3) = 1.3 \\ H(q_4) = 0.94 \end{array} \quad \text{with } D = \{a, ba, b\}$$

state	input d^n		
	a	ba	b
q_1	0	-0.36	0
q_2	-0.36	0	0
q_3	-0.36	-0.36	0
q_4	+0.36	0	+0.36
L_1	+0.36	-0.72	+0.36

In regard of L_1, the choice is $d^n = ba$.

$\Rightarrow$ **Criterion** $\left[L_2(q_i^n, d^{n+1}) \right]$:

A length c_{ij} is defined for each arc $\left(q_i, q_j \right)$ of the transition graph such as :

$$c_{ij} = \begin{vmatrix} 1 & if & \exists e_h \in E : \delta(q_i, e_h) = q_j (j \neq i) \\ 0 & if & i = j \\ \infty & otherwise \end{vmatrix}$$

The Bellmann-Kalaba's algorithm enables to calculate a matrix V, the value of its element v_{ij} being the length of the shortest path which leads from q_i to q_* through q_j. For a command d^{n+1} such as $q_j^{n+1} = \delta(q_i^n, d^{n+1})$, the criterion L_2 is then defined by :

$$L_2(q_i^n, d^{n+1}) = v_{ii} - v_{ij}$$

Example 4 : From the transition graph of the example 1 and $q_* = q_4$

$$V = \begin{pmatrix} 2 & 2 & 2 & 2 \\ 3 & 1 & 3 & 1 \\ 4 & 2 & 1 & 1 \\ 4 & 2 & 2 & 0 \end{pmatrix}; \text{ with } \alpha = \frac{1}{5}$$

state	input d^n		
	a	ba	b
q_1	0	0	0
q_2	0	2	2
q_3	0	1	1
q_4	2	0	2
αL_2	0.4	0.6	1
$L_1 + \alpha L_2$	-0.04	-0.12	+1.36

The best choice in regard of the Bayesian strategy is $d^n = ba$. The weight α can be modified during the control process in order to favour either the discrimination of the states if the knowledge of the situation of the automaton is not sufficient, or the shortest displacement in the transition graph.

$\Rightarrow$ **Taken into account different word orders with its intelligibility index**

All state $q_u \in Q$ can be the one of q_*, and each is associated with its intelligibility index I_u. We have now a criterion L_2 composed of partial sub-criteria $L_{2_u}\left(q_i^n, d^{n+1}\right)$ calculated by mean the Bellmann-Kalaba's algorithm. The value of its element v_{uij} is the length of the shortest path which leads from q_i to $q_* = q_u$ through q_j.

So, $L_{2_u}(q_i^n, d^{n+1}) = v_{uii} - v_{uij}$, and $L_2\left(q_i^n, d^{n+1}\right) = \sum_{u=1}^{t} I_u L_{2_u}\left(q_i^n, d^{n+1}\right)$.

D is composed of the t x t commands that enable to go from $q_i\left(\forall i\right)$ to q_*. For the set of commands D, all the elements L_1 and $L_{2_u}\left(q_i^n, d^{n+1}\right)$ of the matrix $L = \left\{L\left(q_i, d_l\right)\right\}$ can be calculated in advance. At the step n+1, the only computation necessary to determine the command $d^{n+1} = d_a$ is the matrix product

$$(L_1 + \alpha(I^t L_2)).P\left[Q^n / x^n\right] = L' = \left\{L'\left(d_l\right)\right\} \text{ with}$$

$$L_1 = \left(L_1\left(q_i^n, d^{n+1}\right) \quad \cdots \quad L_1\left(q_t^n, d^{n+1}\right)\right);$$

$$I^t = \left(I_1 \quad \cdots \quad I_t\right)^t;$$

$$L_2 = \left(L_{2_1}\left(q_1^n, d^{n+1}\right) \quad \cdots \quad L_{2_t}\left(q_t^n, d^{n+1}\right)\right);$$

$$P\left[Q^n / x^n\right] = \begin{vmatrix} P\left[q_1^n / x^n\right] \\ \cdots \\ P\left[q_t^n / x^n\right] \end{vmatrix}$$

d_a is the chosen command if $L'\left(d_a\right) \le L'\left(d_l\right)\forall d_l \in D$. The decision unit corresponding to this strategy is simple and thus, the control of the automaton can be very fast.

The weight α that occurs in the global criterion L has to be defined in advance but can be modified during the control process in order to favour either the discrimination of the states if the knowledge of the situation of the automaton is not sufficient, or the shortest displacement in the transition graph. That is the case when the automaton is close of a final state ordered by voice having a good intelligibility index.

5. Conclusion

We have presented a method which allows to introduce a vocal control system in the production units.

From a speech signal of a word, it is possible to recognise the right control order and label a intelligibility index, providing the intelligibility level of the transmission system. The time/frequency procedure using the loudness calculation allows to take the psychoacoustic characteristics of hearing into account. The LabView software has revealed itself excellent for the realisation of the time/frequency device. We have used the very complete and documented Stuttgart Neural Network Simulator where the Time Delay Neural Network has permitted to solve two important problems in Speech Recognition : taking into account the temporal relations between input events and giving them invariant in temporal translations.

In a very noisy acoustic environment, the speech intelligibility considerably decreases . At some stage and due to the noise of transmission system that throws back a different perceived order from the vocal order, several compatible orders with the present state of the machine controlled are plausible. We have elaborated a new automaton control problem where the speech processing module is put in the feedback loop of the machine. The human - machine interface provides orders taken about information from the past and present state of the surrounding environment. We have the essential elements of an Anticipatory System and we have provided a decision module control which chooses the right order for the convergence of the machine. This problem has been solved here by defining a criteria that depends on an information function which measure the probabilistic knowledge of the state of the automaton, and on shortest path length function, and by using a Bayesian decision strategy.

References

Lang K. and Hinton G.E. (1988). A time Delay Neural Network Architecture for Speech Recognition, TR CMU-CS n° 88-152, Carnegie Mellon University.

Naranjo M., Richetin M. and Rives G. (1980). Study and realisation of a computer aided diagnosis and control system for a rolling mill. DGRST Contract Report 79-7-0299, edited by LASMEA, Blaise Pascal University of Clermont Ferrand (in French).

Naranjo M. and Tsirigotis G. (1997). Time Delay Neural Network in a Psychoacoustic Model of Hearing, SCI'97, World Multiconference on Systemics, Cybernetics and Informatics, Caracas (Venezuela), July 7-11.

Waibel A., Hanazawa T., Hinton G.E., Shikano K. and Lang K. (1987). Phoneme Recognition Using Time Delay Neural Networks, Technical Report TR-1-0006, edited by Advanced Telecommunication Research Institute, Japan.

Zwicker E. and Fast H. (1991). Psychoacoustics Facts and Models, Springer Verlag.

The role of anticipation in multilevel biological systems

V.Csányi
Department of Ethology
ELTE University, Budapest
Hungary

Abstract

First I want to illustrate the ability of animals to anticipate by the case of the paradise fish. We observed that the paradise fish assume intentionality of any object which has a visible mass and two horizontal eye-like spots. Assuming intentionality was always followed by exploratory behavior during which the paradise fish built up a predictive model of the behavior of the object assumed to be intentional in their minds. After a short experience the paradise fish were able to predict certain behavior of an "eye"- carrying object. The anticipatory system of the paradise fish includes both inherited and learned elements.

The second task of my discussion is to show the relation of the ability of anticipation to the properties of the higher system in which the anticipator is a component. I will discuss the levels of the biological organization and the possible function of anticipation which depend entirely on a given level of the whole organization. On a higher level of organization where the organism is only a component the anticipated events are regular parts of cyclic processes which constitute the higher systems. Anticipation is bound to lower organizational levels.

The third part of the paper is devoted to evolution. I want to show how the incessant struggle of the evolutionary systems toward stability created anticipation at the lower levels and how the same process will eliminate anticipation entirely from the behavior of the whole system.

Finally I deal with anticipation and society. I discuss tradition, rule and algorithm based forms of anticipation.

Keywords: anticipation, evolution, model, biological systems, system's level

1. Introduction

It is a quite recent view in biology to accept anticipation, a property to predict future events and react to them properly, as characteristic trait of an animal. Earlier the reactive "brute" concept of animals of the behaviorists, or "the properties are mapped in to the blue print of the DNA" idea of the molecular biologists left no room for intelligence and prediction or anticipation by any organism except human.

Anticipation has a simple cybernetic explanation if we consider the organism as a system able to model its environment. A modeling apparatus can be a relatively simple mechanism with a number of feedback loops, capable of changing its own parameters. Craik (1943) and MacKay (1951) were the first proponents of the concept that the nervous system of the animal is a device for modeling of the environment. Basic system theoretical treatment of this idea by Rosen (1985) is just a little older than a decade.

CP437, *Computing Anticipatory Systems: CAYS--First International Conference*
edited by Daniel M. Dubois © 1998 The American Institute of Physics 1-56396-827-4/98/$15.00

2. Anticipation by brain-models

The heart of the idea is goal directed behavior for which MacKay gave an exact definition. Definition of goal "X" pursued by a system "A" would be as follows. Let "Y" represent "A" and its present environment, and let "X" equal "A" in a state in which it has reached its goal. Then "A" shows goal-directed behavior if it performs internal or external movement that minimizes the difference between "X" and "Y" (Fig.1).

Goal directed behavior in animals

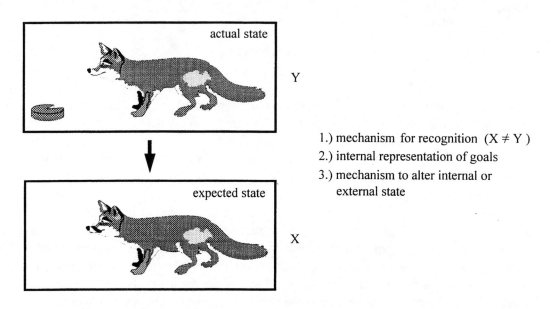

actual state Y

expected state X

1.) mechanism for recognition ($X \neq Y$)
2.) internal representation of goals
3.) mechanism to alter internal or external state

Fig. 1 Schemata of goal directed behavior

This definition of goal-directed behavior is valid not only for living organism but for artificial cybernetic constructions as well. It follows from the definition that a system capable of showing goal-directed behavior has to possess certain mechanisms. It is necessary that the system be able to distinguish between states "X" and "Y" that is to have some kind of *recognition sub-system*, which, in case of artificial systems, can be very simple (e.g. a heat sensor of a thermostat). Furthermore it is necessary that it be able *to change its own state*, based upon this recognition process. Finally it is necessary that the system carry *internal representation* of the possible goals in order to recognize differences and change.

Subsystems of recognition (perception) and change (behavior) are well-known in living organisms. Internal representations are made possible by mechanisms of genetic and neural memory. The question of the organization of the internal representation is, however, more complex. The adaptive "goal" of an organism is to survive in its surroundings, and the internal representation must, therefore be of an environment that in some cases includes the organism itself.

The internal representation of the environment can be considered as a construction, which in essence is a model (Craik 1943), more precisely, a dynamic model of the environment. The "model" is meant in the system science sense: a model of a complex system is always a simpler system whose components and their interactions correspond in some way to components and some interactions of the complex system (Mesarovic 1964). Model building is always a simplification involving a special identification between two different systems, of which one is the model and the other is the system being modeled. The effects of operations on the model are used to predict the behavior of the system being modeled.

The most important biological function of the animal brain is the construction of such a dynamic model of the environment. This model includes environmental factors and interactions most important for the survival and reproduction of the animals, the continuous maintenance and operation of the model, and the use of the obtained data for prediction or *anticipation* of events in the interest of the survival and reproduction of the animal (Csányi 1986b, 1988, 1993).

It is possible to model a complex system very simply but also in more complicated ways. Simple models can anticipate some simple but important events for adaptation. For example, an animal may anticipate that dark periods are followed by daylight. This knowledge is in the form of a model of the dark-light rhythm of the environment that establishes the diurnal cycle in the nervous system of animals. The mechanisms are basically genetic. Different species may be active during the dark or during the light period, but all have the inherited "genetic memory" of the periodicity of this environment. Both genetic and neural memory contribute to the final performance.

More complex phenomena can also be modeled and complex events anticipated. A wolf's brain presumably models not only diurnal cycles but the behavior of prey, fellow members of the pack, order of dominance within the pack, effects of past events, and many other things.

According to MacKay, the neural model is not only a simple projection, but a kind of complex reconstruction, which also contains instruction of the possible behavior of the organism in response to the stimuli of the external world. The animal's activity is not simply organized by responses to the external stimuli, but also by anticipations and analyses of situation based on the internal analysis of the model formed in the nervous system (Gallistel 1980). Information analyzed in this way probably includes that involved in eliciting fear (Hebb 1946), orientation (Sokolov 1960), attack and defense (Archer 1976) and avoidance of predators (Csányi 1985, 1986a, 1993)

2.2 Anticipation by a fish

I want to illustrate the ability of animals to anticipate by the case of the paradise fish based upon ethological observations in our laboratory. We studied avoidance behavior of the paradise fish in the presence of a living pike, a living goldfish, and various fish-like dummies. They were in semi-natural environments or aquatic shuttle boxes. Detailed description of the experiments and their analysis are in the above-cited articles.

First we examined the reaction of naive paradise fish to their first exposure to an individual of another species of fish (a goldfish or a satiated pike) (Csányi 1985).

During the encounters the paradise fish always approached the other fish and examined it thoroughly, swimming around it several times and carefully avoiding its head. The other fish did not react to the exploration of the paradise fish: Goldfish continued their slow swimming, with pikes usually rested under the surface of the water.

Further experiments were carried out to examine the reaction of the paradise fish to attack by a predator. Ten to twelve paradise fish were placed for a short time in an aquarium containing a hungry pike which made several attempts before it was able to catch one paradise fish. During the attacks the paradise fish were chased and some were even bitten by the pike. After the first catch the remaining paradise fish were rescued and put into another aquarium. The following day the paradise fish that had been chased were tested individually in the aquarium of a satiated pike.

We found that in the case of the chased group a well pronounced *avoidance* were observed. The conclusion was that the first encounter with a hungry pike initiated a learning process to recognize and avoid a threatening predator.

This conclusion was given further support, when we successfully transformed a harmless goldfish into an object to be avoided using electric shocks. We thought that if the main motivating factors for avoidance were pain and fear then mild electric shocks applied in the presence of a goldfish would make it fear-inducing and it would be avoided on the next encounter, and this assumption was proved in experiments.

An important finding was that only members of other species can be transformed in this way. No treatment could transform a conspecific to be an avoided object.

In a further experiment we investigated whether there were key stimuli connected to predators important in predator recognition and in the avoidance learning process in the paradise fish.

We experimented with a living goldfish and a series of dummies which looked more and more like a fish. We carried out the usual conditioning experiment in the smaller shuttle tank, except that after the habituation period (6 trials) we put a dummy into the dark compartment as well. In some of the experimental groups we applied electric shocks (50mA) as punishment in addition to the dummy (Csányi 1986a). We completed the experiment with groups where the paradise fish found a <u>live</u> goldfish instead of a dummy in the dark compartment and got mild 50mA shocks in the presence of the goldfish during conditioning. Of the results of this complicated experiment the most striking, the change in entrance latency, is illustrated in Fig. 2.

Increase in latency in groups treated with various dummies (Csányi, 1985)

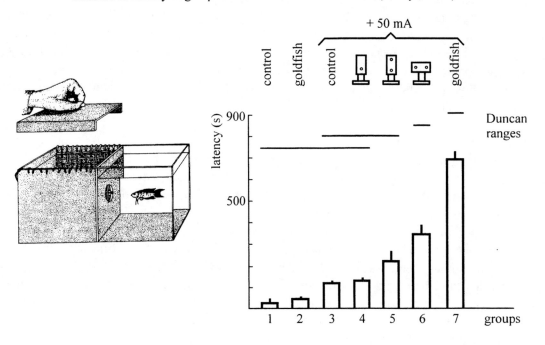

Fig. 2 Increase in latency in groups treated with various dummies. Bars represent means ± SE. group sizes varied between 18 and 79. Duncan ranges are also shown.

Neither the dummies nor the presence of a goldfish nor the electric shocks alone caused significant changes in latency.

Latency significantly increased, however, if punishment was applied in the presence of other dummies. We got the largest increase in the group conditioned in the presence of a live goldfish, and latency also increased remarkably, although to a somewhat lesser extent, in the groups where the paradise fish met a dummy with two lateral eye-like spots. The increase in latency was smaller in the presence of a dummy with two vertically arranged lamps.

Besides recording the data, we also watched the extremely interesting behavior of the paradise fish in this experiment. The paradise fish that were given shocks in the presence of the goldfish or the dummies with "eyes" quickly left the dark compartment, but they usually turned round immediately, swam back to the gate and looked inside from the light area. No behavior like this could be observed in the case of the other dummies (Csányi and Dóka 1992).

Judging from the results of this experiment we can maintain that eyes or lateral eye-like spots on a body have a <u>key stimulus</u> character used in learning for the paradise fish. A remarkable amount of learning can be observed in the presence of the dummies with the key stimulus and if the dummy is "moving", like the goldfish for example, a single trial is virtually sufficient to make the paradise fish avoid the dark compartment.

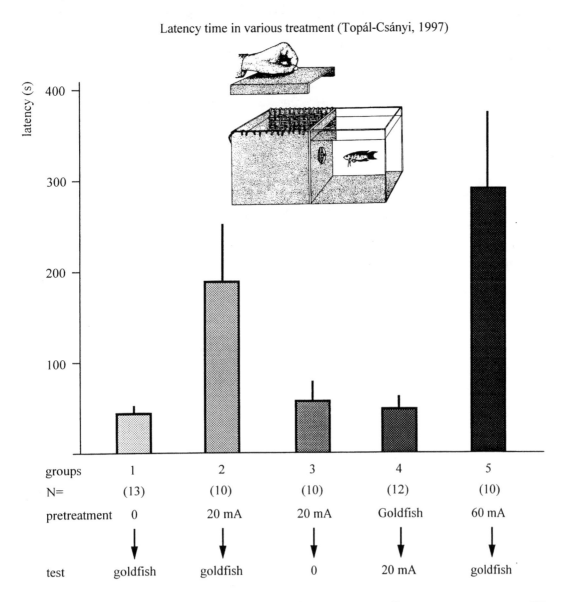

Fig. 3 Changes in latency in groups with various treatments. Bars represent means ± SE.

In the next experiments we made the dynamics of semi-natural environment of the paradise fish more complex. After a series of habituation trials in which the animals let explore the shuttle tanks for several minutes we performed an experiments with two main trials. During the first trial the various groups of paradise fish encountered in the dark compartment either goldfish, or shocks of different intensity. During the second main trial some of the groups found a goldfish restricted by a perforated piece of Plexiglas in the dark compartment and only the group which met goldfish at the previous trial were shocked with 20mA current. The effect of the treatment is shown on Fig.3. Regarding the latency time in boxes where the goldfish were significant effect was found (F(5,86)=5.26, p<0.001).

The results can be summarized as all the fish which were shocked on the first main trial in an empty compartment avoided goldfish in the second one. There was no any effect found if the encounters were in reverse that is goldfish at the first and shocks at the second trial (Topál and Csányi 1997).

The most simple explanation of these experiments is that the paradise fish have *intentionality hipotheses* about other living organisms. We observed that the paradise fish assume intentionality of any object which has a visible mass and two horizontal eye-like spots. Assuming intentionality was always followed by exploratory behavior during which the paradise fish built up a predictive model of the behavior of the object assumed to be intentional in their minds. After a short experience the paradise fish were able to predict certain behavior of an "eye"- carrying object. They completely ignore it if it proved to be harmless during the learning period, but carefully avoid it if their experience were pain- or fearful. The anticipatory system of the paradise fish includes both inherited and learned elements and helps it to adapt into a complex ecosystem.

2.3 Anticipation and system levels

If we examine the relation of the ability of anticipation by an organism to the properties of the higher system in which the anticipator is only a component we found that on a higher level of organization the anticipated events are *regular parts* of cyclic processes which constitute the higher systems. Predators of the paradise fish are regularly appears during the replication of the ecosystem in which paradise fish are components. The exact time of the predator appearance or the exact form of its body is not determined, but the regular appearance of harmful swimming creatures with two eyes is a constant feature of the water environment.

Mechanisms of anticipation in living organisms

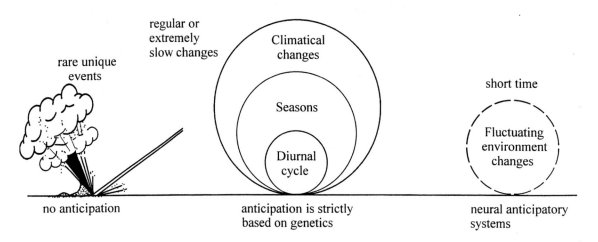

Fig. 4 Various anticipatory mechanisms are bound to the cyclical changes of the biosphere

Anticipation evolved in the paradise fish and in other livings as well is mere a reflection of a certain feature of the higher level organization.

Appearance of a predator, an event which belongs to a cyclic process of a system, to the reproduction of predators in the ecosystem, appear as random at the organizational level of the prey component, but it's probability is one at the higher system level (Csányi 1989a). The ability to predict seems to be a highly evaluated property at the level of the prey component but it is only part of a predetermined dynamic cyclic process at the higher level. The "future" which is anticipated by the prey is the "present" for the ecosystem (Fig. 4).

Predator-prey interaction in which not only the prey animal is able to anticipate but the predator as well is evolved to maintain the stability of the ecosystem. Without anticipation of either side, the fine balance between predators and prey would be harmed with the breakdown of the ecosystem.

All evolutionary system can be characterized by the constant replication of components (Csányi 1989b). Each component produced in surplus and disappears in some breakdown process like predation. During evolution there is a constant tendency to increase the stability of components at every level. Information acquired about the next higher level processes is the essence of anticipation, that is anticipation is a form of knowledge about the higher system by which its stability is maintained. In the same time anticipation as a mechanism disappears at the highest level, there is no such trait of the whole system, the biosphere could be found. Students of evolution maintain that evolution is blind, there is no goal or definite direction of the whole process. Therefore anticipation has no any roles at the highest level, it is simply a mechanism which uses information about the existing high level processes to maintain stability at the lower levels.

Finally I deal with anticipation and society. I discuss tradition, rule and algorithm based forms of anticipation.

3. Anticipation in human societies

Evolutionary process led to the emergence of humans and society in which new mechanisms of anticipation appeared but the basic relation of this trait to the levels of organization has not been changed (Csányi 1992a, 1996).

Evolutionary emergence of the conceptual thought based on language of man resulted in a fundamentally new way of model making (Csányi 1992b). Mental superstructures built up from linguistic concepts also reflect experiences, and therefore they can be regarded as models of the outer world. And their values are also dependent solely on their predictive and explanatory power (Csányi 1996).

The medium of human culture is language which is based upon the biological linguistic competence of the species. By linguistic means the past and the present can be created, natural phenomena can be decomposed into their constituents and new virtual structures can be created out of them.

Constant interaction among the members of the early group-cultures allowed the construction of a complex *supermodel* of the individual and its environment based on a large amount of evidence collected by the whole group (Csányi 1988, 1989a, 1992a). This supermodel can be regarded as a type of *social mind*, which is made up of the experiences and thought processes of the individuals, but which has far greater autonomy than has any one of the individuals. Through stories and myths it enables the past to be recalled, providing a mental and behavioral framework drawing on ancestral experiences.

3.1 Anticipation based on tradition and rules

Predicting and anticipation have a firm role in the social supermodel. In animals only the individual experiences provide information about the environment, about the complex nature of the next higher level of biological organization. In society tradition, experience of past individuals, experience of the previous generations are available for constructing the structure of the environment, this enhances the power of anticipation enormously but it never makes it independent of the social supermodel.

This enhanced form of anticipation is based on tradition but there is an other aspect of society which makes it very different from any biological system.

Rule following behavior is also a species-specific feature of man, which can be defined as a tendency to obey rules extracted from social interactions. Language is one of the most manifest forms of it. However, rules are not limited to language. The kinship system for an example is, in fact, social affection regulated in a defined way, with family relations being manifested in rules of behavior and in respecting the rules. Making of artifacts also depends on following rules, activated by the fondness for objects. This rule following is also manifested in elaborating ideas. A habit or a technique may be characterized by a description or through a fixed system of given rules, which means that the shaping of an idea, in our mind, or in our behavior, or in any other way, is reflected just like the shaping of an object in the successive application of a set of rules. Obeying the rules is the most essential biological feature of man.

Human cultures, in their essence, display systems of rules operated continuously by social affection and the fondness for artifacts. Systems of rules concerning human relations are understood within the cultural group, as well as the rules concerning the preparation, the use, the exchange, and the production of objects and rules relating to the genesis, the values, the functioning, and the history of the culture. Language is the general communication system which, once again, consists of a system of rules which intervenes in shaping and transmitting the various rules of the culture, and thus reflects the fullness of the given culture.

Anticipation inside in such a rule govern system became very peculiar because the existing social structure, the environment for the individual, is determined by and large the network of the rules. Prediction and anticipation in society is successful only if it is based on the recognition of these rules.

The role of anticipation either based on tradition or based on rule recognition is the same as in case of model making animals, it transfer information from the higher organizational level to the lower and helps to maintain stability.

3.2 Limitless anticipation by the use of algorithms

Language permits a particularly precise form of conceptual thinking to develop. Combined with rule following behavior logic and analytic thinking can emerge, which led to the construction of artificial rule systems like the algorithms. With algorithms man could search enormous virtual spaces for usable abstract structures which can be used as models for natural phenomena. A clear example is physics which using mathematical algorithms made the most important discoveries of science. Anticipation by using algorithms entirely different from anticipation based on individual experience. Each algorithm creates a conceptual space which can be searched as a first step of the new type of anticipation, the next step is to find a natural phenomena which have a more or less similar dynamics to the objects find in the algorithm created conceptual space, and then the third step is making a new model of the given part of reality. A well known example of such use of algorithm for new discovery is the application of game theory to animal behavior.

This new form of anticipation is seemingly limitless. Virtual spaces and their abstract objects are also subject of evolution, therefore new and ever newer algorithms emerge offering an endless way to search reality.

To summarize: anticipation is a system property of multilevel systems. It is an important mechanism in biological and social systems which transfer information among organizational levels and help to maintain stability of the whole system. With abstract thinking anticipation provide a limitless searching tool to humanity.

References

Archer, J. (1976). The Organization of Aggression and fear in Vertebrates. In: Bateson P.P.G., Klopfer, P.H. (eds.), Perspectives in Ethology, vol. 2, Plenum, pp. 231-298.

Craik, K.J.W. (1943). The nature of explanation. Cambridge Univ. Press, Cambridge.

Csányi, V. (1985). Ethological Analysis of Predator Avoidance by the Paradise Fish (Macropodus opercularis). I. Recognition and Learning of Predators. Behavior 92:227-240.

Csányi, V. (1986a). Ethological Analysis of Predator Avoidance by the Paradise Fish (Macropodus opercularis). II. Key Stimuli in Avoidance Learning. Anim. Learn. Behav. 14: 101-109.

Csányi, V. (1986b). How is the Brain Modelling the Environment? A Case Study by the Paradise Fish. In Montalenti G and Tecce G (eds.), Variability and Behavioral Evolution, Proceedings, Accademia Nazionale dei Lincei, Roma, 1983, Quaderno No. 259, pp. 142-157.

Csányi V. (1988): Contribution of the Genetical and Neural Memory to Animal Intelligence. in: H.Jerison and Irene Jerison (eds.) "Intelligence and Evolutionary Biology" Springer- Verlag, Berlin pp.299-318

Csányi, V. (1989a): Origin of Complexity and Organizational Levels During Evolution. in: Wake, D.B. and Roth, G. (eds.) "Complex Organizational Functions: Integration and Evolution in Vertebrates", John Wiley & Sons LTD. pp.349-360

Csányi, V. (1989b). Evolutionary Systems and Society:a general theory. Duke University Press, Durham, pp. 304.

Csányi, V. (1992a): Ethology and the Rise of the Conceptual Thoughts. In: J. Deely (ed.) Symbolicity" University Press of America, Lanham, MD. p. 479-484

Csányi, V. (1992b): Nature and Origin of Biological and Social Information. in: K.Haefner (ed.) "Evolution of Information Processing Systems" Springer, Berlin, pp.257-281

Csányi, V. (1993): How genetics and learning make a fish an individual: a case study on the paradise fish. in: P.P.G. Bateson, P.H.Klopfer and N.S.Thompson (eds.) "Perspectives in Ethology" Vol.10. Behaviour and Evolution. Plenum Press, New York, pp. 1-52

Csányi, V. (1996):Organization, function, and creativity in biological and social systems. in: K.E. Boulding and E.L.Khalil (eds.) "Evolution, Order and Complexity", Routledge, London pp.146-181

Csányi, V. and A. Dóka (1992): Learning Interactions Between Prey and Predator Fish. Mar. Behav. Physiol., XXII 63-79

Gallistel, C.R. (1980). The Organization of Action: A New Synthesis. Lawrence Erlbaum, Hillsdale, N.J.

Hebb, D.O. (1946) On the nature of fear. Psychol.Rev. 53:259-276

MacKay, D.M. (1951-52) Mindlike Behavior of Artefacts. Brit. J. Phil. 2:105-121

Mesarovic,M.D.(ed.)(1964) Views on General System Theory. John Wiley, New York, pp.5

Rosen, R. (1985) Anticipatory Systems. Pergamon Press, New York

Sokolov, E.N. (1960) Neuronal Models and the Orienting Reflex. In: Brasier M.A.B.(ed.) "Central Nervous System and Behavior", Macy Foundations, New York

Topál, J. and Csányi, V. (1997) The role of mental construction in avoidance behavior of the paradise fish (Macropodus opercularis L.) (submitted)

Robust Near-Optimal Control Via Unchattering Sliding Mode Control

Giorgio BARTOLINI, Saverio SANNA, Elio USAI

Dipartimento di Ingegneria Elettrica ed Elettronica
Università di Cagliari
Piazza d'Armi, 09123 Cagliari, Italy
tel/fax. +39 (70) 675.5876/5900
e-mails: {giob,sanna,eusai}@diee.unica.it

Abstract

A control scheme for the implementation of an optimal control law for a class of systems in presence of modelling mismatches, actuators nonidealities and external disturbances is proposed. The ideal optimal control law is applied to the system model and the corresponding state evolution is considered as a reference trajectory for the real system. The latter is fed with the sum of the ideal control action and a compensatory control evaluated by a state feedback non-linear controller which resort to the robustness properties of the variable structure control with sliding modes. The sufficient conditions for the effectiveness of the proposed procedure are discussed and a simple example is given.

Keywords: Optimal control, Nonlinear systems, Model tracking, Second order sliding modes, Chattering elimination.

1. Introduction

The control problem of physical, social, or economic systems is usually related with the aim of maximizing the return from, or analogously of minimizing the cost of the whole operation (optimal control problem). Furthermore possible constraints both on the control variables and on the system dynamics must be taken into account. A quite general formulation of the optimal control problem is the following

Problem 1: *Given the system dynamics*

$$\dot{\mathbf{x}}(t) = \mathbf{F}\big[\mathbf{x}(t); \mathbf{u}(t); t\big] \quad t \in \big[t_i; t_f\big] \subseteq \big[T_0; T_1\big] \tag{1}$$

subjected to the following constraints

$$\mathbf{x} \in \Xi \subseteq \mathbf{R}^n \tag{2}$$

$$\mathbf{u} \in \Omega \subseteq \mathbf{R}^m \tag{3}$$

$$\mathbf{x}(t_i) = \mathbf{X}_i \tag{4}$$

$$\mathbf{x}(t_f) \in S \subseteq \Xi \times \big[T_0; T_1\big] \tag{5}$$

find an admissible control law $\mathbf{u}(t)$ *such that the following performance index is minimized*

CP437, *Computing Anticipatory Systems: CAYS--First International Conference*
edited by Daniel M. Dubois © 1998 The American Institute of Physics 1-56396-827-4/98/$15.00

$$J = K\left[\mathbf{x}(t_f); t_f\right] + \int_{t_i}^{t_f} L\left[\mathbf{x}(t); \mathbf{u}(t); t\right] dt \qquad (6)$$

where $\mathbf{x}(t)$ *is the n-dimensional state vector,* $\mathbf{u}(t)$ *is the m-dimensional control vector,* $\mathbf{F}[\cdot]$ *is an n-dimensional vector of smooth functions, and* $L[\cdot]$, $K[\cdot]$ *are positive semi-definite smooth functions all satisfying the usual conditions for the existence of a solution.* □

Optimal control theory is the theoretical basis for managing such class of control problems and it constitutes a well known tool of the modern control theory (Athans, 1966; Kirk, 1970; Leitmann, 1981). The solution of an optimal control problem by means of the optimal control theory is usually not an easy task and it involves the solution of either a two point boundary value differential problem or a dynamic programming problem. This fact implies that the solution of the optimal control problem can be obtained in an explicit form only for some particular cases so that resorting to numerical algorithms for finding the problem solution is necessary, and the optimal control law is known only as a time sequence which must be implemented as a feedforward control. Furthermore, the solution of the optimal control problem can be defined as a state feedback control law only for special classes, e.g., LQR problems, (Athans, 1966; Kirk, 1970; Leitmann, 1981).

Various numerical procedures addressed to solve the mathematical problems arising from the specific optimal control problem are reported in the literature based both on dynamic programming techniques and on shooting methods (McGill and Kenneth, 1964; Larson, 1967; Lastmann, 1978; Shimizu and Ito, 1994; Sanna and Usai, 1996). All cited techniques are based on the knowledge of a model of the dynamical system and the solution of the optimal control problem can be expressed as a time sequence of values of the control input or can be derived by evaluating a proper set of parameters, usually the initial values of the co-state variables, which are used for defining the optimal control law.

From the above considerations it is evident that, usually, the optimal control law constitutes a model based feedforward control strategy depending on future values of the state, i.e., the desired final state vector, and the effectiveness of the computed optimal control law depends on the model accuracy and on the influence of disturbances. In the real word the perfect knowledge of the system dynamics and the absence of disturbances is utopia so that the system behaviour is at most quasi optimal even if the optimal control input is applied.

If the computational time is negligible with respect to the system dynamics, one can solve the optimal control problem on line using the actual values of the system state variables so implementing a feedback control scheme. This procedure cannot be applied in general so that the implementation problem of the optimal control must be considered. Usually the solution of the optimal control problem is considered as a reference trajectory and the aim of the control is to keep the trajectory error as small as possible by using a feedback control strategy.

When considering the implementation problem of a control law, the robustness and disturbance rejection properties of the closed-loop system must be analyzed. In recent years the robustness of classical LQ regulators has been deeply investigated and their characteristics have been enhanced by merging the robustness properties of the H_∞ design (Doyle *and al.*, 1989,1994; Stoorvogel, 1993).

A control methodology that is effective in presence of system uncertainties and external disturbances is the variable structure control with sliding modes; in fact, once in sliding mode, the system behaviour depends only on the definition of the sliding manifold $s(\mathbf{x})=0$, and it is insensitive to disturbances and parameter variations (Utkin, 1992). Some analogies between variable structure and optimal control (Zelikin and Borisov, 1994), e.g., the control law can be discontinuous (time optimal control of systems affine in the control), suggest that these two control methodologies can be used together for solving an optimal control problem, nevertheless the sliding mode theory has been applied only to some limited classes of optimal

270

control problems, e.g., the singular optimal control problem and the time optimal control of linear systems (Newman, 1990; Utkin, 1992; Takahashi and Peres, 1996).

The major drawback to the application of the sliding mode control technique to real systems is the presence of the chattering phenomenon, that is high frequency vibration of the system about the sliding manifold, due to the presence of either non ideal actuators or parasitic fast dynamics.

The usual approaches to reduce the effect of chattering are the so called smooth approximation of the switching devices in a boundary layer of the sliding manifold and the observer based sliding mode control. In the first case the ideal switching device is approximated by a linear high gain inside the boundary layer so that the control law is continuous (Zinober, 1982; Utkin, 1992). Unfortunately the system behaviour inside the boundary layer is unpredictable and it is not possible to guarantee that parasitic dynamics are not excited unless the usual hypothesis of negligible fast dynamics is assured by increasing the dimension of the boundary layer, so decreasing the disturbance rejection properties. In the observer based sliding mode control the chattering is localized inside a high frequency loop that bypasses the plant (Utkin, 1992, Bartolini and Pydynowski, 1996). Recently an unchattering variable structure control based on second order sliding modes has been proposed by Bartolini *and al.* (1997-b).

In this paper the problem of the optimal control implementation for a class of single input dynamical systems affine in the control is faced; in particular the class of systems (1) which can be represented with the following structure of the vector function $\mathbf{F}[\mathbf{x}(t),u(t),t]$ is considered

$$\begin{cases} \dot{x}_i(t) = x_{i+1}(t) & i = 1, 2, \ldots, n-1 \\ \dot{x}_n(t) = f[\mathbf{x}(t),t] + g[\mathbf{x}(t),t]u(t) \end{cases} \tag{7}$$

where $\mathbf{x}(t)$ is the n-dimensional state vector, completely available for measurements, $u(t)$ the scalar control variable and $f[\bullet]$ and $g[\bullet]$ are smooth functions satisfying the usual conditions for the existence of the solution of the optimal control problem, i.e. they are continuous with their first and second derivatives, and furthermore the following constraint holds

$$G_1 \le g[\mathbf{x}(t),t] \le G_2 \quad G_1, G_2 \in R_{>0} \tag{8}$$

Note that (7) represents a large class of invertible stable systems with relative degree equal to n, and that the positiveness of the function $g[\bullet]$ is assumed for the sake of simplicity.

In the next section the unchattering control algorithm based on second order sliding modes will be presented, in Section 3 its effectiveness for a desired optimal trajectory following, in the case of the considered class of dynamical systems, by means of feedback tracking control is shown. In Section 4 two simulation examples are reported, and then, in Section 5 some conclusions are drawn.

2. The Chattering Elimination Problem for Uncertain Nonlinear SISO Systems

When the sliding mode approach for the control of real plants is considered, the chattering problem, arising from nonidealities of real actuators, must be faced. The finite frequency control arising from various kind of nonidealities could excite unmodelled oscillatory modes with unpredictable effects on system behaviour. Any attempt to smooth the discontinuity of the control could even worsen such a situation. One approach to chattering reduction, by maintaining a very high commutation frequency, is based on the use of observers for the modelled part of the system (Utkin, 1992). The sliding mode is attained on the observer state space with a motion which is close to the ideal one. The resulting high frequency control is filtered out by the fast dynamics of the plant so that a practical continuous control is fed to the slow dynamical subsystem. In the case of known nonlinear systems, a general framework has been extended to uncertain systems in the paper of Bartolini and Pydynowski (1996). Recently, the authors presented a chattering free control scheme based on second order sliding modes (Bartolini *and al.*, 1997-b), hereafter reported for the sake of clarity.

Given the system

$$\begin{cases} \dot{z}_i(t) = z_{i+1}(t) \quad i = 1, 2, \dots, n-1 \\ \dot{z}_n(t) = \Phi[z(t),t] + \Gamma[z(t),t]w(t) \end{cases} \tag{9}$$

with $z(t)=[z_1,z_2,\dots,z_n]^T$ representing the completely available state and $\Phi[z(t),t]$ and $\Gamma[z(t),t]$ uncertain smooth functions satisfying the classical condition for the existence and unicity of the solution, and the following inequalities

$$0 < \gamma_1 \leq \Gamma[z(t),t] \leq \gamma_2$$
$$\left|\Phi[z(t),t]\right| \leq \Pi_\Phi + \Theta_\Phi \|z\|$$
$$\left\|\frac{\partial\Phi[z(t),t]}{\partial z}\right\| \leq \Pi_{d\Phi_z} + \Theta_{d\Phi_z}\|z\|$$
$$\left\|\frac{\partial\Phi[z(t),t]}{\partial t}\right\| \leq \Pi_{d\Phi_t} + \Theta_{d\Phi_t}\|z\| \tag{10}$$
$$\left\|\frac{\partial\Gamma[z(t),t]}{\partial z}\right\| \leq \Pi_{d\Gamma_z} + \Theta_{d\Gamma_z}\|z\|$$
$$\left\|\frac{\partial\Gamma[z(t),t]}{\partial t}\right\| \leq \Pi_{d\Gamma_t} + \Theta_{d\Gamma_t}\|z\|$$

with γ_1, γ_2, Π_Φ, Θ_Φ, $\Pi_{d\Phi_z}$, $\Theta_{d\Phi_z}$, $\Pi_{d\Phi_t}$, $\Theta_{d\Phi_t}$, $\Pi_{d\Gamma_z}$, $\Theta_{d\Gamma_z}$, $\Pi_{d\Gamma_t}$, $\Theta_{d\Gamma_t}$, real positive known constants, the problem is to find a continuous control $w(t)$ such that, in spite of the uncertainties (10), the state of (9) is steered to zero exponentially.

To determine the desired continuous control the following procedure needs to be followed (Bartolini and Pydynowski, 1996):

1. differentiate the last equation of (9), setting $z_{n+1}=dz_n/dt$ and consider the augmented order system

$$\begin{cases} \dot{z}_i(t) = z_{i+1}(t) \quad i = 1, 2, \dots, n \\ \dot{z}_{n+1}(t) = \frac{d\Phi[z(t),t]}{dt} + \frac{d\Gamma[z(t),t]}{dt}w(t) + \Gamma[z(t),t]\dot{w}(t) \end{cases} \tag{11}$$

2. Choose an n-th order sliding manifold

$$s[z(t)] = z_n(t) + \sum_{i=1}^{n-1} c_i z_i(t) = 0 \tag{12}$$

with c_i, $i=1,\dots,n-1$, real positive constants such that the characteristic equation $z^n(t)+\sum_{i=1}^{n-1}c_i z^i = 0$ has all roots with negative real part.

3. Consider the first and second time derivative of $s[z(t)]$, namely

$$\dot{s}(t) = \Phi[z(t),t] + \Gamma[z(t),t]w(t) + \sum_{i=1}^{n-1} c_i z_{i+1}(t) \tag{13}$$

$$\ddot{s}(t) = \frac{d\Phi[z(t),t]}{dt} + \frac{d\Gamma[z(t),t]}{dt}w(t) + c_{n-1}\{\Phi[z(t),t] + \Gamma[z(t),t]w(t)\} + \sum_{i=1}^{n-2} c_i z_{i+2}(t) + \Gamma[z(t),t]\dot{w}(t) \tag{14}$$

272

If it is possible to steer s[z(t)] to zero in a finite time by using a discontinuous control signal dw(t)/dt, then, the corresponding w(t) is continuous, thereby eliminating the undesired high frequency oscillations of w(t) (chattering effect) typical of the standard Variable Structure Control (VSC) design. Once on s[z(t)]=0, the system performs like a reduced order linear system with stable transfer function. Assume $y_1(t)=s[z(t)]$ and $y_2(t)=ds(t)/dt$, then, relying on (12), the system dynamics (9) and the relevant uncertain dynamics (13), (14) can be rewritten as

$$\begin{cases} \dot{\hat{z}}(t) = A\hat{z}(t) + By_1(t) \\ z_n(t) = -C\hat{z}(t) + y_1(t) \\ \dot{y}_1(t) = y_2(t) \\ \dot{y}_2(t) = \Psi[z(t), w(t), t] + \Gamma[z(t), t]v(t) \end{cases} \tag{15}$$

where $\hat{z}(t) = [z_1, z_2, ..., z_{n-1}]^T$, $C=[c_1, c_2, ..., c_{n-1}]$, A is a $(n-1)\times(n-1)$-matrix in companion form whose last row coincides with vector $-C$, $B=[0, ... , 0, 1]^T \in R^{n-1}$, $v(t)=dw/dt$, and $\Psi[\bullet]$ collects all the uncertainties not involving $v(t)$. The first two lines of (15) correspond to a linear system controlled by $y_1(t)$, and this system is stable by assumption. The second two equations of (15) correspond to a nonlinear uncertain second order system ($y_2(t)$ is not available for measurement) with control $v(t)$. If the control $v(t)$ steers to zero both $y_1(t)$ and $y_2(t)$, then the linear system becomes an autonomous one evolving on the manifold defined by (12). Note that the last two equations of (15) are coupled with the previous ones through the uncertainties $\Psi[z(t), w(t), t]$, $\Gamma[z(t), t]$.

2.1 The Auxiliary Problem

As a preliminary step of our treatment, we assume that, instead of bounds (10), the following particular bounds

$$\begin{aligned} 0 < \gamma_1 &\leq \Gamma[z(t), t] \leq \gamma_2 \\ |\Psi[z(t), w(t), t]| &\leq \Pi \end{aligned} \tag{16}$$

are considered.

With this assumption, which will be dispensed with in the next section, the dynamics relevant to $y_1(t)$ and $y_1(t)$ can be isolated and the following auxiliary problem can be solved separately.

Problem 2: *Given a second order system*

$$\begin{cases} \dot{y}_1(t) = y_2(t) \\ \dot{y}_2(t) = \Psi[y(t), t] + \Gamma[y(t), t]v(t) \end{cases} \tag{17}$$

with $y = [y_1, y_2]^T$, $y_2(t)$ *unmeasurable, and bounds as in (16), find a control law* $v(t)$ *such that* $y_1(t)$, $y_1(t)$ *are steered to zero in a finite time in spite of the uncertainties.*

Since $y_1(t)$ is not available and $\Psi[y(t), t]$, $\Gamma[y(t), t]$ are uncertain, this problem is not easily solvable by consolidated theory.

A possible solution is derived from a sub-optimal version of the well-known bang-bang time optimal control for a double integrator in which the switching line (Athans, 1966; Kirk, 1970; Leitmann, 1981) is equivalently defined as the line in which the difference between the $y_1(t)$ current value and one half of its last extremal value y_{1M} changes its sign. The corresponding sub-optimal control algorithm can be obtained by setting $\alpha^*=1$ in the following

Algorithm 1:

1. Set $\alpha^* \in\ ]0;1] \cap\ ]0;\ 3\gamma_1/\gamma_2[$

2. Set $y_{1M} = y_1(0)$

3. Repeat, for any $t > 0$, the following steps

    a) If $[y_1(t) - 0.5\ y_{1M}][y_{1M} - y_1(t)] > 0$ then set $\alpha = \alpha^*$ else set $\alpha = 1$

    b) If $y_1(t)$ is extremal then set $y_{1M} = y_1(t)$.

    c) Apply the control law

$$v(t) = -\alpha \cdot V_M \cdot \text{sign}\left[y_1(t) - \tfrac{1}{2}y_{1M}\right] \tag{18}$$

    Until the end of the control time interval

This algorithm is equivalent to the traditional one if $y_1(0)y_2(0) \geq 0$, while it has only one more commutation if $y_1(0)y_2(0) < 0$, in the case when a double integrator is considered. In the case of uncertain second order systems in question, it is still possible to reach in a finite time the origin of the y_1Oy_2 plane provided that some slight modifications to the algorithm are introduced. To this end, the following theorem has been proved (Bartolini *and al.*, 1997-a)

Theorem 1: *Given the state equation* (17) *with bounds as in* (16) *and* $y_2(t)$ *not available for measurement, then, if the extremal value of* $y_1(t)$ *is evaluated with ideal precision, for any* $y_1(0)$, $y_2(0)$, *the sub-optimal control strategy defined by* Algorithm 1 *with the additional constraints*

$$\begin{aligned} &\alpha^* \in\ ]0;1] \cap\ ]0;\tfrac{3\gamma_1}{\gamma_2}[\\ &V_M > \max\left\{\tfrac{\Pi}{\alpha^*\gamma_1};\tfrac{4\Pi}{3\gamma_1 - \alpha^*\gamma_2}\right\} \end{aligned} \tag{19}$$

causes the generation of a sequence of states with coordinates (y_{1Mi} ; 0) *featuring the following contraction property*

$$\left|y_{1M_{i+1}}\right| < \left|y_{1M_i}\right| \quad i = 1, 2, \ldots \tag{20}$$

Moreover, the convergence of the system trajectory to the origin of the error state plane takes place in a finite time. □

Algorithm 1 and the related Theorem 1 are effective in the ideal case, that is the extremal values (local maxima, minima and horizontal flex points) of the variable $y_1(t)$ are evaluated with ideal precision by means of an ideal infinite bandwidth peak detector. In the real case the estimates of the extremal values are affected by an error that can be associated to an equivalent time delay δ, and only the finite time reaching of a δ-vicinity of the origin of the state plane can be assured (Bartolini *and al.*, 1997-a). Previous results cover a large class of uncertain systems and are semiglobal; the extension of the proposed control algorithm to a larger class of system with bounded dynamics has been presented in (Bartolini *and al.*, 1996).

Note that Algorithm 1 performs a second order sliding mode, in the sense defined in the work of Fridman and Levant (1996), for the system (17) if the sliding manifold $s[z(t)] = z_1(t) = 0$ is considered.

2.2 Chattering Elimination

Now, it will be shown that it is possible to solve the chattering elimination problem by relying on the results obtained with reference to the auxiliary problem in the previous section. To this end, consider equation (15) which can be viewed as the connection of two systems coupled through the signal $y_1(t)$ and the nonlinear term $\Psi[z(t), w(t),t]+\Gamma[z(t), t]v(t)$. Simply assume that (10) holds, which is reasonable in many practical situations, while $\Psi[z(t), w(t),t]$, though bounded in any bounded domain, cannot be a-priori assumed to be bounded, since proving the boundness of its arguments is a goal of this treatment. Thus, the aim of the following analysis is to prove that, after an initialization phase, the state trajectories reach regions of the state space including the origin. Once such regions are reached, the application of Algorithm 1, with minor modifications, leads to a contractive process steering $y_1(t)$, $y_2(t)$ to zero in a finite time. After that time, the further evolution of the system states is that of an autonomous linear exponentially stable system. In order to formally describe this procedure, the following proposition has been proved (Bartolini *and al.*, 1996, 1997-b).

Proposition 1: *Given the state vector* $z(t)$, *its norm can be bounded, in any bounded interval, by a function of its initial values and of the maximum value assumed by* $y_1(t)$, *whose evolution is represented by* (17), *in such interval, that is*

$$\|z(t)\| \le \Pi_z \|z(t_i)\| + \Theta_z Y_{1_{(t_i;t_f)}}$$
$$Y_{1_{(t_i;t_f)}} = \max_{[t_i;t_f]} |y_1(t)| \tag{21}$$

Furthermore, the following bound for the uncertain function $\Psi[z(t), w(t),t]$ *holds*

$$|\Psi[z(t), w(t),t]| \le \Theta_0[z(t)] + \Theta_1[z(t)]y_1(t) + \Theta_2[z(t)]y_2(t) + \Gamma[z(t),t]\tfrac{\partial\Gamma}{\partial z_n}w^2(t)$$
$$\Theta_0[z(t)] \le a_{0,0} + a_{1,0}\|z(t)\| + a_{2,0}\|z(t)\|^2 + a_{3,0}\|z(t)\|^3 \tag{22}$$
$$\Theta_i[z(t)] \le a_{0,i} + a_{1,i}\|z(t)\| + a_{2,i}\|z(t)\|^2 \quad (i=1,2)$$

with $a_{i,j}$ (i,j=0,1,2) *and* $a_{3,0}$ *real non negative constants.* □

Note that the term depending on $w^2(t)$ would not appear in case $\Gamma[z(t), t]$ were not dependent on $z_n(t)$, as in the case of mechanical systems.

From previous analysis it can be proved that the chattering elimination problem, after an easily implementable initialization procedure, has the same feature of Problem 1 provided that the control $v(t)$ is modified according to the following theorem (Bartolini *and al.*, 1996, 1997-b).

Theorem 2: *Given the system* (15), *provided that for* $t \in [t_{M_i}, t_{M_{i+1}}]$, *(i=1,2,...), where* t_{M_i} *and* $t_{M_{i+1}}$ *are the time instants corresponding to two subsequent extremal values of* $y_1(t)$, y_{1M_i} *and* $y_{1M_{i+1}}$, *the control signal* $v(t)$ *is chosen as*

$$v(t) = -\alpha V_{M_i} \text{sign}\left\{y_1(t) - \tfrac{1}{2}y_{1_{M_i}}\right\} + H_i \text{sign}\left\{y_{1_{M_i}}\right\} \tag{23}$$

where α *is defined according to* Algorithm 1, *and* V_M *is chosen as in* (19) *with* Π *satisfying*

$$\Pi \ge \sup_{[t_{M_i};t_{M_{i+1}}]}\left\{\Theta_0[z(t)] + \Theta_1[z(t)]y_1(t) + \Theta_2[z(t)]y_2(t)\right\} + \gamma_2\left\{\Pi_{d\Gamma_z} + \Theta_{d\Gamma_z}\left[\Pi_z\|z(t_i)\| + \Theta_z y_{1_{M_i}}\right]\right\}w^2(t_{M_i}) \tag{24}$$

and H_i *such that the following inequality holds*

$$H_i \ge \left\{\Pi_{d\Gamma_z} + \Theta_{d\Gamma_z}\left[\Pi_z\|z(t_i)\| + \Theta_z y_{1_{M_i}}\right]\right\}|w^2(t) - w^2(t_{M_i})| \tag{25}$$

275

then, the trajectories of $y_1(t)$ *and* $y_2(t)$, *in the considered time interval, lie between the abscissa axis and the "external" limiting curve defined by* Algorithm 1. □

The previous theorems assure the finite time convergence of system (9) to the sliding manifold (12) and the maintenance of the sliding behaviour by means of a continuous control so avoiding the chattering phenomenon.

Remark 1: previous theorems define sufficient but non necessary conditions guaranteeing the finite time convergence to the sliding manifold, this means that it is possible to apply successfully the proposed control algorithm with a control amplitude not satisfying conditions (19), (23)-(25).

3. The Near-Optimal Control

Consider the following mathematical model

$$\begin{cases} \dot{x}_i(t) = x_{i+1}(t) & i = 1, 2, \ldots, n-1 \\ \dot{x}_n(t) = f_m\big[x(t), t\big] + g_m\big[x(t), t\big]u(t) \end{cases} \tag{26}$$

which constitutes an approximation of the real plant dynamics (7).

Let $u_m(t)$ be the solution of the optimal control problem, namely Problem 1, when the plant model dynamics (26) is considered, and $x_m(t)$ is its corresponding state trajectory. Obviously if the ideal optimal control law $u_m(t)$ is applied to the real plant the associated state evolution cannot correspond to the optimal one, because of the model uncertainties $\Delta f[x(t),t]$ and $\Delta g[x(t),t]$. Furthermore, due to the non-idealities of the actuators, the real control input is different from the optimal one, that is $u(t)=u_m(t)+\Delta u(t)$. From these considerations, it is straightforward that an additional control effort to constrain the real state trajectory to the optimal one is needed, so that

$$u(t) = u_m(t) + \Delta u(t) + w(t) \tag{27}$$

where $\Delta u(t)$ is the above error input and $w(t)$ is the compensatory input.

Assume that the real plant dynamics is such that the following relationships hold

$$\begin{aligned} f\big[x(t), t\big] &= f_m\big[x_m(t), t\big] + \Delta f\big[x(t), t\big] \\ g\big[x(t), t\big] &= g_m\big[x_m(t), t\big] + \Delta g\big[x(t), t\big] \end{aligned} \tag{28}$$

then if the error state is defined as the difference between the real and the optimal state trajectory, i.e., $e(t)=x(t)-x_m(t)$, by means of (7) and (26)-(28), the error state dynamics is defined as follows

$$\begin{cases} \dot{e}_i(t) = e_{i+1}(t) & i = 1, 2, \ldots, n-1 \\ \dot{e}_n(t) = \Phi\big[x(t), u_m(t), \Delta u(t), t\big] + g\big[x(t), t\big]w(t) \end{cases} \tag{29}$$

where

$$\Phi\big[x(t), u_m(t), \Delta u(t), t\big] = \Delta f\big[x(t), t\big] + \Delta g\big[x(t), t\big]u_m(t) + g\big[x(t), t\big]\Delta u(t) \tag{30}$$

collect all the uncertainties in the real system behaviour.

It is evident that the error state dynamics could belong to the class of uncertain nonlinear dynamical ones considered in the previous section if uncertainties $\Phi[x(t),u_m(t),\Delta u(t),t]$ and $g[x(t),t]$ are bounded with their derivatives, so that conditions equivalent to (10) are satisfied. In This case the compensatory control $w(t)$, defined according to the previously proposed unchattering variable structure control, is able to steer in finite time the error dynamics to a proper sliding manifold, so that the real system behaviour follows the desired optimal one asymptotically, with arbitrary dynamics.

3.1 Continuous Optimal Control

Consider the case of continuous solution of the optimal control problem, that is the control $u_m(t)$ is represented by a continuous function of time, this means that, on the basis of the necessary conditions on the functions $F[\cdot]$, $L[\cdot]$ and $K[\cdot]$ for the existence of a solution of Problem 1, also the corresponding optimal state trajectory $x_m(t)$ is continuous with its first derivatives. Furthermore the control error due to the actuators nonidealities can be sensibly assumed continuous.

The above considerations suggest that a large class of dynamical systems can satisfy the following constraints

$$
\begin{aligned}
&\left| \Delta f\left[x(t),t \right] \right| \leq F \\
&\left\| \frac{\partial \Delta f\left[x(t),t \right]}{\partial x} \right\| \leq F_{d\Phi_z} \\
&0 < G_1 \leq g\left[z(t),t \right] \leq G_2 \\
&\left| u_m(t) \right| \leq U \\
&\left| \Delta u(t) \right| \leq Ue \\
&\left\| \frac{\partial \Delta f\left[x(t),t \right]}{\partial t} \right\| \leq F_{d\Phi_t} \qquad t \in [t_i;t_f] \\
&\left\| \frac{\partial \Delta g\left[x(t),t \right]}{\partial x} \right\| \leq G_{d\Phi_z} \\
&\left\| \frac{\partial \Delta g\left[x(t),t \right]}{\partial t} \right\| \leq G_{d\Gamma_t} \\
&\left| \frac{du_m(t)}{dt} \right| \leq U_{dt} \\
&\left| \frac{d\Delta u(t)}{dt} \right| \leq Ue_{dt}
\end{aligned}
\tag{31}
$$

which guarantee that the implementation problem of an optimal control can be reduced to the problem of the unchattering variable structure control of uncertain system treated in Section 2.

In fact if the sliding manifold

$$
s\left[e(t) \right] = e_n(t) + \sum_{i=1}^{n-1} c_i e_i(t) = 0
\tag{32}
$$

is properly chosen, the application of the procedure proposed in the paper of Bartolini and Pydynowski (1996), previously reported, reduces the uncertain system (29) to the two following coupled dynamics

$$
\begin{cases}
\dot{\hat{e}}(t) = A\hat{e}(t) + By_1(t) \\
e_n(t) = -C\hat{e}(t) + y_1(t)
\end{cases}
\tag{33}
$$

$$
\begin{cases}
\dot{y}_1(t) = y_2(t) \\
\dot{y}_2(t) = \left\{ \frac{d\Phi}{dt} + \frac{dg}{dt} w(t) + \sum_{i=1}^{n-1} c_i e_{i+1} \right\} + g\left[x(t),t \right] \frac{dw(t)}{dt}
\end{cases}
$$

where $\hat{\mathbf{e}}(t) = [e_1, e_2, ..., e_{n-1}]^T$, $\mathbf{C} = [c_1, c_2, ..., c_{n-1}]$, $\mathbf{A}$ is a (n-1)x(n-1)-matrix in companion form whose last row coincides with vector -$\mathbf{C}$, $\mathbf{B} = [0, ... , 0, 1]^T \in \mathbb{R}^{n-1}$, $y_1(t) = s(t)$, and $y_2(t) = ds/dt$.

By choosing, after a proper initialization phase, the derivative of the compensatory control w(t), dw/dt, according to Theorem 2, the finite time reaching of the sliding manifold is assured, i.e. s(t)=0 for any $t \geq t_1$, so that, from that time instant on, the reduced linear system evolves as an autonomous stable system with arbitrary dynamics, and the asymptotic following of the optimal state trajectory is guaranteed.

3.2 Discontinuous Optimal Control

When the Pontryagin's function is linear with respect to the control variable it is well known that the optimal control is discontinuous and commutes among finite values (Athans, 1966; Kirk, 1970; Leitmann, 1981). The implementation of this type of control law is performed by real relays which usually introduce a time delay in the control action so that also the control error $\Delta u(t)$ is a time discontinuous function. On the contrary the state trajectory, due to the conditions on the vector function $\mathbf{F}[\cdot]$ for the existence of a solution of the optimal control problem, is continuous but with discontinuous bounded time derivatives.

Let $t_{c_{km}}$ (k=1,2,...,r) and t_{c_k} (k=1,2,...,r) be the optimal and the corresponding real switching time instants respectively; as the time derivatives of both the ideal and real control input evaluated at the switching time instants are impulsive, the last two inequalities in (31) are not satisfied and the following inequalities hold

$$\left| \frac{du_m(t)}{dt} \right| \leq U_{dt} \qquad t \in \left] t_{c_{k_m}} ; t_{c_{k+1_m}} \right[\qquad k = 0, 1, 2, ..., r \quad t_{c_{0_m}} = t_i \; ; \; t_{c_{r+1_m}} = t_f \qquad (34)$$
$$\left| \frac{d\Delta u(t)}{dt} \right| \leq Ue_{dt} \qquad t \in \left] t_{c_{k_m}} ; t_{c_k} \right[\bigcup \left] t_{c_k} ; t_{c_{k+1_m}} \right[$$

The above inequalities (34) imply that the sufficient conditions for reaching the sliding manifold (32) and the maintenance of the sliding behaviour for the error state $\mathbf{e}(t)$ are verified only within bounded time intervals. In fact at any switching time instant the derivative of the sliding variable

$$\frac{ds(t)}{dt} = \Delta f[\mathbf{x}(t), t] + \sum_{i=1}^{n-1} c_i e_{i+1}(t) + \Delta g[\mathbf{x}(t), t] u_m(t) + g[\mathbf{x}(t), t] \Delta u(t) + g[\mathbf{x}(t), t] w(t) \qquad (35)$$

changes its value instantaneously and the condition for the sliding motion to occur, i.e. $s[\mathbf{e}(t)] = ds[\mathbf{e}(t)]/dt = 0$, is verified no more.

Nevertheless the proposed control strategy could be applied also in this case but a sequence of reaching phases and sliding motions will be obtained instead of an asymptotic convergence of the real state to the optimal one. It must be pointed out that in this case the definition of the sliding manifold is crucial since the settling time of the sliding mode, and of its correspondent reaching phase, must be negligible with respect the time interval between two subsequent switchings in order to have an acceptable behaviour of the system.

The considerations previously reported allow one to consider the proposed control scheme as a near optimal implementation of an optimal control law. The real control law is near optimal because an error with respect to the optimal trajectory is always present, and the presence of the term w(t) in (27) could imply that the actual control law u(t) does not belong to the set of admissible control, e.g. the maximum amplitude of the actual control is greater then that considered in the corresponding optimal control problem. Both the error $\mathbf{e}(t)$ and the added control w(t) depend only on the system uncertainties and they are always null if there is no uncertainty. Possible external disturbances can be considered equivalent to the control error $\Delta u(t)$ so that the previous treatment still holds.

4. Example

Let us consider the second order system of example n°1 in Lastman (1978),

$$\begin{cases} \dot{x}_1(t) = x_2(t) \\ \dot{x}_2(t) = -x_1(t) + u(t) \end{cases} \tag{36}$$

which can be considered as the model of a simple mechanical system constituted by a mass and a spring subjected to an external force. The real system is characterized by a nonlinear spring and an uncertain mass, so that the actual system dynamics is the following

$$\begin{cases} \dot{x}_1(t) = x_2(t) \\ \dot{x}_2(t) = \frac{1}{m}\left[-x_1(t) + 10^{-2}x_1^3(t) + u(t)\right] \end{cases} \quad 0.9 \le m \le 1.1 \tag{37}$$

4.1 Continuous Control

As a first case, the control aim is steering the system to a neighbourhood of the origin of the state plane, starting from the point (3;0.5), while minimizing the following performance index

$$J = 5\left[x_1^2(4) + x_2^2(4)\right] + \int_0^4 \left[1 + u^2(t)\right]dt \tag{38}$$

The final state must lie within a circle centred in the origin of the state plane and with radius equal to 0.25, i.e.,

$$x_1^2(4) + x_2^2(4) \le \frac{1}{16} \tag{39}$$

By means of the variational approach to the optimal control problem, it can be shown that the optimal control law which minimizes the performance index (38), when model (36) is considered, is defined by the following

$$u_m(t) = -\lambda_2(t) \tag{40}$$

where $\lambda_2(t)$ is the second element of the co-state vector $\lambda(t) = [\lambda_1(t); \lambda_2(t)]^T$, whose dynamics is defined by the Eulero-Lagrange necessary conditions for the optimality, but with "a piori" unknown initial values.

By applying the algorithm proposed by Sanna and Usai (1996), the initial co-state vector which corresponds to the minimum of the performance index, J=4.6306, has been found, i.e., $\lambda(0) = [1.68397514527; 0.41845177092]^T$. The optimal control law and the corresponding state trajectory is evaluated by integrating the augmented differential system according to the Eulero-Lagrange necessary conditions.

The proposed control scheme has been applied considering the actual value of the uncertain mass m=1.1, with the sliding manifold (32) characterized by c_1=5. A digital implementation of the control Algorithm 1 with an equivalent delay δ=2 *ms*, V_M=7, and α*=1 has been used. The real and optimal state trajectories are practically the same as reported in Fig.2, while the actual plant input (27) differs from the optimal one because of the system uncertainties (Fig.1). The performance index, evaluated for the real system, is equal to 4.9050.

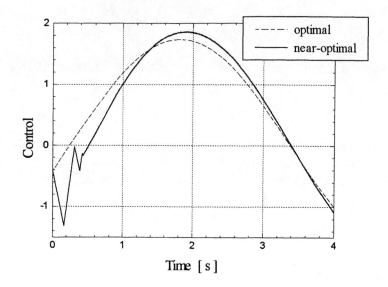

Fig.1. continuous control inputs

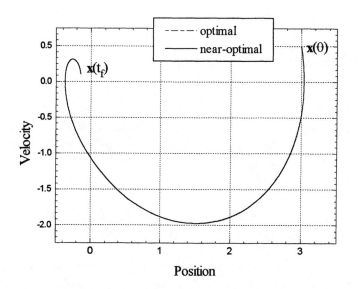

Fig.2. state trajectories in case of continuous input

4.2 Discontinuous Control

Consider the time optimal control problem for system (36) with the condition on the control amplitude $|u(t)| \leq 1$; the optimal control law is

$$u_m(t) = \begin{cases} +1 & \lambda_2 < 0 \\ -1 & \lambda_2 > 0 \end{cases} \tag{41}$$

where $\lambda_2(t)$ is the second element of the co-state vector $\lambda(t)=[l_1(t); l_2(t)]^T$. By applying the algorithm proposed by Sanna and Usai (1996), the initial co-state vector which corresponds to the minimum time interval, $t_f = 4.9924154$, has been found, i.e., $\lambda(0) = [0.9350228009; 0.3668778501]^T$. In this case it has been supposed that the real relay causes a constant delay in the control switching equal to 10 *ms*, and if the optimal control (41) alone is applied the goal of steering the system to the origin of the state plane is not attained. The proposed control scheme has been applied with the sliding manifold (32) characterized by $c_1=10$. A digital implementation of the control Algorithm 1 with an equivalent delay $\delta=1ms$, $V_M=25$, and $\alpha^*=1$ has been used. The real and optimal state trajectories differs only in the vicinity of the initial and switching time instants as reported in Fig.4. In Fig.3 it is evident the effect of the switching delays due to the actuator nonideality which causes a transient phase at each commutation. These transient phases are the consequence of the control discontinuity which causes the instantaneous variation of the derivative of the sliding variable. It must be pointed out that the compensated control input doesn't satisfy the control constraints which are considered for solving the time optimal control problem.

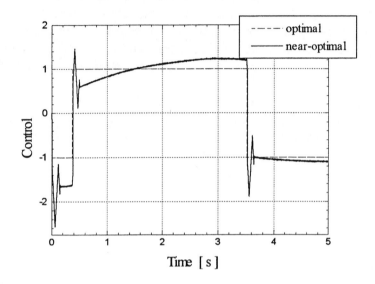

Fig.3. time optimal control inputs

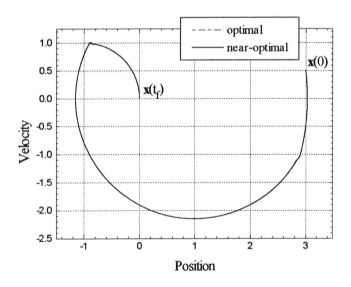

Fig.4. time optimal state trajectories

5. Conclusions

The problem of the optimal control implementation in presence of modelling approximations, actuators nonidealities and external disturbance has been considered for a class of single input dynamical systems. A mixed feedforward/feedback control scheme has been presented, in which either an on-line or an off-line optimal control and trajectory evaluation, and an on-line nonlinear controller are implemented.

The knowledge of a model of the controlled system allows one to define a solution of the considered optimal control problem which is used for applying the ideal optimal control law to the plant by means of a feedforward controller as well as to define the reference optimal state trajectory. The error between the real and reference trajectories constitutes the input of a nonlinear feedback controller which defines a compensatory input.

The nonlinear controller is based on a recently proposed control algorithm which is able to perform a second order sliding mode control. The use of such method extends the robustness properties of sliding mode control to the implementation of optimal control to the detriment of an increase of the performance index. In fact uncertainties due to modelling errors, actuators nonidealities, and external disturbances affect the real behaviour of the plant so that the optimal value of the considered performance index cannot be reached even if the theoretic optimal control is applied.

The proposed control scheme is defined near optimal as the difference between the real and the ideal state trajectory is steered to zero asymptotically and the control effort differs from the optimal one due to uncertainties. Of course if no uncertainty is present the feedback loop doesn't produce any action and the system behaviour is optimal.

References

Athans M. and P.L. Falb (1966). *Optimal Control: An Introduction to the Theory and Its Applications*. McGraw Hill, New York

Bartolini G. and P. Pydynowski (1996). "An improved, Chattering Free, V.S.C. scheme for Uncertain Dynamical Systems". *IEEE Transaction on Automatic Control*, vol. AC-41, n°8, pp. 1220÷1226

Bartolini G., A. Ferrara and E. Usai (1996). "Second order VSC for non linear systems subjected to a wide class of uncertainty conditions". *Proceedings of 1996 IEEE Int. Workshop on Variable Structure Systems - VSS'96, Seiken Symposium n°19*, Tokyo, Japan, pp. 49÷54

Bartolini G., A. Ferrara and E. Usai (1997-a). "Application of a Sub-Optimal Discontinuous Control Algorithm for Uncertain Second Order Systems". *Int. J. of Robust and Nonlinear Control*, vol.7, pp. 299-319

Bartolini G., A. Ferrara and E. Usai (1997-b). "Chattering avoidance by second order sliding mode control". *IEEE Trans. Aut. Control*, to appear

Doyle J.C., K. Glover, P.P. Khargonear and B.C. Francis (1989). "State-Space Solutions to Standard H_2 and H_∞ Control problems". *IEEE Trans. Aut. Cont.*, vol. AC-34, n° 8, pp. 831-847

Doyle J.C., K. Zhou, K. Glover and B. Bodenheimer (1994). "Mixed H_2 and H_∞ Performance Objectives II: Optimal Control", *IEEE Trans. Aut. Cont.*, vol. AC-39, n° 8, pp. 1575-1587

Fridman L. and A. Levant (1996). "Higher Order Sliding Modes as a Natural Phenomenon in Control Theory". in *Robust control via variable structure and Lyapunov techniques-* (Lectures

Notes in Control and Information Science 217), F. Garofalo & L. Glielmo editors, Springer-Verlag, London

Kirk D.E. (1970). *Optimal Control Theory - An Introduction*. Prentice Hall, Englewood Cliffs

Larson R.E. (1967). "A Survey of Dynamic Programming Computational Procedures", *IEEE Trans. Automatic Control*, vol. AC-12, n° 6, pp.767-773

Lastman G.J. (1978). "A shooting method for solving two point boundary value problems arising from non singular bang-bang optimal control problems". *Int. J. Control*, vol.27, n°4, pp.513-524

Leitmann G. (1981). *The Calculus of Variations and Optimal Control - An Introduction*. Plenum Press, New York

McGill R. and P. Kenneth (1964). "Solution of Variational Problems by Means of a Generalized Newton-Raphson Operator". *AIAA Journal*, vol.2, n°10, pp.1761-1766

Newman W.S. (1990). "Robust Near Time-Optimal Control". *IEEE Trans. Aut. Control*, vol. AC-35, n° 7, pp. 841-844

Sanna S. and E. Usai (1996). "A general algorithm for the formulation and solution of non singular optimal control problems". *Proceedings of CESA 1996, Symposium on Control, Optimization and Supervision*, pp. 1294÷1298, Lille, France

Shimizu K. and S. Ito (1994). "Constrained Optimization in Hilbert Space and a Generalized Dual Quasi-Newton Algorithm for State-Constrained Optimal Control Problems". *IEEE Trans. Automat. Contr.*, vol. AC-39, n°5, pp.982-986

Stoorvogel A.A. (1993). "The Robust H_2 Conrol Problem: A Worst-Case Design". *IEEE Trans. Aut. Cont.*, vol. AC-38, n° 9, pp. 1358-1370

Takahashi R.H.C. and P.L.D. Peres (1996). "Sliding Modes Solution for the H2 Singular Problem". *Proceedings of the 35th IEEE Conference on Decision and Control*, Kobe, Japan*l*, pp. 243-248

Utkin V.I. (1992). *Sliding Modes in Control Optimization*. Springer-Verlag, Berlin

Zelikin M.I. and V.F. Borisov (1994). *Theory of Chattering Control with applications to Astronautics, Robotics, Economics, and Engineering*, Birkhäuser, Boston

Zinober A.S.I., O.M.E. El-Ghezawi and S.A. Billing (1982). "Multivariable structure adaptive model following control systems". *IEE Proceedings*, vol. 5, pp. 6-12

Social Autopoiesis: A Concept in Search of a Theory

Jean Paul BROONEN
Université de Liège
Service d'Orientation Universitaire
Bld. du Rectorat, B33
B-4000 LIEGE 1 BELGIUM
Fax: + 32 4 366 29 88
E-mail: JP.Broonen@ulg.ac.be

Abstract

This paper is a brief report on the issue of extension of tne concept of autopoiesis to social systems. The arguments developped by four groups of authors to bring a response to that issue are summarized: Maturana and Varela, the fathers of the concept of autopoieis; Zeleny & Hufford who proposed a simple extension of the concept to social systems; Luhmann and Hejel with two different transformations of the concept; Morgan and his metaphorical perspective. The determinist vs teleological conception of (social) autopoiesis explicitly or implicitly sustained by several authors is emphasized.

Keywords: social systems, autopoiesis, teleology, functionalism.

Introduction

The issue of if and how the concept of autopoiesis can or should be extended to social systems has produced a good deal of discussion. This paper summarizes in the light of Mingers' (1995) review (see also Mingers, 1992) and Whitaker's (1995) as well as Teubner's (1989) views some important arguments developped until now by several authors to bring responses to that issue. Epistemological aspects implied by the various solutions will be emphasized.

The extension of the concept of autopoiesis to social systems brings up fundamental questions of ontology, epistemology, and methodology. The concept crosses topics of several disciplines. That makes its richness and the difficulty to understand what it means precisely when one goes from a specific discipline to another.

In a history of the concept of autopoiesis Mingers (1995) emphasized that the concept of autopoiesis was originally developed about what distinguishes living entities or systems from nonliving. Responses varied from Bergson's vitalism (1907). The first systemic, developping, about complex, goal-seeking behaviors of organisms, concepts such as feedback, homeostasis, and open systems, is purely mechanistic (e.g. Cannon, 1939). But these concepts are also applicable to machines which could not be qualified as living systems. In a more recent approach Miller (1978) and Bunge (1979) specified the necessary characteristics of a living organism such as reproductive ability, information-processing capabilities, etc. The problem is, Mingers (1995) noticed, that the list of these characteristics working by noting some of the common characteristics of observed systems that are accepted as living "this tactic assumes precisely that which is in need of explanation - the distinction between the living and the nonliving" (p. 10). Besides, there is no agreement on the elements of a specific list and on the number of the necessary elements.

Contrasting with these views, Maturana and Varela pointed "the single, biological individual (for instance, a single celled creature such as an amoeba) as the central example of a living system. " (Mingers, *ibid.*). They sustained that autopoiesis is its essential character. Maturana (1980) proposed a basic definition of an autopoietic system:

CP437, *Computing Anticipatory Systems: CAYS--First International Conference*
edited by Daniel M. Dubois © 1998 The American Institute of Physics 1-56396-827-4/98/$15.00

"A dynamic system that is defined as a composite unity as a network of productions of components that, a) through their interactions recursively regenerate the network of productions that produced them, and b) realize this network as a unity in the space in which they exist by constituting and specifying its boundaries as surfaces of cleavage from the background through their preferential interactions within the network, is an autopoietic system ". (p. 29)

Mingers (1995) considered two critical points when one tries to extend the concept of autopoiesis outside biology, especially to the social domains: (1) the nature of the components, and (2) the nature of the boundaries of the system. These two elements are not specified in the definition. In particular, are they physical? And if they are not, what is their domain of existence? Put in another way, is it necessary to conceive of social autopoiesis from its biolological roots or do we need an autonomous concept of social autopoiesis?

In the the following lines I will essentially consider four groups of authors as far as the response they do to these questions:

- the fathers of the biological autopoiesis: Maturana and Varela;
- Zeleny & Hufford who proposed a simple extension of the concept to social systems;
- Luhmann and Hejel with two different transformations of the concept;
- Morgan for a metaphorical perspective.

1. Maturana and Varela

Maturana and Varela developped the concept of autopoiesis to explain the nature of the phenomena of a living being, that is the molecular processes at the level of cellular activity and neural processes at the level of the nervous system.

Varela (1981) did never develop any social theory. He claimed that the level of autopoiesis concerns only biological cells, the only systems where components are produced in a space. Rather, social systems are autonomous in the sense they maintain their organization, but they are not autopoietic since no spatial boundary is discernable (for an observer) and they do not necessarily regenerate the network of interactions that produced them. Therefore, operations other than productions capture better relations that define units like a firm.

Maturana distinguishes the autopoietic unities and the social systems. Autopoietic unities are cells - that he calls *first-order* autopoietic systems - and multicellular systems or *second-order* autopoietic systems: " A multicellular living system is accomplished through the autopoiesis of its cellular components and, through its own fulfillment as a multicellular totality, makes possible the autopoiesis of them ". (Maturana *et al.*, 1995). The social systems - e.g. social insects or human societies (families, clubs, political parties, and the like that the author calls natural) are *third-order* systems and autopoietic only "in an apparent manner". They constitute the *medium* in which these autopoietic unities interact:

"(...) a collection of interacting living systems that, in the realization of their autopoiesis through the actual operation of their properties as autopoietic unities, constitute a system that as a network of interactions and relations operates with respect to them as a medium in which they realize their autopoiesis while integrating it, is indistinguishable from a natural social system and is, in fact, one such system." (Maturana, 1980, p. 11)

Thus:

" A living system exists in 2 domains: a) in the domain in which its components realize it as a first or second order autopoietic entity, and b) in the domain in which it interacts with the medium that contains it as a totality, namely in the relational or behavioral domain." (Maturana *et al.*, 1995)

An important concept in the theory is the structure-determined character of a system:

"As a structure-determined composite entity, a system is such that all that happens in it and to it is determined in its structure (...). Nothing external to a structure determined system (SDS) can specify what happens in it. The structural changes that an SDS undergoes arise either as a result of its internal dynamics, or triggered through the encounter of the properties of the elements that compose it with the elements that compose the medium. The external elements (...) do not determine the structural changes that arise in it, they only trigger them. We call perturbation the interactions that trigger in a SDS a change of state, and destructive interactions those interactions that trigger its disintegration." (Maturana *et al.*, 1995).

The system and its circumstances change congruently or the system disintegrates. The concept of structural coupling refers to that process by which, in a medium where recurring states happen, autopoiesis (constant internal biochemical correlations) continuously leads to select in the organism a structure "suitable for" (Mingers, 1995, p. 35) that medium (which can be other organisms). Among several states made possible at any given moment by the system's structure, the environnement selects which maintains autopoiesis. As Mingers (*Ibid.*) comments: "Structural coupling is a reformulation of the idea of adaptation, but with the important proviso that *the environment does not specify the adaptative changes that will occur.*" No information is transfered across the boundaries of the systems engaged in that reciprocal adaptation which results in structural changes in both.

Let us now emphasize that Maturana and Varela always strongly objected to a teleological conception of autopoiesis, i.e. to a recourse to ideas of function and purpose.

First, there is no other world than the world of the observer. This is a position of radical constructivism. The description of the "real world" is impossible because the stucture-determined nature of the observer is such as the resulting experience reflects necessarily the observer. Scientific statements are only validated by the scientific method, "defined and constituted only in terms of the operational coherences of the domain of experiences of standard observers" (Maturana, 1990, p. 20), not by correspondence to an external world, "which is not" (Maturana and Varela, 1974, p. 121). That *constituted objectivity* as opposed to a *transcendental objectivity* (the "realist" view of reality) recognizes even the structural determinism as an abstraction of the experiential coherence of the observer. This philosophical position (highly contestable - see for example Mingers, 1995, *passim* - in its ultimate extension) is quite opposed to a teleological one.

Secondly:

"If living systems are physical autopoietic machines, teleonomy becomes only an artifice of their description, which does not reveal any feature of their organization, but which reveals the consistency in their operation within the domain of observation. Living systems, as physical autopoietic machine, are purposeless." (Maturana and Varela, 1980, p. 86)

The basic activity of a living system is maintaining autopoiesis and the fact that its activity leads to an appropiate behavior is only a post hoc effect of the structural coupling.

The nervous system is not in itself autopoietic but organizationally closed:

"The nervous system does not operate making representations of the medium in which the living system that it integrates exists, representations that would serve after as the basis for some processing in order to arrive at the appropriate behaviour. Nevertheless, it has a plastic structure that changes following the contingencies of the living system while this system maintains its autopoietic organization in a medium." (Maturana *et al.*, 1995*)*

The way of treating intentionality in animals is a good example of Maturana' approach:

"Intentionality is a commentary that an observer makes about the flow of the behaviour of an animal, as he or she relates the present behavior with the outcome that it may have, and does so, speaking as if the outcome were an argument in the generation of the behavior that gives rise to it. Intentionality is not a feature of the operation of the nervous system because the nervous system does not and cannot do so. What the nervous system does while the animal is "thinking", is to operate in its internal dynamics according to the structure it has at that moment as a result of the structural changes it has undergone contingently to the living of the animal." (Maturana et al., 1995).

But what about human beings where intentions, purposes are evident? On that point, Maturana and Varela's conceptions about language are crucial. A basic property of the nervous system is that the state of its relative activity becomes itself an object of interaction for it. And the human brain is much more responsive to its own internal structures than to its sensory/effector surfaces. This is the basis for the emergence of symbolic domain, of language and, consequently, by a series of recursions, of self-consciousness (the observation of the observer), responsability as self-awareness and freedom as self-awareness of self-awareness.

From a phylogenetic perspective:

" a first recursion in the linguistic behavioral domain, as it becomes part of the manner of living of such an organism, will constitute language and 'languaging', in terms of consensual coordination of consensual coordinations of behavior (Maturana, 1978). At the same time, as the circular processes of the brain become coupled to the linear flow of 'languaging', that brain becomes a 'languaging' brain (...). A second recursion gives rise to observing, that is, the distinction of the operation of distinction of an object. A third recursion gives rise to the observer, in the distinction of observing that localizes observing. Self-consciousness, that is, the observing of the observer, will arise in the fourth recursion of the coordination of coordinations of consensual behaviour. The fifth recursion gives rise to the experience of responsibility as self-awareness, and the sixth gives rise to the experience of freedom as self-awareness of self-awareness. All these operations are operations in language, that is, features of the operation of the organism (i.e., human being) in its relational space, and although they require the nervous system to take place, do not take place in it. Or, in other words, recursive linguistic behavior is not an operation of the brain, and is not determined by a particular feature of the nervous system." (Maturana et al., op. cit..)

We see here the possibility of escaping from predetermination, though the status of autopoiesis remains absolutely amentalistic as it appears in the following lines:

" (...) as intentionality becomes part of the manner of living of the observer, as he or she lives in conversations of intentionality, the structure of the nervous system of the observer changes in a manner contingent to that sensory/effector correlations that entail intentions. The nervous system of the observer becomes an intentional "languaging" brain, but intentionality, as a relational feature of the flow of behavior, remains a feature of the relational space in which the observer lives" (Maturana et al., 1995, passim).

This does not simply means that there is no location of intentionality or self-consciousness in the brain nor specific intentional or self-consciousness neurones. This means that intentionality or self-consciousness, though generated through the participation of a brain activity, take place as operations in the relational domain of the organism, and they have consequences in the body dynamics. Therefore, the real problem is to understand how relational phenomena have consequences in the body dynamics.

By the way it is interesting to establish the link between the role of recursive operation in the domain of linguistic behavior which gives rise to new phenomenal dimensions and Skinner's distinction between rule-gouverned behavior - by-product of language - and contingencies-modeled behavior. In Skinner's view, intentions for example are nothing else than verbal - eventually inner - precursor behaviors as discriminative stimuli which do not cancel the action of immediate contingencies. And the origin of these precursor behaviors has to be found in interindividual exchanges of the linguistic community.

Now to come back to social systems in the Maturana's view, they are the medium for interaction and structural couplings realized through networks of recurring conversations (i.e. a braiding of language stricto sensu, emotions, and bodyhood). This is a first characteristic by which social domain is distinct from other domains of interaction.

As a second characteristic, a social system is a network of consensual (rather than explicit consensus) co-ordinations of action, through which its members realize their biological autopoiesis. Membership is determined by implicit acceptance (or rejection) -which forms an implicit boundary for the system - : becoming a member involves the organism's acquiring the behaviors and conversations that are appropiate to that domain. "Consensual" refers to a domain of mutual acceptance. In a paper of 1988, Maturana (p. 64) writes:

"The emotion that makes possible recurrent interactions in mutual acceptance is that which we connote in daily life with the word love."

As Mingers comments, social systems are thus restricted to those organisms with a high degree of flexibility with respect to their behavior. Also mutual acceptance - and a fortiori love - implies that other forms of recurrent interactions as military forces based on obedience or work communities based on task fulfillment have to be excluded, which is difficult to admit.

A third distinct characteristic of social system is that it is an emergent domain.

"Particular members may join or leave, but the social organization continues. The relationship between people and the social system is circular. The participants , as structure-determined entities, have properties and behaviors determined by their structure. These properties and behaviors realize the particular social systems to which they belong. But this , in turn, selects particular structural states within the participants (...). In other words, a social system inevitably selects or reinforces behaviors that confirm it and deselects those which deny it." (Mingers, 1995, p.131)

Fourth characteristic: a social system is conservative. Change cannot be imposed by the system but arrives only because members modify their behaviors, despite the fact that the social system maintains its homeostasis:

"An individual may enter a social system and not become structurally coupled to it (...) or already existing members can reflect upon their experiences in other domains and choose to modify their own behaviors, thus realizing an altered social system." (Mingers, *op. cit.,* p. 132)

In summary, a social system is a network of consensual co-ordinations of action, through which its members realize their biological autopoiesis.

2. Zeleny and Hufford

Let us now move to some authors who simply tranfered the concept of autopoiesis from the biological to the social domain. Mingers (1992, 1995) provided a radical criticism of Zeleny and Hufford's (1992) thesis who claimed in an attempt to apply autopoietic theory to social systems that family, as an example of a natural social system, is autopoietic, that all natural social systems are autopoietic, and that all autopoietic systems "both 'organic' and possibly 'inorganic' are necessarily social" (p. 156).

This last contention is the more easy to criticize, Mingers (1995) argues, as an unwarranted transference of the social to physics and chemistry:

"(...) there are at least two elements that must be common to all definitions of social: first it relates to the activity of groups of entities rather than single individuals; second it concerns rule-based behavior rather than physical cause and effect. Thus the behavior of billard balls on a table could not in any sense be termed social. While physical systems do consist of groups of components, it cannot be said that they follow rules. The essence of rule-governed as opposed to cause-and-effect behavior is that a rule can be broken, or followed in right or wrong ways (Winch, 1958; Wittgenstein, 1978). This is clearly not the case for physico-chemical interactions (...). Molecules cannot choose to interact or not: their behavior is determined (...)." (pp.125-126)

In applying autopoiesis to natural social systems, three problems have to be analyzied: the boundary, the nature of the components and the process of production of the components.

First problem. Contrary to Zeleny and Hufford's assertion, the supposed boundary of the family is not clear because the notion of membership is not. Mingers asks how to classify the cousin who lives abroad and is never seen or the *au pair* in the family for three months, or the boy friend, or the divorced wife etc. And what about families in other cultures than in the USA? Will everybody agree on what it is to be a member or a non-member of the family?

Second problem. Components of the boundary must be identifiable. For Zeleny and Hufford, the components of a family are the father, the mother, children aso. But, as Mingers emphasizes, this is a confusion between the biological domain and the roles belonging to the social domain: a "father" may be the biological farher, an adoptive father, a foster father...

Third problem. Following the theory, the components participate in processes of production that define an autopoietic system. But it remains here unclear if the processes of production are biological, social, or both. The actual production of biological organisms is a biological process quite independent of whether such organisms then participate in a social family.

To summarize:

"crucial contentious terms are not clearly defined, (...) the authors do not acknowledge the variety of views concerning the nature of the social world and the complex relations between social structure and individual action" (...) and " appear to subscribe to a very functionalist view, quite against Maturana's own prescription." (Mingers, 1992, pp. 229-230).

This last comment is quite interesting: one of the crucial issues of the debate on autopoiesis is the ontological question.

3. Modification of the definition of autopoiesis

A third group of authors modify the definition of autopoisesis, while applying the concept to social domain.

As Whitaker (1995b) argued, Maturana and Varela's views on social systems are difficult to apply for those who want to employ the concept of autopoiesis, but consider an organization (in the usual sense) as a unit distinct from its participants. "If human beings are the constituent elements of a social system, how can the social system (itself) be considered self-reproducing in terms of the humans?" asks Whitaker (1995b).

"Luhmann's solution", Whitaker (1995b) notices, "is to search for some constituent element other than humans. The rationale is that '...social processes must correspondingly produce social components if the concept of autopoiesis is to be extended to the social domain with any validity.' (Bednarz, 1988, p. 61)". Luhmann redefined social systems as being reproduced by always self-referential 'communications'. The required conditions for autopoiesis are met in terms of such communications, in the sense of communicative activities consisting in three elements: information, utterance, and comprehension. (see Mingers,1995, p. 139 *sqq.*, for a clear overview of Luhmann's autopoietic society based on communications). The system itself determines (it is the autopoietic closure) what is information for it; the utterance is the "why now", the "how", and the "who" of the communication, and comprehension how the utterance may be interpreted. A communicative act finds a sense if it generates some understanding by somebody else and communicative autopoiesis is not a physical self-reproducing of its own structure, but a network of events which are different along the line of the time. Autopoietic communication is seen as a 'meaning-processing'.

Society differentiates itself into subsystems (education, science, etc.). Each subsystem is an autopoietic network of recursive communications. Society as a whole is also autopoietic composed of all communications generated in subsystems plus all others which are not involved in particular subsystems. This has to be distinguished from the environment: the physical environment and people with their consciousnesses. Events that occur in the physical world (e.g. pollution) do not affect society until they enter the network of communication. "Society cannot communicate with but only about its environment according to its capacities for information processing" (Luhmann, 1986, p. 177). Society or subsystems are not isolated; they are organizationally closed but interactively open to triggering. People in particular can irritate society, and society may in response generate a communication, but the type of communication will be determined by society, not by the environmental triggering.

The concept of structural coupling was recently inserted in the theory (Luhmann, 1992, commented by Mingers,1995). In the social case, the concept refers to expectations of communications. This has the advantage of describing the system in terms of its operational characteristics, independent from the specific participants in that system at any given time.

In general, it can be concluded with Whitaker that Luhmann has developed a theory of the society as autopoietic where confusion between the biological and the social domains is avoided making the human participants peripheral components "at most".

His approach is radical in the sense that it treats social systems solely in terms of 'communications'. His ideas are developed in details in specific application of the principles to the field of law (see for a critical and extended discussion, Teubner, 1988, 1989 for the French edition). Teubner's own theses on autopoisis applied to law are outside my competence.

Thereupon, the Mingers' essential criticism is that firstly the problem of boundaries is not resolved. Boundary consists of particular components, which is not the same thing as simple differentiation of systems between themselves by defining their own comunications. That is why Mingers prefers to speak of a case of organizational closure rather than autopoiesis.

Secondly "there is no significant attempt to show how societal communication, as an independant phenomenal domain, emerges from the interactions of the human beings who ultimately underpin it. Without human activity there would be no communications" (Mingers, 1995, p. 149). And "It is one thing to say analytically that communications generate communications, but operationally they require people to undertake specific actions" (*Op. cit.,* pp. 148-149). Mingers sustains, correctly I think, that if a communication may be stimulated by another, it cannot be produced by it.

But what is considered as a weakeness by Mingers is precisely for Teubner (1989) a way of treating the problematic nature of an autopoietic higher-order system whose components are themselves autopoietic at their own level. From this point of view, whether or not social systems are organisms is not relevant since social autopoiesis has to be conceptualized apart from the autopoiesis of living beings. Emerging unities in the second-order autopoietic system are produced communications which constitue the elements of the third-order autopoietic system. Society is not a biosystem, but a system of meanings. It can be seen that nevertheless the issue of the relationship between the psychic individual systems and the social system of communication remains unclarified.

Mingers also emphasizes that Luhmann's overt functionalism had to be clarified (Luhmann uses the expression 'functionally differentiated society'), but he considers that it is not impossible to reformulate the theory of structural coupling in order to show nonteleologically the emergence of the society's particular structure.

A very consistent approach with Maturana and Varela's statements on social systems is represented by Hejel's social theory. A clear overview of the salient points is given by Whitaker (1995a, 1995b).

Hejel (e.g.1984) treats social systems as emergent from interactivity among their participants. He defined a social system as "...a group of living systems which are characterized by a parallelization of one or several of their cognitive states and which interact with respect to these cognitive states." (p. 70)

As far as autopoiesis is concerned, social systems cannot be characterized neither by self-organization, neither by self-maintenance, neither by self-reference.

Hejl provides precise definitions of these terms:

- Self-organizing systems are those "(...) which, due to certain initial and limiting conditions arise spontaneously as specific states or as sequences of states." (*Op.cit.,* pp. 62-63).
- Self-maintaining systems are a series of " (...) systems in which self-organizing systems 'produce' each other in an operationally closed way." (*Op.cit.,* p.63).
- Self-referential systems " (...) organize the states of their components in an operationally closed way." (*Ibid.)*

From these basic definitions, Hejel sustained that social systems cannot be described as

- strictly self-organizing, because they lack spontaneity;
- self-maintaining, because they do not directly generate the components which realize themselves;
- even self-referential, because the components may participate in several social systems in the same time and they have the possibility to withdraw.

Rather, social systems are defined in terms of an intersection between their composite identity and the individual participants. Hejel characterizes such phenomena as syn-referential. This concept refers to:

"components, i.e., living systems, that interact with respect to a social domain. Thus the components of a syn-referential system are necessarily individual living systems, but they are components only inasmuch as they modulate one another's parallelized states through their interactions in an operationally closed way." (1984, p. 75)

If the Hejl' s exploratory analysis is considered by Whitaker as "an example of how autopoietic theory can be reasonably applied to interpersonal networks" and "a guide for future applications", Teubner (1989) sees syn-referentiality as a vague concept and of limited application.

4. A metaphorical conception of social autopoiesis

The last group of authors to examine is composed of scholars who prefer to admit that autopoiesis is only a practical metaphore. For example, following Morgan's approach (1989), the Organizations' own internal self-images or identity are more important in the relations they entertain with the environnement than the environnement itself: the Organizations try to maintain (cf. autopoiesis) their image and identity by projecting themselves onto the environment.

Mingers (1995) comments that "using autopoieisis metaphorically is reasonably unproblematic (...). However, the results are merely metaphoric" (pp. 151-152). Is this position not too strict? As Simon reminded to us in his Nobel conference, one might not reject a metaphore if it is impossible to oppose an alternative. And in a logic of action, metaphores are useful. Morgan (1989) provided a useful guide for using metaphores to understand organizations.

5. Discussion

Interestingly in the context of this congress, it seems to me, a dividing line can be drawn between those who as Maturana defend a constructivist but anti-finalist position and others who propose a resolutely finalist-constructivist view. For instance, Le Moigne, claiming that "Monod, compelled by an unconscious disciplinary imperialism, restricted to living beings (the) fundamental property of teleology" (1990a, p. 136, my translation), wrote:

"to recognize that the models of the object have a privilege that the physicians devoted only to the Systems of Representations, the wright of fancy, becomes to-day unavoidable if we want to explain through our models behaviours of many objects we observe (human and social objects, of course, but also objects-artefacts, and even, why not, natural objects)" (1990a, p.144, my translation)

Of course, it is by virtue of the methodological rejection of disjunction between the object and the subject that the ideal systemic modelisation is conceptualized as projectivity, that is "the capacity of the modelisator to explain his/her 'projects of modelisation, i.e. goals he/she proposes to the model of a complex system that he/she sees as a priori goal-oriented and goal-orienting (without being constrained to know a priori in a sure and demonstrable way goals of the system, which are hypotheticaly constructed)" (Le Moigne, 1990b, p. 65, my translation).

Between these extremes, Barel (1979), maintaining that 'self-reproduction', i.e. the property of a system by which it actively takes part in its own reproduction, is the basic feature by which a complex and differenciated set can be analyzed as a system, sustained that an essential component of self-reproduction is self-finalisation, in the following precise sense that a system determines itself, though partly, its own goals, which need not to be conscious. This last feature is important. Barel emphasized that, *stricto sensu*, refering to self-reproduction condemns to consider only living systems, "i.e. biological or ecological systems (...), and (...) social systems if one admits (...) that self-reproduction has a real existence, and not the status of a simple metaphore" (p. 220). But Barel saw an open field outside an anthropomorphic conceptualization of self-reproduction that "could take as a starting point the fact that some non-living systems (...) can take part in their own production and reproduction, insofar as they have a "surplus" and can use it to self-"finalizing" (*in the cybernetic meaning, which does not imply any consciousness, any project, any final causality[1]*). The notion of surplus " requires nothing else than an *uncertainty* in the manner a system, for example physical, will react to a given situation. Any system that 'has the choice' (i.e. in which 'choice' is partly determined from its internal conditions) draws into the way of self-reproduction which, of course, does not more delete the boundary between living and non living than the social self-reproduction deletes what separes itself from biological reproduction" (*loc. cit.*).

Another position is developped by Teubner (1989) about the nature of the evolution of law (I only summarize what is understandable by a non specialist of law). Rejecting the "syndrom of the old evolutionnist concept ", Teubner does not agree with the idea of a modification oriented toward a precise goal. Rather, evolution is "simply 'teleonomic'. It pursues a construction on the basis of its fundamental architectonic, following its own rules (...)" (p. 77), combining programs that brought results and excluding others that brought nothing. "But though this progression is irreversible, no step is 'better', nor more 'viable' nor even more 'sure' than the previous one" (*Ibidem*).

---

[1] My italics.

292

As a general conclusion, this brief report tried to show that the extension of autopoiesis to social systems is problematic though "social autopoiesis remains a highly debatable possibility" (Mingers, 1995, p.152).

Probably, as Varela argues, applying to social systems autopoiesis in the biological sense, which includes boundaries *stricto sensu*, is difficult to sustain.

One way (see Luhmann) is to think out social autopoiesis independently of the autopoieis of the living systems. Social systems (and psychic systems) which are considered as closed, self-referential, without having physical production as their mode of operation, are communications. The richness of the basic concepts are maintained, but the cut is questionable.

Another way (see Hejel) consists in restricting autopoieis to biological systems but, in a bottom-up application, in developing an analysis of social systems from its human composites as autopoietic living systems. Here, we get rid of the controversial aspects of the basic theory, but at the expense of a shinkage of the basic autopoietic concepts. This is not necessarily detrimental. As Whitaker noticed, "Autopoietic theory provides ample concepts and principles to keep us busy with its applications for many years to come".

Finally, metaphorical use of autopoiesis, which has already proved its richness, is not to be despised.

References

Barel, Y. (1979). Essai sur le paradoxe et le système le fantastique social. Presses universitaires de Grenoble.

Bergson, H. (1907). L'évolution créatrice. Paris: Presses Universitaires de France.

Bunge, M. (1979). Ontology II: A world of Systems. Dordrecht: Reidel.

Cannon, W. (1939). The wisdom of the body. New York: Norton.

Hejel, P. M. (1984). Towards a theory of social systems: Self-organization and self-maintenance, self-reference and syn-reference. In H. Ulrich and G. J. B. Probst (Eds.), Self-organization and management of social systems. Insights, promises, doubts and questions (pp. 60-78). Berlin: Springer.

Le Moigne, J.-L. (1990a). La théorie du système général. Théorie de la modélisation. Paris: Presses Universitaires de France.

Le Moigne, J.-L. (1990b). La modélisation des systèmes complexes. Paris: Dunod.

Luhmann, N. (1986). The autopoiesis of social systems. In F. Geyer & J. van der Zouwen (Eds.), Sociocybernetic paradoxes (pp. 172-192). London: Sage Publications.

Maturana, H. (1978). Biology of language: The epistemology of reality. In G. Millar & E. Lenneberg (Eds.), Psychology and biology of language and thought: Essays in honour of Eric Lenneberg (pp. 27-63). New York: Academic Press.

Maturana, H. (1974). Cognitive strategies. In H. von Foerster (Ed.) Cybernetics of Cybernetics (pp. 457-469), Biological Computer Laboratory, University of Illinois.

Maturana, H. (1980). Man and society. In F. Benseler, P. Hejl, & M. Kock (Eds.), Autopoiesis, communication, and society: The theory of autopoietic systems in the social sciences (pp. 11-32). Frankfurt: Campus Verlag.

Maturana, H. (1988). Reality: The search for objectivity or the quest for a compelling argument. Irish Journal of Psychology, 9, 25-82.

Maturana, H. (1990). Science and daily life: The ontology of scientific explanations. In W. Krohn, G. Kuppers, and H. Nowotny (Eds.), Self-organizations: Portrait of a Scientific Revolution (pp. 12-35), Dordrecht: Kluwer Academic Publishers.

Maturana, H., Mpodozis. J. Letelier, J. C. (1995). Brain, Language, and the Origin of Human Mental Functions. Biological Research, 28, 15-26. [Quotations are made from the electronic publishing].

Maturana, H., & Varela, F. (1980). Autopoiesis and cognition: The realization of the living. Dordrecht: Reidel.

Maturana, H., & Varela, F. (1987). The tree of knowledge. Boston: Shambhala.

Miller, J. (1978). Living systems. New York: McGraw Hill.

Mingers, J. (1992). The problems of social autopoiesis. International Journal of General Systems, 21, 229-236.

Mingers, J. (1995). Self-producing systems: Implications and applications of autopoiesis. New York, London: Plenum Press.

Morgan, G. (1989). Images de l'organisation. Toronto: Presses de l'Université Laval et Editions ESKA.

Teubner, G. (1989). Le droit, un système autopoiétique. Frankfurt and Main: Suhrkamp Verlag.

Varela, F. (1979). Principles of biological autonomy. New York: Elsevier-North Holland.

Varela, F. (1981). Describing the logic of the living: The adequacy and limitations of the idea of autopoiesis. In M. Zeleny (Ed.), Autopoiesis: A theory of living organization (pp. 36-48). New York: North Holland.

Whitaker, R. (1995a, April). An electronic forum for autopoiesis and enactive cognitive science. The Observer, 10.

Whitaker, R. (1995b, December). Whitaker's autopoiesis. Athena, W3 Researches and Education, Manheim, Furtwangen, Heilderberg: Electronic publishing.

Winch, P. (1958). The idea of social Science. London: Routledge and Keegan Paul.

Wittgenstein, L. (1978). Philosophical investigations. Oxford: Blackwell.

Zeleny, M., & Hufford, K. D. (1992). The application of autopoiesis in systems analysis: Are autopoietic systems also social systems?. International Journal of General Systems, 21, 145-160.

Morphogenesis by Diffusive Chaos in Epidemiological Systems

Daniel M. DUBOIS
Université de Liège,
Institut de Mathématique
12, Grande Traverse
B-4000 LIEGE 1, BELGIUM
Fax: + 32 4 366 94 89,
E-mail: Daniel.Dubois@ulg.ac.be

Philippe SABATIER
Ecole Vétérinaire de Lyon
Unité Bio-Informatique
1, Avenue Bourgelat,
F-69280 MARCY L'ETOILE, FRANCE
Fax: */33/(16)/78872667,
e-mail: sabatier@clermont.inra.fr

Abstract

This paper deals with a discrete mathematical model of epidemiological systems based on the SIRS model. A population of susceptible individuals can be infected by contact to infective individuals which recover. The first part of this paper deals with temporal discrete models. Three threshold values of a control parameter depending on the transmission rate and the total population share the behaviour of the epidemic in four zones: no infection, increase of infection, bifurcation and then chaos. It is shown that for the same transmission rate, a critical total population shares the dynamics of the infection: for low population, infection disappears and for high population, the infection increases but the number of susceptible individuals remains at a positive constant value.

In the second part of this paper, the SIRS model is generalised in taking into account the spatial diffusion of the populations. Classically space-time models are based on the reaction-diffusion equation. There is a problem with such a parabolic equation because this predicts an infinite speed of propagation for the population disturbances. To avoid this biologically incorrect feature, a non-stationary term is introduced. Numerical simulations show the propagation of the infection with chaotic spatial heterogeneities, called morphogenesis by diffusive chaos. In spatial region with low population, no infection propagates.

Keywords: Epidemiological systems, deterministic chaos, diffusive chaos, morphogenesis.

1 Introduction

In mathematical models for the spread of infections the force of infection is defined as the probability per unit of time that a given susceptible individual becomes infected (Capasso, 1993). The force of infection can be split in three factors : the contact rate the average number of relevant contacts with others individuals per unit of time ; the probability that such a contact occurs with an infectious individuals ; the probability that such a contact results in transmission of the infection. In this paper we are concerned with possible effects of population size on this crucial concept, and consequently on the transmission of diseases.

The infection process is generally derived from the mass-action assumption, by analogy to chemical reaction kinetics, which consider that the density of relevant contacts per unit of time is proportional both to the density of susceptible individuals and to the density of infectious individuals.

CP437, *Computing Anticipatory Systems: CAYS--First International Conference*
edited by Daniel M. Dubois © 1998 The American Institute of Physics 1-56396-827-4/98/$15.00

The classical model of epidemic (Kermack and McKendrick, 1927) is given by the SIRS differential equations:

(1a) $dS/dt = -\beta IS + \rho R$
(1b) $dI/dt = +\beta IS - \gamma I$
(1c) $dR/dt = +\gamma I - \rho R$

where S, I and R are the susceptible to a disease, infective and recovered populations; β is the rate of transmission of the infection, γ is the rate of decay of infective individuals and ρ the rate of decay of the recovered individuals.
The variation of the total population $N = S + I + R$ can be found in adding the eqs. 1abc

(1d) $dN/dt = 0$

so, the population is time invariant and can be determined by initial conditions. Different initial conditions give rise to different values of the total population defined by the parameter C

(1e) $N = C$

This model has two steady states : (1) absence of infection, when $I^* = R^* = 0$ and the population size $N^* = S^*$, and (2) presence of infections, when $N^* = S^* = \gamma/\beta$. The first stationary state is locally asymptotically stable when $R_{0,N} < 1$, and the endemic steady state is stable when $R_{0,N} > 1$ where : $R_{0,N} = N^*\beta / \gamma$.

Anderson and May (1979) formulate transmission in terms of numbers of individuals instead of densities : they assumes that the number of infections per unit of time is proportional to the product of the number of susceptible individuals and the number of infectious individuals. In this formulation, the transmission rate is independent on N.

The dependence of the transmission rate on N in the true mass-action, opposed to the pseudo mass-action, reflect the fact that whereas for constant density and increasing population size, the number of individuals encountered par individual does not change, the probability of encountering any particular individual decreases. On the original meaning of the mass-action, the transmission rate is expressed in densities.

The total population is normalised to the unity, $C = 1$, which means that the S, I and R populations must be interpreted as densities independent of the total area and, hence, of the total population. It is likely that the mass-action is not an accurate description of contact rates will occur for high densities of individuals (Mollison, 1985). It is possible to develop a model which takes into account an increasing of the contact rate with the density (Heesterbeek, Metz, 1993).

However in what is to follow we will analyse the dynamics of an epidemic in function of different size of the population $N = C$, with constant density, so C becomes a control parameter. With the same values of all the parameters β, γ and ρ, threshold values of C will define different classes of behaviour of the epidemic.

2 Temporal Discrete Epidemiological Systems

In view of simulating this system on computer, the eqs. 1abc are transformed to finite difference equations (Sabatier, Guigal, Dubois, 1995, 1996):

(2a) $S(t + \Delta t) = S(t) - \beta \, \Delta t \, I(t)S(t) + \rho \, \Delta t \, R(t)$

(2b) $I(t + \Delta t) = I(t) + \beta \, \Delta t \, I(t)S(t) - \gamma \, \Delta t \, I(t)$

(2c) $R(t + \Delta t) = R(t) + \gamma \, \Delta t \, I(t) - \rho \, \Delta t \, R(t)$

where we have taken a discrete time step Δt. The conservation of the total population, eqs. 1de, is given in the discrete equation

(2d) $N(t + \Delta t) = S(t + \Delta t) + I(t + \Delta t) + R(t + \Delta t) = N(t) = S(t) + I(t) + R(t) = C$

where C is a control parameter fixed by the initial condition of the total population $N(t)$ at time $t = 0$..

With $C = 1$, these eqs. 2abc give rise to different types of solution depending on the values of the parameters β, γ and ρ. Sabatier, Guigal, and Dubois (1995, 1996) showed the emergence of deterministic chaos from these discrete eqs. 2abc and incursive control of such a chaos.

This paper will analyse these eqs. 2abc in function of the value of the total population N from the control parameter C. The fact to consider the total population as a control parameter means that the epidemic is a fast dynamics compared with the slow dynamics of the growth of the population.

2.1 Anticipation of Epidemic by Population Thresholds

In view of demonstrating the influence of the total population on the behaviour of the epidemic, let us consider the simplest model which can give rise to a very complex dynamics. For that, let us derive finite difference equation from the SIS model of Bailey (1957):

(3a) $S(t + \Delta t) = S(t) - \beta \, \Delta t \, I(t)S(t) + \gamma \, \Delta t \, I(t)$

(3b) $I(t + 1) = I(t) + \beta \, \Delta t \, I(t)S(t) - \gamma \, \Delta t \, I(t)$

This model means that the infective population $I(t)$ will become a susceptible population with the rate γ. There is no intermediate recovered population:

(3c) $R(t + \Delta t) = R(t) = 0$

so $\rho R(t)$ is replaced by $\gamma I(t)$ in eq. 2a and eqs. 3ab are obtained.
The total population given by eq. 2d is then replaced by

(3d) $N(t + \Delta t) = S(t + \Delta t) + I(t + \Delta t) = N(t) = S(t) + I(t) = C$

From this eq. 3d, the susceptible population is given by

(3e) $S(t) = C - I(t)$

In replacing eq. 3e in eq. 3b, the evolution of the infective population is obtained in function of itself and three parameters β, γ and C:

(3f) $I(t + \Delta t) = I(t) + \beta \, \Delta t \, CI(t)[1 - I(t)/C] - \gamma \, \Delta t \, I(t)$

which is similar to the Pearl-Verhulst map giving rise to chaos (May, 1976; Feigenbaum, 1978).
In considering a time duration $\Delta t = 1/\gamma$ related to the duration of the infection, this eq. 3f becomes

(3g) $I(t + 1/\gamma) = (\beta/\gamma) \, CI(t)[1 - I(t)/C]$

With a normalisation of the total population, C = 1, the parameter $\sigma = \beta/\gamma$ describes the potential number of infectious contacts which might be made by an infected individual (Kemper, 1980).
In defining the proportions of the populations by $i(t) = I(t)/C$ and $s(t) = S(t)/C$, eq. 3g becomes

(4) $i(t + 1/\gamma) = 4 \, \mu \, i(t)[1 - i(t)]$

in defining the parameter μ

(5) $\mu = (\beta/\gamma) \, C / 4 = \sigma \, C / 4$

This parameter μ can be interpreted as a probability: indeed, $0 \le i(t + 1/\gamma) \le 1$ and the maximum of $i(t)[1 - i(t)] = 1/4$ for $i(t) = 1/2$, so $0 \le \mu \le 1$.
Three threshold values of μ define four different classes of behaviour related to the theory of deterministic chaos. For $\mu \le 1/4$, the solution is a null fixed point I = 0. For $1/4 < \mu \le 3/4$, a fixed point $I = C (1 - 1/4\mu)$. For $3/4 < \mu \le 0.89...$, a diagram of bifurcation with period doubling. For $0.89... < \mu \le 1$, the deterministic chaos with a sensitivity to initial conditions.

Figures 1ab give the bifurcation diagram of eqs. 3ab where the susceptible S(t) and infected populations I(t) are a function of σ with C = 1. Three threshold values of $\sigma = \sigma_1, \sigma_2, \sigma_3$ share four classes of behaviour of the infection:
1. Constant susceptible and no infective populations, 2. Non-linear decrease of susceptible and non-linear increase of the infective populations, 3. Bifurcations, and 4. Chaos in both populations.

Figures 2ab give the bifurcation diagram of the same eqs. 3ab where the susceptible S(t) and infective I(t) populations are a function of the total population given by the parameter C with $\sigma = 1$. In this case, the three threshold values deal with the total population $C = C_1 , C_2 , C_3$ which share four classes of behaviour of the infection, with a novel interpretation of the epidemic:
1. Linear increase of the susceptible and no infective populations, 2. Constant susceptible and linear increase of infective populations, 3. Bifurcations, and 4. Chaos in both populations.

It is interesting to point out two important results related to the critical threshold $T = \sigma\, C = 1$ which defines two main zones. Firstly, below this threshold, no infection can propagates and secondly, above this threshold, the averaged susceptible population remains constant (even in the bifurcation and chaos regions). It exists thus, for a given rate of transmission by contacts, a critical total population for the propagation of the infection.

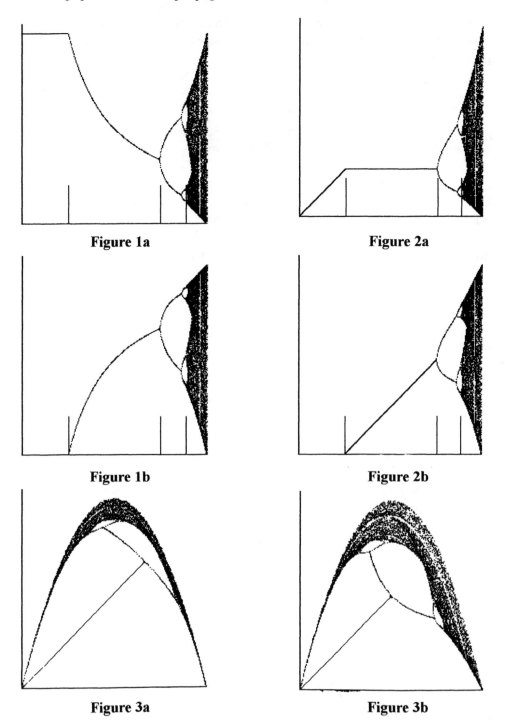

Figure 1a

Figure 2a

Figure 1b

Figure 2b

Figure 3a

Figure 3b

It would be interesting to validate this first anticipation of epidemic with actual data. The same infection would propagate in regions of high density of individuals and would decay in regions of low density of individuals. In a population with infective individuals, there would always exists a constant susceptible population. Experimental data could validate this second anticipation of epidemic.

Indeed, it is possible to find experimentally the chaotic law of transmission of a disease in plotting the values of the infective population $I(t)$ at time t as a function of the same infective population $I(t - \Delta t)$ at time $t - \Delta t$.

Figures 3ab give $I(t + \Delta t)$ as a function of $I(t)$ for the simulations shown in Figs. 1ab and 2ab, respectively. This corresponds to a parabola given by eq. 4. In the theory of deterministic chaos, for the first return map, it was demonstrated that any function similar to the parabola gives rise to the same qualitative bifurcation diagram and chaos.

In 1980, Collet, Eckmann and Lanford demonstrated the universality of the Feigenbaum constants for any iterative law $x(t+1) = f(x(t))$ under sufficiently broad conditions: f is continuous and differentiable, it presents a quadratic extremum, and a condition of convexity.

2 Spatio-Temporal Discrete Epidemiological Systems

To take into account the spatial spread of population, a diffusion factor can be added to the SIRS model 1abc in the following way:

(6a) $\partial S/\partial t = - \beta IS + \rho R + D \nabla^2 S$

(6b) $\partial I/\partial t = + \beta IS - \gamma I + D \nabla^2 I$

(6c) $\partial R/\partial t = + \gamma I - \rho R + D \nabla^2 R$

where D is the diffusion coefficient and ∇^2 is the classical Laplacian operator

(7) $\nabla^2 F(x,y,z,t) = \partial^2 F / \partial^2 x + \partial^2 F / \partial^2 y + \partial^2 F / \partial^2 z$

which gives the second derivatives related to the x, y, z spatial co-ordinates.
The space-time dynamics of the total population $N(x, y, z, t)$

(8) $N(x, y, z, t) = S(x, y, z, t) + I(x,, y, z, t) + R(x, y, z, t)$

is obtained in adding the three eqs. 6abc as follows:

(9) $\partial N(x, y, z, t) / \partial t = D \nabla^2 N(x, y, z, t)$

which is similar to the heat conduction equation.

There is a problem with such a parabolic equation because this predicts an infinite speed of propagation for the population disturbances.
To avoid this biologically incorrect feature, let us introduce a non-stationary term similarly to the term added in the Fourier's law for heat conduction by some authors (Maxwell, 1867,

Cattanbo, 1958, Green and Laws, 1972, Vernotte, 1958, Chester, 1963, Kranys, 1966, Müller, 1967, Lambermont and Lebon, 1973, 1976).

In introducing such a non-stationary term, eq. 9 can be written in the following way

(10) $\partial N(x, y, z, t) / \partial t = -\tau \partial^2 N(x, y, z, t) / \partial t^2 + D \nabla^2 N(x, y, z, t)$

where τ is a relaxation time. With τ and D with the same sign, eq. 10 is a hyperbolic equation exhibiting a finite speed of propagation given by

(10a) $v = (D / \tau)^{1/2}$

Eq. 10 can be discretized in one space dimension as follows

(11) $N(x, t + \Delta t) = N(x, t-\Delta t) - 2 \Delta t \tau [N(x, t + \Delta t) - 2 N(x,t) + N(x, t - \Delta t)] / \Delta t^2$
        $+ 2 \Delta t D [N(x + \Delta x,t) - 2 N(x, t) + N(x - \Delta x, t)] / \Delta x^2$

In choosing

(12) $\Delta t = 2\tau$

eq. 11 becomes

(13) $N(x, t + 2\tau) = N(x, t) + 2 \tau D [N(x + \Delta x, t) - 2N(x, t) + N(x - \Delta x, t)]/\Delta x^2$

This discrete eq. 13a is numerically stable. For example, in choosing

(14) $\Delta x^2 = (2\tau).(4D)$

eq. 13a becomes

(13b) $N(x, t + 2\tau) = N(x, t) + [N(x + \Delta x, t) - 2N(x, t) + N(x - \Delta x, t)]/4$

With $2\tau = 1$ and $4D = 1$ with an initial condition $N(4, 0) = 1$, the spatial evolution of $N(x, t)$ in function of time gives the spread of the population with time as shown in the following figure

x =	0	1	2	3	4	5	6	7	8
t = 0	0	0	0	0	1	0	0	0	0
t = 1	0	0	0	1/4	1/2	1/4	0	0	0
t = 2	0	0	1/16	1/4	3/8	1/4	1/16	0	0
t = 3	0	1/64	3/32	15/64	5/16	15/64	3/32	1/64	0

with a correct conservation of the total population equal to 1 at each time step.

Eqs. 6abc can be discretized by the same discretization as in eq.13a.

The space-time discrete model SIRS can be thus written as follows

(14a) $S(x, t + 2\tau) = S(x, t) + 2\tau[- \beta IS + \rho R] + 2\tau D[S(x + \Delta x, t) - 2S(x, t) + S(x - \Delta x, t)]/\Delta x^2$

(14b) $I(x, t + 2\tau) = I(x, t) + 2\tau[+ \beta IS - \gamma I] + 2\tau D[I(x + \Delta x, t) - 2I(x, t) + I(x - \Delta x, t)]/\Delta x^2$

(14c) $R(x, t + 2\tau) = R(x, t) + 2\tau[+ \gamma I - \rho R] + 2\tau D[R(x + \Delta x, t) - 2R(x, t) + R(x - \Delta x, t)]/\Delta x^2$

The numerical simulation of these eqs. 14abc shows morphogenesis by diffusive chaos as introduced by Dubois(1996) for space-time Pearl-Verhulst and Lotka-Volterra systems.

2.1 Morphogenesis by Diffusive Chaos in SIRS model

Numerical simulations of eqs. 14abc were performed with the following values of the parameters:: $\Delta t = 2 \tau = 1$, $\Delta x = 1$ and $D = 1/16$.
Eqs. 14abc are then considered as cellular automata: The space variable is represented by 200 cellular automata: $x = 1, 2, 3, ... 200$ by step $\Delta x = 1$.
The time variable varies by integer step $\Delta t = 1$: $t = 0, 1, 2, 3, ...$
The boundaries conditions are considered as reflecting walls:
$S(1, t) = S(2, t)$, $I(1, t) = I(2, t)$, $R(1, t) = R(2, t)$
$S(200, t) = S(199, t)$, $I(200, t) = I(199, t)$, $R(200, t) = R(199, t)$
Two simulations were considered with different initial conditions.

On one hand, the initial total population $N(x, 0) = S(x, 0) + I(x, 0) + R(x, 0) = C = 1$ is constant and uniform in the space, but the initial distribution of infection is not uniform in the space. The initial condition are
$S(x, 0) = 0.8 + 0.2 \sin(2 \pi x / 400)$,
$I(x, 0) = 1 - S(x, 0)$ and $R(x, 0) = 0$ for x varying from 0 to 200
Figures 4abvdefghij show the spatial distribution of $I(x, t)$ at successive times $\Delta t = 1$. Morphogenesis by diffusive chaos is represented essentially by travelling phase waves of epidemic because in this simulation the diffusion coefficient is small.

On the other hand, the initial total population $N(x, 0) = S(x, 0) + I(x, 0) + R(x, 0)$ varies with the space variable and is not uniform in the space. A very small initial infection is localised in a single automaton $I(100, 0) = 0.01$ at the centre of the space.
The initial conditions are
$S(x, 0) = \sin(2 \pi x / 200)$ for x varying from 50 to 99 and from 101 to 150,
$S(x, 0) = 0$ for x varying from 1 to 49 and from 151 to 200, and $S(100, 0) = 0.99$
$I(x, 0) = 0$ for x varying from 1 to 99 and 101 to 200, and $I(100, 0) = 0.01$
$R(x, 0) = 0$ for x varying from 0 to 200
Figures 5abcdefghij show the spatial distribution of $S(x, t)$ at successive times $10 \Delta t = 10$. The surprising phenomenon is the high speed of the propagation of infection by diffusive chaos by comparison to the spread of the population by diffusion. The threshold effect, explained in the first part of this paper is well-seen.

Finally, Figure 6a shows the first return map of $I(x, t + 1)$ as a function of $I(x, t)$ for all the 200 automata $x = 0$ to 200, during a time interval $t = 0$ to $t = 200$ in the same conditions as in Figures 4. The parabolic picture confirms the presence of chaos. Figure 6b shows the diagram of $I(x, t)$ as a function of $S(x, t)$ for the same condition as in Figure 6a.

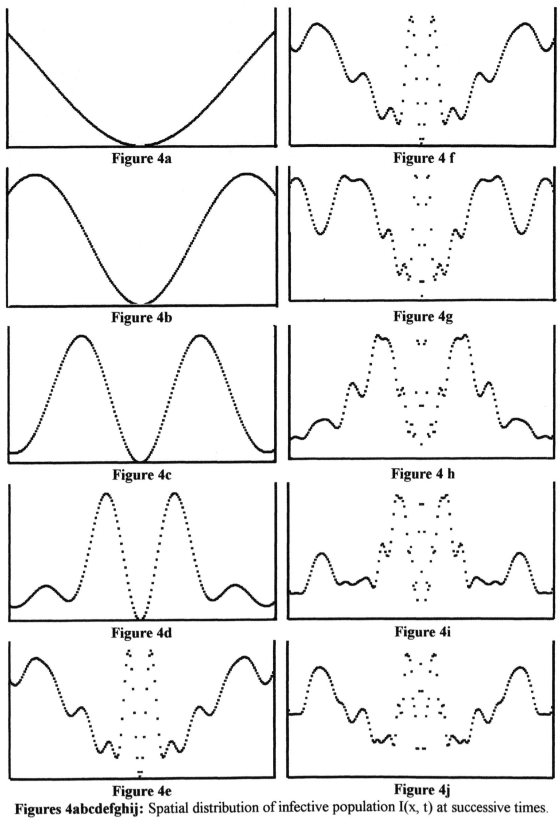

Figure 4a

Figure 4b

Figure 4c

Figure 4d

Figure 4e

Figure 4 f

Figure 4g

Figure 4 h

Figure 4i

Figure 4j

Figures 4abcdefghij: Spatial distribution of infective population I(x, t) at successive times.

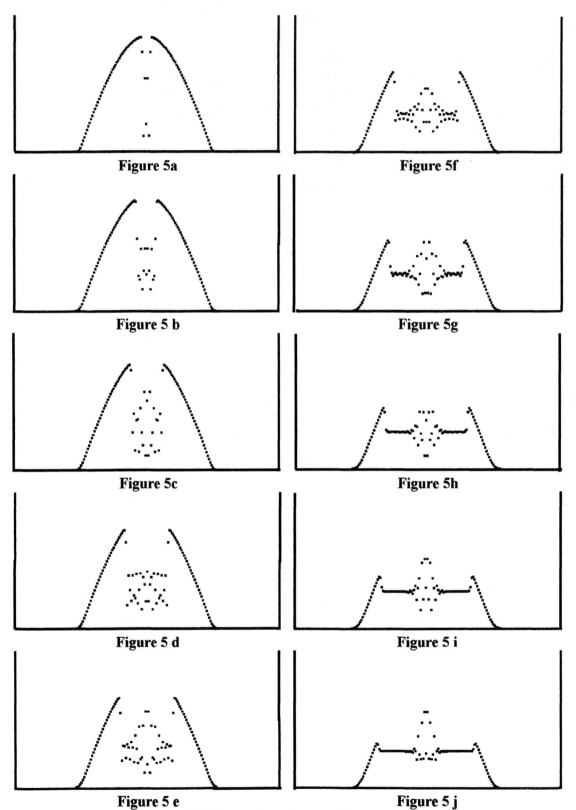

Figure 5a

Figure 5 b

Figure 5c

Figure 5 d

Figure 5 e

Figure 5f

Figure 5g

Figure 5h

Figure 5 i

Figure 5 j

Figures 5abcdefghij: Spatial distribution of susceptible population $S(x, t)$ at successive times. The propagation of epidemic is speeder than the spread of the population by diffusion.

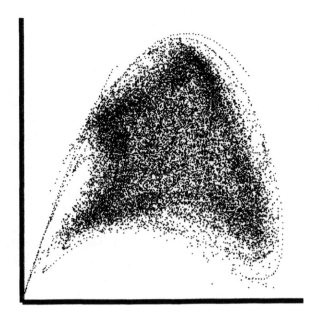

Figure 6a: First return map of I(x, t + 1) in function of I(x, t) for the conditions given at Figures 4. The parabolic picture confirms the diffusive chaos.

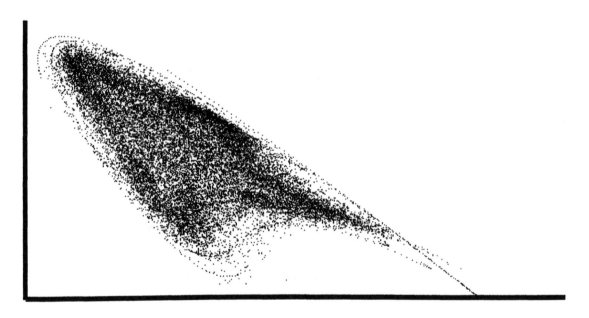

Figure 6b: Diagram of I(x, t) in function of S(x, t) for the conditions given at Figure 6a.

Conclusion

Classically, the total population is normalised to unity N = 1, and the master control parameter r is the number of contacts per unit of time. We have show that some anticipation of the evolution of an epidemic can be made from the actual value of the population. Several threshold values of the total population N define different classes of the propagation of the infection: 1. Decay of the infection, 2. Stabilisation of the infective population, 3. Bifurcation of populations and 4. Deterministic chaos with sensitivity to initial conditions. This is an important result for three reasons. First, it is easier to measure the values of the populations than the number of contacts. Second, the critical thresholds are related to the product of the total population by the number of contacts by unit of time. Third, with the same number of contacts per unit of time, the evolution of an infection is directly related to some critical thresholds in the total populations. In region with low density of population, an infection disappears and in region with high density of population, chaotic propagation of epidemic occurs.

The space-time SIRS model shows the importance of space in the understanding of epidemic propagation. Infective population behave as travelling phase waves of high speed in comparison to the speed of the spread of the populations by spatial diffusion.

It is thus wrong to think that the propagation of an epidemic is only due to the spatial diffusion of populations. The threshold effect is well confirmed as well as the role of chaos.

The morphogenesis by diffusive chaos was mathematically demonstrated for the simple Pearl-Verhulst model (Dubois, 1996). Research is in progress in this direction.

References

Anderson, R. M., May, R.M. (1979). Population biology of infectious diseases I." Nature Vol. **280**: pp. 361-367.

Bailey, N. T. J. (1957). *The Mathematical Theory of Infectious Diseases and its Applications*. London: Charles Griffin & co. Ltd. 413 p.

Capasso, V. (1993). Mathematical Structures of Epidemic Systems, Lecture Notes in Biomathematics, Vol 97, Springer-Verlag : Berlin, Heidelberg. 283 p.

Cattaneo, C. (1958), Sur une forme de l'équation éliminant le paradoxe d'une propagation instentanée." C.R. Acad. Sc. Vol. T 247: pp. 431-433.

Chester, M. (1963). Second Sound in Solids." Phys. Rev. Vol. **131**: pp. 2013-2015.

Collet P., Eckmann J. P., Landford, O. E. (1980). Universal Properties of Maps on an Interval, Communications in Mathematical Physics, Vol. 76, pp. 211-254.

Derrick, W.R., Van den Driessche, P. (1992), A disease transmission model in a nonconstant population. J. of Math. Biol., Vol. 31, pp. 495-512.

Dubois D. M.(1996). *Emergence of Space-Time Structures from Diffusive Chaos in Pearl-Verhulst and Lotka-Volterra Cellular Automata*, in Cybernetics and Systems'96, volume 1, Edited by R. Trappl, published by the Austrian Society for Cybernetic Studies, Vienna, pp. 100-105.

Dubois D. M., Sabatier Ph. (1996). "Qualitative Anticipation of Deterministic Chaos in Epidemiological Systems". Advances in Modeling of Anticipative Systems. Edited by G. E. Lasker, D. M. Dubois, B. Teiling. Published by The International Institute for Advanced Studies in Systems Research and Cybernetics, pp. 51-55.

Feigenbaum M. J. (1978), "Quantitative universality for a class of non-linear transformations", Journal of Statistical Physics, Vol. 19, pp. 25.

Green, A. E., Laws, N. (1972). On the Entropy Production Inequality." <u>Arch. Rat. Mech. Anal.</u> Vol. **45**: pp. 47-53.

Heesterbeek, J. A. P., Metz, J.A.J. (1993). The saturating contact rate in marriage and epidemic models." <u>J. Math. Biol.</u> Vol. **31**: pp. 529-539.

Kemper, J. T. (1980). On the identification of the superspreaders for infectious diseases. *Math. Biosc.* Vol. 48, p. 111-127.

Kermack, W.O., Mac Kendrick, A.G. (1927). Contribution to the mathematical theory of epidemics. Proc. Roy. Soc. A, Vol. 115, pp. 700-721.

Kolmogorov, A., Petrovski, I., Piscounov, N. (1937). Etude de l'équation de diffusion avec croissance de la quantité de matière et son application à un problème biologique." <u>Bull. Univ. Mosc. Ser. Int.</u> Vol. **A1**: pp. 141-158.

Kranys, M. (1966). The Postulate of the Relaxation of Non-Stationary Processes Caused by the Finite Transport Velocity of Interaction in the Matter." <u>Phys. Letters,</u> Vol. **22**: pp. 285-286.

Lambermont, J., Lebon, G. (1973). On a Generalization of the Gibbs Equation for Heat Conduction." <u>Phys. Lett.</u> Vol. **42A**: pp. 499-500.

Maxwell, J. C. (1867). On the Dynamical Theory of Gases." <u>Phil. Trans. Roy. Soc.</u> Vol. **157**: pp. 49-88.

May R. M. (1976). "Simple mathematical models with very complicated dynamics", Nature, 261, p. 459.

Mena-Lorca, J., Hethcote, H.W. (1992), Dynamic models of infectious diseases as regulators of population size, J. Math. Biol., Vol. 30, pp. 693-716

Mollison, D. (1985). Sensivity analysis of simple epidemic models. <u>Population Dynamics of Rabies in Wildlife</u>. B. P.J. London, Academic Press: pp. 224-234.

Müller, I. (1967). Zum Paradoxen der Wärmeleitungstheorie." <u>Zeitschrift für Phys.</u> Vol. **198**: pp. 329-344.

Müller, I. (1971). The Coldness, a Universal Function in Thermoelastic Bodies." <u>Arch. Tat. Mech. Anal.</u> Vol. **41**: pp. 319-322.

Sabatier, P., Guigal, P. M., Dubois, D. M. (1995). "Stabilising the Dynamical Behaviour of an Infection: Anticipatory Regulation by Incursive Control in a Discrete Equation". Advances in Interdisciplinary Studies in Systems Research and Cybernetics, vol. III, Edited by G. Lasker, published by The International Institute for Advanced Studies in Systems Research and Cybernetics, pp. 31-35.

Sabatier P., Guigal P. M., Dubois, D. M. (1996). "Emergence of Chaos in the Classical SIRS Epidemiological Model". Cybernetics and Systems'96, Vol. 1, Edited by R. Trappl, Austrian Society for Cybernetic Studies, pp. 618-623.

Schenzle, D., Dietz, K. (1987). Critical population sizes for endemic virus transmission. <u>Raumliche Persistenz und Diffusion von Krankheiten</u>. W. a. H. Fricke, E. Heidelberg, Heidelberger geographische. Vol. **83**: pp. 31-42.

Thieme, H. R. (1992). Epidemic and demographic interaction in the spread of potentially fatal disease in growing populations." <u>Math. Biosci.</u> Vol. **111**: pp. 99-130.
Vernotte, P. (1958). Les paradoxes de la théorie continue de l'équation de la chaleur." <u>C.R. Acad. Sc.</u> Vol. **T 247**: pp. 3154-3155.

Competing Technologies: Lock-ins and Lock-outs

Loet Leydesdorff
Department of Science & Technology Dynamics,
Nieuwe Achtergracht 166, 1018 WV Amsterdam, The Netherlands;
<l.leydesdorff@mail.uva.nl>
&
Peter Van den Besselaar
Department of Social Science Informatics,
Roetersstraat 15, 1018 WB Amsterdam, The Netherlands;
<peter@swi.psy.uva.nl>

Abstract

W. Brian Arthur's model for competing technologies is discussed from the perspective of evolution theory. Using Arthur's own model for the simulation, we show that 'lock-ins' can be suppressed by adding reflexivity or uncertainty on the side of consumers. Competing technologies then tend to remain in competition. From an evolutionary perspective, lock-ins and prevailing equilibrium can be considered as different trajectories of the techno-economic systems under study.

Our simulation results suggest that technological developments which affect the natural preferences of consumers do not induce changes in trajectory, while changes in network parameters of a technology sometimes induce ordered substitution processes. These substitution processes have been shown empirically (e.g., Fisher & Prey, 1971), but hitherto they have been insufficiently understood from the perspective of evolutionary modelling. The *dynamics* of the selective network determine the possible dissolution of a lock-in given the presence of competing technologies.

Keywords: competition, lock-in, alternative technologies, substitution, trajectory

1. Introduction

Arthur (1988a, 1988b, 1989) specified why one expects a lock-in in the case of competing technologies, randomness in initial purchasing behaviour, and absorbing barriers because of network effects. A random walk will necessarily reach one of the barriers. In an evolutionary model, the absorbing barriers can also be considered as selectors on the variation. Since Arthur's model was specified with reference to competing technologies, the selection environments are then market segments.

In this study, we use Arthur's model for a discussion of lock-ins from the perspective of evolution theory. Using comparable simulations, we first show that one does not need the assumption of absorbing barriers in order to find lock-ins as a result of random walks. If a network effect (that is, a path-dependency upon previous adopters) is assumed, lock-ins can be generated endogenously.

Second, we discuss the economic interpretation of these effects in terms of consumer behaviour on markets. We shall show that under realistic assumptions, continuing competition is a third trajectory, in addition to lock-in on either side. Third, we focus on 'lock-out': under which conditions will technologies that have been locked-in be able to leap out? In a final section, our results are interpreted in terms of non-linear dynamics.

CP437, *Computing Anticipatory Systems: CAYS--First International Conference*
edited by Daniel M. Dubois © 1998 The American Institute of Physics 1-56396-827-4/98/$15.00

2. Arthur's (1988) model

Arthur generalized his model from so-called Pólya-urn models for the purposes of studying economic processes like standardization, network effects, and so-called 'increasing returns' (David 1985; cf. Arthur 1990). As in the case of path-dependency in the probabilities of drawings in urn models, one can assume that the choices made by previous adopters from among competing technologies will matter for individual consumers (Leydesdorff 1992; 1995).

While an adopter may have a given ('natural') inclination toward one technology, this preference can be modified by information about the market shares of various technologies. For example, even if one wished nowadays to buy a video cassette recorder using a system other than VHS, one would soon realize the need to conform to this standard in order to profit from the apparatus given current market conditions. Standardization tends to lock a prevailing technology and a market into each other (David & Foray 1994). This lock-in is caused by 'network externalities' in the diffusion phase like technological interrelatedness, learning by doing, or economies of scale and scope.

Arthur (1988a, 1988b, 1989) distinguished two technologies which he labeled A and B. Additionally, he assumed two types of agents, R and S, with different 'natural inclinations' towards A and B. R-agents have a natural inclination a_R towards type A technology, and a lower inclination b_R towards B. Similarly, one can attribute parameters a_S and b_S to S-agents ($b_S > a_S$).

The network effects are modelled as independent terms, again differently for R-type agents and S-type agents. The appeal of a technology is increased by previous adopters with a term r for R-type agents, and s for S-type agents. These assumptions lead to the following model:

	Technology A	Technology B
R-agent	$a_R + rn_A$	$b_R + rn_B$
S-agent	$a_S + sn_A$	$b_S + sn_B$

Table 1. *Returns to adopting A or B, given n_A and n_B previous adopters of A and B. (The model assumes that $a_R > b_R$ and that $b_S > a_S$. Both r and s are positive.)*

If R-type and S-type agents arrive on the market randomly, the well-worked-out theory of random walks predicts that this competition will necessarily lock-in on either side (A or B). *Table 2* provides code in BASIC for simulating this model, and *Figure 1* shows the results of ten runs in a population of 10,000 adopters. The line in the middle corresponds to a 50% market share for each technology. As predicted, lock-in occurs in all cases although not necessarily before the end of a simulation.

The network effects are not a consequence of hypothesizing absorbing barriers. They are generated *endogenously* as a direct consequence of the positive values of parameters r and s. The absorbing barriers are not caused by (static) market conditions, but structural in a random walk with path-dependent feedback. However, if we reduce both r and s by 50% to 0.005 in the above model, lock-in will often not occur in a population of this size (10,000). Although a lock-in will occur in principle given path-dependency, in practice the system may thus remain in equilibrium for extensive periods (e.g., 100,000 adopters).

310

```
10 INPUT N
20 SCREEN 11: WINDOW (-2, 0)-(N, 100): CLS
25 FOR J = 1 TO 25
30 LINE (-2, 50)-(N, 50)
40 AR = .8: BR = .2: SA = .2: BS = .8: NA = 1: NB = 1: S = .01: R = .01
45 RANDOMIZE TIMER
50 FOR I = 1 TO N
70 XX = RND
90 IF XX < .5 GOTO 100 ELSE GOTO 125
100 RETURNA = AR + R * NA: RETURNB = BR + R * NB
110 IF RETURNA > RETURNB THEN NA = NA + 1 ELSE NB = NB + 1
120 GOTO 140
125 RETURNA = SA + S * NA: RETURNB = BS + S * NB
130 IF RETURNA > RETURNB THEN NA = NA + 1 ELSE NB = NB + 1
140 Y = NA + NB: Z = 100 * NA / Y
150 PSET (Y, Z)
160 NEXT I
165 NEXT J
170 END
```

Table 2
Code for the simulation of Arthur's (1988) model

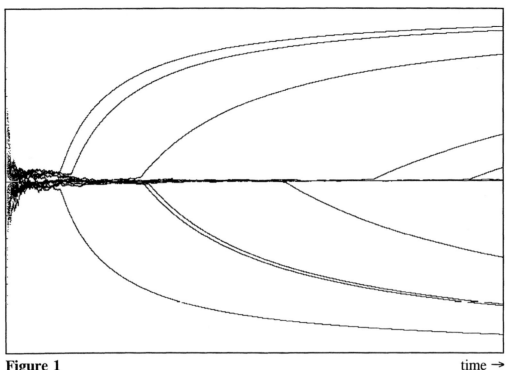

Figure 1   time →

Arthur's model as specified in Table 2; after 10 simulation runs

2.1.   Uncertainties about Relative Market Shares

Visual inspection of *Figure 1* suggests that the points representing the occurrence of a lock-in are closer to the 50%-line when this event happens at a later stage. In this section, we shall show that lock-in is dependent on the absolute difference in the number of adopters of technology *A* and technology *B*. This absolute number declines as a percentage of market share with increasing adoption.

Lock-in into technology *A*, for example, occurs when it becomes more attractive for *S*-agents to buy this technology despite their natural preference for technology *B*. From *Table 1*, we can see that this is the case when:

$$a_S + sn_A > b_S + sn_B$$

thus:

$$sn_A - sn_B > b_S - a_S$$

$$(n_A - n_B) > (b_S - a_S)/s$$

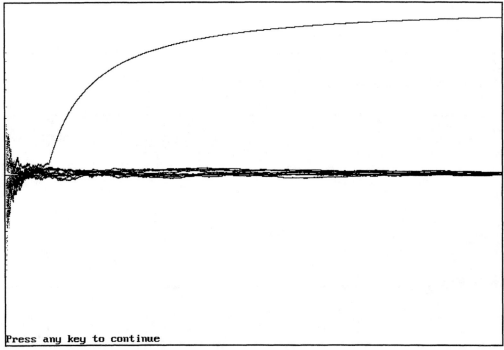

Figure 2   time $\rightarrow$

Uncertainty about a difference in market shares smaller than 5% of the market leads to suppression of lock-in in nine out of ten cases

In short: given the values for various parameters in the model, the lock-in depends only on the absolute difference in the number of previous adopters $| n_A - n_B |$.[1]   When this difference surpasses a critical value, the system is locked-in. With increasing diffusion, this

---

[1] In the case of a lock-in in technology *B*, the analogon of the above inequality is: $(n_B - n_A) > (a_R - b_R)/r$.

absolute difference ($|n_A - n_B|$) becomes smaller as a percentage of the total number of adopters ($n_A + n_B$).

This relative decrease of the percentage has an economic interpretation. While consumers may be able to distinguish large differences in adoption between two technologies, decreasingly smaller percentages may become difficult to perceive. Let us, for example, assume that consumers can appreciate the difference in market penetration of technologies *A* and *B* as long as this difference is larger than 5% of the market. When the difference becomes smaller than 5%, adopters become uncertain, and one may assume that adopters would hold to their 'natural inclination' under this condition. *Table 3* shows how the insertion of lines 80, 100, and 160 in the code specifies this effect. *Figure 2* illustrates that lock-ins virtually disappear in this case.

```
10 INPUT N
20 SCREEN 11: WINDOW (-2, 0)-(N, 100): CLS
25 FOR J = 1 TO 10
30 LINE (-2, 50)-(N, 50)
40 AR = .8: BR = .2: SA = .2: BS = .8: NA = 1: NB = 1: S = .01: R = .01
45 RANDOMIZE TIMER
50 FOR I = 1 TO N
70 XX = RND
80 IF (NA - NB) > 0 THEN M = (NA - NB) ELSE M = (NB - NA)
90 IF XX < .5 GOTO 100 ELSE GOTO 160
100 IF M * 20 < (NA + NB) GOTO 110 ELSE GOTO 130
110 NA = NA + 1
120 GOTO 210
130 RETURNA = AR + R * NA: RETURNB = BR + R * NB
140 IF RETURNA > RETURNB THEN NA = NA + 1 ELSE NB = NB + 1
150 GOTO 210
160 IF M * 20 < (NA + NB) GOTO 170 ELSE GOTO 190
170 NB = NB + 1
180 GOTO 210
190 RETURNA = SA + S * NA: RETURNB = BS + S * NB
200 IF RETURNA > RETURNB THEN NA = NA + 1 ELSE NB = NB + 1
210 Y = NA + NB: Z = 100 * NA / Y
220 PSET (Y, Z)
230 NEXT I
240 NEXT J
250 END
```

Table 3
Revised model with uncertainty at 5% level

2.2. Reflexivity on the Side of Consumers

The Arthur-models assume that adopters are passive followers of market forces. However, people have a tendency to maintain their preferences even if they have to pay a price for them. In this section, we assume that switching to a technology other than the one 'naturally' preferred, can only be induced by the expectation of a net profit of 5% or larger.

Thus, consumers will estimate their profits reflexively, and no longer react immediately to marginal profits.

This model implies changing only the statements in lines 110 and 130 of the original programme (*Table 1*) from 'IF RETURNA > RETURNB THEN' into 'IF RETURNA > 0.95 * RETURNB THEN', and *vice versa*. The results of ten rounds of simulation under these conditions are exhibited in *Figure 3*.

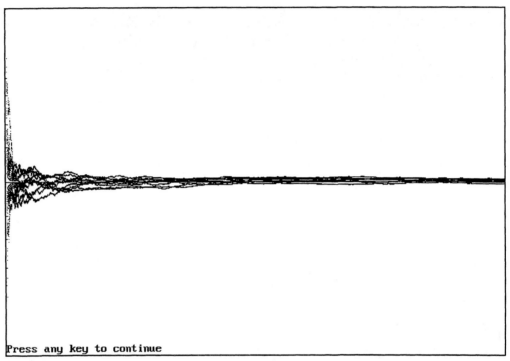

Press any key to continue

Figure 3 time →

Consumers keep to their original inclinations when the expected profit is smaller than 5%

The lock-ins tend to disappear. As in the case of uncertainty above, lock-ins may occur incidentally as a consequence of swings in market shares before equilibrium is achieved.

2.3. Technological leap-frogging

Is a technology that has lost out in the competition for lock-in able to leap-frog back in at a later date? Let us, for example, assume that a technological breakthrough is achieved in technology B after a lock-in in technology A. If the market is sufficiently large (e.g., $n_A + n_B > 2000$), it can become attractive to the suppliers of technology B to invest in recapturing this market. The breakthrough is modelled in the next program (*Table 3*) given the above market size and the condition that technology A has become dominant to the extent of capturing two-thirds of the market. What are the chances for technology B?

Remember that we have two types of parameters: one set like a_R and a_S that model the natural, that is, intrinsic, inclinations of R- and S-type agents to choose technology A, and another set $\{r, s\}$ that models the network effects of previous adopters. The technological breakthrough first operates on intrinsic inclinations because it changes the functional characteristics of the technology (for example, the price/performance ratio), and only upon

314

diffusion can there be a network effect. These network effects will be discussed in the next section.

Under the specified conditions (line 80 in *Table 4*), we simulate the breakthrough in technology *B* by resetting (in an additional line 81) the 'natural inclination' towards technology *B* for *R*-type adopters to 2.0 (instead of 0.2 as before), while their 'natural inclination' for technology *A* is reduced to 0.08 (versus 0.8 before). Analogously, the natural inclination of *S*-type adopters towards technology *A* is reduced with an order of magnitude to 0.02, and the inclination to the already preferred technology *B* is increased to 8.0.

The results of this simulation are exhibited in *Figure 4*. The lock-ins are never affected by these dramatic changes in the parameters. Thus, a lock-in indeed prevents technological leap-frogging of a superior technology after the fact.

```
10 INPUT N
20 SCREEN 11: WINDOW (-2, 0)-(N, 100): CLS
25 FOR J = 1 TO 10
30 LINE (-2, 50)-(N, 50)
40 AR = .8: BR = .2: SA = .2: BS = .8: NA = 1: NB = 1: S = .01: R = .01
45 RANDOMIZE TIMER
50 FOR I = 1 TO N
70 XX = RND
80 IF (NA + NB) > 2000 AND NA / NB > 2 GOTO 81 ELSE GOTO 90
81 AR = .08: BR = 2: SA = .02: BS = 8
90 IF XX < .5 GOTO 100 ELSE GOTO 125
100 RETURNA = AR + R * NA: RETURNB = BR + R * NB
110 IF RETURNA > RETURNB THEN NA = NA + 1 ELSE NB = NB + 1
120 GOTO 140
125 RETURNA = SA + S * NA: RETURNB = BS + S * NB
130 IF RETURNA > RETURNB THEN NA = NA + 1 ELSE NB = NB + 1
140 Y = NA + NB: Z = 100 * NA / Y
150 PSET (Y, Z)
160 NEXT I
165 NEXT J
170 END
```

Table 4
Model with (failure of) technological leap-frogging

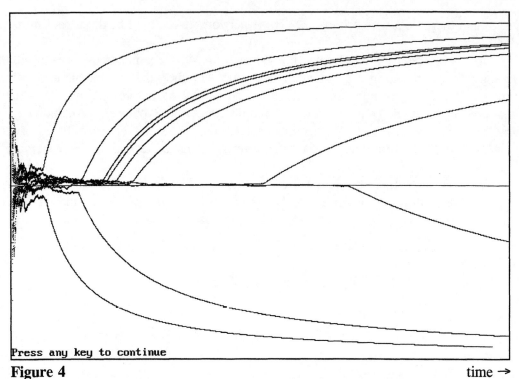

Press any key to continue

Figure 4   time →

Technological breakthrough with major effects on 'natural inclinations'

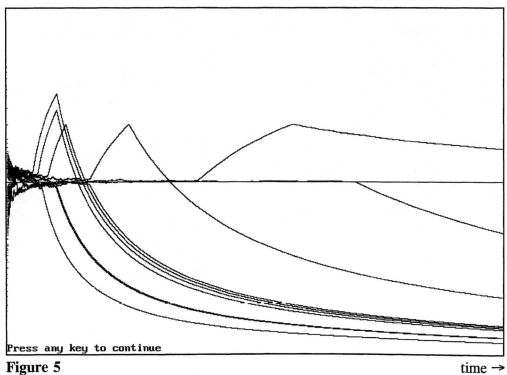

Press any key to continue

Figure 5   time →

Forcing technological 'lock-out' and possible return to equilibrium
(20,000 adopters)

316

If we increase and decrease all parameters indicating inclinations with yet another order of magnitude, we force a reversal of the lock-in that we shall call a 'lock-out'. Interestingly enough, an in-between trajectory is visible when the market is sufficiently large. The system then returns to equilibrium instead of overshooting into a lock-in of technology *B* (*Figure 5*; based on 20,000 adopters).

2.4 Changes in Network Parameters

If, under the conditions specified above where technology *A* has captured two-thirds of a market with more than 2000 adopters, the network effect *r* (associated with *R*, and therefore with a preference for *A*) is reduced with three orders of magnitude, and the network effect *s* is increased with a factor of one thousand, the lock-ins into technology *A* are *not* affected. Thus, a lock-in is also robust against dramatic changes in the network parameters.

However, if *s* vanishes at these conditions (*s* = 0) so that network effects disappear for *S*-type agents, equilibrium tends to be restored (*Figure 6*). This is an analytical consequence of the model specified in *Table 1*. Given the fact of a lock-in, one should invest in technologies that counter-act network externalities rather than in technological breakthroughs that may affect natural inclinations.

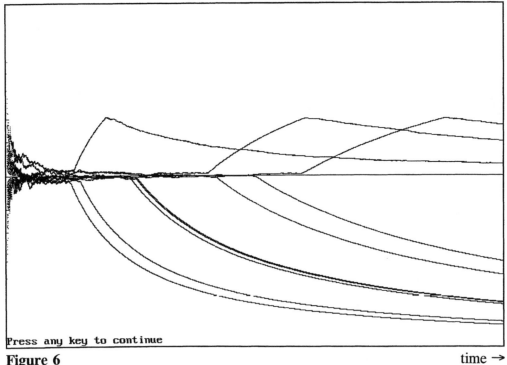

Press any key to continue

Figure 6   time →
S-type agents no longer profit from the network externalities (s = 0)

3. Locked-in versus Locking-in

These results raise the question of whether network externalities should be attributed to choices by previous adopters or to the technologies involved. If we associate *r* (along the column in *Table 1*) with technology *A*, and correspondingly *s* with technology *B*, we obtain a model that is highly sensitive to changes in both *s* and *r* (*Table 4*).

	Technology A	Technology B
R-agent	$a_R + rn_A$	$b_R + sn_B$
S-agent	$a_S + rn_A$	$b_S + sn_B$

Table 4. *Networks effects are an attribute of technologies*
(as opposed to number of previous adopters as in Table 1)

For example, if this network effect r is reduced by only 50% (to 0.005) under the conditions above—where technology A has captured two-thirds of a market with more than 2000 adopters—the lock-in for technology A revert to lock-ins for technology B, unless the initial lock-in was so early that technology B could not penetrate the market at all (*Figure 7*).

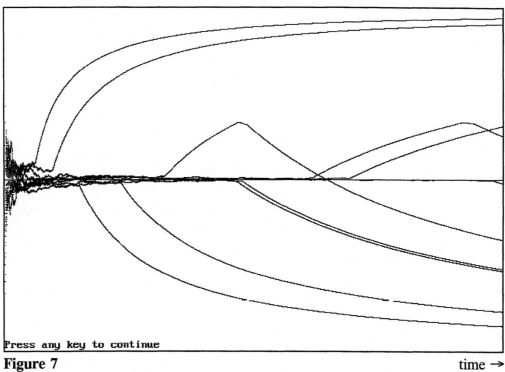

Press any key to continue

Figure 7   time $\rightarrow$
Threshold effect on r-parameter

Interestingly enough, not only a decrease in r leads to this effect, but also an increase in s, that is, if the network effects of the non-dominant technology are gradually strengthened, for example, because of generational drift in the population. Thus, the lock-out in this case is an effect of changes in the network parameters of the technology which at that moment is no longer being traded. S-type agents (which arrive randomly) change their purchasing behaviour, and thereby force a change in lock-in.

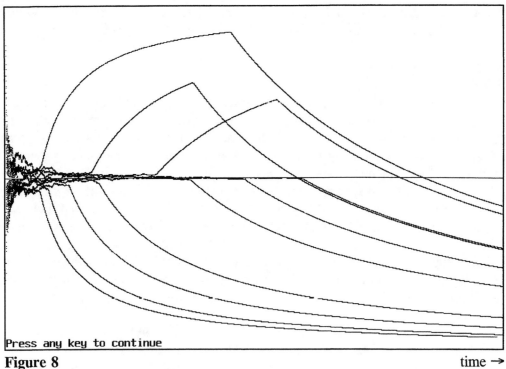

Press any key to continue

Figure 8   time →

Gradual increase of network effect s, favouring technology B

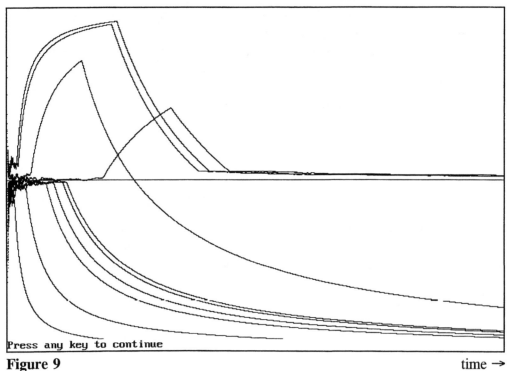

Press any key to continue

Figure 9   time →

Lock-in, lock-out, and return to equilibrium in a population of 30,000 adopters

Note that equilibrium cannot be restored by changing only the network parameters. One needs another mechanism like uncertainty among adopters to induce a return to equilibrium. For example, if we combine a simulation with the above condition of keeping

319

to one's original preference in the case of less than 5% difference in market shares (given a market of more than 10,000 adopters),[2] the system obtains a window for returning to equilibrium. This is demonstrated in *Figure 9* for a population of 30,000.

4. Summary and Discussion

Arthur (1988b) noted that lock-ins can again disappear if the network effects of previous adopters are attenuated, for example, along a learning curve (cf. Kauffman 1988; Frenken & Verbart 1998). In this study, we showed that the network effects can also be suppressed *a priori*, for example, by reducing the relevant parameters or by counter-acting them. The competing technologies then tend to remain in competition. One might wish to save Arthur's theory by stating that the system of competing technologies is in this case 'locked into an equilibrium.' However, this begs the question of the necessity of an eventual lock-in.

Lock-ins can be considered as discrete long-term expectations of network effects. Selectors at the network level are a consequence of the relational factors in the variation. (In the model, these parameters are distinguished from adopters' 'natural inclinations' for one or another technology.) In other words, the lock-in in one technology, or in the other, or in an equilibrium between the two can be considered as three different trajectories of the techno-economic systems under study.

It was shown that a lock-in as defined by Arthur (1988a) is dependent on the *difference* between the numbers of adopters of the two technologies. In Arthur's model, however, adopters were considered passive followers of market forces. If one assumes that adopters have a reflexive capacity to estimate uncertainties, categorical lock-ins tend to disappear in favour of a long-term trajectory of competitive equilibrium.

Thus, a co-evolution between competing technologies may not be exceptional in view of the prevailing reflexivity and uncertainty at the actor level (cf. Alchian 1950). When only two technologies are assumed, a competitive balance between these dynamics is not expected to be stable given endogenous reinforcement mechanisms. A third subdynamic, like a degree of freedom on the consumer side, enables the system to stabilize into a steady state.

When a fourth degree of freedom can be specified, this potentially globalizes the stabilization as a next-order recursion of the selection (Leydesdorff 1994). The obvious candidate for this degree of freedom is 'technological change' during the adoption process. Technological developments may affect both parameters in the model: (a) the natural inclinations of consumers, and (b) the positive network effects from the number of previous adopters. For example, if in the future it becomes possible to download movies from the Internet, the advantages of using one type of VCR or another might disappear.

Furthermore, our simulations show that trajectory transitions can indeed be induced by changing network parameters. However, the emergence of a new technology affecting the natural inclinations of consumers does not lead to a transition in this model if there is already a lock-in. In their study of hyper-selection in innovation processes, Bruckner *et al.* (1994) have shown that only within niches can the separatrix between the two basins of attraction sometimes be tunneled given a stochastic model. This conclusion fully accords with our results: if sufficiently present, network parameters become dominant.

Finally, we could show that changes in the network effect can sometimes lead to *ordered* substitution processes, notably following the lock-in line of the substituting technology. These rapid, but ordered substitution processes have been demonstrated

---

[2] An additional threshold of 10,000 adopters for making uncertainty relevant is necessary, since otherwise no lock-ins may occur, as demonstrated in *Figure 2*.

empirically (e.g., Fisher & Prey 1971). Hitherto, they have been insufficiently understood from the perspective of evolutionary modelling. In our opinion, evolutionary networks should be considered as hyper-networks that are able to suppress relevant subdynamics when this suppression appears to be functional for further development (cf. Bruckner *et al.* 1996). Not the emergence of a new technology, but structural adjustments in the dynamics of the network seem to determine the dissolution of one lock-in or another given a choice between competing technologies (cf. Freeman & Perez 1988; David & Foray 1994).

	a_R	b_R	a_S	b_S	r	s
Figure 1 (Table 2)	0.8	0.2	0.2	0.8	0.01	0.01
Figure 2	0.8	0.2	0.2	0.8	0.01	0.01
Figure 3	0.8	0.2	0.2	0.8	0.01	0.01
Figure 4 *after technological breakthrough*	0.8 *0.08*	0.2 *2.0*	0.2 *0.02*	0.8 *8*	0.01 0.01	0.01 0.01
Figure 5 *after technological breakthrough*	0.8 *0.008*	0.2 *20*	0.2 *0.002*	0.8 *80*	0.01 0.01	0.01 0.01
Figure 6 *after threshold*	0.8	0.2	0.2	0.8	0.01	0.01 *s = 0*
Figure 7 *after threshold*	0.8 0.8	0.2 0.2	0.2 0.2	0.8 0.2	0.01 *0.005*	0.01 0.01
Figure 8 *after threshold*	0.8	0.2	0.2	0.8	0.01	0.01 $s_{n+1} = s_n * 1.001$
Figure 9 (N = 30,000) *after threshold*	0.8	0.2	0.2	0.8	0.01	0.01 $s_{n+1} = s_n * 1.001$

Table 5

Summary of parameter values and changes in parameter values for different figures

Acknowledgement

We wish to thank Koen Frenken for stimulating discussions and valuable suggestions (cf. Frenken & Verbart 1998). We also acknowledge partial funding by the European Commission, TSER project PL97-1296.

References

Alchian, A. A. (1950). 'Uncertainty, Evolution, and Economic Theory,' *Journal of Political Economy* 58, 211-22.

Arthur, W. Brian (1988a). 'Competing technologies,' in: Dosi, G., C. Freeman, R. Nelson, G. Silverberg, and L. Soete (eds.), *Technical Change and Economic Theory*. London: Pinter, pp. 590-607.

Arthur, W. Brian (1988b). 'Self-Reinforcing Mechanisms in Economics,' in: Anderson, Ph. W. *et al.* (Eds.), *The Economy as an Evolving Complex System*. Redwood City, CA: Addison-Wesley, pp. 9-32.

Arthur, W. Brian (1989). 'Competing Technologies, Increasing Returns, and Lock-In by Historical Events,' *Economic Journal* 99, 116-31.

Arthur, William B. (1990). 'Positive Feedbacks in the Economy,' *Scientific American*, February 1990, 80-85.

Bruckner, Eberhard, Werner Ebeling, Miguel A. Jiménez Montaño, & Andrea Scharnhorst (1994). 'Hyperselection and Innovation Described by a Stochastic Model of Technological Evolution,' in: Leydesdorff, Loet & Van den Besselaar, Peter (Eds.), *Evolutionary Economics and Chaos Theory: New directions in technology studies*. London: Pinter, pp. 79-90

Bruckner, Eberhard, Werner Ebeling, Miguel A. Jiménez Montaño, & Andrea Scharnhorst (1996). 'Nonlinar stochastic effects of substitution — an evolutionary approach,' *Journal of Evolutionary Economics* 6, 1-30.

David, Paul A. (1985). 'Clio and the Economics of QWERTY,' *American Economic Review* 75, 332-7.

David, Paul A. & Dominique Foray (1994). 'Dynamics of Competitive Technology Diffusion Through Local Network Structures: The Case of EDI Document Standards,' in: Leydesdorff, Loet & Van den Besselaar, Peter (Eds.), *Evolutionary Economics and Chaos Theory: New directions in technology studies*. London: Pinter, pp. 63-78.

Fisher, J. C., and R. H. Pry (1971). 'A Simple Substitution Model of Technological Change,' *Technological Forecasting and Social Change* 3, 75-88.

Freeman, Chris and Perez, Carlota (1988). 'Structural crises of adjustment, business cycles and investment behaviour,' in: Dosi, G., C. Freeman, R. Nelson, G. Silverberg, and L. Soete (eds.), *Technical Change and Economic Theory*. London: Pinter, pp. 38-66.

Frenken, Koen and Okke Verbart (1998). 'Simulating paradigm shifts using a lock-in model,' in: Petra Ahrweiler and Nigel Gilbert (eds.), *Simulations in Science and Technology Studies*. Berlin: Springer Verlag (forthcoming).

Kauffman, Stuart A. (1988). 'The Evolution of Economic Webs,' in: Anderson, Ph. W. *et al.* (Eds.), *The Economy as an Evolving Complex System*. Redwood City, CA: Addison-Wesley, pp. 125-46

Leydesdorff, Loet (1992). 'Irreversibilities in science and technology networks: an empirical and analytical approach,' *Scientometrics* 24 (2), 321-57.

Leydesdorff, Loet (1994). 'The Evolution of Communication Systems,' *Int. J. Systems Research and Information Science* 6, 219-30.

Leydesdorff, Loet (1995). *The Challenge of Scientometrics: The Development, Measurement, and Self-Organization of Scientific Communications*. Leiden: DSWO Press, Leiden University.

CAUSALITY, OPERATORS AND FINALITY IN SYSTEMS THEORY

Robert Vallée
WOSC
2, rue de Vouillé
75015 Paris, France

Causality, eigen-behavior, epistemo-praxiology, incursivity, system

Abstract

Causal operators acting, by definition, on the past and present functions of time are first presented. Then are described dynamic systems involving a causal operator in their evolution equation, both in the cases of a continous and a discrete time.

A first example of causal dynamic system is given by the modelling of a subject observing an object (Von Förster 1976). The operator involved acts in fact only on the present. In case of convergence it gives rise to an "eigen-behavior" representing the stabilized state of what concerns the modes of action of the subject and of what concerns the object.

Then is presented the concept of "epistemo-praxiology" (Vallée, 1974, 1995) involving an "observation operator'" (Vallée, 1951), then decision and effect operators. All these operators are causal (acting upon past and present). They explain the subjective limitations of the observing and deciding subject in front of an object (pragmatic indiscursibility and pragmatic inverse transfer of decisional structures). They show also the coevolution of the subject and the object as a fixed point of a composite operator having observation, decision and effection traits.

The notions of "incursivity" and "hyperincursivity" (Dubois and Resconi, 1992 ; Dubois, 1995) involve operators acting on the past, present and future of functions of time and considerations linked to Aristotle's final causes. They give tools for solving problems of stability of certain systems.

1. CAUSAL OPERATORS

A mathematical operator is said to be causal when, acting on an input function of time $x(u)$, it gives as an output a function of time whose value at instant t depends only upon the past of $x(u)$, $u \leq t$, or $x/^{t}_{-\infty}$. If the operator acts upon an input function $x(u,m)$ of both time and space, giving an output function of time and space also, it is said to be causal under the same condition : the value of the output function, at instant t and point m, depends only upon the past of the values of $x(u,n)$ whatever be the points considered $(u \leq t)$. In other words the (u,n) to be considered constitute the half time-space[1] defined by $u \leq t$.

---

[1] In the framework of special relativity (u,n) must belong to the anterior part of the light cone of summit (t,m) which reduces to the half time-space defined by $u \leq t$ when the velocity of light is infinite.

CP437, *Computing Anticipatory Systems: CAYS--First International Conference*
edited by Daniel M. Dubois © 1998 The American Institute of Physics 1-56396-827-4/98/$15.00

If we consider a dynamical system whose "state" $x(t)$, at instant t, belong to $\mathbb{R}^n$, we have two main possibilities : the "continous" case where $t \in \mathbb{R}$ and the "discrete" case where $t \in \mathbb{Z}$. In the first one the equation of evolution involves a derivative. We have

$$dx(t)/dt = f(x \,/^t_{-\infty}, v \,/^t_{-\infty}, t), \, t \in \mathbb{R},$$

where f is an operator, depending generally upon instant t, and v being a function of t representing the influence of the environment. Operator f is causal as regards the pair (x,v), that is to say that it acts upon the past of $(x(u), v(u))$, $u \leq t$. The initial condition at t_0 is the past of $x(u)$ for $u \leq t_0$ (or a portion of it). In the second case, the equation of evolution involves the difference $Dx(t) = x(t+1) - x(t)$, which plays the part of a derivative, and

$$Dx(t) = f(x \,/^t_{-\infty}, v \,/^t_{-\infty}, t), \, t \in \mathbb{Z},$$

with the same remarks as above about operator f.

When the "state" of the dynamical system, at instant t, is given by $x(t,m)$ where $m \in \mathbb{R}^3$ and $x(t,m) \in \mathbb{R}^n$, outside the case where the evolution equation is a partial derivatives equation or a partial differences equation, we have

$$dx(t,m) \,/dt = f(x \,/^t_{-\infty}, v \,/^t_{-\infty}, t), \, t \in \mathbb{R},$$

or

$$Dx(t,m) = f(x \,/^t_{-\infty}, v \,/^t_{-\infty}, t), \, t \in \mathbb{Z},$$

with the same remarks as above.

For the sake of simplicity let us consider the discrete case with a state depending only upon time. We may write

$$Dx(t) = x(t+1) - x(t) = \varphi(x(t), x(t-1), \ldots; v(t), v(t-1), \ldots; t)$$

or

$$x(t+1) = \psi(x(t), x(t-1), \ldots; v(t), v(t-1), \ldots; t)$$

since the pasts of x and v are given respectively by the sequence $x(t), x(t-1), \ldots$ and $v(t), v(t-1),$ $\ldots$ The initial condition $x(t_0), x(t_0-1), \ldots$ being given, the whole future of $x(u)$, for $u > t_0$, is defined, provided that function v is known. But what about the so called shock of the future, that is to say the influence of the future of $x(u)$, $u > t$, upon x itself ? If we introduce in ψ, or φ, $\ldots x(t+3), x(t+2), x(t+1)$, these values must be evaluations based on the part of $x(u)$ (regularities for example) or on a predictive model.

2. SECOND ORDER CYBERNETICS

In the frame of his second order cybernetics, Heinz Von Förster (1976), inspired by the works of Jean Pioget, presented in an original way the problem of a subject S in front of an object O (or of several objects). Using a discrete time $(t \in \mathbb{Z})$, he represented by $obs(t)$ the observables, for a given observer, concerning the actions of subject S and also concerning object O (or the objects). The evolution of $obs(t)$ is then given by the recurrence equation

$$obs(t+1) = Coord \, obs(t), \, t \in \mathbb{Z},$$

325

where *Coord* is an operator acting upon *obs(t)*, *obs(t₀)* being given. Operator *Coord* coordinates both what concerns the actions of *S* and what concerns *O*, giving rise to *obs(t+1)*. It is obviously a causal operator acting meraly upon the present of function *obs*.

If the recurrent process is convergent, that is to say if *obs(t)* → *obs(∞)* when *t* → ∞, we have

$$obs(\infty) = Coord \; obs(\infty).$$

Then *obs(∞)*, which is a root of the above equation, is called by Heinz Von Förster eigen behavior or eigen value. This last expression comes from quantum mutanios and, more generally, from the theory of linear operators. Since operator *Coord* is generally non linear we may use, for *obs(∞)*, the expression fixed point[1] of operator *Coord*. This fixed point emuges asymptotically from the evolution of system *(S,O)*, depending on the basis of attraction to which belongs *obs(t₀)*. According to Heinz Von Förster, *obs(∞)* represents the stabilized state of what concerns the modes of action of *S* and of what concerns *O*.

3. EPISTEMO-PRAXIOLOGY

We consider a system *S* able to observe, decide and act. Let functions *x* and *v* represent, respectively, the evolution of system *S* and its environment *E*. So the pair $\xi = (x,v)$ represents the evolution of the whole universe *U* to be considered here. Function ξ is what *S* observes through its "observation operator" *O* (Vallée, 1951), giving function η representing the evolution of its perceptions

$$\eta = O(\xi) = O(x,v).$$

Obviously *O* is a causal operator, it acts only upon the part of $\xi(u)$, $u \leq t$. In other terms

$$\eta(t) = O(x \, |^{t}_{-\infty}, \; v \, |^{t}_{-\infty}) = O(\xi \, |^{t}_{-\infty}), \; t \in \mathbb{R} \text{ or } \mathbb{Z}.$$

Moreover, generally if not always, *O* has no inverse, a circunstance which generates many subjective limitations such as indiscernibility and "inverse transfers" (Vallée, 1974).

The "Observation operator" *O* is followed by a decision operator *D*, also non inversible and causal. We call the product $P = DO$ the "pragmatical operator". It is, of course, non inversible and causal. It generates specific indiscernabilities and "inverse tranfers". We have

$$\zeta = D(\eta) = DO(\xi) = P(\xi),$$

where function ζ represents the evolution of the decision of *S*.

If we represent by A_S the action operator of *S* we may write

$$A_S(\zeta,x,v) = A_S(DO(\xi),x,v) = A_S(DO(x,v),x,v) = x,$$

because, influenced by ζ, and also by *v*, A_S gives rise to the evolution (described by *x*) of *S* itself.

---

[1] By definition $\bar{x}$ is a fixed point of operator *h* if $h(\bar{x}) = \bar{x}$, *h* being a mapping of a set of *E* on itself. It is a root of equation $n(x) = x$. If *h* is linear, $\bar{x}$ is an eigen vector of operator *h* with eigen value 1.

If A_E is the evolution operator of E

$$A_E (v,x) = v.$$

The two above equations may be written synthetically

$$A (x,v) = (x,v)$$

or

$$A (\xi) = \xi.$$

The global action operator A is also causal, as well as A_S and A_E. It concerns the super-system (S,E) or universe U. We see that function ξ, which describes the co-evolution of system S and its environment E, or the evolution of universe U, is a fixed point of operator A. This description of the evolution of both system S and its environment E (which may be considered as an object for S) involves observation, decision and action, taking into account many aspects of the subjectivity of S. It gives an introduction to what we call "epistemo-praxiology" (Vallée, 1995).

4. INCURSIVITY AND HYPERINCURSIVITY

Here we shall consider only a discrete time $(t \in \mathbb{Z})$ and, in some cases, also a discrete space (the physical space or a parameters space). We have a recurrence equation

$$Dx(t) = x(t+1) - x(t) = \varphi (x(t), x(t-1), ...; v(t), v(t-1), ... ; t)$$

defining a recursion. $Dx(t)$, involving $x(t+1)$, may be called the foward difference or "forward derivative" $D_f x(t)$, in the same fashion $x(t) - x(t-1)$ may be called backward difference or "'backward derivative" $D_b x(t)$ (Dubois, 1995) ; derivative having here a meaning adapted to the discrete case. We have in the second case

$$D_b x(t) = x(t) - x(t-1) = \varphi (x(t), x(t-1), ...; v(t), v(t-1), ... ; t)$$

which gives also, changing t into $t+1$,

$$x(t+1) - x(t) = \varphi (x(t+1), x(t), ...; v(t+1), v(t), ... ; t+1)$$

The fact that $x(t+1)$, and even $v(t+1)$, appear in the second member of the equation, changes the recursion, obtained with D_f, into what Dubois calls an incursion or even an hyperincursion if the above equation has more than one solution in $x(t+1)$. If solved in $x(t+1)$ the equation becomes again a recursion as far as x is concerned but $v(t+1)$ is nevertheless in the second member.

More generally we may consider an equation like

$$x(t+1) - x(t) = \varphi (...x(t+2), x(t+1), x(t), x(t-1), ...; ...v(t+1), v(t), v(t-1), ... ; t)$$

Where $x(t+1)$, if the equation can be solved in $x(t+1)$, depends, in one or several ways, upon the future values... $x(t+3)$, $x(t+2)$. In such cases we have, by definition, an "incursion", if there is one solution in $x(t+1)$, or an "hyperincursion", if there are several solutions in $x(t+1)$

(Dubois, 1995). Both involve considerations linked to Aristotle's final causes, the operator of evolution being obviously non causal.

Il the state of the system depends upon instant $t \in \mathbb{Z}$ and point $m \in \mathbb{Z}^3$ (or parameter $p \in \mathbb{Z}^k$, we may have, in the very much simplified case where $m \in \mathbb{Z}$ and there is no influence of the environment

$$x(t+1,m+1) = \psi \ (...x(t,m+2),\ x(t,m+1),\ x(t,m),\ x(t,m-1),...;$$
$$...x(t-1,m+2),\ x(t-1,m+1),\ x(t-1,m),\ x(t-1,m-1),...;\ etc\ ;\ t)$$

and similar expression for $x(t+1,m-1)$ etc. As far as variable m is considered we have formally an "incursion" but no causal problem since m is a point of the physical space or of a parameters space.

The concepts of "incursion" and "hyperincursion" have been used (Dubois, 1995) for the study of chaos systems and other problems. For example let us consider the recursion

$$x(t+1) = b\ x(t)\ (1-x(t)),\ x \in \mathbb{R},\ b > 0,$$

which implies the Pearl-Verhulst map $y = b\ x(1-x)$. This dynamical system has two points of equilibrium 0 and $\underline{\dfrac{b-1}{b}}$, stable or unstable depending upon the value of b.

For example $\dfrac{b-1}{b}$ is unstable for $b \geq z$. If we may consider the "incursive control".

$$u(t) = b\ x(t)\ (x(t)-x(t+1)),$$

we have

$$x(t+1) = b\ x(t)\ (1-x(t)) + u(t) = b\ x(t)\ (1-x(t+1)).$$

This is an incursive process equivalent to the following recursive one

$$x(t+1) = b\ x(t)\ /\ 1+b\ x(t).$$

This process has the same points of equilibrium as the initial recursion. But now equilibrium $\dfrac{b-1}{b}$ is stable for $b > 1$ and possibilities of chaos, present in the first case, are eliminated.

References

Dubois Daniel M., Resconi G., (1992). Hyperincursion, a New Mathematical Theory. Presses Universitaires de Liège, Liège.

Dubois Daniel M., (1995). Introduction to Aristotle's final causes in CAST : Concept an method of incursion an hyperincursion. In Computer Aided Systems Theory. EUROCAST'95. Edited by F. Pichler, R. Moreno Diaz, R. Albrecht. Lectures Notes in Computer Science, volume 1030, Springer-Verlag, Berlin, Heidelberg, New-York, pp. 477-493.

Vallée Robert, (1951). Sur deux classes d'"opérateurs d'observation". Comptes Rendus de l'Académie des Sciences, tome 233, pp. 1950-1951.

Vallée Robert, (1974). Observation, decision and structure transfers in systems theory. In Progress Cybernetics and Systems Research. Edited by R. Trappl, F. Pichler. Hemisphere Publishing Corporation, Washington, pp. 15-20.

Vallée Robert, (1995). Cognition et Système, Essai d'Epistémo-praxéologie. L'Interdisciplinaire, Lyon-Limonest.

Von Förster Heinz, (1976). Objects, tokens for (eigen)-behaviors. Cybernetics Forum, volume 5, n. 3-4, pp. 91-96.

Anticipatory Autonomous Systems
and Robotics

Evolutionary Learning of Autonomous Agents with anticipatory capabilities

Clemens Pötter*
Institut für Neuroinformatik
Ruhr-Universität Bochum
44780 Bochum, Germany

Abstract

Finding the structure of a control instance for autonomous robots for a class of tasks is a key problem in robotics. In order to extend the classical and behaviour-based paradigms, the use of anticipation is a promising direction. As the basic structure we take neural networks; the connections between the neurons are improved by means of adaptive Evolutionary Algorithms. As an example for a class of tasks, in our model a mobile robot has to reach several goals in different environments with obstacles. For each step, a neural network calculates a new heading direction; in an imaginary step the robot anticipates a collision with an obstacle. In case of collision another direction is calculated. The model delivers good results. Furthermore, it turns out that adding some noise supports generalisation capabilities.

Keywords: Evolution strategies, mobile robot, neural network, imaginary step, noisy fitness function

1. Introduction

We consider anticipating systems from an evolutionary point of view. It is reasonable to assume that anticipation can give an edge to a competitor among others that cannot anticipate. Humans obviously use anticipation very often. Whether or to which extend animals at a earlier evolutionary stage use it, is not clear. Thus, there are drawbacks that we will discuss in the next chapter.

The aim of the paper is to show a way how anticipation can be used in an evolutionary framework; we present a model of a mobile robot which is guided by a neural network. This network is optimised by an evolutionary algorithm. We take neural networks as a basic structure, because they are robust against noise and their natural counterpart is a product of evolution. Due to the bias-variance dilemma the network must not be too small, because the degree of precision of the small network might not suffice, whereas the network with many neurons does not generalise well and learns slowly.

The structure of the paper is the following:
In the next section we discuss the role of anticipation in general. In section 3 we briefly describe some previous models in which mobile robots are optimised by Evolutionary Algorithms, and then go into the details of our model. In section 4 a short introduction to evolutionary algorithms is given and the coding of the parameters in the genome is explained. Section 5 shows the results: the development of the fitness, the influence of noise, some examples of paths and the evolved neural structures. In section 6 we discuss the results, and in the final section we conclude and give a brief outlook on future work.

---

*Clemens.Poetter@neuroinformatik.ruhr-uni-bochum.de

CP437, *Computing Anticipatory Systems: CAYS--First International Conference*
edited by Daniel M. Dubois © 1998 The American Institute of Physics 1-56396-827-4/98/$15.00

2. The Role of Anticipation

According to (Rosen, 1985) an anticipatory system "is a system containing a predictive model of itself and/or its environment, which allows it to change state at an instant in accord with the model's prediction to a latter instant."

What are the advantages? To begin with an example, a dog chasing a rabbit predicts the trajectory and can take a short cut. Simply trying to follow the trajectory more quickly is not a succesful strategy. Another example is a traveller who avoids crossing dangerous regions due to the thought of being robbed. Thus, advantages are:

- more efficient use of energy and time

- protection against dangerous situations.

The drawbacks of anticipation are:

- time, attention, and energy has to be spend on anticipation. In dangerous situations this can be fatal. For that reason the deliberate actions should be complemented by reflexes.

- future states probabilities have to be estimated; if this estimation is not appropriate, strange behaviour can occur.

- This emerging behaviour is more difficult to control and explain. From an engineering point of view the security requirements are more difficult to fulfill.

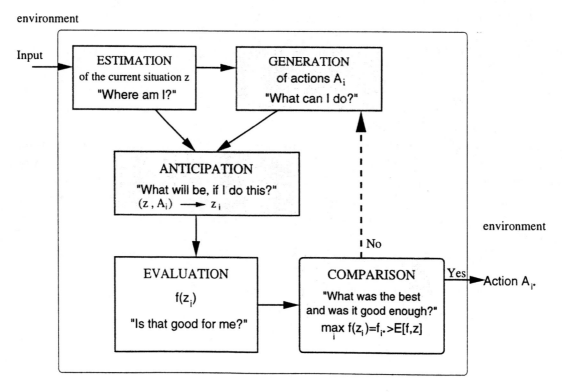

Figure 1: A possible use of anticipation. Explanation given in the text.

In fig. 1 is shown how anticipation can be used. After the system has estimated its situation z, it generates – according to external laws, or with the support of experience –

some possible actions (or sequences of actions) A_i, $i \in I$ (I is the index set of the generated actions); then it predicts by means of a function Z, which it has to learn and is crucial to whole procedure, possible configurations $z_i = Z(z, A_i)$ in the future; the system evaluates them and assigns values $f_i = f(z_i)$ which indicate how desirable the state is. The system determines the action A_{i^*} with $f_{i^*} = \max_i f_i$ and compares it with an expectation value $E[f, z]$ which has to be adapted and requires experience; if an action has to be quick or many actions have already been tested, E is decreased. With this method a trade-off between finding quickly an action and finding action that is improved by anticipation can be realised. If $f_{i^*} < E[f, z]$, the system assumes better alternatives are available, and new actions have to be created. If $f_{i^*} \geq E[f, z]$, the system is "content", and the action A_{i^*} will be performed. This enables the system to behave more suitable and to have an advantage over competitors. In our model only a simple version of this concept is used ($\rightarrow$ section 3.2.).

In strategy games like chess, there is only a finite number of options. This is due to the rules of the game and can be implemented explicitly on a computer. In natural environments, different types of actions (run, jump, fight, eat, etc.) are possible. In a simulation these actions have to be described in a mathematical form. The choice of the describing variables is not clear in general. In our model we chose the variables **heading direction** and **velocity** of the robot.

In fig.1 anticipation is only one of five components, which have to be adapted simultaneously; its influence is fairly indirect and therefore more difficult to optimise for evolutionary algorithms. However, there are more direct methods to train the prediction ability, i.e. evolution of neural networks for time series prediction; these methods are not subject of this paper.

To summarise, on the one hand anticipation has the advantage of providing information for intelligent behaviour, on the other hand it costs time and attention.

3. Model

Before we describe our model we review briefly some previous models of autonomous agents which are interesting in the context of anticipation. In (Brave, 1996) a model with an exploration and a planning phase was studied. While in the exploration phase the robot moves physically, it performs only imaginary steps in the planning phase and anticipates detection of goals. With **Genetic Programming** a robot was evolved that built up a representation of its environment: For each location the robot took a node and the transition from one location to a neighbour location was represented by a labeled edge. In the fitness function the robot had to find the way from every possible start location to every possible goal location. A complete map of the environment could successfully be evolved. As the environment consisted only of a 4×4 toroidal grid without obstacles, it is an open question whether this result can be generalised.

Exploration in mobile robotics was investigated in (Sullivan and Pipe, 1996). A radial basis function network (**RBFN**) represents the environment with their obstacles and goals. Several variants for the next step are produced (direction and step length $\vec{v}_i \triangleq z_i$) and the attractiveness of the step is evaluated by the RBFN, which corresponds to f and is updated by a Temporal Difference learning algorithm. Using the derandomised mutation operator in the evolution strategy they could produce quite good results with small populations.

In (Pötter, 1996) a robot, equipped with a dynamic control, was to find several goals in one environment. It turned out that the evolutionary algorithm could optimise the number of reached goals and the path lengths. The parameters in the dynamic control equation are optimised within a single environment and therefore contain information about it. The generalisation to other environments could not be achieved.

In the present model, which is an extension of that described in (Pötter, 1996), the task remains to find "good" paths from defined starting positions to different goals. We simulate several environments in order to avoid a specialisation of the robot on a single environment. The idea is that evolving the networks in a variety of environments, a class of them can be learned. This is motivated from the biological evolution, where animals have to cope with different areas. Parameters contain information about a class of environments rather than of a single environment.

3.1. Environment

The complete environment consists of 32×32 squares, where some represent obstacles and the others are free. There are several starting positions and several goals for each starting position. In order to confront the robot with a variety of environments different obstacle configurations have been devised. It is shown in fig.2. Some so called chicken cages have been put up to see whether the robot gets trapped in it or finds a way out.

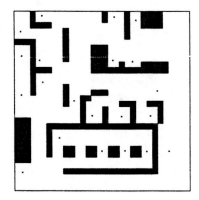

Figure 2: complete environment of the robot with square obstacles, and goals (marked as crosses)

3.2. Robot

The robot is modeled as a disc and has 3 degrees of freedom: 2 dimensions in the plane $\vec{x} = (x, y)$ and a rotational degree of freedom ϕ. It is endowed with a sensory system and a neural net.

The *sensory system* of the robot consists of a number of sensors $n_{sensors}$ which will also be determined by the evolution strategy. The sensors are placed equidistant around the robot, at the angles

$$\phi_k = k \frac{360^0}{n_{sensors}} \quad , k = 0, .., n_{sensors} - 1.$$

The heading direction is $\phi = 0$. The input of the sensor depends on the distance to the obstacle. If no obstacle is visible to the sensor, the input is 0. The closer the obstacle,

334

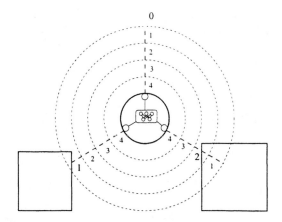

Figure 3: A robot with 3 sensors. The numbers indicate the input, which depends on the distance to the obstacle.

the higher the input:

$$I_k = \begin{cases} \lceil n_{resolution} \left(1 - \frac{r_{obst}}{r_{gaze}} \right) \rceil & \text{for } r_{obst} < r_{gaze} \\ 0 & \text{else} \end{cases}$$

where $\lceil x \rceil$ is the least integer $\geq x$. Additionally, the direction of the goal is given as an input

$$I_{goal} = \phi_{goal} = \arccos \left(\frac{(\vec{x}_{goal} - \vec{x})\dot{\vec{x}}}{||\vec{x}_{goal} - \vec{x}|| \cdot ||\dot{\vec{x}}||} \right).$$

The *neural network* controls the heading direction. The input layer, which is the sensory system, is fully connected to the hidden layer. It consists of a part with sigmoid neurons and a part with Gaussian neurons. We chose to determine their sizes instead of determining the activation function for each neuron independently, because the latter encoding would be very redundant. There are four possible structures within each part, shown in fig.4: no lateral connection, lateral input from the updated neighbour neuron, lateral input from the neighbour neuron that is not yet updated, lateral input from both neighbour neurons.

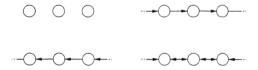

Figure 4: possible structures of the neural net parts

Thus, feedforward as well as recurrent networks are possible. The structure is shown in fig. 5. Noise is added to the output of the neurons.[†]

$$s_i(t) = g \left(w_{0i} + \sum w_{ji} x_j + \tilde{w}_i^l s_{i-1}(t) + \tilde{w}_i^r s_{i+1}(t-1) \right) + \xi$$

with ξ uniformly distributed random number in $[-\xi_{max}, \xi_{max}]$ and $g(x) = (1 + e^{-x})^{-1}$ for sigmoid neurons, and $g(x) = \exp(-x^2)$ for Gaussian-neurons.

[†]The noise makes the system non-deterministic and therefore we do not have the problems of self-referential logical systems (Ekdahl et al., 1995).

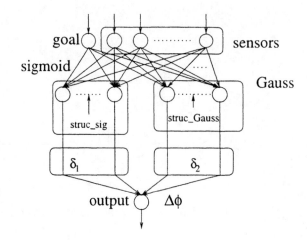

Figure 5: The structure of the neural network which controls the robot's direction

For the velocity we chose a heuristic function: the velocity is constant until the robot is close to the goal; near the goal the robot slows down in order to approach the goal more precisely. Furthermore, the velocity is reduced by a constant factor for each collision in the imaginary steps.

$$v(\vec{x}(t)) = \begin{cases} \|\vec{x}_{goal} - \vec{x}(t)\|v_{near} & \text{for } \|\vec{x}_{goal} - \vec{x}(t)\| < R \\ f_{reduce}^b v_0 & \text{for } b \text{ collisions in the imaginary steps} \end{cases}$$

An energy value is assigned to the robot; this will be used in the fitness function in section 4.2. The start energy $E(t=0) = E_0$ is 10 units; for each physical step it is decreased:

$$E(t+1) = E(t) - \alpha_{rest} - \alpha_{step}v(t)$$

where $\alpha_{rest} = 0.01, \alpha_{step} = 0.01$ are arbitrarily chosen.

A new direction is calculated by

$$\Delta\phi(t) = \delta_1 \sum_{i \in I_{sig}} v_i s_i + \delta_2 \sum_{i \in I_{Gauss}} v_i s_i + \zeta$$

The first term describes the contribution of the sigmoid part, the second of the Gaussian part; $(\delta_1, \delta_2) \in \{(1,0),(0,1),(1,1)\}$ determines whether the sigmoid, the Gaussian or both parts contribute. The third part simulates noise in the rotation angle, which occurs in real systems. ζ is a uniformly distributed random number in $[-\zeta_{max}, \zeta_{max}]$. We change the noise level ζ_{max} to shed some light on the advantage and disadvantage of noise.

Before executing the step the robot performs an imaginary step and anticipates collisions. In case of a collision the direction is calculated anew. The robot has 3 trials to calculate a direction in which there is no obstacle. If it fails 3 times the present goal is aborted, the robot is reset to its starting position and the next goal is given. If the goal is visible to the robot, i.e. the goal is in the gaze range of the robot and no obstacle is between the goal and the robot, it changes the strategy and moves directly towards the goal.

If we set $|I| := 1$, $f(\text{collision}) := -1$, $f(\text{no collision}) := 1$, $E[f,z] := 0$, our model fits in the framework of section 2.

336

4. Evolution Strategy

Evolution Strategies (ES (Schwefel, 1994) and (Rechenberg, 1994)(German)), Genetic Algorithms (GA) and Genetic Programming (GP) belong to the stochastic optimisation methods called Evolutionary Algorithms (EA), which are based on selection, recombination, and mutation. An excellent overview of EA for parameter optimisation is given in (Bäck and Schwefel, 1993). As opposed to GA, which work on bitstrings, or GP, which work on programs, ES use continuous variables in the genome and are more suitable to optimise neural weights. The so-called *genome* contains the object parameters, which describe the individual to be evaluated, and the strategy parameters, which determine the mutation of the object parameters. The mutation of an object parameter is performed by adding a normal distributed random number. As the variance of this distribution depends on the problem, it is encoded in an additional chromosome and therefore also evolved. The mutation of the strategy parameters is performed according to (Schwefel, 1994) by multiplication with a log-normal distributed random number.

The recombination for object parameters is normally performed by choosing a parent for each single parameter independently and by taking the corresponding parent parameter. For the strategy parameters usually the average value of the corresponding parameters in the parent population is taken.

The selection operator choses the best μ individuals of λ offsprings as the next parents. ES are a flexible tool and have proved to avoid local minima. Under weak mathematical conditions, it is proved in (Rudolph, 1994) to converge to the global optimum.

The parameter space in our model has an undefined dimension, because the number of neurons is dynamically determined. Considering the subspaces with a fixed dimension, local minima can exist.

4.1. Encoding

The genome for the robot consists of three chromosomes:

1. continuous robot parameters

2. integer robot parameters

3. mutation parameters

The contents of the chromosomes is described in the tables 1 and 2.

As the number of neurons and sensors is variable, we only use the submatrices of dimension $n_{sensors} \times n_{sig}$ and $n_{sensors} \times n_{Gauss}$ of the complete weight matrices of dimension Max_Neurons$\times$ Max_Sensors. Only weights of existent neurons are mutated, we do not allow mutations of virtual weights. This complies better with the requirement of strong causality (Sendhoff et al., 1997). Hidden mutations occur if a neuron is removed and readded, because its weights remain in the matrix. Initially all weights are set to zero. We chose to group the mutation parameters, because a single mutation parameter would not be appropriate for the different types of parameters and a mutation parameter for each robot parameter would be too high dimensional; as the adaptation of the mutation parameters is of second order[‡], we tried to reduce their number.

[‡]Mutation parameters do not influence the fitness directly; the adaptation of the robot parameters is of first order.

continous robot parameters			
symbol	**description**	**range**	**mutated by**
v_{start}	starting velocity	$[0, v_{max}]$	$+N(0, \sigma_1)$
ϕ_{coll}	rotation angle of the robot if the imaginary step results in a collision	$[-180^0, 180^0]$	
v_0	constant added to the output	$\mathbb{R}$	
ξ_{max}	maximum noise on the hidden neurons	$\mathbb{R}^+$	
r_{close}	distance to the goal which initiates the slowing down of the robot	$\mathbb{R}^+$	
v_{close}	determines the velocity in the neighborhood of the goal $$v_{near} = ||\vec{x}_{goal} - \vec{x}(t)||v_{close}$$	$\mathbb{R}^+$	$+N(0, \sigma_2)$
f_{reduce}	Factor by which the velocity is reduced if the imaginary step results in a collision	$[0, 1]$	
w_{ij}	weight from sensor i to neuron j		
w_{0j}	threshold of neuron j		
$\tilde{w}_i^{r,l}$	lateral weight from neuron i to the right or to the left neuron (i+1,i-1 respectively)	$\mathbb{R}$	$+N(0, \sigma_3)$
v_j	weight from neuron j to the output		

Table 1: First chromosome

4.2. Fitness Function

We start the robot at different points in the environment to simulate different operating areas. The fitness function gives priority to finding the goals, the second criterion is to shorten the path, if the goal is found, and to lengthen the path otherwise. The sum of the number of steps and the total number of imaginary collisions is limited by 200. In our model we define the fitness function as follows: Let $\delta_i' = 1$, if the ith goal is reached while the number of steps plus the number of imaginary collisions is less than 200. and $\delta_i' = 0$ else. Then the fitness function can be written as

$$F = \sum_i (B + E)\delta_i' + (1 - \delta_i')(E_0 - E) \tag{1}$$

with $B = 1000, E_0 = 10$ and the energy E of the robot while reaching the goal i.

The presence of noise implies that the fitness function is not deterministic; we therefore perform several runs over the goals in order to get a more reliable evaluation. The disadvantage of the higher computational cost because of the averaging process is more than compensated by the advantage of a better generalisation to finding untrained goals. The generalisation is tested on a validation set of shifted goals.

We use a pure (10,70)-strategy, i.e. 10 parents and 70 offsprings, and no parent enters the next generation.

338

integer robot parameters			
symbol	description	range	mutated by
$n_{sensors}$	the number of sensors	1,..,16	
$n_{sigmoid}$	the number of sigmoid neurons	0,..,15	p_1
n_{Gauss}	the number of Gaussian-neurons	0,..,15	
$n_{resolution}$	the number of equidistant points on the sensor ray	1,..,5	
struc_{final}	whether the output neuron gets input from the sigmoid part, the Gaussian part or both	0,1,2	
$\text{struc}_{sig-lateral}$	whether there is lateral input in the sigmoid part	0,1,2,3	p_2
$\text{struc}_{Gauss-lateral}$	whether there is lateral input in the Gaussian part	0,1,2,3	

Table 2: Second chromosome

5. Results

In the training set we put 40 goals, and 20 in the validation set. Each set has 4 starting positions and we averaged over 10 runs. In order to decrease the difficulty of finding goals that are far away, we put 20 goals at positions which are easier to approach. This is a "pedagogical" ansatz and effectively smoothes the fitness landscape. This was also used in (Elman, 1993) for the training of networks.

In all simulations "small" systems evolved, i.e. the number of neurons was less than 10 and the number of sensors less than 6.

We compare the development of the fitness of the model without anticipation and the model with anticipation (at noise level 3): Without anticipation only 10 goals on the training set and none of the validation set could be found. With anticipation 36 and 16 goals, respectively, were found.

5.1. Development of the fitness function

In the following subsections we present the results depending on the noise levels $\zeta_{max} = 0,1,3$, and 30. On all noise levels, after about 200 generations the optimum was found. On noise level 0 solutions are found, where 38 out of the 40 training goals are reached, and 12 out of 20 in the validation set. On noise level 1, 37 and 14 goals are found, on level 3 we have 36 and 16, respectively, and on level 30, 35 and 15 goals. The development of the fitness using different noise levels is shown in fig.6. The validation reveals that the ranking of the parents, which is based on the training set, is not always valuable for the validation set. Increasing the noise level implies a decrease in the fitness on the training set, but an improvement on the validation set after 400 generations. If the noise level is too high (e.g. 30), the performance deteriorates on both, the training and the validation set.

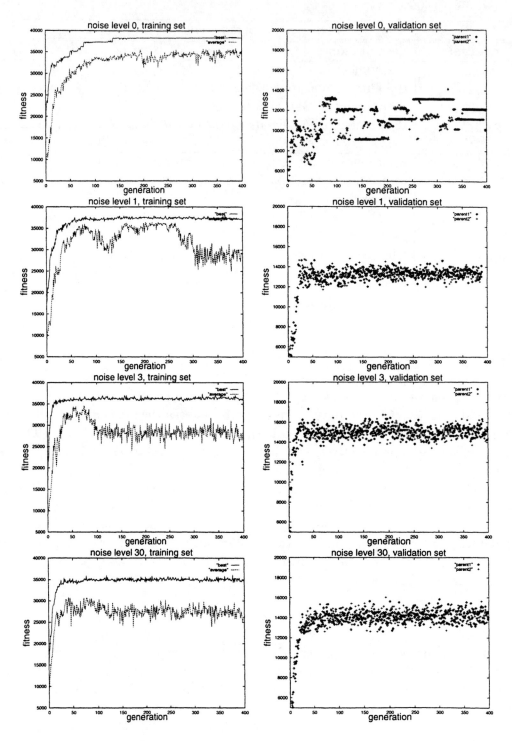

Figure 6: The fitness of the best and the average fitness depend on the generation. On the left: Training set; On the right: Validation set. Noise level 0 (top), 1(2nd row), 3 (3rd row), 30(bottom)

5.2. The Distribution of the evolved structures

For noise level 1, the development of the structure is shown in fig.7.

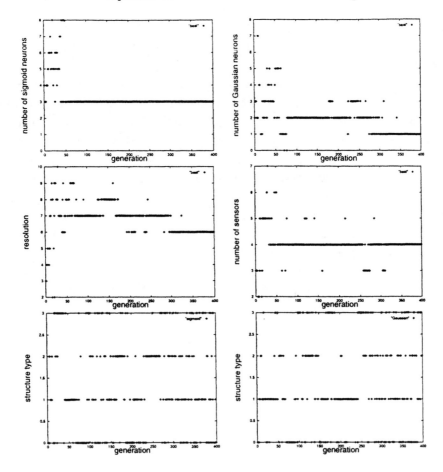

Figure 7: The structure of the best individual depends on the generation.

5.3. Some paths of the evolved robots

For noise level 0, 3, and 5 different paths are shown in fig.8; due to the motoric noise the trajectories from the starting position to the goal can deviate one from another.

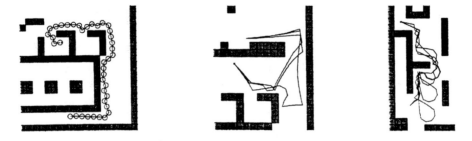

Figure 8: Examples of a pathsfor noise level 0, 3, and 5.

On an ULTRA SPARC workstation for each generation, run, and goal, it takes 1.5s to compute. With 40 goals, 10 runs and 400 generations we need 66 hours.

6. Discussion

In order to optimise a neural network to control a robot's motion, we have to face several problems, especially concerning the encoding in the genome and the evaluation of the robot's performance. In this paper we presented a solution which delivers good results for the restricted model we used in the simulation.

It turns out that anticipation in the form of an imaginary step improves the performance considerably.

As to the role of noise we can summarise that increasing the noise level implies a decrease in the fitness on the training set, but an improvement on the validation set. But if the noise level is too high (e.g. 30), the performance deteriorates on both, the training and the validation set. At motoric noise level 0, the neural noise level converged to zero. We can conclude that determining the appropriate noise level by adding it to the genome does not suuport the generalisation capability. In order to specialise to the given training set the individual noise level decreases.

From the small sizes of the evolved networks we conclude that evolution prefers simple solutions to complicated solutions, if the simple ones achieve nearly the same. This might be surprising with respect to the complexity of natural organisms; a short explanation of this is that in simple sytems only a few parameters have to be adapted, whereas in complicated systems, many parameters have to be chosen simultaneously. It is also described in (Walter, 1961) that evolved creatures are parsimoniously equipped; in a complex environment, however, they can behave in a complicated, or even unpredictable way. On the other hand, we cannot expect the evolution of systems that can solve problems whose degree of complexity is higher than that of the task in the evolution.

Evolution does not produce anticipatory systems, if a simpler, straightforward method or rule can do the same. The task has to be sufficiently complex in order to have the evolution create anticipatory systems.

7. Conclusion and Future Work

In our model we optimised a mobile robot which anticipates its next step and checks for collision. It was shown that the model works quite successful. However, the anticipation was "hard-wired" whereas the emergence of it would be desirable.

We believe that the combination of evolutionary algorithms and neural networks for creating a control instance for autonomous systems provides the right direction for achieving anticipation and "intelligent" behaviour. However, in the present model the anticipation was artificially introduced and we will try to devise a model in which anticipation can emerge more naturally.

The simulation will be implemented on a Cray T3E with 512 parallel processors. If the evaluation of the fitness takes most of the time, by parallelising the calculations an efficient speedup is possible (e.g. with a simple farm-ansatz (Santibáñez-Koref, 1995)). This allows to test multi-population strategies. Subpopulations are isolated for some time, and global selection does not eliminate good structures with accidentally unfitting continuous parameters.

Instead of fixing the weights by evolution, a learning rules will be implemented which finetunes the weights in a learning stage, which precedes the goal finding stage. This is expected to be advantageous to fixed weights, because a quick adaptation to the environment is possible, avoiding another time-consuming evolution strategy.

Acknowledgements. I would like to thank B. Sendhoff and Dr. S. Wacquant for their comments. This work was supported by the DFG grant KOGNET II.

References

Bäck, T. and Schwefel, H.-P. (1993). An overview of evolutionary algorithms for parameter optimization. *Evolutionary Computation*, 1(1):1–23.

Brave, S. (1996). The evolution of memory and mental models using genetic programming. In *International Conference on Genetic Programming '96*, pages 261–266.

Ekdahl, B., Astor, E., and Davidsson, P. (1995). *Intelligent Agents - Theories, Architectures, and Languages*, chapter Towards Anticipatory Agents, pages 191–202. Number 890 in Lecture Notes in Artificial Intelligence. Springer Verlag.

Elman, J. L. (1993). Learning and development in neural networks: the importance of starting small. *Cognition*, 48:71–99.

Pötter, C. (1996). Path planning for autonomous vehicles. In *World Automation Congress'96*, Albuquerque, NM 87191 USA. TSI Press.

Rechenberg, I. (1994). *Evolutionsstrategie '94*. frommann-holzboog.

Rosen, R. (1985). *Anticipatory Systems - Philosophical, Mathematical and Methodological Foundations*. Pergamon Press.

Rudolph, G. (1994). Convergence analysis of canonical genetic algorithms. *IEEE Transactions on Neural Networks*, 5(1):96–101.

Santibáñez-Koref, I. (1995). Parallele Implementation von Multipopulationsstrategien. Interner Bericht 03, Technische Universität Berlin.

Schwefel, H.-P. (1994). *Evolution and Optimum Seeking*. John Wiley.

Sendhoff, B., Kreutz, M., and von Seelen, W. (1997). A condition for the genotype–phenotype mapping: Causality. In Bäck, T., editor, *International Conference on Genetic Algorithms*. Morgan Kaufman.

Sullivan, J. and Pipe, A. (1996). Efficient evolution strategies for exploration in mobile robotics. In *International Conference on Genetic Algorithms*, pages 245–259.

Walter, W. G. (1961). *Das lebende Gehirn*. Kiepenheuer & Witsch. The Living Brain.

The Role of Anticipation in Adaptive Embodied Autonomous Systems

Erich Prem

The Austrian Research Institute for Artificial Intelligence

Schottengasse 3, A-1010 Wien, Austria

erich@ai.univie.ac.at

The aim of this paper is to study the role that anticipation plays in adaptive autonomous systems. The emphasis will be on epistemological consequences of adaptation in practical robotic systems as they are currently developed in the new field of embodied artificial intelligence. The autonomy of physical robots is peculiar, because it consists in behavioral autonomy as well as epistemic autonomy. While the former is a problem that is often addressed, the latter poses difficult foundational questions for the field. We study the role that anticipation plays in this context. It is argued that embodied systems are a particularly interesting case for the study of epistemic autonomy. This is due to the fact that the adaptation process in robots generates a special form of representation that indicates the outcome of interaction and thus can support action selection schemes. The role of these representations and their epistemic and ontological consequences for the system as well as epistemological consequences for system observers are investigated.

Key words: robotics, embodied artificial intelligence, representation, epistemology, ontology

1 Introduction

We share our world with an endless number of other animals. It seems that every possible niche on earth, from hostile deserts to the deep sea, is populated with its own characteristic species. Even more peculiar than the large number of individual types is the seemingly unproblematic nature in which these animals interact with their corresponding environments. From the simplest bacteria to the puzzling behaviors of birds and mammals, all seem most adequately adapted to whatever surrounds them. This immediate way of coping with life has long puzzled scientists, all the more with respect to the way in which humans behave in their environment. In the context of this paper, the term "intelligence" is used to describe exactly this phenomenon of immediate coping with the problems of everyday life.

One problem when trying to understand the richness of behaviors that exist in the animal world is its wide variety of complexity. The behavior of simple unicellular animals can be explained by control systems that exploit stimulus-response mechanisms. The behavior of higher animals, on the other hand, seem to at least sometimes involve deliberation, planning, and forecasting.

CP437, *Computing Anticipatory Systems: CAYS--First International Conference*
edited by Daniel M. Dubois © 1998 The American Institute of Physics 1-56396-827-4/98/$15.00

The aim of this paper is to suggest that this wide variety may indeed possess a common root: anticipation.

Many scientific disciplines are involved in efforts to explain how intelligence functions, what causes intelligent behavior and how evolution provided this richness of living phenomena. Such themes, however, are no longer exclusively reserved to natural sciences. Nowadays technological disciplines try to understand intelligent behavior in order to construct intelligent machines and at the same time construct intelligent machines to understand intelligent behavior. One such discipline is the field of embodied artificial intelligence (AI).

1.1   Embodied Artificial Intelligence

The credo of embodied AI can be succinctly summarized as "Intelligence is determined by the dynamics of the interaction with the world." [Brooks 91] It is based on the idea that in order to build intelligent autonomous systems it is necessary to have it directly and dynamically interact with the world. The departure from traditional approaches is characterized by an increasing importance which the physical structure of the robot body plays. This body is not regarded as a mere box to be moved by the robot's control system. Instead, it is an integrated part of the interaction process between the robot and its environment. To the behavior-based roboticist it is unimportant whether some desired behavior is generated by a computational process in the robot's software or by a physical characteristic of the shape of the robot body. The main effect of this new emphasis on physical aspects is a dramatic departure from the traditional control architecture with the goal to increase the robot's interaction dynamics. Some of these new architectural considerations based on the requirements for creatures are the following (cf. [Brooks 85, Prem 97a]):

- In order to cope quickly with changes in the environment it is necessary to sense the environment often by evaluating rather simple predicates as they are needed by the individual subsystems.

- Robustness is achieved by means of multiple parallel activities. There is no central model of the world, instead individual layers extract only relevant aspects ("projections of a representation into a simple subspace").

- Each layer of control has its own implicit purpose (or goal), "sometimes goals can be abandoned when circumstances seem unpromising".

- The overall system action is driven by the autonomously acting parts, no central (or distributed) process selects from an explicit representation of goals to decide what to do next.

Interestingly, embodied AI did not originate as a theoretical approach to the problem of cognition but as a response to technical problems mainly within the field of robotics. The main difficulty for the robotic engineer was getting the interaction dynamics of the robot right. The traditional approaches to building robot control architectures were highly similar to the method pointed out above. The systems consisted of modules that were based on functional decompositions. Each module served to realize a clearly defined I-O function. The whole process of moving the robot through its environment was supposed to be an information processing task consisting in mapping inputs at data interfaces to outputs at interfaces. Internal system time was not in accord with what was going on outside the robot. For a very long time, roboticists believed that the time constraints could be neglected, since there would be a day when processors would become fast enough for realizing the desired calculations just in time.

The design of a concrete robot control architecture suffers from a lack of information about the physical properties of the robot environment. Exact simulations of these environments are usually impossible due to sensor errors, unpredictable measurement, and peculiar mechanical details of robot-environment interaction. This increases the need for any autonomous robotic system to be adaptive to its environment. This adaptation is formed by goal-driven interaction with the environment. In the next section, the relation between anticipation, representation and adaptation is studied in more detail.

2 Anticipatory adaptive representations

2.1 Representation in artificial systems

For a long time, representation was a (if not *the*) central notion of artificial intelligence. For a system to exhibit intelligent behavior, it seemed necessary to possess representations of knowledge. In an intelligent action, it would then use these representations in a process similar to logical deduction in order to "derive" appropriate actions. However, it soon turned out that such representations were hard to design. How could one formalize all the details of a specific system environment? This problem turned out to be all the more difficult, for the systems themselves were so difficult to describe.

Learning promised to be the solution to such problems. If a system could develop its own view of the world and accordingly construct its representations of the environment, the designer would no longer have to anticipate what the system really needs. In this way, an intelligent system should be able to construct a model of the world, and probably a model of itself in order to be able to choose actions that are appropriate.

Fig. 1 shows a part of a behavior-based control system that reactively transfers inputs to system outputs. It uses an applicability predicate p to decide whether and when an output should be generated. For example, a mobile robot may detect an obstacle to the right and accordingly generate more output on the motor that drives the right wheel. Such a coupling from input to

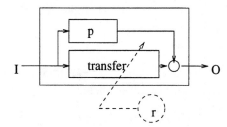

Figure 1: Behavioral module that transfers an input to an output if the applicability predicate p is true. The transfer function is learned based on a training signal r.

output need not be designed in advance. It could also be learned by means of a training signal r. In our example, this could be a signal coming from a bumper that detects collisions with objects (to the right) and changes behavior in the right way.

One possible choice for a learning method are neural networks. These systems became very popular in the field of artificial intelligence, because they promised to solve the above problem of designing representations by simply developing them. However, it soon turned out that the price for this development was that representations within the systems could no longer be explained in any simple way. Consequently, the behavior of systems driven by neural networks was even more difficult to explain. Before we look at a specific example of a neural network let us briefly revisit the framework within which R. Rosen discusses neural networks and anticipation.

Rosen's framework

Let us assume that a training signal (usually based on reinforcement) serves to optimize some criterion that is of importance to the system. It might, for example, assist in the provision of food. Following a description by [Rosen 85, Rosen 91] and Fig. 1, we realize that the system's input is I, while the adaptation is determined by the optimization criterion r. Of course, it is reasonable to believe that there is a linkage between the "predicate" to be learned and the observables of the system environment I. In the course of learning, the system implicitly generates a *model* of the linkage, and also, of the system-environment interaction.

Rosen argues that the result of such a learning or selection mechanism is a transfer function that "predicts" external reinforcement. It drives the system in a way that fitness is optimized before it is evaluated. This is why Rosen calls such systems *"anticipatory."* Representations (data structures, models) in the transfer functions become shaped based on their predictive value with respect to maximizing fitness. As a further consequence, these representations must be properly explained with reference to the future outcome of the system's interaction with its environment. This results in a finalistic or teleological terminology. Let us look at an example of this process (taken from [Prem 95b] and developed there in the context of symbol grounding).

2.2 A neural network example

Fig. 2 shows the basic idea and architecture employed in our experiments. The system sees sequences of patterns that are supposed to be generated by sequences of objects. Two consecutive objects are first categorized by means of a self-organizing network that implements a simple clustering algorithm. This can be best thought of objects measurements (which can be distorted) that map onto different regions of a self-organizing map. These regions can then be labeled through another layer, the symbol layer. Whereas the formation of categories happens without a teacher, the name for a category (i.e. which unit of the symbol layer should represent the category) is defined by a human supervisor and learned by means of an error-minimizing routine.

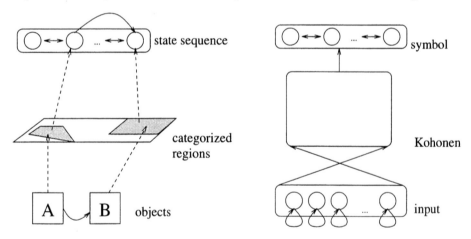

Figure 2: Left: Two consecutive images are categorized in a Kohonen network and generate a state sequence. Right: The network consists of recurrent input units, a Kohonen layer and an interactive activation-output layer.

The input layer which receives signals from the (simulated) measurements consists of self-recurrent units. The weights of the self-recurrent connections range from 0.0 to 0.7 so as to only slowly change their activation through time and to capture different (historical) aspects of the time sequence, cf. [Ulbricht 94]. The training of this architecture with a sequence of input patterns produces a categorical map of the input patterns that can then be summarized by one unit in the output layer. Therefore, what happens in this architecture is that groups of input patterns become represented in one output unit. The environment, if you like, is categorized and named.

However, this is not the usual architecture for a neural network that learns to drive an autonomous agent. There is no external training signal involved in Fig. 2. We now change the architecture to make use of a conventional three-layered perceptron that can learn a specific mapping from inputs to outputs depending on a predefined error function. The new architecture is depicted in Fig. 3.

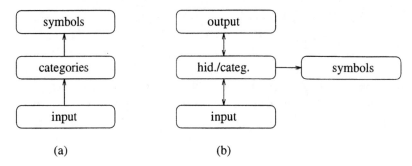

(a)   (b)

Figure 3: Variants of two architectures for labeling inputs: (a) input- and (b) hidden layer categorization.

The new architecture will generate representations in the hidden layer that are much more dependent on the *output* of the multi-layer perceptron than on its input. We can, of course, still use this hidden layer and try to summarize the classes of inputs as in the previous example. However, this will fail, if the hidden layer is sufficiently small. In other terms, in the case of the architecture which categorizes the hidden states of the backpropagation network trying to describe the set of inputs with symbols fails (if the categorization component is forced to find a small set of categories). In this case, only labels for the *output* can be grounded, i.e. only labels for the output of the backpropagation network can be correctly produced.

The practical consequence of this result is that if one tries to use the representations generated in the hidden layer of a backpropagation net which is not auto-associative, arbitrary input descriptions can no longer be produced by mapping the hidden layer to an output. Only those classes can be distinguished, which are in accordance with the goal of the backpropagation network. Without knowing this goal, an explanation of the hidden layer representation is hardly possible.

3   Anticipatory behavior

The example from the previous section is of an inherently static nature. The problem of explaining an interactive adaptive system are more difficult, but we can efficiently use the insights we have gained in our network example.

The representations formed by autonomously acting, adaptive systems have important epistemological characteristics. As Bickhard has argued [Bickhard & Terveen 95, Bickhard 97], these representations serve to indicate interaction outcome. They are used by the system for action selection. As the outcomes are not guaranteed, the system is able to detect representational errors with respect to the "desired" outcomes. Indications of this kind can be considered to constitute a fundamental form of representation. Its content consists in presuppositions about the environment. These implicit "assumptions" concern properties of the environment for which indicated outcomes occur. This naturally leads to a new notion of truth value for these representations. Indicated outcomes can be false, if the presuppositions, i.e. the implicit representational contents, are wrong.

To understand this better, we can compare it with the medieval rule that *veritas* is *adaequatio intellectus rei,* i.e. that truth consists in the correspondence of intellect and object. Finding out what is true can therefore mean several different processes of adaptation. First, the intellect can adapt to the things out there: this is the traditional approach by philosophy, artificial intelligence and cognitive science. Secondly, the things can adapt to the intellect, as Kant proposed. But thirdly, as we have seen now, truth itself can be dependent on what it means to be adequate.

This kind of representations now has important implications for a system's own ontology, i.e. for what there is in the environment of the interactive agent (be it a robot or an animal). To understand how such an agent sees its world, we shall make use of concepts developed by theoretical biology at the beginning of the 20th century.

3.1 Interaction circuits in theoretical biology

A conceptual framework for the analysis of functional circuits dates back to the work of Jakob von Uexküll on theoretical biology. His important contribution in our context is the analysis of "the world according to the animal." A "thing" in the animal's world is only "effector-" and "receptor-bearer". It can be thought of as a generator for signals to receptor organs and as a receptor of manipulations through effectors.

The formation of sensory experience is not only based on inter-*action*. Even more importantly, the interaction has a specific purpose. Such a purpose turns the object from a collection of merely causally operating parts of physical entities into a meaningful assembly of things which are integrated in a purposeful whole. The essential point is *to understand how the thing is embedded in an action and how this action is embedded in a purposeful interaction with the world.* In order to fully understand the system's world, our task consists in the dissection of the functional world (i.e. the whole of the subject's functional circuits).

Such a point of view is surprisingly close to the credo of behavior-based robotics where the descriptive strategy outlined above is turned into a design method [Connell 90]. Starting from functional interaction circuits, the engineer tries to develop a minimalist architecture that fulfills the system requirements. Equipped with such a conceptual framework, we can now study the consequences of adaptation in autonomous systems for the meaning of internal representations.

As innocent as this descriptional framework may look, it has a rather strong influence on the system's representational framework. It is now likely that sensory impressions of the system are categorized in classes that form items of the same usefulness to the system. This is the same process that we have already seen in the static neural network example, but now with action added to the system. This active part is important, for it determines the ways in which the system interacts with its environment. We could even say that it determines what there is in the system's environment.

In the same sense that "chairs" are properly described by their function "for sitting" for humans, objects in an embodied system's environment will now be classified due to their functional properties. It is clear that such a representational frame can be conceptually opaque in relation to human concepts.

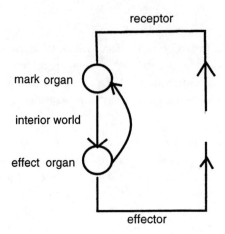

Figure 4: Action circuit as described by von Uexküll.

3.2 Understanding rational behavior

Understanding and explaining the behavior of interactive and adaptive systems thus turns into the problem of understanding a system's goals at the first place. It is only from these goals that we can study the way in which the body and the environment interact to produce internal control systems which make the systems survive in its world.

This view can also explain, why traditional artificial intelligence (and also sometimes psychology) adopted an insufficient picture of how intelligence relates to its environment. In these disciplines it was always assumed that anticipation is a process that operated on a distinguished model of the environment. The idea was that some sort of manipulation of internal image-like representations of the external world was used as a model of the environment in order to forecast it. However, the view proposed in this paper is that the model is much more implicit rather than explicit. It is already the objects themselves which are "anticipations" of possibilities to act.

It follows, for example, that we do not have to deliberately plan our interaction with a glass in order to drink beer. It is quite the other way round: the glass will show up as a possibility to drink beer. Here, the glass is the anticipation of a specialized interaction outcome, it forecasts a special type of satisfaction. Of course, viewed from the outside, the beer-drinking behavior of a human may be described as a rational process, in which the subject searches for a solution to the problem of bringing liquid to its mouth. However, this view implies an ontological reality of epistemic structures that has proofen unhelpful, at least in the field of robotics. The epistemic structures concern image-like representations and rules that an intelligent agent would need to manipulate in order to solve problems. With ontological reality I mean that scientists have in fact been looking for these representations and rules in both real brains and robots.

Recently, empirical results form the cognitive neurosciences suggest that it are exactly the above mentioned finalistic representations that play a major role in the nervous systems of animals. There is evidence at the neurophysiological level that indicative representation plays a major role in sensory-motor body-environment interaction. Groups of cells in the cerebral cortex of monkeys seem to be exclusively related to a sequence of multiple movements performed in a particular order. These cells can contribute a signal about the order forthcoming multiple movements and are useful for planning and coding of several movements ahead [Tanji et al. 94].

The subject centered viewpoint of Uexküll has also been well supported by neurophysiological evidence. Experiments by Graziano show that premotor neurons play a major role in the coding of visual space. The evidence suggests that the encoding of the spatial location of an object happens in arm-centered coordinates rather than using a retinocentric representation. This again points to the way how system-environment interaction of an embodied system creates models that are heavily oriented by the system's functional needs [Graziano et al. 94].

Taking into account these forms of representations and, additionally, embodied systems as the prototypical example for intelligent beings, severly changes our models of cognitive phenomena and intelligent systems [Prem 97b]. In this view, rationality is not some feature added on top of the control system that moves a body through its world. Instead, it is the way in which things around us show up for us in our everyday live.

4 Ontological implications

This finalistic view brings with it the development of a rather distinct system ontology. The tradtional ontology used in the natural sciences consists of substances and their properties (among which we find "form" as the most prominent one [Prem 97d]). The ontology that is used here, however, is a much more teleological one, with purposes and adequateness as the primary categories of what there is in the world.

The ontological position described here is so surprisingly similar to the existential-ontological philosophy of Martin Heidegger that it is worth describing a few points of contact between both ontologies. Our notion of *things* in the animal's world can be best compared to what the philosopher Martin Heidegger calls *equipment* [Heidegger 27]. In the human Being's everyday practices things in our world make sense because we can use them.

The entities that will be encountered this way are not objects in the above sense. We do not simply add a functional predicate to them. *Dealing* with them is our primordial way of having them, not some bare perceptual cognition. To paraphrase Heidegger, "hammering" does not know about this property of being a tool. Instead, the more we are immediately engaged in coping with the problem of fixing something, the less the hammer is taken as an object which can be used in-order-to hammer. Strictly speaking, for Heidegger nothing like one equipment in this sense exists. This is because anything which we are using is embedded in a whole of multiple references to other tools and purposes. The hammer thereby refers to nails, tables, wood, etc.–i.e. a whole world of equipment and also of meaningful coping with the world. As long as we are engaged in "hammering"–in a purposeful dealing with equipment–and this equipment simply is "available", we do not even think about it. In such a situation the tools are simply "ready-at-hand".

In Fig. 5 I tried to depict how Heidegger's account of tools explains our everyday way of coping with the world and how this relates to anticipatory representations in embodied systems. The latter kind of representations are the source for the generation of things and they implicitly *are* our ways of dealing with them. Traditional AI has concentrated on the description of things with (optical, haptic, ...) properties and explicitly listed their functions, e.g. [Minsky 85]. In embodied AI, things are derived from interacting with them *in order to* achieve a pre-defined goal. This makes things the solutions to problems as well as the sensoric impressions they generate while using them.

Very importantly, this approach puts the things encountered by the intelligent system in an immediate context. This means that the pen I hold in my hand while I am writing will be different from the pen that is lying over there on the desk. But while it is in my hand, it simply "is" a way of writing, it is embedded in a whole of functional references to other things like paper, ink, etc. At different times, however, the pen-object may appear as a tool for rather different purposes, e.g. for picking, piercing, etc.

In [Dreyfus 90] it is rightly argued that conventional science has always sought to give descriptions

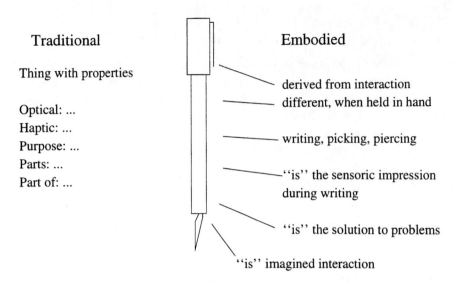

Traditional	Embodied

Thing with properties

Optical: ...
Haptic: ...
Purpose: ...
Parts: ...
Part of: ...

— derived from interaction
— different, when held in hand

— writing, picking, piercing

— "is" the sensoric impression during writing

— "is" the solution to problems

"is" imagined interaction

Figure 5: Things in the world according to an intelligent embodied being. [Prem 97c]
Reprinted with kind permission of the Austrian Computer Society, Proceedings of the First Workshop on Teleoperation and Robotics, 1997.

of objects that do not depend on the context in which an object may be used. This is at the same time the reason, why putting back the context to a thing described as a list of physical properties turned out to be so difficult in AI. The reason lies in the abundance of "facts" that are "known" by such an intelligent being. As Heidegger points out, in the moment where we only look at things in order to describe their pure external features, we strip away context and at the same time significance. We then deprive the objects of their everyday use. In our ontology here, we could even say that we then change the nature of objects to something we have not dealt with so far: merely occurent things that do not show up as possibilities to (inter-)act.

This, however, is the point where embodied Artificial Intelligence finds itself today. While we have a clear idea about how everyday interaction with things can be explained and probably be implemented on a robot, we do not exactly know how to deal with abstract objects and "theoretical" interest in one's environment. However, we believe that this is a step that must be based on first understanding how everyday intelligence works in adaptively coping with the world consisting of (and in) anticipated possibilities to act.

References

[Bickhard & Terveen 95] Bickhard M.H., Terveen L. 1995. *Foundational Issues in Artificial Intelligence and Cognitive Science.* Elsevier Science Publishers.

[Bickhard 97] Bickhard M.H. 1997. The Emergence of Representation in Autonomous Agents, in Prem E. (ed.) *Epistemological Issues of Embodied AI.* Cybernetics & Systems, 28(6), 1997.

[Brooks 85] Brooks R.A. 1985. A Robust Layered Control System for a Mobile Robot. *AI-Memo 864.* Cambridge, MA: AI-Laboratory, Massachusetts Institute of Technology.

[Brooks 91] Brooks R.A. 1991. Intelligence without Representation. In Special Volume: Foundations of Artificial Intelligence, *Artificial Intelligence,* 47(1–3).

[Connell 90] Connell J.H. 1990. *Minimalist Mobile Robotics.* San Diego, C.: Academic Press.

[Dreyfus 90] Dreyfus H.L. 1990. *Being-in-the-world.* Cambridge, MA.: MIT Press.

352

[Graziano et al. 94] Graziano M.S.A., Yap G.S., and Gross C.G. 1994. Coding of Visual Space by Premotor Neurons. *Science,* 11 November 1994, 266, pp. 1054–1057.

[Heidegger 27] Heidegger M. 1927. *Sein und Zeit.* (Being and Time.) Tübingen: Niemayer.

[Minsky 85] Minsky, M. 1985. *The Society of Mind.* New York, NY: Simon & Schuster.

[Prem 95a] Prem E. 1995. Grounding and the Entailment Structure in Robots and Artificial Life, in Moran F., et al.(eds.), *Advances in Artificial Life.* Berlin: Springer, pp. 39–51.

[Prem 95b] Prem E. 1995. Dynamic Symbol Grounding, State Construction, and the Problem of Teleology, in Mira J. et al. (eds.), *From Natural to Artificial Neural Computation.* Proceedings of the International Workshop on Artificial Neural Networks. Berlin: Springer, pp. 619–626.

[Prem 95c] Prem E.: Understanding Complex Systems: What Can the Speaking Lion Tell Us?, in Steels L.(ed.), The Biology and Technology of Autonomous Agents, Springer, Berlin Heidelberg New York, NATO ASI Series F, Vol. 144, 1995.

[Prem 97a] Prem E. 1997. The behavior-based firm. *Applied Artificial Intelligence.* **11** (3), pp. 173–195.

[Prem 97b] Prem E. 1997. The implications of embodiment for cognitive theories. *Austrian Research Institute for Artificial Intelligence* Vienna, TR-97-11. `ftp://www.ai.univie.ac.at/papers/oefai-tr-97-11.ps.Z`

[Prem 97c] Prem E.: The world according to a humanoid robot, in Beneder M.(ed.), *robots@aec.at,* R. Oldenburg, Wien, Schriftenreihe der Oesterreichischen Computergesellschaft, Bd. 101, pp. 17–27, 1997.

[Prem 97d] Prem E.: Epistemic Autonomy in Models of Living Systems, Proc. of the Fourth European Conference on Artificial Life, Brighton, MIT Press/Bradford Books, 1997.

[Rosen 85] Rosen, R. 1985. *Anticipatory Systems.* Oxford, UK: Pergamon.

[Rosen 91] Rosen R. 1991. *Life Itself.* New York: Columbia University Press.

[Tanji et al. 94] Tanji J., and Shima K. 1994. Role for supplementary motor area cells in planning several movements ahead. *Nature,* 371 (6496), pp. 413–416.

[Ulbricht 94] Ulbricht C.: Multi-recurrent Networks for Traffic Forecasting, Proceedings of the AAAI'94 Conference, Seattle, Washington, 883–888, 1994.

[von Uexküll 28] von Uexküll J. 1928. *Theoretische Biologie.* (Theoretical Biology.) Frankfurt/Main: Suhrkamp.

Learning of an Anticipatory World-Model
and the quest for General versus Reinforced Knowledge

Andreas Birk

Artificial Intelligence Laboratory, Vrije Universiteit Brussel

Building G 10, Pleinlaan 2, 1050 Brussels, Belgium

cyrano@arti.vub.ac.be, http://arti.vub.ac.be/~cyrano

Keywords: learning, eye-hand coordination, J.P. Piaget, G.L. Drescher, autonomous mobile robots

1 Introduction

The ultimate aim of AI is to understand intelligence in a constructive way, i.e., to build systems exhibiting intelligent behavior. An anticipatory system, i.e., a system having a model of itself and/or its environment (Rosen 1985), has obviously a "higher cognitive level" than a purely reactive system, as it is capable of planning, imitation, and so on. But from a "designer viewpoint", i.e., as an AI researcher who actually has to build systems, the modeling is a hard and labor-intensive task. For this reason among several others, e.g. to be able to cope with changing environments, it is tempting to propose learning as panacea, i.e., to build up and update the model while interacting with the world. When the world-model is build up a crucial question involved is the balance between general and reinforced knowledge, i.e., information that could be useful and information that has already been indicated as being profitable. A naive solution is to make a complete model, meaning to store as much information about the world as possible. But this is usually infeasible for reasons of computational complexity, especially in respect to memory and speed. Another problem is that some information about the world can only be found in an active way, i.e., through interaction with the world including trial and error manners. But this can be dangerous; harmful situations can be encountered that otherwise would not emerge.

There are several interesting possibilities to do the arbitration between the amount of general and reinforced knowledge. First, some sensor-effector combinations can not be learnable at all, thus reducing computational complexity. For example Gallistel et al. (1991) report that rats rapidly associate poisoned water with smell, but they cannot learn to associate it with visual or auditory information. This restriction of the search-space leaves more room for storing general knowledge. Second, the arbitration can be done in respect to the computational resources available. Starting with a more or less tabula rasa, a learning system can gather as much information as possible in the beginning. But as the model grows information too general has to be deleted to make room for more useful ones. The world-model so to say evolves towards being more and more "greedy", i.e., focusing on usefulness. Starting from storing any experience without getting reinforcement or

CP437, *Computing Anticipatory Systems: CAYS--First International Conference*
edited by Daniel M. Dubois © 1998 The American Institute of Physics 1-56396-827-4/98/$15.00

information from a teacher, proves of usefulness are more and more needed to include or keep something in the world-model as time passes. A third mechanism has to deal with dangerous situations because searching general knowledge can provoke harmful events. So, this kind of undesired strong reinforcement has to override the drives to generate general knowledge as otherwise "curiosity kills the cat".

The rest of the paper is structured as follows. Section 2 gives a conceptual overview over Stimulus Response Learning. This learning approach features the possibility to build an anticipatory world-model from scratch. In section 3 concrete experimental results of learning eye-hand-coordination with Stimulus Response Learning are presented. In doing so, we have a deeper discussion of how the arbitration between storing general and useful knowledge is done. Section 4 features an example of how the "search for knowledge" can lead into dangerous situation. More precisely, we describe experiments where mobile robots can actively destroy themselves when exploring their "complete" range of possible behaviors. In section 5 we conclude the paper and give a brief outlook on future work.

2 Stimulus Response Learning

Stimulus response learning combines an improved version of the schema mechanism of Gary L. Drescher (1991) and the concepts of evolutionary algorithms. It is related to the the common classes of evolutionary algorithms, genetic algorithms (Holland 1975, Goldberg 1989), genetic programming (Koza 1992, 1994), evolutionary strategies (Rechenberg 1973, Schwefel 1977) and evolutionary programming (Fogel et al. 1966), only in respect to the basic inspiration: the principle of evolution in nature. Stimulus response learning permits systems with sensors and effectors, so-called animats (Wilson 1991), to explore and internally model an environment.

The system starts learning with an empty model of the environment (*tabula rasa*). It constructs the model from scratch using primitives which are in no way adapted to the environment and effectors and only in a very mild way adapted to the sensors of the system. In what follows we will call the model of the environment simply the *world model*. The central data structure of the world model is a dynamic directed graph. Its nodes are simple rules for behavior-control, the so-called *stimulus response rules (SRRs)*, which are composed of predicates and actions. In its simplest form such a rule has the form *c/a* which says, that action *a* can be performed, if condition *c* holds. Performing *a* when *c* holds is called *execution* of the rule. In a more elaborated form, the rules have the form *c/a/r* that includes a result *r*. A result *r* describes the state of world in terms of sensor perception after execution of *c/a*.

An edge between two rules stands for possible consecutive execution of the rules. The graph is constructed by an evolutionary algorithm using sets of SRRs and sets of edges as population. The fitness of rules and edges is measured by two simple, universal and purely statistical quality measures, namely *reliability* and *applicability*. Reliability counts how often the prediction of are single rule is right, applicability counts how often a rule is actual used. Reliability is kind of the most general drive in the learning process. In the beginning, every information about constant relations between sensor-input and effector-output is stored. But the mechanism of applicability prunes the model. If some knowledge in the model has not been used for quite some time, then it is removed.

By building up a directed graph, the system constructs a spatial, but in general not Euclidean, model of the world. In the graph we keep track of the SRR executed last. This SRR is called *standpoint*, and models the system's current position in the world. It is introduced for two reasons. First, planning reduces to the search of paths from the standpoint to an SRR with fitting result. Second, we have the possibility to restrict the population of SRRs who participate in a learning step to the neighborhood of the standpoint in the graph theoretic sense. This measure greatly reduces the complexity of learning steps and captures the following intuition. If we want to explain a new phenomenon at some place in the world, modeled by the standpoint in the world model, then we expect existing knowledge about that part of the world, the neighborhood of the standpoint, to be more useful than existing knowledge about remote parts of the world.

3 Learning of Eye-Hand Coordination

We achieved promising results with Stimulus Response Learning in experiments to control a robot-arm with a camera. In one class of experiments the camera picture is sectioned as a grid (figure 1). In each field of the grid the most frequent color in it is determined. With this set-up the system learned to move its hand - a red colored gripper - and to grasp and move building blocks. Gary L. Drescher presents in (Drescher 1991) an own approach using a similar, but to some extent richer,

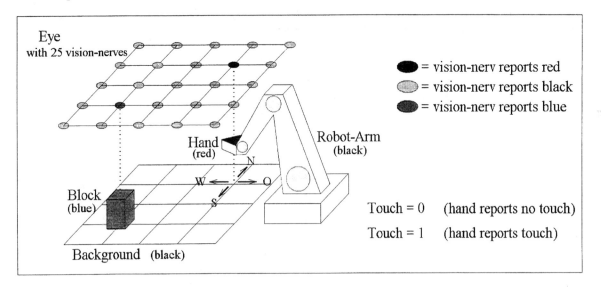

Figure 1: schematic layout of the set-up for learning eye-hand coordination

environment. His best run on a Thinking Machines CM2 (16K processors, 512 Mbyte main memory) ended after two days with memory overflow. His system learned approximately 70% of the desired world-model. A corresponding world-model is found with Stimulus Response Learning in 25 seconds on a SUN Sparc 10 completely. The total amount of memory used is less than 250 Kbyte.

Furthermore, Drescher's experiments are done in simulations only. All experiments with Stimulus Response Learning were done in real world set-ups as well. In doing so, the system was very successful in dealing with noise and errors. In the most challenging class of experiments unprocessed real world images were used (figure 2). The system learned an unpredicted solution

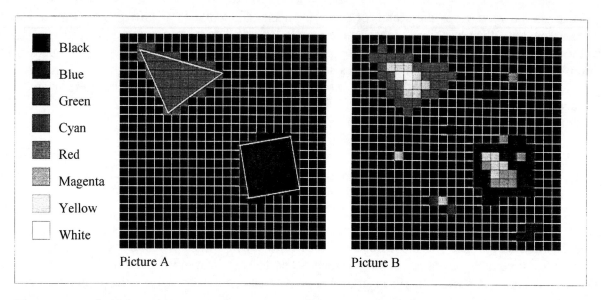

Picture A   Picture B

Figure 2: part of a high-resolution camera picture showing the robot gripper and a building block. Picture B features the raw input. Picture A shows what the system learns to recognize: the triangle of the robot gripper and the rectangle of the building block.

for classifying the gripper and the building blocks by inventing a kind of edge detection. In a real world set-up hand movements and grasping were learned successfully - in every run - in approximately 50 hours on average. In these experiments the run-time was dominated by the speed of the robot arm.

In this paper our main interest is in the question how the arbitration between general and reinforced knowledge can be done, i.e., how to decide what should be included in a world-model. With the experimental results from the learning of eye-hand-coordination we demonstrate here how the mechanisms of *reliability* and *applicability* can be used for this purpose.

When learning eye-hand-coordination the system starts with the "discovery" of rules like

```
red spot at (x,y) / activate motor k / red spot at (x',y')
```

These rules contain the knowledge where the hand (red spot) is moved to when at a certain hand position certain motors are activated. The whole model for moving the hand around consists of a grid-like graph where the rules for moving to position (x,y) are connected to rules which tell where the hand can move to from position (x,y). At first glance, this seems to be simple to achieve but in the real world there are many problems involved. First, there are "errors" due to reflexes, shadows, and so on. So what happens is that the system perceives for example a change in its environment in form of a "popped up" white spot (a reflex) which it "believes" it can influence (as the reflex might

358

disappear due to a motor activation). As a consequence the system stores these "experiences" in rules similar to the above one.

Fortunately the system can find out that these rules are "nonsense" by actively trying to validate the knowledge in the model. This is realized by using alternatively so-called creation- and training-phases during the learning process. In a creation phase, the actual "discoveries" are made, i.e., the rules and edges are formed depending on sensor-changes due to random effector-activation. In a training phase, the system "applies its knowledge" by executing rules. In doing do, it does not need a teacher. It simply can train itself by making kind of random walks through its knowledge learned so far. It consecutively executes rules which are connected by edges and which have fulfilled conditions. In doing so it counts how often the prediction made by a rule holds, i.e., it computes the *reliability*. Only rules with high reliability (close to One) are kept in the model. Therefore, rules describing "perceptual errors" (reflexes, shadows, etc.) are eliminated.

Some "nonsense" rules have conditions corresponding to "perceptual errors" which occur very rarely. Therefore, they are not tested during a training phase (as their conditions are not fulfilled). But these rules are recognized as being "bad" due to their low *applicability*, i.e. the number of executions since creation of that particular rule. In addition to its role in the elimination of "nonsense"-knowledge, *applicability* is useful to generalize knowledge. Take for example the learning of rules for moving the hand (red spot) in the presence of a building block (blue spot). Following two rules are both very *reliable*

```
red at (4,5) + blue at (2,3) / motor 1 / red at (4,5) + blue at (2,3)

             red at (4,5) / motor 1 / red at (4,5)
```

But the second rule is much more general than the first one. The first rule can only be used to move the hand if a building block is present at the right position. This somehow largely restricts hand-movements. But fortunately, the second rule has a much higher *applicability* than the first one as it can be executed much more often. Therefore, it is kept in the world-model whereas rules like the first one disappear.

An important point is that the knowledge is not generalized too much by the mechanism we are using in our system. Note that if the presence of a "blue spot" at a certain fixed position is actually important[1] for hand movements then the second rule will have a low *reliability* as it fails whenever the blue spot is not present but the rule is executed.

One of the most important features of using the combination of *reliability* and *applicability* to guide the process of learning a world-model is the "greediness" involved in these concepts, i.e., starting from including very general knowledge there is a tendency to keep only knowledge in the model that gets a reinforcement in form of usage. As mentioned in the introduction, the learning of a world-model is constrained by computational limits in processing-power and memory-space. Storing general knowledge without revision leads easily to an explosion of the model in size. In Stimulus Response Learning there is kind of a compromise. General knowledge is stored, but after a while it is forgotten again if there was no "reason" to use it.

---

[1] We did experiments where a "magic lamp" indicated if the hand can be moved into a certain direction.

Let us illustrate this in an example. When eye-hand-coordination is learned the system builds a world-model for hand-movements and for gripping and moving of building blocks. Hand-movements are always possible whereas gripping a block is restricted to situations where the hand is by chance above the block (given "random-walks" through the knowledge). Therefore, rules for hand-movements are much more *applicable* than rules for gripping. Though the system learns to grip in our experiments, it "forgets" this again after a while. But if the systems is "told"[2] to grip and move building blocks occasionally then this "reinforcement" causes this part of the model to be kept.

4 When fast Feedback is needed

The disadvantage of the mechanisms presented above is that they rely on repeated "experiences" and active "exploration" of the world. This means that they involve some trial and error components. Therefore they can drive the system into harmful situations and this even several times.

Figure 3: one of the VUB AI-lab mobile robots. The metal antenna on top of it is used for the recharging process.

In the VUB AI-lab, we are working with autonomous mobile robots (figure 3) in a so-called ecosystem setting. The ecosystem (figure 4) includes a charging-station where the robots can re-fill their batteries and so-called competitors. The competitors are boxes housing lamps connected to the same global energy-source as the charging-station. They are therefore "eating up" some of the robots' resources. But when a robot pushes against a box, the lights inside dim for some time and there is more energy in the charging station. From the viewpoint of AI, learning of the crucial behaviors and eventually world-models is an important aim. But there are serious dangers involved when using a system that tries to learn general anticipatory knowledge. So burn the drive-motors for example if the robot is stuck at an obstacle or the batteries explode when they

---

[2] Remember that our system is capable of planning. When we confront it with a goal in the form of "bring the blue spot to position (x,y)", then it can search a rule with corresponding result and do a search for the shortest path from the standpoint to this rule in the world-model-graph. If the system already learned enough, this results in moving the gripper to the building block, gripping, movement to the position (x,y), and release of the block.

are overcharged. It follows that a complete world-model must include this knowledge. But obviously we don't want our robots to learn this from experience.

Figure 4: the VUB ecosystem with a mobile robot (right), the charging station (middle), and a competitor (left).

Figure 5: an exploded robot due to an overcharged battery-pack

In experiments of learning basic behaviors in the ecosystem we therefore used a "pain-like" criterion that gives fast feedback. This criterion is based on short-term monitoring of the internal current of the robot. We claim that this kind of monitoring of the *essential variables* (Ashby 1952) is necessary for any system that learns world-models. In case one (or more) of the variables gets towards the borders of the *viability space* (Mc Farland et al. 1981) fast and effective measures are needed. Otherwise "curiosity kills the cat".

5 Future Work

The obvious next step is to learn an anticipatory world model on the mobile robots in the ecosystem using Stimulus Response Learning and mechanisms handling "emergencies". A very useful and in our opinion feasible to learn model is a kind of map featuring rules that have tests which are capable of "identifying" locations. This can be achieved for example by using information based on touch (e.g. corners), vision, sensing active beacons and so on. It is of course an open question if the accuracy of the model can be made sufficiently high, so that "hardwired" reactive robots can be outperformed. Nevertheless, the complexity of the environment plus the fact that it is embedded in the real world make this task a fascinating goal.

Acknowledgments

Many thanks to all members of the VUB AI-lab who work jointly on maintenance and improvements of the robots and the ecosystem, as well as on the concepts behind it. The robotic agents group of the VUB AI-lab is partially financed by the Belgian Federal government FKFO project on emergent functionality (NFWO contract nr. G.0014.95) and the IUAP project (nr. 20) CONSTRUCT.

References

Ross Ashby (1952). Design for a brain. Chapman and Hall, London.

Gary L. Drescher. (1991) Made-up minds, A constructivist approach to artificial intelligence. The MIT Press, Cambridge.

L.J. Fogel (1966), A.J. Owens, and M.J. Walsh. Artificial Intelligence through Simulated Evolution. Wiley, New York.

Karl Pribam (1960), George Miller, Eugene Galanter. Plans and the structure of behavior. Holt, Rinehart & Winston, New York.

C. Gallistel (1991), A.L. Brown, S. Carey, R. Gelman, F.C. Keil. Lessons From Animal Learning for the Study of Cognitive Development. In The Epigenesis of Mind, S. Carey and R. Gelman, eds, Lawrence Erlbaum, Hillsdale, NJ.

David Goldberg (1989). Genetic Algorithms in Search Optimization and Machine Learning. Addison-Wesley, Reading.

John H. Holland (1975). Adaptation in Natural and Artificial Systems. The University of Michigan Press, Ann Arbor, 1975.

JohnR. Koza (1992). Genetic programming. The MIT Press, Cambridge.

John R. Koza (1991). Genetic programming II. The MIT Press, Cambridge.

D. Mc Farland (1981), A. Houston. Quantitative Ethology: the state-space approach. Pitman Books, Lonodon.

Jean Piaget (1991). Gesammelte Werke. Klett-Cotta, Stuttgart.

Ingo Rechenberg (1973). Evolutionsstrategie: Optimierung technischer Systeme nach Prinzipien der biologischen Evolution. Fromman-Holzboog, Stuttgart.

Robert Rosen (1985). Anticipatory Systems – Philosophical, Mathematical and Methodological Foundations. Pergamon Press.

Hans Paul Schwefel (1977). Numerische Optimierung von Computer-Modellen mittels der Evolutions-Strategie. Birkhaeuser, Basel.

S.W. Wilson (1991). The animat path to AI. In From Animals to Animats. Proc.of the First International Conference on Simulation of Adaptive Behavior. The MIT Press/Bradford Books, Cambridge.

Connectionist Reinforcement Learning
of Robot Control Skills

Rui Araújo, Urbano Nunes, and A. T. de Almeida

Institute of Systems and Robotics (ISR); and
Electrical Engineering Department; University of Coimbra;
Pólo II; Pinhal de Marrocos; 3030 Coimbra - Portugal;
Tel: +351-39-7006200; Fax: +351-39-35672; email: rui@isr.uc.pt

Abstract

Many robot manipulator tasks are difficult to model explicitly and it is difficult to design and program automatic control algorithms for them. The development, improvement, and application of learning techniques taking advantage of sensory information would enable the acquisition of new robot skills and avoid some of the difficulties of explicit programming. In this paper we use a reinforcement learning approach for on-line generation of skills for control of robot manipulator systems. Instead of generating skills by explicit programming of a perception to action mapping they are generated by trial and error learning, guided by a performance evaluation feedback function. The resulting system may be seen as an anticipatory system that constructs an internal representation model of itself and of its environment. This enables it to identify its current situation and to generate corresponding appropriate commands to the system in order to perform the required skill. The method was applied to the problem of learning a force control skill in which the tool-tip of a robot manipulator must be moved from a free space situation, to a contact state with a compliant surface and having a constant interaction force.

Keywords: Self-learning, neural networks, robot, skills.

1. Introduction

There have been developed control theories to successfully deal with a large class of control problems by mathematically modelling the system/plant to be controlled. These analytical models are used to in the controller to generate control actions. Traditional robot programming and control techniques have faced difficulties in contributing to the extension of application domains of robots. In fact, programming based on trajectory interpolation along a prespecified sequence of points has limited ability to cope with complex tasks. The robot programming and control architectures must be equipped to face unstructured environments, which may be partially or totally unknown at programming time, or environments which are time varying and unpredictable. The use of sensory information feedback has been identified as a requirement for achieving flexibility in robotics. However, sensory data is not enough. It is required a set of reflexive *skills* or fine-motion control strategies (e.g. (Lozano-Péres, Mason and Taylor, 1984)) that perform a real-time mapping between the current sensory situation and an appropriate action or actuation to the system. A skill represents, and has associated, knowledge to perform a certain task or operation of the robot. In this article, skills are viewed as an activity for control of some feature characterising the interactions between the a robot manipulator and its environment.

However, the problem arises as of how to generate those skills. Supervised learning methods, the most commonly used in neural networks, require a training data set composed of input vectors and corresponding desired output vectors. If the desired output of the network is not known, these methods can not be applied. An interesting possibil-

CP437, *Computing Anticipatory Systems: CAYS--First International Conference*
edited by Daniel M. Dubois © 1998 The American Institute of Physics 1-56396-827-4/98/$15.00

ity, is *learning* to act over the world by using information received from the sensors for autonomous self-improvement of controller performance. In this article we use a *reinforcement learning* (RL) control system to generate control actions for the robot manipulator. A reinforcement-learning robot (or system) *learns by experimentation* and does not require a teacher for proposing correct actions for all possible situations the robot may find itself in. The robot searches and tries different actions for every situation it encounters and selects the most useful ones. This search/selection process is guided by a *reinforcement* signal that is a performance evaluation feedback function. Thus, unlike supervised learning, the reinforcement learning problem has only very simple "evaluative" or "critic" information instead of "instructive" information available for learning. Reinforcement learning (Barto et al., 1983; Williams, 1992) brings some interesting associated ideas. First, it's based on a concept of *self programming* in which control of a complex system in principle does not need extensive analysis and modelling by human experts. Instead the system discovers it's own abilities. Second the system can cope with disorder. For example, the ability to deal with noisy input data is important for any robot dealing with information close to the raw sensory data. Also the ability to cope with non-stationarity is important for the robot to adapt itself to new or time-varying environments.

There has been effort concerning the application of reinforcement learning methods for learning to control "skills" with mechanical devices. An experiment that is frequently used in reinforcement learning studies (and other) is the problem of controlling an inverted pendulum. Barto et al. (1983) proposed and simulated a method that includes an internal mapping mechanism based on a decoder performing a kind vector quantisation, and producing binary force commands for this problem. On this same problem, Anderson (1989) also uses binary force commands, but uses neural networks for internal mapping. Lin and Lee (1994) use a multilayer neural network implementing a fuzzy logic mapping mechanism, with the overall system producing continuous force commands. Gullapalli (1990) proposes the SRV stochastic neural unit as a basic building block for RL, and in a simulation experiment applies the idea to learning the inverse kinematics function of a planar robot manipulator, with two links and two revolute joints. The same ideas were applied to learning a peg-in-hole insertion operation with a a real robot manipulator, and learning a ball balancing operation with a real device.

The structure of the chapter is as follows. Section 2 presents a reinforcement learning controller architecture. In section 3 we formulate the force control skill to which the controller will be applied. Section 4 presents simulation results. Finally, in section 5 we make some concluding remarks.

2. Learning Architecture

Figure 1 presents the overall architecture of the reinforcement learning control system. The controller is based in reinforcement learning methods (Barto et al., 1983; Williams, 1992). It receives from the environment a time-varying vector of input states, $X(k)$, and a time-varying scalar performance evaluation function, $r(k)$, called the (*external*) *reinforcement* (*signal*) that is generated in an *evaluation module* inside the environment. The objective of learning is for the controller to try to maximise some function of this reinforcement signal, such as the expectation of its value on the upcoming time step or the expectation of some form of accumulation of it's future values.

The precise method for the generation of the reinforcement must be appropriate to both the particular problem being solved, and the type of sensory information available and it's relation to the problem. The generation of the reinforcement is considered to

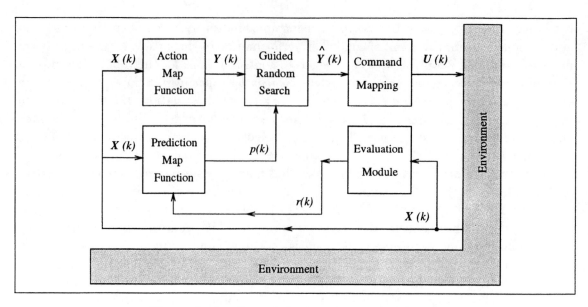

Figure 1. Outline of the learning architecture.

be done in the environment and the method used for this is assumed to be unknown to the learning system. Generally the reinforcement may be function of both the state of the environment and the action command generated by the controller (and possibly some other internal state).

The learning architecture involves two map functions (figure 1): the *action map function* and the *prediction map function*. In general those two map functions are multidimensional and are implemented by two neural networks: the action neural network and the prediction neural network. The *action neural network* learns to select actions as a function of the system states, $X(k)$. The *prediction neural network* or *evaluation neural network* learns to predict the value of the function of reinforcement that we want to maximise. Thus the prediction network has only one output. The inputs to the prediction network are also the system states, $X(k)$. The output of the network, the function $p(k)$, is also called the *expected reinforcement*, or the *internal reinforcement signal*.

The action network decides the best actions to impose to the system in the next time step according to the current environment state that is perceived, $X(k)$. The evaluation network models the environment such that it can perform a prediction of the reinforcement signal that will eventually be obtained from the environment, for the current action that the action network chooses to apply to the system. The prediction network can provide anticipated and more detailed reward/penalty information about the candidate action. This will enable the action network to speed up learning and to decrease the uncertainty it faces. The resulting system may be seen as an anticipatory system that constructs an internal representation model of itself and of its environment. This enables it to identify its current situation and to generate corresponding appropriate commands to the system in order to perform the required skill.

The idea of the learning algorithm is simple. The output of the action network is the expected action vector, $y(k)$. The Guided Random Search module of figure 1 performs an exploration transformation from $y(k)$ to $\hat{y}(k)$. In fact, the expected action vector, $y(k)$, is not applied directly to the system. Instead, it is treated as a mean (expected) action. The actual action, $\hat{y}(k)$, is chosen by performing a random exploration in a neighbourhood around this mean point. This procedure is illustrated in figure 2. The range of exploration is controlled by the variance of a probability density function. In our work, the amount

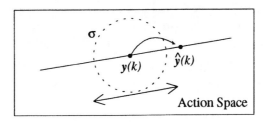

Figure 2. Searching the action space.

of exploration is determined by the following standard deviation, $\sigma(k)$:

$$\sigma(k) = K \left| \frac{x - x_1}{x_0 - x_1} \right|^{2n} \tag{1}$$

where $x = p(k) \in [x_0, x_1]$ is the predicted reinforcement, and $K = \sigma(x_0)$ is a constant that determines the maximum search range that occurs at minimum predicted evaluation.

Equation (1) represents a monotone decreasing function between K and 0, and $\sigma(k)$ can be seen as the extent to which the output node searches for a better action. Since $p(k)$ is the expected reinforcement signal, if $p(k)$ is small, according to (1), the exploratory range will be large, such that the actual action $\hat{y}(k)$ has high probability of being quite different from the mean action $y(k)$. This can be seen as an attempt to discover a better action, since the expected action for the current state, $\hat{y}(k)$, is predicted to give a not very good evaluation. On the contrary, if $p(k)$ is large, the exploratory range will be small, such that the actual action $\hat{y}(k)$ has high probability of being very close to mean action $y(k)$ since it is expected that the mean action is very close to the best action for the current input vector. The amount of search is thus a function of the current predicted quality of the expected action network.

In this work we used a Gaussian probability density function to generate the actual action vector $\hat{y}(k)$ from the expected action vector $y(k)$,

$$\hat{y}(k) = N(y(k), \sigma(k))$$

Thus, $\hat{y}(k)$ is a normal random vector where the density function of each component is given by:

$$f(\hat{y}_i) = \frac{1}{\sigma\sqrt{2\pi}} e^{-\frac{\|\hat{y}_i - y_i\|^2}{2\sigma^2}} \tag{2}$$

The transformation from $y(k)$ to $\hat{y}(k)$ is performed by the Guided Random Search module of figure 1.

In order to perform learning, the weights of both the action network and the prediction network must be adjusted. Learning in this system is performed by error back-propagation (Rumelhart et al., 1986). For simplicity, in the following explanation of the learning algorithm we assume that $p(k)$ is to predict at time step k, just the next value of $r(k)$. However it may be necessary to have $p(k)$ predicting a function of more that one future value of $r(k)$. Also, in the presence of delayed reinforcement, a *temporal credit assignment* problem may appear. In those two cases we can use the *temporal differences method* (Sutton, 1988). In this case the following discussion should be slightly modified. However the main ideas remain valid.

To perform learning in the prediction network, we have the desired output - it is a function of the external reinforcement signal - so that the output gradient can be calculated and the problem reduces to a back-propagation supervised learning problem. For the action network we must take into account the stochastic search module. In this case we use a method closely similar to the *REINFORCE* algorithm (Williams, 1992). For this purpose we first estimate the gradient of the reinforcement signal $r(k)$ with respect to the mean output values. Supposing the function we are wishing to maximise is $r(k)$ itself then we have:

$$\frac{\partial r}{\partial y} \approx [r(k) - p(k)] \left[\frac{\hat{\boldsymbol{y}} - \boldsymbol{y}}{\sigma} \right]_{k-1} \tag{3}$$

The difference to the *REINFORCE* algorithm is only in the calculation of this gradient. Using the *REINFORCE* approach and taking into account the Gaussian probability density function of equation (2) the gradient information would be estimated as (Williams, 1992):

$$\frac{\partial r}{\partial y} \approx [r(k) - p(k)] \left[\frac{\hat{\boldsymbol{y}} - \boldsymbol{y}}{\sigma^2} \right]_{k-1}$$

The time indexes in equation (3) take into account the assumption that the reinforcement signal at time step k depends on the input actions at time step $k - 1$. In equation (3), the term $(\hat{\boldsymbol{y}} - \boldsymbol{y})/\sigma$ is the normalised difference between the actual and expected actions, $r(k)$ is the real reinforcement signal provided by the evaluation module in response to the actual action $\hat{\boldsymbol{y}}(k - 1)$, and $p(k)$ is the predicted reinforcement for the expected action $\boldsymbol{y}(k-1)$. Equation (3) has the following intuitive foundation. If $r(k) > p(k)$, then $\hat{\boldsymbol{y}}(k - 1)$ is a better action than the expected one $\boldsymbol{y}(k - 1)$. In this case $\boldsymbol{y}(k - 1)$ should be moved closer to $\hat{\boldsymbol{y}}(k - 1)$. If $r(k) < p(k)$, then $\hat{\boldsymbol{y}}(k - 1)$ is a worse action than the expected one $\boldsymbol{y}(k-1)$. In this case $\hat{\boldsymbol{y}}(k - 1)$ should be moved farther away from $\hat{\boldsymbol{y}}(k-1)$. After the gradient information for the action network is estimated using equation (3), the reinforcement learning problem becomes a supervised learning problem. At this point we can apply the error back-propagation algorithm (Rumelhart et al., 1986) to adjust the action network weights.

The Command Mapping Module of figure 1 is just an interface module that is used to make some transformation of commands in order to adapt them to the system (the robot system). This module may perform a variety of transformations. It could for example change the frame where the sent commands are expressed, or it could perform a change in the form of representation of the commands (e.g. different representations of rotations transformation).

3. A Force Control Skill

The method of the previous section was applied to the problem of learning a "force control skill" in which the tool-tip of a robot manipulator must be moved from a free space situation, to a contact state with a compliant surface and having a constant interaction force, f_r. The PUMA 560 robot manipulator working in "Real-time Path Control" (RPC) mode (Unimation, 1984) was simulated along one Cartesian degree of freedom (the world Z direction). The contact was against a spring-like compliant surface whose model can be approximated by a stiffness of $K_m = 37\,dN/mm$. Figure 3 illustrates the force control skill. In order to perform the skill, the control system may generate a unidimensional "change of position" command to the environment (thus $\hat{\boldsymbol{y}}(k)$ reduces to a scalar). Due to the compliance of the environment, this action will generate a change in the force state. In this force control skill, the command mapping module of figure 1, just implements a safeguarding saturation function in order to limit the absolute value of the velocity commands sent to the PUMA 560 robot manipulator.

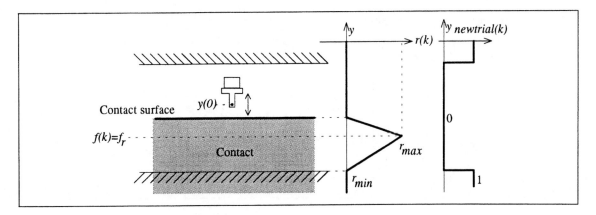

Figure 3. Free-space to contact task.

The function $r(k)$ of equation (4) (illustrated in figure 3),

$$
r(k) = \begin{cases}
r_{min}, & |f(k)| \leq \epsilon_2 \vee |f(k)| > |f_r| + \epsilon_3 \\
r_{max} + \dfrac{(f(k) - f_r)(r_{min} - r_{max})}{(0 - f_r)}, & |f(k)| > \epsilon_2 \wedge |f(k)| < |f_r| \\
r_{max} + \dfrac{(f(k) - f_r)(r_{min} - r_{max})}{(-(|fr| + \epsilon 3) - f_r)}, & |f(k)| > \epsilon_2 \wedge |f_r| < |f(k)| < |f_r| + \epsilon_3
\end{cases} \tag{4}
$$

was used as the external reinforcement to represent the force control skill in the learning algorithm. The contact force between the end-effector and the compliant surface is represented by $f(k)$. In this simulation $f(k) \leq 0$. This reinforcement function is r_{min} in the non-contact situation. When in contact, $r(k)$ decreases linearly from a maximum value of r_{max} to r_{min}, as the absolute value of the error force increases. ϵ_2 and ϵ_3 are positive constants representing forces. ϵ_2 is a small positive constant representing the lowest force at a contact situation, and is used to distinguish between free space and contact situations. ϵ_3 is a constant representing an addition of contact force absolute value that is allowed, above the absolute value of the reference force, before the external reinforcement (evaluation) becomes r_{min}. Thus in this simulation we have $r(k) \in [r_{min}, r_{max}]$, and taking into account equation (1) we have $x_0 = r_{min}$ and $x_1 = r_{max}$. In particular we have used for best and worst evaluations $r_{max} = 1$ and $r_{min} = -1$ respectively.

4. Simulation Experiments

For implementing the prediction and action map functions of figure 1, we used two neural networks with two layers and a feedforward path from inputs to output as illustrated in figure 4. Even though neural networks with two cascaded layers are able to approximate arbitrarily well continuous multivariate functions (Funahashi, 1989), we adopted from (Anderson, 1989) a slightly different architecture, that includes a feedforward path from inputs to outputs.

In order to study the applicability and performance of the learning approach, some simulations were performed. The simulations were organised as follows. Each simulation is called a Run. Each run was composed of trials that were sequentially performed while the cumulative number of simulation time steps in a run remains below some prespecified maximum number. In each trial the end-effector started in a fixed position in free space given by $y(0) = -252\,mm$. The contact between the end-effector and the compliant surface occurs for positions $y \leq y_0 = -253\,mm$. In each trial, the presented learning approach was used but started with the "learned know-how" that was achieved at the end of the previous trial. The desired contact force was $f_r = -200\,dN$.

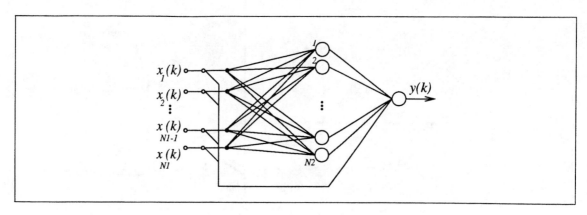

Figure 4. Neural network architecture for implementing the action and the prediction map functions of figure 1.

Each trial in a run was ended when the robot reaches a state that is considered too far from the desired state of having $f(k) = f_r$. More precisely, we defined and used the following end-of-trial function (also depicted in figure 3) to detect when an end of trial occurs, and a new trial must be started.

$$ newtrial(k) = \begin{cases} 1, & (|y(k) - y(0)| > \epsilon_1 \wedge |f(k)| \leq \epsilon_2) \vee |f(k)| > |f_r| + \epsilon_3 \\ 0, & \text{other cases} \end{cases} \qquad (5) $$

This function is one on, and only on, the end of a trial. On a free space situation this function is one, only if the end-effector becomes too far from the compliant surface. On a contact situation, this function becomes one if the contact force increases too much.

The mapping from states $X(k)$, to expected action $y(k)$ and predicted evaluation $p(k)$, was made by two neural networks with two layers composed of hyperbolic tangent neurons.

After some trials in a run, the controller has accumulated some learned experience enabling, to some extent, the control of the force. Figure 5 presents the force response in the first 10000 time steps in a trial, after some other previous learning trials have been made in the same run. Beyond the 10000 steps, the system maintains the same behaviour tendencies. We can see that, on the first part of the response (approximately 4000 time steps), the system performs higher exploration of action-space. In figure 6 we see the corresponding evolution of the expected action applied to the system. This action has the meaning of a velocity that is applied to the robot end-effector. Figure 7 presents the evolution of predicted evaluation $p(k)$. As can be seen, the predicted evaluation is higher when the force is closer to the desired value of $f_r = -200 \ dN$. This is in accordance with its role in the learning algorithm that we have presented. This enables the system to increase exploration in regions of the state space where there is a low predicted evaluation of controller performance. In fact, as can be seen in figure 8, and in accordance with equation (1), the standard deviation of the stochastic search module decreases when the predicted evaluation increases.

After starting completely from scratch, the system has, to some extent, learned to successfully control the force. On steady-state the system exhibits an oscillatory behaviour, and a DC error. Further improvement was not achieved in part because the standard deviation for stochastic search has become too small to allow an exploration of better commands. The learning algorithm requires further enhancements in order to improve its performance. An important aspect is the improvement of coordination between the

370

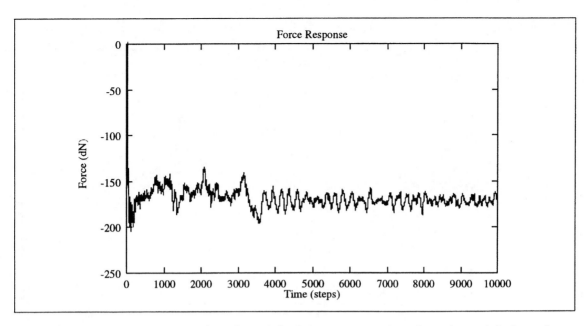

Figure 5. Force response results of a trial after some previous learning trials have been made in the same run.

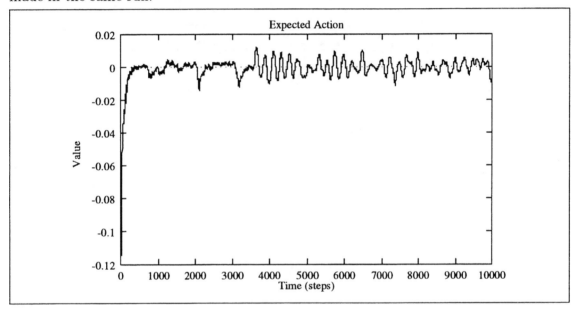

Figure 6. Expected action (velocity), in the same trial of figure 5.

different modules of the learning architecture (figure 1). The development of better functions and/or methods for controlling the standard deviation of stochastic search, and for estimating the gradient information on the action network is also important. Another difficulty is the great amount of design parameters that are involved on the various modules of figure 1. The task of tuning together all those parameters is not trivial. Another issue, is designing the structure of both the prediction and action neural networks.

5. Conclusion

In this paper we have presented an approach for self-learning of robot control skills. The method is based on the application of reinforcement learning techniques. To perform

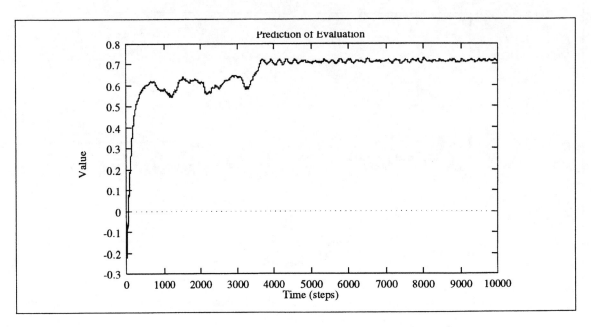

Figure 7. Prediction of evaluation, in the same trial of figure 5.

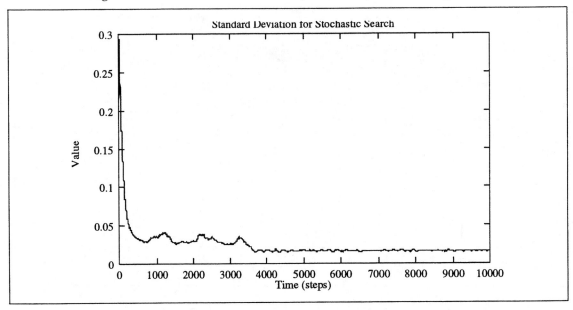

Figure 8. Standard deviation used on the stochastic search module, in the same trial of figure 5.

learning, an action neural network, and a prediction neural network, are automatically trained to respond to the evolution in the system state with appropriate commands to the system. Other techniques can be used for designing force controllers (e.g. (Nunes et al., 1993)). However, this problem is a starting point to explore the application of learning to typical robot tasks. Learning approaches may prove itself useful in more complex tasks where modelling and the designing of the controller are more difficult. Simulation results were presented concerning application to a robot skill consisting on the transition from free space to a contact situation with control force of the interaction force.

References

Anderson, C. W. (1989). Learning to control an inverted pendulum using neural networks, *IEEE Control Systems Mag.* **9**(2): 31–37.

Barto, A. G., Sutton, R. S. and Anderson, C. W. (1983). Neurolike Adaptive Elements That Can Solve Difficult Learning Control Problems, *IEEE Trans. Syst. Man Cybern.* **SMC-13**(5): 834–846.

Funahashi, K. (1989). On the approximate realization of continuous mappings by neural networks, *Neural Networks* **2**: 183–192.

Gullapalli, V. (1990). A stochastic reinforcement learning algorithm for learning real-valued functions, *Neural Networks* **3**: 671–692.

Lin, C.-T. and Lee, C. S. G. (1994). Reinforcement Structure/Parameter Learning for Neural-Network-Based Fuzzy Logic Control Systems, *IEEE Trans. Fuzzy Systems.* **2**(1): 46–63.

Lozano-Péres, T., Mason, M. T. and Taylor, R. H. (1984). Automatic synthesis of fine-motion strategies for robotics, *Int. J. Robotics Research* **1**(3): 3–23.

Nunes, U., Araújo, R. and de Almeida, A. T. (1993). Application of a Robust Indirect Adaptive-Control Method in Task-Space Hybrid Manipulator Control, *Proc. IEEE Int. Conf. Industrial Electronics, Control and Instrumentation (IECON'93)*, pp. 1842–1847.

Rumelhart, D. E., Hinton, G. E. and Williams, R. J. (1986). Learning Internal Representations by Error Propagation, *in* D. E. Rumelhart and J. L. McClelland (eds), *Parallel Distributed Processing: Explorations in the Microstructure of Cognition*, MIT Press/Bradford Books, Cambridge, MA, pp. 318–362.

Sutton, R. S. (1988). Learning to Predict by the Methods of Temporal Differences, *Machine Learning* **3**: 9–44.

Unimation (1984). *Unimate Industrial Robot Programming Manual User's Guide to Val II*, Unimation.

Williams, R. J. (1992). Simple Statistical Gradient-Following Algorithms for Connectionist Reinforcement Learning, *Machine Learning* **8**(3/4): 229–256.

Robot Motion Planning with Virtually Modified Geometry

Boris Baginski
Institut für Informatik
Technische Universität München
Orleansstr. 34, D-81667 München, Germany
e-mail: `baginski@informatik.tu-muenchen.de`

Abstract

We present a novel approach to motion planning for robot manipulators in known environments. The key concept is to evaluate *complete* trajectories between start and goal in the workspace and to *reshape* them incrementally. The evaluation is based on virtual modifications, especially shrinking and expansion, of the geometry model of the robot. The trajectories are bended in space to be improved with respect to the evaluation. We initialize this planning with a possibly colliding connection, evaluate it in our world model and incrementally decrease the *degree of collision* for the whole trajectory. The planning is applicable to realistic robot tasks, even problems with a high number of degrees of freedom can be solved easily. The principle of planning with whole trajectories instead of a moving position along a trajectory can be seen as a shift of perspective that may serve as an example for other planning domains.

Keywords: robotics, motion planning, path planning, computational geometry

1   Introduction

1.1   Motivation

Our overall research target is the development of autonomous robots that are commanded with task level instructions. The robot is only told *what* to do, not *how* to do it. The task has to be decomposed through several layers of intelligent planners. One of the most important *low-level planner* such a system needs is a motion planner that enables the robot to move between arbitrary positions in its working area without colliding with obstacles or with itself. Motion planning based on an internal representation of the environment is inherently anticipatory: the robot decides how to move *now* to avoid *later* collisions.

In the following, we will address this problem for manipulators only, not for mobile robots. Robot arms have a completely different kind of workspace and in general a higher number of degrees of freedom (DOF) than mobile robots. Thus they require specific planning principles.

We are presenting a motion planning algorithm that can take and immediately use the geometry data of the environment without expensive preprocessing. In many realistic cases, this data only gets available when instant planning is needed, and the workspace often changes between the tasks. This is the case for service robots that use sensors to build up a model of the environment. Another example is a mobile manipulator in an industrial production facility. All geometry is known, but the position of the manipulator itself dependends on the position of the platform, and no manipulator-specific preprocessing is possible.

The paper is organized as follows: We first introduce the problem of motion planning and some important terms. We then give a brief overview of other approaches to motion planning.

CP437, *Computing Anticipatory Systems: CAYS--First International Conference*
edited by Daniel M. Dubois © 1998 The American Institute of Physics 1-56396-827-4/98/$15.00

Then, we develop our approach of complete trajectory modification based on the principle of shrinking and expanding geometry models. Its different aspects are analyzed in detail. This is followed by a description of our prototype planner. We first define a useful measure of shrinking for positions and path segments, then we describe how the planner can handle these values successfully. We give some experimental results, achieved with this planner in our simulation environment. In the end, we discuss our ongoing work, other applications of this principle and how it can serve as a general paradigm for motion design.

1.2 Motion Planning / Path Planning Terms

A manipulater is a chain of *n+1 links (0...n)*, connected with *n joints*. The first link (*base*) is fixed in the workspace. The last link carries the *tool* (e.g. gripper and load). The abstract term for this mechanical setup is *open kinematic chain*. The number of *geometric degrees of freedom* of a manipulator is the minimum number of parameters required to describe the position of its tool. A universal robot has 6 geometric degrees of freedom, as an arbitrary point with orientation in cartesian space requires 6 parameters (3 translations and 3 orientations). The number of *kinematic degrees of freedom* is the minimum number of parameters to describe the positions of all links in space. We consider independent joints only, so the number of kinematic degrees of freedom (*DOF*) equals the number of joints. Industrial manipulators usually have the minimum number of 6 joints. For more complex manipulation tasks, robots with more joints (called *redundant* or *hyperredundant* manipulators) are required, e.g. if operations are to be executed through a small hole.

There are two basic types of joints. *Rotational* (or *revolute*) *joints* turn the following link around a fixed axes. *Translational joints* move the following link along an axes. More complex joints can be modeled as a chain of basic type joints.

In the following, we denote the (kinematic) degrees of freedom DOF with *n*. The position of a joint *i*, *i=1...n*, is denoted with q_i. The possible values of a joint are limited in the interval $[q_{imin}, q_{imax}]$. The possible values within these intervals are assumed to be steady. The position of a robot is thus completely described with the vector $q=(q_i, ..., q_n)^T$. All possible robot positions span an *n*-dimensional bounded space, the *configuration space (c-space) C*. A point in c-space describes a posture of the robot in its work space. The work space of the robot contains obstacles. The set of all positions in *C* that result in a collision of any part of the robot with an obstacle are denoted with C_{coll}, all other positions are collision free and lie in C_{free}. For illustration see Fig. 1.

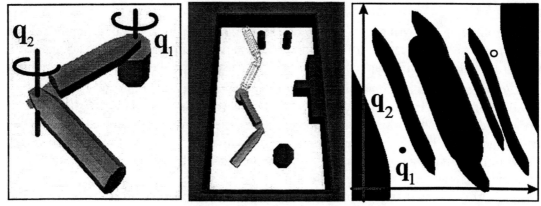

Fig. 1. An example for a two-joint robot [left]. The robot in a demonstration workspace with an example task (solid position: start, wired position: goal) [center]. A map of the respective c-space (black: C_{coll}, white: C_{free}) with the the same task (solid point: start, empty point: goal) [right].

We only consider the case of a moving robot in a static environment (static for one task at least), thus C_{coll} is constant. *Path planning* means to find a collision free path between a start position q_{start} and a goal position q_{goal}. A valid path is any steady curve in C_{free} connecting q_{start} and q_{goal}. *Motion planning* means to consider additional constraints for the shape of the path, arising from dynamic properties of the robot or other constraints set up by the user, e.g minimal path length or minimal energy consumption. C is the search space for planning, thus path planning and motion planning are (at least) *n*-dimensional planning problems.

1.3   Related Work

Path planning is often refered to as *piano movers problem* or *generalized movers problem* in the literature. Some important overview works were published by Latombe (1991) and by Hwang and Ahuja (1992). The complexity was analyzed by Reif (1979). A very rough classification is the following:

Explicit modelling of the c-space. To construct a complete representation of the search space appears promising. But all attempts to build up complete maps of the c-space fail for more than 4-5 degrees of freedom due to the exponential complexity. Even for 3 DOF simplyfied geometry models of the robot and its environment are required to handle such a map, thus this kind of approach is not feasible for manipulators in realistic environments.

Approximative complete modelling of the c-space. The axis of the c-space can be discretized to create a mesh of nodes. All nodes are evaluated by collision detection in the work space, path planning is transformed to graph searching. But again, the effort is of exponential complexity (in computational time and memory), and an upper bound for this approach are 5 DOF (see e.g. Ralli, 1996). The discretization has another negative effect on the planning, as it implies a lower limit for the resolution of the robot's movements.

Approximative incomplete modelling of the c-space. Again, the axes are discretized, but the nodes in the mesh are evaluated only if they are needed (in the graph search), and the mesh can be refined if necessary. This approach has a worst case exponential complexity, and its resolution is limited.

Random modelling of the c-space. Another nowadays popular approach is to build up a randomized graph within C_{free}, see e.g. Kavraki and Latombe (1994). A given number of subgoals are placed randomly and connected in a neighbourhood by simple local search strategies (e.g. check linear connection for collision). Thus the topology of C_{free} can hopefully be covered without exponential complexity. After this preprocessing, motion planning means just finding connections from start and goal to the graph, and graph search. But the success depends on the right choice for the number of subgoals, and the required preprocessing may take very long for complex environments (at least in the order of minutes or hours). The preprocessed c-space map gets invalid if the environment changes, and every object that is moved yields a modification.

Local Planning for positions in work space. The idea is to move from the start position towards the goal, and to pass obstacles along the way, only considering 'local' information. The complexity of this kind of approaches is only linear in the number of degrees of freedom. One possibility is the so-called potential field approach, see e.g. Latombe (1991). The goal implies an attractive force, the obstacles are repulsive. The robot (a point in c-space) moves in this force field, following the gradient. The major drawback of this method is to get stuck in local minima of the potential field, thus requiring random escape or random exploration techniques to avoid the complexity of global search. The *Randomized Path Planner*, developed by Barraquand and Latombe (1990), uses random escape motions of arbitrary length and direction until a position with a better potential value is found. Our Z^3 planner places subgoals randomly in the whole c-space to find a decomposition that can be planned locally with a heuristic search strategy, see Baginski (1996a). Local planning for positions is powerful for simple environments, but usually gets expensive when local minima are to be escaped.

Local Planning for whole trajectories in the workspace. Another local approach is to take a complete initial path (usually with collisions), and to modify it in a way to get it collision free. Therefore, a measure of *collision depth* or *badness* for a whole trajectory must be developed. There are only few publications in this area. Buckley (1989) and Ong/Gilbert (1994) use *minimal directed distances* and the *penetration growth distance* respectively, both rather expensive to calculate, and were not able to apply their schemes to high DOF robots in realistic environments. Improvements of these calculations (reported by Cameron, 1997) were not yet applied to path planning. Quinlan (1995) needs a collision free trajectory to be reshaped under some kind of artificial forces, thus the path planning problem needs to be solved *beforehand*. The approach described in the following is a continuation of (Baginski, 1996b).

2   Local Planning for whole Trajectories

2.1   Principle

Local planning for whole trajectories can be seen from two points of view. One possibility is to look at a curve in the c-space, the other possibility is to consider the volume taken by the robot´s motion in the workspace.

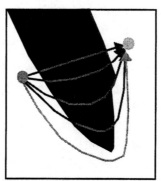

Fig. 2. Planning for a whole trajectory observed in the c-space.
The trajectory is gradually modified until it is free of collision.

Planning principle in the c-space. All possible trajectories (if we ignore the obstacles) between q_{start} and q_{goal} can be considred as an indefinite bundle of curves, occupying the complete c-space. Some of these trajectories pass through obstacles, some are collision free. As we cannot explicitly construct a representation of the complete c-space, we cannot construct *all* trajectories as well. But we can select any one trajectory and examine it, this is a local operation, as we just examine a onedimensional subspace of the c-space. We now modify the trajectory - again a local operation - to a neighboring trajectory in the bundle of all possible trajectories, in a way that the new trajectory is *better* than the old one. This requires a *measure of badness* for the trajectory. This evaluation function has to fulfill several conditions. It has to be unique and it should be strictly monotone (at least locally) to make any planning possible. It would be nice if it would be derivable, but this would limit the trajectory description to parameterized paths with a finite (and constant) number of parameters and seems hardly to be achieved. Thus we have to take some trial modifications of the path and take a better one if it is found (resulting in a numerical gradient descent of unlimited dimension). If there is no local minimum in the evaluation function, the trajectory is pushed out of

378

the c-space-obstacles into the free c-space. We start with an initial trajectory, and within the whole planning process, start and goal are actually connected. We just remove the *badness* by gradually reshaping the trajectory, until the whole trajectory is *good*, i.e. free of collisions. The success depends on the evaluation function. But looking at the c-space with its counter-intuitively shaped obstacles - even in the two dimensional case - no useful evaluation function comes to mind. This changes, if we look at the . . .

Planning principle in the work space. q_{start} and q_{goal} describe two postures of the robot in the work space. Any trajectory between these postures sweeps out a certain volume in the work space. If the trajectory is colliding, a part of the volume is occupied by the work space obstacles. Modifying the trajectory to make it *better* now gets a more obvious meaning. If the trajectory can be reshaped in a way that it intersects *less* with the workspace obstacles, it is most certainly closer to a feasible trajectory than before. So the planning process in the workspace is to *push* the swept volume of the whole trajectory *out* of the obstacles into the free space, see Fig. 3. The evaluation function thus needs to measure the *degree of intersection*, a purely virtual attribute.

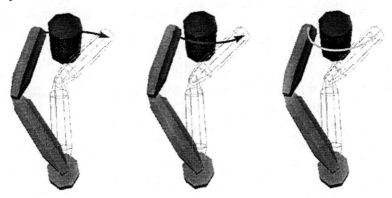

Fig. 3. Modification of the trajectory as it appears in the workspace (same example as in Fig. 2).

2.2 Objectives

The local planning with whole trajectories keeps control of the shape of the trajectory at all time. This allows us to define several objectives for the plannung process.

No collisions. This is, of course, the primary objective. If there are any collisions along the trajectory, the motion is physically obstructed and cannot be used by the robot. All other objectives are secondary.

Keep a safety distance. There are uncertainties between the geometry model of the environment that is used by the robot and the real world. Thus it is desirable to keep the robot away from the obstacles whenever possible. In most realistic tasks, start and goal positions are in contact with the environment (the manipulator transports a load from one position to another, or executes an operation at a certain object, e.g. welding) and cannot be planned with safety distances. But along the way, the robot can be kept clear, if the evaluation function is extended to the free space accordingly.

Find short trajectories. In general, short motions are preferred to longer ones, as they are less time and energy consuming. The planning process should prefer shorter trajectories by including the length in the evaluation function. The length can be measured in a *weighted* c-space, where the axes are streched proportional to the energy costs of the respective joint. As a result, the robot will avoid obstacles by moving *cheaper* joints first. In general, the outermost joints are cheaper to move, as they just carry the load and not (part of) the robot´s arm.

Find smooth trajectories. Any sharp bend along the c-space trajectory requires the robot to come to a complete stop and to start again moving in the new direction. Thus, the planning should return smooth curves that allow continuous motion of the robot. This objective might be in conflict with the previous one and has to be compromised according to the user´s demands.

Respect additional constraints. Additional constraints might be supervised within the planning, especially dynamic constraints, accelleration and speed limits etc. But they require careful consideration. If they are expensive to calculate, they might slow down the overall planning, and if it fails, all efforts were spent in vain. In this case, it might be better to optimize a solution path after the planning.

2.3   Trajectory Modelling

A trajectory consists of an infinite number of points. To be able to handle it within the planning process, we need to model it with a set of supporting points, connected by a certain interpolation function. This interpolation function can be of any kind if it is intersectable, see below. It is possible to use a robot-specific interpolation that might be part of the proprietary control unit. Another possibility is to use a spline function for smooth trajectories, or to ues just straight line segments, that can model arbitrary shaped trajectories, if the supporting points are set close enough.

One condition must be fulfilled by the interpolation function. To adapt to the a priori unknown topology of the free space, it must be possible to insert new supporting points between existing ones, while keeping the overall shape of the trajectory. If this would not be possible, the number of supporting points would have to be set up and kept fixed for the whole planning process, thus limiting the planner in an unacceptable manner.

2.4   Trajectory Evaluation

Evaluation for the primary objective. The work space point of view implies several possibilities to define a useful evaluation function. We suggest to *shrink* the robot if it collides with an obstacle. Shrinking means to reduce the size of the robot´s geometry model until it just fits and the motion becomes possible. This is physically impossible, but very easy to calculate for a geometric data set. The evaluation has to be unique. With the shrinking process, this is achieved if all smaller models are always completely included in all larger models. If this holds, then there is exactly one largest shrunk model that fits for a colliding trajectory. A possibility is just to *shorten* the robot, as a manipulator is usually a lengthy device. The degree of shrinking can serve as evaluation value.

The evaluation we suggest uses the worst position along the trajectory and assigns it to the whole trajectory. There are other possibilities, e.g. integrating some measure along the whole trajectory. But this purely local scheme turns out to be computationally cheap and is sufficient under regard of the concepts set up in section 2.1. If the worst position is improved, the whole trajectory is improved.

Evaluation for secondary objectives. The safety distances can be included in the evaluation by *blowing up* the geometry model beyond its original size. This expansion must assure the same distances all around the robot. Within the planning process, start and goal requires special consideration (they may be *in touch* with the obstacles as mentioned above).

Shortness and smoothness can be integrated in part into the modification process, see below. If the trajectory is in free space and the safety distance is achieved, shortness and smoothness can be evaluated by calculating the length and the derivative/angle of the sharpest

380

bend. Planning becomes pure optimization and gets anytime characteristics, as it can be stopped whenever desired.

2.5   Trajectory Modification

Initialization. Before the planning can start, an initial trajectory must be chosen. It can either be constructed with regards to user constraints and demands, or it can just be the shortest and smoothest trajectory in c-space, the straight line.

Test step generation. Planning means to find a slightly modified trajectory that allows a slightly larger model to be moved along, until finally the whole robot can pass. This stepwise modification is achieved by moving the supporting points. Two major issues arise when moving a supporting point: what is a proper direction and what is a proper distance? As we do not expect our evaluation function to be derivable with respect to the supporting points, heuristic schemes must be employed. They should create a limited number, dependent on the number of degrees of freedom, of possible positions, these positions are evaluated and the best one is chosen. The direction has just one limitation: The supporting point must not be moved *along* the trajectory in a way that this movement does not modify its shape.

Insertion of supporting points. If no supporting point can be moved to improve the overall evaluation, new supporting points have to be inserted to allow a proper adaption of the trajectory´s shape. There must be a heuristic to guide this insertion, a promising idea is to take either the center point or the point of maximum shrinkage. This heuristic will start the planning process, if the straight line was chosen as initial trajectory and collisions were detected.

Dead end. There must be a limit for the insertion of new supporting points to assure termination of the planning. It is sufficient to limit the minimum length between two supporting points to a reasonable short value. This value sets the minimum 'granularity' of the planning and can be set according to the motion capabilities of the robot.

Removing supporting points. Due to the heuristic limitations on step size and direction, parts of the trajectory may be modeled with more supporting points then necessary at a later stage of the planning process. For this reason, and to get shorter and smoother trajectories, it can be tried to improve the evaluation (or keep it constant, at least) by removing selected supporting points.

2.6   Complexity

The complexity of this planning process is linear in the number of degrees of freedom, as only onedimensional subspaces are examined and the modification generates a number of teststeps that linearily depends on the number of degrees of freedom. The overall computational effort of the planning depends on the problem size: as longer the initial/final trajectories are, and as more steps are required to transform the one into the other, the longer the planning takes. But as it assures an improvement in each taken step, the planning is deterministic and will terminate in finite time, for good or for bad (see below).

2.7   Completeness

The planer can get stuck in local minima, as it is working purely locally. If the robot has only few degrees of freedom, obstacles can be constructed, where this problem is obvious (see Fig. 4 for illustration). If a 'global' detour is required and the 'local push' is not sufficient, this principle can not find a path. The situation becomes less obvious (and less often) for robots with more degrees of freedom. If there are some joints 'behind' the link in collision, in most cases a passage can be found. This is a surprising aspect of this planning: it benefits from more degrees of freedom. In contrast to other planners, that can hardly work with more and more joints, our scheme finds more and more solutions in these cases, with just linear complexity.

But none the less, it may terminate in a local minimum without further possible imrovement. To overcome these minima, we embedded the planner into a random subgoal planner (see also Baginski, 1996a). If q_{start} and q_{goal} cannot be connected with local planning, a random subgoal q_{sub} is placed in the configuration space, it is tried to connect q_{start} with q_{sub}, and, on sucess, q_{sub} with q_{goal}. On failure, further subgoals are tested. The maximum number is a user parameter to assure termination. If the planner fails finally, it is assumed that it is *highly improbable* that a trajectory exists. Local planning and random exploration are incomplete by construction, no definite decission can be drawn from failure. But our experiments show that this planning is very reliable and failure is extremely seldom.

Fig. 4. A task that cannot be planned with local trajectory modification

3 Prototype Implementation

Based on the ideas described above, we implemented a prototype planner that is able to do local planning for whole trajectories. There are some limitations, though, but the results achieved are very promising. We just included one secondary objective: find a path with as few supporting points as possible.

3.1 Shrink Measure

Evaluating positions. A robot manipulator is modeled as a chain of bodies (the links), connected with joints. To assign a unique shrink measure to a position, we check the bodies of the chain individually for collision with the environment, beginning at the base of the robot. If all links are free of collision, the shrink measure of this position is set to 1. If a link i collides (and the algorithm finds the first colliding link in the kinematic chain), its size is reduced with respect to the origin of its local coordinate system, located on the joint axis of the link and on the surface of the previous link. Thus, the shrink center is the fixed touching point in the case of a rotational joint, and lies within the current link in the case of a translational joint. A precondition for this shrinking process to yield a unique result is that the link is *convex* (or at least *star-shaped* with respect to its local origin, the shrink center).

The shrink factor s_i of link i is defined as the largest value between 0 and 1 that allows the link to remain free of collisions with the environment. It is calculated approximately with a depth-limited bisection, checking for collision with $s_i=0.5$ at the beginning (we know that the link collides for $s_i=1.0$ and that it does not collide for $s_i=0.0$, as the previous link is free of collision). If it collides, we check again for $s_i=0.25$, else we check $s_i=0.75$ and so on. We limit the number of tests to 8, yielding a precision of $1/256$.

To calculate the shrink measure $s(q)$ for the whole robot model in a certain posture q, each link of the robot is assigned a partial interval between 0 and 1, these parts are ordered in the order of the kinematic chain and together they cover the interval between 0 and 1, without gap and without overlapping. Now the local shrink factor s_i of the shrunk link is projected in its partial interval. Only the first colliding link of a robot is taken into account for the shrink measure calculation. All outer links are ignored, they can be interpreted as shrunk to zero size. The resulting measure is unique and in the interval *[0,1]*. The calculation of the shrink measure is illustrated in Fig. 5.

382

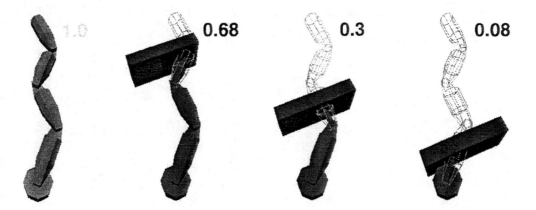

Fig. 5. Illustration of the shrink measure culculation. The size of the first colliding body in the kinematic chain is reduced until it does not collide with the obstacle. The overall shrink measure is calculated by mapping the link´s shrink factor into the interval *[0,1]* for the complete manipulator.

Evaluating trajectories. According to our concept - to modify complete trajectories - we need a shrink measure for complete trajectories and not only for positions. As discribed above, we simply define the minimum shrink measure along a path as the shrink measure of the whole path. For the calculation, the path is discretized into a sequence of single positions. This allows a very cheap computation of the shrink measure, as the collision detection and the bisection can be executed for all discretization points simultaneously. If a link collides for several of the discretization points, all of the collision free discretization points need no further consideration, as we seek just the smallest value. This exclusion continues within the bisection, thus reducing the number of required collision detections drastically.

3.2 Planning Algorithm

Trajectory modelling. To simplify the handling of a trajactory, we model it as a sequence of connected linear segments. The supporting points are the connection points of the linear segments.

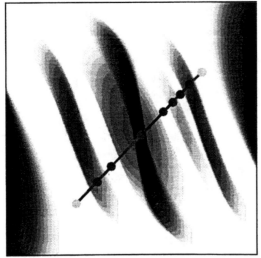

Fig. 6. Initialization of planning for the task shown in Fig. 1. The different brightness of the c-space obstacles reflect the local shrink measure. As darker a region is, as smaller the robot has to be to not collide. The planner evaluates only the straight line between start and goal and has no idea of the surrounding space. Initial supporting points are placed at points of *deepest intersection* and in the center of collision free intervals.

Initialization. Given a start and a goal position, the linear connection is discretized and examined by calculating the shrink value for each discretization point. This is just done once, resulting in a 'shrink profile' of the linear connection between q_{start} and q_{goal}.

If no collision occurs, a solution is found. If there are collisions, initial supporting points are placed along this connection. Nodes are placed at the positions of minimal shrink measure within each collision interval, and in the center of collision free intervals, if there was more than one collision interval detected. All resulting path segments are attributed with their shrink measure, i.e. the minimum value of all their discretization points. The initialization for the sample task shown in Fig. 1 is shown in Fig. 6.

Iterative modification. The main loop of the algorithm tries to move the supporting points adjacent to the segment(s) with the minimal shrink value.

To move the supporting point *away* from the trajectory, we calculate a hyperplane in the *weighted* c-space, orthogonal to the linear connection between the two adjacent nodes. The joints are weighted with their maximum radius that they have if the manipulator is fully stretched. The hyperplane is calculated in the i lower dimensions of the c-space only, as a movement of the upper joints will not improve the placement of the colliding joint. The restriction to orthogonal directions in an i-dimensional subspace is a limitation for the spacial adaption of the trajectory, but this still leaves i-1 base vectors and their respective reverse vectors as test directions. This gives a total number of $2(i$-$1)$ test directions.

The step length is calculated as a quarter of the linear distance between the two adjacent nodes, a heuristic that creates steps in relation to the 'granularity' of the trajectory that is modeled through the supporting points.

The test positions are evaluated by recalculating the shrink measure for the two adjacent linear segments, and the best possible position is taken as the new position for the supporting point. We demand that the shrink measure of at least one of the adjacent line segments is increased. If no point can be found, the step size is cut by half, this may even be repeated. The maximum number of tests executed to improve one supporting point is $6(i$-$1)$.

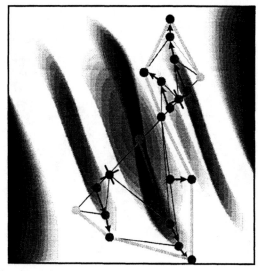

Fig. 7. Simplyfied visualization of planning. The algorithm moves the supporting points out of the obstacles, inserts new ones if necessary and removes nodes if possible.

Insertion of additional supporting points. If, within one iteration, the linear segment with the minimum shrink value cannot be improved (neither of its adjacent supporting points were moved), a new supporting point is inserted in the center to allow a better adaption to the topology of the obstacles. If the distance between two supporting points falls below the discretization distance, the planner terminates with failure and reports a local maximum.

On-line optimization. To achieve the secondary objective, trajectories with few bends, we included an on-line optimization. At the end of each iteration the direct connection between its two adjacent supporting points is evaluated for all unmoved supporting points. If its shrink measure is above the shrink measure of the two segments in between, the node is removed.

The main loop of modification, insertion and optimization is iteratively repeated. The planner terminates with success, if all linear segments of the trajectory have a shrink measure of 1, else it terminates with failure. The planning process for the sample task is visualized in Fig. 7 and Fig. 8.

Fig. 8. The stages of planning shown in Fig. 7 visualized in the robot´s workspace. It can be seen how the trajectory is *pushed* into the free space.

4   Some Experimental Results

We have used the planner in scenarios from 2 to 16 DOF, and in general it is very successful for all kinds of different tasks. The planning time is mostly dependent on the complexity of the geometry models of the robot and the environment. All randomly created tasks in the two dimensional example of the previous chapters are solved without failure and within 1 to 5 sec. If the models are quite simple, we can plan extremely complex motions. An example for an 8 DOF robot is shown in Fig. 9, a 16 DOF robot is shown in Fig. 10. All calculation times refer to simulations on a standard HP Unix workstation.

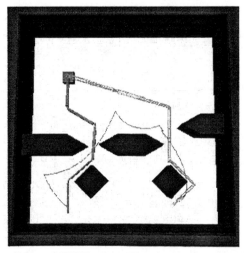

Fig. 9. This solution for an 8 DOF robot (5 rotational and 3 translational joints) is planned in less than 15 seconds. The resulting trajectory consiststs of just five linear segments connecting start and goal via four supporting points.

The examples show that the planner succesfully fulfills the objectives. The trajectories consist out of very few linear pieces (in the c-space). Even very small passages are passed with long straight motions, a capability that no other local planner has.

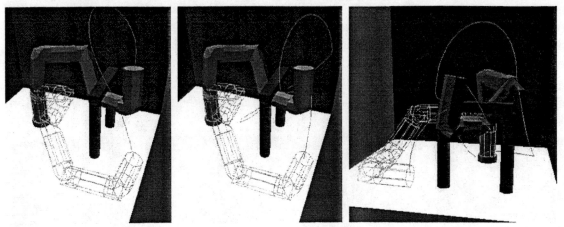

Fig. 10. A very complex task for a 16 DOF robot (all joints rotational), a amall gate has to be escaped. The initial straight trajectory in c-space [left] is transformed into a solution [center and right] in about 9 seconds, consisting of just 7 linear segments.

The success rate of our planner is very high, compared to other local planning algorithms. For robots with a higher number of DOF, especially for hyperredundant robots, the situations of failure are very rare and the critical obstacle configurations are not obvious. In most cases, the failure seems not to be a consequence of the principle itself, but of the heuristics used for the step size and the step directions.

5 Future Work

We want to increase the capabilities of our implemented planning system in several ways by stepwise inclusion of the concepts described in the second chapter.

The planning shall be continued in free space in an analogous fashion, to optimize the trajectory and to ensure safety distances. For this purpose, the robot model has to be expanded beyond ist real size. The problem that arises is to 'summarize' possibly different expansions for the different links into one scalar evaluation value.

We are in principle not bound to trajectories that are made up of straight line segments. The planner is able to work with all kind of shapes that can be modeled with a set of supporting points, e.g. splines. This is going to be integrated to create trajectories that can be executed continously by a real robot in our laboratory.

As the main effort of the planning arises from the collision detection, we want to employ algorithms that simplify the geometry models hierarchically. This would enable us to plan the parts of motion that are close to obstacles with a higher 'resolution', and if we are in free space or in deep collision, we can use much coarser models.

Our planning principle has proven to be very effective in the case of manipulators. But it is applicable in other planning domains as well. We are currently adapting our algorithm to digital mock-up, where a 'free-flying' object must be placed within an assembly. If there is an initial trajectory known that is free of collisions for just a single point, the planner can start with this small 'seed' along the trajectory and gradually increase the size of the object.

To turn our approach into a real anticipatory system as defined by Rosen (1985), time has to be included as a parameter and the gap between model and real world has to be closed. In our current model, the environment is static and the time needed for the motion is not considered. But if there are independently moving obstacles or other robots in the workspace, time has to be modeled to allow planning. This is considered as a future direction of our research.

6   Conclusion

The use of a virtual value, the shrink measure, is the new quality of our concept, because it is cheap to calculate and gives us a matter to manipulate complete trajectories, that are, in the real work space, just colliding. For two fixed, collision free postures of the robot, the planning system incrementally decreases the degree of collision in between.

From a more abstract point of view, the *uncertainty* and *badness* between two *fixed* points is brought under control by evaluating it, to make it comparable to neighbouring paths between the fixed points, thus allowing to finally find a *good* solution. The whole evaluation takes place locally, the system draws all its information from the evaluated paths, it never tries to examine the whole high dimensional space. As an abstract scheme, this may serve as an inspiration for totally different planning domains.

References

Baginski, Boris (1996a). The Z^3-Method for Fast Path Planning in Dynamic Environments. In: Proceedings of IASTED Conference on Applications of Control and Robotics. Orlando, January 1996, pp. 47-52.

Baginski, Boris (1996b). Local Motion Planning for Manipulators Based on Shrinking and Growing Geometry Models. In: Proceedings of the IEEE International Conference on Robotics and Automation, Minneapolis, April 1996, pp. 3303-3308.

Barraquand, J. and Latombe, Jean-Claude (1990). A Monte-Carlo Algorthm for Path Planning with Many Degrees of Freedom. In: Proceedings of the IEEE International Conference on Robotics and Automation, Cincinnati, May 1990, pp. 1712-1717.

Buckley, C. E. (1989). A Foundation for the „Flexible-Trajectory" Approach to Numeric Path Planning. In: The International Journal of Robotics Research, Vol. 8, No. 3, June 1989, pp. 44-64.

Cameron, Stephen (1997). Enhancing GJK: Computing Minimum and Penetration Distances between Convex Polyhedra. In: Proceedings of the IEEE International Conference on Robotics and Automation, Albuquerque, April 1997, pp. 3112-3117.

Hwang, Yong K. and Ahuja, Narendra (1992) Gross Motion Planning - A Suryey. In: ACM Computing Surveys, Volume 4, Number 23, September 1992, pp. 219-291.

Kavraki, Lydia and Latombe, Jean-Claude (1994). Randomized Preprocessing of Configuration Space for Fast Motion Planning. In: Proceedings of the IEEE International Conference on Robotics and Automation, San Diego, May 1994, pp. 2138-2145.

Latombe, Jean-Claude (1991). Robot Motion Planning. Kluver Academic Publishers.

Ong, Chong Jin and Gilbert, Elmar G. (1994). Robot Path Planning with Penetration Growth Distance. In: Proceedings of the IEEE International Conference on Robotics and Automation, San Diego, May 1994, pp. 2146-2152.

Quinlan, Sean (1994). Real-Time Modification of Collision-Free Paths. Ph.D. Thesis, Stanford University, December 1994.

Reif, John H. (1979). Complexity of the Mover´s Problem and Generalizations. In: Proceedings of the 20th IEEE Symposium on Foundation of Computer Science, pp. 421-427.

Ralli, E. and Hirzinger, G. (1996). A Global and Resolution Complete Path Planner for up to 6 DOF Robot Manipulators. In: Proceedings of the IEEE International Conference on Robotics and Automation, Minneapolis, April 1996, pp. 3295-3302.

Rosen, Robert (1985). Anticipatory Systems. Pergamon Press.

Workspace analysis and design of open-chain manipulators

Marco Ceccarelli

Department of Industrial Engineering
University of Cassino
Via Di Biasio 43 Cassino (FR), Italy
fax:+39-776-310812; e-mail: cecca@ing.unicas.it

Abstract - Workspace is a primary characteristic of manipulators both for use and design. A characterisation of workspace can be formulated by means of several approaches. However a numerical efficient method can be proposed in term of an algebraic description of the workspace geometry. This formulation has been useful to deduce an interesting algorithm for workspace analysis by using an evaluation of the workspace boundary. This formulation can be also inverted for synthesis purposes and an example is presented in the paper. Mayor capabilities of the formulation have been used to formulate a modern design approach by means of an optimization problem. The analytical characteristics both of the formulation and the workspace can be used to propose an efficient procedure for an optimum design of manipulators with prescribed workspace. This paper has been written with the aim to give a unified treatment of the workspace formulation, evaluation and design by using and algebraic approach.

Keywords: Robot Kinematics, Computational Geometry, Manipulators, Workspace.

1. Introduction

Manipulator arms are the structural chains which are devoted to the aim to let the end effector reach positions in the space. The orientation capabilities are usually due to wrist devices.

Manipulator workspace is defined as the region of reachable points by a reference point H on the extremity of a manipulator chain and it is usually referenced as a basic characteristic for a successful use and design of robot manipulators. Therefore manipulator arm capabilities can be characterised by workspace characteristics mainly in term of extreme reaches.

The manipulator workspace has been recognised of fundamental importance in robot design as well as in automation applications since the first studies on robot kinematics, (Vertut et al. 1974; Roth 1975), and even from commercial viewpoint, (Nof 1985), as for example in (COMAU 1994; Staubli 1996). Since then numerous investigations have approached the problem of workspace analysis with evaluation and classification purposes more than with the design aims.

The complexity of the workspace analysis problem has been emphasised since the first studies on the topics, (Vertut et al. 1974; Kumar and Waldron 1981; Konstantinov and Genova 1981; Tsai and Soni 1981; Yang and Lee 1983), and even through developments regarding the smallest robot chains as the two-link arms, (Fichter and Hunt 1975; Tsai and Soni 1984).

Different approaches for workspace analysis have been proposed through specific algorithms: by considering the extreme reach configurations in expressing the peculiar geometry, (Derby 1981; Shimano and Roth 1976; Sugimoto and Duffy 1981 a and b), or the mechanical stability, (Kumar and Waldron 1981); by using a matrix formulation in scanning recursively

CP437, *Computing Anticipatory Systems: CAYS--First International Conference*
edited by Daniel M. Dubois © 1998 The American Institute of Physics 1-56396-827-4/98/$15.00

the joint angles domain, (Kumar and Patel 1986; Lee and Yang 1983), or the Cartesian space through inverse kinematics algorithms, (Hansen et al. 1983), and through optimization techniques, (Cwiakala and Lee 1985; Jo and Haug 1989; Lin and Freudenstein 1986; Rastegar and Fardanesh 1990; Rastegar and Perel 1990; Tsai and Soni 1983; Yang and Lee 1984; Haug et al. 1996); by formulating the Jacobian singularities, (Hsu and Kohli 1987; Kohli and Hsu 1987; Liegeois et al. 1986; Oblak and Kohli 1988; Rastegar and Deravi 1987), or the polynomial discriminants of the chains, (Kohli and Spanos 11985; Spanos and Kohli 1985); by considering the geometric process of workspace generation in algebraic formulations, (Ceccarelli 1989, 1996; Fichter and Hunt 1975; Freudenstein and Primrose 1984), giving explicitly the possibility of a synthesis procedure, (Roth 1986; Ceccarelli and Vinciguerra 1990; Ceccarelli 1995a).

In particular, the algebraic geometry has been recognised a powerful means to develop an algebraic treatment of robot workspace which is useful in analysis algorithms and in synthesis criteria, (Freudenstein and Primrose 1984).

The importance of workspace knowledge has been stressed also for design considerations, (Beni and Hackwood 1985; Gupta and Roth 1982), so that it is interesting to develop workspace formulations with synthesis purposes. Workspace characteristics, such as the cross-section shape and the hole and voids geometry, have been studied and discussed for particular robot chains, (Ceccarelli 1989; Fichter and Hunt 1975; Freudenstein and Primrose 1984; Lee and Ting 1991; Rastegar 1988; Tsai and Soni 1984), as well as for general structures, (Ceccarelli 1996; Gupta and Roth 1982; Gupta 1986; When and Jin 1987; Yang and Lee 1983), and even for biomanipulation understanding, (Lenarcic et al. 1990).

However, the design problem by using workspace formulation has been approached only recently and few works have been developed to propose design algorithms. In particular, two main approaches can be recognised: by inverting or manipulating a workspace formulation (Gupta and Roth 1982; Freudenstein and Primrose 1984; Tsai and Soni 1985; Roth 1986; Ceccarelli 1995a), or by formulating a design optimization problem using or including workspace formulation and/or evaluation (Yang and Lee 1984; Gosselin and Guillot 1991; Ceccarelli et al. 1994; Ceccarelli 1995b).

In this paper an attempt is reported to present a unified treatment of a manipulators workspace formulation for both analysis and synthesis purposes. In particular, the problem of a workspace description has been illustrated as a means to characterise manipulators basic capabilities which can be prescribed for design purposes.

2. The problem

A N-revolute manipulator can be sketched as an open chain with N revolute joints. A frame $X_i Y_i Z_i$ can be attached to each link i of the chain in a way that axis Z_i coincides with the joint axis, X_i with the common normal between two consecutive joints axes and Y_i is consequently determined to perform a Cartesian frame. A base frame can be considered to be coincident with $X_1 Y_1 Z_1$ at an initial manipulator configuration. The geometrical parameters of a general N-R manipulator are: the link lengths $a_i \neq 0$ (i=1,...,N), the link offsets d_i (i = 2,...,N), and the twist angles $\alpha_i \neq k \pi/2$ (i=1,...,N-1), (k=0, ...,4). d_1 may not be considered since it shifts the workspace up and down only and therefore it does not affect a workspace evaluation. The joint angles θ_i (i=1,...,N) are defined as the angles between two consecutive X_i axes.

Workspace design data can be prescribed by means of workspace limits in term of radial reach r and axial reach z so that the design problem can be formulated as to find the dimensions of a

389

manipulator arm whose workspace cross-section is within or is delimited by the given axial and radial reaches r_{min}, r_{max}, z_{min}, z_{max}.

An optimum design of manipulator kinematic chains should include the optimum location of the robot base with respect to a fixed frame. Therefore, the design parameters can be considered the link time-independent sizes of the chain, i.e. for N revolute-connected manipulators a_i, d_i, $i = 1,...,N$, and α_i $i = 1,..., N-1$, and the vectors $\underline{b}$ and $\underline{k}$, which represent respectively the position and the orientation of the robot base with respect to a fixed frame XYZ, Fig.1.

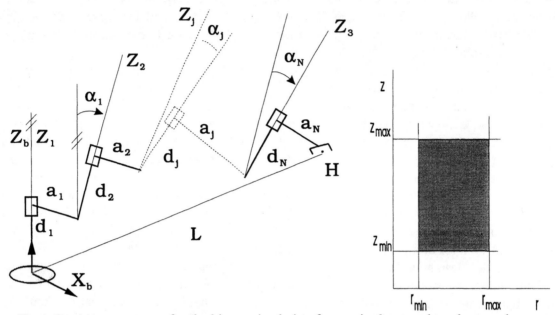

Fig.1. Design parameters for the kinematic chain of a manipulator and workspace data

Since the main task of a manipulator is recognised in moving the end-effector or a grasped object in the space, it is suitable to achieve an optimum design and location of a robot taking into account basically workspace characteristics.

Thus a workspace formulation is of fundamental importance not only for workspace analysis and evaluation but even more to deduce rational design criteria and algorithms for manipulators.

The workspace is generated when a reference point H on the extremity of the manipulator chain is moved to reach all the possible positions because of the mobility ranges of the joints. Basically, workspace determination is a typical problem of direct robot kinematics but the numerical efficiency and the feasibility for design algorithms may require particular attention and advanced modelling and formulation.

3. Workspace characteristics

Workspace geometry of manipulators with revolute joints has been recognised as a ring topology, (Gupta and Roth 1982), and this can be conveniently used to describe analytically the workspace yet.

The workspace is generated when a reference point H on the extremity of the chain is moved to reach all the possible positions because of the movements of the joints.

However, from a descriptive geometry viewpoint the workspace of a telescopic arm with also prismatic joints can be characterised by looking at the figures which are generated by H with successive full movements of the joints starting from the last up to first which is fixed to the manipulator base. Thus, movements of the sliding joint will generate a straight line segment. Then the second revolute pair of the chain let perform a full rotation of the straight line segment which generates a cylindrical cone. Indeed, depending on the orientation of the straight-line segment with respect to the revolute joint axis we may have a cylinder, a cone or generally an hyperboloid. Finally, a full revolution of the first revolute joint will generate, by revolving the hyperboloid, a solid of revolution which can be generally depicted as a "Cylindroid Ring", (Ceccarelli 1996 b), which contains a ring topology yet.

Thus, the ring geometry can be used with the following modelling and reasoning referring to Fig.2 to deduce an algebraic formulation for the manipulators workspace.

A ring is generated by revolving a torus about an axis. Therefore, a ring volume $W_{3R}(H)$ can be thought as the union of the points swept by the revolving torus $T_{R_2R_3}(H)$, due to the mobility in R_2 and R_3 joints, during the θ_1 revolution about Z_1 axis

$$W_{3R}(H) = \bigcup_{\vartheta_1=0}^{2\pi} T_{R_2R_3}(H) \qquad . \qquad (1)$$

Alternatively, a ring $W_{3R}(H)$ can be considered as the union of the tori $T_{R_1R_2}(H)$, which are due to the mobility in R_1 and R_2 joints and are traced by all parallel circles which can be cut on the generating torus $T_{R_2R_3}(H)$ so that

$$W_{3R}(H) = \bigcup_{\vartheta_3=0}^{2\pi} T_{R_1R_2}(H) \qquad . \qquad (2)$$

Thus, the boundary $\partial W_{3R}(H)$ of a ring can be thought as the envelope of torus surfaces generated by revolution of the generating torus or, alternatively, it can be obtained by an envelope of torus surfaces traced from the parallel circles of the generating torus. The latter procedure can be expressed according to Eq.(2) as

$$\partial W_{3R}(H) = \operatorname*{env}_{\vartheta_3=0}^{2\pi} T_{R_1R_2}(H) \qquad (3)$$

where "env" is an envelope operator performing an envelope process.

A ring geometry has been recognised topologically common also to the hyper-rings as a "solid hollow ring" shape on the basis of a consideration of an iterative revolving process for the hyper-ring generation in N-R manipulators. Infact, a (N-j+1)R hyper-ring $W_{(N-j+1)R}(H)$ is generated by revolving a (N-j)R hyper-ring $W_{(N-j)R}(H)$ about Z_j axis. The (N-j+1)R hyper-ring, which is traced by a point H on the last link of an open chain with N-j+1 consecutive revolute pairs, is the volume swept by the (N-j)R hyper-ring during its revolution, Fig.2.

A basic element is a generating parallel circle, which is cut on the revolving $W_{(N-j)R}(H)$ because of a revolution about Z_{J+1} axis and whose a revolution about Z_J generates a torus $T_{R_jR_{j+1}}(H)$. The envelope of the tori of the family will give the boundary $\partial W_{(N-j+1)R}(H)$ of the hyper-ring. The chain dimensions directly involved in the process are the link sizes a_J and α_J and the axial and radial reaches r_J and z_J , whose value can be determined for each parallel circle individuated by a point of the boundary $\partial W_{(N-j)R}(H)$ of the revolving $W_{(N-j)R}(H)$.

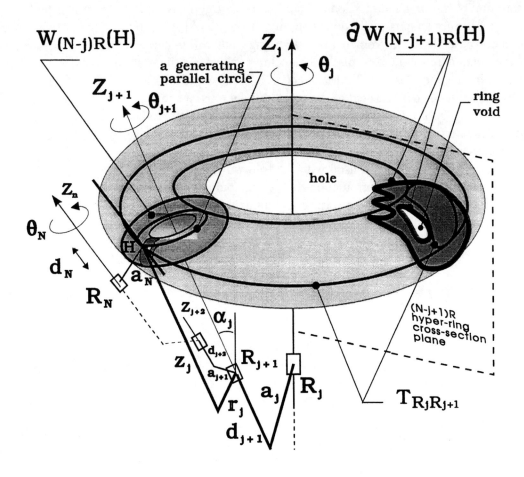

Fig.2 A descriptive view of workspace generation.

The generation process of a hyper-ring is a consecutive revolving process of a circle, a torus, a ring, a 4R hyper-ring, and so on. This can be expressed through a revolution operator Rev in the form

$$W_{(N-j+1)R}(H) = \mathop{\text{Re v}}_{\vartheta_j=0}^{2\pi} W_{(N-j)R}(H) \qquad j = 1,...,N-1 \qquad (4)$$

where $W_{(N-j+1)R}(H)$ represents reachable points, with respect to frame $X_j\,Y_j\,Z_j$ fixed on link j, due to (N-j+1) last revolute joints in the chain; and $W_{(N-j)R}(H)$ is with respect to $X_{j+1}\,Y_{j+1}\,Z_{j+1}$ on link j+1 due to N-j revolute pairs.

Alternatively, $W_{(N-j+1)R}(H)$ can be considered as the union of a suitable torus family which is traced by the boundary points in the revolving torus, ring, 4R hyper-ring and so on, when they are rotated completely about the first two revolute axes in the corresponding generating sub-chain,

$$W_{(N+1-j)R}(H) = \mathop{\text{U}}_{\vartheta_j,\vartheta_{j+1}=0}^{2\pi} T_{R_j R_{j+1}}\left[\partial W_{(N-j)R}(H)\right] \qquad (5)$$

392

where $T_{R_jR_{j+1}}(H)$, represents a torus generated by revolutions θ_j and θ_{j+1} about the joints axes of R_j and R_{j+1}, Fig.2. The revolution in θ_{j+1} generates a parallel circle in $W_{(N-j)R}(H)$ and together with θ_j generates the torus $T_{R_jR_{j+1}}(H)$.

Hence, the boundary $\partial W_{(N-j+1)R}(H)$ of a $(N-j+1)R$ hyper-ring can be described as an envelope of the torus family traced by all the points on $\partial W_{(N-j)R}(H)$, Fig.2, and it can be synthetically expressed as

$$\partial W_{(N-j+1)R}(H) = \mathop{\text{env}}_{\vartheta_j, \vartheta_{j+1}=0}^{2\pi} T_{R_jR_{j+1}}\left[\partial W_{(N-j)R}(H)\right] \quad . \tag{6}$$

Thus, a workspace boundary $\partial W_{NR}(H)$ of a general N-R manipulator can be generated by using recursively Eq.(6), to determine the tori envelopes from the ring up to the NR hyper-ring in the chain from the extremity to the base of the manipulator chain and it has been presented in (Ceccarelli 1996a).

The term hyper-ring has been used to stress revolution operations about more than three axes, which is the case usually referred to the ring geometry.

A "Cylindroid Ring" shows a hollow bulk shape and its cross-section is characterised by straight lines with possible cusps and two circular contours on the top and the bottom, (Ceccarelli 1996b).

The procedure of Eqs.(5) and (6) can be conveniently adjusted for manipulators with prismatic joints too as Fig.2 synthetically illustrates and first results for telescopic arms have been reported in (Ceccarelli 1996b).

A straightforward result of determining a workspace boundary is the individuation of the workspace shape, the existence and extension of hole and voids, which are recognised as basic characteristics for manipulators.

A hole is a region, outside the ring but surrounded by the ring yet, within which it is possible to individuate at least a straight line of points not belonging to the ring. Therefore, a hole is generated when the revolving torus does not intersect neither touches the revolution axis.

A void can be generally identified as an internal region, within the workspace itself, which is not reachable by point H.

Particularly, two branches of envelope contours in a cross-section of a workspace boundary are observable: an external one and an internal ones. The external branch of the torus family envelope determines the external boundary of a bulk hyper-ring and it can be used for the generation of subsequent hyper-rings.

The internal branches of the torus family envelope show several separated and intricate contours in the cross-section. Nevertheless they may indicated one ring void only. Two types of voids have been individuated: a ring void and apple voids, (Ceccarelli 1989 and 1996). A ring void is an internal region not belonging to the ring or hyper-ring and, topologically, it is ring shaped. In fact, a ring void is generated in the revolving workspace by the hole, which is delimited by the smallest parallel circle cut in the generating ring or hyper-ring. Therefore, when this is rotated about an axis, this parallel circle rotates as well so that it generates once more a torus which is the boundary surface of a ring void.

An apple void is a region of points not belonging to the workspace and containing a segment of the axis of revolution. It can be thought to be generated, when no hole exists, by the volume inside a hole on the revolving workspace when the axis of revolution intersects the revolving workspace itself. In a ring it has been detected that more than one and at the most three apple voids may occur, (Ceccarelli 1989).

The common geometrical characteristics in ring and hyper-rings can be recognised in the recursive use of the ring equations. Nevertheless, each additional use introduces a multiplying effect in a sense that, for example, apple voids may occur in a number greater than the three ones forecast for a ring geometry. It seems, indeed, that each revolving operation may double the number of the envelope contours and consequently the number of apple voids may be one more of the double in revolving workspace. Thus, a 4R manipulator may have at the most seven apple voids, a 5R chain my show fifteen apple voids and a 6R hyper-ring may present thirty-one apple voids, and so on, (Ceccarelli 1996).

A general shape of hyper-ring with hole may show in its cross section many inlets looking at the axis of symmetry as many possible apple voids it may have. Indeed, boundary inlets may generate apple voids and vice versa when the hyper-ring is changed to have or not an hole. However, the peculiar shape with boundary inlets occurs when the hyper-ring boundary is near to the first revolution axis and it rapidly goes to a more regular circular shape with the increase of the minimum radial reach, i.e. a larger hole.

On the contrary, the ring void boundary seems to behave invariable geometrical characteristics. This can be explained by taking into account that the generation process of a ring void is produced by the hole in the revolving hyper-ring and it's always due to a revolution of the circle giving the size of the generating hole.

4. A numerical procedure for workspace analysis

It is recognised the importance of a characterisation for the workspace points and thus primary and secondary workspaces have been defined since the first studies on workspace, (Kumar and Waldron, 1981), depending on the reachability and multiple reachability of points, respectively.

The workspace can be described through a binary mapping of the cross section, (Lee and Yang, 1983), according to a numerical algorithm whose basic steps are, (Ceccarelli and Gabriele 1995):

1. Dividing the cross-section plane r, z into I x J small rectangles of width Δi and height Δj, where I and J are the number of divisions along the r axis and the z axis, respectively. Each rectangle is individuate by $f(i, j)$ to provide a binary image of the workspace cross-section. The values of the width Δi and the height Δj can be properly selected as a function of the $\Delta\theta_k$ scanning intervals.

2. Initialisation by setting $f(i, j)=0$ for all i and j.

3. A scanning process for each joint angle θ_k from θ_{kmin} up to θ_{kmax} with step $\Delta\theta_k$ to compute the workspace point coordinates by using a matrix approach in the form

$$x_k = r_k \cdot \cos\theta_k \qquad y_k = r_k \cdot \operatorname{sen}\theta_k \qquad z_k = z^*_k \qquad (7)$$

where $r_k = \sqrt{x^{*2}_k + y^{*2}_k}$ and

$$\begin{bmatrix} x \\ y \\ z \\ 1 \end{bmatrix}^*_k = [T]_{k+1} \cdot \begin{bmatrix} x \\ y \\ z \\ 1 \end{bmatrix}_{k+1} \qquad (8)$$

The $[T]_{k+1}$ transformation matrix between frame k and frame k+1 is given as

$$[T]_{k+1} = \begin{bmatrix} \cos\beta_k & -\mathrm{sen}\,\beta_k & 0 & a_k \\ \mathrm{sen}\,\beta_k \cos\alpha_k & \cos\beta_k \cos\alpha_k & \mathrm{sen}\,\alpha_k & b_k \cdot \mathrm{sen}\,\alpha_k \\ -\mathrm{sen}\,\beta_k \,\mathrm{sen}\,\alpha_k & -\cos\beta_k \,\mathrm{sen}\,\alpha_k & \cos\alpha_k & b_k \cdot \cos\alpha_k \\ 0 & 0 & 0 & 1 \end{bmatrix} \tag{9}$$

where a_k, α_k and b_k are the link length, the twist angle and the link offset, respectively; θ_k takes into account the limits of joints mobility.

4. Construction of the binary map f(i, j) = 1 of the workspace cross-section by determining i and j as

$$i = \mathrm{fix}\left[r_k / \Delta i \right] \quad , \qquad j = \mathrm{fix}\left[z_k / \Delta j \right] \tag{10}$$

using the operator fix to compute the integer value of the abovementioned ratios.

Because a binary image f(i, j) = 1 of the workspace has been given by Eqs.(7)-(10) an efficient algorithm for determining workspace boundary can be developed by using f(i, j) itself. This is based on the geometry of the discretization since a workspace point within the bulk of the reachable area is related to a sum of the f(i, j) values for the points surrounding it which is equal to nine. When the workspace point is on the boundary contour this sum is less than nine, since there are contacting point not belonging to the workspace with f(i, j) = 0. This can be analytically expressed through a variable, named as sum, given by

$$\mathrm{sum} = \sum_{i-1}^{i+1} \sum_{j-1}^{j+1} f(i, j) \tag{11}$$

whose detection can be used to generate a binary mapping g(i, j) = 1 for the boundary points. Two particular cases may arise:

a) a rectangle equal to zero is surrounded by eight rectangles equal to one. This rectangle is not necessary a void, but it may be due to an unsuitable discretization for both $\Delta\theta_k$ or Δi and Δj;

b) a rectangle equal to one is surrounded by seven rectangles equal to one and the eighth equal to zero, determining a gross boundary contour, because of a not fine resolution.

Both ambiguities can be solved by selecting the reference value of the sum variable in Eq.(11) equal to eight or less, instead of nine. It is worth noting that the numerical efficiency of the algorithm increases while the its correctness does not decrease if the rectangles have small enough width Δi and height Δj.

A map of primary workspace can be obtained through the cross-section binary image f(i, j). Secondary workspaces are related to dextrous performances of a robot. Therefore, it is useful to determinate also the dextrous workspace as a measure of the kinematics performance of a robot manipulator. However, it is in general very complex to determine the dextrous workspace, (Kumar and Waldron, 1981). Secondary workspaces can be defined as regions of points that can be reached several times and it is convenient to determine regions with the same number of reaches. Thus, a measure of the reachability fq(i, j) for each rectangle (i, j) of a workspace discretization can be introduced as a function of the number nq of times that a rectangle has been reached during the scanning process of the joints mobility. Because of this numerical definition, fq(i, j) can be generated during the generation of f(i, j) itself.

It is to note that fq(i, j) greatly depends on the size of the rectangles, but it is very interesting since, more than a performance evaluation of points, it gives a practical information on a dextrous performance which in industrial robots can be thought related to small rectangles due to precision and repeatability capabilities. In addition it is to remark that, assuming smaller or larger values for the discretization of both $\Delta\theta_k$ or Δi and Δj, the reachability measure will change in magnitude but the area and the shape of secondary workspaces drawn within two figures, which are distinguished by the mean value of the reachability, will be not modified.

A numerical example for a COMAU robot SMART 6.100A, (Comau, 1994), has been reported in Fig.3

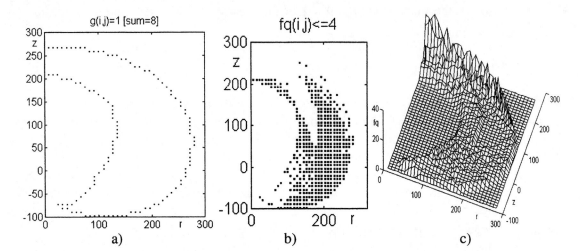

Fig.3. Results from the numerical algorithm for a robot manipulator COMAU SMART 6.100A with $a_1 = 300$ mm, $\alpha_1 = 90$ deg., $a_2 = 1100$ mm, $\alpha_2 = \alpha_3 = 0$ deg., $a_3 = 1625$ mm, and $b_1 = b_2 = b_3 = 0$, (Comau, 1994): a) cross-section workspace contour generated with sum = 8; b) cross-section area of secondary workspace with $f(i, j) \leq 4$; c) a map for the reachability measure.

5. An algebraic formulation for the workspace analysis

The boundary of an (N-j+1)R hyper-ring can be expressed algebraically when it is thought generated by enveloping the torus family traced by the parallel circles in the boundary of the revolving (N-j)R hyper-ring, according to Eq.(6), (Ceccarelli 1996).

An equation for a torus family can be expressed with respect to the j link frame, assuming $C_j \neq 0$ and $\cos\alpha_j \neq 0$, as a function of the radial r_j and axial z_j reaches in the form

$$(r_j^2 + z_j^2 - A_j)^2 + (C_j z_j + D_j)^2 + B_j = 0 \tag{12}$$

where the so-called torus parameters are a_j, α_j, r_{j+1}, z_{j+1}, and the coefficients are given as

$$
\begin{aligned}
A_j &= a_j^2 + r_{j+1}^2 + (z_{j+1} + d_{j+1})^2 \\
B_j &= -4a_j^2 r_{j+1}^2 \\
C_j &= 2a_j / s\alpha_j \\
D_j &= -2a_j(z_{j+1} + d_{j+1})c\alpha_j / s\alpha_j
\end{aligned}
\tag{13}
$$

396

in which $s\alpha_j$ and $c\alpha_j$ designate $\sin\alpha_j$ and $\cos\alpha_j$, respectively.

Particular cases with $C_j = 0$ or $\cos\alpha_j = 0$ are not represented by Eq.(12) and specific formulation can be developed when the torus boundary is not generated as an envelope of the revolving circle: $a_j = 0$, since the torus degenerates in an ellipsoid surface with major axis r_{j+1}; $s\alpha_j = 0$, since the torus degenerates in a circular rim of width r_{j+1}; and $c\alpha_j = 0$, since "common or symmetrical-offset torus" is obtained from the revolving circle of radius r_{j+1}.

Indeed, a torus family can be thought generated by considering r_{j+1} and z_{j+1} as variables, which can be calculated from the boundary points of the revolving $(N-j)R$ hyper-ring. Particularly, r_{j+1} and z_{j+1} can be expressed recursively by using Eq.(12) up to the last revolving torus workspace, or they can be formulated as a function of the revolute joint angles θ_{j+1}, θ_{j+2}, or finally, they can be a function of the last revolute joint angle by using an iterative formulation of the revolving boundary workspace and its torus family up to the extreme ring workspace for the manipulator chain.

r_{j+1}, z_{j+1} can be usefully expressed as a function of the last kinematic joint angle θ_N in the N-R chain to have a single variable formulation. Moreover, θ_N can be indeed considered as a torus family parameter since for each value a torus of the enveloping family can be determined and points of the workspace boundary can be generated.

The envelope equations of a torus family can be obtained from Eq.(12) and its derivative with respect to the torus family parameter θ_N. After some algebra for which $C_j \neq 0$ and $E_j \neq 0$ are needed, the so-called ring boundary equations can be obtained in the form, (Ceccarelli 1996),

$$r_j = \left[A_j - z_j^2 + \frac{(C_j z_j + D_j)G_j + F_j}{E_j} \right]^{1/2}$$

$$z_j = \frac{-F_j G_j \pm Q_j^{1/2}}{(E_j^2 + G_j^2)C_j} - \frac{D_j}{C_j} \tag{14}$$

where the so-called ring coefficients are given as

$$E_j = R_{j+1} + S_{j+1}$$
$$F_j = -2\,a_j^2\,R_{j+1}$$
$$G_j = -2a_j\,z_{j+1}'\,c\alpha_j / s\alpha_j \tag{15}$$
$$Q_j = -E_j^2 \left[F_j^2 + B_j \left(E_j^2 + G_j^2 \right) \right]$$

with

$$R_{j+1} = \left[(C_{j+1}z_{j+1} + D_{j+1})(E_{j+1}G_{j+1}' - G_{j+1}E_{j+1}') + F_{j+1}'E_{j+1} - F_{j+1}E_{j+1}' \right] / E_{j+1}^2$$
$$+ G_{j+1}E_{j+1}(C_{j+1}z_{j+1}' + G_{j+1})/E_{j+1}^2 + E_{j+1} - 2z_{j+1}z_{j+1}' \quad,$$

$$S_{j+1} = 2(z_{j+1} + d_{j+1})\,z_{j+1}' \tag{16}$$

$$z_{j+1}' = \left[\left(\pm 0.5\,Q_{j+1}'Q_{j+1}^{-1/2} - F_{j+1}G_{j+1}' - G_{j+1}F_{j+1}' \right)\left(E_{j+1}^2 + G_{j+1}^2 \right) \right.$$

$$+ 2\left(F_{j+1}G_{j+1}\,m\,Q_{j+1}^{1/2} \right)\left(E_{j+1}E_{j+1}' + G_{j+1}G_{j+1}' \right) \right] / \left(E_{j+1}^2 + G_{j+1}^2 \right)^2 C_{j+1} - G_{j+1}/C_{j+1}$$

The symbol ' represents the derivative operator with respect to the torus family parameter θ_N. Eqs.(16) can be computed through the equation coefficients E_{j+1}, F_{j+1}, G_{j+1}, Q_{j+1} and their derivatives E'_{j+1}, F'_{j+1}, G'_{j+1}, Q'_{j+1}, whose expressions and values can be deduced from r_{j+1} and z_{j+1} formulation. Particularly, the derivatives of the ring coefficients can be calculated, from Eqs.(15), as

$$
\begin{aligned}
E'_j &= R'_{j+1} + S'_{j+1} \\
F'_j &= -2a_j^2 R'_{j+1} \\
G'_j &= -2a_j z''_{j+1} c\alpha_j / s\alpha_j \\
Q'_j &= -2E_j^2 \left[F_j \left(F'_j + E_j^2 + G_j^2 \right) + B_j (E_j E'_j + G_j G'_j) \right] + 2Q_j E'_j / E_j
\end{aligned}
\tag{17}
$$

The derivatives R'_{j+1}, S'_{j+1} and z''_{j+1} can be computed by means of a further derivative operation from Eqs.(16) and by using, additionally, the second derivatives of the ring equation coefficients (15). The radial r_{j+1} and axial z_{j+1} reaches, the ring coefficients and derivatives of the (N-j)R workspace boundary may be expressed iteratively by means of the envelope development from the (N-j-1)R hyper-ring boundary up to the extreme ring workspace due to the last three revolute joints in the chain. Finally, last ring workspace can be expressed in an algebraic form as a function of the torus family parameter θ_N.

The algebraic formulation of Eqs.(13)-(17) can be used to generate numerically the workspace boundary of a N-revolute chain according to Eq.(6), from the extremity to the base of the manipulator when the geometrical sizes of the links are given. It is worth to note that also the generating workspaces can be evaluated for the computations.

The coefficients R_{j+1}, S_{j+1} and z_{j+1} can be calculated through E'_{j+1}, F'_{j+1}, G'_{j+1}, Q'_{j+1}, as well as these can be computed through R'_{j+2}, S'_{j+2} and z''_{j+2} and so on, since the implicit expressions (15) of E_{j+1}, F_{j+1}, G_{j+1}, Q_{j+1}, as far as explicit expressions of E_{n-2}, F_{n-2}, G_{n-2}, Q_{n-2} can be computed. This iterative computation can be expressed, according to Eqs.(17), in a general iterative form

$$
\begin{aligned}
E_{j+1}^k &= R_{j+2}^k + S_{j+2}^k \\
F_{j+1}^k &= -2a_{j+1}^2 R_{j+2}^k \qquad\qquad k = 0, 1,..., j; \; j = 0,1,..., N\text{-}4 \\
G_{j+1}^k &= -2a_{j+1} z_{j+2}^{k+1} c\alpha_{j+1} / s\alpha_{j+1}
\end{aligned}
\tag{18}
$$

and, according to Eqs.(16), as

$$
\begin{aligned}
R_{j+2}^k &= f^k \left(E_{j+2}^{k+1}, F_{j+2}^{k+1}, G_{j+2}^{k+1} \right) \\
S_{j+2}^k &= g^k \left(E_{j+2}^{k+1}, F_{j+2}^{k+1}, G_{j+2}^{k+1} \right) \qquad\qquad k = 0, 1,..., j; \; j = 0,1,..., N\text{-}4 \\
z_{j+2}^{k+1} &= h^k \left(E_{j+2}^{k+1}, F_{j+2}^{k+1}, G_{j+2}^{k+1} \right)
\end{aligned}
\tag{19}
$$

where f^k, g^k, and h^k, represent the k-derivatives of the functions f, g and h which are those expressing R_{j+1}, S_{j+1} and z'_{j+1} in the form of Eqs.(16), respectively.

However, all the abovementioned computations can be numerically evaluated by starting from the computation of derivatives of E_{N-2}, F_{N-2}, G_{N-2}, Q_{N-2} up to (N-3) order. Infact, the ring coefficients can be algebraically expressed from the algebra to deduce the ring equations in the form, (Ceccarelli 1989),

$$E_{N-2} = -2a_N(d_{N-1}s\alpha_{N-1}c\theta_N + a_{N-1}s\theta_N)$$
$$F_{N-2} = 4a_{N-2}^2 a_N(a_N s^2\alpha_{N-1}s\theta_N c\theta_N + a_{N-1}s\theta_N - d_N s\alpha_{N-1}c\alpha_{N-1}c\theta_N) \qquad (20)$$
$$G_{N-2} = 2a_{N-2}a_N c\alpha_{N-2}s\alpha_{N-1}c\theta_N / s\alpha_{N-2}$$

where from the geometry of the manipulator chain it holds

$$r_{N-1} = \left[(a_N c\theta_N + a_{N-1})^2 + (a_N s\theta_N c\alpha_{N-1} + d_N s\alpha_{N-1})^2\right]^{1/2}$$
$$z_{N-1} = d_N c\alpha_{N-1} - a_N s\theta_N s\alpha_{N-1} \qquad (21)$$

An easy numerical algorithm can be developed to compute the radial r_j and axial z_j reaches of all the envelopes for j from N-2 to 1, which is from the extreme workspace ring up to NR hyper-ring. This can be obtained by scanning the joint angle θ_N from 0 to 2π and calculating at each j the coefficients A_j, B_j, C_j, D_j, E_j, F_j, G_j, R_j, S_j and z'_j, and finally r_j, z_j when the j derivatives of E_j, F_j, G_j, are evaluated by using previous calculations for R_{j+1}, S_{j+1} and z'_{j+1}.

6. A manipulator synthesis by inverting a workspace formulation

The algebraic formulation for workspace boundary can be conveniently used for synthesis purposes when the workspace is prescribed by means of workspace limits, Fig. to give precision points for workspace boundary. In this case, in fact, the algebraic formulation can be used to build a system of algebraic design equations where the axial and radial reaches are given by the prescribed workspace data and the design unknowns are the ring or hyper-ring coefficients. once these coefficients are solved the chain parameters can be calculated by using the definitions of the coefficients yet.

To better illustrate the synthesis procedure the case of a three-revolute manipulator has been reported in this paper in the form developed in a previous paper (Ceccarelli 1995a).

The design problem of a general three-revolute open chain manipulator can be formulated through a set of workspace equations whose number is equal to the number of the unknowns which are represented by the structural coefficients A_1, B_1, C_1, D_1, E_1, F_1, G_1, and the manipulator base location vectors $\underline{s}$, $\underline{k}$.

When the manipulator base location is known, assuming seven given workspace boundary points, whose coordinates can be expressed directly with respect to $X_1Y_1Z_1$, the aforementioned unknown structural coefficients can be calculated through the Newton-Raphson technique from a set of equations which express the workspace boundary points. This set is formulated through the second of Eqs.(14) and another one which is deduced from Eqs.(14), with the hypothesis $E_1 \neq 0$, in the form:

$$E_1 K_1 (r_1^2 + z_1^2 - A_1) - (-L_1 \pm Q_1^{1/2}) G_1 - F_1 K_1 = 0 \qquad (22)$$

In a similar way, in the case of the unknown vectors $\underline{s}$ and $\underline{k}$, the Eqs.(14) can be conveniently expressed to obtain two decoupled set of design equations, whose first one can be expresses as

$$E_1 K_1 (\overline{x} \cdot \overline{x} - 2\overline{s} \cdot \overline{x} + \overline{s} \cdot \overline{s} - A_1) - (-L_1 \pm Q_1^{1/2}) G_1 - F_1 K_1 = 0. \qquad (23)$$

By writing Eq.(23) for eight given workspace boundary points the unknown A_1, E_1, F_1, G_1, Q_1 and $\underline{s}$ can be solved when the ambiguity in the sign of the radical Q_1 is resolved taking into account that the upper branch of the envelope boundary is related to the positive sign and the bottom part to the negative sign, (Ceccarelli, 1989). Once the first set of equations is solved, the first of Eq.(14) can be used with the remaining four boundary points to give C_1, D_1 and $\underline{r}_3$ taking into account that $\underline{r}_3$ represents Z axis unit vector with respect to $X_1 Y_1 Z_1$, and therefore its components must satisfy the unit vector condition.

The, assuming $w = \sin^2 \alpha_1$, Eqs.(13) can be inverted to give

$$a_1 = 0.5\, C_1\, w^{1/2}$$
$$r_2^2 = -\frac{B_1}{w\, C_1^2}$$
$$z_2 + d_2 = -\frac{D_1}{C_1 \left(1 - w\right)^{1/2}}$$

(24)

and only the parameter w needs to be solved. Substituting Eqs.(24) into the first of Eqs.(11), with the position $w = y + (1 + 4\, A_1 / C_1^2)\,/\,3$, it yields

$$y^3 + 3\, p\, y + 2\, q = 0 \tag{25}$$

where

$$p = \frac{4}{9\, C_1^2}\left[A_1 - \frac{C_1^2}{4} - \frac{12}{C_1^2}\left(\frac{B_1 + D_1^2}{4} + \frac{A_1^2}{3} \right) \right]$$

(26)

$$q = \frac{1}{27\, C_1^2}\left(1 + 4\frac{A_1}{C_1^2} \right)\left(10\, A_1 - 16\frac{A_1^2}{C_1^2} - 18\frac{B_1 + D_1^2}{C_1^2} - C_1^2 \right) + 2\frac{B_1}{C_1^4}.$$

Eq.(25) can be solved algebraically to give one, two or three solutions depending on the discriminant D by using Eqs.(26). It is observable that Eq.(25) gives one or two significative solutions of w according to the condition $0 < w < 1$, (Ceccarelli 1995a) .

Once Eq.(25) is solved, we can compute the parameters a_1, r_2 and $(z_2 + d_2)$ and the a_1 angle so that each w solution corresponds to two manipulators distinguished at this step by the α_1 sign and to two more manipulators taking into account the supplementary values of α_1.

Successively, with the hypothesis that $\theta_3 = 0$, inverting Eqs.(15), which assume the forms of Eqs.(20), the remaining chain parameters can be obtained as

$$d_2 = -\frac{E_1\, a_1}{G_1 \tan \alpha_1}$$
$$d_3 = \frac{z_2}{\cos \alpha_2}$$
$$a_3 = \frac{G_1 \tan \alpha_1}{2\, a_1 \sin \alpha_2}$$
$$a_2 = \left(r_2^2 - z_2^2 \tan^2 \alpha_2 \right)^{1/2} - a_3$$

(27)

where α_2 needs to be previously solved from

$$\alpha_2 = \pm \sin^{-1}\left[\frac{L^2 + r_2^2 \pm \left[\left(L^2 - r_2^2\right)^2 - 4\,L^2\,z_2^2\right]^{1/2}}{2\left(r_2^2 + z_2^2\right)}\right] \tag{28}$$

where L is a short form for the quantity $(G_1 \tan\alpha_1)\,/(2a_1)$. Eq.(28) provides none, two or four solutions depending on the values of L, r_2 and z_2. Two or four more solutions can be obtained by taking into account supplementary values of the calculated α_2.

It is to note that each numerical solution for the structural coefficients and the manipulator base location corresponds to sixty-four different manipulator parameter sets at the most, depending on the number of solutions for α_1 and α_2. However, since the above mentioned values for the twist angles α_1 and α_2 may give negative values for the length sizes but they do not correspond to alternative design solutions, meaningful solutions can be considered only the sixteen sets which can be synthesised for $-\pi/2 < \alpha_1 < \pi/2$ and $-\pi/2 < \alpha_2 < \pi/2$.

7. An optimum design of manipulators with prescribed workspace

An optimum design of manipulators can be formulated as an optimization problem in the form

$$\min L \tag{29}$$

subject to

$$\underline{x}_j \ge \underline{X}_j \qquad j = 1,...,J \tag{30}$$

$$V \ge V_0 \tag{31}$$

where L is the objective function; $\underline{X}_j$ (j=1,...,J) represent given precision points, with respect to the fixed frame, to be reached as workspace points $\underline{x}_j$, V_0 is a minimum value for a desirable workspace volume.

The robot base location can be included into the design parameters through the $\underline{b}$ vector expressed by means of the radial a_0 and axial d_0 components, Fig.1. Particularly, the angles α_0 and θ_0 can be considered as describing the orientation of the manipulator base as Z_b axis with respect to Z axis and X_b with respect to X, respectively. In addition, the base frame can be conveniently assumed to be parallel with the frame $X_1Y_1Z_1$, fixed on the first link of the manipulator chain, at a starting motion configuration. Therefore, the dimensional parameters for the optimal design problem will also include the parameters α_0, θ_0, a_0 and d_0.

Once the problem of Eqs. (29) to (31) has been solved, optimum design parameters θ_0, a_i, and d_i (i=0,1,...,n) and α_i (i=0,1,...,n-1) will be obtained since they affect L, $\underline{x}_j$, and V.

Several workspace characteristics can be used to formulate the objective function, (Mata and Ceccarelli 1993), but however workspace volume and manipulator length are usually preferred. These characteristics can be also conveniently combined to formulate a performance index for manipulators.

The workspace precision points can be considered inside as well as on the boundary of the workspace volume. Nevertheless, they are assumed, at most, as limiting points for the workspace design capability and, consequently, it is usually convenient to think of them as workspace boundary points. In this case it is also possible to prescribe workspace characteristics as voids and hole, which can be useful in practical applications as save regions for equipment and personnel in the automated environment. Therefore, a workspace description through determination of its boundary is needed and $\underline{x}_j$ is assumed to be determinable from an analytical expression of the workspace boundary. Once workspace boundary points are given, the manipulator solution must satisfy exactly the restriction with its workspace boundary and the sign equal is to be considered. Otherwise, a delimiting region can be assigned within which the workspace may be outlined and a weak restriction can be used to give larger design possibilities.

The constraint on the workspace volume can be of a determinant significance, since this constraint may have an effect on the robot size to counterbalance the minimisation of L in order to ensure a no-degenerated solution of the manipulator chain.

The general optimization formulation of Eqs. (29) to (31) can be better illustrated by referring to a specific case for a manipulator optimum design.

Thus the optimum design of a general three-revolute manipulator deals with the synthesis of the parameters a_1, a_2, a_3, d_2, d_3, α_1, α_2, (d_1 is not meaningful since it shifts up and down the workspace only), Fig.2, when the workspace characteristics are considered to fulfil the design requirements which are expressed in Fig.1. The optimum design can be formulated as an optimization problem in the form

$$\min \; - \; \frac{V^*}{L^3} \tag{32}$$

subject to

$$
\begin{aligned}
&\min (z) \; \geq \; z_{min} \\
&\max (z) \; \leq \; z_{max} \\
&\min (r) \; \geq \; r_{min} \\
&\max (r) \; \leq \; r_{max}
\end{aligned}
\tag{33}
$$

where V* indicates a measure of the workspace volume; L is the total dimension of the manipulator which can be given as

$$L = \sqrt{\left(a_1^2 + d_2^2 + a_2^2 + d_3^2 + a_3^2\right)} \qquad ; \tag{34}$$

z and r are the axial and radial reaches of the boundary points of the manipulator workspace.

Eq.(32) has been used to express the maximisation of manipulator workspace in term of volume V referring to the design dimension L of the manipulator. Indeed, the ratio V* / L gives a measure of the workspace maximisation since the volume V* is compared to a spherical volume achievable with a single link arm O size L with a base spherical joint.

Eqs.(33) express the constraints of Fig.3 in a general way to include several case for which the workspace is only prescribed within a given volume. Indeed the debile disequations permit an easier solution of the design problem that the case for a strict prescribe, since they give a wider field of feasible solutions.

A way to perform numerically in an efficient procedure requires a suitable analytical formulation for the involved quantities of the manipulator workspace. Because of the optimum formulation a formulation for the manipulator workspace can be conveniently expressed in term of the workspace boundary. Thus the algebraic formulation for the ring

402

workspace has been used in the form which has been deduced by the author in a previous paper, (Ceccarelli 1989).

Thus, the volume V can be calculated by summing up, algebraically, the volumes of thin cylinders individuated, each one, by two boundary points k and k+1, in the form,

$$V^* = \frac{\pi}{2} \sum_{k=1}^{k=n} \left(z_{1,k+1} + z_{1,k}\right)\left(r_{1,k}^2 - r_{1,k+1}^2\right) \qquad (35)$$

in which $r_{1,k}$ and $z_{1,k}$ are computed from Eqs.(3)-(5) for n points due to the θ_3 scanning.

This expression for workspace volume evaluation enlightens, moreover, the fact that also a numerical procedure for workspace analysis, like that one expressed by Eqs.(7) to (11), can be used into an optimum design procedure.

The design problem for three-revolute manipulators has been formulated in the form of an optimization problem as a function of seven design variables a_1, a_2, a_3, d_2, d_3, α_1, α_2. The involved expressions are no linear functions of the design variables which have been expressed as algebraic functions of the data because of the algebraic formulation for the workspace. This makes possible and indeed convenient to use available commercial software packages for the calculations of optimisation problems. The numerical procedure, which has been used in (Ceccarelli et al. 1994; Ceccarelli 1995b; Ceccarelli 1997), has taken advantage of the formulation for the workspace boundary and the Sequential Quadratic Programming technique for the non linear optimization problems. Indeed, the Optimisation Toolbox of Matlab (The Math Works 1994) has been used to perform the numerical solution with a Sequential Quadratic Programming technique. This numerical procedure works in such a way that at each step k a solution is found along a search direction δ_k with a variables update ψ_k. The iteration continues until the variables vectors converge. However the procedure has been developed so that the formulation for the workspace has been easily and conveniently included within the solving procedure for the optimization problem by using the facilities of the Optimization Toolbox of Matlab, (The Math Works 1994), which permits an easy arrangement for an optimum design with analytical expressions. Of course the optimum solution is affected by the initial guess manipulator, although the algebraic formulation for the workspace seems to be useful to obtain optimum design also when the initial workspace of the guess manipulator is far away from the prescribed limits or even it violates some of the prescribed requirements.

9. Conclusion

Workspace is a characteristic which can be very useful for an evaluation and a design of manipulators.

In this paper, previous experiences by the author have been summarised to present a unified treatment which shows how an algebraic formulation for manipulator workspace, deduced by a geometric-kinematic modelling, can be used both for analysis and design purposes.

References

Beni G., Hackwood S. (Eds.), (1985). Recent Advances in Robotics. J. Wiley & Sons, New York, Ch.3, pp.71-130.

Ceccarelli M. (1989). On the Workspace of 3R Robot Arms. Proceedings of the Fifth International IFToMM Symposium on Theory and Practice of Mechanisms, Bucharest, Vol.II-1, pp. 37-46.

Ceccarelli M. (1995a). A Synthesis Algorithm for Three-Revolute Manipulators by Using an Algebraic Formulation of Workspace Boundary. Journal of Mechanical Design, Vol. 117, pp.298-302.

Ceccarelli M. (1995b). Optimal Design and Location of Manipulators. In: Computational Dynamics in Multibody Systems, M.F.O.S. Pereira and J.A.C. Ambrosio (Editors), Kluwer, Dordrecht, pp.131-146.

Ceccarelli M. (1996a). A Formulation for the Workspace Boundary of General N-Revolute Manipulators. IFToMM Journal Mechanism and Machine Theory, Vol.31, pp.637-646.

Ceccarelli M. (1996b). A Workspace Analysis for RRP Manipulators", Proc. of 1996 ASME Design Engineering Technical Conferences, 24th Biennial Mechanisms Conference, Irvine, 1996, paper 96DETC-Mech1012.

Ceccarelli M., (1997). Diseño Optimo de Brazos Manipuladores respecto al Espacio de Trabajo. Associazione Spagnola di Ingegneria Meccanica, Bilbao, Año 11, n.1, Vol.3, pp. 243-250.

Ceccarelli M., Vinciguerra A. (1990). A Design Method of Three-Revolute Open Chain Manipulators. Proceedings of VIIth CISM-IFToMM Symposium on Theory and Practice of Robots and Manipulators, Hermes, Paris, pp.318-325.

Ceccarelli M., Mata V., Valero F. (1994). Optimal Synthesis of Three-Revolute Manipulators. International Journal Meccanica, Kluwer, Dordrecht, Vol.29, n.1, pp.95-103.

Ceccarelli M., Gabriele E. (1995). Determining Primary and Secondary Workspaces of Industrial Robots. 4th International Workshop on Robotics in Alpe-Adria Region, Portsach, Vol-II, pp.259-262.

Comau, (1994). Technical Sheet of SMART 6.100A.

Cwiakala M., Lee T.W. (1985). Generation and Evaluation of a Manipulator Workspace Based on Optimum Path Search. Jnl of Mechanisms, Transmissions and Automation in Design, Vol.107, pp.245-255

Derby S. (1981). The Maximum Reach of Revolute Jointed Manipulators. Mechanism and Machine Theory, Vol.16, pp.255-261.

Fichter E.F., Hunt K.H. (1975). The Fecund Torus, its Bitangent-Circles and Derived Linkages. Mechanism and Machine Theory, Vol.10, pp.167-176.

Freudenstein F., Primrose E.J.F. (1984). On the Analysis and Synthesis of the Workspace of a Three-Link Turning-Pair Connected Robot Arm. Jnl of Mechanisms, Transmissions and Automation in Design, Vol.106, pp.365-370.

Gosselin C.M. and Guillot M. (1991). The Synthesis of Manipulators with Prescribed Workspace. Jnl of Mechanical design, Vol.113, pp.451-455.

Gupta K.C., Roth B. (1982). Design Considerations for Manipulator Workspace. Jnl of Mechanical Design, Vol.104, pp.704-711.

Gupta K.C. (1986). On the Nature of Robot Workspace. International Jnl of Robotics Research, Vol.5, n.2, pp.112-121.

Hansen J.A., Gupta K.C., Kazerounian S.M.K. (1983). Generation and Evaluation of the Workspace of a Manipulator. International Jnl of Robotics Research, Vol.2, n.3, pp.22-31.

Haug E.J., Luh C., Adkins F.A., Wang J. (1996). Numerical Algorithms for Mapping Boundaries of Manipulator Workspaces. Jnl of Mechanical Design, Vol.118, pp.228-234.

Hsu M.S., Kohli D. (1987). Boundary Surfaces and Accessibility Regions for Regional Structures of Manipulators. Mechanism and Machine Theory, Vol.22, n.3, pp.277-289.

Jo D. Y. Haug E.J. (1989). Workspace Analysis of Multibody Mechanical Systems Using Continuation Methods. Jnl of Mechanisms, Transmissions and Automation in Design, Vol.111, pp.581-589.

Kumar A., Waldrom K.J. (1981). The Workspace of a Mechanical Manipulator. Jnl of Mechanical Design, Vol.103, pp.665-672.

Kumar A., Patel M.S. (1986). Mapping the Manipulator Workspace Using Interactive Computer Graphics. International Jnl of Robotics Research, Vol.5, n.2, pp.122-130.

Kohli D., Spanos J. (1985). Workspace Analysis of Mechanical Manipulators Using Polynomial Discriminants. Jnl of Mechanisms, Transmissions and Automation in Design, Vol.107, pp.209-215.

Kohli D., Hsu M.S. (1987). The Jacobian Analysis of Workspaces of Mechanical Manipulators. Mechanism and Machine Theory, Vol.22, n.3, pp.265-275.

Konstantinov M.S. and Genova P.I. (1981). Workspace and Maneuvrability Criteria for Robots. Proceedings of IVth CISM-IFToMM Symposium on Theory and Practice of Robots and Manipulators, Warsaw, pp.382-391.

Lee T.W., Yang D.C.H. (1983). On the Evaluation of Manipulator Workspace. Jnl of Mechanisms, Transmissions and Automation in Design, Vol.105, pp.70-77.

Lee N.L., Ting K.L.(1991). Workspace and Sensitive Postures of Planar Open-Loop Manipulators. Mechanism and Machine Theory, Vol.26, n.6, pp. 593- 602.

Lenarcic J., Umek A., Savic S. (1990). Considerations on Human Arm Workspace and Manipulability. Proceedings of the 2nd International Workshop on Advances in Robot Kinematics, Linz.

Liegeois A., Borrel P., Tanner P. (1986). Automatic Modeling of the Workspace for Robots. Proceedings of the Third International Symposium on Robotics Research, M.I.T. Press, pp.205-211.

Lin C.D., Freudenstein F. (1986). Optimization of the Workspace of a Three-Link Turning-Pair Connected Robot Arm. The International Jnl of Robotics Research, Vol.5, n.2, pp.104-111.

Mata A.V. and Ceccarelli M. (1993). Funciones Objectivo para la Optimizaciòn de la Cadena Cinemàtica de Robots. 1° Congresso Iberoamericano di Ingegneria Meccanica, Madrid, Vol.3, pp.47-54.

Nof S.Y. Editor (1985). Hanbook of Industrial Robotics. J. Wiley & Sons, New York.

Oblak D., Kohli D. (1988). Boundary Surfaces, Limit Surfaces, Crossable and Non-Crossable Surfaces in Workspace of Mechanical Manipulators. Jnl of Mechanisms, Transmissions and Automation in Design, Vol.110, pp.389-396.

Rastegar J. (1988). Workspace Analysis of 4R Manipulators with Various Degrees of Dexterity. Jnl of Mechanisms, Transmissions and Automation in Design, Vol.110, pp.42-47.

Rastegar J., Deravi P.(1987). Methods to Determine Workspace, Its Subspaces with Different Numbers of Configurations and All the Possible Configuration of a Manipulator. Mechanism and Machine Theory, Vol.22, n.4, pp.343-350.

Rastegar J., Fardanesh B.(1990). Manipulator Workspace Analysis Using the Monte Carlo Method. Mechanism and Machine Theory, Vol.25, n.2, pp.233-239.

Rastegar J., Perel D. (1990). Generation of Manipulator Workspace Boundary Geometry Using the Monte Carlo Method and Interactive Computer Graphics. Jnl of Mechanical Design, Vol.112, pp.452-454.

Roth B. (1975). Performance Evaluation of Manipulators from a Kinematic Viewpoint. NBS Special Publication on Performance Evaluation of Programmable Robots and Manipulators, pp.39-61.

Roth B. (1986). Analytical Design of Two-Revolute Open Chains. Preprints of 6th CISM-IFToMM Symposium on Theory and Practice of Robots and Manipulators, Cracow, pp.180-187.

Shimano B.E., Roth B. (1976). Ranges of Motion of Manipulators. Proceedings of 2th CISM-IFToMM Symposium on Theory and Practice of Robots and Manipulators, Elsevier, pp.17-26.

Spanos J., Kohli D. (1985). Workspace Analysis of Regional Structures of Manipulators. Jnl of Mechanisms, Transmissions and Automation in Design, Vol.107, pp.216-222.

Staubli (1996). The RX60 is born. Staubli Flash, n.2 pp.1.

Sugimoto K., Duffy J. (1981a). Determination of Extreme Distances of a Robot Hand - Part 1 A General Theory. Jnl of Mechanical Design, Vol.103, pp.631-636.

Sugimoto K., Duffy J. (1981b). Determination of Extreme Distances of a Robot Hand - Part 2 Robot Arms with Special Geometry. Jnl of Mechanical Design, Vol.103, pp.776-783.

The Math Works (1995). Matlab Optimization Toolbox.

Tsai Y.C., Soni A.H. (1981). Accessible Region and Synthesis of Robot Arms. Jnl of Mechanical Design, Vol.103, pp.803-811.

Tsai Y.C., Soni A.H. (1983). An Algorithm for the Workspace of a General n-R Robot. Jnl of Mechanisms, Transmissions and Automation in Design, Vol.105, pp.52-57.

Tsai Y.C., Soni A.H. (1984). The Effect of Link Parameter on the Working Space of General 3R Robot Arms. Mechanism and Machine Theory, Vol.19, n.1, pp.9-16.

Tsai Y.C., Soni A.H. (1985). Workspace Syntheiss of 3R, 4R, 5R and 6R Robots. Mechanism and Machine Theory, Vol.20, n.4, pp.555-563.

Vertut J. et Al. (1974). Contribution to Analyse Manipulator Morphology Coverage and Dexterity. Proceedings of First CISM-IFToMM Symposium on Theory and Practice of Robots and Manipulators", CISM, Udine, pp.277-302.

When M.H., Jin W.M. (1987). On Primary Workspace of Manipulators. Proceedings of 7th IFToMM World Congress, Pergamon Press, Sevilla, Vol.2, pp.1171-1174.

Yang D.C.H., Lee T.W. (1983). On the Workspace of Mechanical Manipulators. Jnl of Mechanisms, Transmissions and Automation in Design, Vol.105, pp.62-69.

Yang D.C.H., Lee T.W. (1984). Heuristic Combinatorial Optimization in the Design of Manipulator Workspace. IEEE Trans. on Systems, Man, and Cyberbetics, Vol. SMC-14, n.4, pp.571-580.

Incursive Anticipatory Control of a Chaotic Robot Arm

Daniel M. DUBOIS
Institute of Mathematics, UNIVERSITY OF LIEGE
Grande Traverse 12, B-4000 LIEGE 1 (Belgium)
Fax + 32 (0)4 366 94 89 - E-mail: Daniel.Dubois@ulg.ac.be
http://www.ulg.ac.be/mathgen/CHAOS/CHAOS.html

Abstract

This paper deals with an innovative mathematical tool for modelling, simulating and controlling systems in automation engineering. Classically, feedback processes are based on recursive loops where the future state of a system is computed from the present and past states. With the new concept of incursion, an inclusive recursion, the future state of a system is taken into account for computing this future state in a self-referential way. The future state is computed from the mathematical model of the system. With incursion, numerical instabilities in the simulation of finite difference equations can be stabilised. The incursive control of systems can also stabilise feedback loops by anticipating the effect of the control what I call a feed-in-time control. In this short paper, the particular case of the modelling, simulation and control of a robot arm in a working space is studied. The highly non-linear model is based on recursive finite difference equations which give rise to instabilities, bifurcations and fractal chaos. The anticipatory control by incursion, called feed-in-time control, of such a robot arm stabilises its trajectory. Numerical simulations show that the robot arm reaches set points in a few steps in any point of the working space.

Keywords: Incursive anticipatory control, robot arm, chaos, feed-in-time, incursion.

1 Introduction

The paper will discuss the potential innovative power of the incursion in automation engineering with the presentation of a typical example in robotics. The incursion is a new concept developed by Dubois (1992, 1995, 1996abc), which deals with an extension of the recursion. A recursive process is always depending on the present or past states of a system. With incursion, the process can depend on past and present states of the system but also to its potential future states.

Classical control (e.g. Craig, 1989) of any system by feedback is based on recursive processes in which the current control function u(t) at the present time step t is a function of the deviation (the error) between the set point x_r and the state x(t) resulting from the preceding control u(t-Δt) at the preceding time step t-Δt, the effect of which giving the new state x(t+Δt) at the next time step t+Δt. So a time delay Δt is present in this feedback loop which can give rise to instabilities or critical oscillations around the set point. Let us remark that numerical instabilities can also appear in the simulation of discretised differential equations when the time step Δt of the discretisation is too large. It was demonstrated that such instabilities can disappear even for large time step Δt with incursion in using backward and forward derivatives, for example, for linear and non-linear discrete oscillating systems

CP437, *Computing Anticipatory Systems: CAYS--First International Conference*
edited by Daniel M. Dubois © 1998 The American Institute of Physics 1-56396-827-4/98/$15.00

(Dubois, 1995, 1996bc; Dubois and Resconi, 1996). This would permit to simulate discretised equation systems in real time for automation engineering. When multiple iterates are generated at each step, the incursion is a hyperincursion (Dubois and Resconi, 1992, 1995; Dubois, 1996a), an extension of hyper recursion.

This paper deals with a feed-in-time control in which the control function $u(t)$ at time step t takes into account of its effect on the state of the system $x(t+\Delta t)$ at the future time step $t+\Delta t$. This future state is estimated from a mathematical model of the dynamics of the system. In other words, the control function is computed in anticipating its effect on the system at the next time step. The formal model of this feed-in-time control is given by the following simplified example. Let us consider the incursive control of a system represented by discrete equations (Dubois, 1995):

$$u(t) = u_r + c(x).(x(t+\Delta t) - x(t))/\Delta t \qquad (1a)$$
$$x(t+dt) = x(t) + \Delta t.[f(x(t),p) + b.u(t)] \qquad (1b)$$

where $u(t)$ is the incursive control function, u_r is a function of the explicit set-point or reference signal $r(t)$ of the variable, $x(t)$ the state variable of the system, f the recursive function of the model of the system depending on the variable $x(t)$ and on the parameter p, b is a known parameter. The function $c(x)$ can be explicitly defined (Dubois, 1996a) as a function of the following derivative of the function $f(x,p)$:

$$c(x) = A + B.df(x,p)/dx \qquad (1c)$$

where A and B are constants. Let us remark that with non-linear functions $f(x,p)$ of at least of degree two in x, the control $u(t)$ is non-linear. Such an incursive control was applied to stabilise the chaos in the Pearl-Verhulst map $f(x,p) = p.x.(1 - x)$ (Dubois, 1992, 1995). Equation 1c gives $c(x)=A-B.(p-2.p.x)$. Without set point ($u_r = 0$), the incursive control can stabilise a system in an unstable regime. Indeed, with unstable or chaotic systems, the incursive control changes the unstable regime to a stable one which becomes the implicit set point (Dubois, 1995, 1996a; Dubois and Resconi, 1994, 1995). The incursive method is general, the variable x and the function f can be given by a vector and a matrix, as demonstrated for non-linear Lotka-Volterra equations and for the inverse cinematic problem for a robot arm (Dubois and Resconi, 1995). This paper will present a new application which will show this stabilisation of the chaos in the case of another model of a robot arm.

The incursive control $u(t)$ is a forward derivative which depends on the measured value of the variable at the present time t and its unknown value at the next future time step $t+\Delta t$. In adaptive control, such a future state is estimated (Portier and Oppenheim, 1993).

The originality of the incursive control deals with the way to compute the future state without estimation. The variable $x(t+\Delta t)$ in eq. 1a can be explicitly replaced by eq. 1b by self-reference as follows (Dubois, 1995, 1996a):

$$u(t) = u_r + c(x).(f(x(t),p)) + b.u(t)) \qquad (2a)$$

and an incursive equation is still obtained because the control $u(t)$ is self-referential. It is important to notice that the incursive control takes into account its action at the future time step and can be transformed to the following recursive control:

$$u(t) = [u_r + c(x).f(x(t),p)]/[1- c(x).b] \qquad (2b)$$

This paper deals with an application in robotics. The arm of a robot is considered in a two-dimensions working space. A mathematical model given by finite difference equations takes into account the angles of the two members of the arm and the set point is defined in polar co-ordinates. The incursive control of such a system stabilises the chaos of the recursive model by a feed-in-time.

2 Controlling the Chaos of a Robot Arm

This paper deals with a special problem in robotics, that is the inverse cinematic problem of computing the iterates t = 1, 2, 3, ... of two angles $\alpha(t)$ and $\beta(t)$ of a robot arm with two members of length 1 from initial conditions $\alpha(0)$ and $\beta(0)$ to a set point given by polar co-ordinates (ρ_r, γ_r) in the working space (x,y) as shown in Figure 1.

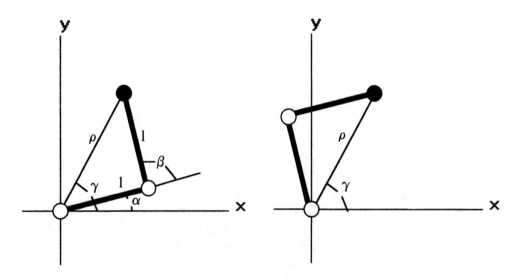

Figures 1a-b: (a) The arm of the robot is constituted of two members of length 1 with angles α and β. The extremity of the arm is defined by the polar co-ordinates (ρ,γ). (b) For the same polar co-ordinates, there are two symmetrical solutions.

From the following discrete recursive equation system (Dubois, 1996a)

$$\alpha(t+\Delta t)=\alpha(t)+\Delta t.[\ \rho_r/2\text{-}1.\cos(\gamma_r\text{-}\alpha(t)]/c \qquad (3a)$$
$$\beta(t+\Delta t)=\beta(t)+\Delta t.[1.\sqrt{(2.(1+\cos(\beta(t))}\text{-}\rho_r]/c \qquad (3b)$$

the iterates $\alpha(t+\Delta t)$ and $\beta(t+\Delta t)$ are computed from their preceding values at time step t, starting with the initial state of the arm $\alpha(0)$ and $\beta(0)$ in function of the final state defined by the polar co-ordinates (ρ_r,γ_r). The value of the constant c depends on the length and time units. The angles are given in radians. The advantage of this model is the fact that the two equations are independent of each other: the two control motors of each member of the arm can move independently. The second equation is only a function of the radial set point which depends only on the angle $\beta(t)$.

Let us remark that the man moves a hand in estimating with the eyes the radial distance of an object to be taken and the rotation angle.

408

When this system reaches at a steady state, we have

$$\alpha(t+\Delta t)=\alpha(t) \text{ and } \beta(t+\Delta t)=\beta(t) \tag{4a}$$

so that
$$\rho_r=2.1.\cos(\gamma_r-\alpha(t)) \text{ and } \rho_r=1.\sqrt{(2.(1+\cos(\beta(t)))} \tag{4b}$$

The criteria of stability of the steady state are given by

$$\left| d\alpha(t+\Delta t)/d\alpha(t) \right| = \left| 1-\Delta t.1.\sin(\gamma_r-\alpha(t))/c \right| < 1 \tag{5a}$$
$$\left| d\beta(t+\Delta t)/d\beta(t) \right| = \left| 1-\Delta t.1.\sin(\beta(t))/[c.\sqrt{(2.(1+\cos(\beta(t)))]} \right| < 1 \tag{5b}$$

For $\Delta t < 2\,c\,/\,1$, the system is only stable for some values of the angles.

There are many set points in the working space which give rise to instabilities, transient bifurcations or chaos for the angles $\alpha(t)$ and $\beta(t)$.

For example, the set point $(\rho_r=0, \gamma_r=0)$ is particularly unstable and gives rise to chaos for $\beta(t)$.

Figure 2a shows the bifurcation diagram of $\beta(t)$ obtained by numerical simulation of eq. 3ab, where $\cos(\beta(t))$ is plotted as a function of the interval of time Δt varying from 0 to $2.c\,/\,1$ (c is taken equal to 1) for the set point $(\rho_r=0, \gamma_r=0)$.

Typical fractal chaos is well-seen for $\Delta t > c\,/\,l$.

The incursive anticipatory control, called feed-in-time control, stabilises such a chaotic behaviour. Indeed, let us introduce the following incursive control for eqs. 3a-b from eqs. 1a-c (Dubois, 1996a):

$$u(t)=[-\sin(\gamma_r-\alpha(t))].[\alpha(t+\Delta t)-\alpha(t)] \tag{6a}$$
$$v(t)=[-\sin(\beta(t)/\sqrt{(1+\cos(\beta(t))}].[\beta(t+\Delta t)-\beta(t)] \tag{6b}$$
$$\alpha(t+\Delta t)=\alpha(t)+\Delta t.[\ \rho_r/2-1.\cos(\gamma_r-\alpha(t)]/c+u(t) \tag{6c}$$
$$\beta(t+\Delta t)=\beta(t)+\Delta t.[1.\sqrt{(2.(1+\cos(\beta(t))-\rho_r]/c+v(t)} \tag{6d}$$

where the control functions u and v depend on the present and future values of the angles at time t and t+Δt. These control functions are discrete forward time derivatives and play the role of an anticipation for the next future step, what is called a feed-in-time. Classically a feedback control takes into account the discrete backward time derivatives which are given by the outputs of the system at times t and t-Δt. It can be remarked that the control functions are equal to zero when the system reaches its steady state. The role of the anticipatory incursion is to increase the stability of the system for larger time steps Δt and to enhance the transient path to the set point which can be quicker.

Figure 2b shows the numerical simulation of eqs. 6abcd in the same conditions as in Figure 2a (cos b(t) as a function of Δt varying from 0 to 2 c / 1): the movement of the robot arm is stabilised at the set point $(\rho_r=0, \gamma_r=0)$.

With incursive anticipatory control no more chaos is present for $\beta(t)$.

Figures 3ab show the movement of the robot arm for the set point $(\rho_r=0, \gamma_r=0)$ with the initial condition $(\rho_r= \sqrt{2}.1, \gamma_r=-\pi/4)$, respectively without and with anticipatory control. **In Fig. 3a, the movement of the angle $\beta(t)$ is chaotic In Fig. 3b the robot arm reaches the set point which is stabilised by incursive anticipatory control.**

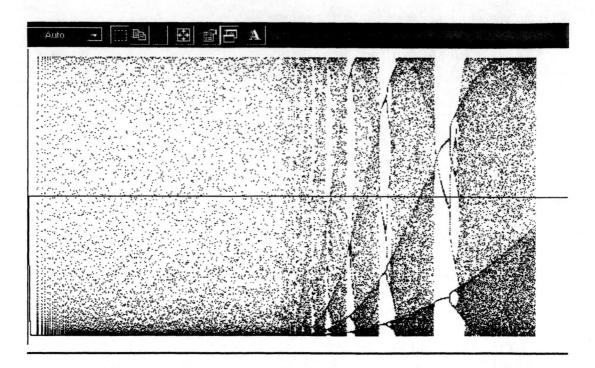

Figure 2a: Bifurcation diagram of cos β(t) (vertically from + 1 to -1) as a function of the value of the time interval Δt (horizontally from 0 to 2 / 1) showing the chaos in the movement of the robot arm.

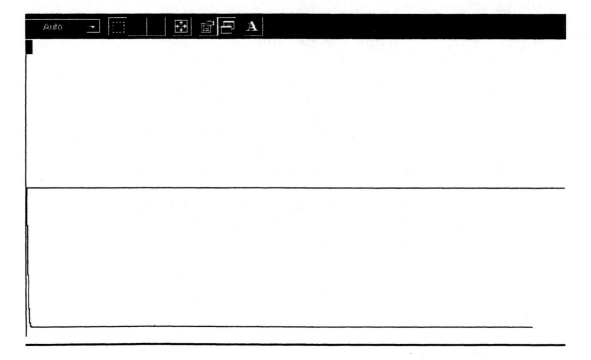

Figure 2b: Suppression of the chaos by incursive anticipatory control of the robot arm in the same condition as in Figure 2a: cos β(t) is plotted as a function of Δt.

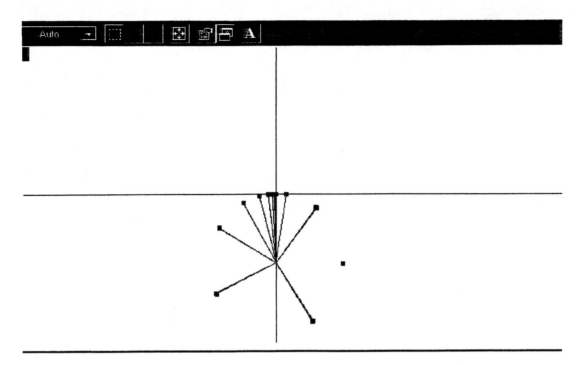

Figure 3a: Chaotic movement of the robot arm at the set point ($\rho_r=0,\gamma_r=0$) starting from the initial condition ($\rho_r= \sqrt{2.1},\gamma_r=-\pi/4$), in the working space.

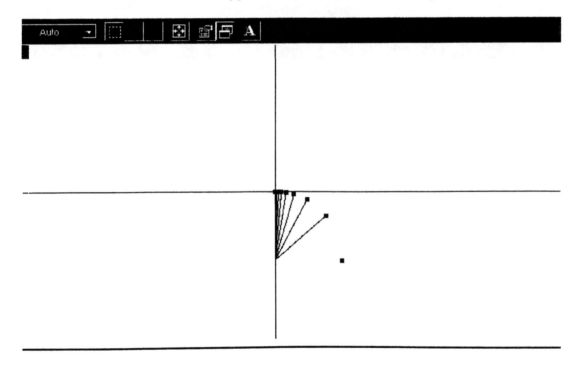

Figure 3b: Suppression of the chaos by incursive anticipatory control of the movement of the robot arm at the set point ($\rho_r=0,\gamma_r=0$) starting from the initial condition ($\rho_r= \sqrt{2.1},\gamma_r=-\pi/4$), in the working space.

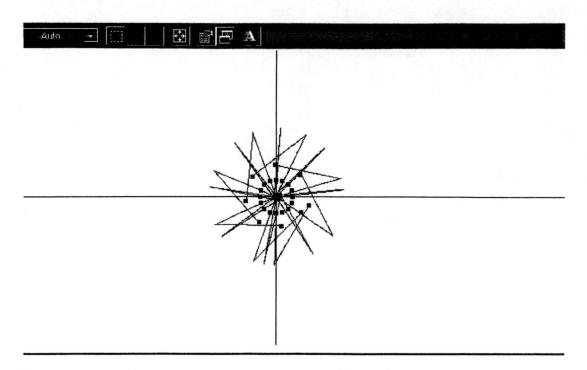

Figure 4a: Instability (bifurcation) of the movement of the robot arm tracking a circular trajectory (16 points) at distance equal to 0.5 l from the centre of the working space, with the initial condition $\alpha = 0$, $\beta = \pi$.

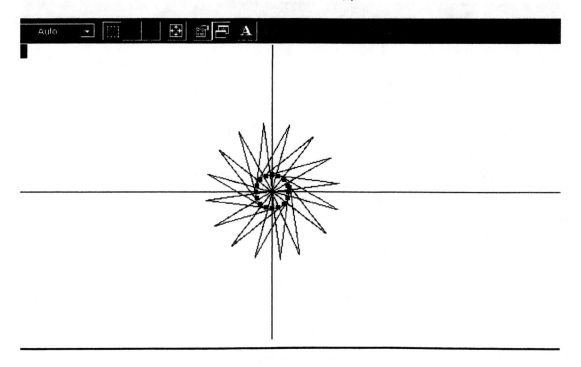

Figure 4b: Stabilisation by incursive anticipatory control of the movement of the robot arm tracking a circular trajectory (16 points) at distance equal to 0.5 l from the centre of the working space, with the initial condition $\alpha = 0$, $\beta = \pi$.

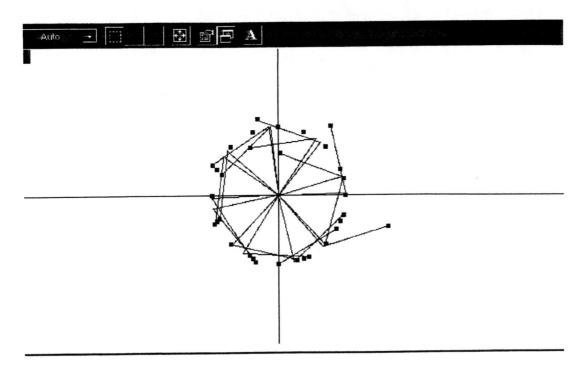

Figure 5a: Instability of the movement of the robot arm tracking a circular trajectory (16 points) at distance equal to 1.0 l from the centre of the working space, with the initial condition α = 0, β = π.

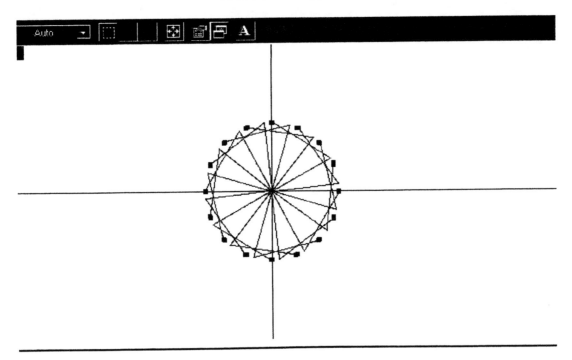

Figure 5a: Stabilisation by incursive anticipatory control of the movement of the robot arm tracking a circular trajectory (16 points) at distance equal to 1.0 l from the centre of the working space, with the initial condition α = 0, β = π.

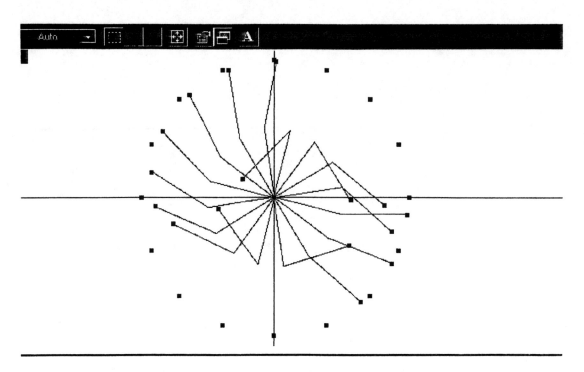

Figure 6a: Instability of the movement of the robot arm tracking a circular trajectory (16 points) at distance equal to 2.0 l from the centre of the working space, with the initial condition $\alpha = 0$, $\beta = \pi$.

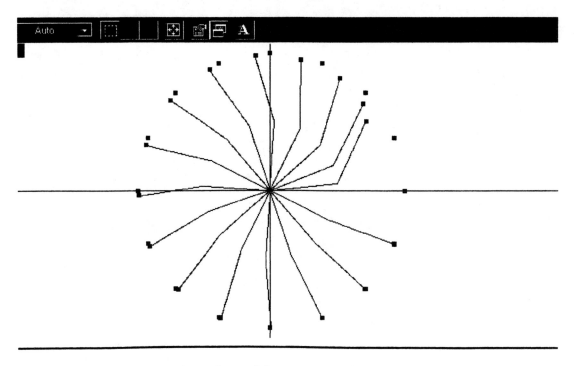

Figure 6a: Stabilisation by incursive anticipatory control of the movement of the robot arm tracking a circular trajectory (16 points) at distance equal to 2.0 l from the centre of the working space, with the initial condition $\alpha = 0$, $\beta = \pi$

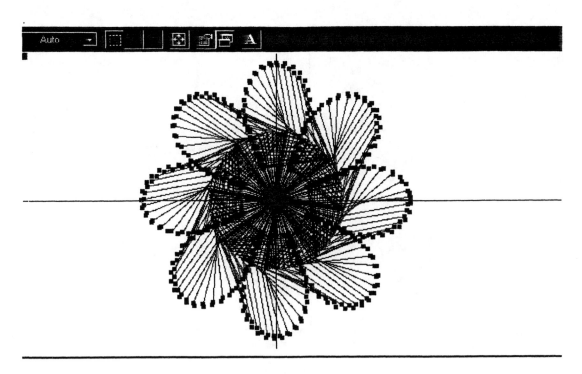

Figure 7: Incursive anticipatory control of the movement of the robot arm tracking a more sophisticated trajectory in the working space starting from the initial condition $\alpha = 0$, $\beta = \pi$.

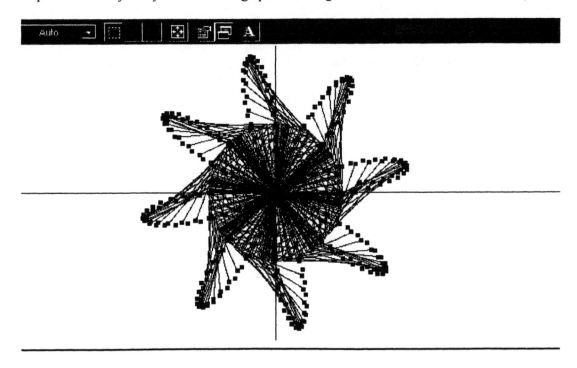

Figure 8: Incursive anticipatory control of the movement of the robot arm tracking a more sophisticated trajectory in the working space starting from the initial condition $\alpha = 0$, $\beta = \pi$.

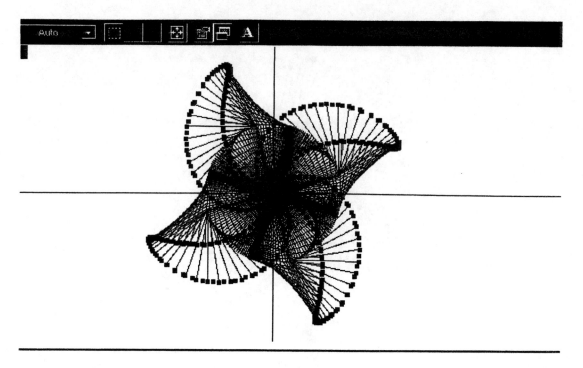

Figure 9: Incursive anticipatory control of the movement of the robot arm tracking a more sophisticated trajectory in the working space starting from the initial condition $\alpha = 0$, $\beta = \pi$.

Figure 4a shows, by numerical simulation of eqs. 3ab, the instability (bifurcation) of the movement of the robot arm tracking a circular trajectory (16 points) at distance equal to $\rho_r = 0.5$ l from the centre of the working space, with the initial condition $\alpha = 0$, $\beta = \pi$.

Figure 4b shows, by numerical simulation of eqs. 6abcd, the stabilisation by incursive anticipatory control of the movement of the robot arm tracking, as in Figure 4a, a circular trajectory (16 points) at distance equal to $\rho_r = 0.5$ l from the centre of the working space, with the initial condition $\alpha = 0$, $\beta = \pi$.

Figures 5a and 6a show the same simulations as in Figure 4a for the robot arm tracking a circular trajectory (16 points) at distances respectively of $\rho_r = 1.0$ l and $\rho_r = 2.0$ l. The instability of the movement of the robot arm is well-seen.

Figures 5b and 6b show the same simulations as in Figure 4b for the robot arm tracking a circular trajectory (16 points) at distances respectively of $\rho_r = 1.0$ l and $\rho_r = 2.0$ l. The stabilisation of the movement of the robot arm is well-seen.

Finally, Figures 7, 8 and 9 show the numerical simulations of eqs. 6abcd of the incursive anticipatory control of the movement of the robot arm tracking different sophisticated trajectories in the working space. Each point of the tracking trajectory is reached by the robot arm by only one computing iterate of eqs. 6abcd.

416

3 Conclusion

In conclusion, the feed-in-time control by incursive anticipation could be a new innovative tool for modelling, simulating and controlling of systems in automation engineering with the introduction of incursion as an extension of the recursion in taking into account the future states in defining self-referential systems.

References

Craig J. J. (1989). *Introduction to Robotics, Mechanics and Control*, Addison-Wesley Publ. Company.

Dubois D. M. (1992). *The Fractal Machine*. Presses Universitaires de Liège, 375 p.

Dubois D. M. (1995). "Total Incursive Control of Linear, Non-linear and Chaotic Systems." *In* G. Lasker (ed.): Advances in Computer Cybernetics. Int. Inst. for Advanced Studies in Syst. Res. and Cybernetics, vol. II, pp. 167-171.

Dubois D. M.(1996a). Feed-In-Time Control by Incursion, in Robotics and Manufacturing: Recent Trends in Research and Applications, volume 6, edited by M. Jamshidi, F. Pin, P. Dauchez, published by the American Society of Mechanical Engineers, ASME Press, New York, pp. 165-170.

Dubois D. M. (1996b). "Introduction of the Aristotle's Final Causation in CAST: Concept and Method of Incursion and Hyperincursion". In F. Pichler, R. Moreno Diaz, R. Albrecht (Eds.): Computer Aided Systems Theory - EUROCAST'95. Lecture Notes in Computer Science, 1030, Springer-Verlag Berlin Heidelberg, pp. 477-493.

Dubois D. M. (1996c). "A Semantic Logic for CAST related to Zuse, Deutsch and McCulloch and Pitts Computing Principles". In F. Pichler, R. Moreno Diaz, R. Albrecht (Eds.): Computer Aided Systems Theory - EUROCAST'95. Lecture Notes in Computer Science, 1030, Springer-Verlag Berlin Heidelberg, 494-510.

Dubois D. M., Resconi G. (1992). *HYPERINCURSIVITY: a new mathematical theory*. Presses Universitaires de Liège, 260 p.

Dubois D. M., Resconi G. (1994). "Holistic Control by Incursion of Feedback Systems, Fractal Chaos and Numerical Instabilities." *CYBERNETICS AND SYSTEMS'94*, edited by R. Trappl, World Scientific, pp. 71-78.

Dubois D. M., Resconi G. (1995). *Advanced Research in Incursion Theory applied to Ecology, Physics and Engineering*. COMETT European Lecture Notes in Incursion. Edited by A.I.Lg., Association des Ingénieurs de l'Université de Liège, D/1995/3603/01, 105 p.

Portier B., Oppenheim G. (1993). "Adaptive Control of Nonlinear Dynamic Systems: Study of a Nonparametric Estimator." *J. Syst. Eng.*, 1, pp. 40-50.

Anticipatory Computing with a Spatio Temporal Fuzzy Model

Stig C Holmberg

Department of Informatics
Mid Sweden University
SE-831 25 ÖSTERSUND, Sweden

Abstract

A cellular automata computer model (STF) for simulation and anticipation of geographical or physical space is constructed. STF has a normalised and continuous, i.e. fuzzy, system variable while both the time and space dimensions take on discrete values. Global rules are employed in STF, i.e. there is a total interdependence among the cells of the automata. Outcomes of the model can be interpreted more as possible future states than exact predictions. Preliminary results seem to be well in line with main characteristics of real geographical spaces.

Key words: Geographical space, Anticipatory Computing, Cellular Automata, Spatio Temporal Fuzzy Model.

1. INTRODUCTION

Physical or Geographical three dimensional space is recognised as a common space, shared by all concrete systems, living and non living (Miller, 1978). Due to this all inclusive property of geographical space, any event taking place somewhere in it may have an impact, sometimes severe, on any individual or group of individuals living within its boundaries. Hence, effective, appropriate and timely anticipation (Rosen, 1985), or early warning, emerges as an urgent and highly desirable task in our common living space. Research on anticipation in geographical space has already been reported, among others, by Phipps and Langlois (1996), Langlois and Phipps (1997), and Holmberg (1997c). However, the research in this field still is in its infancy, the complexity of the research task is considerable, and the set of possible techniques and tools to use in the solutions is vast. In short, the research challenge is far from trivial and most of the work remains to be done. Hence, the purpose of this paper is to launch a research process toward such a system by proposing some new approaches and ideas to anticipation and early warning within geographical space.

2. PROBLEM ANALYSIS AND BASIC APPROACHES

In this section we will start with defining our problem domain, i.e. a geographical space (GS) or territorial concern (TC). Further, the main properties of that space will be analysed and at last, some proper conceptual and scientific tools for its modelling and anticipation will be identified.

2.1 Geographical space as a territorial concern

Geographical space (GS) taken as a territorial concern (TC) is the problem domain addressed in this paper (Figure 1). In this context, a TC being a community based organisation for the design, construction, and maintenance of a territory, i. e. a geographical region (a space) with all the living creatures and non living objects residing within its boundaries. With other words, a TC is a living system or a society with the responsibility (the concern) to establish and maintain a structure and to keep a set of essential variables within critical levels (Holmberg, 1994). Examples of such essential variables may be employment rates, housing standards, education possibilities, health care and other public services, communication and so on. Taken together those variables can be seen as an index of the quality of life. A common political goal may be to provide the same standard or quality of life throughout the whole country, i.e. the whole GS or TC.

Further, the interdependencies among all those variables are very complex and not fully known. In many cases there are also considerable time lags between cause and effects. A common theory, anyhow, is that each area has a direct influence on its neighbours, i.e. the same conditions or states tend to hold for an area and its direct neighbours (Langlois and Phipps, 1997).

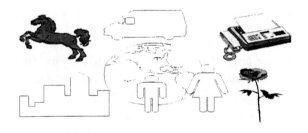

Figure 1. A Territorial Concern is a geographical space with all living and non living objects within its boundaries.

However, within a TC, let it be a city, a nation, or the whole globe, current states or events at one point also seem to have a direct influence on future states at locations further away within the same space, i.e. there will be a global influence. Some examples may help to illustrate this idea. (1) In a city, the separation of housing and business into disjoint areas may increase the risk of riots and social unrest by young people in the city centre. (2) The closing down of a factory in one part of a country will have a positive influence on job opportunities and the same concern's factory in another part of that country. (3) A restrictive attitude to smoking in one part of the world will increase smoking and deteriorate health conditions in other parts. Hence, findings in complexity and chaotic behaviour (Casti, 1994; Prigogine, Stengers, 1984) obviously are relevant also for the TC). Even more, the current trend seems to go toward increasing interdependence. Mulgan (1997) expresses this by saying that the world has become more interdependent and interconnected and that more people depend on more other people than ever before.

420

2.2 The fuzzy nature of the territorial concern

Most entities or aspects of the territorial concern (TC) are continuous. For example, the landscape continuously changes from flats to hills or cultivable land gradually changes into sterile ground. With other words, the reality is continuous (Miller, 1978). Further, if we choose to express the TC as a system model, the system state of that model will be expressed with a set of system variables (Klir, 1991). Hence, in order to express the continuous character of reality those system variables also ought to be continuous. This is a bit contradictory to classical bivalent logic but fits very well with modern fuzzy or multivalent logic (Klir, Yuan, 1995) and the fuzzy principle, i.e. "Everything is a matter of degree" (Kosko, 1993).

2.3 The Space Hypothesis

In the so called Space hypothesis (Holmberg, 1997c), it is concluded that future states of a point or area within a geographical space will be functionally dependant of earlier states at all other points of the same space, i.e., the local future will already be present at the current time, somewhere within the boundaries of the geographical space at hand. The rationale for the hypothesis can be found within Systems Thinking and Systems Research. Firstly, according to Miller (1978) all concrete systems, living and non living alike, share a common space, the physical or geographical one. Secondly, according to Ackoff (1981) a system can be seen as a set of interacting units (variables) with relationships among them, i.e. the behaviour of each element or unit has an effect on the behaviour of all other units, and each element is effected by the behaviour of all the other units in the system. Hence, it may be justified to identify the GS or TC as a system and further to conclude that all the units of that space are interdependent.

2.4 Anticipation for Probability or Possibility

Dealing with the future, but also the past, involves a great deal of uncertainty. Just the actuality, or current state at time t_0, is known up to the degree that variables are identified and their values are measured (Figure 2). Traditionally such uncertainty has been handled with help of classical probability theory.

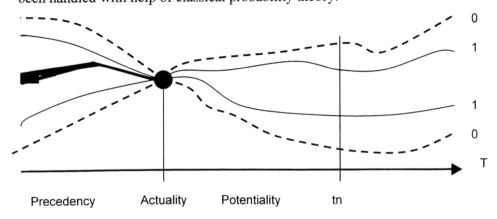

Figure 2. The anticipation situation with the possible states seen as fuzzy super sets.

However, with the growing insights in fuzzy logic, possibility theory (Klir, Yuan, 1995) has established itself as a serious alternative with many compelling properties. Hence, looking backwards or forwards from the current state, anticipation does not only have to be seen as a prediction of the most probable future states (probabilistic anticipation), it can also be seen as a computation of all possible future states (possibility anticipation). With that later approach, The future, or potentiality, becomes a superset of fuzzy sets, one set for every discrete time step t (:t = 0..n). For example, in figure 2 the intersections between the time profile t_n and the boarder lines 0, 1, 1, and 0 constitute the characteristic points of the membership function (Klir, Yuan, 1995) of the fuzzy set of possible system states at time t_n.

2.5 Equifinality, Homeostasis and Autopoiesis in relation to Anticipation

The concept of equifinality (Bertalanffy, 1968) seems to be utterly important in this context. Equifinality says that the steady state of open systems depends more upon system constants or system properties than environmental conditions and foreign disturbances. Hence, if the system's memory fails to recall the exact system trajectory it is always possible to apply reversed anticipation in order to calculate the precedency area in figure 2. That area is enfolding all earlier points which, due to equifinality, can be possible precedence states to the current one.

Homeostasis (Cannon, 1932) and autopoiesis (Varela et al, 1974) may be two other system concepts of great relevance in this context. First, homeostasis is a tendency by virtually any living systems to remain in steady state or to keep its variables within pre-set limits.

> "When a system remains steady it does so because any tendency toward change is met by increased effectiveness of the factor or factors which resist the change" (Cannon, 1932).

The research and writings on homeostasis or steady state are comprehensive and convincing. Hence, it may be assumed that homeostasis is a vital factor also within TC:s.

Second, autopoiesis is a system property making it possible for the system to produce itself. Formally an autopoietic organisation is defined as a unity by a network of productions of components which (i) participate recursively in the same network of productions of components which produced these components, and (ii) realise the network of productions as a unity in the space in which these components exist. (Varela et al, 1974). Hence, to the degree that the TC is an autopoietic system its evolution over time will be inside the system from the very beginning. The problem is just to find that pattern.

2.6 An Ideal Solution of anticipation within a territorial concern

In this context, Ideal Solution or Ideal Design is used in the sense of Ackoff (1981), i.e. not ideal in any absolute sense but ideal as the best solution that the designer can conceptualise at the moment of design. In the next step, a real solution will be realised or implemented as close to the ideal one as possible. The main benefit of this two step approach is that it helps overcome self imposed mental restrictions.

Since their invention by von Neumann, cellular automata (CA) (Burks, 1966) have become well known even in simulation contexts. Further, already they have successfully been used also for simulation of geographical space (Langlois and Phipps, 1997). Therefor, in this first approach it feels natural to apply a CA-solution. The system set will be expressed with a set of system variables (Klir, 1991). In the CA this will be solved by several layers, one layer for each variable.

In their approach however, Langlois and Phipps uses only discrete or crisp values for the system states and their future states are calculated just from the von Neumann or Moore neighbourhoods (Rietman, 1993), i.e. due to local rules (Figure 3). In the ideal solution on the other hand, according to the analysis above new states will be calculated with help of global rules and model outputs will take on continuous values in the interval [0,1]. Those can be interpreted as membership values in the fuzzy set of acceptable system states.

The term global will be applied both in a spatial and a temporal sense. Spatially this means that the values of all the cells in the CA will be used for calculating each new cell value (Holmberg, 1997c). In the temporal dimension it will be possible to apply Incursion and Hyperincursion (Dubois, Resconi, 1992; Dubois, 1995; Dubois, 1996). In this way global rules will be applied also in the temporal dimension.

The rules governing the behaviour of the CA can be deterministic or stochastic. Due to what has already been said about equifinality, autopoiesis and homeostasis it can be assumed that in anticipating a TC deterministic rules will be most realistic and suitable. However, it will also be possible to let a stochastic process influence the behaviour to a certain, but normally rather small, degree.

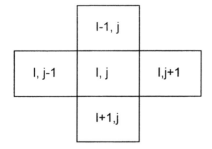

Figure 3. The primary local von Neumann and Moore neighbourhoods for cellular automata.

Further, in the ideal solution the CA is conscious about its own state. Hence, it will also be able to self adjust or to apply a homeostasic behaviour. In this way the rules will be, not only global, but also dynamic, continuously adapting to the actual situation or system state.

The ideal solution will also be learning, i.e. it will be able to identify a goal or goal function and to adapt its behaviour for following the goal as close as possible. At last, it will preserve the topology of the actual territorial concern it is mapping.

3 THE STF MODEL

The Spatio-Temporal Fuzzy Model (STF), as it has been presented by Holmberg (1997c), exhibits some, but not all, of the properties asked for in the above problem analysis and ideal solution approach. Hence, after a short description of the original model (STF-0) an enhanced model (STF-1), which will meet some more of the requirements, will be proposed. However, compared with the ideal solution STF, even in its enhanced version, still represents just a preliminary approach.

3.1 The original version, STF-0

- **The cellular grid**

  In STF-0 the geographical region in focus is modelled as a two dimensional matrix **S** with m x n quadrate cells or sites. In order to fit into a natural region with irregular boundaries, a logical matrix **L** will be used to mask out the parts of **S**, which fall outside of the mapped geographical region.

  In addition to the state matrix S, there are also impact matrices $I_{i, j}$, one for each cell in S. Hence, $I_{i, j}$ is the impact matrix of cell i, j in S. The purpose of $I_{i, j}$ is to express the strength of impact on cell i, j from the other cells in the region according to what will be explained under the heading "transition rules" below.

- **The states**

  In cellular automata (CA) the state variable s will normally take on discrete, or even binary, values (Rietman, 1993, Langlois and Phipps, 1997). However, in STF the state variable will be a normalised continuos variable in the interval [0, 1]. This can be seen as a fuzzy value indicating to what degree the cell belongs to the fuzzy set of acceptable states.

  The original cellular automata, which were invented by von Neumann already in the late 1940's, have been characterised as discrete space-time models (Burks, 1966). Due to this change from a discrete to a continuous state variable it is appropriate to characterise STF as a fuzzy space-time model.

- **The neighbourhood**

  In most CA local neighbourhoods of the von Neumann or Moore type (Figure 3) are employed (Rietman, 1993), i.e. just the four or eight closest cells respectively are taken into account. However, as it has been demonstrated by the above analysis, future states of any cell in the region may be influenced by states, decisions or actions in any other cell. Hence, in order to accommodate to this observation a global neighbourhood will be used in STF, i.e., we will have a transition function of the general form.

$$S_{i, j, t+1} = f(s_{1,1,t} \ldots\ldots\ldots\ldots s_{m,n,t}) \tag{3}$$

That global neighbourhood is expressed in the right side of equation (3) because here all cells, from the first $s_{1,1}$ to the last $s_{m,n}$, take part in the computing of new values. This way of defining the neighbourhood is also in correspondence with main stream findings of system theory (Ackoff, 1981, Miller, 1978).

- **The initial state**

The S-values, i.e. the values of each cell in the space matrix S, can be set manually but in the normal case a random number generator with a rectangular distribution will be used to set values in the interval [0,1]. The same with the impact or I matrices, there is a manual option but normally even here the same random number generator will be used.

The values of the I-matrices can be both positive and negative, indicating both positive and negative influence. Numerically, however, they will normally be rather small, i.e., $-0.3 \leq i \leq 0.3$. The only exception being the element $i_{i,j}$ indicating the influence of cell value $s_{i,j}$ on its next generation. This value being in the interval 0.7 to 0.9 in order to give some inertia to the system.

The normal procedure will be to use the random number generator in most cases. However, before starting the simulation run some values may be set manually, for example in order to test for sensibility to small changes in initial values.

- **The transition rules**

"Transition rules" is used in normal CA vocabulary for denoting the rules governing the calculation the states of new generations. With a continuous state variable, however, it will be no real "transitions" and it becomes more appropriate to speak about "state calculation rules" or something like that. Anyhow, in STF-0 for calculating the next value of cell $s_{i,j}$ the impact matrix $I_{i,j}$ is used as a filter. Hence, by multiplying a cell in S, for example $s_{r,c}$, with its corresponding cell in $I_{i,j}$ the impact or contribution from cell r, c on cell i, j will be given. Further, by calculating those contributions from all the cells and adding them together the new value of $s_{i,j}$ will be obtained.

$$s_{i,j,t+1} = \Sigma i_{i,j,c,r} \, s_{c,r,t} \quad (:c = 1..n, \, :r = 1..m) \qquad (4)$$

- **The frequency vector**

In a crisp CA, a frequency vector is normally used to record the number of cells having the different possible values at each simulation step. This is a good tool for analysing what has happened during the simulation but no such device has been implemented in STF-0.

- **The probabilities**

A normal crisp CA can be calculated due to deterministic or probabilistic rules (Langlois and Phipps, 1997). Even a CA with mixed rules would be possible. In this case part of the new state, perhaps the main part, is calculated with help of a deterministic rule while the rest, normally a minor part, is calculated probabilistically. This later procedure seems to be a realistic modelling of what is happening in many real situations. However, STF-0 has no such probabilistic component.

- **The general procedure**

The simulation develops in the following general steps:
First the initial state, i.e. at time $t = 0$, is set, both for the state matrix S and for the impact matrices $I_{i,j}$ ($:i = 1..m$, $:j = 1..n$). This state is produced by a random number generator, but before the simulation begins, individual cell-values can be set manually to any desired value within permissible limits. Next, new states of S, i.e. S_t ($:t = 1$, tmax), are calculated in succession from its prior state according to the transition rules. Hence, S_{t+1} is in fact a function of S_t and the impact matrices as denoted by (5)

$$S_{t+1} = f(S_t, I_{i,j})\qquad\qquad(5)$$

It is worth noting that the impact matrices are not changing during the simulation. Further, a new state of S is calculated with help of just the prior generation of S, i.e. its immediate predecessor. However, this simplification in temporal dependency is compensated for by the very complicated spatial dependency. Hence, as all cells are related to all other cells there are a very great number of impact routes leading to a cell, some of those are direct one step routes while others can have a length up to the total number of cells in the space. The consequence being that the impact on a cell $s_{i,j}$ is due to values of other cells just one step back up to the total number of cells, due to the length of the impact route.

3.2 The enhanced version, STF-1

During the initial work with STF-0 several improvements of the original design have become obvious. Those enhancements will be discussed here.

- **The cellular grid**

In STF-0 the topological shape of the S-matrix is similar to that of the mapped geographical region. However, by transforming S into a linear one-dimensional vector V, calculations will become more easy and straightforward. The connection to the geographical space can be preserved even in this case, if for each unit in the vector V, a record is kept with the state value s and the geographical coordinates x an y. Further, instead of an impact matrix I for each cell in S, something that quickly will become very cumbersome with increasing number of cells, we will with this approach become just one binary fuzzy relation.

This fuzzy relation is expressing the impact between the elements in V, i.e. we will become a membership function of the form

$$R: V \times V \rightarrow [0,1].$$

For each element i and j in R the membership degree R(i, j) may be interpreted as the degree to which the value of V_i will have an impact on V_j. Hence, as a result of this transformation we will become one state vector V and one relation matrix R. The dimension of R being n rows by n columns, where n is the number of elements in V (Figure 4).

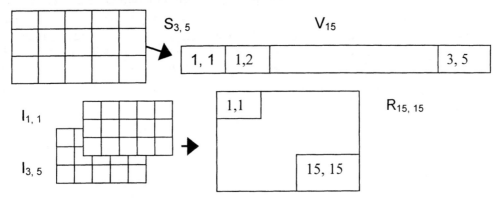

Figure 4. The state matrix S is transformed into a state vector V and the impact matrices I into a quadratic relation matrix R.

- **The states**

  It can be hypothesised that in many cases it is not so much the absolute values of the cells at a certain moment or simulation step, as their trends that will have the greatest influence on the other cells in the space. In order to handle that situation not only the state values s but also their delta values, ds will be part of the model.

- **The transition rules**

  In order to take advantage of the ds-values it will in each simulation be possible to choose to what degree new state values will be computed from s-values or ds-values respectively. Hence the generic formula will be like equation 6.

  $$V_{t+1} = C_1 \, f(VS_t, R) + C_2 \, f(VdS_t, R) \qquad (6)$$

- **The precedency or outcome history**

  A database will in this version be connected to the model. Hence, it will be possible to store the system state at each simulation step, i.e. the whole history or precedency of the model will be accessible for later analysis. Appropriate analysis tools have to be developed in order to take advantage of this history base and to ease analysis and understanding.

- ## The general procedure

The simulation will work the same way as earlier but in this version it will be possible to switch between two modes, simulation mode and analysis mode. It will also be possible to restart the simulation from an arbitrary point, i.e. different alternatives can easily be investigated and analysed.

4 THE STF-0 COMPUTER PROGRAM

The Spatio-Temporal Fuzzy Model (STF-0) has been implemented as a computer program (STFP-0). With this program it is possible to make simulations according to the rules defined in STF-0. A print out of the program's computer screen is shown in figure 5. The program is event driven with command buttons for the available operator actions. The space matrix, in this particular case a 8 x 8 quadratic matrix, is displayed to the left while one impact matrix at a time is displayed to the right. By operating the navigation buttons under the display of the impact matrix it is possible to move between the impact matrices. Under the state matrix some essential system state parameters are displayed.

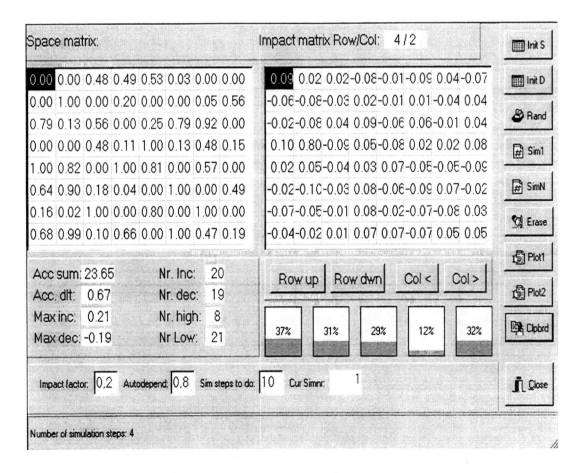

Figure 5. Print out of the STFP-0 computer screen.

428

"Acc sum" displays the accumulated sum of the state values of all cells in the space. "Acc dlt" displays the sum of all the changes during the latest simulation step. "Max inc" and "Max dec" is the greatest positive and negative change in any cell during the latest simulation step. "Nr inc" and "Nr dec" is the number of cells which have increased respectively decreased their value during the latest simulation step. At last, "Nr high" and "Nr low" are the number of cells which have reached the max (1.0) and min (0.0) values. The values of "Acc sum", Nr inc", "Nr dec", "Nr high" and, "Nr low" is also displayed graphically.

In figure 5 the impact factor is set to 0.2. This means that the values in the impact matrices in this case will fall in the interval [-0.1, 0.1]. The impact a cell will have on its own next generation, however, is set to 0.8. The actual simulation step is displayed in the lower left corner of the screen.

The simulations can be stopped and resumed at arbitrary steps. At each step, the values are automatically saved into a data base.

The s-values can be set manually, but will normally be set in the interval [0,1] using a random number generator with a rectangular distribution. The same with the I-matrices, they can be set manually, but normally even here a random number generator with a rectangular distribution will be used. However, before starting the simulation run, it is possible to manually change the generated s- and i-values in order to test for sensibility to initial values.

With the current formula according to equation 4, the s-values can grow outside the normalised interval limits [0,1]. In this version of the program that problem is solved by just clipping the values at the interval limits.

Due to lack of time, so far the program has been used only in a limited number of simulation runs. Those initial simulations have been focused on the following questions:

- Does the results support the space hypothesis?
- Is the model very sensitive to small changes in its initial conditions?
- Are there any parameter settings which make the model's behaviour chaotic?
- Are there any strange or surprising patterns in the model's behaviour?

5   SIMULATION RESULTS

Due to the limited number of simulation data obtained so far, there will still be too early to draw any well grounded or statistically proved conclusions. Anyhow, the current indications point in the following directions:
- There are always winners and losers in the model space and normally there are some more losers then winners. This pattern develops after just a few simulation steps. In following steps interesting and surprising changes will occur in details but the general picture will remain very stable.

429

- In most cases the model seems not to be unstable or overly sensible to initial conditions. However, the possibility of a chaotic behaviour (Peitgen et al, 1992) has to be investigated in more detail.
- The number of cells with intermediate values, i.e., $0.0 < s < 1.0$, will decrease with time, but will eventually stabilise well above zero. However, if the self impact factor is set below a certain limit (often about 0.7), all cells will quickly go down to zero.
- There is an obvious cyclical behaviour in the model, possibly with several superimposed frequencies.
- Compared to cellular automata in their traditional form, STF seems to have introduced several new and insight generating dimensions.
- The hypothesis concerning anticipation in spatial systems seems to be reinforced by the initial simulation outcomes. Hence, the crucial point being if it is possible to find enough empirical data for being able to solve the dependency equations. If that is possible, STF can be applied for anticipating real geographical systems.

To the degree that STF is a true model of a real TC it may also be possible to draw two preliminary and tentative conclusions concerning the real geographical space or territorial concern. First, it may be very difficult to obtain the political goal of equal opportunities in all parts of a country. In all simulation runs great differences have evolved rapidly. Secondly, any country may collapse if it is not managed properly. However, a suitable antcipatory or early warning system may give timely warnings.

In most cases the impact values for a real geographical system or TC will not be known. However, in principle they can be calculated from historical data. Hence, given that those values or system relationships can be found, the model, i.e. STF, can be simulated or evolved in Model-Time in order to anticipate the Real-Time behaviour of the TC.

Compared with an ordinary cellular automata (Rietman, 1993), the STF is based on global rules and the cell states are continuous values in the interval [0,1]. As a consequence, STF will exhibit both positive and negative feedback and the length of the feedback loops will vary from one up to the total number of cells in the model.

Dubois (1995) has presented hyperincursion in which future states are computed from their neighbours' states at past, present but also future time steps. In STF future steps are calculated not only from neighbours but from all cells in the system. By combining the two approaches, a spatial hyperincursion will emerge. The anticipated properties of such an operation are enticing, but still they have to be explored.

Just to mention one obvious extension or improvement of STF. Fuzzy or Soft Computing methods and tools for spatial decision support, for example in finding optimal or suitable locations for different activities, has been developed lately (Holmberg, 1997b). By integrating those tools and methods with STF, it will be possible to test or simulate the global outcome of different localisation alternatives.

6 CONCLUDING REMARKS

It is more or less impossible to make justice to a subject as complex as anticipation in geographical space in just a few printed pages. Specially so in a paper just trying to analyse the basic problems and requirements and to launch a first attempt to a solution approach. Further, it has to be emphasised that anticipation in general, and specially anticipation in geographical space, is a young and rapidly developing scientific branch. Weather it will grow and establish itself as a recognised field or if it will fragment and become submerged in other branches of systems science is still a delicate question to answer. However, the work underlying this paper, irrespective of its initial and tentative character, has indicated that anticipation in geographical space is here to stay. The urgent need for anticipatory, or early warning, systems in the current world is well recognised and hence, hopefully, the coming years will witness both the maturing of the theory and the development, implementation and employment of fully operational anticipatory systems.

References

Ackoff R. L. (1981), Creating the Corporate Future, Wiley, New York.

Bertalanffy von, L. (1968), General System Theory, Foundations, Development, Applications. George Braziller, New York.

Burks A. W. (Ed.) (1966), Theory of Self-Reproducing Automata. John von Neumann. University of Illinois Press, Illinois.

Cannon W. B. (1932), The Wisdom of the Body. W. W. Norton, New York.

Casti J. L. (1994), Complexification; Explaining a Paradoxical World Through the Science of Surprise. Harper Collins, New York.

Dubois D. M. (1995), The Hyperincursive Field and the Aristotelian Final Cause. In Lasker G. E. (Ed), Advances in Computer Cybernetics, Vol III, pp 13-17, The International Institute for Advanced Studies in Systems Research and Cybernetics, Windsor.

Dubois D. M. (1996), Introduction of the Aristotle's Final Causation in CAST, Concep and Method of Incursion and Hyperincursion. In Pichler F., Moreno Díaz R., Albrecht R. (Eds), Computer Aided Systems Theory – EUROCAST'95. Lecture Notes in Computer Science, Springer, Berlin.

Dubois D. M., Resconi G. (1992), Hyperincursivity, a new mathematical theory. Presses Universitaries de Liège, Liège.

Holmberg S. C. (1994), Geoinformatics for urban and regional planning. Environment and Planning B: Planning and Design, Vol. 21, pp 5-19.

Holmberg S. C. (1997a), Fuzzy Anticipation of Complex System Behaviour. In Lasker, G. E., Dubois, D. M. (Eds), Advances in Modeling of Anticipative Systems, pp 36-40. ISBN 092-1836430, International Institute for Advanced Studies in Systems Research and Cybernetics, Windsor.

Holmberg S. C. (1997b), Fuzzy Control for Command, Control and Decision Support in Mobile and Geographically Distributed Operations. SOCO'97, pp 73-77, ICSC Academic Press, Zurich.

Holmberg S. C. (1997c), Anticipation in Spatio-Temporal Systems. Paper presented at the 9th International Conference on Systems Research, Informatics and Cybernetics, Baden-Baden, August 18-23, 1997.

Klir G. J. (1991), Facets of Systems Science. Plenum, New York.

Klir G. J., Yuan B. (1995), Fuzzy Sets and Fuzzy Logic, Theory and Application. Prentice Hall, New York.

Kosko B. (1993), Fuzzy Thinking, The new Science of Fuzzy Logic. Flamingo, Harper Collins Publishers, London.

Langlois A., Phipps M. (1997), Automates cellularies, application à la simulation urbaine. Hermes, Paris.

Miller J. G. (1978), Living Systems. McGraw-Hill, New York.

Mulgan G. (1997), Connexity; How to Live in a Connected World. Chatto & Windus, London.

Peitgen H-O., Jürgens H., Saupe D. (1992), Chaos and Fractals, new frontiers of science. Springer-Verlag, New York.

Phipps M., Langlois A. (1996), Spatial dynamics, cellular automata and parallel processing computers. Environment and Planning B.

Prigogine I., Stengers I. (1984), Order out of Chaos; Man's new Dialogue with Nature. Bantam Books, New York.

Rietman E. (1993), Creating Artificial Life; Self-Organization. Windcrest/McGraw-Hill, New York.

Rosen R. (1985), Anticipatory Systems. Pergamon Press.

Varela F. G., Maturana H. R., Uribe R. (1974), *Autopoiesis: The Organization of Living Systems, its Charecterization on a Model.* Bio. Syst, **5,** 187-197.

Some Issues of the Human Arm Motion
Obtained from its Kinematic Model

Jadran Lenarčič

Department of Automatics, Biocybernetics and Robotics
The Jožef Stefan Institute
Jamova 39, Ljubljana, SLOVENIA
jadran.lenarcic@ijs.si

Abstract - A possible direction in the development of future robot manipulators is in the so-called humanoid robot systems. The objective is to develop robots that behave, work, and communicate like humans. This will open a new era in robot applications, in particular in service robotics, will facilitate robot programming, and create a more effective human-robot interface. This paper presents some potential motion properties of a future humanoid robot manipulator retrieved from kinematical models of the human arm, such as the kinematic redundancy, the workspace properties, and the velocity-torque capabilities. Special attention is given to the utilisation of the kinematic singularities in handling heavy objects. Based on experiments, it is unequivocally demonstrated that the human arm takes advantage of the kinematic singularities in order to compensate weak actuation, while in the actual robotic practice the kinematic singularities are avoided.

Keywords: kinematic model, human arm, robot manipulator, humanoid robot, singularity

1. Introduction

The human arm is a very complex mechanism containing hundreds of degrees of freedom, joints and actuators. The work that has been done in the past to explain the human arm motion has mainly been concentrated on the biomechanical aspects. In this area, the objective has been to develop precise anatomical models which include the complete system of muscles, tendons, and bones. Such models, unfortunately, are too complex to be used in mechanical design of robots and in other applications in engineering. On the other hand, there have been attempts in robotics to develop human-like robot arms or hands, as well as entire human-like bodies containing arms and legs. The "piano-player" robot designed by Kato in Japan had fascinated the visitors of Expo some years ago and has therefore become one of the most popular robots in the whole era of robotics. This kind of robots, however, are only visual copies of humans. They directly (as much as possible) copy human proportions and are usually programmed in the play-back manner.

The performances of humanoid robots are in some sense similar to human's capabilities. In the utilisation of humanoid robots, this fact improves the human-robot interface, facilitates the robot programming, and increases the adaptability of robots to different and more complex

CP437, *Computing Anticipatory Systems: CAYS--First International Conference*
edited by Daniel M. Dubois © 1998 The American Institute of Physics 1-56396-827-4/98/$15.00

applications. A future industrial robot would thus be able to learn different jobs, change its working place under request, communicate with humans or other robots in the working process, and learn from experience. Similar properties are requested for service and personal robots introduced in non-manufacturing applications. So far, the applications of humanoid robots have been seen in different types of entertainment. However, the new technological development, the new types and higher capacity of sensors, control systems, driving systems etc. has opened new possibilities in the development of a new generation of humanoid robots devoted to more "serious" applications, such as medical care systems or personal robots. This direction of research has also been stimulated by the latest achievements in service robotics that include a lot of human-type issues, such as sensing and perception, autonomy, mobility, dual-arm cooperation - all supported by an effective human-robot interface. Some very recent experimental results have announced the new era in robotics. In the last December, the research group of Honda in Japan has presented a prototype of a biped humanoid robot. It has two arms and two legs. The same research direction has clearly been noticed in other laboratories throughout Japan and USA. It seems, however, that the European research community hasn't yet been fully attracted by this topic.

In our investigations of the human arm motion, we have promoted a kinematic model that enables us to simulate and understand a repertoire of the human arm motion abilities. By studying this model, we have found or speculated characteristics and explanations that could give an impact on the design of future robot manipulators, their control and programming in order to increase their adaptability and autonomy. In the present paper, we discuss the human arm proportions in relation to its reachability, we study the amount of the kinematic redundancy, the workspace geometry, and the velocity-torque capabilities. In particular, we are interested in the utilisation of the kinematic singularities in handling heavy objects. By a series of experiments we show that the human arm takes advantage of the kinematic singularities to compensate weak actuation, while in the actual robotic practice the kinematic singularities are avoided because they may cause stability problems in standard control algorithms.

2. Mechanical Portrait

The so-called anthropomorphic mechanisms used in industrial robotics must provide a good dynamic performance. They are, therefore, a very simplified approximation of the human arm. Even the 3R mechanism (if we do not consider the wrist) with successively perpendicular revolute joints, two in the shoulder and one in the elbow, is termed anthropomorphic. Such a structure, however, cannot represent one of the most typical issues in the human arm motion, which is the kinematic redundancy. To represent the spatial motion of the human arm, in our opinion, the simplest still acceptable approximation of the human arm mechanism must contain at least four revolute joints, where three are perpendicular rotations put in the glenohumeral joint (humeral abduction, humeral flexion, and humeral rotation) and one is the rotation representing the elbow flexion-extension.

From the mechanical point of view that considers the number of degrees of freedom, the human arm mechanism, particularly the shoulder joints, are the most complex copy of the human body (Högfors *et al.*, 1987). In addition, the human arm possesses a number of branches of parallel mechanisms (more than one actuators are driving the same joint) whose links are deformable. Nevertheless, in different investigations of the human arm motion capabilities its mechanism usually modelled either by a serial mechanical manipulator consisting of a group of rigid links connected in a special arrangement of one-degree-of-freedom revolute joints as reported by Benati *et al.* (1980), and Engin (1990)., each possibly specifying a selective motion, or by a spherical joint describing the shoulder complex where a selected curve specifies the attainable range of motion as reported by Dvir and Berme (1987).

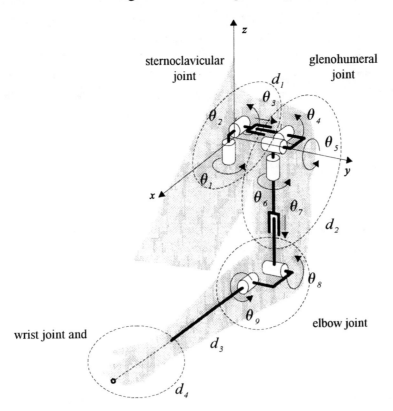

Fig. 1: A serial model of the human arm kinematic structure

Fig. 1 shows a kinematical model of the human arm mechanism as it was proposed by Lenarčič and Umek (1994) and improved by Umek-Venturini and Lenarčič (1996). This model has been obtained based on a number of optical measurements of some selective movements of the arm. The main aspiration of the model has been to compute the reachable workspace of the arm for evaluation and quantification purposes in a rehabilitation procedure but it has turned out that it may have implications in many other areas. This model incorporates a kinematic structure of nine degrees of freedom (without the wrist and the hand). The first group forms the sternoclavicular joint, where θ_1 is the angle of the clavicular flexion/extension, θ_2 is the angle of the clavicular abduction/adduction, θ_3 is the clavicular translation. The second group is the glenohumeral joint, where θ_4 is the angle of the humeral abduction/adduction, θ_5 is the angle of the humeral flexion/extension, θ_6 is the angle of the humeral rotation, and θ_7 is the humeral translation. There are two rotations in the elbow joint, θ_8 is the angle of the elbow flexion/extension, while θ_9 is the angle of the elbow rotation. It is important to stress that the translations along the clavicula and along the humerus (θ_3 and θ_7) are dependent degrees of freedom. Their dependency on the elevation of the arm is demonstrated in Umek-Venturini and Lenarčič (1996). It has been calculated that the upper arm and the shoulder link are shortening as a function of the arm elevation. Other joints can be considered independent. The model contains three rigid links, the shoulder link which is represented by its length d_1, the upper arm link whose length is d_2, and the forearm link whose length is d_3. The length of the palm is d_4. Additional (at least 2) rotations must be included in the wrist, but they are not taken into account in this paper. The basic drawback of the presented human arm model is that it is a purely serial mechanism and that it (since it is serial) contains kinematic singularities inside the workspace. An unavoidable singularity occurs when θ_5 is ± 90 deg. In this singularity the mechanism cannot provide continuos motion in at least one direction inside the workspace. The real human arm does not have any limitations of this kind. Note that, in order to avoid the problem, a model of the human arm should be made as a combination of parallel and serial mechanisms.

3. Workspace Geometry and Proportions

The ranges of motion in joints change from one configuration to another. Some are very large, for instance, the humeral flexion/extension or humeral abduction/adduction, others are quite small. The ranges of motion in the sternoclavicular joint are only few degrees. Even though, a functional analysis has shown that the sternoclavicular joint has an important role which is the collision avoidance between the arm links and the body. If we fix the sternoclavicular joint, the motion of the remaining part of the arm will be troubled by collisions with the body and the resulting reachability of the wrist will become much smaller. It is also possible that the sternoclavicular joint indirectly contributes to avoiding singular configurations of the arm. On the other hand, the glenohumeral joint provides the majority of the motion ability of the arm. Its task is to point the humeral link and therefore contains wide ranges of motion, it covers more than a hemisphere. But it must be pointed out that only the elbow joint, more precisely the elbow flexion-extension, provides the workspace volume. If the elbow-flexion is disturbed, the workspace volume will be severely troubled. Thus, the elbow flexion-extension is one of the most functional degrees of freedom of the human body. From the viewpoint of positioning and orienting the hand, the mechanism of the human arm is kinematically redundant. The arm contains superfluous degrees of freedom. By analysing the model in Fig. 1 (without the sternoclavocilar joint), Lenarčič (1997) has shown that the critical point for the arm redundancy is the humeral rotation. If the humeral rotation is fixed, the mechanism will loose a majority of the kinematic redundancy. Theoretically (without the sternoclavocilar joint), the "size" of redundancy is linearly proportional to the range of the humeral rotation.

One of the basic kinematic issues of an anthropomorphic mechanism is associated with the proportions of the arm, such as the link lengths and the limits in joints. It is interesting, for instance, that the axis of the upper arm, the axis of the forearm, and the axis of the palm (including fingers, as presented in Fig. 2) form a rectangular triangle when the tip of the fingers touches the shoulder's Acromion and the elbow is in its maximum flexion. If the model of the human arm is given as in Fig. 1, the limits in joint angles change from one configuration to another and are strongly dependent on the values of other joint angles. This dependency can be shown either in a (more precise) numerical manner as reported by Umek-Venturini and Lenarčič (1996) or in an analytical manner by some polynomial approximations as reported by Lenarčič and Umek (1994). It is also interesting that the limits in joints do not permit collisions between the links of the arm. This valid in any configuration of the mechanism and in any point of the mechanism's workspace.

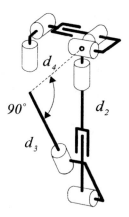

Fig. 2: **Proportions of the human arm kinematic structure**

The reachable workspace of a manipulator is defined (Kumar and Waldron, 1981) as the region within which all points can be reached by the selected reference point on the hand. The workspace geometry represents an important criterion in evaluation, programming, and design of mechanical manipulators and robots. A thorough analysis of the human arm workspace characteristics can therefore provide not only additional information for the design and control of artificial mechanisms but also can help us better understand the human proportions and the human motion performances. It can imply new directions in robotics research, especially in interactions with humans, such as telemanipulation and task description. Some investigators have affirmed (based on extremely simple kinematical models) that the maximum volume of the reachable workspace of an anthropomorphic mechanism is obtained with the proportions similar to the human arm (Yang and Lee, 1984).

In our opinion, however, the workspace volume only partially outlines the mechanism's functionality and does not cover the whole aggregate of the workspace geometry. The other part consists of the form of the workspace. A very practical problem would be to search for an optimum ratio between the lengths of the upper arm and the forearm to obtain a maximum volume of the workspace and a desired form. In this sense, Lenarčič *et al.* (1989) have optimised the workspace of a manipulator's mechanism with respect to the volume of the reachable workspace and with respect to the form of the reachable workspace given in terms of the so-called compactness. The notion of compactness has been introduced in this work to measure the similarity of the workspace form with the form of a sphere. A more compact workspace is more round and may provide more flexibility to solve different tasks by the use of the same mechanism. In a general case, the maximum volume and the maximum compactness of the workspace are obtained with different ratios between the link lengths. It has turned out, as the matter of fact, that in an anthropomorphic kinematic structure(that includes an approximation of joint limits (it has been demonstrated by Lenarčič *et al.*, 1989), the optimum ratio between the lengths of the forearm and the upper arm is practically equal when we maximise the workspace volume and when we maximise the workspace compactness. Moreover, this optimum ratio is very similar to the conventional human arm proportions as presented in Fig. 2. From this point of view, the anthropomorphic manipulator is more universal and flexible in executing spatial paths of different sizes and patterns.

The reachable workspace of a healthy human arm is shown in Fig. 3. On the left, we can see the measured workspace. It was obtained by measuring the reach of the hand step by step. In each step, the subject drew a line with the arm entirely extended on the plane of motion which was physically imposed. The workspace on the right was calculated by using the kinematical model in Fig. 1. The so-called sweeping-type algorithm incorporated a series of nested loops related to each joint coordinate and included a detection of the collisions between the arm and the body. Evidently (see Fig. 3), the computed workspace is quite a good approximation. This fact confirms the quality of the model.

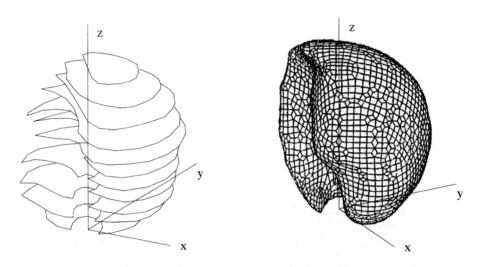

Fig. 3: **The measured (left) and the calculated (right) workspace of the human arm**

The motivation of this investigation has been to produce a figure of the human arm reachable workspace with the scope to quantify a rehabilitation procedure, for instance, of a patient after the stroke. For the calculation of the workspace we must provide only few input parameters, which are the link lengths and the ranges of motion in pre-selected planes, such as frontal, sagittal, and horizontal. This is a very productive advantage of the mathematically obtained model of the reachable workspace since in a great majority of patients the it cannot be measured at all.

4. Amount of Kinematic Redundancy

Redundancy is an immense source of freedom in solving tasks (Khatib, 1996). Redundancy enables to solve an assigned task in an infinite number of ways. Consequently, a redundant robot can operate respecting a given set of performance criteria, such as minimum execution time or minimum energy consumption. Moreover, a redundant robot is more versatile and is therefore capable to solve a task and simultaneously avoid obstacles in the workspace. A mechanism is termed redundant when its number of degrees of freedom is greater than is required by the task. The mechanism of the human arm can thus be considered kinematically redundant when the task of the arm is to move the hand in a selected position and orientation.

This is a standard definition of kinematic redundancy utilised in robotics. Redundancy, as defined in this sense, is a common feature of living mechanisms. Snakes can even be treated as hyper-redundant mechanisms. The most evident property of a redundant mechanism is the self-motion ability. It means that the same task can be solved in an infinity of ways giving to the control system the freedom to choose the best one. For instance, if the hand is fixed in a given position and orientation in the space, the rest of the arm can still move. The self-motion of the human arm manifests itself as a rotation of the centre of the elbow joint about an axis ξ passing through the centre of the glenohumeral joint and the wrist (see Fig. 4). During the self motion the angle γ, measured between this axis and the axis of the forearm, remains constant.

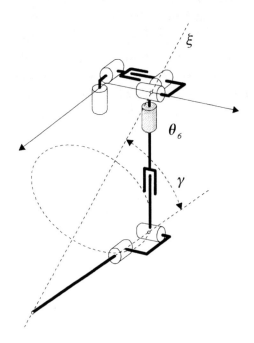

Fig. 4: Self-motion ability of the human arm kinematic structure

By some mathematical manipulations in the kinematic model of the arm that consider the properties of trigonometric functions it is possible to verify that the quantity of the human arm self-motion directly corresponds to the range of motion in the humeral rotation θ_6 - mathematical formulations are reported by Lenarčič (1997). Even though the joint limits in the human arm mechanism depend in a very complex manner on the values of other joint angles and evidently change throughout the workspace, the range of the self-motion practically remains constant and is about 90 deg everywhere in the central region of the workspace (but also close to the inner and outer limits of the workspace, as well as near mechanism's singular configurations).

In contrast to what one could expect, the quantity of the self-motion of the human arm is quite small. Nevertheless, the motion properties (static and dynamic capabilities) of the arm can crucially change by changing the configurations of the arm within the range of the self-motion (with the hand in the same pose of the workspace). An illustrative example is the distribution of the manipulability ellipsoids (Lenarčič, 1994). The manipulability ellipsoid represents the amount of displacements of the hand in the Cartesian space that correspond to the joint displacements that are bounded within a sphere in the space of joint coordinates. The manipulability ellipsoid thus illustrates the velocity capability of the manipulator. The longest principal axis of the manipulability ellipsoid shows the direction and the amplitude of the maximum velocity of the hand. It is also known that the manipulability ellipsoid connects the outside force applied to the hand with the corresponding joint torques. The joint torques are minimised when the force is applied in the direction of the shortest principal axis of the manipulability ellipsoid. In a singular configuration, the manipulability ellipsoid collapses in at least one direction. If the force is applied in that direction, the joint torques will be zero and the manipulator will resist to an infinite outside force. The advantage of a redundant manipulator, which is the property of the human arm, is that it is possible to change the geometry of the manipulability ellipsoid by the self-motion of the mechanism with the hand fixed in a desired pose. The distribution, the shape and size of the manipulability ellipsoids inside the workspace of an anthropomorphic mechanism have been studied by Lenarčič and Umek (1992). The goal of this investigation had been to verify if the arm can produce flat and round manipulability ellipsoids in the same point of the workspace by only changing the configuration within the range of the self-motion. It has been demonstrated, based on the presented model of the human arm, that the arm is capable to simultaneously provide very flat and very round manipulability ellipsoids in the workspace region close to the shoulder complex (as presented in Fig. 5).

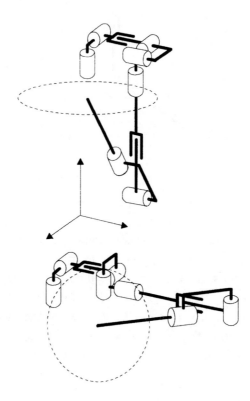

Fig. 5: Manipulability ellipsoid in different configurations

When the hand is put in this region, the manipulability ellipsoids are very flat if the elbow is positioned downward (the upper arm axis is vertical). This is, for example, statically the best configuration for resisting vertical outside forces (weights). When the elbow is upward (the upper arm axis is horizontal in the level of the shoulder), the manipulability ellipsoids become very round. In this configuration, the mechanism can provide uniform velocities of the hand in all directions. Such a property can be well used to execute a variety of tasks that request fine and precise movements of the hand.

5. Exploitation of Kinematic Singularities

As mentioned, the kinematic redundancy of a manipulator enables to optimise the motion with respect to a given set of secondary tasks, such as the minimisation of the joint torques. It has been speculated by Kieffer and Lenarčič (1994) that the humans tend to use singularity configurations of their limbs to gain mechanical advantage and thus decrease the driving forces in joints to minimise the fatigue. It is therefore expected that the minimum joint torque motion will become one of the most requested capabilities of future robots, in particular the humanoid robots, in applications such as lifting weights, pushing and handling heavy objects.

The joint torques that correspond to an outside force applied on the hand in a certain direction become zero in the singularity configuration of the mechanism. Based on this principle, weak actuation can be compensated by more sophisticated control that exploits singularities. However, this may represent significant problems for conventional control schemes where robots are programmed in cartesian space with rigorously prescribed timing that in singularities guides to infinite joint velocities and accelerations. The objective of Kieffer and Lenarčič (1994) had been to study a series of experiments solving human-type tasks, such as weight lifting and drawing a bow. I has been demonstrated that weight lifters are taking advantage of the singularity configurations of the arms. From the floor, the weight is thrown upward using forces generated by leg muscles and transmitted through their arms which are stretched downward in a singular configuration. The weight is then caught with arms folded in a second singularity configuration that is similar to the centre point singularity of a planar 2R manipulator. After a pause, the weight is again thrown upward using leg muscles and caught with arms stretched overhead in a third singularity configuration. Obviously, the weight is too heavy for the arms to manipulate except near these singularities and the lifting procedure is one of transferring the weight between them.

In drawing the bow, archers also make use of singularity configurations to minimise muscle effort in their arms. As an explanation, consider the planar dual arm system shown in Fig. 6 as an approximation of an archer. The bow is modelled as a constant force spring. The goal is to extend the spring from the initial length (Fig. 6 - above) to a given length. The system contains four degrees of freedom is redundant with respect to this task which imposes only one constraint. In the first case (Fig. 6 - middle), the spring is extended based on a usual robotic-type algorithm that takes into account only the positional constraints of the task and minimises the corresponding joint motion. In the second case (Fig. 6 - below), a solution is presented that minimises the joint torques and is obviously similar to human behaviour. It was shown in by Kieffer and Lenarčič (1994) that, if we compare the joint torques versus the spring length for the two solutions, the second solution substantially reduces the effort. A clear tendency of the second solution is to take advantage of kinematic singularities. In the final configurations, the arms ended in (or near, as much as the joint limits permit) the singular configurations.

These examples show that in walking and weightlifting humans use singularity configurations mainly to support heavy loads, rather than to apply forces. Singularities seem to play an important role in the human arm motion in order to minimise fatigue. In contrast, standards control schemes for robots are aimed to avoid singularities because they provoke undesired motion and infinite joint velocities. It must be pointed out that the human arm motion does not suffer from kinematic singularities but it obviously takes advantage from them.

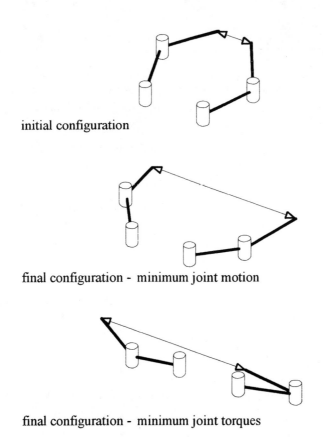

initial configuration

final configuration - minimum joint motion

final configuration - minimum joint torques

Fig. 6: Different solutions in extending a spring to a given length

Lenarčič and Žlajpah (1993, 1994) have studied the operation of weight lifting with a planar 2R manipulator in which the norm of joint torques caused by the gravity forces of the system was minimised based on a local optimisation that follows the steepest gradients of the cost function. The path provided by this optimisation, relative to different ratios between the link lengths and masses of the links and the load, have been compared with a global solution defined as the minimum of the maximum joint torques produced along the operation. In the studied series of examples, it has been confirmed that the difference between the local and the global solution is very small. The worst case is when the mechanism consists of equal links and the mass of the load is very small with respect to the mass of the mechanism's links. One of the findings is that the major part of the load gripped by the manipulator's hand must be carried by the elbow joint. When the weight, represented in Fig. 7 by a force F, is lifted from the bottom (point P_1 - with the arm stretched downward) to the top (point P_2 - with the arm stretched upward), the mechanism must always pass through a configuration in which the forearm link is perpendicular to the applied force causing a abundant torque in the elbow joint. On the other hand, the joint torque in the shoulder joint can be kept small by the control algorithm along the whole path. The optimum path which minimises the sum of the absolute values of joint torques

442

is shown in Fig. 7. The shoulder joint must basically carry only the weight of the links. A similar performance may by found in human weight lifting. Moreover, in the human arm the elbow joint is obviously stronger than the whole shoulder complex. It affirms that humans probably utilise a minimum-joint-torque-type motion in a majority of applications.

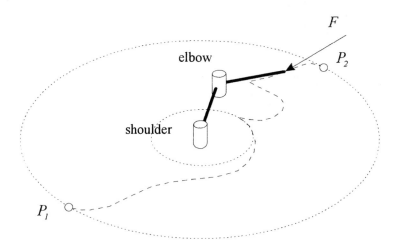

Figure 7: **Optimum path that minimises the static joint torques**

6. Conclusions

In our recent investigations we have developed a kinematic model of the human arm mechanism. In the present paper we utilise this model to discover and discuss a series of the human arm motion characteristics that in our opinion could have a significant impact on the design of advanced mechanical manipulators. In particular, we study the proportions of the arm links in relation to its reachability, we study the amount of the kinematic redundancy, the workspace geometry, and the velocity-torque capabilities of the arm. We are also interested in the utilisation of the kinematic singularities in handling heavy objects. By a series of experiments we show that the human arm takes advantage of the kinematic singularities to compensate weak actuation, while standard control strategies in robotics struggle to avoid singularities because they cause stability problems. The scope of the paper is to demonstrate that the kinematic structure of the human arm offers immense potentials that may be exploited through and alternative approach to the design, control and programming of mechanical manipulators. In doing so we expect to provoke an increased interest of the robotics community in humanoid robotic systems.

References

Benati, M., Gaglio, S., Tagliasco, V., Zaccaria, R. (1980). Anthropomorphic Robotics. Biological Cybern., Vol. 38., pp. 125-140.

Dvir, Z., Berme, N. (1987). The Shoulder Complex in Elevation of the Arm: a Mechanism Approach. J. of Biomechanics, Vol. 11, pp. 219-225.

Engin, A.E. (1990). Kinematics of Human Shouldr Motion. In Biomechanics of Diathrotrial Joints. Edited by Mow, V.C., Rotcliffe, A., Woo, S.L.-Y., Springer-Verlag, New York.

Högfors, C., Sigholm, G., Herberts, P. (1987). Biomechanical Model of Human Shoulder - I. Elements. J. of Biomechanics, Vol. 20, No. 2, pp. 157-166.

Khatib, O. (1996). The Impact of Redundancy on the Dynamic Preformance of Robots. Laboratory Robotics and Automation, Vol. 8, No. 1, pp. 37-48.

Kieffer, J., Lenarčič, J. (1994). On the Exploitation of Mechanical Advantage Near Robot Singularities. Informatica, Vol. 18, pp. 315-323.

Kumar, K., Waldron, K.J. (1981). The Workspace of a Mechanical Manipulator. ASME J. of Mech. Design, Vol. 103, No. 3, pp. 665-672.

Lenarčič, J., Umek, A. (1994). Simple Model of Human Arm Reachable Workspace. IEEE Trans. on Syst, Man and Cybern., Vol. 24, No. 8, pp. 1239-1246.

Lenarčič, J., Stanič, U., Oblak, P. (1989). Some Kinematic Considerations for the Design of Robot Manipulators. Int. J. Robotics & Comp. - Integ. Manufact. Vol. 5, pp., No. 2/3, pp. 235-241.

Lenarčič, J. (1997). The Kinematic Flexibility of Articulated Arms Containing One Degree of Redundancy. Int. Conference on Advanced Robotics, ICAR-97, Monterey (USA), 1997.

Lenarčič, J., A. Umek, A. (1992). Optimum Proportions of Human Arm Mechanism with Respect to its Workspace. In Proceedings VI. Mediterranean Conf. On Medical and Biological Eng., Capri (Italy), pp. 283-286.

Lenarčič, J., Žlajpah, L. (1993). Control Considerations on Minimum Joint Torque Motion. In Proceeding 3rd Int. Symp. on Experimental Robotics, Kyoto (Japan), pp. 211-216.

Lenarčič, J., Žlajpah, L. (1994). Comparison of Local and Global Solution in Optimisation of Joint Torques of n-R Manipulator. In Proceedings 4th IFAC Symp. on Robot Control, Capri (Italy), pp. 447-452.

Umek-Venturini, A., Lenarčič, J. (1996). The Kinematics of the Human Arm Elevation in the Frontal Plane. In Proceedings of the 11th Int. Symp. on Biomedical Engineering'96, Zagreb (Croatia), pp. 105-108.

Yang, D.C.H., Lee, T.W. (1984). Heuristic Combinatorial Optimization in the Design of Manipulator Workspace. IEEE Trans. on Syst. Man and Cybern., Vol. 14, No. 4, pp. xxxxxx.

Computational and
Dynamical Systems

Systems with Topological Structures

Rudolf F. Albrecht

Informatik, Universität Innsbruck

Technikerstr.25, A-6020 Innsbruck, Austria

Dedicated to Prof. L. Vietoris on his 106th birthday

Abstract

The first chapter of this article provides the mathematical prerequisites needed. The second chapter deals with valuated objects and logics. In the third chapter the topological foundations for approximation of objects by other objects are given, e.g. valuated filter and ideal bases on partial ordered sets and uniform measures. Then it is shown, how by these classical topological concepts fuzzy set theory and fuzzy logic can be described. In the fourth chapter applications to models of I/O-systems are made, including systems with hierarchical control, complex and evolutionary systems. The final chapter includes historical remarks and comparisons with related developments.

Keywords: systems theory, topology, approximation theory, fuzzy sets, fuzzy logic

1. Mathematical Prerequisites

1.1. Families, Relations, Functions

Let K be a set of indices, Z a set of objects, $K \neq \varnothing$, $Z \neq \varnothing$, ind: $K \to Z$ an indexing with $\wedge k \in K(k \mapsto z_{[k]} \wedge z_k =_{def} (k, z_{[k]}))$. For example, k can be a name to reference $z_{[k]}$ or a constructive procedure which results in $z_{[k]}$. $r =_{def} (z_k)_{k \in K}$ is a "parameterized family", parameterizing the set $ind(K)$. $(z_k)_{k \in K}$ and the family $(z_{[k]})_{k \in K}$ are considered isomorphic. Let $\wedge k \in K(Z_{[k]} =_{def} Z)$, then $r \in \prod_{k \in K} Z_k$ ($\prod$ the set product) and $Z^K =_{def} \prod_{k \in K} Z_{[k]}$. The canonical indexing of Z is $(z_z)_{z \in Z}$. If L is a non-empty index set with card $L = n$ and $K \subseteq \prod_{l \in L} K_l$ then the family r is also named an "n-dimensional parameterized relation", dim $r = n$. A set $R \subseteq \prod_{k \in K} Z_k$ is an unparameterized relation of parameterized families.

We consider the case $K \subseteq J \times I$, dim $r = 2$. Let $\wedge j \in J(I_{[j]} =_{def} \{i \mid i \in I \wedge z_{ji} \in r\} \wedge r_{j.} =_{def} (z_{ji})_{i \in I_{[j]}} \wedge r_{j.} \neq \varnothing)$ and $\wedge i \in I(J_{[i]} =_{def} \{j \mid j \in J \wedge z_{ji} \in r\} \wedge r_{.i} =_{def} (z_{ji})_{j \in J_{[i]}} \wedge r_{.i} \neq \varnothing)$, then r can be represented by $r_{\to} =_{def} (r_{j.})_{j \in J}$ or by $r_{\uparrow} =_{def} (r_{.i})_{i \in I}$ and concatenation (see **1.2.**) of each of these families of families yields r. According to our definitions, $I = \bigcup_{j \in J} I_{[j]}$ and $J = \bigcup_{i \in I} J_{[i]}$ and $K = \{(j,i) \mid j \in J \wedge i \in I_{[j]}\} = \{(j,i) \mid i \in I \wedge j \in J_{[i]}\}$. The reasoning can be extended to the case dim $r > 2$.

CP437, *Computing Anticipatory Systems: CAYS--First International Conference*
edited by Daniel M. Dubois © 1998 The American Institute of Physics 1-56396-827-4/98/$15.00

Let $\bigwedge j \in J$ $(I_{[j]} = D_{[j]} \cup C_{[j]} \wedge D_{[j]} \cap C_{[j]} = \varnothing)$. A "functional decomposition" Z of $r_\rightarrow$ is given by $(x_j =_{def} (z_{ji})_{i \in D_{[j]}}, \; y_j =_{def} (z_{ji})_{i \in C_{[j]}})_{j \in J}$ if $\bigwedge j, j' \in J(y_{[j]} \neq y_{[j']} \Rightarrow x_{[j]} \neq x_{[j']})$. Then a "function" $f : X =_{def} \{x_{[j]} \mid j \in J\} \rightarrow Y =_{def} \{y_{[j]} \mid j \in J\}$ is defined by $\bigwedge j \in J(x_{[j]} \mapsto y_{[j]})$, also expressed by $f \subset X \times Y$. According to f, $x_{[j]}$ is "valuated" by $y_{[j]}$. $r_\rightarrow$ can have many functional decompositions, each of the resulting functions can have many "arities" card $D_{[j]}$ and card $C_{[j]}$. Functional decompositions Z , Z' of $r_\rightarrow$ can be partially ordered by $Z \leq Z'$ $\Leftrightarrow_{def} \bigwedge j \in J(D_{[j]} \subseteq D'_{[j]})$. An analogue reasoning can be applied to the transposed $r_\uparrow$ of $r_\rightarrow$.

A set/hierarchy of unparameterized/parameterized relations of relations is named a "structure". In the case of parameterization each object is uniquely referenced by its (composite, hierarchically structured) index which gives the "selection path" of the object. A structure on indices induces a structure on the indexed objects.

1.2. Algebra of Relations

Let there be given a set $\mathbf{R}$ of parameterized relations of the form $(z_k)_{k \in K}$ and let $R = \{r_q = (z_{qk})_{k \in K_{[q]}} \mid q \in Q\} \subseteq \mathbf{R}$. A functional assignment $\Gamma : R \rightarrow W \subseteq \mathbf{R}$ is an algebraic composition. Examples for algebraic compositions are:

(1) "projections" for the selection of objects by their indices:
given $J = (J_q)_{q \in Q}$, $pr(J) \; R =_{def} \{(z_{qk})_{k \in K_{[q]} \cap J_{[q]}} \mid q \in Q\}$. For a cartesian indexing we write for example simply $pr_2 \; (x, y, z, \dots) = y$.

(2) "cuts" for the selection of objects by their values:
given $M = (M_q)_{q \in Q}$, $cut(M) \; R =_{def} \{(z_{qk})_{k \in U_{[q]}} \mid q \in Q \wedge U_{[q]} =_{def} \{k \mid k \in K_{[q]} \wedge z_{[qk]} \in M_{[q]}\}\}$.

(3) transformation of indices:
given $\tau = (\tau_q : K_q \rightarrow J_q)_{q \in Q}$, $\tau_{[q]}$ bijective, not necessarily an isomorphism with respect to a structure on $K_{[q]}$, $\bigwedge q \in Q(((z_{qk})_{k \in K_{[q]}} \rightarrow (z_{q\tau(k)})_{k \in K_{[q]}}) \wedge z_{q[k]} = z_{q[\tau(k)]})$.

(4) transformation of values:
given $\nu = (\nu_q : (z_{qk})_{k \in K_{[q]}} \rightarrow (w_{qk})_{k \in K_{[q]}})_{q \in Q}$.

(5) "concatenations" to compose objects. We consider $Q = \{q, p\}$ for the sake of simplicity:
$\mathbf{K}(\{r_q, r_p\}) =_{def} (z_l)_{l \in L}$ with $L = (\{q\} \times K_{[q]}) \cup (\{p\} \times K_{[p]})$;
$\mathbf{K}_{join}(\{r_q, r_p\}) =_{def} (z_l)_{l \in L}$ with $L = K_{[q]} \cup K_{[p]}$ and $z_l = z_{[q]l}$ for $l \in K_{[q]}$, $z_l = z_{[p]l}$ for $l \in K_{[p]}$ if $\bigwedge k \in (K_{[q]} \cap K_{[p]})(z_{[p]k} = z_{[q]k})$, else $\mathbf{K}_{join}(\{r_q, r_p\}) =_{def} \varnothing$;
$\mathbf{K}_\Delta(\{r_q, r_p\}) =_{def} (z_l)_{l \in L}$ with $L = (K_{[q]} \setminus K_{[p]}) \cup (K_{[p]} \setminus K_{[q]})$ and $z_l = z_{[q]l}$ for $l \in K_{[q]}$, $z_l = z_{[p]l}$ for $l \in K_{[p]}$ if $\bigwedge k \in (K_{[q]} \cap K_{[p]})(z_{[p]k} = z_{[q]k})$, else $\mathbf{K}_\Delta(\{r_q, r_p\}) =_{def} \varnothing$ (symmetrical difference);
$\mathbf{K}_\varphi(\{r_q, r_p\}) =_{def} (z_l)_{l \in L}$ with $L = K_{[q]} \cup K_{[p]}$ and $z_l = z_{[q]l}$ for $l \in K_{[q]} \setminus K_{[p]}$, $z_l = z_{[p]l}$ for $l \in K_{[p]} \setminus K_{[q]}$, and $z_l = \varphi(z_{[q]l}, z_{[p]l})$ for $l \in K_{[q]} \cap K_{[p]}$, φ a given function.

Let there be given a "connector" $C \subseteq (\{q\} \times K_{[q]}) \times (\{p\} \times K_{[p]})$, non-empty sets V, V', and a "property" $E(C)$ on C by the relation $E(C) = \{(v_{[s]qk}, v'_{[s]pj})_{s \in S(c)} \mid c = (qk, pj) \in C\}$ and $\bigwedge c = (qk, pj) \in C \wedge s \in S(c)(v_{[sqk]} \in V \wedge v'_{[spj]} \in V')$, the $S(c)$ are given non-empty index sets. If $\bigwedge c = (qk, pj) \in C \vee s \in S(c)((z_{qk}, z_{pj}) = (v_{[s]qk}, v'_{[s]pj}) \in E(C))$, then
$\mathbf{K}(\{r_q, r_p\}, E(C), C) =_{def} (w_l)_{l \in L}$ with $L =_{def} (((\{q\} \times K_{[q]}) \setminus pr_1 C) \cup ((\{p\} \times K_{[p]}) \setminus pr_2 C))$ and $w_l =_{def} z_l$ for $l = (q, k) \in ((\{q\} \times K_{[q]}) \setminus pr_1 C)$ and $w_l =_{def} z_l$ for $l = (p, k) \in ((\{p\} \times K_{[p]}) \setminus$

448

pr_2C), else $\mathbf{K}(\{r_q, r_p\}, E(C), C) =_{\text{def}} \varnothing$. In the simplest case $\wedge c \in C(\text{card } S(c) = 1)$. This concatenation covers for example functional composition if $E(C)$ is an equality relation:

Expl.: Let $r_q = (x_{q1}, x_{q2}, y_{q3}, y_{q4})$, $r_p = (x_{p1}, x_{p2}, y_{p3})$, $C = \{(q3, p1)\}$, $E(C) = \{(y_{q3}, x_{p1})$ with $y_{[q3]} = x_{[p1]}\}$, then $\mathbf{K}(\{r_q, r_p\}, E(C), C) = (x_{q1}, x_{q2}, x_{p2}, y_{q4}, y_{p3})$.

1.3. Variables

We consider $(S_p)_{p \in P}$, $\mathscr{S} = \{S_{[p]} \mid p \in P\}$, $\wedge p \in P(S_{[p]} \neq \varnothing \wedge (S_{[p]}$ a structure fully parameterized by $I_{[p]}) \wedge i \subset I_{[p]})$. i is assumed to be independent of p and may be composite. Let $\wedge p \in P(V_{[p]} =_{\text{def}} pr(I_{[p]} \setminus i)S_{[p]} \wedge (C =_{\text{def}} pr(i)S_{[p]}$ being independent of $p))$ and let $\mathscr{V} =_{\text{def}} \{V_{[p]} \mid p \in P\}$. Then $S_{[p]} = \mathbf{K}_{\text{join}}(\{V_{[p]}, C\})$. To facilitate the representation of $\mathscr{S}$ we introduce objects var V and $S(\text{var } V) = \mathbf{K}_{\text{join}}(\{\text{var } V, C\})$, a function val: $P \times \{\text{var } V\} \times \{S(\text{var } V)\} \to \mathscr{V} \times \mathscr{S}$ with $(p, \text{var } V, S(\text{var } V)) \mapsto (V_{[p]}, S_{[p]})$, and a function val^{-1}: $\{(V_{[p]}, S_{[p]}) \mid p \in P\} \to \{(\text{var } V, S(\text{var } V))\}$. The terminology used is: var V is a "variable" on "variability domain" $\mathscr{V}$ with respect to $\mathscr{S}$, $S(\text{var } V)$ is a "structure variable" on "variability domain" $\mathscr{S}$, val is an "assignment"/"control" function with "assignment"/"control" parameter (or "program") $p \in P$, $V_{[p]}$ is "substitutable" in $S(V_{[p]})$ by any $V_{[q]}$ (application of val^{-1} and re-assignment) to give $S(V_{[q]})$. The notations used are: var $V : \mathscr{V}$, $S(\text{var } V) : \mathscr{S}$ and for assignments var $V :=(p) V_{[p]}$, $S(\text{var } V) :=(p) S_{[p]}$.

We assume $\wedge p \in P(V_{[p]} \in \prod\limits_{j \in J(p)} W_j)$. Then we can consider var $V : \{(\text{var } w_j)_{j \in J(p)} \mid p \in P\}$ and $\wedge p \in P((\text{var } w_j)_{j \in J(p)} : \prod\limits_{j \in J(p)} W_j)$.. However, in general the assignment var $w_j := w_j$ is not free but bounded to such w_j that after all assignments to $(\text{var } w_j)_{j \in J(p)}$ are completed $(w_j)_{j \in J(p)} = V_{[p]}$. With this constraint we can define "partial" assignments to var V, var $V := \mathbf{K}(\{(w_j)_{j \in J' \subseteq J(p)}, (\text{var } w_j)_{j \in J(p) \setminus J'}\})$. Considering the set A of all admissible (partial, total) assignments a to var V, an assignment a to var V denoted by $\text{var}_a V$, we can introduce as new objects the families $\mathbf{V} =_{\text{def}} (\text{var}_a V)_{a \in A}$ and $\mathbf{S} =_{\text{def}} (S(\text{var}_a V))_{a \in A}$ with "states" $\text{var}_a V$ and $S(\text{var}_a V)$.

$V_{[p]}$ can contain variables of lower level ("variables on variables" hierarchy).

Expl.: var $y = \text{var } f (\text{var } x)$, var $f : \{f_{[p]} : X_{[p]} \to Y_{[p]} \mid p \in P\}$ on var-level 2, var $f :=(p) f_{[p]}$, var $x : X_{[p]}$, $f_{[p]}(\text{var } x): f_{[p]}(X_{[p]})$ on var-level 1, var $x := x'$, $f_{[p]}(\text{var } x) := f_{[p]}(x')$.

The structures and operations considered in **1.1.** and **1.2.** can be extended to structures and operations with variables.

2. Valuated Objects, Logics

Applying the concepts of chapter **1.** we define recursively the following structure:

Given the non-empty sets $K^{(0)}$ finite, $Z^{(0)}$, $V^{(0)}$, a bijective indexing ind: $K^{(0)} \to Z^{(0)} \times V^{(0)}$, and $\mathbf{L}^{(0)} = \{(z_k^{(0)}, v_k^{(0)}) \mid k \in K^{(0)}\}$.

For $n, N \in \mathbf{N}$, $1 \leq n \leq N$, let there be given $K^{(n)} \subseteq \text{pow } K^{(n-1)}$, $K^{(n)} \neq \varnothing$, $V^{(n)} \neq \varnothing$, $Z^{(n)} = \{z_k^{(n)} =_{\text{def}} (z_l^{(n-1)}, v_l^{(n-1)})_{l \in k} \mid \wedge l \in k((z_l^{(n-1)}, v_l^{(n-1)}) \in Z^{(n-1)}) \wedge k \in K^{(n)}\}$, a set of functions $\Phi^{(n)} \subseteq$

$\{\varphi_k^{(n)} \mid \varphi_k^{(n)} : (V^{(n-1)})^{card\ k} \to V^{(n)} \wedge k \in K^{(n)}\}$, $\Phi^{(n)} \neq \varnothing$, then we define $\mathbf{L}^{(n)} =_{def} \{(z_k^{(n)}, v_k^{(n)}) \mid v_k^{(n)} = \varphi_k^{(n)} ((v_{[l]}^{(n-1)})_{l \in k}) \wedge \varphi_k^{(n)} \in \Phi^{(n)} \wedge k \in K^{(n)}\}$.

$\mathbf{L}(N) =_{def} \bigcup_{n=1}^{N} \mathbf{L}^{(n)}$ has hierarchical level N. The components of $\mathbf{L}(N)$ can be subject to relations, e.g. orderings, index transformations, associativity, cardinality attributes (a property may hold for "all", "infinite many", "finite many", "exactly n", $0 \leq n$, elements of a component, etc.). Usually $\mathbf{L}(N)$ is represented by means of variables.

Expl.: $Z^{(0)} = \{[x, x+1) \mid [x, x+1) \subset (\mathbf{R}, \leq, +) \wedge x \in \mathbf{I}\}$ ($\mathbf{R}$ reals, $\mathbf{I}$ integers), $V^{(0)} = (\mathbf{I}, +)$, $\mathbf{L}^{(0)} = \{(z, v) \mid z \in \{[0, 1), [1, 2), [2, 3)\} \wedge v \in \{0, 1\}\}$ with canonical indexing, $Z^{(1)} = \{((z, v)), ((z, v), (z', v')), ((z, v), (z', v'), (z'', v'')) \mid z \neq z' \neq z''\}$, $\Phi^{(1)} = \{id, +\}$, $\varphi(v) = v$, $\varphi(v, v') = v + v'$, $\varphi(v, v', v'') = v + v' + v''$, $\mathbf{L}^{(1)} = \{(((z, v)), v), (((z, v), (z', v')), v + v'), (((z, v), (z', v'), (z'', v'')), v + v' + v'')\}$ (example for integration).

In the particular case $\wedge 0 \leq n \leq N((V^{(n)}, \leq^{(n)}, \sqcap^{(n)}, \sqcup^{(n)})$ a lattice), the product sets $(V^{(n)})^{card\ k}$ can be ordered by $\wedge v =_{def} (v_l)_{l \in k}$, $v' =_{def} (v_l')_{l \in k} \in (V^{(n)})^{card\ k}$ $(v <^{(n)} v' \Leftrightarrow_{def} (\wedge l \in k (v_{[l]} \leq v_{[l]'} \wedge \vee l' \in k (v_{[l']} < v_{[l']'}))))$. If in addition all $\varphi_k^{(n)}$ are $\leq$- homomorphisms, i.e. $v \leq^{(n)} v' \Rightarrow \varphi^{(n+1)} (v) \leq^{(n+1)} \varphi^{(n+1)} (v')$, or all $\varphi_k^{(n)}$ are $\geq$- homomorphisms (antimorphisms), i.e. $v \leq^{(n)} v' \Rightarrow \varphi^{(n+1)} (v) \geq^{(n+1)} \varphi^{(n+1)} (v')$, we say $\mathbf{L}(N)$ is a "logic" and $\varphi_k^{(n)}$ is a "logic function". Then $\varphi_k^{(n)} (\sqcap^{(n-1)} (v, v')) \leq^{(n)} \sqcap^{(n)} (\varphi_k^{(n)}(v), \varphi_k^{(n)} (v')) \leq^{(n)} \sqcup^{(n)} (\varphi_k^{(n)} (v), \varphi^{(n)} (v')) \leq^{(n)} \varphi_k^{(n)} (\sqcup^{(n-1)} (v, v'))$

Expl.: $Z^{(0)} = \{z(1), z(2), z(3)\}$, $z \in Z^{(0)}$ a proposition, $V^{(0)} = V^{(1)} = \{t, f\}$, $F: Z^{(0)} \to \{t, f\}$. $\mathbf{L}^{(0)} = \{(z, v) \mid (z, v) \in F\}$ with canonical indexing, $\Phi^{(1)} = \{id, \neg, \vee, \wedge\}$, $Z^{(1)} = \{((z, v)), ((z, v), (z', v')), ((z, v), (z', v'), (z'', v'')) \mid z \neq z' \neq z''\}$, $\mathbf{L}^{(1)} = \{(((z, v)), \varphi(v)), (((z, v), (z', v')), \varphi'(v, v')), (((z, v), (z', v'), (z'', v'')), \varphi''(v, v', v'')) \mid \varphi \in \{id, \neg\} \wedge \varphi', \varphi'' \in \{\wedge, \vee\}\}$. $\varphi \in \{\wedge, \vee\}$ is commutative, associativity on level 2: $\varphi(\varphi(v, v'), v'') = \varphi(v, \varphi(v', v''))$ is usually expressed by concatenation to level 1: $\varphi(v, v', v'')$ (example for boolean algebra).

3. Topological Structures

Classical topology is the well developed mathematical discipline dealing with neighborhoods and approximations and it is to be expected that all "fuzzyness", "vagueness", "softness" theories are covered by it.

3.1. Filters and Ideals

Given a complete lattice $(\mathcal{L}, \leq, \sqcap, \sqcup, \mathbf{O}, \mathbf{E})$ and a subset $\mathcal{B} = \{B_{[k]} \mid k \in K\} \subset \mathcal{L}$, the indexing bijective, with the following properties: $\wedge k \in K (B_{[k]} \neq \mathbf{O}) \wedge \wedge k', k'' \in K (\vee k''' \in K ((B_{[k''']} \leq B_{[k']}) \wedge (B_{[k''']} \leq B_{[k'']}))) \wedge \sqcap \mathcal{B} \neq \mathbf{O}$. Then $\mathcal{B}$ is a proper "filter base" on $\mathcal{L}$. If in addition $\wedge k \in K \wedge L \leq E (B_{[k]} \leq L \Rightarrow L \in \mathcal{B})$ then $\mathcal{B}$ is a "filter". The dual notions to filter base and filter are "ideal base" and "ideal". If S is a non-empty set, then this applies to the complete, atomic, boolean lattice $(pow\ S, \subseteq, \cap, \cup, \varnothing, S)$ which we consider in the following. For filter base $\mathcal{B} = \{B_{[k]} \mid k \in K\}$, $B^* =_{def} \lim \mathcal{B} = \bigcap_{k \in K} B_{[k]}$ (S itself has the discrete topology).

The neighborhood of any $s \in B$, $B =_{def} \bigcup_{k \in K} B_{[k]}$, to the elements of B^* can be expressed by membership or non-membership of s in certain $B_{[k]}$: Let $\wedge s \in B((K(s) =_{def} \{k \mid k \in K \wedge s \in B_{[k]}\}) \wedge \overline{K}(s) =_{def} K \setminus K(s))$, $\mathscr{B}_\cap(s) =_{def} \{B_{[k]} \mid k \in K(s)\}$, $\mathscr{B}_\cup(s) =_{def} \{B_{[k]} \mid k \in \overline{K}(s)\}$. We have $s \in \bigcap_{k \in K(s)} B_{[k]} \cap \bigcap_{k \in \overline{K}(s)} \mathbf{C}B_{[k]}$, $\mathbf{C}$ the complement with respect to B. Let $K_{min}(s) =_{def}$ $\{k \mid k \in K(s) \wedge \neg \vee k' \in K(s)(B_{[k']} \subset B_{[k]})\}$, $\overline{K}_{max}(s) =_{def} \{k \mid k \in \overline{K}(s) \wedge \neg \vee k' \in \overline{K}(s) (B_{[k']} \supset B_{[k]})\}$, then $\mathscr{B}_{\cap min}(s) =_{def} \{B_{[k]} \mid k \in K_{min}(s)\}$, $\mathscr{B}_{\cup max}(s) =_{def} \{B_{[k]} \mid k \in \overline{K}_{max}(s)\}$. General "distance" / "similarity" *relations* of s from / with $s^* \in B^*$ are then given by $\wedge s \in B$ $(D_\cap(s^*, s) =_{def} \mathscr{B}_{\cap min}(s) \wedge D_\cup(s^*, s) =_{def} \mathscr{B}_{\cup max}(s))$. $D_\cap(s^*, s) = D_\cap(s^*, s')$ and $D_\cup(s^*, s) = D_\cup(s^*, s'')$ define equivalence relations $s \sim_\cap s'$ and $s \sim_\cup s''$. In particular, if $\mathscr{B}$ is itself a complete lattice then $d_\cap(s^*, s) =_{def} \bigcap D_\cap(s^*, s) \in \mathscr{B}$ and $d_\cup(s^*, s) =_{def} \bigcup D_\cup(s^*, s) \in \mathscr{B}$ are *functional* in s and $d_\cup(s^*, s) \subset d_\cap(s^*, s)$. Dual results hold for $\mathscr{B}$ an ideal base.

3.2. Comparison and Composition of Bases

Given a non-empty set S and two filter bases $\mathscr{B} = \{B_{[k]} \mid k \in K\} \subset \text{pow } S$ with $B =_{def}$ $\bigcup_{k \in K} B_{[k]}$, $B^* =_{def} \bigcap_{k \in K} B_{[k]} \neq \varnothing$, $\mathscr{C} = \{C_{[l]} \mid l \in L\} \subset \text{pow } S$ with $C =_{def} \bigcup_{l \in L} C_{[l]}$, $C^* =_{def} \bigcap_{l \in L} C_{[l]}$ $\neq \varnothing$, all indexings bijective. We say $\mathscr{B}$ is "finer" than $\mathscr{C}$, $\mathscr{B} \prec \mathscr{C} \Leftrightarrow_{def} \wedge l \in L \vee k \in K(B_{[k]} \subseteq C_{[l]})$, $\mathscr{B}$ is "equivalent" to $\mathscr{C}$, $\mathscr{B} \sim \mathscr{C} \Leftrightarrow_{def} B^* = C^*$, and for finite cardinalities, $\mathscr{B}$ "finer granulated" than $\mathscr{C}$ if card $\mathscr{B} > $ card $\mathscr{C}$.

$\mathscr{S} =_{def} \{B_{[k]} \cup C_{[l]} \mid (k,l) \in K \times L\}$ is a filter base on $S \cup T$ with $S^* =_{def}$ $\bigcap_{kl \in K \times L}(B_{[k]} \cup C_{[l]}) = B^* \cup C^*$, $\mathscr{D} =_{def} \{B_{[k]} \cap C_{[l]} \mid (k,l) \in K \times L\}$ is a filter base on $S \cap T$ only if $\wedge(k,l) \in K \times L((B_{[k]} \cap C_{[l]}) \neq \varnothing)$, $D^* =_{def} \bigcap_{kl \in K \times L} (B_{[k]} \cap C_{[l]}) = B^* \cap C^*$.

Let $\mathbf{B} = \{\mathscr{B}_{[m]} \mid m \in M\}$ be a set of either all being proper filter bases or all being proper ideal bases on pow S. To compare the bases $\mathscr{B}_{[m]}$ by a uniform neighborhood / similarity measure we introduce a filter base $\mathbf{D} =_{def} \{D_{[q]} \mid q \in Q\} \subset \text{pow } (\mathbf{B} \times \mathbf{B})$ with the following properties: $\mathbf{B} \times \mathbf{B} \in \mathbf{D}$, diag $(\mathbf{B} \times \mathbf{B}) =_{def} \{(\mathscr{B}_{[m]}, \mathscr{B}_{[m]}) \mid m \in M\} \subseteq \bigcap_{q \in Q} D_{[q]}$, and $\wedge q \in Q(D_{[q]} =$ $D_{[q]}^{-1})$ (symmetry assumed for simplicity's sake). Then for $F_{[mq]} =_{def}$ cut $(\mathscr{B}_{[m]}) D_{[q]} = \{\mathscr{B}_{[m']} \mid (\mathscr{B}_{[m]}, \mathscr{B}_{[m']}) \in D_{[q]}\}$, $\mathscr{F}_{[m]} =_{def} \{F_{[mq]} \mid q \in Q\}$ is a filter base with $\mathscr{B}_{[m]} \in \bigcap \mathscr{F}_{[m]}$. Consequently, $\mathbf{D}$ defines for all $\mathscr{B}_{[m]}$ a uniform and symmetric neighborhood system. Thus according 3.1., for any pair $(\mathscr{B}_{[m]}, \mathscr{B}_{[m']})$ the neighborhood / similarity measures $D_\cap(\mathscr{B}_{[m]}, \mathscr{B}_{[m']})$ and $D_\cup(\mathscr{B}_{[m]}, \mathscr{B}_{[m']})$ can be applied.

3.3. Valuated Bases

Given a proper filter base $\mathscr{B} = \{B_{[k]} \mid k \in K\}$ on $(\text{pow } B, \subseteq, \cap, \cup)$, a non-empty complete lattice $(V, \leq, \sqcap, \sqcup)$ and a $\leq$- homomorphism $\varphi \colon \text{pow } B \to V$, i.e. $\wedge k, k' \in K((B_{[k]} \subseteq B_{[k']}) \Rightarrow (\varphi(B_{[k]}) \leq \varphi(B_{[k']})))$. With $v_{[k]} = \varphi(B_{[k]})$ it follows from $(B_{[k]} \subseteq B_{[k']}) \wedge (B_{[k]} \subseteq B_{[k'']})$ that $(v_{[k]} \leq v_{[k']}) \wedge (v_{[k]} \leq v_{[k'']})$, hence $\varphi(\mathscr{B})$ is a filter base on V and $\varphi(\lim \mathscr{B}) \leq \lim \varphi(\mathscr{B})$.

We consider $\overset{-1}{\varphi} \colon V \to \text{pow } B$ defined by $\wedge v \in V(\overset{-1}{\varphi}(v) =_{def} \bigcup_{\varphi(U) = v} U)$. Then $\overset{-1}{\varphi}$ is a homomorphism. If $\mathscr{V} = \{v_{[l]} \mid l \in L\}$ is a filter base on V and $\wedge v \in \mathscr{V}(\overset{-1}{\varphi}(v) \neq \varnothing)$, then $\overset{-1}{\varphi}(\mathscr{V})$ is a filter base on pow B. This for example is the case if φ is the set extension of a function f: $B \to C$ and $V = $ pow C. Then we have $\mathscr{B} \prec \overset{-1}{\varphi}(\varphi(\mathscr{B}))$.

φ being a homomorphism corresponds to the "neighborhood to $\lim \mathscr{B}$" interpretation (case (N)). Choosing φ as antimorphism, $\varphi(\mathscr{B})$ is an ideal base, which corresponds to the "similarity to $\lim \mathscr{B}$" interpretation (case (S)).

In the function $(B_{[k]}, v_{[k]})_{k \in K}$ the elements $B_{[k]}$ of a (filter / ideal) base $\mathscr{B}$ with "support" $B = \bigcup \mathscr{B}$ are valuated by $v_{[k]}$ with $v_{[k]} = \varphi(B_{[k]})$. Now we consider a non-empty set $\mathbf{M} = \{\mathscr{B}_{[q]} \mid q \in Q\}$ of bases $\mathscr{B}_{[q]}$, all being either filter or all being ideal bases with functions $(B_{[qk]}, v_{[qk]})_{k \in K_{[q]}}$ and all with the same support B. We define $\wedge q \in Q \wedge x \in B (V_{[q]}(x) =_{def} \{v \mid v = \varphi_{[q]}(B_{[qk]}) \wedge k \in K_{[q]} \wedge x \in B_{[qk]}\}$. Then $\mathbf{M}$ can be partially ordered, in case (N): $\wedge q, q' \in Q(\mathscr{B}_{[q']} \leq_{\mathbf{M}} \mathscr{B}_{[q]} \Leftrightarrow_{def} \wedge x \in B(\sqcap V_{[q']}(x) \leq \sqcap V_{[q]}(x))$, in case (S): $\wedge q, q' \in Q(\mathscr{B}_{[q']} \leq_{\mathbf{M}} \mathscr{B}_{[q]} \Leftrightarrow_{def} \wedge x \in B(\sqcup V_{[q']}(x) \geq \sqcup V_{[q]}(x))$. On $(\mathbf{M}, \leq_{\mathbf{M}})$ filter bases can be considered and on $\mathbf{M} \times \mathbf{M}$ distances $D_\cap(\mathscr{B}_{[q]}, \mathscr{B}_{[q']})$ and $D_\cup(\mathscr{B}_{[q]}, \mathscr{B}_{[q']})$ can be introduced.

Expl.: The general case for any f: $B \to C$, φ the set extension of f, any filter base $\mathscr{B}$ on pow B mapped by $\varphi(\mathscr{B})$, and any filter base $\mathscr{C}$ on pow C mapped by $\overset{-1}{\varphi}(\mathscr{C})$, can be found in text books (e.g. Bourbaki, 1951). The particular case $B = [a,b] \subset \mathbf{R}$, $a < b$, $C = [0,1] \subset \mathbf{R}$, f: $B \to C$, $\vee x \in B(f(x) = 1)$, $\mathscr{C} = \{[\alpha,1] \mid \alpha \in [0,1]\}$, $B_{[\alpha]} =_{def} \overset{-1}{\varphi}([\alpha,1])$ an "α - cut", $\mathscr{B} = \{B_{[\alpha]} \mid \alpha \in [0,1]\}$, was introduced by L.A. Zadeh (1965), B named a "fuzzy set" with "membership function" f. Let f': $[a,b] \to [0,1]$ be another function with $\vee x \in B(f'(x) = 1)$ with corresponding quantities primed. If $\wedge x \in B(f(x) \leq f'(x))$ then $\mathscr{B}' \leq_{\mathbf{M}} \mathscr{B}$. Hereby B_α, B'_α may differ and $\lim \mathscr{B} \subseteq \lim \mathscr{B}'$. If f is integrable, then for example the functional $\int_{x \in B_{[\alpha]}} f(x)dx$ corresponds to a logic function on $(f(x))_{x \in B_{[\alpha]}}$, the functional $\int_\alpha^1 B_{[\beta]}d\beta$ corresponds to a logic function on $(B_{[\beta]})_{\beta \in [\alpha,1]}$, both have equal values, $\int_{x \in B} |f(x) - f'(x)|dx$ is an example of a neighborhood measure for $\mathscr{B}, \mathscr{B}'$.

4. Applications

4.1. Approximation of Functions

Let there be given var y = var f(var x) with var f : $\{f_{[p]} : X_{[p]} \to Y_{[p]} \mid p \in P\}$, var x : $\tilde{X} =_{def} \{x_{[pq]} \mid x_{[pq]} \in X_{[p]} \wedge (p,q) \in R \subseteq P \times Q\}$, var y : $\tilde{Y} =_{def} \{f_{[p]}(x_{[pq]}) \mid (p,q) \in R\}$, and let $pr_1 R = P$, $pr_2 R = Q$. We define $\wedge p \in P((\text{var } x_p : X_{[p]}) \wedge (Q_{[p]} =_{def} \text{cut}(\{p\}) R))$. To model the behavior of physical input/output systems, we introduce a logical time $T = \{t(n) \mid n \in (\mathbf{N_o}, <)\}$, with an ordering induced by $<$ on $\mathbf{N_o}$, taking into account that all physical processes corresponding to assignments and to function evaluations have a non-zero duration in physical time. Starting at t(0), the assignment steps (see example in **1.3.**) are at

t(0): given $\qquad$ $p(t(0)) \in P$,

t(1):	assignments	$f(t(1)) =_{def} val(p(t(0)), var\ f)$, $X(t(1)) =_{def} X_{[p(t(0))]}$,
t(2):	given	$q(t(2)) \in Q_{[p(t(0))]}$,
t(3):	assignment	$x(t(3)) =_{def} val(q(t(2)), var\ x_{p(t(0))})$,
t(4):	evaluation and assignment	$y(t(4)) =_{def} f(t(1))(x(t(3)))$.

This 5-step cycle can be continued whereby we agree that each of the left hand side objects keeps its value constant from assignment until next assignment. Then we consider for logical time $S = \{t(m) \mid m = 5n \wedge n \in [1,2,..N]\} \subset T$, $N \leq \infty$, the control processes $(p(t(m)))_{t(m)\in S}$, $(q(t(m)))_{t(m)\in S}$, and the state processes $(f(t(m)))_{t(m)\in S}$, $(x(t(m)))_{t(m)\in S}$, $(y(t(m)))_{t(m)\in S}$. For suitable control processes and under further (sufficient) assumptions the state processes converge when time increases:

Let $X =_{def} \bigcup_{t(m)\in S} X(t(m))$ with $X(t(m)) =_{def} X_{[p(t(m))]}$ be a uniform space, generated by the filter base $D_X \subset pow(X \times X)$, i.e. $diag(X \times X) \subseteq \bigcap D_X$, $\wedge d\in D_X \vee c\in D_X (c \subseteq \overset{-1}{d})$, $\wedge d\in D_X \vee c\in D_X (\mathbf{K}_{join}(\{c,c\}) \subseteq d)$, and let $\bigcup_{t(m)\in S} f(t(m))(X(t(m))) \subseteq Y$, Y being a complete uniform space, generated by the filter base $D_Y \subset pow(Y \times Y)$. Further, let hold $\wedge t,t'\in S(t < t' \Rightarrow X(t) \subseteq X(t'))$. Then $\wedge t,t'\in S(t < t' \Rightarrow f(t')(X(t)) \subseteq f(t')(X(t')))$ and $\wedge x\in X \vee t(x)\in S \wedge t\in S (t(x) \leq t \Rightarrow x \in X(t))$. We assume

(1) $\wedge d_Y\in D_Y \vee t(d_Y)\in S \wedge t(d_Y) \leq t\in S (x\in X(t) \Rightarrow (f(t(d_Y))(x), f(t)(x)) \in d_Y)$ (uniform Cauchy filter base),

(2) $\wedge d_Y\in D_Y \vee t(d_Y)\in S \wedge t(d_Y) \leq t\in S \vee d_X (d_Y)\in D_X \wedge x,x'\in X(t) ((x,x') \in d_X(d_Y) \Rightarrow (f(t)(x), f(t)(x')) \in d_Y)$ (uniform continuity).

From (1) follows because of the completeness of Y: $\wedge x\in X \vee y(x)\in Y \wedge d_Y\in D_Y \vee t(d_Y,x)\in S \wedge t(d_Y,x) \leq t\in S ((f(t)(x), y(x)) \in d_Y)$ (uniform convergence). If Y is separated (i.e. $diag(Y \times Y) = \bigcap D_Y$), $y(x)$ is unique and consequently a function $F: X \to Y$ is defined by $x \mapsto y(x)$. If Y is not separated, a set $Y_{[x]}$ of many $y(x)$ may exist and all $\widetilde{F} = (x, y(x))_{x\in X} \in \prod_{x\in X} Y_x$ are equivalent with respect to $\bigcap D_Y$. From (2) follows by the axioms of uniform structures and with (1): $\wedge d_Y\in D_Y \vee t(d_Y)\in S \wedge t(d_Y) \leq t\in S \vee \widetilde{d}_Y (d_Y)\in D_Y \wedge x,x'\in X(t) (((f(t)(x'), f(t)(x)) \in \widetilde{d}_Y (d_Y) \wedge (f(t)(x), y(x)) \in \widetilde{d}_Y (d_Y)) \Rightarrow (f(t)(x'), y(x)) \in d_Y)$ and $\wedge x,x'\in X \wedge d_Y\in D_Y \vee d_X(d_Y)\in D_X (((x, x') \in d_X(d_Y)) \Rightarrow (y(x), y(x')) \in d_Y)$ (for separated Y this implies uniform continuity of F).

For $z \in \{p,q,f,x,y\}$ the process (time function) $(z(t(n))_{n\in\{0,1,2,..N\}}$, $N < \infty$, can be described by its "1st order events" (transitions), i.e. pairs $(z(t(n)), z(t(n+1)))$: given $((z(t(n)), z(t(n+1)))_{n\in\{0,1,2,..N-1\}}$ and initial value $z(t(0))$, then the concatenation $\mathbf{K}_\Delta(\{(z(t(n)), z(t(n+1)) \mid n\in\{0,1,2,..N-1\}\}) = (z(t(0)), z(t(N)))$, i.e. the transition from initial to final value ("summation" of events along path $(t(n))_{n\in\{0,1,...N-1\}}$).

4.2. Higher Order Control Functions

The question arises, who supplies the $p(t(5n)) \in P$, $(p(t(5n)), q(t(5n+2))) \in R$, $n = 0,1,2,..$, in the control processes described in **4.1.**. Introducing var $p : \{p_{[u]} \mid p_{[u]} \in P \wedge u \in U\}$, var $q_{var\ p} : \{q_{[v]} \mid q_{[v]} \in Q_{[var\ p]} \wedge v \in V_{[var\ p]}\}$, the higher order val-functions for var p, var $q_{var\ p}$ can depend on the current time point t, on external parameters $u'(t)$, $v'(t)$, and on given "goals", "criteria", "rules", which are (composite) relations $r(t)$, $s(t)$ involving all or part of previous $p(t')$, $q(t')$, $x(t')$, $y(t')$, $t' < t$ (in a physical realization this requires a "memory"). Thus for $n = 1, 2, ..$ at

$t(5n)$: $p(t(5n)) = val(t(5n), u(u'(t(5n)), r((p(t'))_{..\leq t' \leq 5n-5}, (q(t'))_{..\leq t' \leq 5n-3}, (x(t'))_{..\leq t' \leq 5n-2},$
$(y(t'))_{..\leq t' \leq 5n-1}))$, var p),

$t(5n+2)$: $q(t(5n+2)) = val(t(5n+2), v(v'(t(5n+2)), s((p(t'))_{..\leq t' \leq 5n}, (q(t'))_{..\leq t' \leq 5n-3}, (x(t'))_{..}$
$_{..\leq t' \leq 5n-2}, (y(t'))_{..\leq t' \leq 5n-1}))$, var $q_{p(t(5n))}$).

Particular cases of these val-functions are independent of explicit time, or of u', v' (autonomous behavior), or of previous x- and y-values (no feed back, "open loop"). If set extensions of the val-functions and of the relations r, s are considered and topologies are defined on the domains of these, p-, q-values within neighborhood classes on pow P and pow $Q_{[p]}$ can be used (topological control). On a further hierarchical level, variables var u', var v', var r, var s can be considered with val-functions with higher order criteria/rules. On the domains of these variables again topologies can be defined.

4.3. Complex and Evolutionary Systems

Given a set $G = \{g_{[k]} : X_{[k]} \to Y_{[k]} \mid k \in K\}$ of primitive functions (generators) $g_{[k]} : (x_{[k]i})_{i \in I_{[k]}}$

$\mapsto (y_{[k]j})_{j \in J_{[k]}}$, and a family $F = (f_l)_{l \in L}$ with $\wedge l \in L(f_{[l]} \in G)$. If $f_{[l]} = g_{[k]}$, we set $I_{[l]} =_{def} I_{[k]}$, $J_{[l]} =_{def} J_{[k]}$, $X_{[l]} =_{def} X_{[k]}$, $Y_{[l]} =_{def} Y_{[k]}$, $x_{[l]i} =_{def} x_{[k]i}$, $y_{[l]j} =_{def} y_{[k]j}$. Further, let be given a cycle free connector $C \subseteq \bigcup_{l \in L} J_l \times \bigcup_{l \in L} I_l$ being functional from right to left. (F, C) represents a "complex" (composite) system with hierarchical level 1 over G, and for values for which the "equality on C relation" $E(C)$ holds, $\mathbf{K}(F, E(C), C)$ defines a function $f : X \to Y$ with $X \subseteq \{x_{li} \mid x_{[l]i} \in X_{[l]} \wedge (l,i) \in \tilde{I} =_{def} \bigcup_{l \in L} I_l \setminus pr_2 C\}$ and with $Y \subseteq \{y_{lj} \mid y_{[l]j} \in Y_{[l]} \wedge (l,j) \in \tilde{J} =_{def} \bigcup_{l \in L} J_l \setminus pr_1 C\}$ (see eg. Albrecht (1994, 1996)). Generalized to a structure with variables, we have var $F = (var f_l)_{l \in var L}$, var $f_l : G$, var C: set of admitted connectors, depending on the assignment to var F. The domains of the variables can be topologized and for proper assignment sequences the sequence of states can converge to a state, which within the imposed topology satisfies a given criterion or behavior. Given the composite function $f = \mathbf{K}(F, E(C), C)$ and a composite function variable var $\Delta f : \{\Delta f_{[q]} : X'_{[q]} \to Y'_{[q]} \mid q \in Q \wedge X'_{[q]} \subseteq Y\}$ together with a connector variable var $\Delta C : \{\Delta C \subseteq \tilde{J} \times var \Delta I\}$, var $\Delta I: \{I'_{[q]} \mid q \in Q\}$. We consider $\mathbf{K}(\{f, var \Delta f\}, E(var \Delta C), var \Delta C)$. By assignment to the variables, f "evolves" to $\mathbf{K}(\{f, \Delta f_{[q]} \}, E(\Delta C), \Delta C)$.

5. Remarks

The presentation of relations, functions and variables as given in **1.1., 1.2., 1.3.** is not standard and was introduced and applied by Albrecht (1994, 1996). Hereby emphasis is put on parameterization of these objects. Most of the topological concepts used are classical ones: Filter bases were first described and applied by L. Vietoris (1921), in other form introduced by E.H. Moore and H.L. Smith (1922), and later by G. Birkhoff (1935, 1937). The theory of filters was developed by H. Cartan (1937), uniform spaces were investigated by A. Weil (1937-1938), and many others. For topological Banach spaces, **4.1.** expresses the fundamental concepts of "consistency", "convergence" and "stability" of numerical approximations. A formal theory of hierarchical control (**4.2.**) seems not to be existing. The hierarchical construction of I/O systems by means of concatenation of functions by connectors was described and used by Albrecht (1994, 1996).

The basic idea of approximate computation is, not to use point to point functions, but the set extensions of the functions or approximations to these in order to map valuated argument filter bases on valuated result filter bases (topological mappings). A well known and widely exploited example is interval arithmetic and interval analysis in $\mathbf{R}^n$, developed by R.E. Moore (1962), and the "Karlsruhe School" (G. Alefeld, J. Herzberger, U. Kulisch, K. Nickel (1966), and others) in the sixties and seventies: In this case the filter base for a function argument consists of $\mathbf{R}^n$ and an n-dimensional interval only, the members of which are all being $\sim$ equivalent. The result of a constructive function evaluation is enclosed in an interval which is expected to be "small". Another example, related to the preceding, is floating point arithmetic with "optimal rounding": The real numbers are partitioned into (in general not equally long) intervals, each interval is represented by a "machine number" (i.e. a floating point number representable by the computer) contained in it, the other members of the interval are rounded to this machine number (roundings are topological mappings, see Albrecht (1977)). The computations with machine numbers are to be performed such that the resulting machine number coincides with the machine number on which the exact result would be rounded. This arithmetic was introduced by U. Kulisch (1973, 1975) (see also Kulisch and Miranker (1981)) and then extended by him and others to a programming system with automatic result verification (see e.g. Kulisch (1996)). For monotone filter bases, computations with valuated arguments and valuated results are widely practiced in "fuzzy computations" and "fuzzy control" (a contemporary presentation of this subject is for example the book of Klier and Folger (1988)) without making reference to the classical topological background. In an underlying theory, the support of the filter bases may be structured (e.g. a power set of a set, a σ - algebra, a Borel set, a set of intervals on $\mathbf{R}^n$) which influences the choice of the filter base elements, and the choice of their valuation depends in general also on this theory.

References

Parts of this article were presented at the "Second Workshop of the Institute for General Systems Studies", Jan. 9-11, 1997, San Marcos, Texas, and at the workshop "New Trends in Informatics", March 10-14, 1997, TU Budapest.

Albrecht, R. F. (1977). Roundings and Approximations in Ordered Sets. In Computing Suppl. vol. 2.

Albrecht, R. F. (1994). Modelling of Computer Architectures. Proc. Conf. on Massively Parallel Computing Systems, IEEE Computer Society Press, Los Alamitos, California, pp. 434-442.

Albrecht, R. F. (1996). On the Structure of Discrete Systems. In Lect. Notes in Comp.Sc.1030, Comp. Aided Systems Theory - EUROCAST95, Springer Verlag, Berlin, Heidelberg, New York, pp. 3-18

Birkhoff, G. (1935). A new Definition of Limit. Bull. Amer. Math. Soc. 41

Birkhoff, G. (1937). Moore-Smith Convergence in General Topology. Ann. Math. 38

Bourbaki, N.(1951). Topologie Général., Act. Sc. In. 1142. Hermann, Paris, pp. 40, 41

Cartan H. (1937). Théorie des Filtres. Filtres et Ultrafiltres. C.R. Ac. Sc. Paris

Klier G.J., T.A. Folger (1988). Fuzzy Sets, Uncertainty, and Information. Prentice-Hall Int. Ed.

Kulisch U. (1973). Formalization and Implementation of Floating Point Arithmetics. Report RC 4608, IBM Thomas J. Watson Res. Center

Kulisch U. (1975). Formalization and Implementation of Floating Point Arithmetic. Computing 14

Kulisch U., W.L. Miranker (1981). Computer Arithmetic in Theory and Practice. Acad. Press

Kulisch U. (1996). Numerical Algorithms with Automatic Result Verification. Am. Math. Soc. Lect. in Appl. Math., 32

Moore E.H. , H.L. Smith (1922). A General Theory of Limits. Amer. J. Math. 44

Moore R.E. (1962). Intervall Arithmetic and Automatic Error Analysis in Digital Computing. PhD Thesis, Stanford University

Moore R.E. (1966). Intervall Analysis. Prentice Hall

Nickel K. (1966). Über die Notwendigkeit einer Fehlerschrankenarithmetik für Rechen-automaten. Num. Math. 9

Vietoris L. (1921. Stetige Mengen. Monatsh. f. Math. u. Phys. 31 (1921)

Weil A. (1937-1938). Sur les Espaces à Structure Uniforme et sur la Topologie Général. Act. Sc. In. 551. Hermann, Paris

Zadeh L.A (1965). Fuzzy Sets. Information and Control 8

The Consequences of Learnability for the A Priori Knowledge in a World

Hanns Sommer

Control Engineering and Systems Theory Group
Department of Mechanical Engineering (FB 15), University of Kassel (GhK)
Mönchebergstr. 7, 34109 Kassel, Germany
Phone: 0561 8043261
Fax: 0561 8047768
email: sommer@rts-pc1.rts.maschinenbau.uni-kassel.de

Abstract: The precondition for the evolution of intelligent beings in a world (modelled by anticipatory systems) is the learnability of some regularities in that world. The consequences, that can be deduced from the learnability property in a world, form the a priori knowledge. This a priori knowledge is independent of the empirical facts in a special world.

It will be shown that the a priori knowledge consists not only of logical tautologies. Also some well-known non trivial theorems in physics and psychology form part of the a priori knowledge. The a priori knowledge obtained in different worlds is necessarily organised by the same structures.

Examples from physics and psychology show the use of a separation between a priori knowledge and empirical knowledge.

Keywords: Learning, fuzzy information, a priori knowledge, anticipatory systems.

(1) Different kinds of knowledge

Technical systems are constructed by their designers for a fixed known purpose. Engineers assume to know the state of the world and they change that state to make the world more agreeable for human race. The problem in social science is that the incompleteness of the knowledge over the world is a fact that cannot be neglected. Models for social systems have to be open in that sense, that new knowledge can be inserted into the model.

A well-known idea to model social systems is Gidden's process of structuration (see Figure 1, [Mingers 1996]).

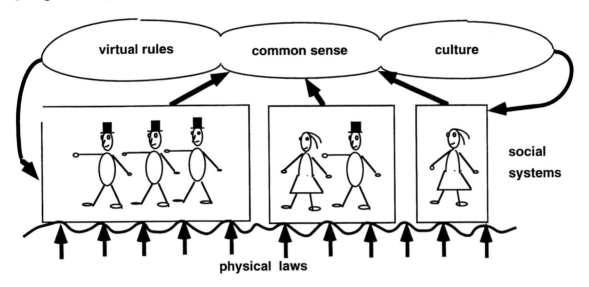

Figure 1: Reality for social systems

CP437, *Computing Anticipatory Systems: CAYS--First International Conference*
edited by Daniel M. Dubois © 1998 The American Institute of Physics 1-56396-827-4/98/$15.00

The elementary social systems (beings) are composed by physical elements and are therefore dominated by the physical laws. But the essential rules in the world are the virtual laws, forming our culture. These rules dominate social systems. The rules of social systems are produced by the social systems itself (men, groups, etc.), but the behaviour of the social systems is dominated by the rules of the social world.

There exist many different possibilities to understand the world. The views of the world postulated by realism and by constructivism provide such frames to understand the world.

Realism postulates: The social systems are consequences of the physical laws. Probabilistic effects create some simple structures that are optimised by Darwin's selection principle. Simple systems grow and form greater more complex entities that are able to create the virtual rules.
The problem of realism: The beings in the world form part of the world. It is a dubious assumption that a part would be able to deduce a model of the whole. The reference systems (called "frames" by Minsky) that comprise for example our language, depend on the beings in the world. But the physical laws cannot be thought independent of these reference systems.

Constructivism: Constructivism claims that the reality for the individual beings is produced by themselves in the process of the production of their virtual rules. (Watzlawik: Reality originates from the conversation of the beings.)
The problem of constructivism: For a mathematical formulation of a theory, models are needed. But it is not possible for us, to make a computer program that constructs ourselves as the observers of reality in the computer world. The observer is not a part of the computer world and therefore the computer world is not a model for Giddens world.

In our world we have:
- **empirical knowledge** (physical knowledge),
- **cultural knowledge** (men made virtual rules),
- **a priori knowledge** (knowledge that is independent of the special world where we live).

These different kinds of knowledge are mingled in the world.

Two questions will be discussed:
(I)   Is the reality detected by the observers in the world completely determined by the empirical facts or exist a priori knowledge independent of the special realisation of the observed world ?
(II)   Is it possible to separate the different kinds of knowledge ?

The method used here is a weak form of Hegel's principle:
"Was vernünftig ist, ist wirklich und was wirklich ist, ist vernünftig"
or translated: "Reality depends on rationality".
Since the idea of rationality cannot easily be formalised we replace rationality by learnability. Reality is our picture learned from the empirical world. Only that part of the world, that can be learned by an observer in the world, can form a part of the observers reality.
Learnabiliy represents only a small part of the concepts included in rationality, but these concepts are accessible to formal definitions. Hegel's principle will be replaced by **the postulate:**

(P)   "Reality depends on learnability".

The consequences of the learnabiliy postulate (P), assumed in our world, form the a priori knowledge in the world.

The a priori knowledge includes logical tautologies but also some theorems that often are falsely considered as a part of the empirical knowledge. The a priori knowledge is independent of our special world, it is true in all possible worlds.

In psychology it is very important, to differentiate between objective knowledge that holds in all possible worlds and the subjective knowledge of one special society. For an individual in a society it is often impossible to make this distinction. He believes in the truth of rules that are limited to his special society. If this society is destroyed, the individual is no longer able to explain or even to understand his foregoing behaviour.

(2) What is optimality in a learning process ?

Definition 1 (Learning): Given a set of learning examples (u_i, o_i), $i = 1, ..., m$ and a space H of hypothesis $h \in H$. A learning process is a search process in the hypothesis space H that finds a hypothesis $h^* \in H$ that is in agreement with all learning examples.

Formal description of a learning process:
u_i is a description of the actual state of the world, a hypothesis $h \in H$ is a description of the laws assumed in the world.

From u_i and $h \in H$ the consequences $o_i' := h(u_i)$ can be deduced.

The error of the selected hypothesis h is proportional to the distance between the calculated consequences o_i' and the real consequences o_i of the state u_i observed in the world.

Let $d(o_i', o_i)$ denote the distance between o_i' and o_i.

The optimal learning process finds:

(EC)
$$h^* := arg \min_{h \in H} \left(\sum_{i=1}^{m} d(h(u_i), o_i) \right)$$

The optimal hypothesis is obtained by minimising an **error criterion**.
If the solution of equation (EC) is not unequivocal, the optimal hypothesis is selected in the set obtained with (EC) by Occam´s Razor Principle ([Li et al 1992]):
(OC)   "Among the several theories (hypothesis) that are all consistent with the observed
          phenomena (examples), one should pick the simplest theory (hypothesis)."

The complexity of a hypothesis is defined as the length of the shortest description of the hypothesis relative to a fixed language.
The optimal hypothesis is obtained by minimising a **complexity criterion**.

The learning process defined by (EC) and (OC) can be practically applied only in very simple situations because of the effect of overfitting. Small perturbations of the data prevent from the selection of the true hypothesis . The process defined by (EC) and (OC) selects a very complex hypothesis that mainly contains the information of the perturbations and does not represent the laws in the world. To avoid this effect, engineers restrict the hypothesis space to a small set of simple hypothesis (for example linear laws [Ljung 1987]) . This method often finds a good hypotheses but it fails in finding the best.

The optimal selection in the hypothesis space is:
Select the hypothesis of the hypothesis space whose certainty is maximally confirmed by the learning examples.

In the following definitions it will be assumed that the hypothesis are descriptions or laws for a part of the events in the world.

Definition 2 (certainty of laws in a world): The certainty of a hypothesis $h \in H$ depends on the confirmation obtained for this hypothesis from the learning examples (u_i, o_i), $i = 1, ..., m$, relative to the complexity of the description of the hypothesis h.

459

The confirmation given by one learning example (u_i, o_i) to the hypothesis h depends on:
- the degree of correspondence of the hypothesis to the example and
- the exactness of the forecast of the hypothesis $h(u_i) =: o_i'$ compared with the exact values o_i found in the world.

The total certainty calculated for the hypothesis has to be distributed among its description parameters. The more complex the description of the hypothesis h is, the greater is number of parts in which the total certainty has to be divided.
The best learning algorithm selects the hypothesis with the highest certainty.

Formal definition of the certainty of a hypothesis:
Let $h \in H$ be defined in the form of a Fuzzy rule IF A   THEN B with Fuzzy predicates $A \in F(M)$ and $B \in F(N)$ over two sets M and N.
$E_\approx : F(M) \times F(M) \to [0,1]$ (or $E_\approx : F(N) \times F(N) \to [0,1]$) denotes an equality measure for Fuzzy predicates (see Appendix (A))
The degree of correspondence of h and u_i is defined: $E_\approx(A, u_i)$,

the exactness of h is defined : $E_\approx(B, o_i)$ and the wrongness of h is defined : $E_\approx(B, \neg o_i)$.
($\neg o_i$ is the symbol for "not o_i ".)
Let $C(h) = (C_+(h), C_-(h))$ denote the certainty of h obtained from former examples. The actualisation of $C(h)$ deduced from the learning example (u_i, o_i) is given by equation (A):

(A+) $\qquad C_+(h) \leftarrow C_+(h)S\big(E_\approx(A, u_i)TE_\approx(B, o_i)\big)$,

(A-) $\qquad C_-(h) \leftarrow C_-(h)S\big(E_\approx(A, u_i)TE_\approx(B, \neg o_i)\big)$

where $r \leftarrow s$ symbolises the operation "r is replaced by s", S denotes a Fuzzy s-Norm ("or"-operator) and T the corresponding Fuzzy t-Norm ("and"-operator) (compare [Pedrycz 1989]).

Let h be defined by the parameters $p_1, p_2, ..., p_{n_h}$ where $\alpha_i \in [0,1]$, $i = 1, ..., n_h$ defines the exactness of the parameter p_i claimed in the description of h. (This means that the parameter p_i in the description of h is given by the Fuzzy predicate $E_{\alpha_i}(p_i, p_i^*)$ with fixed numbers $p_i^* \in \mathbb{R}$, $\alpha_i \in [0,1]$ (see Appendix (A)).)

The **complexity** of h is defined:
$$K(h) := \sum_{i=1}^{n_h} \alpha_i$$
The **confirmation of the parameters of** h is defined:
$$Cp_+(h) := C_+(h)^{1/K(h)T} \ , \ Cp_-(h) := C_-(h)^{1/K(h)T}$$
where $a^{p/qT} = b \ :\Leftrightarrow \underbrace{aTaT...Ta}_{p \text{ factors}} = \underbrace{bTbT...Tb}_{q \text{ factors}}$.

$Cp_+(h)$ is **the certainty for** h **deduced from the learning examples** and $Cp_-(h)$ the **doubt** in h. It depends on the questions if the certainty or the doubt is prevalent in our decision. The stability of a bridge has to be guaranteed without doubt ($Cp_-(h) \approx 0$), but for the decision of a company to search of oil in a region it is sufficient to have a high certainty for finding oil in that region.

The belief function:
From the confirmed hypothesis a belief value for future events can be deduced:
Let $\{h_i\}_{i=1,...,r} \subseteq \mathbb{H}$ denote the set of hypothesis that are confirmed by the learning examples

(u_i, o_i), $i = 1,...,m$, h_i is of the form IF A_i THEN B_i ,

u denotes a description of a situation and $\bar{o}$ a forecast made in this situation.
The belief in $\bar{o}$ is defined:

$$b(\bar{o}) := \overset{r}{\underset{i=1}{S}}\left(C_+(h_i) \, T \, E_\approx(u, A_i) \, T \, E_\subseteq(B_i, \bar{o})\right)$$

Renorming the values $b(\bar{o})$ such that S can be replaced by $+$, provides the well-known Fuzzy Dempster Shafer belief function (with values in $\mathbb{R}_+$) ([Yager 1995],[Mahler 1995]).

$E_\subseteq(\bar{o}, B_i) := max\{\alpha | \bar{o}$ is α finer than $B_i\}$ denotes the degree of the truth of the sentence
"$\bar{o} \subseteq B_i$" and the consequences B_i of the hypothesis h_i are the focal elements of the belief structure.

Applications:

(2.A) Detection of self similarity in measuring data.

It is not a trivial task, to detect chaotic behaviour of a process from measurements. Every curve of discrete values can be approximated as exactly as necessary by a polynomial. The classification of a process (producing values, for example the share values) to be chaotic is often stronger based on the feeling of the experts than on clear mathematical arguments. The following definition provides a mathematical tool, to deduce the property of **self similarity** of a process from its measuring data:
Let T_f denote the transformation of all laws (or hypothesis of the hypothesis space $h \in \mathbb{H}$)
that reduces all parameters p_j and all exactness values α_j by the same factor f:

$$h'(p_j', \alpha_j') = T_f\left(h(p_j, \alpha_j)\right) \text{ where } p_j' = f \cdot p_j \text{ and } \alpha_j' = f \cdot \alpha_j .$$

If the set of laws $h \in \mathbb{H}$ that are confirmed by the measurement data (u_i, o_i), $i = 1,...,m$ of the process is invariant under the transformation T_f, then the process is called **self similar**. The laws of the process are than independent of the scaling used in the observation.

(2.B) The Psycho-Logic of Jan Smedslund

Smedslund defines ([Mock 1996]):
'The common sense of a culture C is equivalent to the system of logical implications taken for guaranteed to be correct by every competent member of C'.
Smedslund postulates the existence of structures of the common sense that are not empirical but formal. An example for formal common sense knowledge is **Smedslunds Theorem** (T):
"If P tries to do A in S and fails, and if she believes she failed because of low ability only, then her belief that she can do A in S will, other things equal, be more weakened than if she believes it was unusual situational circumstances only that led to her failure."

Smedslund proves this Theorem with an empirical method. He offers concretisations of (T) to test persons. For example:
(1) "Jane fails on a mathematical task. Jane's belief that she can solve this type of task is

461

weakened because she believes her failure was due to quite special unfortunate circumstances.

Acceptable explanation ☐

Not acceptable explanation ☐

(2) "Jane fails on a mathematical task. Jane's belief that she can solve this type of task is weakened because she believes she failed because of low ability.

Acceptable explanation ☐

Not acceptable explanation ☐

If a majority of test persons decides for acceptable explanation in the first question and not acceptable explanation in the second question, the Theorem (T) is confirmed by the test.

Using the methods introduced in this article, Theorem (T) is a direct consequence of the learnability assumption in the world. The following sentences have to be examined:

(+) $\qquad$ p and c $\rightarrow$ q

(++) $\qquad$ p and a $\rightarrow$ q

where the meaning of the letters p, c, q, a is defined:

p := "P tries to do A in S and fails"

c := "she believes she failed because of unusual situational circumstances"

a := "she believes she failed because of low ability"

q := "her belief that she can do A in S will be weakened" .

The learning example is: (u,o) with

u := "P tries to do A in S" and o := "P fails".

The certainty of the sentence:

$\bar{o}$:= "P will succeed to do A in S on the next task".

has to be updated.

(u,o) is a negative example for $\bar{o}$. But if condition c holds (+), then the preconditions of $\bar{o}$ are not completely satisfied and therefore the certainty of $\bar{o}$ is not strongly diminished. In contrary, condition a holds also in the future and therefore in the case (++) the preconditions of $\bar{o}$ correspond to the learning example (u,o) and the certainty of $\bar{o}$ is strongly diminished. The difference deduced for the changes of the certainty of $\bar{o}$ is exactly the assertion of Theorem (T).

(3) Learning under restrictions

A learning process is often realised by a very simple algorithm compared to the complexity of the world in which the knowledge should be acquired.

Formal definition of learning under restrictions:

The hypothesis are composed by many description elements $h = h_1, h_2, ..., h_r$ and in one learning step only simple calculations can be executed. For example: a few elements h_i can be modified:

$$h_1, h_2, ..., h_i, ..., h_r \mapsto h_1, h_2, ..., \bar{h}_i, ..., h_r$$

two hypothesis can be combined:

$$\left.\begin{array}{l} h_1, h_2, ..., h_i, h_{i+1}, ..., h_r \\ \bar{h}_1, \bar{h}_2, ..., \bar{h}_i, \bar{h}_{i+1}, ..., \bar{h}_r \end{array}\right\} \mapsto \left\{\begin{array}{l} h_1, h_2, ..., h_i, \bar{h}_{i+1}, ..., \bar{h}_r \\ \bar{h}_1, \bar{h}_2, ..., \bar{h}_i, h_{i+1}, ..., h_r \end{array}\right.$$

or all elements are updated by small changes $\varepsilon_i := \begin{cases} +\delta \\ -\delta: \\ 0 \end{cases}$

$$h_1, h_2, ..., h_i, ..., h_r \mapsto h_1 + \varepsilon_1, h_2 + \varepsilon_2, ..., h_i + \varepsilon_i, ..., h_r + \varepsilon_r$$

The problem of a learning process under restrictions is its instability. Breiman states ([Breiman]):

> "The results of Breiman demonstrate that a significant portion of the classification errors of our learning algorithms are not the result of searching in the wrong hypothesis space - at least in the sense that the space does not contain a good approximation to the target function. Instead, **the errors are the result of instability in the search process.**"

The instability of a learning process is often shown by overtraining ([Sommer 1996]). To avoid overtraining, the learning process has to be simple. Or in a complex world, where a simple learning process cannot be effective, the learning procedure in one level of exactness has to be controlled by a learning procedure in a simplified detection accuracy.

The **coupling of the two learning procedures in different levels of exactness** is realised by the following formalism:

From many learning examples $(u_1, o_1), ..., (u_m, o_m)$ a few simplified examples

$(\tilde{u}_1, \tilde{o}_1), ..., (\tilde{u}_{\tilde{m}}, \tilde{o}_{\tilde{m}})$ (with $\tilde{m} \ll m$) are produced (compare Appendix (B)).

If the changes of the hypothesis does not correspond to the simplified learning examples, the changes will not be executed.

In psychology, the ideas obtained in the strong simplified or generalised detection level are called motivations. The feeling of danger, self-respect, pleasant sensation etc. are examples of this general ideas ([Dörner 1993], [Sommer,Dürrbaum 1994]).

In the first learning step, only with great simplifications or strong restrictions to small parts of the world, knowledge is acquired. But also this part of the knowledge had to be rewarded by the optimality criteria.

Our considerations lead to the following conclusions:

Learnability conditions for learning under restrictions:
(1) A correct description of a simplified part of the world exists. (Existence of a description of the world.)
(2) The correct description is simpler than other descriptions of this part of the world. (Occam Razor principle)
(3) A description of greater parts of the world can be separated into parts that correspond to a simplified view of the world. (Separability of the world.)
    The simples parts that can be represented by a formalism and which are separated from the rest are called elementary elements.
(4) The knowledge in the world is useful for the beings. (Causality)

Applications (Consequences of the learnability conditions):

(3.A) The signification of time deduced from the learnability condition (4):

An order of our pictures from reality in which the complexity of one picture relative to its predecessor is minimised is called time order (compare Figure 2).
The complexity can be defined by the minimal description length of the information (contained in the pictures) relative to a fixed language (the Kolmogoroff complexity measure). The complexity of one picture relative to its predecessor is then the lengths of the description of the changes between the two pictures.

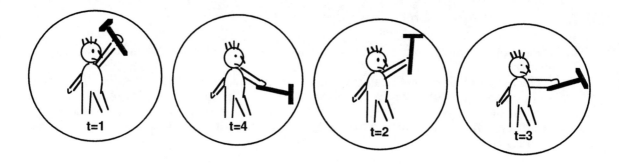

Figure 2: The time order of pictures in our reality.

The time order is the ordering of the events such that the possibility is maximised to make forecasts in our reality.

(3.B) An interpretation of the Schrödinger equation (condition 3):

Learnability condition (3) states that learnability implies the separability of reality into simple parts which can be learned step by step and which are independent of its environment.
This condition provides a reason for Schrödinger equation due to the following facts:

- The exponential function is the description of an undisturbed time development.
- The solution of the Schrödinger equation

$$\frac{d}{dt}\varphi(t,m) = H\varphi(t,m)$$

(where $\varphi(m)$ denotes the state of a quantum mechanical system and $H = \frac{-i}{h} \cdot \hat{H}$ the

Hamilton operator (compare [FICK 1968, p.207]).,)
is an exponential function.

Therefore Schrödinger equation formalises the thesis: Systems that can be recognised, have to be independent from their environment. (Hegel: "Bei sich selbst sein".)

(4) Structure of the knowledge in a learnable world

Learning and forecasting capabilities are preconditions for intelligent beings in a world. Intelligent beings are members of the class of anticipatory systems, which are the systems that compute its present state in function of a prediction of their future knowledge over itself and their environment ([Rosen 1985], [Dubois 1996]).
The structure of the knowledge obtained by learning in a world is therefore an organisation principle that necessarily has to be realised in the knowledge processing unit of anticipatory systems.

Summarising **the results of section (2) and (3)** we obtain:

(1) The knowledge processing unit of an anticipatory system is organised in different levels of exactness. On each level the knowledge is based on elementary description elements defined up to a certain exactitude and combination rules for these elementary elements (compare Figure 3).

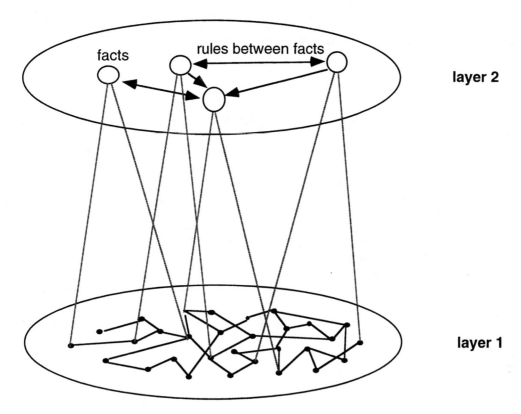

Figure 3: The organisation of the knowledge in various levels of exactness.

(2) The knowledge in each layer can be represented by a Fuzzy Dempster Shafer belief function. It can be shown that in each layer, the aggregation of the knowledge is realised by the Fuzzy operations ([Sommer 1995]) :

> "**and**" realised by a Fuzzy t-norm,
> "**or**" realised by a Fuzzy s-norm,
> "**not a**" realised by the Fuzzy predicate (1-a)   and
> **a combination function for confirmations and veto's**.

(The necessity to introduce a special procedure for the combination of confirmations and veto's to obtain a general knowledge processing system, had been detected by Kiendl ([Kiendl 1993]).)

(3) The upper layers are used to stabilise the knowledge in the lower layers and to prevent from a chaotic behaviour in these layers during the learning phase.

Application:

(4.A) Explication of a fault in the human knowledge processing entity.

The stabilisation of the knowledge in the lower layers by units in higher layers is often responsible for some inflexibility of the knowledge processing system. The knowledge in the lower layer in human beings is organised such that the self respect of the being is maintained. The inflexibility produced by this coupling between the information in different layers is used by gangs of criminals to fix their members to the gang.

The gang requires from a new member to comet a crime. The self respect of the member committing the crime is preserved because he does bad things for the gang leader. The crime is in his opinion a good action for the gang. But if in the future, the member decides to separate himself from the gang, the good reasons for the crime are no longer true and the responsibility for the crime falls back to the member, that cannot preserve his self respect with this blame. Therefore to preserve his self respect, the member cannot broke his connections with the gang.

It is seen, that the discussed fault is a consequence of the necessary and in general effective structure of the knowledge processing unit of anticipatory systems.

Appendix:

(A) GENERALISATION AND SPECIALISATION OF FUZZY PREDICATES

Measures for the similarity of fuzzy sets in $F(X)$ over a ground set X has been introduced and examined by Gottwald [Gottwald 1993] or Cai [Cai 1995]. The disadvantage of this similarity measures in our context is that they are independent of the topology of the ground space . The similarity of two different singletons in $F(X)$ does not depend on the distance between their peaks. But in a similarity theory for fuzzy neural networks, singletons with neighboured peaks has to represent similar predicates.

It is another question, that in biology the topology of the environment has also be learned by the creatures which are living in it. But we assume that this has not been done by an individual. The knowledge of the topology has been acquired by the species and is contained in the a priori knowledge of an individual. It is selected such that the survival of the species is optimised.

The similarity measure used in this article is based on the definition of fuzzy equality on $F(X)$ introduced by Kruse et al. [Klawonn 1994].

DEFINITION 1 (FUZZY EQUALITY OPERATOR ON X): $E_{\approx} : X \times X \to [0,1]$ is called fuzzy equality operator $:\Leftrightarrow$

$\forall x \in X : \ E_{\approx}(x,x) = 1$, $\ \forall x, y \in X : \ E_{\approx}(x,y) = E_{\approx}(y,x)$ and $\ \forall x, y, z \in X : \ E_{\approx}(x,y) T' E_{\approx}(y,z) \leq E_{\approx}(x,z)$,

where T' denotes the Lukasiewicz-t-norm ([Pedrycz 1989]).

An equality operator $E'_{\approx}$ is called finer than $E_{\approx}$ ($E_{\approx}$ is coarser than $E'_{\approx}$) $:\Leftrightarrow$

$\forall x, y \in X : E'_{\approx}(x,y) \leq E_{\approx}(x,y)$

A family $E_{\approx}^{\alpha}$ of equality operators $\alpha \in [0,1]$ is called an ordered family $:\Leftrightarrow$

$\forall \alpha \in [0,1] : E_{\approx}^{\alpha}$ is an equality operator on X , $\quad \forall \alpha, \beta \in [0,1] : \alpha < \beta \Rightarrow E_{\approx}^{\alpha}$ finer than $E_{\approx}^{\beta}$,

$$E_{\approx}^{1}(x,y) = \begin{cases} 1 & \text{for } x = y \\ 0 & \text{else} \end{cases} \qquad \text{and} \qquad E_{\approx}^{0}(x,y) \equiv 1 \text{ for all } x, y \in X .$$

For a fixed equality operator $E_{\approx}$ (with the property $E_{\approx}(x,y) = 1 \Leftrightarrow x = y$) an ordered family of equality operators can be defined by: $E_{\approx}^{\alpha}(x,y) = \left(E_{\approx}(x,y)\right)^{\phi(\alpha)}$,where $\phi : [0,1] \to [0,\infty]$ is surjective and strongly increasing.

(The index α corresponds for α near 1 to the adjective "very" and for α near 0 to the adjective "little".)

The idea is that the membership function of a more general predicate (where many values of X are members) is wider than the membership function of a very special predicate (where only a few values of X are members). This idea leads to the following definition:

DEFINITION 2 (GENERALISATION): Let X denote a finite set, $E_{\approx}^{\alpha}$ an ordered family of equality operators on X. For $\mu \in F(X)$ the α-generalisation is defined by: $\quad \overline{\mu}^{\alpha}(x) := \underset{y \in X}{S} \ E_{\approx}^{\alpha}(x,y) T \mu(y)$,

where T is a t-norm and S the associated s-norm ([Pedrycz 1989]).

(For an infinite set X select a finite set X' whose elements are "equally distributed" in X and restrict the equation to $y \in X'$.) The properties $\overline{\mu}^{1} = \mu$ and $\overline{\mu}^{0} \equiv 1$ are easily deduced from Definition 2.

DEFINITION 3 (SIMILARITY OF FUZZY SETS): With the introduced denotations, for $\mu, \nu \in F(X)$ will be defined: μ is called α-more general than ν (or ν is called α-finer than μ) $:\Leftrightarrow \ \forall x \in X : \overline{\mu}^{\alpha}(x) + \alpha \geq \nu(x)$

The similarity of μ and ν is defined: $\quad E_{\approx}(\mu, \nu) = max \left\{ \alpha \in [0,1] \left| \begin{array}{l} \mu \text{ is } \alpha \text{ finer than } \nu \text{ and} \\ \nu \text{ is } \alpha \text{ finer than } \mu \end{array} \right. \right\}$.

A triangular fuzzy set ν on $F(X)$ is defined by a finite set $X' \subseteq X$ and $\alpha' \in [0,1]$: $\nu(x) := \underset{y \in X'}{\max} E_{\approx}^{\alpha'}(x,y)$

DEFINITION 4 (SPECIALISATION OF FUZZY SETS): With the given notations, for $\mu \in F(X)$ the α-specialisation of μ is defined as the smallest triangular fuzzy set ν with $\overline{\nu}^{\alpha} \geq \mu$.

(B) SIMPLIFICATION OF KNOWLEDGE BY GENERALISATION (FUZZIFICATION)

Fuzzy knowledge is based on predicates (fuzzy sets) and relations between this predicates (rules). The objective of a simplification process is:
- replace all predicates by generalisations,
- collect predicates that are equal in the generalised view and replace them by the same generalised predicate,
- adapt the rules to the new predicates.

A set of fuzzy predicates can be simplified with the procedure:

(1) SIMPLIFICATION PROCEDURE FOR SETS $\left\{o_i\right\}_{i \in I}$ OF PREDICATES: Select $\alpha \in [0,1]$, α near 1.

Find a greatest set $\left\{o_k\right\}_{k \in J} \subseteq \left\{o_i\right\}_{i \in I}$ with $E_\approx\left(o_k, o_l\right) < \alpha$ for all $k, l \in J$.

Find a singleton δ_y $\left(\delta_y(x) := \begin{cases} 1 & \text{for } x = y \\ 0 & \text{for } x \neq y \end{cases}\right)$ such that $\underset{k \in J}{S} s_k T E_\approx\left(\delta_y, o_k\right)$ is maximal,

where s_k denotes the security value of o_k.

Define $\hat{o} := \overline{\delta}_y^\beta$ for a suitable $\beta \in [0,1]$. Each $o_i, i \in I$ with $E_\approx\left(o_i, \hat{o}\right) > \alpha$ will be replaced by $\hat{o}$.

If this procedure is applied to the predicates of the output space $F(Y)$, the predicates of the input space $F(X)$ can be simplified in adaptation to the new predicates over Y.

(2) SIMPLIFICATION PROCEDURE FOR THE PREDICATES OF THE INPUT SPACE: Select all predicates A_{ij} in rules R_i with a consequence B_i that had been replaced by a generalisation in procedure (1).

Simplify this predicates with procedure (1).

Replacing all predicates in the rules by their generalisations, some of the rules will become superfluous. Pruning procedures to simplify the rules are elaborated by Shann and Fu [Shann et al 1995] or Chau and Teng [Chao et al 1995]. A simple procedure works as follows:

(3) PRUNING PROCEDURE FOR THE RULE SET: Let $o \rightarrow \hat{o}$ denote the function from the old predicates o to the new generalised predicates $\hat{o}$. The rule R_i: if u is A_{i1} and....and u is A_{in_i} then o is B_i with

$\tau_i := E_\approx\left(B_i, \hat{B}_i\right)$ and $\tau_{ij} := E_\approx\left(A_{ij}, \hat{A}_{ij}\right)$ is transformed to the rule

R_i: if $\left(\hat{A}_{i1}(u)T\tau_{i1}$ and....and $\hat{A}_{in_i}(u)T\tau_{in_i}\right)T\tau_i$ then o is $\hat{B}_i$.

For $\hat{A}_{ij} = \hat{A}_{ik}$ replace $\hat{A}_{ij}(u)T\tau_{ij}$ and $\hat{A}_{ik}(u)T\tau_{ij}$ by $\hat{A}_{ij}(u)T\tau_{ij}$.

If for some rules R_i, R_j: $\hat{A}_{i1}(u)T\tau_{i1}$ and....and $\hat{A}_{in_i}(u)T\tau_{in_i} \Rightarrow \hat{A}_{j1}(u)T\tau_{j1}$ and....and $\hat{A}_{jn_j}(u)T\tau_{jn_j}$ and

$\hat{B}_i = \hat{B}_j$ holds, then erase rule R_j.

REFERENCES:
Breiman L. (1996). Bagging Predictors. Machine Learning, 24, 123-140.

Cai Kai-Yuan (1995). d-Equalities of fuzzy sets. Fuzzy Sets and Systems 76, 97-112.

Chao Chun-Tang; Teng Ching-Cheng (1995). Imlementation of a Fuzzy inference system using a normalized fuzzy neural network. Fuzzy Sets and Systems 75 17-31.

Dörner D. (1993). Wissen, Emotionen und Handlungsregulation oder die Vernunft der Gefühle. Z. Psychol. 201, 167-202.

Dubois Daniel M. (1996). Introduction of the Aristotle´s Final Causation in CAST: The Concept and Method of Incursion and Hyperincursion. In Compurer Aided Systems Theory - EUROCAST´95. Edited by F. Pichler, R. Moreno Diaz, R. Albrecht. Lecture Notes in Computer Science, Vol. 1030, Springer Verlag, Berlin,Heidelberg,New York, 477-493.

Dubois D. ;Prade H. (1993). Possibility theory, belief revision and non monotonic logic. EUFIT 93,Vol.2,714-719.

Fick E. (1968). Einführung in die Grundlagen der Quantentheorie. Aula-Verlag, 5.Auflage.

Gottwald S. (1993). Defining distances of fuzzy sets through t-norms. EUFIT 93, 351-357.

Hegel G.W.F. (1807). Die Phänomenologie des Geistes. J.A. Goebhardt-Verlag..

Hegel G.W.F. (1812-1816). Wissenschaft der Logik I/II. J.L. Schrag-Verlag.

Kiendl H. (1993). Hyperinferenz, Hyperdefuzzyfizierung und erste Anwendungen. 3. Workshop Fuzzy Control Forschungsbericht Nr. 0293 Universität Dortmund, 82-93.

Kiendl H. (1994). Invarianzforderungen für Inferenzfilter. 4.Workshop Fuzzy Control Forschungsbericht Nr. 0194 Universität Dortmund, 1-12.

Kiendl H. (1997). Fuzzy Control methodenorientiert. Oldenbourg Verlag .

Klawonn F. (1994). Fuzzy sets and vague environments. Fuzzy Sets and Systems 62,2 207-222.

Klir G.J (1995). Principles of uncertainty: What are they? Why do we need them? Fuzzy Sets and Systems 74 15-31.

Kruse R.; Nauck D.; Klawonn F. (1995). Neuronale Fuzzy-Systeme. Spektrum der Wissenschaft,6, 34-41.

Li M.; Vitanyi P.M.B. (1992). Inductive Reasoning and Kolmogorov Complexity. J. of Computer and Systrem Sciences 44 343-384.

Lin C.T.; Lin C.J.; Lee C.S.G. (1995). Fuzzy adaptive learning control network with online neural learning. Fuzzy Sets and Systems 71, 25-45.

Ljung L. (1987). System Identification: Theory for the User. Prentice-Hall, Englewood Cliffs, NJ.

Mahler R.P.S. (1995). Combining ambiguous evidence with respect to ambiguous a priori knowledge. Part II: Fuzzy logic. Fuzzy Sets and Systems 75, 319-354.

Mahler R.P.S. (1996). Combining Ambiguous Evidence with Respect to Ambiguous a priori Knowledge, I: Boolean Logic. IEEE Transactions on System, Man and Cybernetics-Part A, Vol.26, No.1, 27-41.

Mingers John (1996). A Comparison of Maturana´s Autopietic social Theory and Gidden´s Theory of Structuration. Systems Research, Vol. 13, N. 4, 469-482.

Mock V. (1996). Common Sense und Logik in Jan Smedslunds 'Psycho Logik'. Journal for General Philosophy of Science 27; 281-306.

Pedrycz W. (1989). Fuzzy Control and Fuzzy Systems. Research Stud. Press: Taunton and Wiley, New York.

Rosen Robert (1985). Anticipatory Systems. Pergamon Press.

Shann J.J.; Fu H.C. (1995). A fuzzy neural network for rule acquiring on fuzzy control systems. Fuzzy Sets and Systems 71 345-357.

Slotine J.J.E.; Sanner R.M. (1993). Neural Networks for Adaptive Control and Recursive Identification: A Theoretical Framework. ECC`93,Groningen 381-436.

Sommer H.J. (1996). Learning in Complex Environments. EUFIT 705-709.

Sommer H.J. (1995). Zur einfachen Darstellung eines allgemeinen Ansatzes der Wissens-verarbeitung mittels Fuzzy-Logik. 5.Workshop Fuzzy Control Forschungsbericht Nr. 0295 Universität Dortmund, 1-13.

Sommer H.J.; Dürrbaum A. (1996). The use of ignorance in learning systems. FUZZY96 Zittau.

Sommer H.J.; Dürrbaum A. (1994). Übertragung biologischer und psychologischer Prinzipien auf technische Systeme mittels Fuzzy-Logik. Forschungsbericht 194 Uni Dortmund 29-40.

Sun R. (1995). Structuring Knowledge In Vague Domains. IEEE Trans. Knowledge and Data Engineering, 7,1, 120-136.

Sun R. (1996). Commonsense reasoning with rules, cases, and connectionist models: A paradigmatic comparison. Fuzzy Sets and Systems 82, 187-200.

Tiehl R. (1996). Dialectics and mathematical systems theory. FUZZY 96, Zittau, 484-487.

Yager R.R, Filev P.(1995). Including Probabilistic Uncertainty in Fuzzy Logic Controller Modeling Using Dempster-Shafer Theory. IEEE Transactions on System, Man and Cybernetics, Vol.25, No.8, 1221-1230.

Zadeh L.A. (1965). Fuzzy sets. Information and Control 8, 338-353.

Pattern Formation in Spatially Extended Nonlinear Systems: Toward a Foundation for Meaning in Symbolic Forms

David DeMaris

Dept. of Electrical and Computer Engineering

University of Texas at Austin

demaris@ece.utexas.edu

submitted March 1997; Revised January 1998

Abstract

This paper brings together observations from a variety of fields to point toward what the author believes to be the most promising computational approach to the modeling of brain-like symbol formation, unifying perceptual and linguistic domains under a common computational physics. It brings Cassirer's Gestalt era evolutionary theory of language and symbolic thought to the attention of the situated cognition community, and describes how recent observations in experimental brain dynamics and computational approaches can be brought to bear on the problem. Research by the author and others in oscillatory network models of ambiguous perception with an attentional component is emphasised as a starting point for exploring increasingly complex pattern formation processes leading to simple forms of linguistic performance. These forms occupy a space between iconic representations and grammar. The dynamic pattern network framework suggests that to separate perception, representation or models and action in a realistic biophysics of situated organisms may be problematic.

Introduction

The formalism of coupled map lattices (CMLs) allows efficient computer simulation of such oscillating cell assemblies, in which perceptual and cognitive phenomena are embodied through dynamic pattern formation processes. The deterministically fluctuating state variables of each cell recursively executing a nonlinear map correspond to the microscopic pulse populations, while the patterns formed in the lattice as a whole (or interactions between lattices considered as interacting subsystems) correspond to the macroscopic state transitions. In the context of symbolic forms, these macroscopic state transitions might be prototypical spatial organizations such as gradients of oscillation clusters (DeMaris 1997) or spatial patterns (possibly only seen through time averaging) acting as control parameters in the synthesis of

CP437, *Computing Anticipatory Systems: CAYS--First International Conference*
edited by Daniel M. Dubois © 1998 The American Institute of Physics 1-56396-827-4/98/$15.00

imagery and utterance. The constraints governing the construction of macroscopic states parameters and patterns are provided by the perceptual process.

Evidence has accumulated that large scale spatial patterns (affecting many cortical columns) play a hitherto unsuspected role in cognitive processing. Philosopher of language Ernst Cassirer suggested early in the Gestalt psychology era that a single overarching mechanism for symbolic processing did not exist, instead positing a set of basic modalities. We consider several forms of pattern dynamics seen in various neurological syndromes, EEG and MEG brain imaging. This pattern ecology may serve as the basis forms underlying various symbolic modalities. Emergent dynamic spatial patterns may serve as foundations of meaning playing roles both in both perception and the formation of linguistic utterances. In the latter case, a key observation is that patterns of excitation in layers coding for semantics may influence the spatio-temporal dynamics of motor control and planning layers to produce correlations between aspects of perception and linguistic utterance. These correlations arise due to excitations acting as biases on bifurcation parameters in the output layers, constraining them from reaching synchronization.

Finally we review pattern generation computations and sketch the nature of computational architectures suitable for exploring this more sophisticated symbol formation process. When the spatial patterns influencing linguistic productions are relatively simple, such as monotonic gradients, sound symbolism should dominate in the language. When more complex forms are coupled with these simple forms in output or production layers, we predict that the system will show the hybrid characteristics of sound symbolism mixed with formal grammars characteristic of human language. In this sense, the patterns produce the *modes* of linguistic utterance that embody and communicate meaning. If this view of neural dynamics and sub-symbolic processing is correct, it goes some way towards explaining why the so-called hard problems of consciousness have yielded little. Until the advent of simulation techniques and computational power to constructively model such systems, the interaction with spatial patterns of parameters tend to force models into an oversimplified view. Even the conceptual language of attractors, order parameters and wave solutions may be inadequate to describe the complex interaction of perceptual and memory based spatial forms with local and global bifurcation parameters. Individual lattice sites in some subsystems may operate in continually perturbed transient states, never reaching attractors per se.

Symbolic Form, Perception and Anticipation

The definition of symbol and the binding of meaning to symbol is central to the study of mind. The basis of all predictive or anticipatory behavior which is not obviously seen as mechanistic, even if adaptive control, relies on the creation, relation, and transformation of symbols. This paper will briefly note major trends in this long running debate and focus on an approach rooted in the Gestalt psychology era. We will then treat the relationship of neural models at both micro-level and large scale to this approach to symbol formation. Finally, we consider evidence that oscillating dynamic pattern networks may function in a similar fashion in both perception and in an evolutionarily early component of language known as sound symbolism.

In the recent era of cognitive psychology, the prevailing computational metaphor is often simply described as *symbolic* artificial intelligence. The arbitrary relation between the physical token in a symbol system and real world has been defined by Harnad (1990) as the symbol grounding problem. He suggests that connectionist artificial neural networks, coupled to sensory transducers and feature detectors, are able to produce symbols whose encoding and relations to other symbols are non-arbitrary, having been formed through a bottom up process of iconic and categorical relations. The supervised learning connectionist model has in turn been criticized as a model of perception, given that organisms must survive without a teacher , and for not conforming to the complexities of perceptual phenomena as known behaviorally since the Gestalt psychology era and now subject to detailed measurements at the level of neural signals. A further criticism of the biological realism of such models is the unnatural separation of learning epochs and memory (Tsuda 1992).

The symbolic cognitive model has also been criticized for its remoteness from social embedding (Winograd 1986) and from the lack of any connection to motor behavior and embodiment (Varela, Rosch and Thompson, 1993). Both critiques are aligned with the philosophical school of phenomenology, and begin to address the problem of the location and origin of meaning. This problem is in part due to the assumption which has prevailed since the rise of structural linguistics that the lexicon is arbitrary; in the terms of semiotics, the words of a language are signs defined by social convention rather than coding by an iconic mapping.

Initiating the modern study of symbolic forms, E. Cassirer (1929 / 1955) examined early material from linguistic anthropology, aphasia studies, cultural myths and religious practices, and mathematical and scientific reasoning to deduce the ways in which basic a priori categories (space, time, and number) manifest in the symbolic forms of language, myth, art, and scientific thought. Cassirer arrived at an evolutionary conception of symbolism as emergent in language and co-evolving with the fundamental subject - object distinction in perception. In this view, expression (or expressive perception) is prior to the segmentation or distinction of objects and their attributes, and thus subsumes or underlies conventional categories of metonymy (context sensitive) and metaphor (context free) known to later structural linguistics. He argued that expression emerges with perception from the organisms living in a world with a basic orientation toward action, anticipating the contemporary inactive or situated cognition approach. In contrast to the ecological or functional orientation of situated cognition, Cassirer emphasized the living presence of *mythical* thought as a prior and coexisting component of language in opposition to discursive or logical thought; he argued that formation of mythic symbols was an expressive process in which heightened affect towards some natural 'momentary god' created a focus of the sensory and affective field to create a verbal or gestural symbol. Expression enters into the formation (more dynamic than grounding) of symbols when mimesis, in imitative hand or vocal gestures, is replaced by vocal utterances expressing aspects of feeling, and as body-derived spatial designators begin to indicate an awareness of space and objects. This becomes captured in language as sound symbolism or phonetic symbolism (Hinton et al. 1994). Within a culture phonetic patterns become identified with expressive effects. In Indo-European languages, for example there are trends in associating vowels (a, o, u) with distant objects, while vowels (e and i) are associated with near to the speaker. In some existing languages (Nuckolls, 1996) similar effects can become the basis of grammar and of a variety of performance effects. In this sense, the underlying expressive character serves as a framework to anticipate both the micro-level unfolding of individual sentences and as a larger culture -bound frame of values and relationships to the natural environment.

Cassirer's work was contemporary with and strongly influenced by Gestalt psychology. Gestalt researchers (Kohler 1926) also explored the relationship between sound patterns and visual patterns by making subjects assign correspondences between invented visual forms and words. He found that a large majority of subjects would assign "takete" to a more angular shape, "maluma" to a rounded shape. Allot (1981, 1995) suggests that correspondences between the neural responses of the kinesthetic and visual systems give rise to "natural" word forms. The late writings of Whorf (1956) should also be considered a historical precedent for the theoretical approach expressed here. Whorf suggested that Sanskrit terms from Indian philosophy of symbolism be adopted. He associated *Nama* and *Rupa* with lexical processes and spatial binding respectively as formative processes of phonetic symbolism, while associating the *Arupa* construct as a non-temporoal, non-spatial pattern level governing linguistic processes of morphological patterning and grammar

In reviewing these investigations, we conclude that simple iconic mapping (onomatopeic words for example) is rarely the case. The mappings of phonetic, pitch or other performance codes to symbols capture affect and intention toward the world in ways that existing grounded connectionist models do not. Oahu (1995) turns the usual assumptions of causal flow in expressive speech when he suggests that facial expressions associated with emotional display actually regulate large scale neurodynamics by changing cortical blood flow patterns.

Emergent dynamic spatial patterns may serve as foundations of meaning playing roles both in both perception and the formation of linguistic utterances. In the latter case, a key observation is that patterns of excitation in layers coding for semantics may influence the spatio-temporal dynamics of motor control and planning layers to produce correlations between aspects of perception and linguistic utterance. These correlations arise due to excitations acting as biases on bifurcation parameters in the output layers, constraining them from reaching synchronization. When the spatial patterns influencing linguistic productions are relatively simple, such as monotonic gradients, sound symbolism should dominate in the language. When more complex forms are coupled with these simple forms in output or production layers, we predict that the system will show the hybrid characteristics of sound symbolism mixed with formal grammars characteristic of human language. In this sense, the patterns embody and produce the *modes* of linguistic utterance that capture and communicate meaning.

Anticipation in language enters at micro-levels, when what we designate as grammar - the contextual modification of lexical objects - is operative.The anticipatory nature of language in performance is has been described by Friedrich:

... The relation of linear to nonlinear is thus quite complicated and paradoxical. In the audible chain of conversation, for example, some sounds will be imagined before others are being articulated, and this anticipatory imagining will affect what is being said. In the same way, the short-term memory of what has just been articulated will affect the present act. This high-speed cybernetic scanning up ahead and backwards along the imagined sound stream presupposes both the (linear) stream and also an essentially nonlinear or alinear monitoring and synthesis of that stream as a field or simultaneously apprehended structure. (Friedrich, 1986)

Neural Networks and Dynamic Patterns

While debates on the nature of symbolism have raged in the various communities for years, neuroscience and neurology have had to directly confront the complexities of brain organization. The conceptual and measurement tools of signal processing and information theory have seen the most widespread use, but alternatives in the form of nonlinear dynamics and dynamic pattern networks have emerged as viable alternatives. In the latter paradigms, modeling typically occurs in terms of coupled oscillators, where the state variables are *macrostate* variables corresponding to quantities such as average ensemble frequency, rather than single neuron dynamics. We consider some problems and considerations in neural modeling before returning to the problem of symbolism and language. Multistable percepts, such as depth reversals in the Necker Cube, pose challenges for any theory of neural coding due to the complexity of phenomena and apparent correlations with attention (as evidenced by eye fixations).

A Very Brief History of the Cell Assembly Concept

Hebbian cell assemblies are traditionally defined as functional units (i.e. smallest 'meaningful' neural substrates) comprising neurons that are strongly interconnected by excitatory synapses such that activation of a subset leads to 'ignition' of the whole assembly as soon as a given threshold is passed (Abeles, 1991). The constituent neurons may be spatially distributed, and any neuron may participate in many assemblies. While simultaneous activity was considered as the primary characteristic for some time, more recent emphasis has been on phase-locked oscillations of groups of neurons. (Eckhorn et. al 1992; Grossberg 1991). The emphasis on synchronized oscillations has been challenged in favor of chaotic oscillations, and further that spatial patterns of such oscillations are the carriers of perceptual codes (Skarda and Freeman 1987; Yao and Freeman 1990).

The next level of organization - how neural coding is organized into functional modules such as those hypothesized by lesion studies in neurology - proves an even more daunting

473

problem. Local circuits operating as feature detectors in primary sensory areas have typically been assumed to serve various architectures: higher order networks of combined feature detectors, connectionist assembly dynamics (with fixed point attractor basins), or more complex adaptive resonance networks for gestalt formation and memory. Connectionist networks in which the global state vector evolves to a static condition might be termed convergent or equilibrium, in the sense that during recognition of learned categories they remain in the same convergent phase regime due to stationary parameters in the network dynamics. The Hopfield network with symmetric connections and multilayer back propagation networks are examples of equilibrium networks in this sense, with the weights stabilized by training to a stationary state. In order to process continuing input, such networks must jump out of the equilibrium state encoding a recognized memory, that is to say they must be reset. Typically some supervisory process is proposed. In contrast, biological networks exhibit continual non-stationary dynamical activity, with intrinsic resetting or cyclic behavior of dynamical control parameters governed by perception, attention, mood, and intrinsic cycles such as breathing (Elbert et al, 1994). Even when dynamical control parameters are stationary, much of the work on such dynamic pattern networks explores the chaotic and intermittent dynamical regimes to explain cognitive and perceptual phenomena. In such paradigms the traditional concept of cell assemblies breaks down; it may be better to speak of spatio-temporal patterns emerging as *coherence assemblies*, in recognition of the fact that oscillations may be occurring at all points in space in chaotic or intermittency modes, but coherently oscillating only in certain spatial regions.

In a recent review of the role of deterministic chaos in cell assemblies, Elbert and colleagues question the traditional separation of cortex into functional modules:

> ... A cell assembly includes sometimes widespread cortical neurons including sensory, cognitive (meaning) and motor functions. Any restrictive separation into highly specialized "modules" as is fashionable in present day neuropsychology, is obsolete; vis-à-vis the fact that every sufficiently large pool of neurons of the cortex is connected to every other neuronal pool, forming the anatomical basis of our illusion of a unified consciousness. The meaning and qualitative nature of an event, an idea, an emotion, or a percept is reflected in the local topography of its connections and firing patterns, so to speak in the topographical "gestalt" of an assembly in its phase space

The present paper suggests that firing patterns are mediated to a greater extent than previously suspected by emergent spatial patterns in oscillating networks. These patterns affect the "local" dynamics observed in different modular regions by virtue of long distance connections. They may thus serve as selective bifurcation parameters in more loosely coupled regions which, if disconnected, would have different "local" dynamics. These bifurcation controls may give rise to further spatio-temporal patterns of synchronized activity localized in time (in the case of motor control) or distributed over space (in the case of imagery). The emergent form of the patterns is intimately tied to behavioral and emotional aspects of organismic being. In a model, these must be handled by including them in the form of order parameters - *coupling* and *bifurcation* parameters governing the general evolution of patterns as the model states are perturbed by sensory input and memory. Patterns emerge anticipating future actions or sensory input in both organism and model, but assigning a *causal* status to the patterns in an unfolding process is problematic. That the patterns are themselves structurally and in part genetically tuned possibilities means that they are, at that phylogenetic level as well, anticipatory of environmentally fit perceptual and behavioral interactions.

Pyschophysics of the Necker Cube

The Necker cube has elicited a large literature, with new and interesting variants introduced regularly. The following discussion may seem a slight digression; the chief reason for introducing it here is to consider in some detail the way in which details of a pattern - in this case, the retinal image of a cube - influence both attentional forms and temporal flows. Considering the psychophysics and dynamical models of the problem in some detail may serve to help the reader visualize how such dynamics could play a role in cognitive processes normally thought of as symbolic (in the conventional computational sense).

Over time, explanations of the involuntary perceptual state changes involving satiation (Attneave, 1971) gave way to those involving attention and interaction with top-down cognitive components. Cognitive explanations, in which a spatially global percept enforces a top-down schema, lost credibility when demonstrations involving multiple cubes which showed that multiple conflicting perceptions could exist in the same visual field (Long and Toppino 1981). Size effects on reversal rate means and distributions have been studied over a wide range by Borsellino et al. (1982).

The interaction of gestalt formation with attention is well documented in the case of reversible depth perception with the Necker cube. Eye fixations were also observed to correlate with reversals. In an attempt to resolve a long standing dispute on whether eye movements were an effect or cause of reversals, Ellis and Stark (1978) undertook a study of location and durations of fixations during reversals; subjects fixations were allowed to range freely over a fairly large (12 degrees) cube while recording fixation points, duration, and reversal times. They demonstrated that fixations are attracted along diagonals in the cube, had longer duration during reversal events (600-800 ms vs. 350-700 ms), and that fixations near (but not on) a central vertex forced its interpretation as part of the near face; they summarize scanning behavior as "back and forth between temporally changing externally appearing (i.e. nearest) corners". Experiments indicating a strong causal relationship of fixation point on reversal are reported by Kawabata et al. (1978). When fixation is near a certain vertex prior to flashing the cube on a screen, that vertex is perceived as nearer and the residency time is strongly biased towards.

Synergetic Modeling and Order Parameters

The author has previously argued (Demaris, 1997) that it may be helpful to consider depth perception and attention as a unified field constructed from the interaction of visual forms and gradients in a spatial distribution of oscillation clusters. This approach may provide insight into the complex interaction of switching and residency times with attention (as evidenced by eye fixation), depth gestalt, and spatial scale dependencies in multistable phenomena. This approach is close in some respects to that advanced in the Ditzinger and Haken (1986) model for figure ground reversal, but attempts to introduce a spatial dimension and focus to account for the known size effects and the links between depth organization and eye movement biasing effects. In their model, relationships between attention parameters and an order parameter are defined such that there is a saturation nonlinearity of an attention parameter when an order parameter (corresponding to a particular percept of figure) increases, forcing a "reset" in the attention state variable. This in turn triggers a dynamical reorganization and emergence of the alternate attractor in the order parameter. They note that a time constant in the equation linking order parameters and attention controls the reversal times between percepts. In an extention of this model (Ditzinger and Haken 1996) they acknowledge the dependence of the distribution on the visual angle of the cube and suggest that the visual angle is correlated with a parameter governing fluctuation strength, but give no deterministic derivation for this fact.

A more recent nonlinear oscillator formulation is proposed by Kelso et al (1995). They propose that changes in the shape of the distribution of switching time observed with rotation of a Necker cube can be modeled by a system of oscillators coupled through a phase relation operating in the intermittency regime near tangent or saddle-node bifurcations. The system is governed by a scalar order parameter which is iterated in the course of a model, occupying a space bounded by coupling. While such a model does produce distribution changes similar to observed psychophysics, the process by which the spatial form itself leads to changes in coupling is not made explicit. There is also no suggestion of how the resulting percept of space is coded which would further support motor actions or speech acts which are consistent with a percept, and no explanation of the observed interaction of attention with switching states. Since multiple cubes can sustain conflicting orientations are observable (von Grunau et al. 1984), it has been asserted that the character of perceptual organization is local, which is not yet addressed by the synergetics models.

The model (Fig. 1) outlined in DeMaris (1997) attempts to address the latter deficiencies. Depth perception and attention are treated as a unified composition of fields (2 dimensional arrays of coupled maps) computed from the interaction of visual forms and gradients in a spatially distributed bifurcation parameter. The dynamics (detailed in the Appendix) in which the bifurcation state is continually updated by the local oscillations toward a quasi stable fluctuation around the transition to chaos might better account for the association of transition time and eye movements with the spatial scale of the cube.

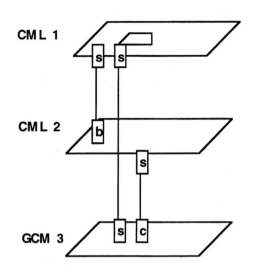

CML 1

CML 2

GCM 3

Input layer
Extracted features create oscillations distributions for binding (Binding layer not shown.)

Attention layer:
Form mediated fluctuations between chaotic and periodic generate attention flow and control coupling in depth field layer 3.

Gestalt (depth field) layer:
Coupling modified during attention events, influencing cluster distribution.

Fig 1. Three layer coupled map architecture for simulating interactive formation of attention foci and depth gestalt formation.

The model is outlined In Figure 1, The first layer represents the shape processing reaction diffusion layer described in equations (2-4, in Appendix).   The second layer represents the emerging attention foci when the fluctuations of equation 4 fall below a threshold indicating periodic oscillations. This layer drives both layer 1 and controls the coupling parameter in layer 3 (equation 7). A third layer and set of equations is used to encode the actual depth interpretation, by the spatial distributions of clusters (particular oscillation modes) when the entire lattice is oscillating near the edge of its coherent regime, supporting only a few oscillations modes. Links between layers indicate the nature of the causality or parametric role played by the same state value in different layers; note that oscillation state in layer 1 influences bifurcation in layer 2, while layer 2 state influences mean field coupling in layer three.

The next section will consider evidence that large scale spatial forms generated by spatially distributed oscillating systems play a significant role in cognition and behavior. Some theoretical paradigms relating cognition and emotional states are reviewed and contrasted briefly with computational and signal processing approaches.   It is suggested that underlying perceptual-motor form generation processes underlie language and perception in spatially distributed systems, and that the dynamics leading to utterance may be similar to those described above for the interactions between Necker Cube dynamics, depth gestalt and attention. The brain regions responsible may in fact have a common evolutionary origin, which has been modified in the language system to interact more strongly with memory and emotional content from the limbic system.

Pollack, working on the problem of machine induction from positive exemplars, demonstrates that bifurcation in a learning process on recurrent higher order Sigma-Pi networks corresponds to the ability to generalize; he hypothesizes that by operating in the complex or chaotic regimes, a system may handle non-regular languages, perhaps extending to natural languages. The system generates finite state automata which can then accept strings as belonging to the formal language. He concludes with a hypothesis relating the complexity of languages generated by the dynamical recognizer:

The state-space of a dynamical recognizer is an attractor which is cut by a threshold (or similar) decision function. The complexity of the generated language is regular if the cut falls between disjoint limit points or cycles; context free if it cuts a self-similar recursive region, and context sensitive if it cuts a chaotic region (Pollack 1991).

No studies of dimensionality have been performed yet on model time series with coupled layers of lattices (see Appendix), but this approach may be worth pursuing. The dynamics of controlling the coupling of the global map through the attention layer are subject to a threshold affect, suggesting that the pattern recognition capability may be tunable to suit environmental conditions through small parameter changes in spatially distributed systems.

Basis Forms

Evidence has accumulated that large scale spatial patterns (affecting many cortical columns) play a hitherto unsuspected role in cognitive processing. Cassirer suggested that a single overarching mechanism for symbolic processing did not exist, instead positing a set of basic modalities. We consider several forms of pattern dynamics seen in various neurological syndromes, EEG and MEG brain imaging. This set of "pattern alphabets" operating at various spatial and temporal scales, may serve as the *basis forms* underlying various symbolic modalities. Finally we review pattern generation computations and sketch the nature of computational architectures to explore a more sophisticated symbol formation.

Phenomenelogical Reports of Patterns

The patterns known as *form constants* in psychological literature, frequently associated with visual hallucinations, seem implicated in the phenomena of synesthesia. This generic syndrome of cross-modal sensory interaction can manifest in many ways, but at least one characteristic mode is that overlaid forms from the repertoire of form constants appear in normal visual space. The forms are transformed in systematic ways for a single subject, but there is great inter-subject variability. Complexity of an auditory (musical stimulus) may be reflected in the complexity of the experienced form. The syndrome has been associated with unusual patterns of cortical blood flow (Cytowic 1987).

Elecrophysiological Reports of Patterns

At the spatial scale of hemispheric dynamics, spanning what have historically been considered as separate functional/computational modules, large scale patterns in alpha band EEG signals have been found which vary according to the percept within a task (Fuchs et al 1991; Kelso et al 1995). These patterns may be decomposed to linear combinations of amplitude modulated spatial forms or modes. Lehman and coworkers (1995) have also detected large scale spatial patterns in EEG and identified these "microstates", with sub-second duration, with specific cognitive modes. Microstates corresponding to production of visual imagery and of abstract thought were distinguishable on the basis of angles between voltage centroids and the scalp axis, and front-rear and left-right centroid distances. Kovacs and Julesz (1994), working in smaller scales (a single cortical region), report that the visual system extracts 'skeletons' derived from global shapes as an intermediate stage of object representation. The physiological origin and basis of the patterns is unknown; they are detected through their modification of the contrast sensitivity of local feature detectors.

Computational Basis for Patterns

Various computational models have also been shown to generate patterns resembling form constants. R. Abraham and colleagues have demonstrated similar pattern evaluations in toroidal topology locally-coupled map lattices; characteristic forms emerge from other simple forms undergoing dynamic evolution (Abraham 91). Abraham notes the similarity of the forms to the Chladni patterns (Jenny 74) which arise in sand, water, etc. subject to periodic forcing through vibrating plates. Many of these patterns bear a strong resemblance to the form constants documented by Cytowic in synesthetic perceptions. A model utilizing a two-dimensional sheet of pulse coupled nonlinear oscillators driven by uncorrelated external noise has been shown to spontaneously break translational symmetry and to generate quasi-hexagonal clusters (Usher et al, 1995). Glass patterns (saddle, spiral, vortex, and degenerate node archetypes) are realizable from random input by parameter changes in numerical relations of partial differential equations (Kelso et al, 1995), which can be converted to the discrete coupled map formalism. A few numerical experiments noted in the authors thesis (DeMaris 1995) suggest that alternating chaotic and periodic regimes in a reaction diffusion model may produce skeletal forms similar to those noted by Kovacs.

The temporal aspect of pattern formation in computational systems is obviously of great importance in assesing their biological plausibility. Pattern generation is often determined by local interactions and formation of stationary waves in diffusive processes - the time scale of the process, in terms of iterations, may be infeasable for neurodynamic pattern generation. In a biological system these iteration epochs may correspond to significant frequencies in EEG at which coupling between local oscillators occurs (Baird 1990). One possible solution (not explored to the author's knowledge) are hybrid locally -globally coupled systems, such as the depth attention model below, with re-entrant connections from global lattices rapidly influencing the pattern formation in the local layers corresponding to sensori-motor maps.

Microgenesis and Convergence Zones

The existence of spatial patterns in brain dynamics is indicated by a variety of phenomenological evidence and more recently by imaging studies. At this point, we must ask what role such patterns might play in functional or behavioral terms? What implications do they have for traditional assumptions of regional modularity? Two theoretical positions regarding modularity which seem well aligned with these observation on pattern dynamics are microgenetic theory and the theory of convergence zones.

Microgenetics, derived from evolutionary theory and observations in clinical neurology, claims to explain imagery and perceptual functions (Brown, 1977), and considers them intimately related. Cytowic (1989), summarizes the theory with respect to perception:

" ... a perception consists of a series of levels of space representation leading from a two dimensional space map with the body organized around the brainstem-tectal system ... In the perception of an object, there is an emergence of an object from a system of limbic structures where objects are selected out of experientially, symbolically, and effectively related objects toward a final stage of feature analysis. This is different from the conventional direction of visual perception. ... Normally, afferents come into visual cortex and relay to inferotemporal cortex for matching to memory images or to parietal lobe for updating with the environment. That is, there are successive levels of shape construction. Conventionally, we think that an object is constructed and then matched to memory. The reverse is true in microgenesis. The object unfolds out of memory toward its analysis in the external world. It unfolds from within the observer, within the subject, from a primitive archaic space through a dream-like system of experiential, symbolic stages toward a representation as a holistic object and object relations in a three-dimensional Euclidean space. It unfolds out of a volumetric or egocentric space within the body, a space of hallucination, out toward the space of the external world.

For microgenetic theory, the relevant modularization of the brain is the evolutionary division between limbic, parietal, and occipital, each involved in a separate "pipeline" of constructive stages in image formation. Microgenetics grows out of an attempt to explain aphasias and other syndromes resulting from lesions or organic disturbances. Parietal lesions, for example, involve spatial distortions in perception, which are in turn linked with distortions in language.

Damasio & Damasio (1994) discuss the importance of 'convergence zones' of association cortex (especially anterior temporal and prefrontal) with convergent afferent and divergent feedback connections to many sensorimotor areas. In this view, activity patterns distributed over 'lower' sensory cortices are recorded by convergence zones and can be reconstructed by them via the divergent feedback projections. Convergence zones do not store 're-representations' of sensorimotor encodings which are combined to percepts, concepts etc., but only 'codes' capable of 'reconstructing' these fragments. This concept of *dispositional memories* seems closely related to microgenesis, and recalls Whorf's disccusions of pattern formation in thought. The atemporal, aspatial *Arupa* patterns postulated by Whorf might be considered as dynamical parameters governing the dynamical emergence of a gross spatial forms which in turn allow construction of finer scale behavioral, imagery, and linquistic utterance through dynamic processes similar to those introduced here.

Language Considered as Emergence from Basis Forms

In contrast to connectionist training approaches, language development would operate through the gradual *constraint* of pattern generators in convergence zones to produce only specific spatial forms or combinations, which, after dynamic evolution in other coupled modules towards synchronized oscillations to motor regions, result in activation of grammatical or morphologically correct utterances.

Cytowic relates the visual phenomena of synesthesia (cross-modal sensory interactions) to microgenetic theory, on the basis of subjects' reports positioning their synesthetic percepts as "close in personal space", and undergoing spatial distortion and transformation according to the character of the incoming sensory experience. Music translated into visual form constants, for example, are more or less stable for one subject based on the musical complexity of the piece generating the overlaid form constants. If these forms are engaged, as imcrogenesis suggests, as intermediate stages of production whose generative parameters would be coded in convergence zones, then the forms must evolve or constrain the dynamics of connected neural regions in the production of the percept.

Conversion of Spatial Forms to Ensemble Density Distributions

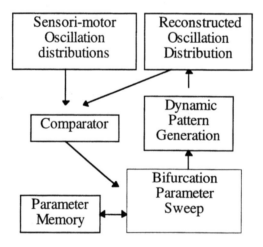

Fig 2 Architectural diagram illustrating a microgenetic approach to memory.

Figure 2 illustrates the architecture for microgenetic construction of a pattern matching diverse features coded by an oscillations distribution. When a pattern or combination from the repertoire of basis forms is found which generates sufficiently similar oscillation distributions, perhaps through dynamics similar to those described in DeMaris (1995), the parameter memory can recall the generative parameters by creating a recallable attractor. Similarity in a computational system may be a simple distance function between instantaneous histograms or time averaged spectra. The pattern, capturing significant features of an environmental image, can then be used as the basis for motor or vocal gestures which are not simply iconic, but are *natural* in the sense used by Alott (1981).

Conversion of Spatial forms to Temporal Dynamics

To the extent that form constants are patterns of large scale interactions coupled to other systems, these spatial structures may influence the time course of other brain subsystems by modulating the operating parameters of another region. The time taken by an iterative system to reach a stable oscillation depends on the distance from a bifurcation point; in a spatially extended system, this might be controlled by modifying excitatory - inhibitory ratios at a local point where the constant forms "overlay" a sensory or re-synthesized imagery field. The imposition of form constants may then impose a natural temporal organization. A hypothesis worthy of exploration is whether particular constants are associated with particular temporal organizations. Spiral or radiating spatial organizations might be associated with cyclic temporal forms for example; lattice forms with gradients might be expected to produce one-time event forms such as single morpheme, statement, or musical phrase. Linear combinations of modulated forms may serve as parametric controls in synthetic or motor layers to produce more complex gestures.

Describing a case of motor aphasia, Merleau-Ponty (1962) writes

> The patient is unable to pronounce, in isolation, a letter or word within a familiar motor series, through being incapable of differentiating between the 'figure' and 'background' and freely conferring upon a certain word or letter the value of a figure. Articulatory and syntactical accuracy always stand in inverse ratio to each other, which shows that the articulation of a word is not a merely motor phenomenon, but that it draws upon the same energies which organize the syntactical order. When certain disturbances of verbal intention are present, as in the case of literal paraphasia in which letters are omitted, displaced, or added, and in which the rhythm of the word is changed, it is, a fortiori, clearly not a question of a destruction of engrams (memory traces), but of the reduction to a common level of figure and background, of a powerlessness to structurize the word and grasp its articulatory physiognomy.

Cytowic describes perceptual motor distortions resulting from brain injuries:

> (An) object becoming smaller or larger; acceleration of time, inversion phenomena ... micropsia and macropsia. All these are collectively known as metamorphasia. ... In addition they (lesions) involve defects in spatial and motor action in the space of the limb perimeter, such as misreaching, inability to draw; the spatial distortions in perception are intimately linked with relationships to limb action.

Such statements suggest that perceptual Gestalt processes such as figure ground separation and depth organization play a constructive role in generation of utterances, which has been asserted by linguists concerned with sound symbolism and the motor theory of lanuage

482

production (Allot, 1981). Recalling the hypothesized dynamics for depth construction outlined above, it is suggested that the same processes which result in transient coherent oscillations in response to an *external* form extracted by the early visual layers may result in the flow of motor activation mediated by *internal* forms stabilized during the microgenetic memory formation process. I emphasize mediated, as the generative process is closer to the character of a reaction-diffusion or spreading activation process rather than the conventional signal decoding or computational recall processes. The process of generating images and utterances in this view is closer to a rapid morphogenetic process, with the basis forms serving as boundary conditions for the emergence the output layers. Brown notes that brainstem hallucinocis typically results in reported imagery such as faces, insects, and animals. This phenomena and the universality of such imagery in hypnagogic imagery suggest that the boundary conditions to generate such images from an oscillating reaction-diffusion process may function in ways similar to those in which boundary conditions cooperatively generate forms in morphogenesis (Kaufmann, 1993).

In this sense, the idea of basis forms can be contrasted with the prevailing use of "basis functions" used in engineering signal compression and synthesis. Basis functions are used in Fourier and wavelet signal analysis to reconstruct a signal instantaneously (limited only by parallel computational resources), by series expansion (addition) of a family of prototypical functions with weighting coefficients. Temporal or spatial extension is inherent in the protoype. In contrast, a basis form undergoes deformation in a parallel spatio-temporal process which is computationally irreducible (i.e. it cannot be accelerated by additional computing resources to produce a mental image or utterance). Basis forms may be considered a subset of the class of order parameters as known in the synergetics and statistical mechanics community, but are considered primarily for their effect in networks or cascades of loosely coupled spatially extended systems. The loose coupling insures that the forms can persist long enough to be effective in mediating between brain regions which are primarily sensory, motor, memory, or logical - discursive.

A particular unfolding sequence of forms and the derived transient synchronization can capture personal and cultural valuations, through the coupling of affective state and patterns. The reciprocal relations between affective state, patterns, and utterances are taken up in more detail in the next sections.

Anxiety and Silence

Attention, seen as a transient coherent spatial oscillation emerging fromns driven by an external or internal form is very much the type of perceptual focus or condensation to physically the mythic symbolic modality of Cassirer. In this prototypical symbolic formation (See Figure 2), diverse sensory or memory component oscillations are combined and linked along with the order parameter code which determines limbic or brainstem derived patterns coding for their meaning.

The normal production of utterances, like the normal flow of attention over a visual scene, may depend on specific spatio-temporal patterns. Since emotional or affective state presumably plays the role of a bifurcation or order parameter in the formation of patterns, some disorders (aphonia) related to extremes of anxiety may be most directly caused by cessation of formation of synchronized focus spots when their source patterns increase in spatial frequency or fractal dimension. Zimmer, in his discussion of the broader importance of multistability, constructs tilings of the Necker cube with increasing spatial frequency and space filling character. It is evident that multistability, with attendant bifurcations and attention flows only occurs when the patterns consist of low spatial frequencies. We conjecture that similar distortions of internally generated basis forms may lead to cessation or distortions in output. Neurologically such distortions can be produced by lesions or electrochemical disruption in either the immediate field where patterns are produced, in connecting pathways, or in areas which constrain the formation of patterns by "embodying" bifurcation parameters governing the spatial frequency of patterns.

In Brown's microgenetic theory, aphasias are considered as stages in normal language production which do not complete the normal "developmental" course of a linguistic utterance. His discussion of neologistic jargon seems to indicate that this phenomena may have similar dynamics to *glossolalic utterance*; the interaction between sounds of the emerging utterance stream results in so-called "klang effects". Perhaps such effects result from unusual degrees of symmetry in basis forms due to decoupling of gradients associated with normal grammatical production. Sass (1991) has suggested that one pathway to schizophrenia is through a loss of cognitive grounding in body image and kinesthetic sense. This would be a kind of disassociative state in which, perhaps due to disturbances in transmitter production or uptake, the gradients underlying both image in perception and in normal grammatical production and modulation are deactivated or have less influence in the production of the flow of utterance. Such hypothesis should be testable with the models proposed here.

Sass, Brown and others have also reviewed the distortions in the style of visual artists undergoing schizophrenic episodes. Distortions of spatial placement are seen, as well as trends towards increasing spatial frequency, symmetry and fractal dimension. Sass discusses the phenomenological reports of a physician Schreber whose world involves a gradual immersion in a world of "nerves and rays" and comments on their relation to inner speech. Shreber's hallucinations were apparently synesthetic, given his explanation that the rays "do not understand the meaning of the words they speak, but apparently they have a natural sensitivity for similarity of sounds". (Readers troubled by the consideration of such experiential reports should consult Laughlin et. al (1990) for an extensive treatment of the problems of admitting, or not admitting, such data in cognitive neuroscience. Cytowic makes similar arguments regarding synesthesia)

Cultural Interaction and Values

One strength of such an acccount is that it may allow for a larger role for social and cultural imbedding, by allowing the entrainment or constraint process of basis form creation to be modified by the cultural values placed on aspects of situated experience. These may be expressed ultimately in language and other cultural practices. Bateson , in his description of style in Balinese painting (1971), suggests that an art style and individual work should serve to impart "a correction, in the direction of systematic wisdom" of natural oversimplifying or polarizing tendencies. Viewed in light of some of the other work surveyed here what might be considered "ornamental" aspects of art such as the elaboration of space, the tendency of forms towards space filling motifs in contrast to sketches, may be considered as indicators of relationships to body centered or enactive universals. Perhaps Bateson's larger theme of a corrective wisdom is active in this sense; a culture or individuals may attempt to balance or even oppose certain valuations or polarities in spatial perceptions by reducing or reversing the gradients in form which might be implied by this *embodied component of their ethos*. The Balinese imagery schema of a turbulent lower field opposed to a serene upper field is contrasted with the Western tendency (in Bateson's view) toward the reverse.

Such a systematic metaphorical strategy integrated with a cosmology (Laughlin et al, 1990) may serve to modify inherent gradients in perceptual - motor systems and limbic / brainstem tuning, arrived at through ecological and cultural interactions during individual development. If motor action is built up first through *axial* and then *radial* body centered coordinate systems as suggested by Brown, a differing balance between these systems will have deep implications for the way object and conceptual formation are entwined with this aspect of ethos.

Foundations of Dualism and Coding in Sound Symbolism

If such composite nonlinear field networks can adequately model the simultaneous cooperative emergence of depth percepts and attentional flows, a natural extension is to project this encoding an additional level, coding for a simple sound symbolic presentation . Distance from the observer, coded as an oscillation cluster gradient, may be the prototype for dualistic conceptions. Kaneko indicates that the dynamics of globally coupled maps can be adjusted to exist in a range which supports only two coexisting clusters. If we suppose that the oscillation states of these clusters serve as linear components of a bifurcation field in another module serving as motor planning, either transients or a spatially global envelope lowering the bifurcation parameter will produce transients in a fluctuation layer which may serve as the action triggers to homuncularly mapped motor cortex. Temporal differences arise from the cluster coding differences when translated into the control layer bifurcation field. Memory or intention may overlay a "masking" form to select only a part of the field for production. The diagram below (Figure 3) indicates the outline of such an architecture, which is ultimately formalized in equations similar to those in Appendix 1.

485

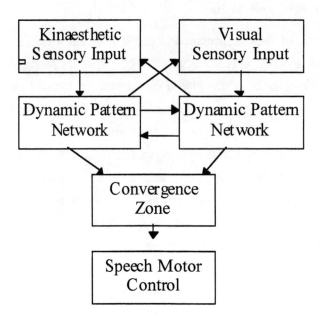

Fig 3. Architecture of layers for a higher level network translating visual percepts into utterances exhibiting phonetic symbolism

Modeling and Complexity Considerations

Synchronization, Assemblies and Causal Ambiguity

As noted earlier, the notion that spatial patterns of coherent oscillations, of a transient, dynamically evolving, or relatively stable character play important processing roles by serving as mediating forces in neurodynamics is relatively new and raises some problematic issues in modeling and presentation of results. The patterns themselves, which may be important in discussions on attributing or partitioning causality in a complex system, do not appear explicitly in the model. The transient fluctuations in the attentional parameter of models can be consider as efficient causes of transitions in the global layer, but they are pattern dependent and also dependent on the ongoing memory and intentional states of the organism. Rocha (1996) discusses the applicability of model-relative causality to pattern formation processes, which may be applicable in this case.

Order Parameters may Anticipating Post-Bifurcation Dynamics

If an anticipatory system is one that takes present action based on an expectation of a future state, perhaps the concept of order parameters in synergetic computers (Ditzinger and Haken, 1995; Mainzer 1994) may be an important part of a robust computational approach underlying action. In this particular case the order parameter is a measure of the overlap between a prototype pattern and a test pattern, ultimately minimizing a potential function. As statistical mechanical systems subject to phase transitions approach critical points, they anticipate new behavior by making microscopic adjustments in parameters such as fluctuations in density parameters (Reichi 1980). If phase transitions and subsequent reorganizations of pattern dynamics underly grammatical transformations in language, more subtle deformations (i.e. context dependence of phoneme duration and frequency) may be governed by these dynamic, anticipatory effects in advance of a bifurcation.

Situated Cognition and Anticipatory Systems

Another aspect of anticipatory systems is the interaction of models of the system and models of the environment. In models of situated or enactive cognition, the closed loop between behavior, kinesthetic and sensory feedback from the body is necessary to define the appropriate perturbations and structural coupling to ensure the organisms fit to the environment during development. Is there a clear answer for which category to assign, say, the sensory feedback from an animal (or robot, or simulacra animal/ environment model) and kinesthetic feedback? In modular signal processing -computational hybrids models, at least "feature detector" visual feedback would probably be assigned to the environment model category. In a theory like microgenesis, and given the evidence that dynamic patterns encode perception rather than sensory input, the patterns are not easily assigned to causal categories. Microgenesis suggests that even the experienced*present* is anticipated; that it is a construction constrained by environmental perturbation*at least* "feature detector" visual feedback would probably be assigned to the environment model category. In a theory like microgenesis, and given the evidence for dynamic patterns representing perception rather than sensory input. Microgenesis suggests that event the present is anticipated, is a construction constrained by environmental perturbation.

Conclusion

Regarding the emergent modes of knowledge he derived over the course of four volumes, Cassirer (1996) states

> None of these termini - understanding, reason, judgment - are to be understood in terms of a faculty psychology but only in a strictly transcendental sense ... They cannot be understood in any way as "things", neither in external nor in internal experience, in the "outer world" or "inner world". They must always be thought of as having the character of pure "conditions of experience".

Indeed, given the emergent and transient nature of patterns in spatially extended nonlinear systems, it is difficult to partition any intermediate state leading to such a terminus as having the character of a thing *in the sense that it explicitly appears as a state variable in a formal model.*. The historically arrived at conceptions of representations and models are difficult to tease apart of such a pattern dynamics. Yet cultures, pre-scientific and scientific alike do *produce* and *perform* linguistic symbols for emergent and transient phenomena in the world, often productively in the sense that correlations are made with ecologically and culturally relevant behaviors. Bridging these two worlds, seeking a complex dynamics which both resembles the reductions of linguistics and transcends them to approach the living character of performed language, is a grand challenge in cognitive science and linguistics.

Acknowledgements

This work was supported in part by the IBM graduate work study program. Discussions with I.Tsuda, K. Kaneko, J. Sherzer , F. Abraham, O. Roessler, K. Pribram, R. Picard, B. Goertzel, B. Womack, and M. Jacome have been helpful and encouraging in the pursuit of this work.

Appendix

Terminology and Definitions for Coupled Map Lattices

Discrete maps of the form $f_{t+1} = f(f_t)$ can be extended to networks of coupled nodes, where each node has a real valued state which iterates to perform as a discrete oscillator. Such models are known as *coupled map lattices* (Kaneko 83). The network attractor is then a vector or array of the states of all nodes. For such dynamical networks, a spatial bifurcation behavior at the level of the entire network is evident, emerging from the interaction of excitatory and inhibitory connections between the elements. This network-level bifurcation may be tuned by controlling the phase regime of the individual nodes, the number of connections between nodes (neighborhood size), the ratio of excitatory to inhibitory connections, or the coupling strength between nodes (Kaneko 89). The term attractor is sometimes invoked at the network or lattice level, but is generally less useful at the lattice level, where oscillations, cooperative and competitive interaction between clusters (oscillation modes or nodes oscillating in the same attractor basin) form *dynamic patterns*. The *evolution* of a network from an initial state under relatively low coupling results in an organization in which patterns of continuing activity between interacting cells are spatially bounded by "frozen" areas, in which the neighborhood interactions reach a stable state. The local active areas are referred to as *domains*, while the frozen separating regions are domain boundaries. In a sense, the network organizes itself into sub-networks, with the activity pattern in a domain more properly related to the definition of an attractor. [1] *Intermittency*, or *chaotic itinerancy*, is a phenomenon appearing in a small, weakly chaotic region of the parameter space, in which the dynamic behavior is a blending or linking of periodic attractors existing in the more ordered regions of parameter space. Individual periodic

---

[1] The concept of a domain is really only useful when the lattice has been iterated for many (> 1000) generations, so that transients have died out. In most of the perceptual simulations only a few iterations are run.

attractors are linked by intermittent chaotic transitions. Various network or neighborhood topologies have been explored for spatio-temporal chaotic systems. Network nodes may be *locally coupled* to adjacent nodes, *diffusively coupled* to a small region of the lattice, *globally coupled* to every node, coupled to a random set of neighbors, or some blending of these conditions.

When modeling physical or psychological field phenomena with nonlinear dynamics, it is common for each variable in a field (each 'unit' in a coupled map lattice) to represent a quantity associated with an aggregate of microscopic units. This kind of representation, common in statistical mechanics or fluid dynamics, is known as a *macrostate* variable. Temperature or instantaneous velocity of a fluid, for example, might be macrostate variables in a fluid study. In neural modeling the macrostate variables will be quantities like ensemble activation (average spike train frequency), coherence (pulse train density or spikes / unit time), neurotransmitter fluxes, and excitatory - inhibitory ratios of synapses.

Modeling of Depth Gestalt and Attention Phenomena

The mathematical model of coupled map lattice layers which simultaneously result in evolution of attentional transients and depth gestalts in a globally coupled layer is given below, following the presentation in (DeMaris 1997).

The equation for the logistic map is

Eq. 1

$$S_{t+1} = 1 - bS_t^2 \quad \text{subject to the constraints } -1 < S < 1, 0 < b < 2;$$

where b is a bifurcation parameter; changing the parameter forces a structured transition between phases following the sequence {fixed point : limit cycles: intermittency (unstable quasi-periodic motion): chaos}. Typically the transition points between phase regimes are visualized by bifurcation trees for systems with one bifurcation parameter, or phase space plots for spatially extended systems with multiple parameters. The initial state S is defined by perturbation from early sensory inputs; in simulations to date edge extraction is assumed, so that initial states are from the set {0.0, 1.0)

When a physical system is simulated with a coupled map model, a sequence of processes is decomposed into simple parallel dynamics at each lattice point, with each process carried out successively. In the present model, this means that at each iteration, a diffusion step is performed modeling lateral entrainment of cell assemblies, then a reaction step representing local evolution within each assembly. The bifurcation parameter for the logistic map at each site is also a variable parameter in this system and is updated at a slower rate based on the local neighborhood evolution.

The entire diffusion step can be expressed as:

Eq. 2

$$S_d(x,y) = (1-c)S_t(x,y) +$$

$$\frac{c}{4}\left[S_t(x,y+1) + S_t(x,y-1) + S_t(x+1,y) + S_t(x-1,y)\right]$$

where S_d is the intermediate diffusion array, t is the current time step, x, y are the spatial indices of the pixel array S at the center of the diffusion neighborhood, S is the state variable at each pixel of the array, and c is the coupling constant restricted to the range (0.0 to 1.0). (1-c) scales the state at each node prior to summing the neighbor states, to insure that the states remain within bounds. In practice the step is implemented by a diffusion / scaling step followed by the application of the map.

The second computational unit applied in each time step is the logistic map:

Eq. 3

$$S_{t+1}(x,y) = 1 - b_t(x,y)S_d(x,y)^2$$

where S, t, x, and y are as above and where b is the bifurcation parameter, restricted to the range (0.0 < b < 2.0). S is restricted to the range (-1.0 < S < 1.0). In Kaneko's studies the lattice dynamics are characterized in the space of bifurcation and coupling, but with each held constant in a particular simulation. In the model here b (x,y) is itself allowed to vary anisotropically in space, computed from the local evolution of the map according to the following equation:

Eq. 4

$$b_{t+1} = b_t * (1-c)\frac{c}{4}\left[\left\{S_t(x,y+1) + S_t(x,y-1) + S_t(x+1,y) + S_t(x-1,y)\right\}+1\right]$$

$$* S_t(2-b_t)$$

where b, c, S, x, and y are as previously defined. The second factor is simply the local neighborhood average previously computed for this iteration, now used in a multiplicative fashion to influence the bifurcation state in the next cycle, with the addition of 1 causing this to be an excitatory factor, balanced by the inhibitory factor (2-b). The combination of the two terms results in *spatially mediated fluctuations* around an unstable periodic mean value. From a programming and visualization standpoint, the S and b evolution equations are implemented as two separate two-dimensional lattices. These are represented as modular layers in the system architecture in Figure 1.

The equations governing the globally coupled map (GCM) depth encoding layer and interactions with the second (bifurcation fluctuation attention layer) are given below

Eq. 5

$$g_m(x,y) = 1 - b_g\left\{g_t(x,y)\right\}^2$$

where g_m is an intermediate product, representing the nonlinear map evolution prior to mean-field coupling. much like the S_d term in equation 2. The term b_g is a separate bifurcation parameter active at all sites, and may be considered a kind of arousal state of the network. It is currently set at 1.54, near the transition to chaos for the map.

A set of sites in the global layer is now selected by *attentional masking* by setting the coupling coefficients based on the form mediated bifurcation layer. The following constants were chosen based on analysis of the phase transition behavior of globally coupled maps.

Eq. 6

$$gc_t(x,y) = .37 \text{ if } b(x,y) < b_{thresh}, 0.1 \text{ if } b(x,y) > b_{thresh}$$

where gc_t is the global coupling term which is allowed to vary over space, under the control of the b layer. A partial mean field (mean of the active, most coherent sites) is computed:

Eq. 7

$$gmean_t = \frac{\left\{\sum_{i=1}^{N}\left(g_{sub-thresh(x,y))}\right)\right\}}{N}$$

where N is the number of sites in the b lattice which are below the threshold as defined in equation 5. The sites in g whose x,y coordinates in the b lattice are below the threshold are the only ones used in computing the mean, hence those sites tend to pull others into a collective basin of attraction based on this coherent cell assembly.

Finally the next state g is computed by:

Eq. 8

$$g_{t+1}(x,y) = (1 - gc_t(x,y)) * g_m(x,y) + gc_t(x,y) * gmean_t$$

References

Allot R. (1981) Lexicon and surface syntactic structure of languages as societal but not arbitrary selections from a range of potential physiologically determined "natural" word-forms and syntactic processes. UNESCO Symposium on Glossogenetics. Paris.

Allot R. (1995) Sound Symbolism, in Language in the Ice Age. U. Figge and W. Koch (eds.) Bochum: Brockmeyer.

Abeles M. (1991) *Corticonics*, Cambridge UK: Cambridge University Press.

Abraham R. H., Corliss J., and Dorband J. E. (1991) Order and Chaos in the Toral Logistic Lattice, J. of Bifurcations and Chaos 1:1, pp.227-234.

Attneave F. (1971) Multistability in Perception, Scientific American 225:6, pp.62-71.

Baird B. (1990) Bifurcation and Category Learning in Network Models of Oscillating Cortex, Physica D 42, pp. 365-384.

Bateson, G. (1972) Style, Grace, and Information in Primitive Art, in Steps to an Ecology of Mind, New York : Ballantine Books.

Borsellino, A., Carlini F., Riani, M., Tuccio, M.T., De Marco, A., Penengo, P., & Trabucco, A. (1982). Effects of visual angle on perspective reversal for ambiguous patterns. *Perception.* 11, pp. 263-273.

Brown J. W. (1977) *Mind, Brain, and Consciousness.* New York: Academic Press.

Cassirer, E. (1946) *Language and Myth* , New York, Dover.

Cassirer, E. (1955), *The Philosophy of Symbolic Forms* (Vols. 1-3), New Haven, Yale University Press.

Cassirer, E. (1996), *The Philosophy of Symbolic Forms* (Vol 4), New Haven, Yale University Press.

Cytowic R. (1989) *Synesthesia: a union of the senses*, New York: Springer Verlag.

Damasio A.R. and Damasio H. Cortical systems for the retrieval of concrete knowledge: the convergence zone framework, in Koch (ed.), Large Scale Neuronal Theories of the Brain, Cambridge MA:MIT Press, 1994

DeMaris, D. (in press, 1997). Attention, Depth Gestalts, and Chaos in the Perception of Ambiguous Figures, in Levine, D., Brown V., Shirey T. (eds.) Oscillations in Neural Systems, Hillsdale NJ, Lawrence Erlbaum Associates

DeMaris, D. (1995). Computing shape similarity with chaotic reaction diffusion spectra. In *Proceedings of 1995 World Congress on Neural Networks.* 270-273.

DeMaris, D. (1995) Spatially Extended Chaos and the Perception of Form, MS thesis, Dept. of Electrical and Computer Engineering, University of Texas, Austin.

Ditzinger T. and Haken H. (1986) Oscillations in the perception of ambiguous figures, Biological Cybernetics V61 pp. 279-287, 1986

Ditzinger T. and Haken H. (1995) A Synergetic Model of Multistability in Perception, in Kruse P. and Stadler M. (eds.) Ambiguity in Mind and Nature, Berlin: Springer-Verlag.

Eckhorn R., Schanze T., Brosch M., Salem W., Bauer R. (1992) Stimulus specific synchronizations in cat visual cortex : multiple electrode and correlation studies from several cortical areas, in *Induced Rhythms in the Brain*, Basar E. and Bullock T. H. (eds.) Boston: Birkhauser.

Elbert, T., Ray W.J., Wowalik Z.J., Skinner J.E., Graf K.E., & Birbaumer N. (1994). Chaos and physiology: deterministic chaos in excitable cell assemblies. Physiological Reviews. 74:1, 1-40.

Ellis S.R. and Stark L. (1978) Eye movements during the viewing of Necker cubes, Perception 7, pp. 575-561.

Freeman W. J. (1986) Simulation of chaotic EEG patterns with a dynamic model of the olfactory system. Biological Cybernetics 56, pp. 458-459.

Freeman W. J. and Van Djik, B.W. (1987) Spatial patterns of visual cortical fast EEG during reflex in a rhesus monkey, Brain Research 422, pp. 267-276.

Freeman W.J. (1995) *Societies of Brains: A study in the neuroscience of love and hate.* Hillsdale NJ: Lawrence Erlbaum.

Friedrich P. (1986) *The Language Parallax: Linguistic Relativism and Poetic Indeterminacy,* Austin, University of Texas Press.

Fuchs A., Kelso J.A.S., Haken H. (1992) Phase Transitions in the Human Brain: Spatial Mode Dynamics, Int. Journal of Bifurcations and Chaos, 2: 917-939.

Grossberg S. and Somers D. (1991), Synchronized Oscillations During Cooperative Feature Linking in a Cortical Model of Visual Perception, Neural Networks 4, pp. 453-460.

Harnad, S. (1990) The symbol grounding problem, Physica D 42, pp. 335-346.

Hinton L., Nichols J., Ohala, J.J. (Eds.).(1994*) Sound Symbolism* , Cambridge: Cambridge University Press.

Jenny H. (1974) *Cymatics: The Structure and Dynamics of Waves and Vibrations,* Basel: Basilius Press.

Kaneko K. (1993) Overview of coupled map lattices, Chaos 2.3, pp. 279-282

Kaneko K. (1989) Spatiotemporal chaos in one and two dimensional coupled map lattices, Physica D 37 pp. 1-41.

Kaneko K. (1990) Clustering, coding, switching, hierarchical ordering and control in network of chaotic elements, Physica D 41 pp.137-142

Kaneko, K. and Tsuda, I (1994). Constructive complexity and artificial reality: an introduction, Physica D 75, 1, 1994.

Kelso, J.A.S, Case P., Holroyd T., Horvath E. , Raczaszek J., Tuller B. and Ding M. (1995) Multistability and Metastability in Perceptual and Brain Dynamics, in Kruse P. and Stadler M. (eds.) Ambiguity in Mind and Nature, Berlin: Springer-Verlag

Katchalsky A.K, Rowland V., and Blumenthal R. (1974) *Dynamic Patterns of Brain Cell Assemblies,* Cambridge MA: MIT Press.

Kauffman S. (1993) *The Origins of Order: Self-organization and Selection in Evolution,* New York: Oxford University Press, 1993

Kawabata N., Ymagami K., and Noaki M., (1978) Visual fixation points and depth perception, Vision Research 18, pp. 853-854.

Kohler W. (1929) *Gestalt Psychology.* New York: Liveright.

Kovacs I. and Julesz B. (1994) Perceptual sensitivity maps within globally defined visual shapes. Nature, London, v 370, pp. 644-646.

Laughlin, Jr. C.D., McManus J., d'Aquili E.G. (1990) Brain, symbol & experience : toward a neurophenomenology of human consciousness, Boston : New Science Library.

493

Lehmann D. (1995) Brain Electric Microstates, and Cognitive and Perceptual Modes, in Kruse P. and Stadler M. (eds.) Ambiguity in Mind and Nature, Berlin: Springer-Verlag.

Long G. M., and Toppino, T.C. (1981) Multiple representations of the same reversible figure: implications for cognitive decisional interpretations, Perception 10 pp. 231-234.

Mainzer K. (1994) *Thinking in Complexity: The complex dynamics of matter, mind, and mankind*, New York: Springer-Verlag.

Merleau-Ponty M. (1962) *Phenomenology of Perception,* Smith C. (trans.), New York: Humanities Press.

Nuckolls, J. B. (1996) Sounds like Life: Sound-symbolic grammar, performance, and cognition in Pastaza Quecha , Oxford: Oxford University Press.

Ohala, J.J. (1994) The frequency code underlies the sound symbolic use of voice pitch, in Hinton L., Nichols J., Ohala, J.J. (Eds.). Sound symbolism , Cambridge: Cambridge University Press.

Pollack J. B. (1991) The Induction of Dynamical Recognizers, Machine Learning 7, pp. 227-252.

Rocha L.M. (1996) Eigenhebavior and Symbols In: Systems Research, Vol. 12 No. 3, pp. 371-384, 1996, Glanville R. (Ed.) Special Issue Heinz von Foerster Festschrift

Rosen Robert (1985). Anticipatory Systems. Pergamon Press.

Sass, L. (1992) *Madness and Modernism: Madness in the light of modern art, literature and thought.* New York, Basic Books.

Skarda C., Freeman, W. J. (1987) How brains make chaos in order to make sense of the world, Behavioral and Brain Sciences 10, pp. 161-195.

Tsuda, I. (1992) Dynamic Link of Memory. Chaotic Memory Map in Nonequilibrium Neural Networks, Neural Networks 5, pp. 313-326.

Varela F. J., Thompson E., and Rosch E. (1991)*The Embodied Mind: Cognitive Science and Human Experience*, Cambridge MA: MIT Press.

Usher, M., Stemmler, M., Olami, Z. (1995) Dynamic Pattern Formation Leads to 1/f Noise in Neural Populations. Physical Review Letters, 74(2) pp. 326-329.

von Grunau M. W., Wiggin S. and Reed M. (1984) The local character of perspective organization, Perception and Psychophysics 35 pp. 319-324.

Whorf (1941) Language, Mind, and Reality, in Language, Thought and Reality: Selected Writings of Benjamin Lee Whorf, Cambridge: MIT Press.

Winograd T., and Flores F. (1986) Understanding Computers and Cognition: a new foundation for design, Norwood NJ, Ablex Publishing Corp.

Yao Y., Freeman W.J. (1990) Model of Biological Pattern Recognition with Spatially Chaotic Dynamics, Neural Networks 3, pp.153-170.

A. Zimmer (1995) Multistability - More than just a Freak Phenomena, in Ambiguity in Mind and Nature, P.Kruse and M. Stadler (eds.) Berlin: Springer-Verlag.

494

The hyperfinite signal : a new concept for modelling dynamic systems.

Jean-Paul Frachet

Equipe Ingénierie des Produits et des Systèmes de Production,
ENSAM : Ecole Nationale Supérieure d'Arts et Métiers
8, boulevard Louis XIV   59 046 LILLE Cedex - FRANCE
Phone: (03 or 33.3).20.62.22.10 - Fax: (03 or 33.3). 20.53.55.93
E-mail: jpfrachet@lille.ensam.fr

ABSTRACT. The notion of hyperfinite signal enables to represent the information we can get from an intelligent observation of the temporal evolution of physical parameters characterizing the behaviour of a dynamic - or reactive, or anticipatory - system. In spite of its totally discrete type, we gain a formalism similar to the one of continuous values evolving in a continuous time, so that the behaviour of the system can be modelled by algebraic or differential equations. The hyperfinite signal theory relies on an original model of time and discrete values, using non standard analysis and finite field theory. In this paper, we focus on a subtype allowing the modelling of discrete events systems.

KEYWORDS : Non standard analysis, finite fields, discrete events systems, time model, behaviour.

RÉSUMÉ. La notion de signal hyperfini permet de représenter l'information que l'on peut tirer d'une observation intelligente de l'évolution temporelle des grandeurs physiques caractérisant le comportement d'un système dynamique, ou encore réactif, ou encore anticipatif. En dépit de son caractère totalement discret, nous obtenons un formalisme tout à fait similaire à celui des systèmes continus, évoluant dans un temps continu, de sorte que le comportement peut-être modélisé par des équations algébriques ou différentielles. La théorie du signal hyperfini s'appuie sur un modèle original du temps et des grandeurs discrètes, exploitant l'Analyse Non Standard et la Théorie des Corps Finis. Dans ce papier, nous développons particulièrement une sous-classe permettant la modélisation des systèmes à événements discrets.

MOTS-CLÉS : Analyse non standard, corps finis, systèmes à évènements discrets, modèle du temps, comportement.

1. Basic principles leading to the concept of Hyperfinite Signal.

1.1 Limits of the classical modelling of temporal evolution of a phenomenon

In the natural sciences such as physics or mechanics, we use to build models of temporal evolution of a phenomenon *in itself*, that is to say independently of its observation : a stone falls, either you look at it or not ... In this paradigm, the time evolution of physical values can be described, formally or informally, by a function $x = f(t)$ where x is the modelling of a value and t is the modelling of the time. That is the way the models ("the physical laws") in which x and t are continuous ($\in \Re$, for instance) are considered as superior since they allow more calculations (especially, $x = f(t)$ can be the solution of a differential equation).

The problems begin when phenomenon is considered in a **real context**, where the physical values, $x(t)$, $y(t)$ and so on, are observed (or actuated) through an instrumentation which gives an altered image of this values introducing a significant bias. In this context, which is the only one we are physically, an infinitely small or a too important variation of x or y or t will of course *never* be observable. More, the finite resolution of the instrumentation (or actuation),

CP437, *Computing Anticipatory Systems: CAYS--First International Conference*
edited by Daniel M. Dubois © 1998 The American Institute of Physics 1-56396-827-4/98/$15.00

independently of the ingenuity of the scientists to correct the errors, gives definitively an impassable difference between the continuous world and the finite one. Trying to overpass this gap involves lot of numerical problems and a lack of confidence in the final results.

1.2 The basic principles of the proposed approach.

In the proposed approach, we don't worry if the value and its evolution in time are discrete or not *in itself*. We interest in the model built **thanks to the observation of the value**. Thus the *x* value is supposed to be seen in the informational space, that is through a sensor (or an actuator) : it is a *piece of information* or a *signal*. As the sensor which gives this signal is, actually or potentially, material, the number of distinct values it may send is finite and even quite small.

The time can not be refined infinitely: an infinite number of dates can not be inserted in between two dates, as it can be done in $\Re$. The most important is to insert a sufficient number in order to "see" the cause to effect chain between two signals for example. The set of the dates when a value can be observed has necessary a finite cardinal.

To characterise this approach, we propose the *Hyperfinite Signal (HS)* term, underlying the fact that the characteristic of the modelling of this value evolution is **informational** and **finite**, but very large (more precisely, large enough). The term **Hyperfinite** comes from **non-standard analysis (NSA)** (Diener-Reeb 1989, in french) but we have not enough room in this paper to say more about the filiation from the NSA to HS.

This approach is different from the classical modelling of discrete systems, the main drawback of which is to forbid the algebraic or differential formalism so familiar, powerful and expressive in the continuous systems. As seen latter, this approach tends to provide discrete systems with calculus laws similar to continuous ones. In fact, there is no more difference between these two classes of systems : it depends only on the glance you give to them.

2. The Hyperfinite Signal

2.1 The set X of the possible values of the signal

We consider the signal can only have a finite number of predefined values which build a set *X*. We usually suppose this finite set *X* owns at least an internal associative law (addition), a neutral element and for each of its elements an opposite (it is a **group**).

We can often suppose that the cardinal of *X* looks like p^n (p represents a prime number and n an integer) in order that *X* can have a field structure (finite field or Galois' field), as E Galois demonstrated in the 1830's (Lidl, Niederreiter 1983). Practically, we consider the common case $p = 2$, which is found for example when the value is observed by a sensor giving information through an analog-to-digital converter (or digital-to-analog converter, for an actuator) represented by a **binary word** whose length is *n* bits (usually, *n* is from 8 to 16, leading card(X) from 256 to 65536, which is very small in itself and very far from an quite good approximation of a real in a computer : 64 bits leading card(X)$\cong 10^{20}$).

In the case n = 1, the set *X* owns only two values, named **0** and **1**. In this article, we shall focus on this particular case.

2.2 The basic postulates

The continuous reference model principle: the time reference model remains the line of reals oriented towards the future. A *date* is an abscissa on this line considered from an origin on, whereas a *duration* is a segment of this line. These two notions imply an arbitrary *unity of time*.

The sampling principle: although the time flows conceptually in a continuous way (stopping it is impossible ...), we suppose it is physically possible to frame a date with a duration both sufficiently long to measure or calculate the value (and thus to get a value of the signal) and sufficiently small so that the variation of the signal is not significant during this length. This principle, coming from the NSA, enables to give a physical sense to the sentence "the value of the signal at a given time".

The principle of the fact that the observation (or the action) is to be finite: in practical applications, the signal can only be observed at a finite number of dates supposed to be placed between a "beginning of working" (we postulate it always exists) and an end of working, or at least of observation, chosen sufficiently in a long time so that every interesting phenomenon is completed.

The principle of a shared clock: we suppose only one line of time exits considered as a ***common referential*** for the modelling of all the signals of a single system. If two distinct signals representing the same phenomenon have a same date, their values are simultaneous. This constraint is strong and difficult to follow in the case of distributed systems.

2.3 The set PD of Pertinent Dates involves a Pertinent Scale of Time (PST)

According to these principles, the time aspect of the signals set concerning a single phenomenon can be modelled by one finite set of dates when the values are observed at. Since those dates are to be chosen in order to obtain the more information on these values, this set of dates is called *finite subset (of $\Re$) of the* **Pertinent Dates PD**, the elements of which are called τ. This set, the cardinal of which is N, is plainly ordered (since all the dates are different) and thus forms an **increasing series,** the index **t** of which belongs to the set T_N (the set of the N first integers, starting from 0). It is called a **Pertinent Scale of Time (PST)** Knowing an integer $t < N$ enables to reach the corresponding date τ of PD : $\tau = $ **PST(t)**. Knowing a date $\tau \in$ PD enables to reach the corresponding index t : $t = $ **PST^{-1}(τ)**. The notations **t + 1** or **t - 1** refer respectively to the dates following or preceding this date τ in **PD**, that's why t is called a **date**, by abuse of language.

2.4 Definition of the Hyperfinite Signal

> A **Hyperfinite Signal** is defined as a **mapping** *valuation* $T_N \xrightarrow{\text{valuation}} X$ associated to a **Pertinent Scale of Time (PST)**, defined as an ordered series of dates.

In what follows, the elements of T_N will be called *dates* (abuse of language). The value of the HS ***a*** at the date t will be called ***a(t)***.

2.5 Notion of Homogeneous Set of HS (HSHS)

In order to establish relationship between several HS concerning the same system, for instance between the *HS-input* and the *HS-output* of a given system, it seems easier to assume that their values are extracted from the same set *X* and given at the same date. So, several HS are said ***homogeneous*** if they have the same set *X* and the same Pertinent Scale of Time. All the HS of a given HSHS forms a **finite set**, the cardinal of which is quite big ($\text{card}(X)^N$) but which can be **structured** in order to permit the calculation. In the following, we shall consider the Pertinent Scale of Time as *implicit* (because it is the same for all the HS belonging to the given HSHS) and we shall only consider the mapping *valuation*, $T_N \xrightarrow{\text{valuation}} X$, characterizing each of them.

2.6 Structuration of an Homogeneous Set of HS

To structure such a set, a method is to define some properties on the values at a date t and then to extend these properties to all the dates. Some basic types, an internal linking law and the derivation and integration operators can be defined that way.

2.6.1 Basic types

If t_0 is an integer $< N$ (element of T_N),

> ***constant HS:*** $\forall t, c(t) = c(t_0)$. Let us denote them a*, where, $\forall t, a^*(t) = a^*(t_0)$
> For instance, 0* and 1* are constant HS

HS steps: called $S_{t0, s}$, they are worth 0 when $t < t_0$ and are worth s ($\in$ X) when $t \geq t_0$. A step is completely defined by the knowing of its date of state modification t_0 and its value s.

HS impulsions: called $I_{t0, i}$, they are worth 0 when $t \neq t_0$ and are worth i ($\in$ X) when $t = t_0$. An impulsion is completely defined by the knowing of ***the*** single date t_0 when it is worth i and its value i.

2.6.2 Internal linking law on an Homogeneous Set of HS

We define the internal linking laws on the signals by using the internal linking laws defined on the set X of the values of the signal. (Recall : X is a least a group) Let us call it *additive law*, denoted $+$: $c = a + b \Leftrightarrow \forall t, c(t) = a(t) + b(t)$.

As the additive law (on X) has a neutral element called **0** and every element has an opposite, we conclude that it exits a constant HS called 0* and an opposite HS b to each HS a, with $a + b = 0*$.

An HSHS is closed for this law and for any other law deduced by adding linking laws in X.

2.6.3 Particular case if X is a FIELD

If X is a field (which is the case if its cardinal is p^n), that is to say it have at least two laws, one denoted additively and the other multiplicatively, having each a neutral element, respectively denoted 0 and 1, and each element a having an opposite, $-a$ and an inverse $1/a$ (except 0), it is easy to show that a given HSHS (built on this X) is a **commutative ring, having divisors of 0.**

0*, 1* are respectively the neutral elements of the additive and the multiplying laws. ($a+0* = a$ and $a.1* = a$) but **there is no inverse** for every element a, such as $a.b = 1*$. More, it have divisors of 0, since $a.b = 0*$ does not imply $a = 0*$ or $b = 0*$

2.7 Definition of the derivation operator

The derivative of an HS a is the HS a', the value of which is for each date the value of the derivative of the HS a: $\forall t, a'(t) = [a(t)]'$.

The value of the derivative of an HS a, at the date t, is by definition:

$$a'(t) = {}^\circ(a(t{}^\circ 1) - a(t))$$
$$\text{with } {}^\circ \text{ for either } + \text{ or } -$$

The evaluation of this formulation requires to suppose:

a(N) = a(N-1): N is long enough, so that the signal is no longer modified.

a(-1) = a(0): the signal has not yet moved.

The sign ${}^\circ$ represents either the sign - or the sign + (left or right derivative). The reason why is that, even if the reference model is continuous and derivable, ***a'*** does not have the same value in these two cases because X and AD are discrete.

In reference to Daniel DUBOIS's works (Dubois 1996), we call these HS ***recursive derivative*** (left derivative (${}^\circ$ means -): its value depends on the previous values) and ***incursive derivative*** (right derivative (${}^\circ$ means +): its value depends on the next values). In this article, the used derivative is the recursive one, unless it is mentioned.

Physical meaning: we can easily check that the derivative of an HS is except the sign the difference between itself and the HS deducted from itself by a temporal step of one date.

The ***HS derived from a constant HS is*** the constant HS 0*.

The ***HS derived from the HS step $S_{t, s}$*** is the HS impulsion $I_{t, s}$.

The derivative of a sum of HS is the sum of the derivatives.

2.8 Definition of the integration operator

The primitive HS of an HS a is the HS which is worth the integrative of a at each date. We define the integrative of an HS a in a date as the sum of the value of this HS from all the dates since the "beginning of working" (j = 0) up to the considered date t (included).

$$I(a(t)) = \sum_{j=0}^{t} a(j)$$

We can check that the primitive of a sum is the sum of the primitives .

Primitive of the impulsion $I_{t, i}$: up to the date t (not included), the integrative is null. At the date t, it is worth i. At the next dates, it is still worth the value i because the HS is null. So the HS primitive of the impulsion $I_{t, i}$ is the HS step $S_{t, i}$.

In the second part of this article, the HS are supposed to take their values in {0, 1}.

3. A particular case: the Scalar Binary Signal (SBS)

3.1 The notion of SBS and its interest

The Scalar Binary Signal (SBS) is a particular hyperfinite signal being worth in X = {0, 1}. It enables the modelling of the function describing the evolution in time of a Boolean variable.

The mathematical study of dynamic discrete systems (combinatorial or sequential) is indeed based on the use of variables being worth in B = {0, 1} and modelling the state *false* or *true* of a logical phrase dealing with an aspect of the currently studied system. The relationships between these variables are often Boolean expressions.

Furthermore, in Dynamic Discrete Events Systems (DDES), these variables can evolve in time and in this case, the Boolean expressions are to be evaluated for the values these variables have at a date. For instance, the "edge" of a logical variable has no sense and the modelling framework has to be extended to give it a meaning.

As written upper, the linking laws of the set $X = \{0, 1\}$ provide the structuration of the set of the SBS. These linking laws are the mappings from $\{0, 1\}^n$ in {0, 1}. These mappings are usually modelled with Boolean functions of n variables. As for us, we prefer a *Galoisian modelling* based on a stronger structuration than the Boolean lattice: the field structure, as said upper.

3.2 Modelling of Boolean systems on $(GF2)^n$

In the proposed approach, as 2 is a prime number, {0, 1} can be given a field structure. The Boolean variables can be considered as the elements of the finite field {0, 1}, called *GF2*. Its two elements are called 0 and 1. Two laws can be defined:

⇒ an *additive law* + (commutative and associative) defined by the truth table:
$0 + 0 = 0, 1 + 0 = 0 + 1 = 1, 1 + 1 = 0$.
0 is the neutral element: $a + 0 = a$.
As each element is its own opposite (a + a = 0), the sign "+" is not different from the sign "-": one or the other can be indifferently chosen.
This law corresponds to the operator "EXCLUSIVE OR" in the Boolean space **B**.

⇒ a *multiplicative law* (commutative and associative), written • (or nothing if there is no ambiguity) defined by the truth table:
$0 \cdot 0 = 0, 1 \cdot 0 = 0 \cdot 1 = 0, 1 \cdot 1 = 1$.
1 is the neutral element: $a \cdot 1 = a$.
By definition 0 has no opposite, 1 is its own opposite.
This law corresponds to the operator "AND" in **B**.

the multiplication is distributive against the addition:
$a_{\bullet}(b + c) = a_{\bullet}b + a_{\bullet}c$.

The mappings from $\{0, 1\}^n$ in $\{0, 1\}$:

In $[GF(p)]^n$ ($\forall$ p prime), we can prove that *every function of* n variables $y = f(x_0, x_1, ..., x_{n-1})$, in which $y, x_0, x_1, ..., x_{n-1}$ are variables being worth in GF(p), can always be written *in an unique way* as a polynomial the maximum degree of which is p - 1 and the coefficients of which are some elements of the field.

In the case p = 2, we conclude that *every function* (mapping from $\{0, 1\}^n$ in $\{0, 1\}$) *can be written as an unique polynomial of n variables* in which each variable only appears with a degree 0 or 1. These polynomials have 2^n terms the coefficients of elements of the field (0 or 1).

Relationship between Boolean space and Galoisian space:

A consequence of the previous paragraph is that *each* Boolean function can be written as a Galoisian *polynomial*, the converse is true, what can be proved easily if we suppose that the operators *NOT* and *AND* build up a complete system of Boolean operators (each Boolean function can be written with these only two operators). We only need to express these two operators in Galoisian, what is evident:

Operator *NOT*: $/a \Leftrightarrow 1 + a$. Operator *AND*: $a \wedge b \Leftrightarrow a.b$.

We can then deduce the other main Boolean operators :

operator **OR** $a \# b = /(/a./b) \Leftrightarrow 1 + (1 + a).(1 + b) = a + b + a.b$,

operator **NOR** $= /OR \Leftrightarrow (1 + a).(1 + b)$,

operator **NAND** $= /AND \Leftrightarrow 1 + a.b$, and so on.

Example of translation : $y_{Bool} = a \wedge /b \# c$. Let's set down $d = a \wedge /b$, $y = d \# c = d + c + d.c$, $d = a.(1 + b)$ then $y_{Gal} = a + a.b + c + a.c + a.b.c = 1.a.b.c + 1.a.b + 1.a.c + 1.a + 0.b.c + 0.b + 1.c + 0$.

3.3   Structure of an Homogeneous Set of SBS: HSSBS

3.3.1 Basic types of SBS

constant SBS: only two exist: 0^* and 1^*. $\forall t$, $0^*(t) = 0$, $1^*(t) = 1$.

SBS steps: called $S_{t0,}$, they are worth 0 when $t < t_0$ and 1 when $t \geq t_0$. A step is completely defined by the knowing of its date of state modification t_0.

SBS impulsions: called I_{t0}, they are worth 0 when $t \neq t_0$ and 1 when $t = t_0$. An impulsion is completely defined by *the single* date when it is worth 1.

3.3.2 Linking laws on an homogeneous set of SBS

Additive law: $c = a + b \Leftrightarrow \forall t, c(t) = a(t) + b(t)$.

Multiplicative law: $c = a.b \Leftrightarrow \forall t, c(t) = a(t).b(t)$.

0^* is the neutral element for addition: $a + 0^* = a$. Each SBS is its own opposite : $a + a = 0^*$. 1^* is the neutral element for multiplication: $a.1^* = a$.

There is no opposite (or inverse) to an SBS.

These laws are commutative and associative. They can be extended to any number (≥ 2) of operated SBS. The multiplication is distributive against the addition: $a.(b + c) = a.b + a.c$. The multiplication is idempotent: $a.a = a$. Then an integer power of an SBS is worth this SBS: $a^n = a$.

In relationship with Boolean algebra, the following further laws are introduced:

Law "complementation to 1^*": $/a = 1^* + a$ then $a + /a = 1^*$

(involutive law: $/(/a) = a$).

500

Law OR #: *c = a # b* ⟺ ∀t, c(t) = a(t) + b(t) + a(t).b(t) ⟹ *a + b + a.b*.

This law can be extended to any number of variables a, b, c, ... by summing all the monomials of 1, 2, 3, ..., n variables. *For example: a # b # c = a + b + c + a.b + a.c + b.c + a.b.c.*

The law OR is commutative and associative: (a # b) # c = a # (b # c) = a # b # c.

The multiplication is distributive against the law OR and mutually: a # b.c = (a # b).(a # c), a.(b # c) = a.b # a.c, yet there is no distributive relationship between + and #. The well-known properties are true : a # (/a) = 1* and a.(/a) = 0*,

the De Morgan theorems too : /(a # b) = /a./b and /(a.b) = /a # /b.

External multiplication by a Galoisian: the external multiplication of an SBS *a* by a Galoisian α is an SBS *b* defined by: ∀t, b(t) = α.a(t). Then : 0.a = 0*, 1.a = a.

3.3.3 Partial order relation on an HSSBS

An SBS a is said to be ≥ b if at each date when b is worth 1, a is worth 1.

Then we find a.b = b or b.(a + 1*) = 0* or /a.b = 0*.

$$a \geq b \leftrightarrow a.b = b$$

So:

$$\forall a, b \in HSSBS, a \geq a.b \text{ and } b \geq a.b$$

3.3.4 Algebraic structure of HSSBS

As a HSHS, an HSSBS is a commutative ring, having divisors of zero. More, it is called *Boolean*, because $a^2 = a$.

3.4 A representation of an SBS

An SBS *a* can be defined by summing the steps the dates t_0 of which are the dates of state modification of the SBS *a*.

$$a = \Sigma a(t).S_t + a(0).1*$$
the sum is extended to all the dates of state modification of the SBS a

(we can noticed that a(0).1* is the integration constant).

Note on the notation:

The sign + has 3 meanings. It may be the addition of two Galoisians, the addition of two SBS or the addition of two integers. The context gives the right meaning:

- a(t) + b(t) is the Galoisian addition.

- a + b is the addition of 2 SBS.

- t + 1 is the addition of 2 integers in a(t + 1).

3.5 Definition of the derivative operator

3.5.1 Derivative at a date of an SBS

As seen upper, the derivative of the SBS **a** is the SBS **a'** the value of which is for each date the value of the derivative of the SBS **a**:

left (recursive) derivation: a'(t) = a(t) + a(t - 1)
right (incursive) derivation: a'(t) = a(t + 1) + a(t)

(the operator + and the operator - are the same in the GF2 algebra).

3.5.2 Derivative of the basic SBS

We could easily prove that the **SBS derivative of a constant SBS** is the **constant SBS 0***:
$(0^*)' = (1^*)' = 0^*$. The **SBS derivative of an SBS step S_t is the SBS impulsion I_t.**

3.5.3 SBS derivative of a composition of SBS

Let a and b, 2 SBS:

The derivative of a sum is the sum of the derivatives: $(a + b)' = a' + b'$.

The derivative of the complement is the derivative: $(/a)' = a'$.

The derivative of a product is: $(a.b)' = a'.b + a.b' + a'.b'$.

The derivative of a composition of SBS using the law OR (a # b = a + b + a.b) is :
$(a \# b)' = a'./b + b'./a + a'.b'$.

Some examples can be found on Figure 1.

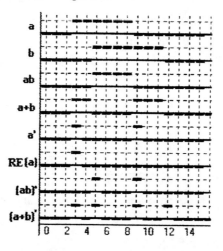

Figure 1. *Some SBS deducted by applying operators of two non-correlated SBS **a** and **b***

3.5.4 Independence principle: notion of non-correled SBS

An SBS often characterises practically the state of a physical object independent of the other physical objects the state of which is modelled by other SBS. What is defined here is the independence principle (or the non-correlation principle), it means two non-correled SBS can not both have a state modification at the same date. This principle is based on characteristics of the information on the SBS. Some more details can be found in Frachet (1993).

$$\forall a, b \in \text{HSSBS and non-correled, } a'.b' = 0^*$$

The results on the derivation of a composition of SBS become: $(a.b)' = a'.b + a.b'$ and $(a \# b)' = a'./b + b'./a = a'./b \# b'./a$ (because the 2 terms can not both be worth 1 simultaneously: a'./b.b'./a $\equiv 0^*$).

3.5.5 Generalisation to n SBS

A generalisation of the properties of the derivation operator to n SBS can be done.
If $a_1, a_2, ..., a_n$, are n SBS, we can prove the following properties:

$(\Sigma a_i)' = \Sigma a'_i$.

$(\Pi a_i)' = \Pi(a_i + a'_i) - \Pi a_i$.

$(\Pi a_i)' = \Sigma(a'_i.\Pi a_{j \neq i})$ if a_i *are non-correled*.

$(\#(a_i))' = \#(a_i + a_i') - \#(a_i)$.

3.6 Definition of the integral operator

An integral operator can be defined as: the integral at a time t of an SBS *a* is the sum of the values of the SBS for all the dates since "the beginning of working" to the given time t, *included*.

$$\boxed{\text{I}(a(t)) = \Sigma a(t) \text{ from the initial time to the time t}}$$

So, the *primitive of an SBS a* is the SBS which is always equal to the *integral of a*. The following proprieties are true : the primitive of a sum is the sum of the primitives and the primitive of the product by a scalar is the product by this scalar of the primitive. The *primitive of an I_t impulsion* is the S_t *step function.* The *primitive of the 0* constant SBS* is the *constant SBS 0**. The *primitive of the constant SBS 1** is the SBS the graph of which is a square function the period of which is 2 and the initial value of which is 1.

4. Some applications to logical and sequential systems

4.1 The edges and their calculus

We can associate to each SBS *a* a "rising edge" SBS *RE(a)* (respectively "falling" *FE(a)*) which represents the time diagram of the rising edges (respectively falling) of this SBS *a*.

According to the convention which is that when the status changes, the function has already its new status, we have: RE(a) = ↑a = a.a' and FE(a) = ↓a = /a.a' = (1* + a).a'. RE(a) and FE(a) are SBS, because they result from a composition of SBS. They are a sum of *pulses,* completely defined by the dates of occurrence of the different edges

We can deduce the following properties:

a' = ↑a + ↓a, so ↑a = a' + ↓a and ↓a = a' + ↑a.

a' = ↑a # ↓a.

Applying these properties to the Boolean expression edges leads (Frachet (1995))] to find easily the properties given by Roussel (1993) :

a # ↑a = a.   a. ↑a = ↑a.   (/a). ↑a = 0*.   ↑(/a) = ↓a.

↑(Πa_j) = (Πa_j).($\Pi(a_i + a_i')$) - Πa_j. = #(↑$a_i \Pi a_j$) for non-correlated SBS.

↑(#(a_j)) = #(↑a_j.Π(↑a_j # /a_j(/↓a_j))).

↑(↑a) = ↑a.   ↑(↓a) = ↓a.   ↓(↑a) = (↑a)' + (↑a) which is the SBS pulse ↑a moved forward 1.

Similar properties can be demonstrated on falling edges.

The examples of Guillaume (1995, 1996) can also be demonstrated.

4.2 Modelling with and solving algebraic equations

We agree upon to call algebraic equation in which only SBS appear, thus in which unknowns are SBS. For instance, let *a* and *b* be given SBS and *x* an unknown one. The equation *a + x = b* has one solution (*x = a + b*) completely-defined solution, whatever the values of *a* and *b* are. On the opposite, for the equation *a.x + b = 0**, which is equivalent to a.x = b, a.x is always ≤ a, so that b has to be ≤ a in order to find a solution ; if b ≤ a, we have b = a.b the equation becomes a.x = a.b or *x = b*.

A significant application can be found when researching some SBS local configurations, with no need to determine globally the SBS involved in the given configuration. The result appears as a constraint on his SBS. Especially, this allows to determine whether a simultaneous occurrence of 2 events is possible or not.

An example of practical application is the checking of the fourth rule of the GRAFCET (IEC (1991)) of simultaneously clearing two transitions (see Figure 2). Let's examine a concrete example : in which case the t11 and t12 transitions, which both precede the same #10 step, are simultaneously cleared. Both transitions have associated transitions defined as edges:

$$R11 = \uparrow(a./c \# b) = \uparrow(r11) = (r11).(r11)'.$$
$$R12 = \downarrow(a.b \# c) = \downarrow(r12) = /(r12).(r12)'.$$

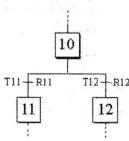

Figure 2. *Example of grafcet for the checking of the fourth rule*

In this problem described by Roussel (1993), a, b and c are **unknown SBS**, supposed **non-correlated**. The problem is to determine the configuration of a, b and c which, for an undetermined time t, could lead to a simultaneous clearing of both transitions, that is solving the equation $R11.R12 \neq 0*$, which becomes $(a./c \# b).(/(a.b \# c)).(r11').(r12') \neq 0*$, so :

$(1* + c).(a + b).(a'.b.c + a.c' + a.b.c') \neq 0*$ and finally $a.c'.(1* + c).(1* + b) \neq 0*$, so that the searched result is **a./b. $\downarrow$c $\neq$ 0***. This result can be checked by examining R11 and R12 :

if $\downarrow$c occurs when a = 1 and b = 0, we have a rising edge for R11 (which can be reduced to /c) and a falling one for R12 (which can be reduced to c).

4.3 Modelling with differential equations

The equation is a differential one since it contains the differentiated SBS to the unknown SBS. Following the example of classical differential equations, it can be solved by numerical or analytical integration. Here are some examples:

Modelling a divisor by two:

Let *e* and *s* be respectively the SBS applied to the input and the SBS given by a divisor by two. We can easily check that s is a solution of the differential equation $s' - e.e'$. The integration constant is the initial value of *s*

Modelling an elementary memory (1 bit):

Let *r* and *s* be the SBS reset and set input of an elementary memory and let q be the SBS showing the status of that memory. We can easily check that q is a solution of this differential non-linear equation (if we assume there is no conflict : $s.r = 0*$) :

$q' = q.(r+s) + s$, which can be written : $q' = s.(1* + q) + r.q$

Note that the terms $s.(1* + q)$ and $r.q$ are pulses as shown on figure 3

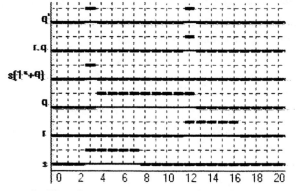

Figure 3. *The chronogram coming from the equation of the memory.*

Note : One can see that it is necessary to use the **incursive** derivative for q' because it is an **actuation** (and not an observation) : q is an output, the state of which is **forced** by the state of the inputs.

In case of conflict ($s.r \neq 0^*$, meaning that it exists at least a date t where $r(t) = s(t) = 1$), the final equation of the memory depends on the behaviour of the memory when a conflict occurs :

q(t) unchanged : $q' = q.(r + s) + s + r.s.$
q(t) = 1 (priority for 1) : $q' = q.(r \# s) + s.$
q(t) = 0 (priority for 0) : $q' - q.(r \# s) + s + r.s.$
q(t) = 1 + q(t - 1) (JK flip-flop) : $q' = q.(r + s) + s = s.(1 + q) + r.q.$

Application to differential modelling of the GRAFCET:

We can consider that the associated transition conditions are in fact a SBS (logical functions evolving with time). Consequently, the activity of the steps can be interpreted as a SBS too.

Starting with the algebraic modelling of GRAFCET which assumes that each step can be modelled by a 1-bit 1-priority memory, we simply have to take the differential equations of this kind of memory for each step and to consider that the SBS r and s are the unions of the Conditions for Overstepping a transition (CO_i) respectively following and preceding the step.

Thus, we obtain a system of n differential equations (where n is the number of steps in the Grafcet).

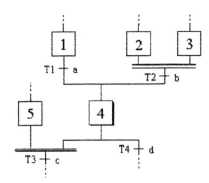

Figure 4. *An extract of a Grafcet in which the Step #4 illustrates the most general case for activating and deactivating a step*

This property applied to the Figure 4 (for the activation of step #4) leads to:

$$CO_1 = a.X_1 \; ; \; CO_2 = b.X_2.X_3 \; ; \; CO_3 = c.X_4.X_5 \; ; \; CO_4 = d.X_4$$

So that $r = CO_3 \# CO_4$ and $s = CO_1 \# CO_2$. The differential equation giving the activity of step # 4 is: $X'_4 = X_4.(r \# s) + s.$

Future of the occurrence of an event:

We can check that the SBS *f*, solution of $f' = a.(1^* + f)$ is worth 0 while *a* is worth 0 and then constantly 1 during and after the first date where *a* is worth 1, whatever are the values taken by *a* in the future. So, f is the SBS future of the event « first occurrence of $\uparrow a$ ». This simple equation can be used to modellize the order of occurrence of events.

5. Conclusion and perspectives

The Hyperfinite Signal notion is strongly inspired by the concepts of Non Standard Analysis which were used (Frachet, Colombari (1992)) in order to link a synchronous (Benveniste, Berry (1991)) temporal semantics to the GRAFCET. Our research topics is based on the applications of SBS and the validation and implantation of logical systems. We also work on the extension to the Scalar Galoisian Signals (SGS) (card(X) = 2^n)

We thank in advance any reader who would communicate his advices and opinions on this paper.

6. References

A. BENVENISTE, G. BERRY, (1991) "The synchronous approach to reactive and real time system", In *Proceedings IEEE*, vol. 79 n° 9, September 1991.

F. DIENER, G. REEB, (1989) "Analyse non standard", *Hermann éditeur*, 1989. (in french)

D. M. DUBOIS (1996). « Introduction of the Aristotle's Final Causation in CAST: Concept and Method of Incursion and Hyperincursion ». In *Computer Aided Systems Theory - EUROCAST'95. Edited by F. Pichler, R. Moreno Diaz, R. Albrecht. Lecture Notes in Computer Science*, volume 1030, Springer-Verlag, Berlin, Heidelberg, New York, pp. 477-493

.J.-P. FRACHET, G. COLOMBARI, (1993) "Elements for a semantics of the time in GRAFCET and dynamic systems using Non-Standard Analysis", in *Automatique Productique Informatique Industrielle Review*, vol. 27 n° 1, January 1993, pp. 107-125.

J.-P. FRACHET et al, (1995) "Modèles comportementaux pour les Systèmes Dynamiques à Evénements Discrets", *Polycopié du Diplôme d'Etudes Approfondies de Production Automatisée de l'Université de Nancy I et de l'Ecole Normale Supérieure de Cachan*, 1995. (in french)

J.P. FRACHET et al, (1997) "Modelling discrete event systems behaviour using the Hyperfinite Signal", in *RAIRO-APII-JESA* Vol. 31 n°3/1997, pp. 453 à 470

M. GUILLAUME, (1995)"D'une théorie des systèmes séquentiels à la notion de codeur : validation globale du Grafcet", *Thèse de doctorat de l'Ecole Supérieure d'Electricité, Université de Rennes*, December 1995.

M. GUILLAUME, J.-M. GRAVE, P. CHLIQUE, (1996)"Formalization of edges for the Grafcet state machine", in *Proceedings of the CESA'96 IMACS Multiconference - Computational Engineering in Systems Applications*, Lille (France), July 9-12, 1996.

IEC (1991) "Preparation of function charts for control systems", *International Standard*, IEC 848, Second edition, 1991.

R. LIDL, H. NIEDERREITER, (1983)"Encyclopaedia of mathematics and its applications", Section: "Algebra", Volume 20: "Finite fields", *Addison-Wesley Publishing Company*, 1983.

J.-M. ROUSSEL, J.-J. LESAGE, (1993)"Une algèbre de Boole pour l'approche événementielle des systèmes logiques", *Automatique Productique Informatique Industrielle Review*, vol. 27 n° 5, 1993, pp. 541-560.

Anticipatory Programming
– Higher-Order Dynamics, Fault-Tolerance, Composition –

F. Geurts   L. Onana Alima

Département d'Ingénierie Informatique, Université catholique de Louvain

Place Sainte Barbe 2, B-1348 Louvain-la-Neuve, Belgium

{gf,lo}@info.ucl.ac.be

Abstract. Fault-tolerance is often formalized in terms of systems adapting their behavior to their eventually stable environment. Anticipatory systems are based on predictive models of their environment. Learning provides the higher-order dynamics leading to "good" models. Using anticipation, we extend adaptive programming concepts to stronger forms of interactions with the environment leading, among others, to fault-containment. Inference rules and composition laws are given, defining how anticipation propagates from individual systems to concurrent ones.

Keywords: dynamical systems, anticipation, learning, fault-tolerance, composition.

1   Introduction

Context. In the context of (distributed) programs, fault-tolerance is often defined in the uniform formal framework of self-stabilization after transient perturbation and, more generally speaking, adaptation (Burns, Gouda, Miller, 1993; Gouda, 1995; Gouda, Herman, 1991). The system eventually stabilizes to "healthy" states, which means that it recovers a controlled behavior in finite, maybe arbitrarily large, time by successively adapting its transitions to the source of faults, that is, its environment. After each transition, the system agrees with the previous state of its environment, which eventually leads to a stable agreement.

Following the same idea, fault-containment strengthens fault-tolerance by restricting the recovering time or the amplitude of the damage caused by transient faults, in order to self-stabilize its execution to correct states (Ghosh, Gupta, *et al.*, 1996).

Aim. The aim of this paper is to show that anticipation (Rosen, 1985) is an adequate tool to build a uniform framework for evolutionary fault-tolerant systems. More precisely, we introduce anticipation in the transitions of a system, i.e., the ability to use a model of its environment in order to adapt its own states to future states of this environment. Better than recovering from transient faults, the idea is to predict, adapt to, and contain them before they occur.

CP437, *Computing Anticipatory Systems: CAYS--First International Conference*
edited by Daniel M. Dubois © 1998 The American Institute of Physics 1-56396-827-4/98/$15.00

Outline. After defining anticipation from the notion of adaptation (§2), we will show that learning permits a stepwise refinement of the predictive model used by the system, in terms of higher-order or evolutionary dynamics (§3). Then, we will add a notion of environment invariance to adaptation and anticipation, that we will compare to self-stabilization (§4). Different levels of fault-tolerance will be defined according to the degree of adaptation or anticipation a system can use (§5). Laws of anticipation, i.e., inference rules, will be given (§6), as well as composition laws allowing the propagation of eventual anticipation from individual systems to their composition (§7). Finally, we will draw some conclusions and highlight some aspects of our work devoting further investigation (§8).

2 From adaptation to anticipation

In order to analyze interactions between systems and some environment, we have to define how these components evolve in time, and how they agree on some subsets of their state space.

Environment/system. Let us first define the *environment*, given by the pair (E, g), where E represents the state space and $g : E \mapsto E \in E^E$ is the function describing its transitions: $\forall e \in E$,

$$
\begin{aligned}
g^0(e) &= e \\
g^{t+1}(e) &= g(g^t(e));
\end{aligned}
$$

thus, for any initial state $e^0 \in E$, we have

$$
\forall t \geq 0, e^{t+1} = g(e^t)
$$

where e^t represents the environment state at time t.

Our intention here is not to enter the discussion as to whether Nature is deterministic or not. For the sake of simplicity, we consider functional environments. Of course, this includes nondeterministic relations by considering them as set-to-set functions.

Second, a *(dynamical) system* that can be influenced by its environment is defined as (S, f), where S represents the state space and the function $f : S \times E \mapsto S \in S^{S \times E}$ describes its transitions:

$$
\forall (s^0, e^0) \in S \times E, t \geq 0, s^{t+1} = f(s^t, e^t).
$$

This kind of dynamics finds its foundations in the reactive paradigm, as every single state is just obtained by reaction to a previous state of itself and of the environment.

Note the asymmetry of the previous two definitions: environment states are explicitly present in the dynamics of the system, while the opposite is not true. Indeed, to keep the presentation clear, we choose the dynamical effect to be one-way: only the system can be affected by the environment.

Agreement. Given a starting state $(s^0, e^0) \in S \times E$, the dynamics of the pair system-environment is fully defined, mathematically speaking. However, from a program-synthesis point of view, nothing is said about the goal the system should reach with respect to its environment. Specifying an *agreement*, that is, an objective of this coupled dynamics, amounts to giving a relation $A \subseteq S \times E$ that both systems should verify, be it always or from some moment on.

Adaptation. Let us now define *adaptation*. The system f adapts to its environment g according to the agreement A from T on if, after each transition following T, the system agrees with the previous environment state, that is, the one it could observe:

$$f \hookrightarrow_A g \text{ iff } \forall t \geq T, (s^{t+1}, e^t) \in A.$$

Prediction. Instead of observing the environment and adapting itself to this state, the system could try to predict, and adapt to, the next environment state. This precisely corresponds to anticipation according to a predictive *environment model*. Such a model is a function $\tilde{g}$ that is "very close" to g with respect to a given distance d in the functional space E^E where g is defined, say smaller than a given (small) value ε:

$$d(g, \tilde{g}) \leq \varepsilon.$$

In order to emphasize the use of this predictive model in the estimation of the next environment state, we rewrite $f(s, e)$ as $f(s, \tilde{g}(e))$. Thus:

$$\forall (s^0, e^0) \in S \times E, t \geq 0, s^{t+1} = f(s^t, \tilde{g}(e^t)).$$

Anticipation. We can now define *anticipation*. The system f anticipates its environment g according to the agreement A from T on if, after each transition following T, the system agrees with the environment state, that is, the one it could predict using its model $\tilde{g}$ of g:

$$(f, \tilde{g}) \rightsquigarrow_A g \text{ iff } \forall t \geq T, (s^{t+1}, e^{t+1}) \in A.$$

Note the similarity between adaptation and anticipation: the latter is nothing but the former with respect to future states of the environnment.

3 Learning and higher-order dynamics

Usually, a "good" environment model is not necessarily known at the beginning of the history of the pair system-environment. Incremental learning permits to refine the model by successive approximations. For instance, neural-based time-series prediction and function approximation make extensive use of incremental algorithms based on steepest gradient descent (Anderson, 1989; Blum, Li, 1991; Haykin, 1994): each new value in the series entails slight modifications of the network weights in order to better fit the underlying unknown generating function.

Higher-order dynamics. At each point in time, a model $\tilde{g}^t : E \mapsto E$ is made available by some learning algorithm that realizes a higher-order or evolutionary dynamics

$$G : E^E \times E^* \mapsto E^E$$

(where E^* denotes the set of finite sequences on E), starting from an initial model $\tilde{g}^0$:

$$\forall t, \tilde{g}^{t+1} = G(\tilde{g}^t, (e^0, e^1, \cdots, e^t)).$$

The functional G is stated here in its most general form, taking as arguments the previous model and all environment states since the beginning of the evolution.

Learning quality. For the evolutionary dynamics to be useful, the learning algorithm has to be such that the distance $\varepsilon^t = d(g, \tilde{g}^t)$ decreases with time, at least while ε^t remains greater than a given limit ε. If there exists a function h such that

$$\forall t, \varepsilon^{t+1} = h(\varepsilon^t)$$

then its derivative h' must be negative above the fixed limit:

$$\forall \gamma > \varepsilon, h'(\gamma) < 0.$$

Building a predictive model. In (Snyers, 1996), the author defines a hierarchy of agents interacting with each other and within an environment: reactive, hedonic and eductive agents. By simple reinforcement learning based on notions of good and ill, hedonic agents anticipate their close future. On the other hand, eductive agents are able to build models of their own sensori-motor perceptions and, thereby, they anticipate further ahead. This means that they possess a learning algorithm which helps building a right model. In (Crutchfield, 1994), the author proposes a general model construction, called the ε-machine, based on Chomsky's language-theoretic hierarchy of grammars. This allows him to relate the complexity of the object being modeled to the language complexity of its observable dynamics, i.e. the sequences (or words) it generates.

We do not exploit learning algorithms in the following, although this topic in itself would deserve an entire paper.

4 Eventual adaptation/anticipation

In this section, we slightly modify the definitions given in §2, in order to make them more realistic. The idea is to restrict the environment dynamics by adding an invariant in which it must stay indefinitely in order to lead to states where an adaptive or anticipative agreement with the system can happen.

Invariance. More precisely, an environment *invariant* P is a g-closed subset of E, i.e. $g(P) \subseteq P$:

$$\forall e \in P, g(e) \in P.$$

510

Eventual adaptation. Let us first define *eventual adaptation*: the system f eventually adapts to its environment g according to the agreement A from the invariant P,

$$f_P\Diamond\!\!\hookrightarrow_A g \text{ iff } (e^0 \in P) \Rightarrow (\exists T, \forall t \geq T, (s^{t+1}, e^t) \in A).$$

Self-stabilization. Usually (Burns, Gouda, Miller, 1993; Schneider, 1993), *self-stabilization* to a set $Q \subseteq S$ is defined by the existence of some point in time, T, such that

$$(\forall t < T, s^t \notin Q) \wedge (\forall t \geq T, s^t \in Q).$$

It is thus related to eventual adaptation by the following implication:

$$f \text{ self-stabilizes to } A \Rightarrow f_E\Diamond\!\!\hookrightarrow_A g.$$

The converse does not hold in general.

Eventual anticipation. In the same fashion, we define *eventual anticipation*: the system f eventually anticipates its environment g according to the agreement A from the invariant P,

$$(f, \tilde{g})_P\Diamond\!\!\rightsquigarrow_A g \text{ iff } (e^0 \in P) \Rightarrow (\exists T, \forall t \geq T, (s^{t+1}, e^{t+1}) \in A).$$

Note that, in both definitions, T is "some point in time" in the joint history of the system and its environment: it is not precisely defined but it exists.

5 Fault-tolerance

In a pair system-environment, the latter often represents a source of possible faults. The system interacts with it provided no fault occurs. When things go wrong, the system enters a transient sequence of meaningless states until the environment stabilizes back to normal states, after which it resumes its correct execution by self-stabilization. Most of the fault-tolerant models are based on combinations of such transient phenomena with self-stabilization. Fault-containing systems are able to treat some limited transient faults without undergoing too much perturbation. For instance, they recover in constant time or, in case of distributed systems, they restrict the amount of infected parts to constant-size neighborhoods.

Both models can be related to each other using three embedded sets of states:

$$H \subseteq C \subseteq T.$$

Let us consider H as the set of normal states of the system, C as the set of states the system reaches after contained faults, and T as the set of states reached after tolerated faults. A fault-tolerant system immediately reaches T from H, and takes the direction of H as soon as faults disappear. A fault-containing system is able to treat some faults by jumping to C and taking the direction of H faster than from tolerated faulty states.

In the rest of this section, we will successively review conditions of decreasing strength that can be assumed on the environment, and still guarantee the correct execution of the system. These increasing forms of fault-tolerance will be related to adaptation and anticipation.

Fault-intolerance. If the healthy behavior of the system relies on the permanence of the environment, that is,

$$\forall e \in E, g(e) = e,$$

then it is simply *fault-intolerant*.

Fault-tolerance. If transient faults are accepted, which means that

$$\exists e \in E, \forall t > T, e^t = e,$$

then the system is *fault-tolerant*. This notion can be implemented using adaptation in the following way: after time T, the environment does not change anymore; thus, by successive adaptations, the system ends up agreeing with it.

Fault-containment. If faults occur in a predictible way, the system can adapt its behavior to them in advance and reduce their global impact. The system is *fault-containing*. Here, we see that adaptation alone is insufficient to achieve this goal. Actually, anticipation seems to be the right tool to prevent strong perturbations due to some faults. It is thus of utmost importance in the context of fault-tolerance, and should be investigated deeply with this objective in mind.

Different degrees of prediction could be examined: knowing g precisely, i.e.

$$d(g, \tilde{g}) = 0;$$

having a good model of g, i.e.

$$d(g, \tilde{g}) \le \varepsilon;$$

having a probabilistic model of g, i.e.

$$\forall x, P(d(g(x), \tilde{g}(x)) > \varepsilon | x) < \varepsilon.$$

Robustness. Finally, if no information is known about the environment, i.e. faults can occur in a totally random way, the system is strongly tolerant or *robust*. Or, it does not depend on its environment!

6 Laws of anticipation

The two laws we present below are derived using the previous definitions. They, among many others, can be used in reasoning about system anticipation.

In the following theorem, we use the same notational conventions as before for f and g; $\tilde{g}$ is supposed to be a good model of g; P and Q are environment invariants; A and B are subsets of $S \times E$.

The result shows how to use two instances of anticipation to deduce two more instances of the same property.

512

Theorem 1

If the system f eventually anticipates its environment g from two different invariants P and Q according to the agreements A and B:

$$(f, \tilde{g})_P \diamondsuit \rightsquigarrow_A g \text{ and } (f, \tilde{g})_Q \diamondsuit \rightsquigarrow_B g$$

then the following two anticipation properties hold from their intersection and union, respectively:

$$(f, \tilde{g})_{P \cap Q} \diamondsuit \rightsquigarrow_{A \cap B} g \text{ and } (f, \tilde{g})_{P \cup Q} \diamondsuit \rightsquigarrow_{A \cup B} g.$$

PROOF. Let us start with the first property. Since $P \cap Q \subseteq P$ entails the eventual agreement on A from T_1 on, and $P \cap Q \subseteq Q$ the eventual agreement on B from T_2 on, the agreement happens on $A \cap B$ after $T = \max\{T_1, T_2\}$.

The second property is straightforward, too. At the very beginning, either $e^0 \in P \backslash Q$, $e^0 \in Q \backslash P$, or $e^0 \in P \cap Q$. The first case leads to A, the second one leads to B, and the third one to $A \cap B$. Thus, after some time, the system and its environment agrees on $A \cup B$ if the environment initially belongs to $P \cup Q$.

$\square$

7 Composition

In this section, we show that the composition of several anticipatory systems preserves the property under the assumption of non interference between the constituents. The result presented below is an example of what can be achieved by composition, but it is by no mean the most general form (Gouda, 1995).

Constituents. First, let us introduce the set of constituents:

$$\{f_{i,n_i} : S_i \times \mathcal{E}^{n_i} \times E \mapsto S_i \mid i \in I \text{ and } n_i \in \mathbb{N}\}$$

where $\cup_i S_i \subseteq \mathcal{E}$ and $\mathcal{E}^{n_i}$ denotes the n_ith Cartesian power of $\mathcal{E}$. In the rest of this section, we remove n_i to keep notations clear.

This way of abstractly extending the domain of constituents emphasizes the possibility, for each one of them, to be influenced by others. If f_i is executed alone, no assumption whatsoever can be made on its central argument.

In order to obtain a composed dynamical system, two complementary questions must be tackled: How are constituents related to each other? How does the composite system evolve? Answers are presented in the same order.

Interdependence. When different systems are composed together, some input-output links must be established between them. For example, one of f_i's inputs could be f_j's state.

In the definition of each constituent f_i, the parameter n_i partially plays this role as it determines the potential dependence between f_i and other systems through $\mathcal{E}^{n_i}$. Globally, a *dependence function* must be given to fix the relationships between all constituents of the composite system:

$$\mathcal{D} \;:\; I \mapsto I_\omega^*$$
$$\text{s.t.} \quad \mathcal{D}(i) \in I_\omega^{n_i}$$

where $I_\omega = I \cup \{\omega\}$ and ω means "undefined".

This function defines a *dependence relation*: f_i depends on f_j,

$$i \heartsuit j \quad \text{iff} \quad \exists k \in \{1, \cdots, n_i\}, \mathcal{D}(i)_k = j$$
$$\wedge \quad \exists s \neq s' \in S_j, (f_i(s_i, (\cdots, s, \cdots), e) \neq f_i(s_i, (\cdots, s', \cdots), e)).$$

A *dependence path* is a sequence $(i_1, \cdots, i_n) \in I^*$ such that

$$\forall k \in \{1, \cdots, n-1\}, i_k \heartsuit i_{k+1}.$$

Finally, a *dependence cycle* is a dependence path $(i_1, \cdots, i_n)$ such that $n > 2$ and $i_1 = i_n$.

Dynamics. A priori, two choices seem relevant to dynamically compose systems: a synchronous model, or an asynchronous one (Hoare, 1985; Milner, 1989). We adopt the second case: at each step in the composed system's history, an arbitrary constituent is chosen and enabled to progress. This model somehow interleaves the individual dynamics into a single global one. Usually, an important constraint is added to the way successive arbitrary choices are made along infinite histories of the composed system: fairness, that is, no constituent is disabled forever (Francez, 1986).

The effect of the *composition operator* $\otimes$ is defined as follows:

$$\otimes_\mathcal{D} f_i(s_1, \cdots, s_{|I|}, e) \in \{(s_1, \cdots, f_k(s_k, (s_{\mathcal{D}(k)_j})_{j \in \{1, \cdots, n_k\}}, \tilde{g}_k), \cdots, s_{|I|}) \mid k \in \{1, \cdots, |I|\}\}.$$

Thus, the *composite dynamics* is expressed as:

$$\forall(s^0, e^0) \in \times_i S_i \times E, t \geq 0, s^{t+1} = \otimes_\mathcal{D} f_i(s^t, e^t)$$

where s^0, s^t and s^{t+1} denote state vectors of $\times_i S_i$.

Anticipation. In the next theorem, we suppose that each subsystem f_i has a good model $\tilde{g}_i$ of its environment.

Theorem 2
If all constituents eventually anticipate their environment as follows:

$$\forall i, (f_i, \tilde{g}_i)_{P_i} \Diamond\!\rightsquigarrow_{A_i} g$$

and their dependence function $\mathcal{D}$ does not introduce any cycle, then the composite system eventually anticipates its environment:

$$(\otimes_\mathcal{D} f_i, \{\tilde{g}_i\})_{\cap_i P_i} \Diamond\!\rightsquigarrow_{\cap_i \overline{A_i}} g$$

where $\overline{A_i} = \{(s_1, \cdots, s_{|I|}, e) \mid (s_i, e) \in A_i \text{ and } \forall j \neq i, s_j \in S_j\}$.

PROOF. Each constituent eventually anticipates the global environment according to the agreement A_i, whatever the values of its central argument. This remains true when the constituents are connected to each other via some dependence function.

Thus, if all individual invariants are initially valid, the composite system eventually anticipates the environment according to the conjunction of the individual agreements.

The time needed to reach the global agreement is at most the maximum of all individual ones.

□

Note that the absence of cycle permits to avoid oscillations that would otherwise break the convergence to a global agreement. Moreover, the strong non interference between constituents on which the previous result is based could be relaxed following the arguments of (Gouda, 1995).

8 Conclusion

Summary. We have defined anticipatory systems from the well-known notion of adaptive systems. Then, we have related different levels of fault-tolerance to these forms of interaction between systems and their environment. Finally, we have shown that logic inference rules can be stated, as well as composition laws allowing to build anticipatory systems from smaller ones provided they individually anticipate the same environment.

Discussion. Although we just have defined some notions and stated a few results, this work shows that anticipation can strengthen systems in their reaction against perturbations coming from their environment. For instance, the field of distributed systems constitutes an important area and a source of challenging developments for fault-tolerance. Of course, the theory of incremental approximation of dynamical systems as well as classical techniques of artificial intelligence have to be deeply studied in order to facilitate the learning phase.

Further work. Some aspects of the work devote further investigation. In particular, other inference rules and system compositions should be studied, following, e.g., (Geurts, 1996; Gouda, 1995). Moreover, anticipation should be used in its different forms of predictive models making systems fault-containing. Finally, probabilistic fault-tolerant and fault-containing anticipatory systems could be constructed and their correctness proved in the framework developed in (Morgan, 1995; Morgan, McIver, et al., 1995).

Acknowledgements. The authors would like to thank Daniel Dubois for inviting us to this first International Conference on Computing Anticipatory Systems, and for the enlightening discussions we had on hyperincursion. Furthermore, the first author (FG) is supported by the Belgian FNRS as a "Chargé de Recherches".

References

Anderson C. W. (1989). Learning to control an inverted pendulum using neural networks. *IEEE Control Systems Magazine*, **9**(3):31–37.

Blum E. K., and Li L. K. (1991). Approximation theory and feedforward networks. *Neural Networks*, **4**:511–515.

Burns J. E., Gouda M. G., and Miller R. E. (1993). Stabilization and pseudo-stabilization. *Distributed Computing*, **7**:35–42.

Crutchfield J. P. (1994). The calculi of emergence: Computation, dynamics, and induction. Tech. Rep. 94-03-016, Santa Fe Institute.

Francez N. (1986). "Fairness". Springer-Verlag.

Geurts F. (1996). "Compositional Analysis of Iterated Relations: Dynamics and Computations". PhD thesis, Département d'Informatique, Université catholique de Louvain. To appear as Lecture Notes in Computer Science, Springer-Verlag.

Ghosh S., Gupta A., Herman T., and Pemmaraju S. V. (1996). Fault-containing self-stabilizing algorithms. Tech. rep., Univesrity of Iowa.

Gouda M. G. (1995). The triumph and tribulation of system stabilization. *In* J. M. Hélary and M. Raynal (eds.), "Proc. of the 9th International Workshop on Distributed Algorithms", LNCS 972, pp. 1–18. Springer-Verlag.

Gouda M. G., and Herman T. (1991). Adaptive programming. *IEEE Transactions on Software Engineering*, **17**(9):911–921.

Haykin S. (1994). "Neural Networks. A Comprehensive Foundations". IEEE Computer Society Press.

Hoare C. A. R. (1985). "Communicating Sequential Processes". Prentice Hall.

Milner R. (1989). "Communication and Concurrency". Prentice Hall.

Morgan C. C. (1995). Proof rules for probabilistic loops. Tech. Rep. PRG-TR-25-95, Prog. Res. Group, Oxford U.

Morgan C., McIver A., Seidel K., and Sanders J. W. (1995). Probabilistic predicate transformers. Tech. Rep. PRG-TR-4-95, Programming Research Group, Oxford U.

Rosen R. (1985). "Anticipatory Systems". Pergamon Press.

Schneider M. (1993). Self-stabilization. *ACM Computing Surveys*, **25**(1):45–67.

Snyers D. (1996). "Dynamiques coévolutionistes dans les sociétés d'agents autonomes". PhD thesis, Université de Caen.

Anticipation versus Adaptation in Evolutionary Algorithms: The case of Non-Stationary Clustering

A. I. González, M. Graña, A. D'Anjou, F.J. Torrealdea
Dept. CCIA, Univ. Pais Vasco/EHU[0], Aptdo 649, 20080 San Sebastián, España
e-mail: ccpgrrom@si.ehu.es

Abstract

From the technological point of view is usually more important to ensure the ability to react promptly to changing environmental conditions than to try to forecast them. Evolution Algorithms were proposed initially to drive the adaptation of complex systems to varying or uncertain environments. In the general setting, the adaptive-anticipatory dilemma reduces itself to the placement of the interaction with the environment in the computational schema. Adaptation consists of the estimation of the proper parameters from present data in order to react to a present environment situation. Anticipation consists of the estimation from present data in order to react to a future environment situation. This duality is expressed in the Evolutionary Computation paradigm by the precise location of the consideration of present data in the computation of the individuals fitness function. In this paper we consider several instances of Evolutionary Algorithms applied to precise problem and perform an experiment that test their response as anticipative and adaptive mechanisms. The non stationary problem considered is that of Non Stationary Clustering, more precisely the adaptive Color Quantization of image sequences. The experiment illustrates our ideas and gives some quantitative results that may support the proposition of the Evolutionary Computation paradigm for other tasks that require the interaction with a Non-Stationary environment.

Keywords: Non-Stationary Clustering, Evolutionary Algorithms, Color Quantization.

0. Introduction

The original proposition of Holland (1975) of the Evolution algorithms was intended to provide a means for improving the response of complex systems to changing and uncertain environments. The Evolution algorithms were to be able to adapt to the new environment and to discover new ways to deal with it. The discovering was to be the product of crossover (sometimes called recombination) and mutation operations that will produce spontaneously new information material (rules) that anticipate the environment changes. The adaptation was also included in the scheme through the application of a selection function to the potential parents, that acts as a kind of filter for the generation process. The duality of anticipation and adaptation has been present from the very start of the modern research on Evolutionary Computation algorithms.

However, much of the work reported in the literature is referred to optimization problems (Bäck, Schwefel, 1993) that can be conceptualized as autonomous systems: closed systems without any kind of interaction with the environment. In this setting only the convergence question is of interest, and much effort is being devoted to analyze the convergence to the optima of the fitness function of several kinds of Evolutionary Algorithms. Some fundamental questions related with the interaction with the environment have been oversighted in this line of work.

First of all, in the general case of an open system whose control is somehow directed by an evolutionary mechanism, the fitness function is no longer a time invariant function. The fitness function is in this case a measure of the goodness of the response of the system to environment events. Therefore, a given individual may change its fitness as time goes on and the environment changes. Optimal solutions at a given time can become suboptimal or even very

[0]This work is being supported by a research grant from the Dpto de Economía of the Excma. Diputación de Guipuzcoa, and a predoctoral grant and projects PI94/78, PI97/12 and UE96/9 of the Dept. Educación, Univ. e Inv. of the Gobierno Vasco

bad solutions in the future, and the other way round. Time dependent fitness functions must not be confused with noisy fitness functions (Fitzpatrick, Grefenstette, 1988; Aizawa, Wah, 1994), unless the signal to noise ratio is below one, and the noise is non stationary. The sources of the noise may lie in the communication channels of the system with the environment, in the computation mechanisms or in the environment. In the case of noisy fitness functions the optimal solutions remain the same, partially hidden by the noise. In this case a single value of the fitness function is of no use, and the proper evaluation of the fitness function involves some sampling and averaging process. The noisy fitness function is a special case of the closed systems referred above, that reduce the operation of the Evolutionary Algorithm to the role of a global random search (optimization) strategy.

Confusing the Non-Stationary case, characterized by a time varying fitness function (that can also be corrupted by noise) with the noisy Stationary case can be dangerous. The resampling of the fitness function that is very appropriate to estimate the average fitness of the individual as an indication of its true fitness can be very misleading. In the Non-Stationary case the true fitness is changing in time, and the resampling and averaging mechanism may produce very inconsistent results.

A fundamental problem in dealing with Non-Stationary environments is that of synchronization. There is a strong need to relate the inner clock of the Evolutionary Algorithm with the external clock of the environment in order to guarantee the meaningful operation of the whole. A very slow internal clock will produce a bad sampling of the environment, and, therefore, systematic inconsistencies of the responses given to external events. This a phenomenon already observed by researchers that have tried to apply Evolutionary Computation approaches to real time control of processes. Dealing with Non-Stationary environments involves considerations of response time and sampling frequencies that do not arise in the context of closed systems (pure optimization problems). The solution adopted by the research community is to reduce the application of Evolutionary Algorithms to control problems that can be stated as (closed systems) pure parameter optimization problems. We think that the synchronization problem is still waiting for a proper treatment in the Evolutionary Computation literature.

From the perspective of open systems, in which the role of the Evolutionary Algorithm is that of looking for improvements of the response of the system to its environment, fast convergence to good suboptimal solutions is more important than slow convergence to global optimal solutions. Also the computational considerations take an important position as far as the system is supposed to react in real time to real life situations. Again, these kind of considerations appear in the application of Evolutionary Algorithms to the design of control systems. As far as we know, no real-life on-line control Evolutionary Algorithm application has been realized successfully up to date. Most of the literature deal with experiments performed on simulated systems without any regard for time constraints.

The population diversity is closely related to the anticipation abilities of the Evolutionary Algorithm. There have been some works on this topic, mostly on Genetic Algorithms. However again the emphasis in these works is under the closed system perspective. Population diversity is usually considered as a measure of the Genetic Algorithm potential to escape from uniform populations (all individuals equal) lying in a local extreme. In other words, population diversity is sought as a guarantee against premature convergence to suboptimal solutions. From the perspective of open systems, the greater the population diversity the greater the possibility that an unknown environment configuration can be dealt with at first shot, without need for adaptation. However, preserving meaningful diversity involves the need to forecast the environment, because of the time varying nature of the fitness function in a Non-Stationary environment. We think that the study of the ability of the Evolutionary Computation paradigm to preserve meaningful diversity in a Non-Stationary setting, without explicit forecasting procedures, is a really interesting line of research.

The structure of the paper is as follows. First we will comment on the general statement of the adaptive-anticipative duality in very general terms. Second we will comment how this duality is realized in the setting of Evolutionary Algorithms dealing with Non-Stationary environments and the precise instances of Evolutionary Algorithms applied. Third we will describe our experiments . Fourth we will discuss the results obtained, and finally we will give some conclusions.

1. Anticipative versus adaptive computational systems

From the perspective of a computational system interacting with an environment, the adaptive-anticipative duality is summarized in Figure 1. The kind of systems we are thinking of are processing and control systems that must produce a response to external events or must process information coming from the environment. The environment is changing (Non-Stationary) or even reacting to the responses of the system. Therefore the system must observe the environment to obtain information about its present state (external state) in order to compute its responses. These responses are function also of the system's own internal state.

The adaptive behaviour consists in essence in the generation of responses based on internal states updated according to the present observation of the environment. This implies that the computation of the internal state based on actual information from the environment is previous to the generation of the response.

The anticipative behaviour consists in the generation of responses as soon as the environment information is received. That means that the internal state is not updated previously to the generation of the response, but afterwards. In other words, responses are computed from past internal information. Anticipation implies forecasting the environment in the sense that potential new external states must be predicted and the potential responses precomputed.

Initialization of internal state
Observe environment
while not terminate interaction do
Compute next internal state
Generate response to the
environment
Observe environment
end while

Initialization of internal state
Observe environment
while not terminate interaction do
Generate response to the
environment
Compute next internal state
Observe environment
end while

(a)   (b)

Figure 1 General structure of an Open System as an Adaptive algorithm (**a**) and as an Anticipative algorithm (**b**)

In Figure 1 adaptive and anticipative behaviours are reflected in the relative positions of the **generate** and **compute** statements. From a technical point of view, the distinction between these two cases is not trivial. All the operations (observation, computation, generation) consume time and the interaction between the system and the environment can be highly time-critical. Time-critical means that there is a tradeoff between the optimality of the responses given and the time at which these responses are received. There is a time limit after which even the true optimal response is useless, and the system is doomed. For very time-critical systems, this time limit can be so tight that there is no room for the computation of an updated internal state to generate an optimal response. The response must be generated instantly and the reaction time of the environment employed in updating the internal state. This after-computation of the internal state must therefore anticipate the next environment event to, somehow, precompute the responses. Less time-critical systems allow for a time to adapt the internal state to the new environment information, previous to the generation of the response.

The dominant computational paradigm in technological systems is adaptation, because it is the sure way to go. The critical requirements for technological systems are, therefore, the time constraints imposed on the adaptation. For example, Artificial Neural Networks are adaptive mechanisms that are actually being applied with success to a variety of technical problems. As we will discuss in the results section their performance is greatly reduced when applied in an anticipative setting. On the contrary, Evolutionary Algorithms posses both anticipative and adaptive components, so it can be hypothesized that they can perform similarly as adaptive or anticipative systems. Even it can be though that the variety of solutions represented by a population can produce spontaneously anticipative behaviors. The aim of this paper is to explore experimentally this possibility and to motivate its formal study.

2. Anticipative versus Adaptive Evolutionary Algorithms

The main point in this paper is that of the adaptation versus anticipation issue. Figure 2 shows the adaptive and anticipatory versions of an Evolutionary Algorithm. Readers already familiar with the Evolutionary Computation literature (Bäck, Schwefel, 1996) may notice at first sight the time variables t_{int} and t_{ext} that symbolize the internal and external clocks, respectively. Increasing the external clock means observing the environment to gather new data that will serve to compute up to date fitness values for the individuals. The **actuate** operator involves the generation of actions or responses towards the environment that result from the internal state given by the actual population at the precise instant of the execution of this operator. In both figures $P(t_{int})$ denotes the population at internal time t_{int}, and $E(t_{ext})$ denotes the environment variables at external time t_{ext}.

$t_{int}:= 0$; $t_{ext}:= 1$; frec=0
 initialize $P(t_{int})$
 $F(t_{int}) =$ **evaluate** $(P(t_{int}), E(t_{ext}))$
 while not terminate interaction do
     $P'(t_{int}):=$ **recombine** $P(t_{int})$
     $P''(t_{int}):=$ **mutate** $P'(t_{int})$
     $F''(t_{int}):=$ **evaluate** $(P''(t_{int}), E(t_{ext}))$
     $P(t_{int}+1):=$ **select** $(F(t_{int}), F''(t_{int}))$
     frec:=frec+1;
 	
     if frec $= \tau$ then
         actuate $(P(t_{int}+1), E(t_{ext}))$
         frec=0;
         $t_{ext}:= t_{ext}+1$;
     endif
     $t_{int}:= t_{int}+1$;
 end while

(a)

$t_{int}:= 0$; $t_{ext}:= 0$; frec=0
 initialize $P(t_{int})$
 $F(t_{int}) =$ **evaluate** $(P(t_{int}), E(t_{ext}))$
 while not terminate interaction do
     $P'(t_{int}):=$ **recombine** $P(t_{int})$
     $P''(t_{int}):=$ **mutate** $P'(t_{int})$
     frec=frec+1
 	
     if frec $= \tau$ then
         $t_{ext}:= t_{ext}+1$;
     endif
     $F''(t_{int}):=$ **evaluate** $(P''(t_{int}), E(t_{ext}))$
     $P(t_{int}+1):=$ **select** $(F(t_{int}), F''(t_{int}))$
 	
     if frec $= \tau$ then
         actuate $(P(t_{int}+1), E(t_{ext}))$
         frec=0
     endif
     $t_{int}:= t_{int}+1$;
 end while

(b)

Figure 2 General structure of an Evolutionary Algorithm as an adaptive algorithm (**a**) and an anticipative algorithm (**b**) with a fixed number of generations between interactions with the environment

The value of τ determines the frequency of interaction with the environment specified as the number of generations computed between observations. This value and the objective duration of the computations for each generation determines the synchronization of the internal and external clocks. Note that as the frequency of interaction grows the system becomes a closed system.

The remaining notation used in figure 1 is rather classical. **initialize** denotes the generation of the initial population, **recombine** denotes the application of a recombination or crossover operators, **mutation** denotes the application of a mutation operator, **select** denotes the selection operator. Note that we are assuming an elitist strategy as the general selection strategy. The next generation is somehow built up from the best parents and children. This is the only strategy that guarantees convergence to the optimal values of the fitness function, and is the reasonable choice if fast convergence to (sub)optimal individuals is desired.

The Evolutionary Schemas proposed in figure 2 cover a wide variety of more concrete propositions of Evolutionary Algorithms. In fact we have tested in our works several algorithms that come from the Genetic Algorithm and the Evolution Strategy approaches. Genetic Algorithms may be characterized by the binary nature of the individuals, the relative importance of the crossover operator as global search operator and the use of mutation as the mean to introduce new genetic material that can permit to scape from bad regions of the search space. On the other hand, Evolution Strategies are characterized by real valued individuals, the use of random mutation as the main search operator and its elitist nature.

We will discuss briefly the features that define the approach as adaptive or anticipatory. Notice that in figure 2 the most salient difference between the adaptive and anticipatory definitions is the actual position of the external clock increment in the sequence of operations (t_{ext}:= t_{ext}+1;). The adaptive approach will perform all the mutation and selection operations based on the fitness computed over the present environmental data. The adaptive algorithm gathers the response of the environment and proposes new solutions that may be better adapted to it. In the anticipatory case, the algorithm computes new solutions (through recombination and mutation) as potential solutions for the next external time instant, and they are selected depending on their fitness to the new environment. The algorithm tries to anticipate solutions for the next time instant. This simple change may involve many changes in real life application. In the adaptive approach the system must give answers to situations already present, in the anticipatory approach the system is trying to figure out new solutions for a future situation. Adaptive solutions may be much more time critical than anticipatory ones. Notice that the anticipatory approach performs an optimization on a present data set, whose final result will be evaluated on the future data set.

A key parameter in the behaviour of both adaptive and anticipative algorithm is the interaction frequency τ. This parameter controls the reaction time allowed in the case of the adaptive strategy, and the processing for forecasting time in the anticipative strategy. The very stochastic nature of the Evolutionary approaches implies a great deal of uncertainty about the expected time to convergence. In practical systems, this uncertainty can be unbearable. There is always a time limit either to produce a response (adaptive case) or to process the past information before a new event takes place (anticipative case). This temporal restriction implies, at least, that an upper limit on the number of generations computed must be set. Ideally the system would work on a one to one basis, that is τ=1. We have tested this situation and some less stringent conditions, allowing for the computation of several generations between environment interactions.

The Evolutionary Algorithms tested, in their adaptive and anticipative versions, are:

1- An Evolution Strategy proposed by Babu and Murty (1994) for the case of Stationary Clustering. Individuals represent clustering solutions given by the cluster representatives. Recombination operators and mutations are used and the selection strategy is elitist. We have tested over the Non-Stationary data both the hard and fuzzy fitness functions.

2- An Evolution Strategy proposed by our group for the case of Non-Stationary Clustering (González *et alt*., 1997, 1998). Individuals represent cluster centers, so that the clustering is given by the whole population. Several mutation and selection strategies were tested in other papers. In this paper we apply a deterministic mutation operator and a suboptimal selection. We do not employ any recombination operator. We have found our design to be very fast in convergence to good suboptimal solutions, to the point that we have had success applying it to the image sequence in a one to one time basis: only one generation is computed between images of the sequence.

3- A Genetic Algorithm based on the representation of clusters as partitions of the data sets proposed by Bezdek *et alt.* (1994). Each individual is a clustering solution specified by the matrix of (hard) cluster membership coefficients. Crossover and mutation are rather classic. We have changed the selection strategy to an elitist one (the original proposition of Bezdek does not converge to optimal solutions even in the Stationary case). For the application to the Non Stationary case a reordering step of the sample was needed before the application of the Genetic Algorithm, in order to ensure the consistency of the data and the internal representation.

4- A Genetic Algorithm proposed by Alippi and Cucchiara (1992) again based on partition representations. Individuals are similar to the previous algorithms. Crossover is asexual, based on a single individual. Mutation is classic. Again we enforce an elitist selection strategy. Again a reordering of the sample was needed in the case of the application to the Non-Stationary case.

3. The experiments on Non-Stationary Clustering

In our works we have focused on a specific problem that possess many of the complexities inherent to open systems with time varying environments. The problem is the computation of clusters for a sequence of data sets that show enough variation in their statistical characteristics so as to consider them as samples of a non-stationary process. We call this problem Non Stationary Clustering, and it is closely related to Adaptive Vector Quantization (Gersho, Gray, 1992; Chen *et alt.*, 1994). In practice, the experimental data comes from a sequence of images, and the practical application is that of Color Quantization of an image sequence (Chen *et alt.*, 1995; Goldberg, Sun, 1986). Each data set is a sample of the pixels of each image in the sequence. The solution sought is the optimal set of color representatives for each image, so that the fitness function is the distortion of the Color Quantization. For the image sequence used, this fitness is highly time variant, because of the color variations introduced from one image to the next (it is not a typical talking head sequence). In this paper we will not deal very much on the technical details, because our focus is the relative responses obtained through adaptive and anticipative strategies of Evolutionary Algorithms as discussed above.

In the setting of Non-Stationary Clustering, increasing the external clock means taking into account the data set that describes the environment in the next time instant (we assume a discrete time axis). In the precise instance of Color Quantization of image sequences, the data sample at each time instant is a set of pixels of the image obtained at this moment. Each data point is a point in the RGB unit cube, so that our data form clusters in the tridimensional space whose axes are the basic color components (Red, Green, Blue). Adaptation means that we recompute the color palette using pixels from present image before performing the Color Quantization of the image. Anticipation means that we use the pixel color information of the present image to compute the color palette that will be used in the Color Quantization of the **next** image. The anticipative strategy must predict color variations.

In other reports (González *et alt.* 1997, 1998) we have used as reference responses the results of the application of the Heckbert (1980) algorithm as implemented in MATLAB (Wu, 1991). However, we will not use these benchmark here, because our emphasis is on the duality of anticipative-adaptive responses. But, for further illustration of our ideas we introduce the results obtained with several neural network architectures for Vector Quantization: the Simple Competitive Learning (Ahalt *et alt.*, 1990; Yair *et alt.*, 1992), the Self Organizing Map (Kohonen, 1989), the Fuzzy Learning Vector Quantization (Bezdek, Pal, 1995) and the Soft Competition (Yair *et alt.*, 1992). As Neural Network Architectures are designed as adaptive algorithms, their response in the adaptive setting is quite good. However, as will be shown in the section, their behavior in the anticipative case is very poor.

4. Discussion of the experimental results

We have gathered in figure 3 sample distortion curves of the best individuals computed by the Evolution Algorithms tested under the adaptive and anticipative strategies. These curves show the differences of response obtained in both situations. The first conclusion is that the Evolution Algorithms perform similarly in the anticipative and adaptive cases, showing their inherent abilities as both adaptive and predictive mechanisms. This is rather good news for the real time community, because it opens the door for the application of Evolution Algorithms in an anticipative ways which is much less time critical than the adaptive computation. The main response divergences appear in the middle of the image sequence which corresponds to the images that show the sharpest color variations. That means that the anticipative response degrades relative to the adaptive response when the environment changes are very strong (fast).

Among the responses in figure 3, the best relation of anticipatory and adaptive behaviors is given by the Alippi algorithm . The worst is given by our Evolution Strategy. This last result is not surprising because it was designed as an adaptive algorithm. Nevertheless, the divergence of behaviors in our Evolution Strategy is very localized in the pairs of images that show the strongest color variations.

For the sake of comparison, in figure 4 we show the relative responses of the adaptive and anticipative applications of several Neural Network Architectures. As said before, Neural Network are designed as adaptive algorithms, and their response degrades when applied as anticipative mechanisms. The Neural Networks were trained in a one-pass schedule over the sample, which is equivalent to the one generation interaction frequency of the Evolution Algorithms.

Finally, to summarize the experimental results we present the global distortion along the image sequence for the algorithms tested (the addition of the distortions for each image). In table 1 we show the results of the Evolutionary Algorithms with several parameter settings: population size, number of offsprings allowed and frequency of interaction (number of generations computed between environment interactions). Table 2 shows the results for the inherently adaptive algorithms: the Artificial Neural Networks and our Evolution Strategy. In the case of the Artificial Neural Networks we have applied them using the Euclidean distance and a penalized distance to compute the winner. These tables are given to give a more quantitative evaluation of the relative behavior, besides the qualitative impression given by figures 3 and 4.

The inspection of tables 1 and 2 shows that the Artificial Neural Network and our Evolution Strategy perform better in the adaptive mode than the Evolution Algorithms, even when several generations between interactions are allowed. However, the degradation of the response of the Artificial Neural Networks when applied as anticipative mechanisms is greater than for Evolution Algorithms. When the penalized distance is applied this divergence in behaviors increases as the adaptive nature seems to improve. We remind the reader that our interest lies in the observation of the divergence of responses between the adaptive and anticipative applications of the algorithms. The results of the Artificial Neural Networks were obtained after a careful exploration of the control parameters of the algorithms, and are very good by themselves. We have not performed an equivalent work on the Evolution Algorithms side, so we do not give conclusions based on the magnitudes of the results. It is possible that careful exploration of the Evolution Algorithms control parameters could lead to results similar in magnitude to the ones found with Artificial Neural Networks.

The interesting fact is that we have found that Evolution Algorithms show a strong insensitivity to their application as adaptive or anticipative algorithms, and that in some cases their anticipative responses improve over the adaptive ones (i.e. in the Alippi algorithm).

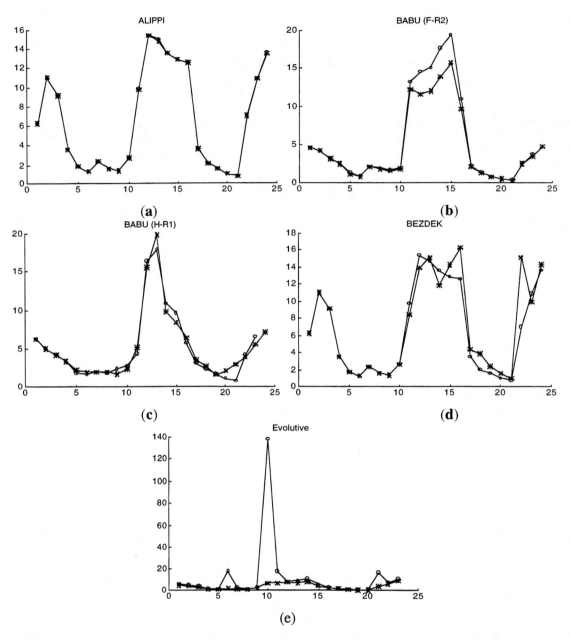

Figure 3. Distortion results on the experimental data sequence from the anticipatory (o) and adaptive (*) versions the Evolution Algorithms: (**a**) Allipi, (**b**) Babu with fuzzy fitness, (**c**) Babu with hard fitness, (**d**) Bezdek and (**e**) our Evolution Strategy

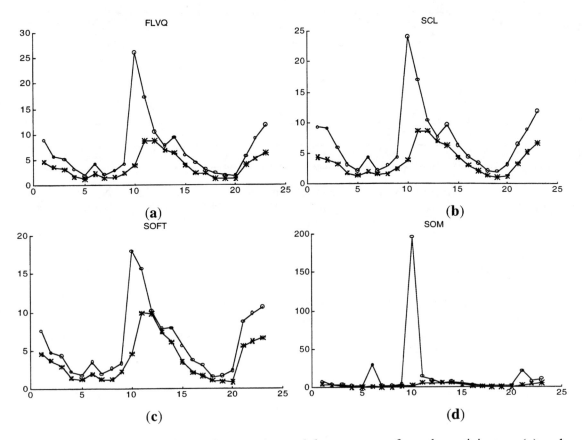

Figure 4. Distortion results on the experimental data sequence from the anticipatory (o) and adaptive (*) versions the Neural Network Architectures: (**a**) FLVQ, (**b**) Simple Competitive Learning, (**c**) Soft Competition, and (**d**) Self Organizing Map

Frequency of interaction	Algorithm	Adaptive		Anticipative	
		pop. size 10 offsprings 6	pop. size 16 offsprings 8	pop. size 10 offsprings 6	pop. size 16 offsprings 8
1 g	Alippi	160,7	150,3	161,4	158,2
	Bezdek	173,7	152,8	160,9	152,7
	Babu HR1	126	129,2	116,4	121,1
	" FR2	114,9	114,1	126,9	117,9
8 g	Alippi	155,2	157,9	159,4	157
	Bezdek	157,4	145,9	209,6	214,2
	Babu HR1	98,13	115,2	130,6	122,3
	" FR2	134,5	111,7	127,6	111,5
16 g	Bezdek	160,1	140,5	166,1	156,8
	Alippi	156,8	150,8	281,3	165,2
	Babu HR1	96,75	102	114,7	113,9
	" FR2	108,1	121,8	118,8	126,3

Table 1: Global distortion along the image sequence for the Evolution Algorithms in the anticipative and adaptive cases. Best individuals results.

Neural Network	Adaptive		Anticipative	
	Euclidean distance	Penalized distance	Euclidean distance	Penalized distance
FLVQ	85,52	76,59	156,5	330,3
Soft Competition	81,14	80,05	139,7	349,3
SCL	85,46	75,57	160,2	315,8
SOM	70,15	71,19	349,5	339,6
Evolution Strategy	88,43		276,99	

Table 2. Global distortion along the image sequence for the Artificial Neural Network in the anticipative and adaptive cases. Our Evolution Strategy is included due to its inherent adaptive design.

5. Conclusions

We have explored the response of various Evolution Algorithms to their application as anticipative or adaptive algorithms to a Non-Stationary problem, that of Non-Stationary Clustering. We have found that they perform similarly in both cases, even that they can show better response as anticipative systems.

This responses are obtained without performing any specific adjustment or introducing any computational trick that could allow some hidden forecasting. Neither we have employed any mean of preserving meaningful population diversity. Therefore we conclude that evolutionary algorithms posses an intrinsic anticipative nature.

Previous applications of Evolution Algorithms to open system problems such as control problems seem to fail due to their inability to cope with the time constraints imposed by the real time interaction with the environment. We hypothesize that this failure is due to their application as adaptive algorithms. The time required to update the internal state and generate the response falls outside the time requirements in most cases. However, we think that their application as anticipative mechanisms could both meet the time performance constraints. When applied in an anticipative way, the time to produce a response would be several orders of magnitude shorter than in the adaptive case. The updating of the internal states, the most computationally intensive task could be overlapped in time with the environment reaction time.

We expect that the anticipative application of Evolution Algorithms will open new venues for practical applications and for interesting research problems.

References

Ahalt S.C., Krishnamurthy A.K., Chen P., Melton D.E., (1990). "Competitive Learning Algorithms for Vector Quantization". In Neural Networks 3, pp. 277-290.

Aizawa A.N., Wah B.W. (1994). "Scheduling of genetic algorithms in a Noisy Environment". In Evolutionary Computation 2(2), pp. 97-122.

Alippi C., Cucchiara R., (1992). "Cluster partitioning in image analysis classification: a genetic algorithm approach". In Proc IEEE COMPEURO, The Hague, pp. 139-144.

Babu G. P., Murty N.M., (1994). "Clustering with evolution strategies". In Pattern Recognition 27(2), pp321-329.

Bäck T., Schwefel H.P., (1993). "An overview of Evolutionary Algorithms for parameter optimization". In Evolutionary Computation 1, pp. 1-24.

Bäck T., Schwefel H.P., (1996). "Evolutionary computation: an overview". In IEEE ICEC'96, pp. 20-29.

Bezdek J.C., Boggavaparu S., Hall L.O., Bensaid A., (1994). " Genetic Algorithm guided Clustering". In Proc 1st IEEE Conf. Evol. Comp., pp. 34-39.

Bezdek J.C., Pal N.R., (1995). "Two soft relatives of Learning Vector Quantization". In Neural Networks 8(5), pp. 729-743.

Chen H.H., Chen Y.S., Hsu W.H., (1995). "Low rate sequence image coding via vector quantization". In Signal Processing 26, pp. 265-283.

Chen O.T., Chen B.J., Zhang Z., (1994). "An adaptive vector quantization based on the gold-washing method for image compression". In IEEE trans cirsuits & systems for video techn. 4(2), pp. 143-156.

Fitzpatrick J.M., Grefenstette J.J., (1988). "Genetic algorithms in Noisy Environments". In Machine Learning 3, pp. 101-120.

Gersho A., Gray R.M., (1992). "Vector Quantization and signal compression". Kluwer Acad. Pub.

Goldberg M., Sun H., (1986). "Image sequence coding using vector quantization". In IEEE Trans. Communications 34, pp. 703-710.

González A.I., Graña M., D'Anjou A., Albizuri F.X., (1997). "A near real-time Evolution Strategy for adaptive Color Quantization of image sequences". In Joint Conference of Information Sciences, pp. 69-72.

Gonzalez A.I., Graña M., D'Anjou A., Albizuri F.X., Torrealdea F.J., (1998). "A comparison of experimental results with a Evolution Strategy and Competitive Neural Networks for near real-time Color Quantization of image sequences". Accepted in Applied Intelligence special issue on Evolutionary Learning.

Heckbert P., (1980). "Color image quantization for frame-buffer display". In Computer Graphics 16(3), pp.297-307.

Holland J.H. (1975). "Adaptation in Natural and Artificial Systems". The Univ. of Michigan Press.

Kohonen T., (1989). "Self Organization and Associative memory". Springer Verlag.

Wu X., (1991). "Efficient Statistical Computations for Optimal Color Quantization". In J. Arvo (ed). Graphics Gems II, Academic Press Professional, pp. 126-133.

Yair E., Zeger K., Gersho A., (1992). "Competitive Learning and Soft Competition for Vector Quantization". In IEEE Trans. Sign. Proc. 40(2), pp.294-308.

Temporal Logics for Analyzing the Behavior of Systems

Mireille LARNAC, Janine MAGNIER, Vincent CHAPURLAT
LGI2P
EMA/EERIE
Parc Scientifique Georges Besse
30000 - Nîmes
France
Phone : +33 (0) 4 66 38 70 26 - Fax : +33 (0) 4 66 38 70 74
email : {larnac,magnier,chapurla}@eerie.fr

Abstract

The task of anticipating the behavior of a system requires to have a model as a basis for reasoning. This statement raises several questions and problems, especially concerning the knowledge the user has got of the system, and his ability to describe the aspects of the system he wants to study. This point also involves that some models are available, and consequently, that some associated methods for predicting the future evolution exist. Furthermore, the description or modeling is sometimes reduced to an observation of the reactions of outputs when inputs are applied, or can be more detailed. Moreover, there exists a large range of ways to handle and represent the "time" factor. The choice can be determining for the possible reasoning processes. This paper discusses these aspects and presents a particular solution for systems described from a discrete time point of view.

Keywords

Modeling, Formal Verification, State Models, Interpreted Sequential Machine, Proof of Properties

1. Introduction

Anticipating the behavior of a system, should it be organizational, industrial, socioeconomic, etc., requires some means to represent either its structure or its dynamics. This obviously constitutes the role of models, which can furthermore be used for performing some analysis. The two major classes of verification and validation are based either on simulation (execution of the model for a given set of input values), or on formal proof of properties performed without any execution of the model.

This paper discusses the modeling problem, and the various aspects (pros and cons, feasibility, complexity) of simulation and formal verification. The second part presents a state of the art of modeling and verification methods which use some formal logic frameworks, the attention being stressed on the way the "time" factor is considered, represented and handled. Then, the last part shows that it is possible, with some classical modeling and formal verification approaches, to prove some properties which highlight anticipative behavior of the system.

CP437, *Computing Anticipatory Systems: CAYS--First International Conference*
edited by Daniel M. Dubois © 1998 The American Institute of Physics 1-56396-827-4/98/$15.00

2. Modeling of Systems

2.1. General considerations on modeling

Modeling a real system comes down to represent one or several of its aspects (structure, behavior, function, etc.) within a given formalism, from a specific point of view. Obviously, there does not exist any universal modeling framework which supports any kind of system representation; conversely, every modeling tool is dedicated to some particular application field. Indeed, the system to be modeled will not be represented within the same formalism whether its nature is technical, socioeconomic, organizational, ecological, and so on. Moreover, the modeling possibilities depend on the knowledge the user has got of the system: it can be only an input/output relationship which is based on an observation of the behavior of the system, or expressed as an evolution between states, or a description of the structure, or a functional and data-driven approach.

It follows that the choice of a model tremendously depends on the possibility for an human operator to describe the modeling concepts which are required for representing a given view of the system, and the expressiveness power of the model also depends on the description mechanisms such as hierarchical decomposition, interconnections, etc. The second criterion for choosing a model is the point of view the user wants to have of the system (structure, behavior, etc.). Furthermore, this process strongly depends on the goal which is pursued: this comes down to answering at least the two following questions: what is this modeling for (is this for making some analysis, and in this case, is it simulation or formal analysis, is this for documentation purpose, etc.) and what is the context (in which life cycle phase)? Last, a very important feature of modeling concerns the «time» factor. Indeed, a key point is to determine if the modeling task requires some time expression, and if needed, what representation of this time factor will be handled. It can be implicit or explicit, and viewed in a continuous, discrete or hybrid manner.

2.2. Modeling of anticipative systems

As stated by Robert Rosen (1985), an anticipatory system is «a system containing a predictive model of itself and/or of its environment in view of computing its present state as a function of the prediction of the model».

Then considering the field of anticipative (or anticipatory) systems, it appears that it is necessary to model the system from a behavioral point of view (since the dynamic evolution of the system is the foundation of the analysis), and if follows that the time factor must appear.

The technique for modeling the system can start at a very low level of knowledge, when the inner structure and transfer functions of the system are totally or partially unknown. In this case, a black box approach for modeling will be chosen, since it is only possible to describe an input/output relationship. The difficulty in this class of approaches lies in selecting the accurate data (inputs and outputs) of the real system to consider into the model. Indeed, the environment of systems is very often constituted of several factors, and it is usually worthless to consider all of them. The two main approaches in this area are the well-known Neural Networks (Hay Kin, 1994), and fuzzy logic-based models (Zadeh, 1965; Jang, 1993; Chiu, 1994)

These two classes of models are based on a first step of learning input/output relationships and therefore automatically estimating and configuring their inner parameters. Here, time remains implicit.

On the other hand, when the user has got a full knowledge of the system, it may be possible to establish an exact model of any point of view.

We stress our attention here on state models which, as stated before, require that it is possible for an human user to identify the states of the system, and the conditions which make it evolve. The basement of all the state models is the Finite State Machine (FSM) model in which the inputs and outputs are of Boolean type, and no environmental influence is considered (Kohavi, 1978; Hartmanis and Stearns, 1966). Several versions of this model exist, especially when considering the evolution rules; indeed, synchronous, asynchronous or event-driven models have been defined, and consequently, their formal basis and properties differ. Nevertheless, the semantic interpretation considers that the temporal evolution is implicitly contained into state (or transition) sequences from the initial state.

Several extensions of the FSM model have been defined, either to refine the temporal behavior by adding duration on transitions and/or states (Alur and Dill, 1990, 1994), or by taking into account and representing the influence of the environment (Cheng and Krishnakumar, 1993; Ostroff, 1989; Larnac et al, 1997). Then, a model of the system being established, one of the major tasks on may wish to perform consists in making some analysis of its behavior.

3. Analysis of the behavior of systems

3.1. The three ways for analyzing

In order to better understand the behavior of a system, three possibilities may be viewed.

The first one, which is the simplest, can be applied only if the system already exists; indeed, it consists in running tests on the real system by configuring it, if possible, into a given state, letting it evolve and then examining the characteristics which are considered as «outputs». Obviously, this approach is very often not applicable, and does not provide any proof that the same experience will give the same result another time. Moreover, the analysis is not supported by any tool, so it is up to the user to make the interpretation of the results.

The second, well-known manner, is based on simulation. This requires to have a model which expresses the behavior of the system. The technique consists in executing the model for a given input vector (i.e. a value is assigned to each input) and examining the obtained output values. A very important point worth stressing is that simulation is definitely not the analysis method by itself, but it constitutes a necessary step for providing data to be studied; in other words, the analysis consists in examining the simulation results. Currently, in most cases, this task is performed manually by the user who just «looks if the simulation results are satisfying or not». Nevertheless, some applications involve an automatic interpretation of the results, based on the use of some formal tool.

To summarize, the pros of simulation are that it is a widely spread method, so several efficient industrial tools, provided with user-friendly interfaces exist on the market. The cons concern the fact that the analysis is not embedded into the simulation principle, that in the case of complex systems, it is almost impossible to test all the possibilities, so the coverage is incomplete. Further, in order to test a given situation, it is necessary to put into play some input generation mechanism, and this task remains of huge difficulty.

Last, the third way to analyze the behavior of a system is to use formal tools which permit to prove properties of the system. This approach requires a logic framework (a formal system and an associated interpretation) and a demonstration tool for first of all describing the behavior of the system into a set a formulae; this constitutes the axiomatic of the system. Then, in order to study a property, it is necessary to express it into a formula which will be proven of being a theorem. It follows that the analysis method is based on reasoning and does not involve any execution of the model.

The use of such methods mainly permits to check that the system fulfills a given property in all cases, or to express into a single formula all the possibilities to obtain a given criterion (e.g. evolution into a given state, production of an output, etc.). The advantages of formal analysis is that the proof of a property is sure (an exhaustive search has been carried out) since it is based on the use of formal tools. On the other hand, only few applications were developed, and the manipulation of this theory still remains delicate.

3.2. Formal verification methods for anticipation?

The question addressed in this paper is to study if formal methods can be applied for describing and analyzing anticipative systems, and hence provide a formal basis for modeling and study.

The requirements for this task is that the formal model is able to handle the time factor, preferably in an explicit way. Moreover, it may be necessary to consider both future and past times. The following of this article explains how a finite state model and its associated verification tool can be used for studying some classes of anticipative systems.

4. Modeling and verification of Finite State Machines

4.1. Definition

The Finite State Machine (FSM) model is classically defined by the 5-tuple (Hartmanis and Stearns, 1966; Kohavi, 1978):

$M = < S, I, O, \delta, \lambda >$ where:

- S is a finite, non empty set of states
- I is a finite, non empty set of inputs
- O is a finite set of outputs
- δ is the transition (next state) function: $\quad \delta : I \times S \to S$
- λ is the transition output function: $\quad \lambda : I \times S \to O$

We denote by #S, #I and #O the cardinalities of S, I and O respectively.

We consider only deterministic machines, which obey the following rules:

- each state has one and only one following state for each relevant input
- no two distinct inputs can be applied simultaneously
- no two distinct outputs can be output simultaneously

Also, we consider machines which are completely specified. This means that for all states, the next state and output are specified for all inputs. It is possible to study machines which are incompletely specified within the framework of our model, by defining a new type of variable for representing the unspecified inputs or outputs. However, we concentrate on completely specified machines in this article.

4.2. Expression in Temporal Logic

We define the following:

Three sets **S**, **X** and **Z** each representing propositions of the same type.

- **S** is the set of state type propositions:

    $s_i \in S$, $\forall s_i \in S$, $i = 0, ..., \#S - 1$, s_i is TRUE when the state of M is s_i.

- **X** is the set of input type propositions:

    $x_j \in X$, $\forall i_j \in I$, $j = 0, ..., \#I - 1$, x_j is TRUE when the present input of M is i_j.

- **Z** is the set of output type propositions:

    $z_k \in Z$, $\forall o_k \in O$, $k = 0, ..., \#O - 1$, z_k is TRUE when the present output of M is o_k.

These definitions allow us to describe the temporal evolution of M (by expressing the behavior of the transitions of M) into the following notation (based on the DUX temporal logic (Gabbay et al, 1980; Manna and Pnueli, 1982)), called Elementary Valid Formula (EVF):

Let us suppose that we have $\delta(s_i, i_j) = s_k$ and $\lambda(s_i, i_j) = o_l$; it follows

$$EVF ::= \Box(s_i \wedge x_j \supset O s_k \wedge z_l)$$

531

whose interpretation is the following: « it is always true ($\Box$ operator) that if s_i is the current state (and therefore s_i is true) and i_j is the current input (x_j is true), then the next state ($\bigcirc$ operator) will be s_k (s_k will be true), and the current output is o_l (z_l becomes true) ».

It follows that the set of all the EVF's (each of which expresses the existence of a transition of the FSM) provides an equivalent representation of the behavior of the FSM model. This statement is true if we also take into account the set of formulae which represent the determinism constraints. The first set contains the state determinism concept, which says that at a given time step, there is one and only one current state. This determinism formula can be written, using the DUX formalism:

$$\text{DF1} ::= \Box[s_i \supset \neg\ s_j]\ \forall\ j \neq i,\ i, j \in \{0,...,\#S - 1\}$$

Similarly, DF2 and DF3 express that at a given time step, the machine cannot have two different inputs, and cannot produce two different outputs:

$$\text{DF2} ::= \Box[x_i \supset \neg\ x_j]\ \forall\ j \neq i,\ i, j \in \{0,...,\#X - 1\}$$
$$\text{DF3} ::= \Box[z_i \supset \neg\ z_j]\ \forall\ j \neq i,\ i, j \in \{0,...,\#Z - 1\}$$

Within the framework of a verification process, it is often necessary to consider time intervals. This leads to the definition of a state, input and output sequences, which are noted, respectively:

$$s_i^n ::= s_{i1} \wedge \bigcirc s_{i2} \wedge \bigcirc^2 s_{i3} \wedge ... \wedge \bigcirc^{n-1} s_{in}$$
$$x_j^n ::= x_{j1} \wedge \bigcirc x_{j2} \wedge \bigcirc^2 x_{j3} \wedge ... \wedge \bigcirc^{n-1} x_{jn}$$
$$z_k^n ::= z_{k1} \wedge \bigcirc z_{k2} \wedge \bigcirc^2 z_{k3} \wedge ... \wedge \bigcirc^{n-1} z_{kn}$$

Then, in order to provide the user with a more global view of the system evolution, first the concept of Temporal Event (Et) which represents the possible effects of the machine functioning has been defined, and then all the conditions which lead to obtaining a given temporal event are gathered into one single formula, called Unified Valid Formula (UVF). A temporal event will be a future state (Et = $\bigcirc s_i$), a future state within n time steps (Et = $\bigcirc^n s_i$), a state sequence (Et = s_i^n), a present output (Et = z_k), a n-future output (Et = $\bigcirc^n z_k$), or an output sequence (Et = z_k^n). The Unified Valid Formula associated with a temporal event Et is thus:

$$\text{UVF(Et)} ::= \bigvee_{(p,q)\ /\ s_p \wedge x_q^n \supset \text{Et}} (s_p \wedge x_q^n)$$

It appears that the calculation of an UVF only consists in manipulating the set of EVF's.

4.3. Verification of properties

Let us come back to the goal of this work; it addresses the formal verification of properties of systems which are represented thanks to a FSM model. The first question to answer is: « what kind of properties can be proven? ». To start with, the structure of the FSM can be exploited, and properties of some states can be of great interest. For instance, it is worth knowing if two states are equivalent, or if a state is a source or a sink. More sophisticated properties consist in establishing the conditions (on input sequences) to make a state being a « functional » sink, even though it is not a structural one, or to generate input sequences to resynchronize the machine into a given state.

In most cases, the state evolution of the machine is not available, and the only means for the user to get some information on the machine evolution is to examine the outputs. So the analysis process of output sequences is very important, and a tool for generating input sequences in order to obtain outputs, or to distinguish internal states must be provided.

Last, it seems very important to be able to formally establish the influence of a current factor (input or state) on the future evolution. This relates to the « sensitivity » of the future with respect to a present situation or decision.

The verification method is based on two approaches:

- in some cases (for some properties), it is sufficient to analyze the EVF's and UVF's (either search if a given formula exists, or what its form is). For example, if s_p is a sink state, it means that all the transitions which leave s_p go back to this state. It follows that all the EVF's which contain s_p in their left part must be of the form: EVF $::= \Box(s_p \wedge x_j \supset Os_p \wedge z_l)$, for all x_j. Similarly, if s_p is a source state, it means that if there exists some transitions whose destination is s_p, they come from s_p. The consequence is that either UVF(Os_p) is empty, or that it has the following form: UVF(Os_p) = $\vee (s_p \wedge x_j)$.

- Unfortunately, this approach of verification based on the study of EVF's and UVF's is not sufficient for analyzing the influence of the present on the future. This is the reason why a formal tool for analyzing the sensitivity has been defined: the Temporal Boolean Difference.

4.4. The Temporal Boolean Difference

The Temporal Boolean Difference (TBD) is the extension of the classical Boolean Difference (Kohavi, 1978) defined on propositional logic, to temporal logic, especially the DUX system. The formal definition of TBD of $f(x_1, ..., x_n)$ with respect to a variable x_i is the following:

$$\frac{\partial f(x_1,...,x_n)}{\partial x_i} = f(x_1,...,x_i = \text{False},...,x_n) \oplus f(x_1,...,x_i = \text{True},...,x_n)$$

The important point is that there is a strong analogy with the Boolean Difference of Boolean functions, but the fundamental differences are that the formulae are expressed in Temporal Logic, and that the variables which are manipulated are typed (states, inputs, outputs) and non independent (because of the determinism properties).

For the verification of properties of a FSM model, the TBD is applied on an UVF for obtaining a Temporal Event which concerns future states or outputs (cf. the definition of Temporal Event), with respect either to a current state or to an input. The formula is called the Derived Valid Formula (DVF) and is the following:

$$DVF(Et,q) = \frac{\partial UVF(Et)}{\partial q} = UVF(Et_{|q}) \oplus UVF(Et_{|\neg q})$$

The result of the calculation of DVF(Et, q) can be:

- False. This means that UVF(Et) is independent of q; in other words, the fact that q changes value has no influence on the fact that Et will occur or not.

- not False. In this case, we obtain a Temporal Logic formula which expresses the sensitivity of UVF(Et) to changes in q, i.e. the conditions for UVF(Et) to pass from True to False (or conversely) when q changes value.

The applications of TBD are various. It permits to generate input sequences for resynchronizing the machine into a given « initial » state, or for distinguishing the inner states (which are unknown) through the generation of distinct output sequences. Moreover, a very wide field of application is the study of the impact of a decision (or a current event) on the future evolution of the system. Further, even though the user has got no possibility to change the present, he knows all the conditions which, when made True, make the system evolve into a given way. It is then up to him to choose his strategy for modifying some parameters and then determine what he wants to get into the future.

In conclusion, a formal method has been defined which provides the user with an equivalent symbolic representation of the behavior of the Finite State Machine model, and then with a tool which supports proof of properties and formal analysis. In order to do this, it has been necessary to define the concept of Temporal Boolean Difference for evaluating the sensitivity of the evolution of a system with respect to some variable change. The details of the modeling and verification approach, as well as the demonstrations of all the theorems can be found in (Magnier, 1990).

The limitations of this approach are linked to the weak expressiveness of the FSM model, which only handles Boolean data, and which needs to express any data influencing the system through inputs or states. It follows that the number of states or transitions tends to increase exponentially as soon as new data have to be taken into account. This is the reason why we have defined an extension of the FSM, called the Interpreted Sequential Machine model.

5. The Interpreted Sequential Machine Model

5.1. Definition

The Interpreted Sequential Machine (ISM) is a state model for the formal representation of the behavior of discrete time systems. It is an extension of the classical Sequential Machine ; the underlying concepts were adapted from the Extended Finite State Machine (EFSM), defined in (Cheng and Krishnakumar, 1993), for functional testing purpose.

The main characteristics of this model are the following:

- the core of the model is a state system (extension of state machine)
- the inputs and outputs can be of any type (Boolean, integer, real, data, event, etc.)
- the data which constitute the environment of the system and which influence the functioning of the sequential part of the system are represented separately

As any sequential model, the state diagram, called Control Graph, is composed of an alternation of states and transitions. Informally, a transition between two states is activated if first an event appears on the inputs, and then if some conditions on the environment (data) are fulfilled; then the effects of this transition firing appear on the outputs and on the data of the environment which all change value.

Structurally, the ISM model owns inputs and outputs, and is made up of two parts (Figure 1):

- the Control Part (CP) contains the Control Graph and some necessary interpreters
- the Data Part (DP) is made up of the set of data which represent the environment of the system, and of the operations on these data

It is possible to partition the set I of inputs into two disjoint sets: the set I_C of the Control Part inputs and the set I_D of the Data Part inputs. Similarly, the set O of outputs is split up into Control Part outputs (O_C) and Data Part outputs (O_D).

The formal model of the ISM is thus the following: ISM = < I, O, CP , DP > with:

- CP = <I_C, CII, **E**, **F**, CG, U, COI, **Z**, O_C> where:
  - I_C is the set of the Control Part inputs
  - **E** is the set of propositional input variables of the Control Graph
  - CII is the Control Input Interpreter. Its role is to evaluate some conditions on the Control Part inputs of I_C, and therefore to give some propositional (Boolean) values to the elements of **E**
  - **F** is the set of propositional enabling variables
  - CG is the Control Graph: CG = <**S**, **E**, **F**, **Z**, U, δ, λ, β> where:
    - **S** is the set of propositional symbolic state variables
    - δ is the propositional transition function (next state): $\delta : \mathbf{S} \times \mathbf{E} \times \mathbf{F} \to \mathbf{S}$
    - λ is the propositional output function: $\lambda : \mathbf{S} \times \mathbf{E} \times \mathbf{F} \to \mathbf{Z}$
    - β is the propositional updating function: $\beta : \mathbf{S} \times \mathbf{E} \times \mathbf{F} \to U$
    A transition t of CG is defined as follows: $t : (s_i, e_j, f_k) \to (s_l, z_m, u_n)$ where:
    $$(s_i, s_l) \in \mathbf{S}^2, e_j \in \mathbf{E}, f_k \in \mathbf{F}, z_m \in \mathbf{Z}, u_n \in U$$
    $$s_l \in \delta(s_i, e_j, f_k), z_m \in \lambda(s_i, e_j, f_k), u_n \in \beta(s_i, e_j, f_k)$$

534

- **U** is the set of propositional updating variables (assigned when β is evaluated)
- **Z** is the set of propositional output variables of the Control Graph (assigned when λ is evaluated)
- COI is the Control Output Interpreter. Its role is to give values to the Control Part outputs from the values of the propositional variables of the variables of **Z**.
- O_C is the set of the Control Part outputs
- DP = <I_D, D, P, EFI, UFI, **F**, **U**, O_D> where:
  - I_D is the set of the Data Part inputs
  - D is the set of internal variables
  - P is the set of parameters (fixed characteristics of the system)
  - EFI is the Enabling Function Interpreter. Its role is to evaluate some conditions on the Data Part inputs, on the internal variables and on the parameters and therefore to give some propositional (Boolean) values to the elements of **F**
  - **F** is the set of propositional enabling variables
  - **U** is the set of propositional updating variables
  - UFI is the Updating Function Interpreter. It contains all the functions that are used for calculating the new values of the Data Part outputs and of the internal variables of D. these updating functions are activated by the variables of **U**
  - O_D is the set of the Data Part outputs

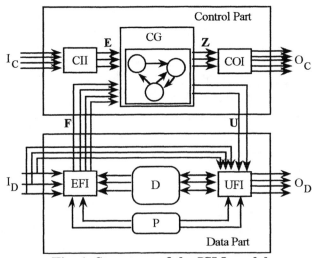

Fig. 1. Structure of the ISM model

5.2. Dynamic evolution of an ISM model

The behavior of an ISM model is described by the dynamic evolution both of the Control Graph, and of the internal variables of D.

The behavior of CG is expressed by the sequence of fired transitions.

The interpretation of a transition t (Figure 2), where t: (s_i, e_j, f_k) → (s_l, z_m, u_n) is the following: if s_i is true (s_i denotes the current state of the Control Graph), then if e_j is true (the value of the inputs of CP verify the conditions which make e_j be true), then, if f_k is true (the value of the inputs of DP, of the internal variables in D and of the parameters of P verify the conditions which make f_k be true), then:

- at the same time, z_m and u_n become true until the next transition firing on CG,
  - the functions associated with z_m in COI update the Control Part outputs

535

- the functions associated with $\mathbf{u_n}$ in UFI update both the internal variables and the Data Part outputs
- the next state will be $\mathbf{s_l}$ ($\mathbf{s_l}$ will be true, while $\mathbf{s_i}$ will become false).

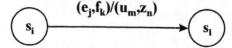

Fig. 2. A transition of the Control Graph

5.3. Verification of properties

Given an ISM model of a real system, it is possible to carry out formal verification by proof of properties.

Informally, the properties which can be proven highlight some characteristics of the system such as:

- Does there exist source states, sink states ? What are the conditions to make a source a "functional" sink ?
- What are the conditions to reach a given state in the future ? How many time steps are necessary ?
- What happened in the past which lead the system into the present state ?
- What are the conditions which will make the system evolve from one state to another ?
- What is the influence of a change of value of a factor on the evolution of the system ?

Similarly to the case of the FSM model, the verification process consists in translating the behavior of the model into a set of temporal logic formulae. This provides the basis for proving formulae (which express properties) to be theorems or not. Here, the formulae refer to the behavior of the Control Graph. Indeed, the variables which are involved into the structure of CG are propositional, and the dynamic evolution of this graph is easily translated into DUX formulae. For instance, the behavior of a transition t: $(\mathbf{s_i}, \mathbf{e_j}, \mathbf{f_k}) \rightarrow (\mathbf{s_l}, \mathbf{z_m}, \mathbf{u_n})$ is the following EVF : EVF ::= $\Box(\mathbf{s_p} \wedge \mathbf{e_j} \wedge \mathbf{f_k} \supset O\mathbf{s_l} \wedge \mathbf{z_m} \wedge \mathbf{u_n})$, and the process for constructing UVF's is still valid. The main difference is that the inputs which were supposed to be independent in the case of the FSM are now linked, since they represent non disjoint conditions on the real inputs. It is then necessary to perform some variable changes in order to fulfill this requirement.

Last, the application of the Temporal Boolean Difference has been defined within the framework of ISM models, and this allows us to perform verifications on the behavior of the Control Graph.

6. Conclusion

Choosing a model for managing the complexity of a real system strongly depends on the knowledge the user has got and on the kind of verification he wants to carry out. We have presented in this paper a formal proof process of the Sequential Machine, and then an extension of this model, called the Interpreted Sequential Machine. This state model permits to represent the behavior of a system from state evolution point of view, taking into account the interactions with the environment. Furthermore, the ISM handles any kind of data; this makes the model more accurate for representing real systems. A formal proof process is associated with the model, allowing us to study the validity of properties of the system.

Within the framework of anticipative systems, this class of modeling and verification methods may be used for aiding in designing models of systems since they permit to better understand their behavior and get the proof of their temporal evolution (in the past as well as in the future).

Bibliography

Alur, R. and Dill, D.L. (1990); Automata for Modeling Real-Time Systems; Proceedings of ICALP'90, LNCS 443, Springer Verlag, 322-335

Alur, R. and Dill, D.L. (1994); A Theory of Timed Automata for modeling; Theoretical Computer Science 126, 183-235

Cheng, K.T. and Krishnakumar, A.S. (1993); Automatic Functional test Generation using the Extended Finite State Machine Model; 30th ACM/IEEE Design Automation Conference

Chiu, S.L. (1994); Fuzzy Model Identification based on Cluster Estimation; Journal of Intelligent and Fuzzy Systems, vol. 2, 267-278

Gabbay, D., Pnueli, A., Shelah, S., Stavi, J. (1980); On the temporal analysis of fairness; 7th ACM symposium on Principles of Programming Languages

Hartmanis, J. and Stearns, R.E. (1966); Algebraic Structure Theory of Sequential Machines; Prentice Hall, Englewood Cliffs, N.J.

Hay Kin, S. (1994); Neural Networks: a Comprehensive Foundation; Mac Millan College Publishing Company, New York

Kohavi, Z. (1978); Switching and Finite Automata Theory; Tata McGraw Hill, Computer Science Series

Jang, J.S.R. (1993); ANFIS: Adaptative Network-Based Fuzzy Inference Systems; IEEE Transactions on Systems, Man and Cybernetics, vol 23, N° 3, May-June 1993

Larnac, M., Chapurlat, V., Magnier, J. and Chenot, B. (1997); Formal Representation and Proof of the Interpreted Sequential Machine; Eurocast'97, LNCS 1333, Springer Verlag, 95-107

Magnier, J. (1990); Représentation symbolique et vérification formelle de machines séquentielles, Thèse d'Etat, University of Montpellier II, France

Manna, Z. and Pnueli, A (1982); How to cook a temporal proof system for your pet language; report No STAN-CS-82-954; Department of Computer Science, Stanford University

Ostroff, J.S. (1989) ; Temporal Logic for Real-Time Systems, Research Studies Press Ltd.

Rosen, R. (1985); Anticipatory Systems; Pergamon Press

Zadeh, L.A. (1965); Fuzzy Sets; Information and Control 8, 338-353

Kolmogorov Systems: Internal Time, Irreversibility and Cryptographic Applications

Josef Scharinger

Institute of Systems Sciences
Johannes Kepler University
Altenbergerstr. 69, 4040-Linz, Austria
e-mail:js@cast.uni-linz.ac.at

Abstract. This contribution describes how conservative time reversible Kolmogorov systems which are some of the most unstable systems currently known in chaos theory allow for the construction of equivalent irreversible dissipative systems by means of a necessarily non-unitary but invertible transformation. This transformation is derived as an operator based on the notion of internal time of the system under consideration which relates entropy to age and essentially implements a convolution operation in an adequately chosen set of basis functions.

As an application we mention the field of cryptography. It will be shown how Kolmogorov systems can be utilized to implement highly efficient computationally secure ciphers for bulk encryption applications. The highly unstable dynamics associated with Kolmogorov systems is thereby taken as a chaotic nonlinear permutation operator, while substitution is implemented using an adaption of a standard shift register based pseudo random number generator.

Keywords: Kolmogorov systems, dissipative systems, irreversibility, internal time, cryptography.

1 Introduction

In chapter 4 of his exciting book *Anticipatory Systems* (Rosen 1985), R. Rosen elaborates in considerable detail on the quality of time. It admits a multitude of different kinds of encoding, which differ vastly from one another. We can consider reversible Hamiltonian time, irreversible dynamical time, thermodynamic time, probabilistic time and sequential or logic time. Each of these capture some particular aspects of our time sense, at least as these aspects are manifested in particular kinds of situations. The conclusion drawn by R. Rosen is that, while certain formal relations could be established between these various kinds of time, none of them could be reduced to any of the others.

This contribution takes a somewhat different standpoint. Using the example of Kolmogorov systems (Arnold & Avez 1968, Shields 1973, Moser 1973) which are some of the most unstable systems currently known in chaos theory, fulfilling special conditions like ergodicity and the mixing property, these reversible conservative systems exhibit additional important properties which allow for the construction of equivalent irreversible dissipative systems by means of a necessarily non-unitary but invertible transformation. This transformation is derived as an operator based on the notion of internal time of the system under consideration which relates entropy to age and essentially implements a convolution operation in an adequately chosen set of basis functions.

As an application we mention the field of cryptography. It will be shown how Kolmogorov systems can be utilized to implement highly efficient computationally secure ciphers for bulk encryption applications. The highly unstable dynamics associated with Kolmogorov systems is thereby taken as a chaotic nonlinear permutation operator, while substitution is implemented using an adaption of a standard shift register based pseudo random number generator. In this way we obtain secure product ciphers which are firmly

CP437, *Computing Anticipatory Systems: CAYS--First International Conference*
edited by Daniel M. Dubois © 1998 The American Institute of Physics 1-56396-827-4/98/$15.00

grounded on systems theoretic concepts, offering many features that should make them superior to contemporary bulk encryption systems, especially in applications where an appropriate combination of efficiency and security plays a major role.

2 Equivalence of conservative Kolmogorov systems and irreversible dissipative systems

With the Second Law of thermodynamics statistics has established itself as an adequate formalism for the description of physical processes which is applied as a supplement to the laws of classical mechanics whenever a system is so complex that an analysis is not possible without the introduction of some sort of approximation. The irreversible phenomena which are found at the macroscopic level in statistical mechanics and thermodynamics are in this interpretation just a consequence of coarse graining and, therefore, no contradiction to the assumption that, on the microscopic level of individual particles, all laws are reversible and time-symmetric. Irreversible behavior is just the result of approximations based on the ignorance of the observer who is unable to precisely capture all individual elements of a complex system.

A fundamentally different way for the generation of time-directed behavior was proposed by I. Prigogine (Nicholis & Prigogine 1989, Prigogine & Stengers 1990, Prigogine 1980, Prigogine 1991) who did most of his research in Brussels, Belgium, and was awarded the Nobel price in chemistry in 1979. In certain cases dissipative behavior can be derived directly from a conservative systems dynamics without introducing any sort of approximation under the precondition that the system under consideration is a sufficiently unstable (chaotic, ergodic, mixing) one. Then it may be possible to reconcile reversible and irreversible systems by establishing an equivalence between conservative and dissipative systems accomplished via a necessarily non-unitary but invertible transformation. Therefore, time-directed phenomena are not generated artificially, but are based on an exact systems analysis.

In Prigogines interpretation the Second Law implies that irreversible processes introduce an arrow of time whereby a positive direction of time is linked to an increase of entropy. According to this formulation this leads to the existence of a very specific function (Ljapunow function) which can only increase (decrease) for an isolated system. Therefore, when interpreted as a dynamical principle, the Second Law has a number of important consequences like irreversibility, arrow of time, instability, non locality and nondeterminism.

2.1 Kolmogorov Systems

Kolmogorov systems are chaotic systems of extraordinary instability. First of all, they fulfill the ergodic hypothesis which roughly means (according to Maxwell) that any trajectory approaches any point on the energy surface arbitrarily close. Next, they are mixing. Let ν be a normalized measure for subsets A of the systems phase space Ω:

$$\nu(A) = \frac{\int_A d\mu}{\int_\Omega d\mu}.$$ (1)

Then the corresponding system is mixing (Misra 1978), iff the following holds for arbitrary sets A and B:

$$\nu(A_t \cap B) \xrightarrow{t \to \pm\infty} \nu(A)\nu(B).$$ (2)

Intuitively this expresses the fact that during the systems evolution in time any subset of the phase space Ω is spread out uniformly over the whole phase space.

Finally, if $\mathcal{B}$ denotes the set of all measurable subspaces of Ω and T_t defines the phase transition function, there exists a generating partition $\mathcal{F}_0 \subseteq \mathcal{B}$ fulfilling:

$$T_t\mathcal{F}_0 = \mathcal{F}_t \subseteq T_s\mathcal{F}_0 = \mathcal{F}_s, t \leq s, \tag{3}$$

$$\bigvee_{\lambda=-\infty}^{+\infty} \mathcal{F}_\lambda = \mathcal{B} \tag{4}$$

$$\bigcap_{\lambda=-\infty}^{+\infty} \mathcal{F}_\lambda = \mathcal{F}_{-\infty} \text{ only contains sets of measure 0.} \tag{5}$$

To study the various aspects of Kolmogorov systems, it is useful to consider the following realization. Let the unit square $\mathbb{E}$ denote the phase space and π the generating partition. Then the following interpretation of the transformation T_π specified by $\pi = (p_1, p_2, \ldots, p_k)$ is valid. For points $(x, y) \in \mathbb{E}$, the point $(x', y') = T_\pi(x, y)$ is defined as:

$$x' = \frac{x - F_s}{p_s}, \tag{6}$$

$$y' = F_s + y * p_s. \tag{7}$$

In this equation we use $F_s = \sum_{i=1}^{s-1} p_i$ and s is the index of the vertical rectangle containing the point (x, y).

Obviously T_π is invertible since the unit square is mapped onto itself. Therefore the inverse transformation T_π^{-1} can be defined which derives the original values $(x, y) = T_\pi^{-1}(x', y')$ from the transformed ones by simply exchanging the coordinates in x and y direction.

$$x = G_s + x' * p_s, \tag{8}$$

$$y = \frac{y' - G_s}{p_s}. \tag{9}$$

In that equation we use $G_s = \sum_{i=1}^{s-1} p_i$, but this time s denotes the index of the horizontal rectangle containing the point (x', y').

Based on the transformation T_π which maps points $(x, y) \in \mathbb{E}$ onto other points in $\mathbb{E}$, a transformation B_π operating on functions $f \in L^2(\mathbb{E})$ can be defined.

$$B_\pi(f)(u, v) = f(T_\pi^{-1}(u, v)), \tag{10}$$

$$B_\pi^{-1}(f)(u, v) = f(T_\pi(u, v)). \tag{11}$$

It is plain to see that the following linear relations hold:

$$B_\pi(af)(u, v) = aB_\pi(f)(u, v) = af(T_\pi^{-1}(u, v)), \tag{12}$$

$$B_\pi^{-1}(af)(u, v) = aB_\pi^{-1}(f)(u, v) = af(T_\pi(u, v)), \tag{13}$$

and

$$B_\pi(f + g)(u, v) = B_\pi(f)(u, v) + B_\pi(g)(u, v) \tag{14}$$
$$= f(T_\pi^{-1}(u, v)) + g(T_\pi^{-1}(u, v)),$$

$$B_\pi^{-1}(f + g)(u, v) = B_\pi^{-1}(f)(u, v) + B_\pi^{-1}(g)(u, v) \tag{15}$$
$$= f(T_\pi(u, v)) + g(T_\pi(u, v)).$$

541

The transformation T_π gets particularly simple if we focus on partitions π where all p_i are equal. Then every p_i can be expressed in the form $\frac{1}{p}$ ($p \in \mathbb{N}$) and we obtain an implementation of the corresponding Kolmogorov system as a p-adic shift as follows:

Let $x, y \in \mathbb{IU}$ be given by their p-adic representations ($x_i, y_i \in \{0, 1, \ldots, p-1\}$):

$$x = \sum_{i=1}^{\infty} x_i p^{-i} \text{ and } y = \sum_{i=1}^{\infty} y_i p^{-i} \tag{16}$$

Then the point $(x, y) \in \mathbb{E}$ becomes a doubly infinite sequence $(u_i)_{i=-\infty}^{\infty}$ with $u_i \in \{0, 1, \ldots, p-1\}$ and

$$u_i = \begin{cases} x_{1-i}, & \text{if } i \le 0, \\ y_i, & \text{if } i > 0 \end{cases} \tag{17}$$

This correspondence has to be read like

$$\ldots u_{-2} u_{-1} u_0 \cdot u_1 u_2 u_3 \ldots \equiv \ldots x_3 x_2 x_1 \cdot y_1 y_2 y_3 \ldots . \tag{18}$$

Now $T_{(\frac{1}{p}, \ldots, \frac{1}{p})}(u)$ can be realized through a simple shift to the right by one position:

$$T_{(\frac{1}{p}, \ldots, \frac{1}{p})}(u) = \ldots u_{-3} u_{-2} u_{-1} \cdot u_0 u_1 u_2 \ldots \tag{19}$$
$$\equiv \ldots x_4 x_3 x_2 \cdot x_1 y_1 y_2 \ldots,$$

and the inverse transformation $T^{-1}_{(\frac{1}{p}, \ldots, \frac{1}{p})}(u)$ can be implemented by a single shift to the left:

$$T^{-1}_{(\frac{1}{p}, \ldots, \frac{1}{p})}(u) = \ldots u_{-1} u_0 u_1 \cdot u_2 u_3 u_4 \ldots \tag{20}$$
$$\equiv \ldots x_2 x_1 y_1 \cdot y_2 y_3 y_4 \ldots .$$

Since $T_{(\frac{1}{p}, \ldots, \frac{1}{p})}(u)$ is uniquely specified by the value p, we will use the shorthand notations T_p and B_p. Finally, if $\to n$ denotes a shift to the right and $\leftarrow n$ a shift to the left by n positions, this results in:

$$T_p(u) = u \to 1, \tag{21}$$
$$T_p^{-1}(u) = u \leftarrow 1, \tag{22}$$
$$B_p(f)(u) = f(T_p^{-1}(u)) = f(u \leftarrow 1), \tag{23}$$
$$B_p^{-1}(f)(u) = f(T_p(u)) = f(u \to 1). \tag{24}$$

2.2 Chrestenson Functions

Chrestenson functions (Chrestenson 1955) are intimately related with the p-adic group. Therefore it seems quite natural to utilize them for the analysis of Kolmogorov systems possessing a realization as a p-adic shift. We will introduce them as complete discrete products of finite cyclic groups similar to Vilenkin systems (Vilenkin 1947).

As introduced in the preceding section, let two points x and y in $\mathbb{IU}$ be specified by their p-adic representations:

$$x = \sum_{i=0}^{\infty} x_i p^{-i-1}, \, y = \sum_{i=0}^{\infty} y_i p^{-i-1}, \tag{25}$$

with $x_i, y_i \in \{0, 1, \ldots, p-1\}$. Additionally there are two indices $h, v \in \mathbb{N}_0$ with $N(h)+1$ and $N(v)+1$ relevant digits, respectively:

$$h = \sum_{i=0}^{N(h)} h_i p^i, \, v = \sum_{i=0}^{N(v)} v_i p^i. \tag{26}$$

Then the values of the one dimensional Chrestenson functions $_p\psi_h$ and $_p\psi_v$ with $h, v \in \mathbb{N}_0$ are defined as follows:

$$_p\psi_h(x) = e^{-\frac{2\pi i}{p} \sum_{i=0}^{N(h)} h_i x_i}, \, _p\psi_v(y) = e^{-\frac{2\pi i}{p} \sum_{i=0}^{N(v)} v_i y_i}, \tag{27}$$

and

$$_p\psi_0(x) = {_p\psi_0(y)} = 1. \tag{28}$$

Based on the one dimensional Chrestenson functions which constitute a basis for the square integrable functions over the unit interval, a basis for functions $f \in L^2(\mathbb{E}))$ can be found. To this end products of one dimensional Chrestenson functions are taken and in this way the two dimensional Chrestenson function $_p\phi_{h,v}$ with horizontal index $h \in \mathbb{N}_0$ and vertical index $v \in \mathbb{N}_0$ at the point $(x, y) \in \mathbb{E}$ reads as:

$$_p\phi_{h,v}(x, y) = {_p\psi_h(x)}_p\psi_v(y) = e^{-\frac{2\pi i}{p} \sum_{i=0}^{\max(N(h),N(v))} (h_i x_i \oplus_p v_i y_i)}. \tag{29}$$

Using the orthonormality of Chrestenson functions, we get the relation

$$\int_0^1 \int_0^1 {_p\phi_{h,v}(x, y)} * \overline{_p\phi_{h',v'}}(x, y) dx dy = \begin{cases} 1, \text{ if } h = h' \wedge v = v' \\ 0, \text{ else} \end{cases} \tag{30}$$

and using their completeness, we know that any function $f \in L^2(E)$ can be represented as a Walsh-Fourier series

$$f(x, y) = \sum_{h=0}^{\infty} \sum_{v=0}^{\infty} F(h, v)_p\phi_{h,v}(x, y) \tag{31}$$

whereby the Walsh-Fourier coefficients are computed as follows:

$$F(h, v) = \int_0^1 \int_0^1 f(x, y)_p\phi_{h,v}(x, y) dx dy. \tag{32}$$

By means of simple mathematical considerations it can be verified that the application of the dynamics of some Kolmogorov systems to any two dimensional Chrestenson function results once again in a Chrestenson function. More explicitly:

$$B_p(_p\phi_{h,v}) = {_p\phi_{h \div p, v*p+h \bmod p}}. \tag{33}$$

and similarly

$$B_p^{-1}(_p\phi_{h,v}) = {_p\phi_{h*p+v \bmod p, v \div p}}. \tag{34}$$

These relations emphasizing the fact that Chrestenson functions are essentially eigenfunctions of Kolmogorov systems will enable us to establish the equivalence between conservative time reversible Kolmogorov systems and dissipative time irreversible counterparts, as will be elaborated in the next section.

2.3 The Λ Transformation

To prove the existence of equivalent transitions between conservative reversible and dissipative irreversible systems, Prigogine et. al. have elaborated a canonical procedure how the invertible transformation implementing the transition (Λ transformation) can be realized. First of all, an adequate orthogonal set of eigenfunctions has to be found. Based on this set, an internal time operator is introduced which has to fulfill the condition that its application to each eigenfunction equals that eigenfunction multiplied by the corresponding eigenvalue. Provided that such an internal time operator can be found, it is very easy to derive an adequate Λ transformation as a monotonically decreasing (increasing) operator of the internal time operator. Taking e.g. the square of the Λ operator is then always guaranteed to be a Ljapunow function implying the existence of a physically observable quantity which can only decrease (increase) in time.

This program has also been applied to a concrete example, the so-called baker system. Pichler (Pichler 1992) has recognized that this method is essentially equivalent to dyadic convolution in terms of Walsh functions (Walsh 1923) and has suggested to construct Λ transformations for more general systems than the baker system by means of generalized Walsh functions. Here we revisit this idea and study Kolmogorov systems, including the baker system as a special case.

General Procedure To construct the Ljapunow function M it makes sense to introduce the internal time operator T which fulfills

$$U^{-n}TU^n = T + nI, T = \sum_{n=-\infty}^{\infty} nE_n \tag{35}$$

Thereby E_n denotes the eigenfunction of T corresponding to the eigenvalue n

$$TE_n = nE_n. \tag{36}$$

The task is to find a suitable set of eigenfunctions for the system under consideration which cope with the following additional requirements:

$$E_n E_m = \delta_{n,m} E_n, \ \sum_{n=-\infty}^{\infty} E_n = I, U_t E_n = E_{n+t}. \tag{37}$$

Let λ_n be a monotonically decreasing sequence starting with 1 and tending to zero as $n \to +\infty$. Then M can directly be realized as monotonically decreasing operator of T.

$$M = \sum_{n=-\infty}^{\infty} \lambda_n^2 E_n, \tag{38}$$

and $\Lambda \equiv M^{\frac{1}{2}}$ equals

$$\Lambda = \sum_{n=-\infty}^{\infty} \lambda_n E_n. \tag{39}$$

Since $\{E_n\}$ is an orthonormal basis for $f \in L^2(\mathbb{E})$, any distribution ρ can be expressed as follows:

$$\rho = \sum_{n=-\infty}^{\infty} c_n E_n. \tag{40}$$

544

$\tilde{\rho} = \Lambda\rho$ thus finally reduces to

$$\tilde{\rho} = \sum_{n=-\infty}^{\infty} c_n \lambda_n E_n. \tag{41}$$

If $1 \geq \lambda_n \geq \lambda_{n+1} > 0$ is a monotonically decreasing sequence, then this procedure ensures that $\mathcal{H}_{\tilde{\rho}} = \int \tilde{\rho}^2 d\omega$ is indeed a Ljapunow function.

Internal Time Operator We use the set of Chrestenson functions and want to define their age in accordance with some basic requirements:

$$T(_p\phi_{h,v}) = n \, _p\phi_{h,v} \Rightarrow \tag{42}$$

$$T(B_p(_p\phi_{h,v})) = T(_p\phi_{h\div p, v*p+h \bmod p}) = (n+1) \, _p\phi_{h\div p, v*p+h \bmod p},$$
$$T(B_p^{-1}(_p\phi_{h,v})) = T(_p\phi_{h*p+v \bmod p, v\div p}) = (n-1) \, _p\phi_{h*p+v \bmod p, v\div p}.$$

This is best done revisiting the generalized Rademacher functions where the assignment shall be as follows $(n > 0)$

$$T(_p\phi_{p^n,0}) = -n \, _p\phi_{p^n,0} \quad T(_p\phi_{0,p^n}) = (n+1) \, _p\phi_{0,p^n}. \tag{43}$$

Consequently, if $_p\phi_{h,v}$ is the product of generalized Rademacher functions which are assigned the eigenvalues n_1, n_2, $\ldots$ by T, then the age of $_p\phi_{h,v}$ becomes:

$$T(_p\phi_{h,v}) = \max\{n_1, n_2, \ldots\}_p\phi_{h,v}, \tag{44}$$

$$T(_p\phi_{0,0}) = 0_p\phi_{0,0}. \tag{45}$$

Entropy Operator Using the assumptions made about the internal time operator

$$T = \sum_{h=0}^{\infty}\sum_{v=0}^{\infty} n(h,v)_p\phi_{h,v} \tag{46}$$

it is easy to construct the Λ operator

$$\Lambda = \sum_{h=0}^{\infty}\sum_{v=0}^{\infty} \lambda(h,v)_p\phi_{h,v} \tag{47}$$

as desired. To this end we only require that the coefficients $\lambda(h,v)$ adhere to

$$n(h,v) < n(h',v') \Rightarrow \lambda(h,v) \geq \lambda(h',v'). \tag{48}$$

This is best achieved if the coefficients $\lambda(h,v)$ are derived as a normalized monotonically decreasing sequence from the integer coefficients $n(h,v)$ in T:

$$\lambda(h,v) = f(n(h,v)), 1 \geq f(n) \geq f(n+1) \geq 0. \tag{49}$$

For instance, the function

$$f(n) = \frac{1}{1 + e^{n\ln p}} = \frac{1}{1 + p^n} \tag{50}$$

would be a valid example.

It is well known that any distribution $\rho \in L^2(E)$ can be expressed as an infinite series

$$\rho(x, y) = \sum_{h=0}^{\infty} \sum_{v=0}^{\infty} F(h, v) \, _p\phi_{h,v}(x, y) \tag{51}$$

Then $\tilde{\rho} = \Lambda\rho$ equals

$$\tilde{\rho}(x, y) = \sum_{h=0}^{\infty} \sum_{v=0}^{\infty} F(h, v)\lambda(h, v) \, _p\phi_{h,v}(x, y) \tag{52}$$

and the function $\mathcal{H} = \int_E \tilde{\rho}\bar{\tilde{\rho}}$ is a Ljapunow function, indeed.

Proof of Monotonicity We begin with the Walsh-Fourier representation of an initial distribution at an arbitrary, but fixed point in time

$$\rho_0(x, y) = \sum_{h=0}^{\infty} \sum_{v=0}^{\infty} F(h, v) \, _p\phi_{h,v}(x, y), \tag{53}$$

Since B_p is a linear operator, it follows

$$\rho_1(x, y) = \sum_{h=0}^{\infty} \sum_{v=0}^{\infty} F(h, v) \, _p\phi_{h \div p, v*p+h \bmod p}(x, y), \tag{54}$$

$$\tilde{\rho}_0(x, y) = \sum_{h=0}^{\infty} \sum_{v=0}^{\infty} F(h, v)\lambda(h, v) \, _p\phi_{h,v}(x, y) \tag{55}$$

and

$$\tilde{\rho}_1(x, y) = \sum_{h=0}^{\infty} \sum_{v=0}^{\infty} F(h, v)\lambda(h \div p, v * p + h \bmod p) \, _p\phi_{h \div p, v*p+h \bmod p}(x, y). \tag{56}$$

The corresponding $\mathcal{H}$ functions are accordingly

$$\mathcal{H}_{\rho_0} = \int_0^1 \int_0^1 \sum_{h=0}^{\infty} \sum_{v=0}^{\infty} F(h, v) \, _p\phi_{h,v}(x, y) \tag{57}$$

$$* \sum_{h=0}^{\infty} \sum_{v=0}^{\infty} \overline{F(h, v) \, _p\phi_{h,v}}(x, y)dxdy,$$

$$\mathcal{H}_{\rho_1} = \int_0^1 \int_0^1 \sum_{h=0}^{\infty} \sum_{v=0}^{\infty} F(h, v) \, _p\phi_{h \div p, v*p+h \bmod p}(x, y) \tag{58}$$

$$* \sum_{h=0}^{\infty} \sum_{v=0}^{\infty} \overline{F(h, v) \, _p\phi_{h \div p, v*p+h \bmod p}}(x, y)dxdy,$$

$$\mathcal{H}_{\tilde{\rho}_0} = \int_0^1 \int_0^1 \sum_{h=0}^{\infty} \sum_{v=0}^{\infty} F(h, v)\lambda(h, v) \, _p\phi_{h,v}(x, y) \tag{59}$$

546

$$* \sum_{h=0}^{\infty} \sum_{v=0}^{\infty} \overline{F(h,v)\lambda(h,v)\,_p\phi_{h,v}}(x,y)dxdy$$

and

$$\mathcal{H}_{\tilde{\rho}_1} = \int_0^1 \int_0^1 \sum_{h=0}^{\infty} \sum_{v=0}^{\infty} F(h,v)\lambda_{h\div p,v*p+h \bmod p}\,_p\phi_{h\div p,v*p+h \bmod p}(x,y) \tag{60}$$

$$* \sum_{h=0}^{\infty} \sum_{v=0}^{\infty} \overline{F(h,v)\lambda_{h\div p,v*p+h \bmod p}\,_p\phi_{h\div p,v*p+h \bmod p}}(x,y)dxdy.$$

These expressions simplify considerably, if we take into account the orthonormality of Chrestenson functions. Using this fact, most of the terms in the product of the two double sums disappear:

$$\mathcal{H}_{\rho_0} = \sum_{h=0}^{\infty} \sum_{v=0}^{\infty} F(h,v)^2, \tag{61}$$

$$\mathcal{H}_{\rho_1} = \sum_{h=0}^{\infty} \sum_{v=0}^{\infty} F(h,v)^2, \tag{62}$$

$$\mathcal{H}_{\tilde{\rho}_0} = \sum_{h=0}^{\infty} \sum_{v=0}^{\infty} F(h,v)^2 \lambda_{h,v}^2 \tag{63}$$

and

$$\mathcal{H}_{\tilde{\rho}_1} = \sum_{h=0}^{\infty} \sum_{v=0}^{\infty} F(h,v)^2 \lambda(h \div p, v*p+h \bmod p)^2. \tag{64}$$

The differences of values taken by the $\mathcal{H}$-function for subsequent densities are therefore

$$\mathcal{H}_{\rho_1} - \mathcal{H}_{\rho_0} = \sum_{h=0}^{\infty} \sum_{v=0}^{\infty} F(h,v)^2(1-1) = 0, \tag{65}$$

$$\mathcal{H}_{\tilde{\rho}_1} - \mathcal{H}_{\tilde{\rho}_0} = \sum_{h=0}^{\infty} \sum_{v=0}^{\infty} F(h,v)^2(\lambda(h \div p, v*p+h \bmod p)^2 - \lambda_{h,v}^2). \tag{66}$$

Using $\lambda(h,v) = f(n(h,v))$, $n(h \div p, v*p+h \bmod p) = n(h,v)+1$ and $f(n) \geq f(n+1)$, finally delivers what we were looking for

$$\mathcal{H}_{\tilde{\rho}_1} - \mathcal{H}_{\tilde{\rho}_0} \leq 0 \tag{67}$$

3 Efficient Data Encryption Using Chaotic Kolmogorov Systems

Since chaotic Kolmogorov systems are so closely related to irreversible systems, it is quite natural to assume that they can be of considerable interest for cryptographic applications where the primary aim is that a non-legitimate user is not able to recover the original plaintext from an observed ciphertext. We have elaborated on this idea and constructed a new symmetric product cipher based on chaotic Kolmogorov systems. Due to the limited space available, the approach will only be outlined. Detail can be found in our various publications on the subject (Scharinger & Pichler 1994, Pichler & Scharinger 1995a, Pichler & Scharinger 1995b, Scharinger, Pichler, Kozek & Feichtinger 1996, Scharinger & Pichler 1996, Scharinger 1997).

3.1 Symmetric Product Ciphers

Symmetric product ciphers (DES (National Technical Information Service 1977), IDEA (Lai & Massey 1990), SKIPJACK (Clipper) (Brickell 1993, Hoffman 1993)) which perform a block-wise encryption of the plaintext input to the system currently offer unparalleled advantages for bulk encryption applications. They are much faster then public-key systems (RSA (Rivest, Shamir & Adleman 1978), ElGamal (ElGamal 1985), Diffie-Hellman (Diffie & Hellman 1976)) and offer a higher degree of security than simple substitution or permutation systems (see e.g. (Schneier 1993, Simmons 1991)) do.

Input to a r-round product cipher is a block of plaintext P_0 and a user-supplied passphrase. From this passphrase the internal key management derives the individual round keys K_i $(1 \leq i \leq r)$ and supplies them to the various encryption rounds. Every round i applies one permutation (rearrange input symbols based on symbol position) and one substitution operation (replace input symbols based on symbol value) to P_{i-1} (keyed by $K_{i,Perm}$ and $K_{i,Subst}$, respectively) and thus computes the ciphertext block C_i. After executing r rounds C_r gives the ciphertext output by the system.

There are many conditions required for a secure cipher. The following ones are probably most important:

Confusion: Ensures that the (statistical) properties of plaintext blocks are not reflected in the corresponding cipher-text blocks. Instead, every cipher-text has to have a pseudo-random appearance to any observer or standard statistical test.

Diffusion in terms of **plaintexts:** Demands that (statistically) similar plaintexts do result in completely different cipher-texts even when encrypted with the same key. In particular this requires that any element of the input block influences every element of the output block in a complex irregular fashion.

Diffusion in terms of **passphrases:** Demands that similar passphrases do result in completely different cipher-texts even when used for encrypting the same block of plaintext. This requires that any element of the passphrase influences every element of the output block in a complex irregular fashion. Additionally this property must also be valid for the decryption process because otherwise an intruder might recover parts of the input block from an observed output by a partly correct guess of the passphrase used during encryption.

3.2 Permutations Based on Chaotic Kolmogorov Systems

Our goal was to find permutation operations where the actual rearrangement (scrambling) of the input block can be determined by a single parameter. To this end we use the most unstable systems known in chaos theory: Kolmogorov systems T_π, where each element of the class is uniquely characterized by a parameter π. To give a notion why Kolmogorov systems are so especially suitable for implementing permutations in the context of a product cipher consider the following properties which give evidence that Kolmogorov systems are perfectly suited for the purpose:

Ergodicity: The trajectory connecting the iterates of almost every starting point approaches any other point arbitrarily close. In cryptographic terms ergodicity means that it is very hard to predict the final position of a point from its initial position (*confusion*).

Mixing-property: Any measurable subset is spread out uniformly. In cryptographic terms this implies that any regular structures contained in the input image never prevail but dissipate all over the corresponding output image, instead (*confusion and diffusion*).

Exponential divergence: Neighboring points diverge exponentially. This is typical for chaotic systems where small variations amplify and change the overall systems behavior in a very complex manner. In cryptographic terms: small modifications in inputs cause huge changes in the corresponding outputs (*diffusion*).

Summing up Kolmogorov flows do precisely exhibit the properties required for a good permutation operator in the context of product cipher. However, to utilize continuous Kolmogorov flows for permuting blocks of plaintext, discrete counterparts must be found.

Let the numbers q_s ($s = 1, 2, \ldots, k$) be specified by $q_s = \frac{n}{n_s}$. Due to the constraint that n_s must divide n, all q_s are positive integers and we have the correspondence to the continuous case: $q_s \triangleq \frac{1}{p_s}$. Additionally N_s shall denote the left border of the vertical strip which contains the point (x, y) that is to be transformed. Obviously the following relation holds for N_s :

$$N_s = \begin{cases} 0 & \text{for } s = 1 \\ n_1 + \ldots + n_{s-1} & \text{for } s = 2, \ldots, k \end{cases} \tag{68}$$

Then for $(x, y) \in [N_s, N_s + n_s) \times [0, n)$ the definition $T_{n,\delta} : [0, n)^2 \to [0, n)^2$

$$T_{n,\delta}(x, y) = (q_s(x - N_s) + (y \bmod q_s), (y \div q_s) + N_s) \tag{69}$$

(mod , $\div$ ... see [1]) constitutes the best possible discrete approximation to the corresponding chaotic continuous transform in the sense that deviations only occur at the least-significant position.

$T_{n,\delta}$ is a bijective map. The inverse $T_{n,\delta}^{-1}$ is easily found to be given by

$$T_{n,\delta}^{-1}(x, y) = ((x \div q_s) + M_s, q_s(y - M_s) + (x \bmod q_s)) \tag{70}$$

for $(x, y) \in [0, n) \times [M_s, M_s + n_s)$. In contrast to N_s which marks the *left* border of the *vertical* strip s, M_s denotes the *lower* border of the *horizontal* strip where the point (x, y) lies in and it can be computed as follows:

$$M_s = \begin{cases} 0 & \text{for } s = 1 \\ n_1 + \ldots + n_{s-1} & \text{for } s = 2, \ldots, k \end{cases} \tag{71}$$

For n an integral power of 2 the calculation of $T_{n,\delta}$ (and also the corresponding inversion formula $T_{n,\delta}^{-1}$) can be done by just using additions, subtractions and bit-shift operations which allows for very efficient SW/HW implementations of the permutation and inverse permutation operator, respectively.

3.3 Substitution Operator

In a product cipher permutations are complemented by substitution operations. Besides other tasks, the substitution has to ensure that the output of the overall encryption process

---

[1] Let the positive integer a be represented by other positive integers in the form $a = b * d + r$ ($r < b$). Then the integer division $a \div b$ gives d and the corresponding remainder $a \bmod b$ is r.

is always a block of uniformly distributed pseudo-random data regardless of the distribution of data in the input block. Our choice for implementing a suitable substitution operator is based on a maximum-period shift register pseudo-random number generator (PRNG). Due to its outstanding efficiency, its superb period and the sufficient statistical properties, we decided to use an adaption of the PRNG called "R250". It is worth noting that randomness in cryptography is different from the ordinary standards of randomness in the sense that, for cryptographic purposes, the matter of predictability is exceedingly important. We take account of this additional requirement by combining random number generation (substitution) in an intertwined and iterated manner with permutation operations, as proposed by the general framework of a product cipher system.

References

Arnold, V. I. & Avez, A. (1968), *Ergodic Problems of Classical Mechanics*, W. A. Benjamin, New York.

Brickell, E. (1993), 'SKIPJACK Review Interim Report: The SKIPJACK Algorithm', posted on sci.crypt; available from NIST.

Chrestenson, H. E. (1955), 'A class of generalized walsh functions', *Pacific Journal of Mathematics* **5**, 17–31.

Diffie, W. & Hellman, M. (1976), 'New directions in cryptography', *IEEE Transactions on Information Theory* **22**, 644–654.

ElGamal, T. (1985), 'A public-key cryptosystem and a signature scheme based on discrete logarithms', *IEEE Transactions on Information Theory* **31**, 469–472.

Hoffman, L. (1993), 'Clipping Clipper', *Communications of the ACM* **36**(9), 15–17.

Lai, X. & Massey, J. (1990), 'A proposal for a new block encryption standard', *EURO-CRYPT 90* pp. 389–404.

Misra, B. (1978), 'Nonequilibrium entropy, lyapounov variables, and ergodic properties of classical systems', *Proc. Natl. Acad. Sci. USA* **75**(4), 1627–1631.

Moser, J. (1973), *Stable and Random Motions in Dynamical Systems*, Princeton University Press, Princeton.

National Technical Information Service (1977), Data Encryption Standard, Technical report, National Bureau of Standards, Federal Information Processing Standards Publication, Springfield VA. FIPS PUB 46.

Nicholis, G. & Prigogine, I. (1989), *Exploring Complexity*, Freeman and Co., New York.

Pichler, F. (1992), 'Realisierung der Λ-Transformation von Prigogine mittels dyadischer Faltungsoperatoren', *Austrian Society for Cybernetic Studies* . ISBN 385206127X.

Pichler, F. & Scharinger, J. (1995*a*), Ciphering by bernoulli–shifts in finite abelian groups, *in* H. Kaiser, W. Müller & G. Pilz, eds, 'Contributions to General Algebra 9', pp. 249–256.

Pichler, F. & Scharinger, J. (1995*b*), 'Finite dimensional generalized baker dynamical systems for cryptographic applications', *Lecture Notes in Computer Science* **1030**, 465–476.

Prigogine, I. (1980), *From Being to Becoming*, Freeman and Co., San Francisco.

Prigogine, I. (1991), New perspectives on complexity, *in* G. J. Klir, ed., 'Facets of Systems Science', Plenum Press, New York, pp. 483–492.

Prigogine, I. & Stengers, I. (1990), *Dialog mit der Natur*, Piper, München, New York.

Rivest, R., Shamir, A. & Adleman, L. (1978), 'A method for obtaining signatures and public-key cryptosystems', *Communications of the ACM* **21**(2), 120–126.

Rosen, R. (1985), *Anticipatory Systems*, Pergamon Press.

Scharinger, J. (1997), Fast encryption of image data using chaotic kolmogoroff-flows, *in* 'Storage and Retrieval for Image and Video Databases V', SPIE Proceedings Volume 3022, pp. 278–289.

Scharinger, J. & Pichler, F. (1994), 'Bernoulli Chiffren', *Elektrotechnik und Informationstechnik* **11**. (in German).

Scharinger, J. & Pichler, F. (1996), Efficient image encryption based on chaotic maps, *in* A. Pinz, ed., 'Pattern Recognition 1996, Proceedings of the 20th Workshop of the AAPR', R. Oldenbourg, Wien, München, pp. 159–170.

Scharinger, J., Pichler, F., Kozek, W. & Feichtinger, H. (1996), Chaotic kolmogorov flows for image encryption, *in* R. Trappl, ed., 'Cybernetics and Systems '96', Vol. 1, Austrian Society for Cybernetic Studies, pp. 111–116.

Schneier, B. (1993), *Applied Cryptograpy, Protocols, Algorithms and Source Code in C*, John Wiley and Sons, New York.

Shields, P. (1973), *The Theory of Bernoulli Shifts*, The University of Chicago Press, Chicago.

Simmons, J. (1991), *Contemporary Cryptography*, IEEE Press.

Vilenkin, N. Y. (1947), 'A class of complete orthonormal series', *Izv. Akad. Nauk SSSR, Ser. Mat* **11**, 363–400.

Walsh, J. L. (1923), 'A closed set of orthogonal functions', *Americal Journal of Mathematics* **45**, 5–24.

On Some Methods Of Discrete Systems Behaviour Simulation

Alexander A.Sytnik, professor, doctor of technical sciences,
E-mail: Sytnik@scnit.saratov.su, Sytnik@sgu.ssu.runnet.ru

Natalia I. Posohina, post-graduate student
Posohina@sgu.ssu.runnet.ru

410071, Russian Federation, Saratov, Astrakhanskaya str., 83
Saratov State University,
TEL: 7-(8452) 51-55-31, 7-(8452) 51-14-39 FAX: 7-(8452) 24-04-46

Abstract

The project is solving one of the fundamental problems of mathematical cybernetics and discrete mathematics, the one connected with synthesis and analysis of managing systems, depending on the research of their functional opportunities and reliable behaviour.

This work deals with the case of finite-state machine behaviour restoration when the structural redundancy is not available and the direct updating of current behaviour is impossible.

The discribed below method, uses number theory to build a special model of finite-state machine, it is simulating the transition between the states of the finite-state machine using specially defined functions of exponential type with the help of several methods of number theory and algebra it is easy to determine, whether there is an opportunity to restore the behaviour (whith the help of this method) in the given case or not and also derive the class of finite-state machines, admitting such restoration.

Keywords: restoring behaviour, modelling, finite-state machines, number theory.

Introduction

The project is solving one of the fundamental problems of mathematical cybernetics and discrete mathematics, the one connected with synthesis and analysis of managing systems, depending on the research of their functional opportunities and reliable behaviour.

Nowdays, two basic types of redundancies are generally in use, they are: structural redundancies and temporary redundancies. In the first case, additional backup copies are inserted in the structure of the system . As soon as the basic part fails or the system needs some updatings of behaviour, the existing structural reserve is assigned the task of realizing the given behaviour. In the second case, an available or specially created reserve of time (temporary redundancy) is used to repeat or restart the operation, previously failed as a result of infringement.

CP437, *Computing Anticipatory Systems: CAYS--First International Conference*
edited by Daniel M. Dubois © 1998 The American Institute of Physics 1-56396-827-4/98/$15.00

The structural reservation absense or failure derives a question: "Whether it is possible or not to use current automatic device functioning properties to create the required (demanded) set of reactions?" The answer assumes the study of system functional redundancy, being available at the moment, and also the study of possible (or probable) variants of its creation. We'll call this kind of restoring behaviour a functional one.

The traditional mathematical model of discrete systems with memory is (despite different lacks) a finite-state machine.

This work deals with the case of finite-state machine behaviour restoration when the structural redundancy is not available and the direct updating of current behaviour is impossible.

The question is, what are the methods that can be used to restore behaviour of the given finite-state machine and which methods prove to be most efficient in each special case. One of the methods, discribed below, uses number theory to build a special model of finite-state machine, it is simulating the transition between the states of the finite-state machine using specially defined functions of exponential type with the help of several methods of number theory and algebra it is easy to determine, whether there is an opportunity to restore the behaviour (whith the help of this method) in the given case or not and also derive the class of finite-state machines, admitting such restoration.

Finite-state machine model is widely used in research of complex discrete systems. As a rule, finite-state machines are considered as transformers, namely, properties are studied through consideration of the way input strings are transformed into output strings. However, sometimes it is even more important to determine the feedback - the group of input strings transformed into the given final string, especially in the synthesis problem solving, that is in constructing a finite-state machine, satisfying given conditions. It is a so-called inverse problem, and it is even more complex in congruence with the direct problem. Formally, the direct problem can be defined as determination of an output string for the given input one - that is the transformer form of behaviour. In case the finite-state machine is represented as a set of final strings, which it generates, the machine is considered to be an enumerator (and it has an enumerator form of behaviour). Actually, the transition from the transformer to enumerator within the framework of the finite-state machines theory is complicated and time-consuming. This work repesents one approach to tie together some elements of finite-state machines and number theories, aiming to construct a so-called "numerical" model of behaviour of a finite-state machine, and then to carry out transition from transformer form of behaviour to enumerator form, using several methods of algebra and number theory.

Constructing the Model of Behaviour

The finite-state machine (FSM) $A=(S,X,Y,\delta,\lambda)$ is given, with the number of current internal condition of FSM as an output, so it is possible to consider $S=Y$ and $\delta\equiv\lambda$, and the initial finite-state machine is brought to the form $A=(S,X,\delta)$. Internal conditions of the finite-state machine should be enumerated with integers from 0 up to m-1, so that $S = \{0,1,...,m-1\} = GL(m)$ (it means, that S coincides with the semigroup of remainders modulo m).

Let $\{h_s\}_{s\in S}$ denote the set of finite-state mappings of the form $h_s:X^* \to Y^*$, where X^*,Y^* are sets of FSM input and output strings, respectively. According to Kurosh (1975), it is possible (without any information loss) to bring it to the set of finite-state mappings $\{g_s\}_{s\in S}$

of the form $g_s : X^* \to Y$, where input strings are associated with the last symbol of the appropriate output string: $x_1 x_2 ... x_n \to y_n$ for $\{g_s\}_{s \in S}$ (instead of $x_1 x_2 ... x_n \to y_1 y_2 ... y_n$ for $\{h_s\}_{s \in S}$).

The transformer form of behaviour of the finite-state machine A is represented with its set of finite-state mappings $\{g_s\}_{s \in S}$. The enumerator form of behaviour of the finite-state machine A is represented with the set of output strings $L(X^*) = \{y_1 y_2 ... y_n \in Y^* | (\forall y_i \in Y)(\exists s_i \in S)(\exists \alpha_i \in X^*): g_{s_i}(\alpha_i) = y_i \}$, generated by A.

The problem of synthesing FSM-enumerator can be defined as constructing a finite-state machine, realizing the given set of finite-state mappings. Determining the set of strings $L(X^*)$ for the given set $\{g_s\}_{s \in S}$ needs a special model of the finite-state machine behaviour, constructed according to some restrictions on the form of elements $\{g_s\}_{s \in S}$.

When the variable x is fixed, the transition function (of the given finite-state machine can be considered as a substitution of the form:

$$\delta_x : \begin{pmatrix} 0 & 1 & ... & m-1 \\ s_0 & s_1 & ... & s_{m-1} \end{pmatrix}. \tag{1}$$

Modeling by functions of exponential types is assumed only by finite-state machines that for any input symbol x possible, the δ_x can be represented by some function of exponential type f_x (with fixed factors from the set $\{0,1,...m-1\}$) of the form:

$$f_x(s) = a_0 + a_1 s + a_2 s^2 + ... + a_l s^l \pmod{m}, s \in S \tag{2}$$

Several important facts concerning the set of functions of exponential type, constructed for the given set $\{f_x\}_{x \in X}$:

Lemma.

The set of functions of exponential type $\{f_x\}_{x \in X}$ simulates the behaviour of the finite-state machine A (it means that any finite-state mapping g from $\{g_s\}_{s \in S}$ can be represented as a composition of functions from $\{f_x\}_{x \in X}$.

Proof:

Let's define the function of two variables (on the Cartesian product of the set of internal states and the set of input strings $\tilde{\delta} : S \times X^* \to S$ as follows:

$$\begin{cases} \tilde{\delta}(s, \Lambda) = s \\ \tilde{\delta}(s, x\alpha) = \tilde{\delta}(\delta(s,x), \alpha) \end{cases}, \forall s \in S, \forall x \in X, \forall \alpha \in X^*,$$ where Λ is an empty string in X^*.

Then $g_s(\alpha) = \tilde{\delta}(s, \alpha), \forall \alpha \in X^*$.

At the same time, $f_x(s) = \delta(s, x)$

Hence, if the identity mapping is defined as follows $\Delta(s) = s \pmod{m}$, then

$$\tilde{\delta}(s, \Lambda) = \Delta(s)$$

$$\tilde{\delta}(s, x_1 x_2 ... x_n) = \tilde{\delta}(\delta(s, x_1), x_2 ... x_n) = \delta(\delta(...(\delta(s, x_1), ...), x_{n-1}), x_n) =$$

$$= f_{x_n}(f_{x_{n-1}}(...f_{x_1}(s)...)) = f_{x_n} \circ f_{x_{n-1}} \circ ... \circ f_{x_1}(s), \forall s \in S, \forall x \in X, \forall \alpha \in X^*$$

Thus, $g_{s(}x_1 x_2 ... x_n) = f_{x_n} \circ f_{x_{n-1}} \circ ... \circ f_{x_1}(s)$

Q.E.D.

The set $\{f_x\}_{x \in X}$ must have a basis, moreover, a finite basis (because both f_x and its initial transition function (, are defined on the finite set $\{0,1,\ldots m-1\}$), and this fact is obvious from the formula (2), since the number l - senior exponent of any function f_x - is already finite. Quite naturally, the basis considered for the set of functions of exponential type is $\{x^0, x^1, x^2, \ldots x^l\}$. Defining the actual number of non-identical functions of the form x^k means determining the senior exponent l for any possible function of the form (2), calculated modulo m, namely: the senior exponent of any function from the set $\{f_x\}_{x \in X}$ constructed for the finite-state machine with m internal conditions does not exceed l.

The functions $\{x^0, x^1, x^2, \ldots x^l\}$ constitute a semigroup of mappings (including a so-called constant mapping $x^0 = 1 (\mod m)$). Its subsemigroup $\{x^1, x^2, \ldots x^l\}$ is a periodic semigroup, generated by x. By definition, the index of a semigroup is the least positive integer r_0 such, that $x^{r_0} = x^{r_0+n}$ for some positive integer n, and the period of semigroup is the least positive integer m_0 of all possible n. The given r_0 and m_0 correspond to a unique semigroup of mappings $\{x, x^2, \ldots x^{r_0+m_0-1}\}$, consequently, $l = r_0 + m_0 - 1$. As far as x belongs to $\{0,1,\ldots m-1\}$, the semigroup of mappings $\{x, x^2, \ldots x^{r_0+m_0-1}\}$ can be considered to be a semigroup of m-tuples of the form $(0^k, 1^k, 2^k, \ldots, (m-1)^k)$ (where x^k is an arbitrary mapping from the set $\{x, x^2, \ldots x^{r_0+m_0-1}\}$). In fact, these m-tuples are the bottom lines of appropriate substitutions of the form (1).

Observation

In the semigroup of m-tuples the multiplication is defined componentwise, therefore all i-th components of m-tuples of the semigroup constitute a subsemigroup of the semigroup $\{0,1,\ldots m-1\}$, generated by some i. Hence, speaking of relations between indexes and periods of m-tuple and componentwise semigroups, it must be necessarily taken into account, that the period of m-tuple semigroup should contain all periods of componentwise semigroups, thus it should be the least common multiple of their periods. The index of multiple semigroup should be the greatest of all indexes of componenwise semigroups. The relations between m and r_0, m_0 are determined by the following theorem:

Theorem 1.

Let $m = 2^{\alpha_0} p_1^{\alpha_1} p_2^{\alpha_2} \ldots p_k^{\alpha_k}$ 　　　　　　　　　　　(3)

be the prime factorization of m, where $\alpha_0 \geq 0, \alpha_i > 0, i = \overline{1,k}$. Then the index and period of semigroup of mappings $\{x, x^2, \ldots x^{r_0+m_0-1}\}$ are:

$$r_0 = \max(\alpha_0, \alpha_1, \ldots, \alpha_k),$$
$$m_0 = p_1^{\alpha_1-1} \cdot \ldots \cdot p_k^{\alpha_k-1} \cdot HOK([2^{\alpha-2}], p_1 - 1, \ldots, p_k - 1)$$

Proof:

By the definition of r_0 and m_0 (mentioned above), r_0 and m_0 - are the least positive integers, satisfying the congruence

$$x^{r_0} = x^{r_0+m_0} (\mod m), \forall x \in \{0,1,\ldots,m-1\} \qquad (4)$$

This congruence can be considered as a system of m congruences. It can be split into three subsystems: the first one contains all the congruences with the values of x, relatively prime with the modulus m, that is $x \in I = \{a \mid HOD(a,m) = 1\}$; the second one contains all the congruences with the values of x, containing all the prime factors of the modulus m, that is x

belongs to the set $\{a\,|\,a=2^{\beta_0}p_1^{\beta_1}p_2^{\beta_2}\ldots p_k^{\beta_k},0<\beta_i\le\alpha_i,i=\overline{0,k}\}$; the third one contains all the remaining congruences, that is x and the modulus m have some common divisor different from 1 and $2p_1p_2\ldots p_k$ (or $p_1p_2\ldots p_k$, in case of odd m), that is $x\in III=\{0,1,\ldots,m-1\}\setminus(I\cup II)$. Each of the subsystems must be solved separately.

I. In the first subsystem (it is a reduced system of remainders modulo m) each congruence can be divided by the integer, relatively prime with the module m, namely by x^{r_0}. We'll have $x^{m_0}=1(\mathrm{mod}\,m),\forall x\in I$. Afterwards, every congruence is split into $k+1$ congruences as follows:

$$\begin{cases} x^{m_0}=1(\mathrm{mod}\,2^{a_0})\\ x^{m_0}=1(\mathrm{mod}\,p_1^{a_1})\\ \ldots\ldots\ldots\ldots\ldots\ldots\\ x^{m_0}=1(\mathrm{mod}\,p_k^{a_k}) \end{cases}$$

Further on, all the congruences derived by now, are split into $k+1$ subsystem (modulo $2^{\alpha_0},p_1^{\alpha_1},p_2^{\alpha_2},\ldots,p_k^{\alpha_k}$ respectively), and x runs throughout the reduced system of remainders in each. The least possible solution for the first subsystem $x^{m_0}=1(\mathrm{mod}\,2^{\alpha_0}),\forall x\in I$ will be $m_0=\begin{cases}1,\alpha_0=0,1\\2^{\alpha_0-2},\alpha_0\ge 2\end{cases}$. Let $[m_0]$ denote the integer part of m_0, then $m_0=[2^{\alpha-2}]$. According to Euler's theorem, the solution of subsystem of congruences of the form $x^{m_0}=1(\mathrm{mod}\,p_i^{\alpha_i}),\forall x\in I,i=\overline{1,k}$, will be $m_0=\varphi(p_i^{\alpha_i})=p_i^{\alpha_i}-p_i^{\alpha_i-1}=(p_i-1)p_i^{\alpha_i-1}$. The required m_0 should be divisible by all the integers $[2^{\alpha-2}],(p_1-1)p_1^{\alpha_1-1},\ldots,(p_k-1)p_k^{\alpha_k-1}$, thus, it should be their least common multiple, and whereas all the $p_i^{\alpha_i-1}$ are relatively prime, they can be carried in front of the lcm mark, so we'll have $m_0=p_1^{\alpha_1-1}\cdot\ldots\cdot p_k^{\alpha_k-1}\cdot HOK([2^{\alpha-2}],p_1-1,\ldots,p_k-1)$. Each integer from the resulted system of remainders, being raised in the power, divisible by its Euler's function, makes 1, thus in this case the index $r_0=1$.

II. Any integer from the set II contains all the prime factors of the prime factorization of m. So, any integer, being raised in some definite power r^*, makes 0, and raising in the greater powers makes also 0. The integer $2p_1p_2\ldots p_k$ (or $p_1p_2\ldots p_k$, in case of odd m), will be distinct from zero, while $r^*<\max(\alpha_0,\alpha_1,\ldots,\alpha_k)$, hence r^* can't be less than the maximum power in the prime factorization of m, at the same time non of the integers from the set II^* can stay distinct from zero, being raised in the power equal or greater than this maximum. Hence, in the worst case, an integer from the II set will stay nonzero, being raised into the power of $r^*<\max(\alpha_0,\alpha_1,\ldots,\alpha_k)$, but any integer in the power of this maximum will make 0. Consequently, according to this observation, in this case $r_0=\max(\alpha_0,\alpha_1,\ldots,\alpha_k)$ and $m_0=1$.

III. All the integers from the IIId set must have at least one common divisor with m, not equal to $2p_1p_2\ldots p_k$ (or $p_1p_2\ldots p_k$, in case of odd m). An arbitrary integer a is a product of an element p from the subsemigroup, containing 1 (wich is a resulted semigroup of remainders) and an element q from the subsemigroup with some other idempotent. The semigroup, generated by a is: $\{pq,p^2q^2,\ldots,p^{r_0}q^{r_0},\ldots,p^{r_0+m_0-1}q^{r_0+m_0-1}\}$. Its period should be divisible by all the periods of semigroups, generated by p and q, and its index, quite naturally, can't be less than their indexes. Hence, we are interested in the index and period of semigroup, generated by q. We'll split the initial system of congruences into $k+1$ subsystems (modulo $2^{\alpha_0}, p_1^{\alpha_1}, p_2^{\alpha_2}, ..., p_k^{\alpha_k}$, respectively), where x runs through the resulted system of remainders.

Let $d = HOD(m, a^{r_0}, a^{r_0+m_0})$ be the greatest common divisor of both parts of the congruence and its modulus (one of $2^{\alpha_0}, p_1^{\alpha_1}, p_2^{\alpha_2}, ..., p_k^{\alpha_k}$). Then, according to Vinogradov (1965), we can divide both parts of congruence and modulus by d, and the equivalent congruence will be:

$$x^* = \bar{x}^*(\operatorname{mod} m^*), x^*, \bar{x}^* \in I^*$$

Now both parts of congruence are relatively prime with the new modulus, therefore we can multiply both parts of the congruence on the converce to x^* modulo m^*, and then solve the congruence according to the method, introduced in the first part of the proof. Besides, the obtained index r_0^* modulo m^* is less or equal to the index obtained modulo m_0, because the powers of the prime factorization of m^* are less or equal to the powers of the prime factorization of m_0, and thus modulus m^* divides modulus m_0.

Choosing lcm of moduli, obtained in each of the three possible cases, we'll get $m_0 = p_1^{\alpha_1-1} \cdot ... \cdot p_k^{\alpha_k-1} \cdot HOK([2^{\alpha-2}], p_1 - 1, ..., p_k - 1)$, and maximum of the three indexes is equal to $r_0 = \max(\alpha_0, \alpha_1, ..., \alpha_k)$.

Q.E.D.

After the leading power of the function of exponential type (2) is determined, it is necessary to find its coefficients $a_0, a_1, ..., a_l$. All the consecutive integers $0, 1, ...m-1$ are substituted in f_x and the results are equated to the integers $s_0, s_1, ..., s_{m-1}$ from the bottom line of the appropriate substitution of the form (1). Thus, the resulting system of linear congruences modulo m is:

$$\begin{cases} s_0 = a_0 (\operatorname{mod} m) \\ s_1 = a_0 + a_1 + a_2 + ... + a_l (\operatorname{mod} m) \\ s_2 = a_0 + a_1 2 + a_2 2^2 + ... + a_l 2^l (\operatorname{mod} m) \\ .... \\ s_{m-1} = a_0 + a_1(m-1) + a_2(m-1)^2 + ... + a_l(m-1)^l (\operatorname{mod} m) \end{cases}$$

Excluding the first congruence (since in is trivial) and transforming the others eliminating variable s_0:

$$\begin{bmatrix} 1 & 1 & ... & 1 \\ 2 & 2^2 & ... & 2^l \\ ... & ... & ... & ... \\ (m-1) & (m-1)^2 & ... & (m-1)^l \end{bmatrix} \times \begin{bmatrix} a_1 \\ a_2 \\ ... \\ a_l \end{bmatrix} = \begin{bmatrix} s_1 - s_0 \\ s_2 - s_0 \\ ... \\ s_{m-1} - s_0 \end{bmatrix} (\operatorname{mod} m)$$

In brief, $Ma=s$.

The solvation of this system helps extracting the criterion of finite-state machine behaviour simulation with a set of functions of exponential type, based on the properties of integer m and the form of substitutions of the form (1).

Theorem 2.

The behaviour of the finite-state machine A can be simulated by a set of functions of exponential type $\{f_x\}_{x \in X}$ if and only if the rank of an extended matrix $[M|s]$ is equal to the rank of a matrix M and each element of the column of constant terms is divisible by the greatest common divisor of all the elements of the appropriate line of a matrix M.

In fact it means, that the matrix system of congruences, having at least one solution, should be solvable over a field of real numbers (by Kurosh (1975) it demands the rank of matrix M to equal the rank of extended matrix $[M|s]$) and besides, it should be solvable over the field of integer numbers as a system of Diophantine equations (it demands each constant term to be divisible by the greatest common divisor of coefficients of all the other terms of the equation). Briefly, the given theorem unites two different criteria.

Consequence

In case the number of states m is prime, the simulating set of functions of exponential type $\{f_x\}_{x \in X}$ can be constructed for an arbitrary finite-state machine A (that is, the behaviour of all finite-state machines with prime numbers of state can be simulated by an appropriate set of functions of exponential type).

If the modulus of system of congruences is a prime, the matrix M will be square, its rank will be equal to the rank of an extended matrix $[M|s]$ exactly. Thus, at least one (the last) element in each line of the matrix will equal 1, so, the gcd of all the elements of the line is also equal 1, and the element of the column of constant terms can be an arbitrary integer from $\{0, 1, ..., m-1\}$. Hence, this special behaviour simulating model can be constructed for an arbitrary column of constant terms.

Finite-state Machine Model

Let's turn back to final state machines. Let n be the rank of matrix M. In $\{f_x\}_{x \in X}$ is chosen a subset of n linearly independent functions $\{e_i\}_{i=\overline{1,n}}$, each of them corresponds to a definite letter z_i of the alphabet Z. There can be made a transition from the basis $\{x^0, x^1, x^2, ... x^l\}$ (standard basis for all the finite-state machines with m states) to the new basis $\{e_i\}_{i=\overline{1,n}}$ (made of functions of exponential type associated with mappings of the given finite-state machine). Each function f_x, not included into the new basis, corresponds to some word β_x, obtained as discribed below: function f_x is divided by functions e_i of the new basis, until it is factorized into the product of functions e_i-s and a constant c from the set $\{0, 1, ..., m-1\}$, and the word $z_{i_1} z_{i_2} ... z_{i_n} = \beta$ consisting of the letters z_j, corresponding to appropriate factors e_j of f_x's factorization (adding, if necessary, the letter z^c, corresponding to the constant c). The question of one-to-one correspondence between f_x and β_x is the same as the question of one-to-one correspondence between the integer and its prime factorization.

The obtained set $\{\beta_x\}_{x \in X}$ is, in fact, the generative set of the set of finite-state mappings. Hence, we've got the generative set $L(X^*)$ for all the words, enumerated by the finite-state machine A.

References

1. Arbib M.A. (1975) The algebraic theory of finite-state machines, languages and semigroups. Moscow: "Statistics"

2. Vinogradov I.M. (1965) Bases of number theory Moscow: "Science".

3. Kudriavtsev V.B., Alioshin S.V., Podkolzin A.S. (1985) Introduction in the theory of finite-state machines. Moscow: "Science".

4. Kurosh A.G. (1975) A course of supreme algebra. Moscow: "Science".

5. Sytnik A.A. Methods and Models for Restoration on Automata Behaviour // Automation and remote control. Consultants Bureau. New York. 1993. P.1781-1790

Neuronal and Cognitive Systems

Learning to Understand —
General Aspects of Using Self-Organizing Maps in Natural Language Processing

Timo Honkela

Helsinki University of Technology

Neural Networks Research Centre

P.O.Box 2200, FIN-02015 HUT, Finland

tel: +358 0 451 3275, fax: +358 0 451 3277

e-mail: Timo.Honkela@hut.fi

WWW: http://nucleus.hut.fi/~tho/

Abstract

The Self-Organizing Map (SOM) is an artificial neural network model based on unsupervised learning. In this paper, the use of the SOM in natural language processing is considered. The main emphasis is on natural features of natural language including contextuality of interpretation, and the communicative and social aspects of natural language learning and usage. The SOM is introduced as a general method for the analysis and visualization of complex, multidimensional input data. The approach of how to process natural language input is presented. Some epistemological underpinnings are outlined, including the creation of emergent and implicit categories by SOM, intersubjectivity and relativity of interpretation, and the relation between discrete symbols and continuous variables. Finally, the use of SOM as a component in an anticipatory system is presented, and the relation between anticipation and self-organization is discussed.

Keywords: natural language processing, self-organizing maps, semantics, epistemology, neural networks.

1. Introduction

Traditionally the formal study of language has centered around structural and static aspects. The automatic analysis and generation of syntactic structures has mainly been based on explicit, hand-written, and symbolic representations. In semantics the main focus has been on propositional structures, quantifiers, connectives, and other phenomena that match well the apparatus of predicate logic. This paper aims at widening the scope to include many more natural features of natural language including contextuality of interpretation, and the communicative and social aspects of natural language learning and usage. The principles of formalizing specific aspects of these phenomena are considered in this paper including the following:

In traditional study of language fixed categorizations are normally used. Lexical items, words and phrases, are positioned into categories such as verbs and nouns, and these categories are used in abstract rules, e.g., of the type "S → NP VP", i.e., a sentence consists of a nominal phrase and a verb phrase. It may seem that the abstract rules are precise, but when they are applied, discrepancies exist between the rules and the actual

CP437, *Computing Anticipatory Systems: CAYS--First International Conference*
edited by Daniel M. Dubois © 1998 The American Institute of Physics 1-56396-827-4/98/$15.00

use of language. A rule may be incorrect in various ways. For instance, a rule may be overtly general and should be refined. Refining the rule may be based on adding extra restrictions on its use, or creating more fine grained categories that divide the feature space into smaller areas. When this refinement process is continued into the extreme it may appear that each word has a category of its own. At least, it seems that a natural grammar has a fractal structure.

- The use and interpretation of language is adaptive and context-sensitive. One can, of course, find the most usual patterns and make definitions based on them, but in the actual discussions and writings, words are often used creatively based on the particular situation. The well-known ambiguity "problem" highlights the context-sensitivity: there may be multiple interpretations for a word or a phrase but in the context the desired interpretation can be understood. Most often human listeners or readers do not even notice the potential alternative readings of distinct words. The preceding text and the overall context supports an anticipatory process that blocks effectively incorrect interpretations.

- The context-sensitivity of interpretation is also relevant when one considers the more fine-grained structure of semantic and pragmatic level. The traditional, logic-based ontology of natural language interpretation is based on the idea that the world consists of distinct objects, their properties, and the relationships between the objects. Such a view neglects the fact that the propositional level of sentences does not have a simple one-to-one counterpart in the reality. The reality is highly complex, apparently high-dimensional in the perceptional level, changing, consisting of non-linear and continuous processes. Thus, studying the epistemological level having basically "names" and "objects" as their referents, may be considered to be far too simplistic. One should, for instance, take into account the relation between discrete symbols and the continuous spaces which the symbols refer to.

- Human understanding of natural language is based on the long individual experience. Inevitably the differences in personal histories cause differences in the way humans interpret natural language expressions. In the light of the previous discussion, it should be clear how approaching this kind of phenomenon is difficult using the apparatus of symbolic logic without considering more refined mathematical tools of algebra. The subjectivity of interpretation is apparent when thoughtfully. The communication is enabled by the intersubjectivity based on the learning process in which the interpretations are adapted to match well enough so that meaningful exchange of thoughts becomes possible. Possibility of fine-grained differences in interpretation become understandable when continuous variables and spaces are considered as the counterparts for the disctinct symbols that are used in communication. For instance, if person A has a prototypical idea of a specific color having the value [0.234 0.004 0.678] in a color coding scheme, and for the person B the corresponding vector is [0.232 0.002 0.677], it is clear that the communication based on the symbol is successful in spite of the small difference, error, in the interpretation. Actually, it may be still more fruitful to consider a reference relation as a distribution rather than a relation between a symbol and a numerical value or vector. Such an approach enriches strongly the possibility to study fine-grained phenomena of natural language interpretation as opposed to the model theoretical approach.

In the following, Kohonen's Self-Organizing Maps (SOMs) (Kohonen, 1982, 1995) are introduced. The SOMs may provide a sound basis for modeling the general underlying principles of natural language learning and interpretation. The motivation for such a claim is presented in the rest of the paper.

2. Self-Organizing Maps

The SOM is a widely-used artificial neural network model in which learning is unsupervised: no a priori classifications for the input examples are needed. In comparison, the backpropagation algorithm, for instance, requires the examples to consist of input-output pairs. The network architecture of the SOM consists of a set of laterally interacting adaptive processing elements, nodes, usually arranged as a two-dimensional grid called the map. All the map nodes are connected to a common set of inputs. Any activity pattern on the input gives rise to excitation of some local group of map nodes. After learning, the spatial positions of the excited groups specify a mapping of the input onto the map. The learning process is based on similarity comparisons in a continuous space. The result is a system that map similar inputs close to each other in the resulting map. The input may be highly complex multidimensional data like in the real-life applications in speech recognition, image analysis, and process monitoring. (Kohonen, 1995)

2.1. Learning Process in SOM

Assume that some sample data sets have to be mapped onto the array depicted in Fig. 1; a sample set is described by a real vector $x(t) \in R^n$ where t is the index of the sample, or the discrete-time coordinate. In setting up the neural-network model called the Self-Organizing Map, we first assign to each unit in the array a parameter vector $m_i(t) \in R^n$ called the codebook vector, which has the same number of elements as the input vector $x(t)$. The initial values of the parameters, components of $m_i(t)$, can be selected at random.

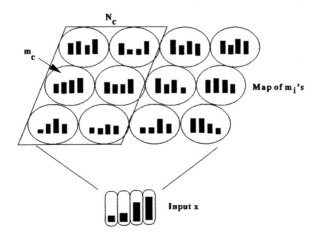

Fig. 1. The basic architecture of the self-organizing map.

565

The "image" of an input item on the map is defined to be in the location, the $m_i(t)$ of which matches best with $x(t)$ in some metric. The self-organizing algorithm that creates the ordered mapping can be described as a repetition of the following basic tasks:

1. An input vector $x(t)$ (like one row of Table 1) is compared with all the codebook vectors $m_i(t)$. The best-matching unit on the map, i.e., the unit where the parameter vector is most similar to the input vector in some metric, called the winner, is identified.

2. The codebook vectors of the winner and a number of its neighboring units in the array are changed incrementally according to the learning principle specified below.

The basic idea in the SOM is that, for each input sample vector $x(t)$, the parameters of the winner and units in its neighborhood are changed closer to $x(t)$. For different $x(t)$ these changes may be contradictory, but the net outcome in the process is that ordered values for the $m_i(t)$ are finally obtained over the array. If the number of input vectors is not large compared with the number of codebook vectors (map units), the set of input vectors must be presented reiteratively many times. As mentioned above, the codebook vectors may initially have random values, but they can also be selected in an ordered way. Adaptation of the codebook vectors in the learning process takes place according to the following equation:

$$mi(t+1) = m_i(t) + \alpha(t)[x(t)-m_i(t)] \quad \text{for each } i \in N_c(t),$$

where t is the discrete-time index of the variables, the factor $\alpha(t) \in [0,1]$ is a scalar that defines the relative size of the learning step, and $N_c(t)$ specifies the neighborhood around the winner in the map array. At the beginning of the learning process the radius of the neighborhood is fairly large, but it shrinks during learning. This ensures that the global order is obtained already at the beginning, whereas towards the end, as the radius gets smaller, the local corrections of the codebook vectors in the map will be more specific. The factor $\alpha(t)$ decreases during learning. Details about the selection of parameters, variants of the map, and thousands of application examples can be found in (Kohonen, 1995). A recent work describing the SOM in data analysis and exploration is (Kaski, 1997).

2.2. Introductory example: from a table to a map

As a very simple illustrative example we demonstrate how different languages can be identified from their written forms by the SOM. Consider Table 1 that is an excerpt from a larger table used in the experiment. It gives the relative frequencies of characters in a small text corpus, selected from each of the languages considered.

Table 1. Part of a table containing relative frequencies of characters.

Language	'a'	'b'	'c'	'd'	'e'	'f'	'g'	'h'	
English	0.078	0.016	0.019	0.053	0.128	0.018	0.024	0.078	...
Estonian	0.118	0.008	0.000	0.055	0.127	0.002	0.016	0.019	...
Finnish	0.119	0.000	0.000	0.007	0.083	0.000	0.000	0.022	...
French	0.069	0.005	0.035	0.042	0.139	0.010	0.015	0.009	...
Hungarian	0.088	0.020	0.009	0.023	0.102	0.007	0.034	0.021	...
...		...	...	...	...	...	...	...	...

Mutual similarities of the different languages are very hard to deduce from this kind of a table. However, when the row vectors of the table are applied as inputs to a self-organizing map, an outcome of the learning process is that statistically similar vectors (rows) become mapped close to each other in the display. In Fig. 2, similar statistics of characters in different languages are reflected in their "images" being correspondingly close on the map: the nearer the "images" of two languages (like *Danish* and *Norwegian*) are, the more similarities they have in their written features.

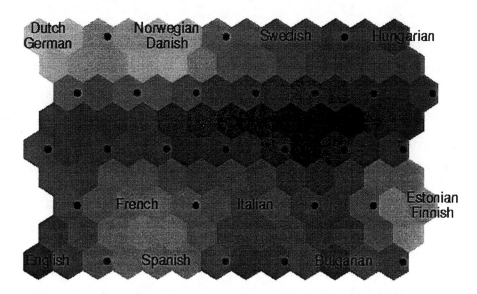

Fig. 2. *A self-organized hexagonal map based on the relative frequencies of characters in some languages. The map tends to preserve the topology of the original input space.*

The comparison was based on the character set which was used in the original material obtained from the Internet. The aim was not to provide a detailed linguistic analysis and, thus, the input material was randomly selected. Simplifying transliterations of many accented characters had been used the original texts. The phonetic resemblances were not considered. However, although only such a set of crude features derived from a small corpus was used as input data, some basic relationships can be seen on the map. The Romance and Germanic languages fall into areas of their own. Finnish and Estonian have been positioned into the same map node. The dark color denotes a larger distance between the codebook vectors. Thus, one can also find meaningful clustering structure on the map while the previously mentioned areas are separated by the "borderlines".

3. Maps of Natural Language

In languages we use signs and conventions to present a kind of terrain. Like maps, language can open up new worlds: we can innovate through language. We can work out problems, understand and communicate complex ideas by unpacking them in words. But like maps, language also distorts. In using language, we necessarily reduce and group and select. This is exactly what self-organizing maps also do: generalize and organize information, as well speech signal, pictorial images as symbolic input if the context is present.

In most practical applications of the SOM, the input to the map algorithm is derived from some measurements, usually after preprocessing. In such cases, the input vectors are supposed to have metric relations. Interpretation of languages, in their written form, is based on the processing of sequences of discrete symbols. To create a map of discrete symbols that occur within the sentences, each symbol must be presented in the due context.

3.1. Basic Principles of Using SOM in Natural Language Interpretation

The Self-Organizing Map, SOM (Kohonen, 1995) is well suited to serve as the central processing element in modeling natural language interpretation because of the following reasons:

- The SOM algorithm modifies its internal presentation, i.e., the map node vectors according to the external input which enables the adaptation.

- The SOM is able to process natural language input to form "semantic maps" (see Ritter and Kohonen, 1989). Natural language interpretation using the SOM has further been examined, e.g., in (Miikkulainen, 1993; Scholtes, 1993; Honkela, 1995).

- Symbols and continuous variables may be combined in the input, and they are associated by the SOM (see, e.g., Honkela, 1991). Continuous variables may be quantized, and a symbolic interpretation can be given for each section in the possibly very high-dimensional space of perceptional variables.

- Because the SOM is based on unsupervised learning, processing external input without any prior classifications is possible (Kohonen, 1995). The map is an "individual" model of the environment and of the relation between the expressions of the language and the environment.

- The SOM enables creating a model of the relation between the environment and the expressions of the language used by the others. In addition, generalizations of this relations can be formed (Honkela, 1993).

3.2. Creating Maps of Words

It has earlier been shown that the SOM can be applied to the analysis and visualization of contextual roles of words, i.e., similarities in their usage in short contexts formed of adjacent words (Ritter and Kohonen, 1989). In the unsupervised formation of the of map of words, each input x consists of an encoded word and its averaged context. Each word in the vocabulary is encoded as a n-dimensional random vector. In our experiments (e.g., Honkela et al. 1995, Kaski et al. 1996, Lagus et al. 1996) n has usually been 90. A more straightforward way to encode each word would be to reserve one component for each distinct word but then, especially with a large vocabulary, the dimensionality of the vector would be computationally intractable. A mathematical analysis of the dimensionality reduction based on random encoding is presented by Ritter and Kohonen (1989).

The basic steps in forming a map of words are listed in the following (see also Fig. 3):

1. Create a unique random vector for each word in the vocabulary.

2. Find all the instances of each word to be mapped in the input text collection. Such words will be called key words in the following. Calculate the average over the contexts of each key word. The random codes formed in step 1 are used in the calculation. The

context may consist of, e.g., the preceding and the succeeding word, or some other window over the context. As a result each key word is associated with a contextual fingerprint. If the original random encoding for a single word is, for instance, 90-dimensional, the resulting vector of key word with context is 270-dimensional if one neighboring word from both sides is included. The classifying information is in the context. Therefore the key word part of the vector is multiplied by a small scalar, ε. In our experiments ε has usually been 0.2.

3. Each vector formed in step 2 is input to the SOM. The resulting map is labeled after the training process by inputting the input vectors once again and by naming the best-matching neurons according to the key word part of the input vector.

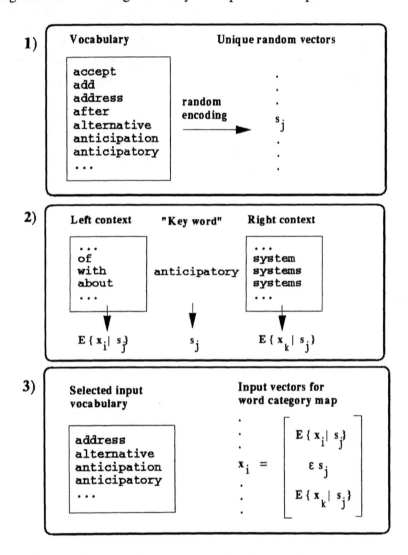

Fig. 3. Creation of input vectors for the word category map.

3.3. Comparisons of Artificial Neural Networks Models

Artificial neural-network models may be classified according to their specific features, such as structure, dynamics, and type of adaptation. The nature and source of the input material and its preprocessing should also be specified. The input data material may be symbolic, numerical,

569

or a combination of the both. The speech signal is a typical example of continuous-valued, non-symbolic input material. On the other hand, written texts are symbolic, although they may be transformed into vectorial form in some way. One method of such a transformation was presented in the previous chapter.

The appearance of natural-language expressions, written or spoken, is sequential. This inherent property raises the question of how to handle time. Elman (1991) presents different ways of representing time. One possibility is to concatenate subsequent "moments" into single input vectors like in the current study. Another possibility is to use networks with recurrent connections to make them operate sequentially. The learning principle of the SOM, however, is not suitable for such recursive operations. On the other hand, the self-organizing map is the only neural-network architecture in which the spatial order between representations emerges automatically. Such a self-organized order is also one of the main properties of the neural structures in the brain. Neurophysiological studies supporting the existence of SOM like processing principles in the brain have been reviewed in Kohonen (1995).

The basic learning strategies of adaptive systems can be categorized into supervised, reinforced, and unsupervised. In supervised learning, the system is given input-output pairs: for each input there must also exist the "right answer" to be enforced at the output. The system then learns these input-output pairs. The task is not trivial, however, and after the learning period the network is also able to deal with inputs that were not present in the learning phase. This property ensues from the generalizing capabilities of the network. The drawback of supervised learning is the need for correct output in each input example. In some cases, obtaining the output for each input case is a very laborious task especially if a large source material is used.

Whereas supervised learning models are suitable for classification, unsupervised learning can be used for abstraction. The self-organizing map, considered in this article, enables autonomous processing of linguistic input. The map forms a structured statistical description of the material, allowing very nonlinear dependencies between the items. Application of the self-organizing maps to natural language processing has been described, e.g., by Ritter and Kohonen (1989), Scholtes (1993), Miikkulainen (1993), Honkela et al. (1995).

4. Epistemological Considerations

In the following, the problem areas presented in the introduction and the methodological tools provided by the self-organizing maps and the underlying principles, are tied together.

4.1. Emergent Categories

Conceptually interrelated words tend to fall into the same or neighboring node in the word category map (see, e.g., Kaski et al., 1996; Kohonen et al., 1996). Fig. 4 shows the results of a study in which texts from the Usenet newsgroup sci.lang were used as input for the SOM (Lagus et al., 1996). The overall organization of a word category map reflects the syntactic categorization of the words. In the study by Honkela, Pulkki, and Kohonen (1995) the input for the map was the English translation of Grimm fairy tales. In the resulting map, in which 150 most common words of the tales were included, the verbs formed an area of their own in the top of the map whereas the nouns could be found in the opposite corner. The modal verbs were in one area. Semantically oriented ordering could also be found: for instance, the inanimate and animate nouns formed separate clusters.

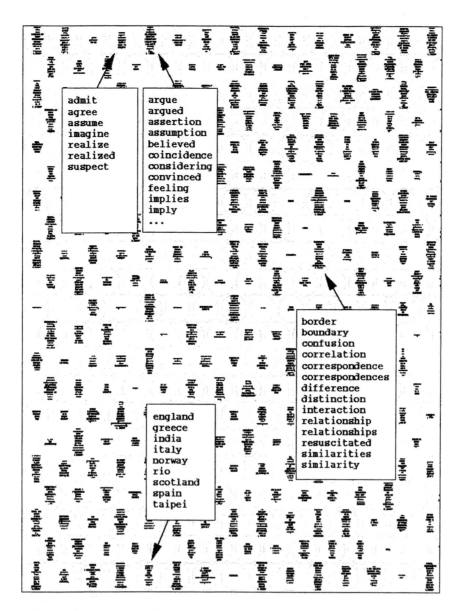

Fig. 4. *An illustration of a word category map based on the texts of the Usenet newsgroup sci.lang. The map consists of 15 x 21 nodes most of which contain several words.*

An important consideration is that in the experiments the input for the SOM did not contain any predetermined classifications. The results indicate that the text input as such, with the statistical properties of the contextual relations, is sufficient for automatical creation of meaningful implicit categories. The categories emerge during the learning process. The symbol grounding task would, of course, be more realistic and complete if it is were possible to provide also other modalities such as pictorial images as a part of the context information.

4.2. Intersubjectivity and Relativity: Relation Between Discrete and Continuous

Subjectivity is inherent in human natural language interpretation. The nature and the level of the subjectivity has been subject to several debates. For instance, the Chomskian tradition of linguistics as well as the philosophy of language based on predicate logic seem clearly to undermine the subjective component of language processing. In them, the relation between the "names" of the language and the "objects" or "entities" may be taken as granted and to be unproblematic.

Consider now that we are about to denote an interval of a single continuous parameter using a limited number of symbols. These symbols are then used in the communication between two subjects (human or artificial). In a trivial case two subjects would have same denotations for the symbols, i.e. the limits of the intervals corresponding to each symbol would be identical. If the "experience" of the subjects is acquired from differing sources, the conceptualization may very well differ. This kind of difference in the density pattern is illustrated in Fig. 4a. In Fig 4b, the interval from x_0 to x_4 is divided into smaller intervals according to the patterns. The first subject uses two symbols, A and B, whereas the second subject also utilizes a third once, namely C for the interval between x_1 and x_3 Thus, if the context (the parameter value in this simplified illustration of Fig. 5) is not present, a communicated symbol may lead to an erroneous interpretation.

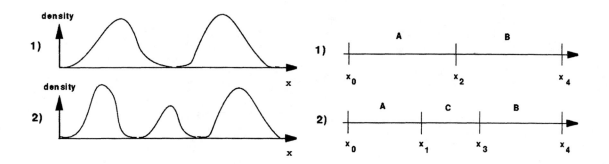

Fig. 5. Illustration of symbols associated to a continuous parameter by two subjects (A and B). In the left, the varying density patterns show the motivation for different conceptualizations of the same phenomenon. The conceptualization, i.e., naming the space, is presented in the right.

One may then ask how to deal with this kind of discrepancies. A propositional level is not sufficient. The key idea is to provide the means for a system to associate continuous-valued parameter spaces to sets of symbols, and furthermore, to "be aware" of the differences in this association and to learn those differences explicitly. These kinds of abilities are especially required by highly autonomous systems that need to communicate using an open set of symbols or constructs of a natural language. This kind of association of set of symbols and a set of continuous parameters is a natural extension or modification of the word category maps (see Honkela, 1993; Honkela and Vepsäläinen, 1991). An augmented input consists of three main parts: the encoded symbol, the context which is the parameter vector in this case, and identification of the utterer or source of the symbol being used. The map nodes associate symbols with the continuous parameters. One node corresponds to an area in the multidimensional space, i.e., a Voronoi tessellation determined by the codebook vector

572

associated with the map node and its neighboring nodes. The relation is one-to-many: one symbol is associated with infinitive number of points.

In this kind of mapping, the error (cf. Rosen, 1985), or a kind of relativity is a necessity in communication. One can define the exact reference of a symbol in a continuous space only to a limit that is restricted by several issues, for instance, the limited time available for communication in which the level of intersubjectivity is raised. A common source of context is often not available either. Von Foerster (1972b) has outlined the very basic epistemological questions that are closely related to the topics of the present discussion. He states, among other things, that by keeping track of the computational pathways for establishing equivalence, "objects" and "events" emerge as consequences of branches of computation which are identified as the processes of abstraction and memorization. In the realm of symbolic logic the invariance and change are paradoxical: "the distinct being the same", and "the same being distinct". In a model that includes both the symbolic description as well as the continuous counterpart, there is no paradox, and the relationship may be computed, e.g., by the self-organizing map.

The previously presented framework also provides a means to consider the relationship between language and thoughts. In the case of colors, one may hypothesize that the perceptual input in the human experience is overwhelming when compared with the symbolic descriptions. Thus, the "color map" is based on the physical properties of the input. On the other hand, abstract concepts are based on the cultural "agreements" and they are communicated symbolically so that the relation to external, physical counterparts is less straightforward. A natural result would be that such concepts are much more prone to subjective differences based on the cultural environment. Even if the original perceptual input is available but it is constantly associated with a systematic classifying symbolic input, the result deviates strongly compared with the case in which the latter information is not available. Von Foerster (1972a) has described the phenomenon and its consequences in the following way: "We seem to be brought up in a world seen through descriptions by others rather than through our own perceptions. This has the consequence that instead of using language as a tool with which to express thoughts and experience, we accept language as a tool that determines our thoughts and experience." In linguistics this kind of idea is referred to as the Sapir-Whorf hypothesis.

4.3. Anticipatory Systems

Rosen (1985) has described anticipatory behavior to be one in which a change of state in the present occurs as a function of some predicted future state. In other words, an anticipatory system contains a predictive model of itself and/or its environment, which allows it to change state at an instant in accord with the model's predictions pertaining to a later instant.

Music involves the expectation and anticipation of situations on the one hand, and confirmation or disconfirmation of them on the other. Kaipainen (1994) has studied the use of SOMs in modeling musical perception. In addition to the basic SOM, Kaipainen uses a a list of lateral connections that record the transition probabilities from one map node to another. The model is based on the specific use of the SOM in which the time dynamics of a process are characterized by the trajectory path on the map. This aspect has been important already in the first application area of the SOMs, namely speech recognition, and more recently in process monitoring. The model of musical perception was tested in three modes called "Gibsonian", "autistic", and "Neisserian". The Gibsonian and autistic are the two extremes regarding the use of anticipation recorded in the trajectory memory: the first model was designed so that it did not use the trajectory information at all. The result was that continuity from one variation to another could not be maintained. On the other extreme, the autistic model was parameterized so that it developed a deterministically strong schematic drive. It began to use its internal representational states, eventually becoming ignorant of the input flow of musical patterns. The intermediate model, denoted as Neisserian, was the one that performed best in musical terms. It was open to to the input having at the same time an internal schematic drive, anticipation, which intentionally actualized musical situations rather than just recognizing them as given.

When the relationship between Rosen's formulations and Kohonen's self-organizing maps is considered the following quote may be of interest (Rosen, 1985): "Briefly, we believe that one of the primary functions of the mind is precisely to *organize* percepts. That is, the mind is not merely a passive receiver of perceptual images, but rather takes an active role in processing them and ultimately in responding to them through effector mechanism. The *organization* of percepts means precisely the establishment of relations between them. But we then must admit such relations reflect the properties of the active mind as much as they do the percepts which the mind organizes." It seems that the SOM concretizes this idea. In general, Rosen's point of view may be characterized as physical and biological whereas Kohonen's main results are related to computational, epistemological, neurophysiological, and cognitive aspects. Many basic issues are interrelated, though, including those of error, order and disorder, similarity, and encoding.

5. Conclusions

The SOM-based approach seems to be well in line with the systemic and holistic principles widely adopted in the cybernetic research community while it concentrates on the interaction between processing elements, studies the effects of interactions, is especially suited to study of nonlinear phenomena with mutually dependent variables, and the validation of the results is based on comparison of the behavior of the model with reality.

In natural language processing new promising application areas are arising. It may be concluded that we are gradually learning to understand how we learn to understand.

References

Elman Jeffrey (1991). Finding Structures in Time. Cognitive Science, 16, pp. 96-132.

von Foerster Heinz (1972a). Perception of the Future and the Future of Perception. Instructional Science 1, 1, R.W. Smith, and G.F. Brieske (eds.), Elsevier/North-Holland, New York/Amsterdam, pp. 31-43. (Also appeared in von Foerster, H.: Observing Systems, Intersystems Publications, Seaside, CA, 1981, pp. 189-204.)

von Foerster Heinz (1972b). Notes on an Epistemology for Living Things. BCL Report No. 9.3, Biological Computer Laboratory, Department of Electrical Engineering, University of Illinois, Urbana, 22 p. (Also appeared in von Foerster, H.: Observing Systems, Intersystems Publications, Seaside, CA, 1981, pp. 258-271.)

Honkela Timo and Vepsäläinen Ari M. (1991). Interpreting Imprecise Expressions: Experiments with Kohonen's Self-Organizing Maps and Associative Memory. Artificial Neural Networks, T. Kohonen and K. Mäkisara (eds.), vol. I, 897-902.

Honkela Timo (1993). Neural Nets that Discuss: A General Model of Communication Based on Self-Organizing Maps. Proc. ICANN'93, Int. Conf. on Artificial Neural Networks, S. Gielen and B. Kappen, Springer, London, 408-411.

Honkela Timo, Pulkki Ville, and Kohonen Teuvo (1995). Contextual relations of words in Grimm tales analyzed by self-organizing map. In F. Fogelman-Soulie and P. Gallinari (eds.) ICANN-95, Proceedings of International Conference on Artificial Neural Networks, vol. 2, pp. 3-7. EC2 et Cie, Paris.

Honkela Timo, Kaski Samuel, Lagus Krista, and Kohonen Teuvo. Newsgroup Exploration with WEBSOM Method and Browsing Interface. Report A32, Helsinki University of Technology, Laboratory of Computer and Information Science, January, 1996.

Honkela Timo (1997). Self-Organizing Maps of Words for Natural Language Processing Applications. Proceedings of Soft Computing '97, in print, September 17-19, 1997, 7 p.

Honkela Timo (1997). Emerging categories and adaptive prototypes: Self-organizing maps for cognitive linguistics. Extended abstract, accepted to be presented in the International Cognitive Linguistics Conference, Amsterdam, July 14-19, 1997.

Kaipainen Mauri (1994). Dynamics of Musical Knowledge Ecology - Knowing-What and Knowing-How in the World of Sounds. PhD thesis, University of Helsinki, Helsinki, Finland, Acta Musicologica Fennica 19.

Kaski Samuel (1997). Data Exploration Using Self-Organizing Maps. Dr.Tech thesis. Helsinki University of Technology, Espoo, Finland, Acta Polytechnica Scandinavica, no. 82.

Kaski Samuel, Honkela Timo, Lagus Krista, and Kohonen Teuvo (1996). Creating an order in digital libraries with self-organizing maps. Proceedings of WCNN'96, World Congress on Neural Networks, Lawrence Erlbaum and INNS Press, Mahwah, NJ, pp. 814-817.

Kohonen Teuvo (1982). Self-organized formation of topologically correct feature maps. Biological Cybernetics, 43, pp. 59-69.

Kohonen Teuvo (1995). Self-Organizing Maps. Springer-Verlag.

Kohonen Teuvo, Kaski Samuel, Lagus Krista, and Honkela Timo (1996). Very large two-level SOM for the browsing of newsgroups. Proceedings of ICANN'96, International Conference on Artificial Neural Networks.

Lagus Krista, Honkela Timo, Kaski Samuel, and Kohonen Teuvo (1996). Self-organizing maps of document collections: a new approach to interactive exploration. E. Simoudis, J. Han, and U. Fayyad (eds.), Proceedings of the Second International Conference on Knowledge Discovery & Data Mining, AAAI Press, Menlo Park, CA, pp. 238-243.

Miikkulainen Risto (1993). Subsymbolic Natural Language Processing: An Integrated Model of Scripts, Lexicon, and Memory. MIT Press, Cambridge, MA.

Ritter Helge and Kohonen Teuvo (1989). Self-organizing semantic maps. Biological Cybernetics, vol. 61, no. 4, pp. 241-254.

Rosen Robert (1985). Anticipatory Systems. Pergamon Press.

Scholtes Jan C. (1993). Neural Networks in Natural Language Processing and Information Retrieval. PhD thesis, University of Amsterdam, Amsterdam.

Emergent Dynamical Properties of Biological Neuronal Ensembles and their Theoretical Interpretation and Significance

Guenter W. Gross[1] and Jacek M. Kowalski[2]
Center for Network Neuroscience
Departments of Biological Sciences (1) and Physics (2)
University of North Texas, Denton Texas 76203

Abstract

New culturing and recording techniques allow long-term, real-time, and multisite monitoring of spontaneous and evoked spike activity from small neuronal networks grown from dissociated embryonic mouse CNS tissue. These systems, typically composed of several hundred neurons, generate complex spatio-temporal spike and burst patterns and exhibit interesting dynamical properties that are generally non-linear and non-stationary with a multitude of inter-state transitions. We have observed network self-activation (or "ignition") to bursting states; complex oscillatory bursting as a basic activity mode of a cellular group; simple, regular oscillations during disinhibition; transient synchronization or approximate phase-locking within active groups; dynamical recruitment with spike frequency-dependent activation of additional neurons and their incorporation into a common network pattern; and dynamical competition among different active cell groups.

In addition to providing a rich test-bed for the application of new data processing and simulation methods, the observed dynamics may reflect some of the behaviors of neuronal networks *in vivo*. They also provide insight into the fault tolerance and packing density problems in neuronal systems. Because individual neurons are relatively unreliable units, a high degree of reliability can be achieved only in a group environment. Therefore, it is reasonable to assume that ensemble properties play a significant role in information processing. Furthermore, if ensemble properties dominate computational events, then the modeling of ensemble dynamics, rather than that of individual neurons, holds the potential for simplifying the ensuing mathematical descriptions to a degree at which realistic emulations may actually be possible.

Key Words

Network dynamics, neurons in culture; self-organization; spatio-temporal spike patterns; burst synchronization.

1. Introduction

Information processing in animals requires a multitude of dynamical events and strategies at several organizational levels. Presently, many of these events are unknown. Although the fundamental neural mechanisms involved have molecular and cellular origins, perception and behavior emerge from sub- and suprathreshold electrophysiological patterns that begin to organize on the level of the neuronal group or neuronal network. In higher animals, behavior is the expression of incredibly complex electrophysiological dynamics that involve many thousands of such networks. The dynamics generated are both non-linear and non-stationary, so that simple neuronal ensembles are inherently attractive, because their organizational principles may be understood in the near future and because some of these principles may be repeated at higher levels. It also is hoped that organizational principles of small neuronal ensembles may lead to a new family of mathematical models that can represent biological systems and, therewith, take us beyond the present level of artificial neural networks (ANN). It is important to recognize that

CP437, *Computing Anticipatory Systems: CAYS--First International Conference*
edited by Daniel M. Dubois © 1998 The American Institute of Physics 1-56396-827-4/98/$15.00

studies of biological network dynamics are in their early stages and could not be incorporated into the existing ANN schemes. Network behavior in ANN is theoretically inferred from highly reductionist approaches and is not likely to represent the actual functioning of the biological system. We still have much to learn from biology. A bumblebee with a total neural mass smaller than the head of a pin can perform incredible feats of pattern recognition and sensory-motor integration. How this is done should not remain a mystery forever!

It is generally accepted that not single neurons but networks represent the fundamental processing units in the brain (Edelman, 1987; Sporns et al., 1991; Levitan and Kaczmarek, 1991; Freeman, 1992; MacGregor, 1993; Douglas et al., 1995). Pattern generation, recognition, manipulation, and storage are all accomplished in a distributive manner by a large number of nerve cells and an even larger number of synapses. Hence, the structural organization, dynamic capabilities, and transient response characteristics of networks are vital investigative domains in neuroscience. It may be argued that we will not understand information processing in the central nervous system until we understand pattern processing in networks. Yet, because of methodological difficulties associated with the simultaneous monitoring of a large sample of neurons in a network, this research area so far has received limited attention.

A basic organizational principle of neocortex architecture is that neurons with similar functions are aggregated together in columns or slab-like arrangements (Mountcastle, 1982; Yuste, 1994; Douglas et al., 1995). Although lateral connections must exist, most connections are made locally within 1mm columns (Hubel and Wiesel, 1962). Theoretical considerations have supported these observations by suggesting the majority of synapses must be made close to their source neurons so as to minimize the volume of axonal wiring and prevent an explosive growth of cortical volume with increasing cell number (Mead, 1990; Mitchison, 1992). It also has been suggested that approximately 80% of the synapses in gray matter are excitatory, and most originate from other cortical neurons (Braitenberg and Schütz, 1991). From these observations, it may be inferred that excitatory cortico-cortical synapses must contribute strongly to pattern generation and processing and that a basic feature of cortical function is massive local excitation. It also is evident that the first important organizational steps are taken within these small, directly-interconnected "processing units".

Studies of the function of such small, densely packed processing units *in vivo* is difficult, because conventional micropipette or wire bundle microelectrodes, if introduced to such small volumes, destroy much of the tissue to be studied. Even sophisticated microelectronic electrodes with linear arrays of 10 to 20 photoetched microelectrodes (Wise, 1975; Kuperstein and Whittington, 1981) are not able to monitor a sufficiently large number of constituent neurons in parallel to provide data on the internal dynamics of such ensembles. Optical methods using voltage-sensitive dyes cannot penetrate effectively into the cell columns and could not be used for long-term (days to weeks) analyses without toxic effects on the tissue. Given these methodological challenges, the cell culture option, when combined with array recording techniques, becomes attractive, because it allows the growth of tissue on recording arrays. These new models of network development and function offer several important advantages: (1) The networks grow on the electrode array insulation layer and develop a highly stable cell-electrode coupling (Gross and Schwalm, 1994). Networks are not injured or changed in any way during preparation for recording. (2) The networks develop spontaneous activity and remain active for many months (maximum to date: 312 days). (3) The networks are histiotypic in their pharmacological responses (Gross and Kowalski, 1991; Gross, 1994; Gopal and Gross, 1996b). Sensitivities to some compounds are in the low nanomolar range (e.g. strychnine, Gross et al., 1991; 1995). (4) The present 64-electrode array yields spatial and temporal statistical data, with multiple chances to select high signal-to noise-ratio samples of representative spike (action potential) activity. This approach provides data on both single cell output and population responses. (5) The *in vitro* environment allows highly reproducible pharmacological and

578

chemical manipulations with no homeostatic interference from other organs. Also, networks *in vitro* are isolated and do not receive undefined input from other neural tissue. (6) In special chambers, the monolayer networks allow long-term (days to weeks) optical, photometric, and electrophysiological monitoring, as well as laser microbeam cell surgery. Major structural changes can be correlated with electrophysiological changes, and network development or deterioration can be followed with time-lapse photography.

2. Experimental Environment

Neuronal networks *in vitro* develop spontaneous activity approximately one week after seeding. This consists mostly of spiking with some weak bursting that is rarely coordinated among channels. At two weeks of age, bursting is stronger, and more recruitment into common patterns is seen. Four weeks after seeding, the networks display a complex repertoire of spatio-temporal spike patterns (Gross and Kowalski, 1991). That level of complexity seems to remain stable for up to six months. Beyond this time, not enough data are available for comparison. Fig. 1 shows the recording matrix of a 64-microelectrode array fabricated in this laboratory. A low density network has developed on the array consisting of four rows of 16 microelectrodes each. Indium-tin oxide conductors are 10 μm wide in the recording matrix and terminate in a shallow, 20 μm diameter recording crater. Neuronal cell bodies (darkly stained cells) reside on a thin glial carpet; however, neurites (especially axons) may be found both on top and underneath the carpet. The glial carpet is confluent, which inhibits glial movement and stabilizes the entire network. Oval, lightly stained structures are glial nuclei. Table 1 summarizes the support environment.

TABLE 1. Support Environment

EXPERIMENTAL SYSTEM....	Monolayer networks growing on thin-film electrode arrays.
TISSUE SOURCE......................	Mouse embryo CNS tissue (14.5 days gestation) dissociated and seeded at 500k on 600 mm^2 surface areas.
CULTURE PLATES.................	5 x 5 cm x 1.2 mm soda-lime barrier glass (1,200Å layer of SiO_2);
ELECTRODE ARRAY..............	64 ITO thin film microelectrodes terminating in a 0.8 x 0.8 mm recording area.
INSULATION............................	Polysiloxane (DC 648, Dow Corning)
DE-INSULATION.....................	Pulsed nitrogen laser , (λ=337nm, E.D.= 2 μJ/μm^2, single shot)
SURFACE PREPARATION....	Oxidation of CH_3 groups to OH- groups via flaming (1 s exposure to propane or butane flame); addition of poly-D-lysine & laminin.
NETWORKS..............................	Randomized cell body adhesion; spontaneously active; minimum size: 1 mm^2 (50 neurons); maximum size: 600 mm^2 (30k neurons)

2.1 Cell Culture

Dissociated spinal and cortical tissue from 14.5-day-old mouse embryos is cultured according to the method of Ransom et al. (1977) with minor modifications that include the use of papain and DNAse for tissue dissociation (Huettner and Baughman, 1986). This procedure produces isolated cells with retracted processes. Cells are seeded on multimicro-electrode plates (MMEPs) containing 64 thin-film electrodes in a 1 mm^2 area (Gross et al., 1985). Poly-D-lysine (25 μg/ml; 30-70kD, Sigma) plus laminin (16 μg/ml) are used for substrate preparation. Neurons are maintained for one week in minimal essential medium (MEM, GIBCO, Gaithersburg, MD) containing 10% fetal bovine serum and 10% horse serum (HyClone Lab, Logan, Utah). Thereafter, cells are fed two times per week with MEM containing 10% horse serum. For spinal cultures, when the glial carpet becomes confluent, glial growth is inhibited by the addition of 7 μM cytosine arabinofuranoside (Sigma). The cultures are maintained at 37°C in an atmosphere of 90% air and 10% CO_2.

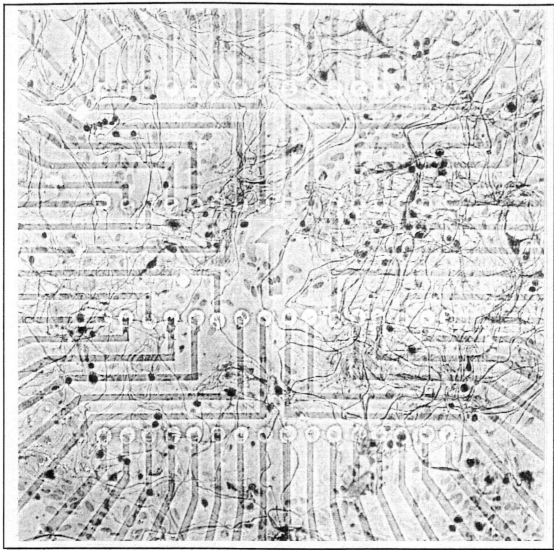

Fig. 1. Recording matrix of microelectrode array consisting of 64 indium-tin oxide conductors terminating in an 0.8 x 0.8 mm region. In the center, conductors are 10 μm wide and spaced 40 μm between columns and 200 μm between rows. The polysiloxane insulation surface is activated and decorated with polylysine and laminin to allow cell adhesion and growth. Most recordings are obtained from axons crossing the recording craters.

2.2 Fabrication of Microelectrode Arrays

The techniques involved in the preparation of photoetched multimicroelectrode plates (MMEPs) and the concomitant culture methods have been described in previous publications (Gross, 1979; Gross et al., 1985; Gross and Kowalski, 1991; Gross, 1994). Commercially available, sputtered ITO plates (1 mm thick) are photoetched, cut into 5cm x 5cm wafers, spin-insulated with polysiloxane, cured, de-insulated at the electrode tips with single laser shots, and electrolytically gold-plated to adjust the interface impedance of the exposed ITO to approximately 1 Mohm. The MMEP insulation material (Dow Corning DC648 polysiloxane resin) is hydrophobic. The surface can be activated by a brief pulse from a propane flame (Lucas et al., 1986). Flaming through masks allows the creation of coarse adhesion patterns such as 1-4 mm diameter islands centered on the electrode array. Poly-D-lysine and laminin are added to the activated surface regions. The insulation has proven remarkably stable despite repeated sterilization via autoclaving and months under warm saline.

2.3 Life-Support, Recording, and Observation

Both open and closed recording chambers have been developed that allow, respectively, network maintenance in a constant bath of 1 to 2 ml medium, or in a much smaller volume of 0.3 ml under a constant medium flow at 10-40 µl/min. The former design is well-suited to rapid medium changes and short-term pharmacological studies. The latter provides a more controlled environment in a flow chamber that allows high power microscopy with upright and inverted microscopes. The performance of both chambers has been described in detail (Gross and Schwalm, 1994; Gross, 1994). The chamber consists of an aluminum base holding the MMEP, a stainless steel chamber cover with the glass microscope port, and medium supply lines. Two zebra strips (carbon-filled silicone elastomere, Fujipoly, Cranford, N.J.) are pressed between the amplifier circuit board and the MMEP to provide electrical contact with the recording matrix.

Multielectrode recording is performed with a computer-controlled, 64-channel amplifier system commercially available from Plexon, Inc., Dallas. VLSI multichannel preamplifiers (Dept. Elect. Eng., Southern Methodist Univ., Dallas) are positioned on the microscope stage to either side of the recording chamber. The preamps allow computer selection of any combination of channels for electrical stimulation with 4 different patterns. The amplifier bandwidth is usually set at 500Hz to 6kHz. Activity is displayed on oscilloscopes and recorded on a 14-channel Racal direct tape recorder. Spike data from active channels also is integrated (rectification followed by RC integration with a resulting time constant of 500ms) and displayed on a 12-channel Graphtek strip chart recorder. The reference electrode is attached to the amplifier ground, which is connected to the stainless steel chamber holding the culture medium.

3. Characteristics of Networks in Culture

3.1 Morphological Self-Assembly

The dissociation of embryonic tissue produces isolated nerve and glia cells. Both are seeded together and begin to interact with the polylysine and laminin surface at the same time. Initial stages of organization seem to be determined by strength of adhesion to the substrate. Axons and dendrites are more tightly adhered than are cell bodies (or somata). Consequently, a neuron-glia competition for the surface usually results in the glia cells growing below the somata. The opposite happens with dendrites or axons that already are established on the substrate. In that situation, an approaching glia cell often will climb over the process . These simple factors, together with more complicated and still unidentified organizational principles, produce a non-homogeneous distributions of neurons (Fig. 2). Neurons, especially those with small somata, frequently are found in clusters. Larger neurons often are found in such proximity that extensive dendritic contact is possible. Consequently, it is difficult to ignore the possibility that extensive subthreshold activity (i.e. membrane potential fluctuations below the action potential level) may play a powerful role in these cultured networks. The systems begin to stabilize as soon as extensive bursting develops in the network. It is known that action potentials inhibit growth cone motility (Cohan and Kater 1986; Fields and Nelson, 1994), and it may be assumed that one role of the spontaneous network activity is to keep the neurons from extensive "hunting" for new targets. Although many synaptic connections may be in a constant state of flux, we do not see obvious axonal growth after the third week in culture.

3.2 Spontaneous Network Activity

The slightly recessed craters surrounding the exposed electrode tip often are invaded by axons with glia above the axonal bundle. Under those conditions, signals that approach one millivolt in amplitude are seen often, whereas signals from axons on top of a glial-filled craters are usually close to the noise line (40 μV). The spontaneous activity of cultured networks is complex. For both the dissociated spinal cord tissue and tissue from the auditory cortex, the predominate feature of the activity is bursting (Fig. 3). A burst is a sudden increase and decrease in firing frequency that allows measurements of burst duration, interval, and period. Maximum spike frequencies, although not visible from these panels, also can be determined in a burst. Hence, the first convenient step of feature extraction is the quantification of the spatio-temporal burst patterns. Thereafter, more complicated spike pattern analyses can be added to the data processing effort. On a single channel, burst patterns can range from random to periodic. However, identifiable patterns are transient and can be locked into a stationary pattern only pharmacologically, especially by the blocking of inhibitory synapses (panel C). Native oscillatory states in cultured networks generally are short-lived and typically damped in period. Only a few experiments

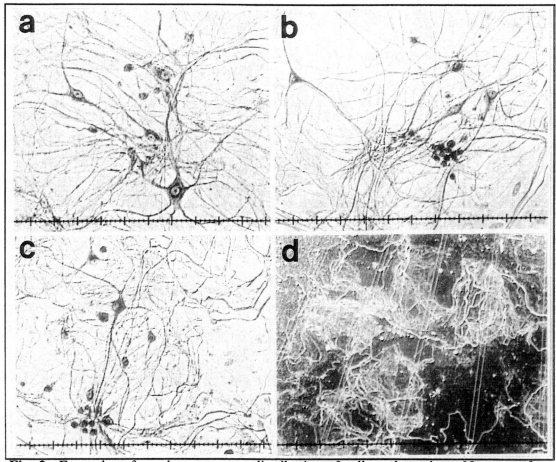

Fig. 2. Examples of non-homogeneous distribution of cells and neurites. Neurons often form small clusters with extensive cell body or dendritic contact. (a) Large multipolar neuron in extensive dendritic contact with three large and five smaller neurons. (b) Cluster of small neurons in field of distributed large neurons. (c) Large multipolar neuron in dendritic contact with cluster of 7 small neurons. (d) Dark field photograph of a 2.5 mm wide region showing clustering of neurites. The neurite clusters form in response to a higher density of cells in those locations (cells not visible). Diagonal lines delineate insulated conductors (50 μm wide) outside the recording matrix. Scale for a, b, and c: one large division equals 40 μm.

have shown long episodes of native oscillations, usually when the cultures have been left undisturbed for long periods of time. When long periods of spontaneous oscillations occur, the bursting always is approximately synchronized.

Fig. 3 shows three panels, each with 16 channels of discriminated spike activity from the same network under different conditions. Shortly after assembly (A), the culture reveals bursting and considerable spiking between bursts. One hour after assembly (B) there is less interburst spiking but similar complex temporal bursting. Finally, the activity after addition of 40 μM bicuculline is shown in (C). Bicuculline is a blocker of the inhibitory transmitter gamma amino butyric acid (GABA). Hence, it represents a "disinhibition" of the network. However, in spinal cord tissue, inhibitory receptors using glycine as a transmitter still remain functional. Nevertheless, this partial disinhibition results in remarkably regular burst oscillations with little spiking between bursts. Each channel represents a different active unit in the network.

Spike and burst patterns are difficult to quantify because repetitions, if they occur, are never exactly the same. However, it is possible to extract major pattern features by visual inspection. For example, panel (A) of Fig. 3 demonstrates the existence of 3 patterns: 14 channels are coordinated into one pattern with only channels 4, 11, and 13 showing obvious

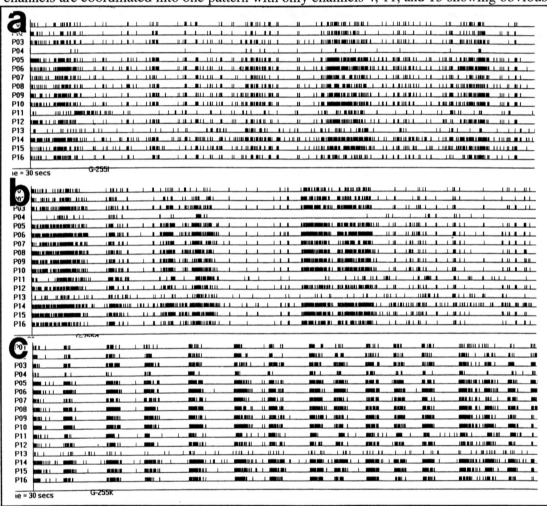

Fig. 3. Typical network recording showing 16 channels of discriminated spike activity for a 30 sec period. (A) 15 minutes after chamber assembly, (B) one hour after assembly, (C) activity after addition of 40 μM bicuculline. The channels, representing different active units in the network, show activity coordination, especially during high frequency spiking and quasi-synchronization after addition of bicuculline (C).

differences. After disinhibition (C), all channels except 13 reflect similar patterns.. As channel 13 does not show strong features of a totally different pattern, we consider this culture to be one network because it recruited all its representative neurons into one pattern. More quantitative descriptions using cluster plots have been described previously (Gross, 1994).

The complexity of typical bursts is shown in Fig. 4 with a simultaneous two channel printout from a digital oscilloscope. Signal-to-noise ratios ranging from 3:1 to 10:1 are typically obtained with this experimental system. Although spike frequencies usually decay with time, there exists a remarkable burst fine structure that has not yet been analyzed. This figure also shows silent periods before and after the burst. Such an absence of spiking after intense bursting is seen often across all channels, which suggests that inhibitory influences are not required to generate such burst patterns.

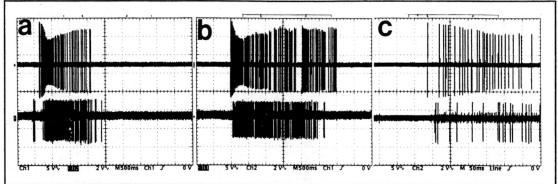

Fig. 4. Typical bursts and their fine structure shown with a two-channel recording from a digital oscilloscope. High frequency activity in the top trace exceeded 200 Hz for (a) and (b). Note decrease of AP amplitude during high frequency spiking. Sweep speeds: (a) and (b) = 500 ms/div; c = 50 ms/div.

3.3 Coexistence of Temporal Patterns

Figure 5A represents complex spontaneous burst activity with at least 3 different patterns coexisting simultaneously, but with the synchronization of one burst (arrow) on all 14 channels. Usually, several electrodes reflect the same pattern, indicating such patterns originate from groups of cells and not single neurons. This activity is not unusual and demonstrates the coexistence of several temporal patterns in a culture. As mentioned in reference to Fig. 3, such independent patterns often merge into one common pattern during disinhibition. When this occurs, we consider the culture to consists of one network. The individual patterns in the native state are assumed to originate in network subdivisions that we call "circuits".

Fig. 5B presents 14-channel integrated spike data displayed on a chart recorder. High amplitude integrated profiles reflect high firing frequencies. It can be seen clearly that intense bursts are synchronized, whereas smaller bursts often maintain their own patterns. Intense bursting leads to synchronization and often engages the entire network in a common global pattern. The transition to such states has been termed "intensity recruitment" (Gross and Kowalski, 1991). Such approximate burst synchronization (with small phase differences) is transient and may occur for simple and complex temporal patterns. Preliminary data from a correlation of network size with the number of major patterns is shown in Fig. 6. Despite substantial fluctuation, these data indicate that larger networks generate a greater number of patterns and demonstrate that disinhibition reduces their number.

Even small monolayer networks, 2-4mm in diameter consisting of less than 500 neurons, may simultaneously display multiple burst patterns, which often are merged into one pattern by disinhibition with GABA receptor antagonists. These data are consistent with a picture in which single cells or, most likely, cell clusters generate suprathreshold spiking from subthreshold oscillations. This spiking, in turn, ignites a local region into bursting that spreads through the network and transiently entrains other regions. Such a

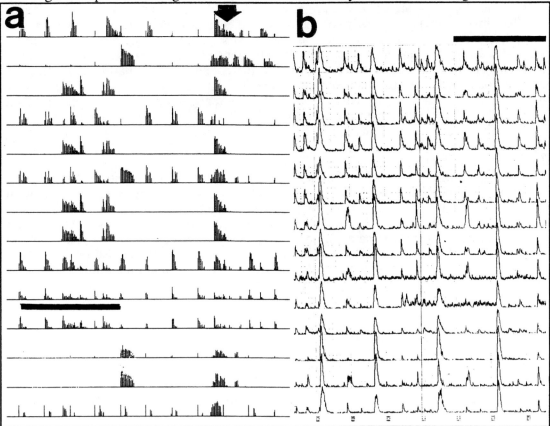

Fig. 5. Coexisting simultaneous patterns in culture. (A) Discriminated spikes per 1 s bin on 14 channels revealing three general burst patterns. (B) Analog integrated burst data showing burst synchronization for the large amplitude integrated profiles. Bar = 1 min.

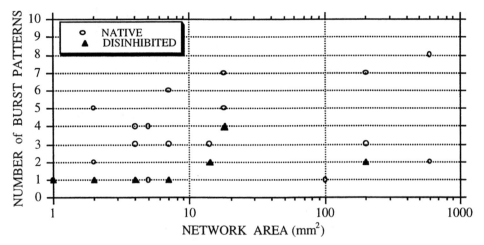

Fig. 6. The number of distinguishable burst patterns increases with network area (number of neurons) and decreases with the blocking of inhibitory synapses (disinhibition).

view suggests that networks are dominated by sensitive loci with high excitatory synaptic density, that there are several of such regions, that they compete for dominance, and that complex network patterns are as much a function of the superposition of circuit patterns and collision phenomena as they are shaped by inhibitory influences.

3. 3 Emergent Properties of Networks

Individual neurons are considered relatively unreliable units that achieve a high degree of reliability only in a group environment. Hence, it is reasonable to assume that ensemble properties play a significant role in information processing. Furthermore, if ensemble properties dominate computational events, then the modeling of ensemble dynamics, rather than the dynamics of individual neurons, holds the potential for simplifying the ensuing mathematical descriptions to a degree where a realistic emulation in electronic or optoelectronic devices may actually be possible. The recognition of higher organizational levels and properties (emergent properties) obviously is essential to our understanding of network dynamics. The exploitation of emergent properties may provide a breakthrough in modeling.

There are four basic dynamical features of the system under study that can be considered as emergent properties of the neuronal ensemble: (1) network self-activation (or "ignition") to bursting states; (2) complex oscillatory bursting as a basic activity mode of a cellular group; (3) simple, regular oscillations during disinhibition; (4) transient synchronization or approximate phase-locking within active groups; (5) dynamical recruitment with spike frequency-dependent activation of additional neurons and their incorporation into a common pattern; (6) dynamical competition among different active cell groups with different dynamics that may lead to transient pattern dominance.

4. Theoretical Considerations

4.1 Model of a Generic Network in Culture

The discussion so far can be summarized in a simple schematic for a generic network in culture (Fig. 7). The model must reflect the tight cell clusters seen in all monolayer cultures, as well as the fact that different patterns can coexist but can be made to merge into a single pattern after disinhibition. As a morphological cell cluster may or may not have any dynamical capabilities, we suggest the term "nacelle" to represent a cell cluster consisting of a small number of neurons with close somatic and/or dendritic contact to allow for constant subthreshold interactions. Nacelles are, therefore, morphologically determined with a stable cell number (for a specific nacelle) and with limited dynamical organization through recruitment. For the next organizational step, we recommend the term "circuits". These represent a closely interconnected group of nacelles that communicate primarily via action potentials and are generally in the same firing mode. In the burst regime, circuits are expected to reflect the same temporal pattern with only small phase delays between bursts. Finally, a group of circuits, interconnected in such a way that entrainment to a common pattern is possible, are designated networks. This is an operational definition for "network" suggested by Gross (1994). In this scheme, cell cultures that unite to a major common pattern under conditions of disinhibition are considered as consisting of one network. Patterns that do not merge during disinhibition reflect the existence of two or more networks.

Given these definitions, several properties should be highlighted. Both circuits and networks have fuzzy boundaries and can change in size with changes in synaptic weights and/or changes in internal dynamics. Circuit size (i.e. cell number) may depend on which nacelle triggers or dominates activity. Especially the circuits are highly flexible and may expand or shrink in cell (or nacelle) number dynamically. The switching of cells between circuits is not new and has been observed in invertebrate ganglia (Hooper and Moulins,

586

1989; Simmers et al., 1995). We would like to support views that consider this phenomenon a common occurrence and a generic behavior of circuits and networks.

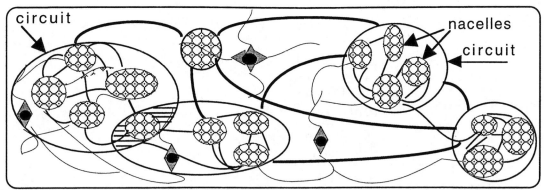

Fig. 7. Generic network in culture. The network consists of several circuits which, in turn, consist of a variable number of nacelles and distributed large neurons. Nacelles represent small clusters of neurons with extensive dendro-dendritic, dendrosomatic, or somato-somatic contact that allow interaction on the subthreshold level. Circuit size may vary dynamically via recruitment of nacelles and single neurons.

4.2 Formation of Spike Patterns in Cultured Networks

Our view of the formation of spatio-temporal patterns in the network is represented schematically in Fig. 8. In the absence of action potential input (in culture that can come only from electrical stimulation), the system is dependent on noise for activation. This is thought to occur in nacelles in which some membrane potential oscillations reach threshold and initiate spiking or bursting. As a second step, a circuit is activated by recruitment of the respective nacelles. Already at this level, the proper conditions must prevail; most nacelles must be close to their thresholds. Also, the emergence of many complex patterns is possible at this level, albeit without reproducibility and pattern fault tolerance. However, the use of many nacelles does provide a fault tolerant ignition mechanism. Equipped with a large number of subthreshold oscillators, the network never dies electrically and "plays with patterns" unless strongly inhibited. The activity now spreads through the network and stimulates other circuits. Without entrainment, each circuit may produce its own (damped) patterns. This type of activity is the actual "resting" or "idling" state of the network. Networks in culture are never quiet.

The presence of idling states may have several important biological implications. They serve as "ready states" that allow the network to respond rapidly to input patterns, and perform as complex, nonlinear filters. It is not unreasonable to suggest that these phenomena may be part of the mechanisms used by anticipatory systems. In many regions of the brain, ready states are most likely a function of the past experience of a network. Hence, a "resonance" between one of the ready states and highly probable incoming patterns may be visualized as underlying unusually rapid, specific responses to stimuli.

Circuits do not exist in isolation but influence each other. The various patterns seen are, therefore, as much a function of the circuit properties as of circuit-circuit interactions. At some point, which is most likely a function of high spike frequencies in bursts, an entrainment phenomenon is triggered that begins to recruit circuits into a common pattern. This is observable in almost every experiment. Weak bursts (low firing frequency and short duration) often are not coordinated among channels, whereas intense bursting with high firing frequencies and relatively long burst durations to enhance burst overlap with other units usually results in the coordination of bursting, development of coarse-grain synchronization, and expression of a common pattern.

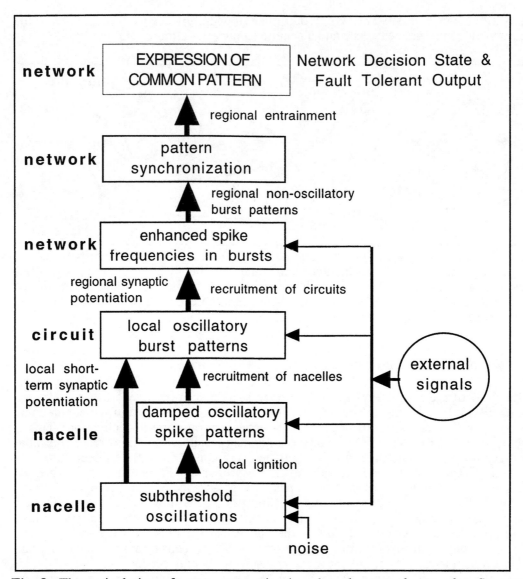

Fig. 8. Theoretical view of pattern generation in cultured neuronal networks. Spontaneous activity is triggered by nacelles as a result of subthreshold oscillations. The recruitment of nacelles with strong excitatory connections leads to the production of spike and burst patterns in circuits. A subsequent recruitment of circuits leads to network activity. Under conditions of strong entrainment, all circuits are drawn into a common pattern.

In summary, networks in culture seem to be systems of smaller, more strongly interconnected local "sub-networks" with highly fuzzy boundaries. As isolated systems, they are influenced only by environmental and neural noise and are assumed to reside in a weakly stable state, close to possible bifurcations that manifest themselves as transitions to subgroup stationary activity modes in terms of nearly synchronized bursting. Previous activity and synaptic potentiation mechanisms may determine the nature of the pattern and the intensity of spiking. Above a certain level of spike frequency (the recruitment threshold), one subgroup becomes dominant and entrains other groups to a common pattern. Under this condition, existing observable burst patterns merge into a global network pattern. Correspondingly, small perturbations received from electrical stimulation (or sensory input in the case of CNS networks *in situ*) will result in the selection of specific patterns and/or the reinforcement of existing patterns (resonance) that lead to rapid recruitment and entrainment to a global pattern.

4.3 The Challenge to Theory

The challenge to theory now is concentrated around the following basic questions: (1) What is a minimal biologically realistic model of a neuronal network with the behaviors described? (2) What are the basic mechanisms of local ensemble formation and local activity synchronization? (3) What are the basic mechanisms of global (network) ensemble formation and network activity synchronization? (4) What is the purpose and biological significance of synchronized bursting modes in contexts other than pacemaker functions? (5) Why is this mode often observed at the network developmental stage? (6) What role does it play in "information processing"? (7) When does a synchronized periodic mode represent pathological (epileptiform) activity?

Some of these questions were addressed by Kowalski et al. (1992) with a model of (1) "point-like" modified Hodgkin-Huxley type neurons, which are not endogenous bursters but may display such behavior when externally driven by other units and/or injected currents, and (2) a simplified phenomenological model of chemical synapses in which significant post-synaptic currents are generated only if rapid (spike-related) changes in the presynaptic membrane potential are present. It was demonstrated that such simplified models typically undergo transitions from quiescent to periodic synchronized bursting states if the interneuronal coupling parameters (synaptic strengths) exceed some critical value. In the simplest case, the dynamics of such a transition can be considered a Hopf bifurcation when the ensemble equilibrium state becomes unstable above some critical coupling strengths and a new, stable periodic state emerges. Much more complex activity modes also are possible in the framework of such models, network "splitting" into synchronized subensembles with a significant phase shift between them, more than one stable activity state, period doubling, and synchronized chaotic modes.

These systems, composed of several model neurons, can be considered a reasonable implementation of a nacelle or a local circuit, depending on the character of the intercellular connnections. Computer simulations of the model show that synchronous, but not necessarily periodic, activity is one of the main modes of a network composed of essentially identical neurons with identical interconnections. However, small changes in the neuronal parameters and/or synaptic efficacies often lead to nearly synchronous activity with small, but fixed, phase shifts between bursts. Our experimental data also show small phase differences during periodic bursting, but the corresponding "phase differences" seem to be random variables with some generic distribution function for a given network (Gross and Kowalski, 1991). This may suggest that parts of real networks displaying nearly synchronous periodic bursting are, on average, structurally homogeneous but undergo small fluctuations of their parameters. In our earlier mathematical model, we did not include noise terms and/or any conduction delays, however we demonstrated that a random reduction of interconnections in an initially homogeneous system leads to desynchronization (Kowalski et al, 1992).

On a more complex level are states when the network is subdivided into two or more nearly synchronized subpopulations with essential phase differences between different groups. This is a "coarse-grain" phase lock that also has been observed in our experiments. The presence of apparently chaotic, but synchronized, modes was one of the most intriguing theoretical observations made by Kowalski et al (1990, 1992). The concept of "synchronous chaos" proved useful in many physical contexts, such as coupled lasers and coding theory. However, the presence of truly chaotic states in biological neuronal networks is still an open question, due to problems with the system stationarity and usually strong noise components.

5. Bridges to Networks *in vivo*

5.1 Differences and Similarities

Recent experiments have revealed that reliable, quantitative information can be obtained from cell culture systems (Gross et al., 1997). The next step must be to demonstrate which aspects of the dynamics mimic *in situ* networks and which are limited primarily to the culture environment. Although cultured networks have reduced synaptic density with a seemingly randomized architecture that lacks normal sensory input, they represent living systems that self-organize into spontaneously active networks with histiotypic pharmacological responses and a complex repertoire of spatio-temporal patterns. The presence of structural and dynamic fluctuations should not prohibit statistical descriptions focusing on highly probable, gross behavioral features of a large family of similar cultures with identical origin, similar neuronal densities, and environmental parameters.

Neurons in primary culture are post-mitotic and retain the differentiated properties of the cell lineage to which they belong *in vivo* (Conn, 1990; Tiffany-Castiglioni, 1993). Motor neurons cultured from embryonic chick (O'Brian and Fischbach, 1986) and from rodent tissue (Schaffner et al., 1987) display dendritic arbors and electrical properties characteristic of motor neurons *in vivo*. Cultures of hippocampal neurons (Goslin and Banker, 1991) and cerebellar Purkinje cells (Linden et al., 1991) resemble their *in situ* counterparts with respect to dendritic arborization and appropriate responses to neurotransmitters. Evidence also exists that cell-cell recognition mechanisms remain operational in culture and that some basic circuitry will self-assemble (Crain and Bornstein, 1972; Fishbach and Dichter, 1974; Bunge et al., 1974; Crain and Peterson, 1976; Camardo et al., 1983; Nelson et al., 1989).

In light of the numerous *in vivo* observations of oscillatory bursting (Freeman and Schneider, 1982; Hatton, 1984; McCormick and Feeser, 1990; Raeva, 1990; Gray, 1994; Jahnsen and Llinas, 1994), it is significant that simple mouse monolayer cultures consisting of 200 to 800 neurons also display such spontaneous, oscillatory burst patterns. Hence, after dissociation of embryonic CNS tissue into its cellular constituents, the self-assembly of networks in culture forms circuits that are, at least in this specific dynamic response, histiotypic (i.e. similar to the parent tissue). It is well established that oscillatory responses can be elicited in practically any region of the brain or spinal cord by disinhibiting agents (Gloor et al, 1977; Alger, 1984; Gutnick and Friedman, 1986). Under the influence of 30 to 120 µM NMDA, rhythmic burst patterns (200 to 800 ms in length) were seen in brainstem slices (Tell and Jean, 1990). Such modes of activity are also seen in culture: the induced oscillatory response is one of the most reproducible of all *in vitro* network behaviors (Droge et al., 1986; Gross and Kowalski, 1991; Gross et al., 1993; Gross, 1994; Gopal and Gross, 1996a,b; Jordan, 1992; Robinson et al., 1993; Rhoades and Gross, 1994; Maeda et al., 1995).

5.2 Network "Decision States"

Networks *in vivo* must be massively multi-tasking and be able to respond to a wide range of inputs with responses that are "situation specific". If cultured networks are even remotely similar to networks in animals, then the latter should also have inherent activity that is not directly input dependent. In more formal language, one might state that networks have many dynamics and each dynamics can generate a large number of activity states. Networks may display several independent states (or patterns) simultaneously or may select a specific state by entraining to a common pattern.

We would like to propose that a basic feature of information processing in animals is the achievement of transient approximate synchronization in neuronal groups. This phenomenon is seen on the levels of the circuit and network in mammalian neuronal cultures and may be a common mechanism on many levels of organization *in vivo*. In the burst regime, this implies the expression of the same (or a very similar) temporal burst pattern in all constituent neurons of a particular organizational structure. This event represents a unique network state that is forged by the excitatory and inhibitory influences of neurons within circuits and the competitive or synergistic interaction of circuit patterns.

From a theoretical point of view, the creation of local, flexible, strongly interconnected neuronal sub-assemblies communicating with each other via long-range, sparse interconnections seems to be a solution for achieving system reliability and fault tolerance in mammalian networks operating with relatively unreliable nodes (or neurons) with variable connectivity. Local synchrony ensures a (necessary) multiplexing of the activity by spreading the pattern over many neurons and automatically duplicating the pattern in all axons leaving the network. This facilitates both local guiding (information) inputs as well as "read-out" by distant parts of the CNS.

Transient synchronization of ensembles at many organizational levels seems to solve another important problem: that of informational capacity and a system's ability to create faithful representations of an ever-changing environment. A system with the ability to transiently assemble a large variety of processing networks no longer would be a "rigid" dynamical system with fixed input and output relationships. Instead, it would have the capability to assemble and disassemble a large family of widely differing dynamical subsystems, each with a large variety of autonomous dynamical behaviors. This represents a step beyond the self-assembling ANN's in which a network after a training phase reaches a single, well-defined architecture representing a single dynamical system. Additionally, in our model, most of the processing is done with self-assembled networks that are each in intrinsically active, autonomous states. In contrast, typical ANN schemes have networks without inputs residing in a single equilibrium state. It is well known that dynamical systems with a single (or even multiple) stable equilibria respond in an "uninteresting way" to weak perturbations because of their stability. However, a system residing on attractors corresponding to autonomously active states typically has many differentiated responses to weak external drives. Such a system also is poised for amplification and resonance phenomena, complex filtering tasks, and complex attractor restructuring. The latter includes period doubling cascades, transition in and out of chaos, and "crises".

5.3 Dynamically Assembled Networks: Mechanisms for Anticipatory Responses

Neural tissue with capability of self-organizing into a large variety of different operational networks may be truly a base for "anticipatory systems" of great flexibility and even predictive powers. Indeed, such tissue is not governed by a single dynamics (no matter how complex) but may transiently organize itself into ever-changing and quite distinct processing networks. In other words, such systems can build "on demand" internal representations of almost any dynamics "observed" by a transient sensory input by assembling a representative active network. The richness of the model is additionally enhanced by the active behavior of such dynamically assembled networks, in which each may have a multitude of attractors. Since dynamically assembled networks are built from the same neuronal tissue that plays a role of an underlying "matrix", the assembling process will be facilitated if such networks had been trained earlier via synaptic potentiation. In this scheme, the system may have a memory of previously assembled networks and dynamical states and may restore them easily under the influence of a proper environmental signal, even if the latter is partly masked or corrupted. These considerations leads to a picture of a neuronal "motherboard" with a history of many competing imprinted attractors that can be reactivated or reinforced by appropriate stimuli.

6. References

Alger, B.E. (1984). Hippocampus: electrophysiological studies of epileptiform activity *in vitro*. In R. Dingledine (ed.), Brain Slices, Plenum Press, N.Y. (pp. 155-199).

Braitenberg, V. and Schütz, A. (1991) Anatomy of the Cortex, Springer.

Bunge R.P., Rees R., Wood O., Burton H., Ko C-P (1974) Anatomical and physiological observations on synapses formed on isolated autonomic neurons in tissue culture. Brain Res. 66:401-412.

Camardo J, Proshansky, Schacter S (1983) Identified aplysia neurons form specific chemical synapses in culture. J. Neurosci. 3: 2621-2629.

Cohan C.S. and Kater S.B. (1986) Suppression of neurite elongation and growth cone motility by electrical activity. Science 232: 1638-160.

Conn PM (1990) Methods in Neurosciences, Vol 2: Cell Culture, Academic Press, N.Y.

Crain SM, Bornstein MB (1972) Organotypic bioelectric activity in cultured reaggregates of dissociated rodent brain cells. Science 176: 182-184.

Crain SM, Peterson ER (1976) Development of specific synaptic networks in cultures on CNS tissues. In: Reviews of Neuroscience (S Ehrenpreis, I Kopin, eds), Raven Press, NY.

Douglas, R.J., Koch, C., Mahowald, M., Martin, C.A.C., Suarez, H.H. (1995) Recurrent excitation in neocortical circuits. Science 269: 981-985.

Droge, M.H., Gross, G.W., Hightower, M.H. and Czisny, L.E. (1986). Multielectrode analysis of coordinated, rhythmic bursting in cultured CNS monolayer networks. Journal of Neuroscience 6: 1583-1592.

Edelman, G.M. (1987). Neural Darwinism. Basic Books Inc., N.Y.

Fischbach, B. D. and M.A. Dichter (1974). Electrophysiologic and morphologic properties of neurons in dissociated chick spinal cord cell cultures. Dev Biol 37: 100-116.

Freeman, W.J. (1992). Neural networks and chaos. INNS Above Threshold 1(3): 8-10.

Fields, R.D. and Nelson, P.G. (1994). The role of electrical activity in formation of neuronal networks. In D.A. Stenger, & T.M. McKenna (Eds.), Enabling Technologies for Cultured Neural Networks. Academic Press, New York (pp. 277-317).

Freeman, W.J. and Schneider, W. (1982). Changes in spatial patterns of rabbit olfactory EEG with conditioning to odors. Psychophysiol. 19: 44-56.

Gloor, P., Quesney, L.F., & Zumstein, H. (1977). Pathophysiology of generalized penicillin epilepsy in the cat: the role of cortical and subcortical structures. II. Topical applications of penicillin to the cerebral cortex and to subcortical structures. Electroencephalog. Clin. Neurophysiol. 43: 79-94.

Goslin, K., and Banker, G. (1991). Rat hippocampal neurons in low-density culture. In: Culturing Nerve cells, G. Banker, K. Goslin, (eds). Cambridge, MA, pp 252-281.

Gopal, K.V. and Gross, G.W. (1996a). Auditory cortical neurons *in vitro*: Cell culture and multichannel extracellular recording. Acta Otolaryngologica 116: 690-696.

Gopal, K.V. and Gross, G.W. (1996b). Auditory cortical neurons *in vitro*: Initial pharmacological studies. Acta Otolaryngologica, 116: 697-704.

Gray, C. (1994) Synchronous oscillations in neuronal systems: Mechanisms and function. J. Comp. Neurosci. 1: 11-38.

Gross, G.W. (1979). Simultaneous single unit recording in vitro with a photoetched, laser-deinsulated, gold multimicroelectrode surface, IEEE Trans Biomed Eng, BME 26: 273-279.

Gross, G.W., Wen, W. and Lin, J. (1985). Transparent indium-tin oxide patterns for extra-cellular, multisite recording in neuronal culture. Journal of Neuroscience Methods 15: 243-252.

Gross, G.W. & Kowalski, J.M. (1991). Experimental and theoretical analysis of random nerve cell network dynamics. In P. Antognetti, & V. Milutinovic (eds.), Neural Networks: Concepts, Applications, and Implementations, Vol. 4 Prentice Hall, Englewood, N.J (pp. 47-110).

Gross, G.W., Rhoades, B.K., and Jordan, R.J. (1992) Neuronal networks for biochemical sensing. Sensors and Actuators 6: 1-8.

Gross, G.W., Rhoades, B.K., Reust, D.L., and Schwalm, F.U. (1993). Stimulation of monolayer networks in culture through thin film indium-tin oxide recording electrodes. J Neurosci Meth 50: 131-143.

Gross, G.W., & Schwalm, F.U. (1994). A closed chamber for long-term electrophysiological and microscopical monitoring of monolayer neuronal networks. J. Neurosci. Meth. 52: 73-85.

Gross, G.W. (1994). Internal dynamics of randomized mammalian neuronal networks in culture. In D.A. Stenger, & T.M. McKenna (eds.) Enabling Technologies for Cultured Neural Networks. Academic Press, N.Y. (pp. 277-317).

Gross, G.W., Azzazy, H.M.E., Wu, M-C, and Rhoades, B.K.(1995) The use of neuronal networks on multielectrode arrays as biosensors. Bisosensors and Bioelectronics 10: 553-567.

Gross, G.W., Norton, S., Gopal, K., Schiffmann, D., and Gramowski, A. (1997). Neuronal networks *in vitro*: Applications to neurotoxicology, drug development, and biosensors. J. Cell. Eng., in press.

Gutnick, M.J., & Friedman, A. (1986). Synaptic and intrinsic mechanisms of synchronization and epileptogenesis in the neocortex. In U. Heinemann, M. Klee, E. Neher, and W. Singer (eds.). Calcium Electrogenesis and Neuronal Functioning, Experimental Brain Research 14: 327-335.

Hatton, G.I. (1994). Hypothalamic neurobiology. In R. Dingledine (ed.) Brain Slices. Plenum Press, N.Y. (pp. 341-374). .

Hooper S.L., Moulins M. (1989). Switching of a neuron from one network to another by sensory-induced changes in membrane properties. Science 244: 187-189.

Huettner, J.E., and Baughman, R.W. (1986). Primary culture of identified neurons from the visual cortex of postnatal rats. J. Neurosci. 6: 3044-3060.

Hubel, D.H. and Wiesel, T.N. (1962). Receptive fields, binocular interaction and functional architecture in the cat's striate cortex. J. Physiol. (London) 160: 106-154.

Jahnsen, H. and Llinas, R. (1994b). Ionic basis for the electroresponsiveness and oscillatory properties of guinea-pig thalamic neurones *in vitro*. J. Physiol. 349: 227-247.

Jordan, R. (1992). Investigation of inhibitory synaptic influences in neuronal monolayer networks cultured from mouse spinal cord. M.S. Thesis, Dept. of Biological Sciences, Univ. of North Texas, Denton.

Kowalski, J.M., Albert, G.L. and Gross, G.W. (1990). On the asymptotically synchronous chaotic orbits in systems of excitable elements. Physical Review A, 42: 6260-6263.

Kowalski, J.M., Albert, G.L., Rhoades, B.K., and Gross, G.W. (1992). Neuronal networks with spontaneous, correlated bursting activity: theory and simulations. Neural Networks, 5: 805-822.

Kuperstein, M. and Whittington, D.A. (1981) A practical 24-channel microelectrode for neural recording *in vivo*. IEEE Transact. Biomed. Eng. BME-28: 288-293.

Levitan, I.B. and Kaczmarek, L.K. (1991) The Neuron: Cell and Molecular Biology. Oxford University Press, N.Y., Ch 16.

Linden, D. J, Dickinson, M.H., Smeyne, M., and Connor, J.A. (1991). A long-term depression of AMPA currents in cultured cerebellar Purkinje neurons. Neuron, 7: 81-89.

Lucas, J.H., Czisny, L.E. and Gross, G.W. (1986). Adhesion of cultured mammalian CNS neurons to flame-modified hydrophobic surfaces. In Vitro 22: 37-43.

MacGregor, R.J. (1993) Theoretical Mechanics of Biological Neural Networks. Academic Press, Boston, pg 206.

Maeda, E., Robinson, H.P.C., & Kawana, A. (1995). The mechanism of generation and propagation of synchronized bursting in developing networks of cortical neurons. J. Neurosci. 15: 6834-6845.

McCormick, D.A., & Feeser, H.R. (1990). Functional implications of burst firing and single spike activity in lateral geniculate relay neurons. Neuroscience 1: 103-113.

Mead, C. (1990). Neuromorphic electronic systems. Proc. IEEE 78: 1629-1636.

Mitchison, G. (1992). Axonal trees and cortical architecture. Trends in Neurosci. 15: 122-126.

Mountcastle, V.B. (1982). An organizing principle for cerebral function: The unit module and the distributed system. In "The Mindful Brain" (H.O. Schmitt, ed.) MIT Press, Cambridge, MA.

Nelson, P.G., Yu, C., Fields, R.D., and Neale, E.A. (1989). Synaptic connections *in vitro*: Modulation of number and efficacy by electrical activity. Science 244: 585-587.

O'Brien, R.J., and Fischbach, G. D. (1986). Modulation of embryonic chick motoneuron glutamate sensitivity by interneurons and agonisists. J. Neuroscience, 6(11): 3290-3296.

Raeva, S. W. N. (1990). Unit activity of the human thalamus during voluntary movements. Stereotactical and Functional Neurosurgery 54-55: 154-158.

Ransom, B.R., Neile, E., Henkart, M., Bullock, P.N., and Nelson, P.G. (1977). Mouse spinal cord in cell culture. J. Neurophysiol. 40, 1132-1150.

Rhoades, B.K., & Gross, G.W. (1994). Potassium and calcium channel dependence of bursting in cultured neuronal networks. Brain Res. 643: 310-318.

Robinson, H.P.C., Kawahara, M., Jimbo, Y., Torimitsu, K., Kuroda, Y., & Kawana, A. (1993). Periodic synchronized bursting and intracelllular calcium transients elicited by low magnesium in cultured cortical neurons. J. Neurophysiol. 70: 1606-1616.

Schaffner, A.E., St. John, P.A., and Barker, J.L. (1987). Flourescence-activated cell sorting of embryonic mouse and rat motoneurons and their long-term survival *in vitro*. J. Neurosci. 7: 3088-3104.

Simmers, J., Meyrand, P., & Moulins, M. (1995). Dynamic network of neurons. American Scientists 83: 262-268.

Sporns, O., Tononi, G., and Edelman, G.M. (1991). Modeling perceptual grouping and figure-ground segregation of means of active re-entrant connections. PNAS 88: 129-33.

Tell, F. and Jean, A. (1990). Rhythmic bursting patterns induced in neurons of the rat nucleus tractus solitarii, *in vitro*, in response to N-methyl-D-aspartate. Brain Res., 533: 152-156.

Tiffany-Castiglioni, E. (1993). Cell culture models for lead toxicity in neuronal and glial cells. Neurotoxicology, 14(4): 513-536.

Wise, K.D. and Angell, J.B. (1975). A low capacitance multielectrode probe for use in extracellular neurophysiology. IEEE Transact. Biomed. Eng., BME 22: 212-219.

Yuste, R. (1994). Calcium imaging of cortical circuits in slices of developing neocortex. In D.A. Stenger, & T.M. McKenna (eds.), Enabling Technologies for Cultured Neural Networks. Academic Press, N.Y. (pp. 207-234).

ACKNOWLEDGMENTS

Supported by the Texas Advanced Research Program and by the Hillcrest Foundation of Dallas, Texas, founded by Mrs. W.W. Caruth, Sr. The authors thank Lori Harvey for her excellent support in cell culture and Todd Hall and Bret Zim for their reliable fabrication of the microelectrode arrays.

UNIVERSALITY OF PASSIVE NEURAL NETWORKS.

A. Łuksza ,W.Citko, W.Sieńko
Department of Electrical Engineering
Maritime Academy
Morska 83
81-225 Gdynia, Poland
e-mail: wcitko@vega.wsm.gdynia.pl

Abstract:

This paper presents some properties and synthesis procedures of nonsymmetric passive neural networks. Universality of these networks follows from the possibility to create neural networks featuring both feedforward and feedback architecture by changing a value of one system parameter. Due to the robustness of obtained structures and their scaleability, they seem to be a real alternative for implementation as Very Large Scale Neural Networks (VLSNN).

Key words: neural networks, chaos, static/oscillatory associative memory, chaotic neural network

1. INTRODUCTION

The development of passive neural networks is motivated by the search for such models of neural nets that could be used for feasible realisation of Very Large Scale Neural Networks (VLSNN). One is convinced that VLSNN can be realised only in feedforward architecture because it is difficult to control stability, attraction basins and positions of equilibrium points in feedback neural networks. A thorough survey and comparison of the feedforward and feedback structures can be found in papers [1, 2]. The purpose of this paper is to present some basic features of passive neural networks and to show their universality. By universality we mean here not only the variety of problems that can be solved by passive neural networks but the fact, that they combine the features of feedforward and feedback neural networks as well.

2. PASSIVITY, PASSIVE NEURON AND PASSIVE NEURAL NETWORKS.

The concept of the passive neural networks has been proposed in the paper [3, 4]. In the classical circuit theory a passive N-port is defined by the sign of energy absorbed by its ports. The concept of the N-port passivity has been extended onto SFN-ports (Signal Flow N-ports). In this way a new design tool and new classes of systems with interesting properties have been created (for example orthogonal filter and wave digital filter theory [5]). A model of SFN-port is shown in Fig.1.

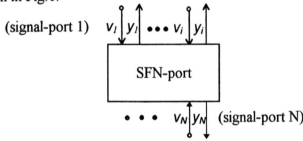

(signal-port 1) v_1 y_1 ••• v_i y_i

SFN-port

••• v_N y_N (signal-port N)

Fig.1. Model of SFN-port.

CP437, *Computing Anticipatory Systems: CAYS--First International Conference*
edited by Daniel M. Dubois © 1998 The American Institute of Physics 1-56396-827-4/98/$15.00

Energy (pseudoenergy) absorbed by the SFN-port is defined as the functional:

$$E(\underline{v},\underline{y}) = \sum_{i=1}^{N} \int_{-\infty}^{t} v_i(\tau)y_i(\tau)d\tau \qquad (1)$$

where: $v_i(t)$, $y_i(t) \in L_2$, $i = 1, 2, ..., N$
SFN-port is passive when energy E is nonnegative:

$$E(\underline{v},\underline{y}) \geq 0, \forall t \qquad (2)$$

There are two basic concepts necessary to develop the theory of passive neural networks. Firstly, it is <u>compatibility of connection</u>:

The connection of a SFK-port and a SFL-port onto SFN-port is compatible if following relation is fulfilled:

$$E(\underline{v},\underline{y}) = E_1 + E_2 \qquad (3)$$

where: E is the energy absorbed at all ports of SFN-port

E_i, $i = 1, 2$ is the energy absorbed at the ports of SFK-port and SFL-port, respectively.

An example of compatible connection is illustrated in Fig.2.

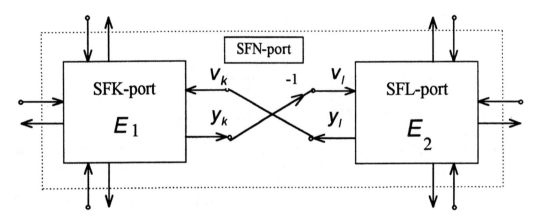

Fig.2. Compatible connection of SFK-port and SFL-port.

It is easy to notice that compatibility of connection preserves the passivity of SFN-port.

The second basic concept is the model of <u>passive neuron</u>. SFN-port shown in Fig.3 is passive under the following conditions:
- transmittance K(s) is a positive real function (p.r.f)
- $\Theta[x]$ belongs to class M nonlinearties i.e.

$$\mu_1 \leq \frac{\Theta[x]}{x} \leq \mu_2; \quad \mu_1,\mu_2 \in (0,\infty) \qquad (3)$$

and it can be considered as a model of passive neuron. It is worth noting that, if K(s) is a transmittance of nonideal integrator:

$$K(s) = \frac{A}{s + \omega_o}; \qquad A, \omega_o \geq 0 \qquad (4)$$

and nonlinearity $\Theta(x)$ is a sigmoidal function, then SFN-port from Fig.3. is McCulloch-Pitts model of neuron. Also selective functions belong to the nonlinearity class given by

596

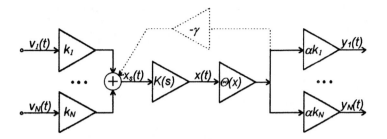

Fig.3. Model of passive neuron

relation (3). A passive neural network is created by the compatible connection of passive neurons. An example of basic 2-neuron network is shown in Fig.4.

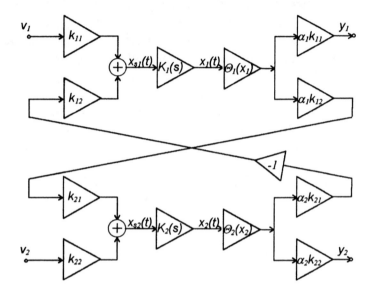

Fig.4. Structure of 2-neuron network.

Scaling this network to connection of N-neurons one obtains a passive neural network described in the state space by:

$$\frac{1}{A_i}\dot{x}_i = \sum_{i=1}^{N} T_{ij}\Theta[x_j(t)] - \frac{\omega_{oi}}{A_i}x_i(t); \qquad i=1, 2,..., N \qquad (5)$$

where:

$$T_{ij} = \alpha_i k_{ij} k_{ji} = -\frac{\alpha_j}{\alpha_i}T_{ji}; \quad i \neq j \qquad (6)$$

For $\alpha_i=1$, i = 1, 2,..., N, connection matrix $[T_{ij}]$ is <u>antisymmetric</u>.
It should be noted that neuron from Fig. 3. is structurally passive i.e. there are finite changes of synapse weights, for which relation (2) is fulfilled. Each neuron can be done passive by connecting a lossy element $\gamma > 0$, as shown by dashing line in Fig.3.

3. STABILITY OF PASSIVE NEURAL NETWORKS.

Passive neural network is a complex dynamical system with number of feedback loops. One of the reasons for introducing the concept of the passivity has been the fact that passivity of network implies its stability. Hence, it can be stated that controlling stability of a large scale neural network means providing passivity of the neurons. From the point of view of nonlinear

597

differential equations theory, the passive neural network has one trivial, globally asymptotical equilibrium point in the state space: $\underline{x} = \underline{0}$. It should be proved under which conditions the passive, asymptotically stable network has the properties of neural net, i.e. whether it is possible to generate $r > 1$ asymptotically stable equilibrium points in the state space having suitable formed region of attraction.

4. SYNTHESIS PROCEDURE OF NEURAL NETWORKS.

Any neural network can be seen as an implementation of a nonlinear mapping with fixed or stationary points. These points create a memory of the neural network. Let us require that $\underline{M} = [\ m_1,\ m_2,...,\ m_r]$ be a set of r bipolar (-1, 1) n-element column vectors being the assumed memory of the network. It is a typical assumption on equilibrium points located in the nodes of a hypercube. In synthesis procedure one determines connection weights $[T_{ij}]$ (connection matrix) in such a way that the network equilibrium points are identical with $\underline{M}$. For the passive neural networks composed of neurons with sigmoidal nonlinear functions $\Theta[\ .\ ]$ the connection matrix $[T_{ij}]$ can be obtained in two following steps [6]:

1. Determining antisymmetric connection matrix $[T_{ij}^a] = \underline{T}^a$.

   Passive neural network with an antisymmetric connection matrix has generally a heterogeneous structure with two layers: input and complementary (hidden). The layers are controlled by two structural neurons. The general schema of such a structure is shown in the Fig.5. Two connection types can be found in the network: local and general being controlled by structural neurons. Hence, these neurons act as general inhibitors (or general excitators when the network is transformed from the static network to an oscillatory one). It is worth noting that the passive neural network structure resembles the double LEGION structure presented in papers [7].

2. Determining symmetric connection matrix $[T_{ij}^s] = \underline{T}^s$.

   Such a matrix can be determined in the following procedure:

   a) $\underline{T}^S = \underline{M}\ \underline{M}^+$

   where: $\underline{M}^+$ is pseudoinverse matrix of $\underline{M}$. Thus, eigenvectors $\underline{T}^s$ are identical with vectors $\underline{m}_i \in \underline{M}$.

   b) $\underline{T}^S = \underline{m}_S\ \underline{m}_S^+$

   where: $\underline{m}_S^+$ is pseudoinverse of subset $\underline{m}_S \subset \underline{M}$. Thus eigenvectors of $\underline{T}^S$ are identical with selected subset of points $\underline{m}_i \in \underline{M}$.

   c) $\underline{T}^S = \underline{\delta} - \underline{D}$

   where: $\underline{\delta} = \mathrm{diag}(\delta_1, \delta_2,..., \delta_n)$; $\delta_i > 0$,

   $\underline{D}$ - symmetric matrix describing a feature of patterns stored in the antysymmetric part of memory.

   Thus, matrix $\underline{T}^S$ selects one pattern $\underline{m}_i$, $i \in [1, ..., r]$ with prescribed feature, i.e.:

   $(\underline{\delta} - \underline{D})\ \underline{m}_i = \mathrm{diag}[k_l]\ \underline{m}_i$; $k_l > 0$.

On contrary to cases a and b, case c can be seen as a difficult minimization problem. Such a problem can be for example encountered at TSP.

In the result of the synthesis procedure the connection matrix of neural passive network is the sum of two matrices:

$$[T_{ij}] = [T_{ij}^a] + \xi[T_{ij}^s] \tag{7}$$

where: $[T_{ij}^a]$ is a composition of two components: local and global one.

$\xi \geq 0$ is a bifurcation parameter.

598

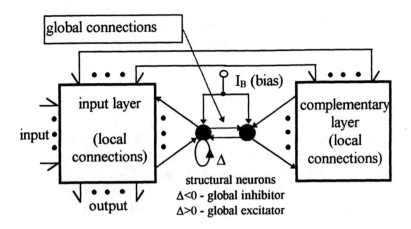

Fig.5. Structure of passive neural network

Let us note that relation (7) gives the possibility to interpret the structure of the passive neural networks in the following way:

- passive neural network structure includes <u>recombination</u> part (antisymmetric part of vector field) and <u>selective</u> part (symmetric part of vector field). Such a model of vector field has been proposed also in genetic models [8].

- passive neural network structure is an implementation of syncretism principle [9], it is a composition of two mechanisms: <u>dumb memory</u> based on antisymmetric connections and learning mechanism based on <u>symmetric selection</u>.

The bifurcation parameter ξ in the relation (7) simulates the mechanism of degradation of passivity by neurons. For given losses of neurons, there is such a bifurcation value ξ_C, that if:

$$0 \le \xi < \xi_C \tag{8}$$

then neurons are passive.

5. PASSIVE NEURAL NETWORK AS NONLINEAR MAPPING

The choice of parameter ξ according to relation (8) preserves the passivity of the network. Hence, this neural network has one trivial equilibrium point but it can be treated as an implementation of nonlinear mapping $F[\ .\ ]$ with given fixed points $\underline{m}_i \in \underline{M}$. Graphical representation of the mapping is shown in Fig.6. It is not difficult to notice that input vector $\underline{I}$ is decomposed into two components:

$$\underline{I} = \underline{v} + \underline{\gamma}\,\underline{y} \tag{9}$$

where: $\underline{\gamma} = \mathrm{diag}(\gamma_1, \gamma_2, ..., \gamma_N)$; $\gamma_i \ge 0$

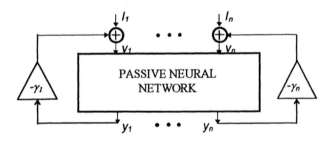

Fig. 6. Model of passive neural network as nonlinear mapping.

Parameter γ_i can be treated as a loss element of passive neuron from Fig. 3. On the other hand, the network from Fig. 6. is an implementation of the nonlinear mapping:

$$\underline{y} = F[\underline{I}] \tag{10}$$

and vectors $\underline{m}_i \in \underline{M}$ are fixed points of F[.]. The relation (9) and (10) let us interpret the acting of the passive neural network in the following way:

- Assuming that model of neuron from Fig.3 is lossless (i.e. $\omega_0 \rightarrow 0$) and $\xi = 0$, relation (9) for $\underline{\gamma} = \underline{1}$ means the orthogonal decomposition of the input vectors $\underline{I}$ (in the space L_2).In this case the neural network can be treated as nonlinear orthogonal filter determining the distance of input signals from the subspace M of the fixed points of the nonlinear mapping F[.]. The output $\underline{v}$ of the novelty filter is the measure of this distance. Hence, passive neural networks can be treated as generalisation of Kohonen's OLAM networks (optimal linear associative memory) [10].

- Assuming the lossy model of neuron ($\omega_0 > 0$) and $0 < \xi < \xi_c$, the network from Fig. 6. is the implementation of associative memory, where the fixed points $\underline{m}_i$ are the centres of the attraction basins. Their sizes are controlled by the value of the bifurcation parameter ξ.

6. PASSIVE NEURAL NETWORK AS A HOPFIELD NETWORK

If the parameter ξ exceeds the bifurcation value ξ_c i.e. $\xi > \xi_c$ passive neural network converts into the Hopfield network - the trivial equilibrium point loses its stability and the points $\underline{m}_i$ become asymptotically stable equilibrium points. The network obtained in this way, still having all functional features of the classical Hopfield network (i.e. with symmetric connection matrix), let us control more effectively the sizes of the attraction regions of the points $\underline{m}_i$ and reduce spurious states. It follows from the other mechanism of the memory activation - stationary states of this memory are selected and cut out in the dumb memory tissue. Spurious states can be eliminated by the complementary layer [11].

7. PASSIVE NEURAL NETWORK AS AN UNIVERSAL NEURAL OSCILLATOR

Recently, oscillatory/chaotic artificial neural networks are being studied as dynamical information processing models [12-18]. Most of chaotic neural structures are based on using a concept of chaotic neuron [19, 20]. Thus, a mechanism of chaos generation is based on a model of 1-D logistic map. The purpose of this paper is to present an oscillatory neural network which structure is derived from a static passive neural network by changing the function of one neuron. Due to such a solution, the known procedures for implementing static, large scale associative memories may provide a tool for designing oscillatory, large scale associative memories.

Assuming, that the memory $\underline{M}$ of neural network is a set of fixed points:
$\underline{M} = [\underline{m}_1, \underline{m}_2,..., \underline{m}_r]$, where $\underline{m}_i$ are column vectors with components (-1, 1), general structure of passive neural network, shown in Fig. 5, consists of two layers: input and complementary (hidden) layer and two structural neurons controlling global connections of network. The neurons in the layers are connected locally. The structural neuron becomes accordingly:

- global inhibitor for $\Delta < 0$ \hfill (11)
- global excitator for $\Delta > 0$ \hfill (12)

where: Δ - transmittance of self-loop.
Neural network from Fig. 5. is described by the following equation:

$$\dot{\underline{x}} = \underline{T}_n \Theta(\underline{x}) - diag[\omega_{oi}] \cdot \underline{x} + \underline{I} \tag{13}$$

where: $\underline{T}_n$ - connection matrix (nonsymmetric)

diag[ω_{oi}], i = 1, 2, ..., N - matrix of dissipative elements of neurons

$\underline{\Theta}(\underline{x})$ - output of network

$\underline{I}$ - input of network (including bias of structural neurons)

The connection matrix $\underline{T}_n$ has the following structure:

$$\underline{T}_n = \underline{T}_g^a + \beta \underline{T}_l^a + diag [0,...,0,\Delta,0,...0] + \xi \underline{T}^s \qquad (14)$$

where: $\underline{T}_g^a$ - antisymmetric matrix describing global connections

$\underline{T}_l^a$ - antisymmetric matrix describing local connections

$\underline{T}^s$ - symmetric matrix controlling the size of attraction basins of points $\underline{m}_i$

β - bifurcation parameter for oscillatory mode

ξ - bifurcation parameter for static mode

Δ - parameter switching the mode of function (static/oscillatory)

Thus the equations (13) and (14) describe an universal neural oscillator:

$$\dot{\underline{x}} = \underline{f}(\underline{x},\Delta,\beta,\xi) \qquad (15)$$

with two modes of function: static mode ($\Delta < 0$) and oscillatory mode ($\Delta > 0$)

The neural network from Fig. 5 may generate chaotic signals in oscillatory mode. To point out this possibility, let us present the general structure of passive neural network in a form of signal flow N-port (SFN-port) (Fig. 7).

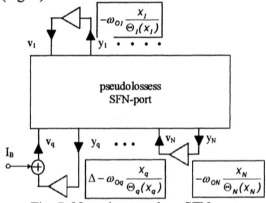

Fig. 7. Neural network as SFN-port

Assuming that sigmoidal nonlinear characteristics of the neuron is approximated by the piecewise-linear characteristics, as shown in Fig. 8.

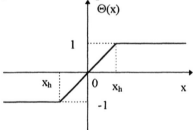

Fig.8. Piecewise-linear characteristics of neuron.

the signal port load is given as:

$$\frac{v_i}{y_i} = \Delta - \omega_{oi}\frac{x_i}{\Theta_i(x_i)} \quad ; \quad \Delta \geq 0. \qquad (16)$$

601

Relation (16) is presented in Fig. 9.

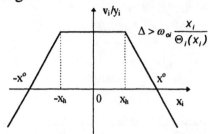

Fig.9. Load of signal-port attached to structural neuron.

Hence, it can be noted that:

a)   condition of oscillation is given by:

$$\Delta > \sum_i \frac{\omega_{ol}}{\Theta_i'(0)} > 0; \quad \text{where: } \Theta_i'(0) = \frac{d\Theta}{dx_i}\bigg|_{x_i=0}, \quad i = 1, ..., N. \qquad (17)$$

b)   there is a trapping region in the space phase of the equation (13) limited by the surface:

$$Q = \frac{1}{2}\sum_i \Theta_i^2(x_i) \qquad (18)$$

Indeed, if Q is taken large enough, then the flow is directed inward i.e. dQ/dt < 0 everywhere on the surface defined by Q.

c) $x_q(t) \in (-x^o, x^o)$ \qquad (19)

where: $x_q(t)$ -q-th component of $\underline{x}$ attached to structural neuron. Hence, if the initial point is outside of (19), then a relaxation time is needed to start the oscillations.

Mechanism of chaos generation

Given matrix $\underline{M} = [\,\underline{m}_1, \underline{m}_2,..., \underline{m}_r]$, $r \geq 2$ be a set of bipolar column vectors (states) memorised by the neural network (13). Let us assume that two states are degenerate i.e. there two vectors $\underline{m}_k$ and $\underline{m}_l$ at Hamming distance equal 2. Hence, it can be shown that in the structure of neural networks implementing given memory $\underline{M}$, there are two neurons outside of control of global connections i.e. they are split from structural neurons. By changing the value of bifurcation parameter β i.e. by changing the strengths of local connections in model of Fig. 5, one observes a number of period-doubling bifurcations. The direct transition from subharmonics of one frequency to a continuos spectrum requires three bifurcations. The power spectrum in the chaotic regime consists of a set of narrow peaks sitting on top of a broadband noise background. Hence, the dynamics of the oscillator could be modelled by a Feigenbaum or parabolic map.

Example of Design

Let us design a 12-neuron oscillator with the following memory matrix: $\underline{M} = [\,\underline{m}_1, \underline{m}_2, \underline{m}_3, \underline{m}_4\,]$ where:

$$\underline{m}_1 = \begin{bmatrix} -1 & -1 & +1 & -1 & +1 & +1 & -1 & -1 & +1 & -1 & +1 & +1 \end{bmatrix}^T$$

$$\underline{m}_2 = \begin{bmatrix} +1 & +1 & -1 & +1 & +1 & +1 & -1 & -1 & -1 & +1 & -1 & -1 \end{bmatrix}^T$$

$$\underline{m}_3 = \begin{bmatrix} -1 & +1 & +1 & +1 & -1 & +1 & -1 & +1 & -1 & -1 & -1 & +1 \end{bmatrix}^T$$

$$\underline{m}_4 = \begin{bmatrix} -1 & -1 & +1 & +1 & +1 & +1 & -1 & -1 & -1 & -1 & +1 & +1 \end{bmatrix}^T$$

602

The neurons with indices 1 to 5 and 8 to 12 belong to input layer and complementary layer, respectively. The structural neurons are indexed as 6 and 7. The connection matrices of this oscillator are given by [6]:

$$
\underline{T}_g^a = \begin{bmatrix}
0 & 0 & 0 & 0 & 0 & -1.429 & -1.429 & 0 & 0 & 0 & 0 & 0 \\
0 & 0 & 0 & 0 & 0 & 2.857 & 2.857 & 0 & 0 & 0 & 0 & 0 \\
0 & 0 & 0 & 0 & 0 & 1.429 & 1.429 & 0 & 0 & 0 & 0 & 0 \\
0 & 0 & 0 & 0 & 0 & 0 & 0 & 0 & 0 & 0 & 0 & 0 \\
0 & 0 & 0 & 0 & 0 & 2.857 & 2.857 & 0 & 0 & 0 & 0 & 0 \\
1.429 & -2.857 & -1.429 & 0.000 & -2.857 & 0 & 14.28 & 2.857 & 0.000 & 1.429 & 2.857 & -1.429 \\
1.429 & -2.857 & -1.429 & 0.000 & -2.857 & -14.28 & 0 & 2.857 & 0.000 & 1.429 & 2.857 & -1.429 \\
0 & 0 & 0 & 0 & 0 & -2.857 & -2.857 & 0 & 0 & 0 & 0 & 0 \\
0 & 0 & 0 & 0 & 0 & 0 & 0 & 0 & 0 & 0 & 0 & 0 \\
0 & 0 & 0 & 0 & 0 & -1.428 & -1.429 & 0 & 0 & 0 & 0 & 0 \\
0 & 0 & 0 & 0 & 0 & -2.857 & -2.857 & 0 & 0 & 0 & 0 & 0 \\
0 & 0 & 0 & 0 & 0 & 1.429 & 1.429 & 0 & 0 & 0 & 0 & 0
\end{bmatrix}
$$

$$
\underline{T}_l^a = \begin{bmatrix}
0 & 2 & 1 & 2 & 2 & 2 & 2 & 2 & 2 & 4 & 2 & 3 \\
-2 & 0 & 0 & 1 & 1 & 1 & 1 & 1 & 1 & 2 & 0 & 0 \\
-1 & 0 & 0 & 2 & 2 & 2 & 2 & 2 & 2 & 3 & 0 & 2 \\
-2 & -1 & -2 & 0 & 1 & 1 & 1 & 1 & 0 & 0 & -1 & 0 \\
-2 & -1 & -2 & -1 & 0 & 1 & 1 & 0 & -1 & 0 & -1 & 0 \\
-2 & -1 & -2 & -1 & -1 & 0 & 0 & -1 & -1 & 0 & -1 & 0 \\
-2 & -1 & -2 & -1 & -1 & 0 & 0 & -1 & -1 & 0 & -1 & 0 \\
-2 & -1 & -2 & -1 & 0 & 1 & 1 & 0 & -1 & 0 & -1 & 0 \\
-2 & -1 & -2 & 0 & 1 & 1 & 1 & 1 & 0 & 0 & -1 & 0 \\
-4 & -2 & -3 & 0 & 0 & 0 & 0 & 0 & 0 & 0 & -2 & -1 \\
-2 & 0 & 0 & 1 & 1 & 1 & 1 & 1 & 1 & 2 & 0 & 0 \\
-3 & 0 & -2 & 0 & 0 & 0 & 0 & 0 & 0 & 1 & 0 & 0
\end{bmatrix}
$$

It can bee seen that states $\underline{m}_1$ and $\underline{m}_4$ are degenerate since:

$\underline{m}_1 - \underline{m}_4 = [0, 0, 0, -2, 0, 0, 0, 0, 2, 0, 0, 0]^T$

The points $\underline{m}_i$, $i = 1, 2, 3, 4$, are solutions of the following equation:

$$(\underline{T}_g^a + \underline{T}_l^a)\Theta(\underline{x}) + \underline{l} = \underline{0} \tag{20}$$

where: $\underline{l} = [0, 0, 0, 0, 0, I_B, I_B, 0, 0, 0, 0, 0]^T$; ($I_B = 20$)

Since, however the neurons with number 4 and 9 are split from structural neurons, there are infinite number of solutions of Eq. (20). Assuming the neuron characteristics as shown in Fig.8 with value of $x_h = 0{,}01$ and $\omega_{oi} = 1$; $i = 1, \ldots, 12$, the considered oscillator is described by:

$$\dot{\underline{x}} = (\underline{T}_g^a + \beta \underline{T}_l^a + diag[0, \ldots, 0, \Delta, 0, \ldots 0])\Theta(\underline{x}) - \underline{x} + \underline{l} \tag{21}$$

where: $\underline{l} = [0, \ldots, 0, I_B - \Delta, I_B, 0, \ldots, 0]^T$; $\Delta = 4$.

The power spectrum of the oscillations is shown in Fig.10 for different value of parameter β.

This spectrum exhibits the features typical for a parabolic map i.e. a set of narrow peaks filled by noise in background. The p-p value of the oscillations depends on parameter Δ. (acc. to Fig.9). It is worth noting that the energy landscape of the phase-space can be changed by the parameter ξ, i.e. by the symmetric part of the vector field (acc.to Eq.(14)). For example, there is such a critical value ξ_C that for $\xi > \xi_C$ the state $\underline{x}(t)$, $t \to \infty$ is attracted by one of the centres $\underline{m}_i$. The proposed model of neural oscillator can be seen as an elementary block for implementing large scale associative oscillatory/static memories. One of the most interesting application field for such networks is digital, spread-spectrum communication. It is clear, that due to the spectral features of the examined here chaotic oscillations (high resolution), such a neural oscillator can be used as a chaotic encoder in CDMA/SSMA (Code Division/Spread Spectrum) communication systems. One of the problem being under examination can be formulated as follows: how many symbols can by encoded by the noise generated in this oscillatory neural network by changing the discrete values of bifurcation parameter β.

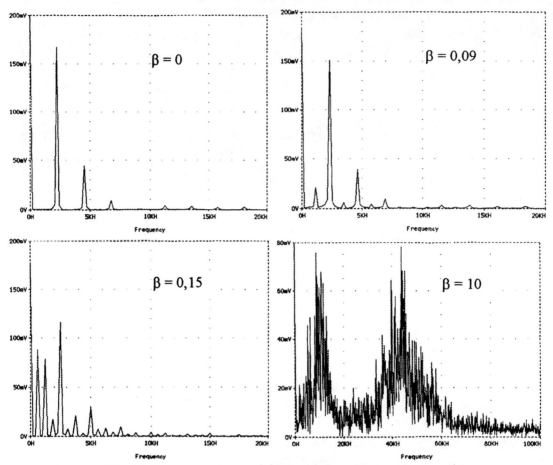

Fig.10. Power spectrum of the oscillations in neural network.

8. SUMMARY

The universality of passive neural networks structure follows from the possibility to create neural networks featuring both feedforward and feedback architecture by changing the value of one system parameter (bifurcation parameter). From the view point of the functions being realized by the passive neural networks, they can be a base for creating the large scale associative memories. Moreover, they can also be a base for large scale oscillatory (also chaotic) memories by changing the general inhibitor function to the function of the general excitator.

Due to the robustness of obtained structures and their scaleability, they seem to be real alternative for implementation of VLSNN. This paper does not deal with specific applications. However, it is worth noting that number of problems concerning the pattern recognition and feature extractions have been simulated by using passive neural networks.

REFERENCES

[1]   B. Kosko, Fuzzy Associative Memory Systems, in A. Kandel Ed. , Fuzzy Experts Systems, CRC Press, (1992).

[2]   W. Duch, Feature, Space Mapping as a Universal Adaptive System, Comp. Physics Comm. 87, (1995).

[3] W. Sieńko, J. Zurada, Neural Signal-Flow Networks, Proc. Neuronet'90 Conf. , Prague, Czech. , (1990).

[4] W. Sieńko, J. Zurada, Pseudopassive Neural Networks, Proc. ISCAS 1991, Singapore, (1991).

[5] A. Fettweis, Some General Properties of Signal-Flow Networks, Proc. NATO Advanced Study Institute, Bournemouth, (1972).

[6] W. Sieńko, A. Luksza, A Novel Deign Method of Nonsymmetric Neural Networks, Proc. XV KKTO i UE, Szczyrk, (1992)

[7] D. Wang, D Terman, Locally Excitatory Globally Inhibitory Oscillator Networks, IEEE Tr. on N.N, vol. 6, No 1, (1995).

[8] E. Akin, Hopf Bifurcation in the Two Locus Genetic Model, Memoirs of the AMS, vol. 44, No 284, (1983).

[9] W. Sieńko, A. Łuksza, W. Citko, Passivity vs Syncretism of Neural Networks, Proc. of EUFIT'96, Aachen, (1996).

[10] T. Kohonen, Self-organization and Associative Memory, Springer, New York, (1989).

[11] W.Łuksza, W.Sieńko, A. Suppression of Spurious States in Passive Neural Networks, Proc. XVII KKTO i UE, Wrocław, (1994).

[12] B. Baird, M. W. Hirsch, F Eeckman, A Neural Networks Associative Memory for Handwrittten Character Recognition Using Multiple Chua Characters, IEEE Trans. on Circuits and Systems-II: Analog and Digital Signal Processing, vol. 40, No. 10, October 1993.

[13] Saito, M. Oikawa, Chaos and Fractals from a Forced Artificial Neural Cell, IEEE Trans. on Neural Networks, vol. 4. No. 1, January 1993.

[14] Y. Choi, B. A. Huberman, Dynamic behaviour of nonlinear networks, Physical Review A, vol. 28, No. 2, August 1983.

[15] Aihara, T. Takabe, M. Toyoda., Chaotic Neural Networks, Physical Review A, vol. 144, No. 6, 7, March 1990.

[16] Yu. M. Sandler, Model of Neural Networks with Selective Memorization and Chaotic Behaviour, Physical Review A, vol. 144, No. 8, 9, March 1990.

[17] Nozawa, A Neural Network Model as a Globally Coupled Map and Applications Based on Chaos, Chaos 2 (3),1992.

[18] Zou, J. A. Nossek, Bifurcation and Chaos in Cellular Networks, IEEE tr. on CAS, vol. 40, No. 3, March 1993.

[19] C. C. Hsu, D. Gobovic, M. E. Zaghloul, H. H. Szu, Chaotic Neuron Models and Their VLSI Implementations, IEEE Tr. on Neural Network, vol. 7, No 6, November 1996.

[20] L. Wang, Oscillatory and Chaotic Dynamics in Neural Networks Under Varing Operating Conditions, IEEE Tr. on Neural Network, vol. 7, No 6, November 1996.

[21] D. Wang, D Terman, Locally Excitatory Globally Inhibitory Oscillator Networks, IEEE Tr. on Neural Network, vol. 6, No 1, 1995.

[22] D. Terman, D. Wang, Global competition and local cooperation in a network of neural oscillators, Physica D 81 (1995) 148-176.

[23] D. Frey, Chaotic Digital Encoding: An Approach to Secure Communication, IEEE Tr. on Circuits and Systems, vol. 2, No. 10, 1993

Unsupervised Hebbian Learning in Neural Networks

Bernd Freisleben[1] and Claudia Hagen[2]

[1]Department of Electrical Engineering and Computer Science, University of Siegen
Hölderlinstr. 3, D–57068 Siegen, Germany
E-Mail: freisleb@informatik.uni-siegen.de

[2]Department of Computer Science, University of Darmstadt
Julius-Reiber-Str. 17, D–64293 Darmstadt, Germany
E-Mail: hagen@iti.informatik.th-darmstadt.de

Abstract. In this paper, a survey of a particular class of unsupervised learning rules for neural networks is presented. These learning rules are based on variants of Hebbian correlation learning to update the connection weights of two-layer network architectures consisting of an input layer with n units and an output layer with m units. It will be demonstrated that the networks are able to perform a variety of important data analysis tasks, including *Principal Component Analysis* (PCA), *Minor Component Analysis* (MCA) and *Independent Component Analysis* (ICA).

Keywords: unsupervised hebbian learning, principal component analysis, minor component analysis, independent component analysis

1 Introduction

Learning algorithms for neural networks can be divided into *supervised* and *unsupervised* algorithms. Supervised learning algorithms rely on an external teacher providing information about the correctness of the output produced by a network upon presentation of an input pattern. In unsupervised learning, the learning goal is not defined in terms of a teacher signal; the only available information is the input data and the information built into the network topology and the learning rules. Thus, an unsupervised learning network exploits the regularities of the input in order to extract statistically relevant features. This principle is believed to exist in cognitive processes [3, 4] and in the preprocessing of sensory inputs in the mammalian primary visual cortex [37].

In this paper, a particular class of unsupervised learning rules is considered, namely those in which variants of *Hebbian learning* are employed. Hebb [27] articulated that the weight of a connection between two neurons should be adjusted according to the correlation of the outputs of the two neurons. Formulated mathematically, the modification of the weight w_{ij} of the connection starting at unit j and ending at unit i is proportional to the product of the outputs $x_j \cdot y_i$:

$$w_{ij}(t+1) = w_{ij}(t) + \alpha \cdot x_j(t) \cdot y_i(t) \tag{1}$$

where $\alpha > 0$ is a learning parameter. However, there is one problem with this learning rule: the weights keep on growing without bound, and learning never stops. Several

CP437, *Computing Anticipatory Systems: CAYS--First International Conference*
edited by Daniel M. Dubois © 1998 The American Institute of Physics 1-56396-827-4/98/$15.00

proposals have been made to constrain the growth of the weights. Oja [43] was the first to observe that the modified Hebbian learning rule

$$\mathbf{w}(t+1) = \mathbf{w}(t) + \alpha \left(\mathbf{x}(t)y(t) - y^2(t)\mathbf{w}(t) \right) \qquad (2)$$

applied to the weight vector $\mathbf{w}$ of a simple linear network model with n input units with input vector $\mathbf{x}$ and a single output unit with output y not only keeps the weights bounded, but also normalizes the weight vector to unit length. More importantly, Oja showed that starting from random initial weights, the weight vector converges to a vector which is identical to the eigenvector belonging to the largest eigenvalue of the covariance matrix of the input distribution. This result has fostered the development of several learning rules which elucidate the relationship between unsupervised Hebbian learning and statistical data analysis.

In this paper, we survey learning rules that are able to perform a variety of important data analysis tasks, including *Principal Component Analysis* (PCA), *Minor Component Analysis* (MCA) and *Independent Component Analysis* (ICA) [31, 9, 44]. All of these learning rules operate on two-layer network architectures consisting of an input layer with n units and an output layer with m units. In some proposals, there are additional trainable lateral connections between the units in the output layer. The architectures of the networks proposed to perform the data analysis tasks mentioned above are shown in Fig. 1.

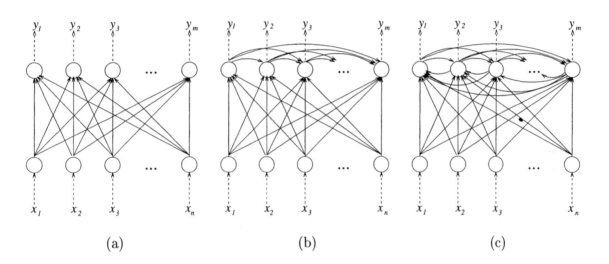

Figure 1: Network Architectures

The motivation for developing neural alternatives for traditional data analysis tasks is that the networks significantly reduce the computational effort otherwise required to perform these tasks. For example, PCA networks adaptively extract the principal components from a given data set (i.e. the eigenvectors belonging to the largest eigenvalues of the covariance matrix of the input distribution) without explicity determining the covariance matrix and performing the time–consuming eigenvector computations. Similarly, MCA networks automatically extract the minor components (i.e. the eigenvectors belonging to the smallest eigenvalues of the covariance matrix of the input distribution) from a given input pattern set. It will be shown that the PCA networks can be used for image compression, and that the MCA networks can be applied to particular feature extraction tasks.

607

Many unsupervised learning proposals to ICA and its central part, *Blind Source Separation* (BSS), are modifications of the neural networks performing PCA. The difference between the BSS networks and the PCA networks is the nonlinear transformation of the outputs in the learning rules of the BSS networks instead of the identity function used in the PCA networks. The motivation for introducing a nonlinear function is to approximate and implicitly optimize the higher order moments of the components of the output vectors during the training phase. It is worth mentioning that the transformation performed by the network between inputs and outputs is still linear. Under special assumptions on the input vectors, these modified PCA networks perform BSS. It will be demonstrated that the networks are capable of performing BSS for a set of acoustic source signals.

Apart from their application potential in data analysis and signal processing, unsupervised networks based on Hebbian learning rules are interesting objects of study with respect to information theory [12, 49], analysis of dynamical systems [26] and self-organization in biological neural systems [58]. Due to their solid mathematical foundations, such network models are also useful for studying various phenomena of anticipation; Rosen has shown that learning in neural networks is an anticipatory process [50].

The paper is organized as follows. Section 2 surveys networks for performing PCA, section 3 networks for performing MCA, and section 4 networks for performing ICA (in particular, BSS). In each section, the data analysis problem is described, the features of the networks are presented, and an application example is given. Section 5 concludes the paper and outlines areas for future research.

2 Principal Component Analysis

PCA is a well known statistical method for analyzing data [31]. The aim is to find a set of m orthogonal vectors in the original n–dimensional data space that account for as much as possible of the data's variance and then project the data onto the M–dimensional subspace spanned by these vectors. For $m < n$, the vectors yielding the smallest mean–squared reconstruction error of the transformed input are the eigenvectors belonging to the m largest eigenvalues of the covariance matrix of the input data. If the data to be analyzed has a Gaussian distribution along each input channel, then these eigenvectors, called the *principal components*, lie in the directions with the maximum variances in each of the perpendicularly arranged subspaces, which in information theoretic terms corresponds to the maximization of the output entropy. PCA is useful in many applications, such as signal filtering and restoration, pattern recognition and classification, and image compression and encoding. Mathematically, PCA can be stated as follows:

Given is a random vector $\mathbf{x} \in R^n$ with an expected value $E(\mathbf{x}) = 0$ and its covariance matrix $E((\mathbf{x} - E(\mathbf{x}))(\mathbf{x} - E(\mathbf{x}))^T) = E(\mathbf{x}\mathbf{x}^T) = \mathbf{C}_{xx}$. We want to find a linear transformation of $\mathbf{x}$ to a vector $\mathbf{y} \in R^m$, $m \leq n$, whose components y_i are uncorrelated and whose variances $E(y_i^2)$ are maximal.

The desired components y_i are the i-th *principal components* of $\mathbf{x}$. Thus, the following steps are required to perform PCA:

1. compute the covariance matrix $\mathbf{C}_{xx}$

2. compute the eigenvectors $\mathbf{v}_1, \cdots, \mathbf{v}_m$ of $\mathbf{C}_{xx}$ belonging to the m largest eigenvalues

3. project $\mathbf{x}$ with $\mathbf{V} = (\mathbf{v}_1, \cdots, \mathbf{v}_m)^T$ onto $\mathbf{y} = \mathbf{V}\mathbf{x}$

The motivation for investigating novel ways for performing PCA is to avoid the need for the time–consuming eigenvector computations required in the conventional PCA scheme. As already mentioned, the possibility of using neural networks for the parallel on–line computation of the PCA expansion was first indicated by Oja [43], who observed that his modified Hebbian learning rule applied to a network with a single linear output unit converges to a weight vector which is identical to the first principal component axis of the input distribution.

Several proposals have been made to extend Oja's work in developing networks that perform a complete PCA [13]. In general, the activation functions of the output units and the learning rules of the proposals made in the literature can be written as follows:

Activation Function:

$$y = f_1(Q)y + f_2(Q)Wx$$

Learning Rules:

$$\triangle W = \alpha \left(f_3(Q, w, y)yx^T - g_1(y, W)W \right)$$
$$\triangle Q = \beta \left(f_4(Q) + g_2(yy^T) \right)$$

In these equations, x is the input vector, y is the output vector produced by the network, W is the weight matrix for the connections between input and output layer, Q is the weight matrix for the lateral connections between the output units, and $f_1, f_2, f_3, f_4, g_1, g_2$ are different functions.

Fig. 2 summarizes the activation functions and learning rules of PCA networks. The networks types given correspond to those shown in Fig. 1, and there are several linear operators on quadratic matrices, such as yy^T:

- DD, which sets all off-diagonal elements to zero

- OD, which sets all diagonal elements to zero

- UOD, which sets all elements above and on the diagonal to zero

- UDD, which sets all elements above the diagonal to zero

- ID, which does not change the matrix

The numbers # represent the following networks:

1. Network (1) of Oja [44]

2. Network of Oja, Ogawa und Wangviwattana [47]

3. Network of Liu und Ligomenides [38]

4. Network of Peper und Noda [48]

5. Network of Sanger [52]

6. Network(2) of Oja [44]

Net #	Type	Activation of $\mathbf{y}$		Learning Rule of $\mathbf{W}$		Learning Rule of $\mathbf{Q}$	
		$\mathbf{f_1(Q)}$	$\mathbf{f_2(Q)}$	$\mathbf{f_3(Q,W,y)}$	$\mathbf{g_1(y,W)}$	$\mathbf{f_4(Q)}$	$\mathbf{g_2(yy^T)}$
1.	(a)	$\mathbf{0}$	$\mathbf{I}$	$\mathbf{I}$	$\mathbf{yy}^T$	$--$	$--$
2.	(a)	$\mathbf{0}$	$\mathbf{I}$	$\mathbf{I}$	$\boldsymbol{\beta}^T \cdot \mathbf{yy}^T$ [1]	$--$	$--$
3.	(a)	$\mathbf{0}$	$\mathbf{I}$	$2\mathbf{I} - \mathbf{WW}^T$	$\mathbf{yy}^T$	$--$	$--$
4.	(a)	$\mathbf{0}$	$\mathbf{I}$	$\mathbf{I}$	$\mathbf{PWW}^T$ [2]	$--$	$--$
5.	(a)	$\mathbf{0}$	$\mathbf{I}$	$\mathbf{I}$	$UDD(\mathbf{yy}^T)$	$--$	$--$
6.	(a)	$\mathbf{0}$	$\mathbf{I}$	$\mathbf{I}$	$\beta \cdot UDD(\mathbf{yy}^T)$	$--$	$--$
7.	(a)	$\mathbf{0}$	$\mathbf{I}$	$\mathbf{I} - \gamma\mathbf{H(y)}$ [3]	$\mathbf{I} - DD(\mathbf{yy}^T)$	$--$	$--$
8.	(b)	$\mathbf{0}$	$\mathbf{I+Q}$	$\mathbf{I}$	$DD(\mathbf{yy}^T)$	$\mathbf{0}$	$-UOD$
9.	(b)	$UOD(\mathbf{Q})$	$\mathbf{I}$	$\mathbf{I}$	$DD(\mathbf{yy}^T)$	$-DD$	$-UOD$
10.	(c)	$\mathbf{Q}^k$	$\mathbf{I}+\cdots+\mathbf{Q}^{k-1}$	$\mathbf{I}$	$DD(\mathbf{yy}^T)$	$\mathbf{0}$	$-OD$
11.	(c)	$\mathbf{0}$	$(\mathbf{I-Q})^{-1}$	$\mathbf{I}$	$DD(\mathbf{yy}^T)$	$\mathbf{0}$	$-OD$
12.	(c)	$\mathbf{0}$	$\mathbf{I}$	$\mathbf{I}+OD(\mathbf{Q})$	$DD(\mathbf{yy}^T)$	$-\mathbf{I}$	$-OD$
13.	(c)	$\mathbf{Q}$	$\mathbf{I}$	$\mathbf{I}$	$DD(\mathbf{yy}^T)$	$-\delta\mathbf{I}$	$-\gamma OD$
14.	(c)	$\mathbf{Q}$	$\mathbf{I}$	$\mathbf{I}+\gamma OD(\mathbf{Q})$	$\mathbf{0}$ [4]	$-\mathbf{I}$	$-OD$
15.	(c)	$\mathbf{Q}$	$\mathbf{I}$	$\mathbf{I}+\gamma OD(\mathbf{Q})$	$DD(\mathbf{yy}^T)$	$-\mathbf{I}$	$-OD$

Table heading: **PCA Network Overview**

Figure 2: Overview of Activation Functions and Learning Rules of PCA Networks

7. Network of Hyvärinen [29]

8. Network of Rubner und Tavan [51]

9. Network of Kung und Diamantaras[35]

10. Network(1) of Földiak [14]

11. Network(2) of Földiak [14]

12. Network(1) of Leen [36]

13. Network(2) of Leen [36]

14. Network of Brause [7]

15. Network of Freisleben [16, 17]

An important application of PCA networks is image compression. The general idea is to exploit the fact that nearby pixels in images are often highly correlated. Therefore, the image to be compressed is divided into equally sized blocks of pixels, and each of these blocks, treated as an n–dimensional input vector, is linearly transformed into a m–dimensional output vector whose components are mutually uncorrelated. The image is scanned from left to right and top to bottom, and the blocks are consecutively presented to the input units of the network, as shown in Fig. 3.

---

[1] $\boldsymbol{\beta}^T = (\beta_1, \cdots, \beta_m)^T$ determines the different biases of y_i

[2] $\mathbf{P} = \text{Diag}(\|\mathbf{w}_1\|, \cdots, \|\mathbf{w}_m\|)^{-1}$

[3] $\mathbf{H(y)} = \text{Diag}(h_1(\mathbf{y}), \cdots, h_m(\mathbf{y}))$ with $h_i(\mathbf{y}) = \prod_{j=1}^{i-1} y_j^2$

[4] instead of a *decay* term in the learning rule, an explicit normalization is performed.

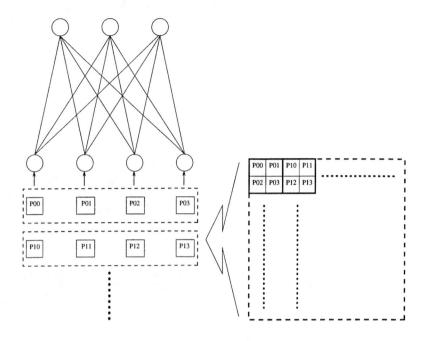

Figure 3: Image Compression with PCA Networks

In the learning mode, the input vectors representing an image are propagated through the network by computing the output activations and performing the corresponding weight updates (the input/output connections are initialized to small random values, whereas the lateral connections, if applicable, are initially set to zero). The whole image is presented to the network several times to give the network time to converge. After the training phase is finished, the image is again processed by the network, however without any connection updates. The image is compressed by multiplying each block represented as an n–dimensional vector by each of the m weight vectors obtained after training to generate m coefficients for coding the block. To decompress it, each block of the image is reconstructed by adding together all the weight vectors multiplied by their coefficients.

It has been shown [30] that the overall average mean-squared reconstruction error is minimized when the rows of the transformation matrix used for decorrelation are the orthonormalized eigenvectors of the covariance matrix of the input vectors. The corresponding optimal transform under this error citerion is called *Karhunen–Loève Transform (KLT)* [30], which is equivalent to PCA. Since the KLT is quite time–consuming to compute, it is in practice substituted by a suboptimal but fast transform. An example is the *Discrete Cosine Transform (DCT)* [57] which performs very closely to the KLT and has therefore been selected as the basis of the *JPEG* image compression standard. In contrast to the KLT, the design of the DCT is input–independent, since the same transformation matrix is used for any image that needs to be compressed.

In order to illustrate the performance of a PCA network, a 256×256 8–bit–pixel image with 256 grey levels has been selected [24, 42]. The 256 grey level values are normalized to values between -1 and 1. The image is split into 8×8 pixel blocks, leading to a network with 64 input units. In the experiments conducted, this 64–dimensional input is compressed to a 3–dimensional output, i.e. the networks consist of 3 units in the output layer. If no adaptive quantization scheme is used, this leads to coding each 8×8 block of pixels (512 bits) with a total of 24 bits, so the data rate is 0.375 bits per pixel and the compression factor is 21.33. The blocks do not overlap, and the complete image is presented 8 times, resulting in 8192 input vectors presented to the networks during the training phase.

In order to evaluate the quality of the reconstructed image, the mean squared error (MSE)

$$MSE = \frac{1}{s \cdot t} \cdot \sum_{i,j} (O(i,j) - R(i,j))^2 \tag{3}$$

between the original $O(i,j)$ and the decompressed image $R(i,j)$ is computed, where s, t denote the number of pixels in the x–direction and the y–direction, respectively.

Fig. 4 shows the original image and the image compressed/decompressed by the PCA network proposed by Sanger [52] (network 5 in Fig. 2). The MSE for the KLT is 29.63, for the DCT 32.19, and for Sanger's network 30.40.

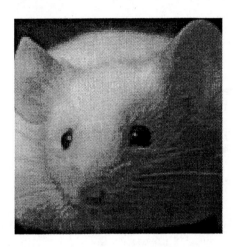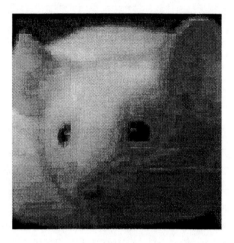

original image   compressed/reconstructed image

Figure 4: Image Compression with Sanger's Network

3 Minor Component Analysis

In several applications, the extraction of features under the variance maximization criterion is not the desired goal. A central problem in pattern recognition is the development of *clustering transformations* which partition the data space into regions each of which contains the sample data points belonging to one class. Such a transformation maximizes the mean–squared distance between data points in different classes and minimizes the distance between points in the same class. In other words, the aim is to extract the most stable features with the *smallest* variances to identify the pure, non–noisy information content of the data. This corresponds to the removal of uncertainty from the data points, and thus to the minimization of the entropy of the clusters under consideration. It has been shown that under Gaussian distribution assumptions the entropy is minimized when the transformation is based on the eigenvectors associated with the m *smallest* eigenvalues of the covariance matrix of the input data. This is equivalent to a reversal of the

PCA method and is therefore called *Minor Component Analysis* (MCA) [44]. Apart from performing clustering transformations, MCA is useful for frequency estimation of signals and curve or surface fitting [55].

Mathematically, MCA can be stated as follows:

Given is a random vector $\mathbf{x} \in R^n$ with an expected value $E(\mathbf{x}) = 0$ and its covariance matrix $E((\mathbf{x} - E(\mathbf{x}))(\mathbf{x} - E(\mathbf{x}))^T) = E(\mathbf{x}\mathbf{x}^T) = \mathbf{C}_{xx}$. We want to find a linear transformation of $\mathbf{x}$ to a vector $\mathbf{y} \in R^m$, $m \leq n$, whose components y_i are uncorrelated and whose variances $E(y_i^2)$ are minimal.

The desired components y_i are the i-th *minor components* of $\mathbf{x}$. Thus, the following steps are required to perform MCA:

1. compute the covariance matrix $\mathbf{C}_{xx}$

2. compute the eigenvectors $\mathbf{v}_1, \cdots, \mathbf{v}_m$ of $\mathbf{C}_{xx}$ belonging to the m smallest eigenvalues

3. project $\mathbf{x}$ with $\mathbf{V} = (\mathbf{v}_1, \cdots, \mathbf{v}_m)^T$ onto $\mathbf{y} = \mathbf{V}\mathbf{x}$

Considering that MCA is a reversal of PCA, it is somewhat surprising that the number of neural PCA proposals is much larger than that of neural proposals for performing MCA. Furthermore, the MCA networks proposed so far are only based on the toplogy (a) shown in Fig. 1. Brause [6], Oja [44], and Xu, Oja und Suen [55] have shown that in a network with a single output unit, the learning rule

$$\mathbf{w}_i(t+1) = \mathbf{w}_i(t) - \alpha \cdot \mathbf{x}(t)y_i(t) \tag{4}$$

converges to a weight vector $\mathbf{w}_i(t)$ which is the smallest eigenvector of the covariance matrix of the input. The above learning is called an *anti-Hebbian* learning rule due to the negative sign used for the Hebbian term.

Anti-Hebbian learning is used in all networks with more than one output unit proposed for performing MCA. In all proposals, the output units have linear activation functions, with the output $y_i(t)$ given by

$$y_i(t) = \mathbf{w}_i(t)^T \mathbf{x}(t), \ i = 1, \cdots, m \tag{5}$$

The learning rules have a similar strucure:

$$
\begin{aligned}
\mathbf{w}_i(t+1) &= \mathbf{w}_i(t) - \alpha(t) \cdot \triangle\mathbf{w}_i(t) \\
\triangle\mathbf{w}_i(t) &= f_2^{(i)}(\mathbf{w}_i(t)) \cdot \mathbf{x}(t)y_i(t) + \mathbf{f}_3^{(i)}(y_1, \cdots, y_{i-1}, \mathbf{w}_1, \cdots, \mathbf{w}_{i-1}) \cdot y_i(t) + \\
&\quad + f_4^{(i)}(y_1, \cdots, y_i, \mathbf{w}_1, \cdots, \mathbf{w}_i) \cdot \mathbf{w}_i(t)
\end{aligned}
\tag{6}
$$

The individual terms of (6) have the following purpose:

The first term determines the strength of the anti-Hebbian term $-\mathbf{x}(t)y_i(t)$ by a scalar function $f_2^{(i)}$, which depends on $\mathbf{w}_i(t)$.

The second term serves to adaptively orthogonalize the current weight vector with respect to the previous weight vectors $\mathbf{w}_1, \cdots, \mathbf{w}_{i-1}$ und for decorrelating the output. The vector function $\mathbf{f}_3^{(i)}$ used for that purpose (which depends on all previous outputs and weight vectors) is multiplied with the output y_i.

The last term is used for adaptively normalizing the current weight vector by multiplying it with the function $f_4^{(i)}$, which may depend on all previous outputs and weight vectors including the i-th, but at least on the i-th output and the i-th weight vector. This term may be omitted, if an explicit normalization step is carried out after each learning update:

$$\mathbf{w}_i(t+1) = \frac{\mathbf{w}_i'(t+1)}{\|\mathbf{w}_i'(t+1)\|} \qquad (7)$$

where $\mathbf{w}_i'(t+1)$ is the weight vector modified by (6).

Fig. 5 summarizes the activation functions and learning rules of MCA networks proposed in the literature.

MCA Network Overview					
Net #	Activation $\mathbf{f}_1^{(i)}$ for y_i	Learning Rule of $\mathbf{w}_i$			expl. Norm.
		$f_2^{(i)}$ for Anti-Hebb	$\mathbf{f}_3^{(i)}$ for Orth./Decorr.	$f_4^{(i)}$ for Norm.	
1.	$\mathbf{0}$	1	$\gamma \sum_{j<i} y_j \mathbf{w}_j$	$y_i^2 + 1 - \mathbf{w}_i^T \mathbf{w}_i$	no
2.	$\beta \sum_{j<i} y_j \mathbf{w}_j$	1	$\beta \sum_{j<i} y_j \mathbf{w}_j$	0	yes
3.	$\mathbf{0}$	$\mathbf{w}_i^T \mathbf{w}_i$	$\sum_{j<i} y_j \mathbf{w}_j$	$y_i^2 + \sum_{j<i} y_i y_j \mathbf{w}_j^T \mathbf{w}_j$	no
4.	$\mathbf{0}$	1	$\gamma \sum_{j<i} y_j \mathbf{w}_j$	0	yes

Figure 5: Overview of Activation Functions and Learning Rules of MCA Networks

The numbers # represent the following networks:

1. Network of Oja [44]

2. Network of Brause [6]

3. Network of Luo und Unbehauen [39]

4. Network of Freisleben and Hagen [15, 18]

In order to visualize the effect of MCA based feature extraction, we have taken a grey–level image and presented it to a network with $N = 16$ input units and $M = 4$ output units. The image has been divided into 4×4 blocks of pixels; each of these blocks represents a 16–dimensional input vector. The image is scanned from left to right and top to bottom, and these vectors are consecutively presented to the input units of the network. The trained network, after several presentations of the whole image, is used to reduce the dimensionality of the input vectors by multiplying each block by each of the 4 weight vectors to generate 4 coefficients for coding the block. Each block of the image is afterwards reconstructed by adding together all the weight vectors multiplied by their coefficients. Fig. 6 shows the original image (a) and the image reconstructed as described above (b), with contrast enhancements to improve the visual presentation of the pure prototype of the object recognized.

4 Independent Component Analysis

Independent Component Analysis (ICA) [9] is a recently developed signal processing technique for representing given noisy mixtures of signals as a linear combination of statistically independent signals. This technique is important for a variety of application areas,

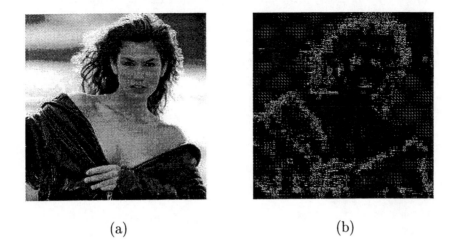

(a)   (b)

Figure 6: Image Clustering with Freisleben's MCA Network

such as antenna array processing, communications, speech processing and medical image processing. The central part of ICA is *Blind Source Separation* (BSS). The problem of BSS is to separate the unknown individual source signals (which are assumed to be statistically independent) out of given linear mixtures of them. From the mathematical point of view, the solution to the BSS problem is a separating matrix which transforms the mixture signals into signals with a maximal degree of independence estimating the original source signals.

Mathematically, the ICA problem is defined as follows:

Assume that there are n source signals $s_1(t), \ldots, s_n(t)$ at discrete time $t = 1, 2, \ldots$ which meet the following conditions:

(i) $s_i(t)$ are the values of a stationary scalar-valued random process at time t, having unit variance and zero mean.

(ii) $s_i(t)$ are statistically independent, i.e. the joint probability density is the product of the marginal densities:

$$p(s_1(t), \ldots, s_n(t)) = \prod_{i=1}^{n} p_i(s_i(t)) \qquad (8)$$

(iii) $s_i(t)$ have non-Gaussian probability density functions, except at most one which may be a Gaussian.

(iv) $s_i(t)$ are not observable.

The strongest condition is (ii), but it is realistic if the source signals come from different physical devices. The condition (i) can always be achieved by normalization. Condition (iii) reflects the fact that a mixture of Gaussians cannot be separated. In many practical situations, the source signals are either *sub-Gaussians* or *super-Gaussians*, i.e. distributions with densities flatter or sharper than that of Gaussians, respectively. Speech signals are typically super-Gaussians, while artificially generated acoustic sounds are often sub-Gaussians [33, 54].

Assume that a linear noisy mixture of source signals $s_i(t)$ at time t satisfying the conditions above is given:

$$\mathbf{u}(t) = \mathbf{M}\,\mathbf{s}(t) + \mathbf{n}(t) \qquad (9)$$

where $\mathbf{s}(t)$ is the n-dimensional source signal vector, $\mathbf{x}(t)$ is the m-dimensional observed signal vector, $m \geq n$, $\mathbf{M}$ is the unknown constant $m \times n$ mixing matrix and $\mathbf{n}(t)$ is a noise vector. The pair $(\mathbf{H}, \mathbf{D})$ is called the ICA of the random vector $\mathbf{x}$ if the following conditions are met:

1. The covariance matrix $\mathbf{C}_{xx} = E(\mathbf{xx}^T)$ can be decomposed into

$$\mathbf{C}_{xx} = \mathbf{H}\,\mathbf{D}^2\,\mathbf{H}^T \qquad (10)$$

   where $\mathbf{H}$ is a $m \times n$ matrix with full column rank n, and $\mathbf{D}$ is a $n \times n$ diagonal matrix, consisting of the positive roots of the eigenvalues of $\mathbf{C}_{xx}$.

2. The input vector $\mathbf{u}$ is related to $\mathbf{H}$ through:

$$\mathbf{u}(t) = \mathbf{H}\,\mathbf{y}(t), \qquad (11)$$

   where the unknown random vector $\mathbf{y} \in \mathbf{R}^n$ has as its covariance matrix $E(\mathbf{yy}^T) = \mathbf{D}^2$ and has a maximal degree of statistical independence, which can be measured by a so called *contrast function*.

The ICA $(\mathbf{H}, \mathbf{D})$ can be uniquely determined under the following additional conditions:

1. The diagonal matrix $\mathbf{D}$ has its elements in decreasing order.

2. The column vectors of the ICA matrix have unit norm.

3. The entry of the largest element in each column of $\mathbf{H}$ is positive.

A contrast function $\mathcal{F}$ is a real valued function defined on a set $\mathcal{X}$ of random vectors and has to satisfy three conditions:

1. $\mathcal{F}(\mathbf{x})$ depends only on the probability distribution of $\mathbf{x}$ for all $\mathbf{x} \in \mathcal{X}$.

2. The value of $\mathcal{F}(\mathbf{x})$ is invariant while scaling the vector $\mathbf{x}$ with a diagonal matrix.

3. If the components of $\mathbf{x}$ are independent non-Gaussians, $\mathcal{F}(\mathbf{x}) \geq \mathcal{F}(\mathbf{A}\,\mathbf{x})$ for any invertible matrix $\mathbf{A}$ and equal if $\mathbf{A}$ is a diagonal matrix.

In several approaches [9, 33, 54], the contrast function is assumed to depend on the fourth-order cumulants of the components x_i of the vector $\mathbf{x}$. In [9, 33], different contrast functions are defined in order to approximate the fourth-order cumulants on one hand and to have a formula as simple as possible on the other hand.

Fig. 7 illustrates the steps of a complete ICA algorithm.

The first box represents the unobservable (linear) transformation of the unknown source signal vector $\mathbf{s}$ into the observed signal vector $\mathbf{u} = \mathbf{M} \cdot \mathbf{s}$. The $m \times n$-mixing matrix $\mathbf{M}$ is assumed to have full column rank, i.e. $\mathbf{u}$ has the dimension m, with $m \geq n$.

The second box represents a *prewhitening* step which transforms $\mathbf{u}$ into $\mathbf{x} = \mathbf{D}^{-1} \cdot \mathbf{F}^T \cdot \mathbf{u}$, such that $\mathbf{x}$ has as its covariance matrix $\mathbf{C}_{xx} = E(\mathbf{xx}^T)$ the identity matrix $\mathbf{I}_n$. Prewhitening simplifies the BSS task and can be achieved by any non-neural or neural

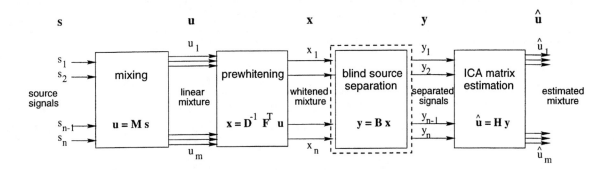

Figure 7: The Steps of an ICA Algorithm

PCA algorithm; the columns of the matrix $\mathbf{F}$ are the n principal eigenvectors of $\mathbf{C}_{xx}$, and the diagonal matrix contains the square roots of the n largest eigenvalues.

The third (dashed) box represents the BSS part: finding an (orthogonal) separation matrix $\mathbf{B}$ to transform $\mathbf{x}$ into an output vector $\mathbf{y}$ whose components have a maximal degree of statistical independence (measured by a contrast function [9]). Each output y_i, $i = 1, \cdots, n$ can be regarded as an approximation of one of the source signals $\pm s_j$, $j = 1, \cdots, n$.

The fourth box represents the last step missing to complete ICA, namely the estimation of the mixing matrix, the ICA matrix $\mathbf{H}$, whose columns form the ICA basis. The output of the last step is the estimated observed mixture vector $\hat{\mathbf{u}} = \mathbf{H} \cdot \mathbf{y}$.

In the following, we concentrate on the networks for performing BSS, because this is the most complicated step. Fig. 8 summarizes the networks proposed for performing BSS. Here, $\mathbf{u}$ denotes the original input and $\mathbf{x}$ denotes the already prewhitened input. Similarly, in the first case there is a weight matrix $\mathbf{B}(t)$ which learns the separation matrix $\mathbf{B}$, and in the second case there is a weight matrix $\mathbf{W}(t)$ which only learns the orthogonal separation matrix $\bar{\mathbf{B}}$. The network architecture used in the majority of BSS proposals is the one shown in Fig. 1 (a). Network 7 is based on an incompletely connected topology, and the networks 8–10 are based on the architecture shown in Fig. 1 (c).

The entries in Fig. 8 have the following meanings:

Networks:
The numbers # represent the following networks:

1. Network of Cardoso and Laheld [8]

2. Network (1) of Yang and Amari [56]

3. Network (2) of Yang and Amari [56]

4. Network of Zhang and Xu [59]

5. Network of Girolami and Fyfe [25]

6. Network of Bell and Sejnowski [2]

7. Network of Deco and Brauer [11]

8. Network of Jutten and Herault [32, 10, 53]

617

<table>
<tr><th colspan="6" align="center">BSS Network Overview</th></tr>
</table>

#	Prerequisites A	B	C	Activation	Learning Rule
1.	fc	n	-	$\mathbf{y} = \mathbf{Bu}$	$\triangle\mathbf{B} = -\alpha\left(\mathbf{yy}^T - \mathbf{I} + \mathbf{g(y)h(y)}^T - \mathbf{h(y)g(y)}^T\right)\mathbf{B}$
2.	r	n	-	$\mathbf{y} = \mathbf{Bu}$	$\triangle\mathbf{B} = \alpha(t)\left(\mathbf{I} - (\mathbf{f}_1(K_3,K_4)\circ\mathbf{y}^2)\mathbf{y}^T - (\mathbf{g}_1(K_3,K_4)\circ\mathbf{y}^3)\mathbf{y}^T\right)\mathbf{B}$
3.	r	n	-	$\mathbf{y} = \mathbf{Bu}$	$\triangle\mathbf{B} = \alpha(t)\left(\mathbf{I} - (\mathbf{f}_2(K_3,K_4)\circ\mathbf{y}^2)\mathbf{y}^T - (\mathbf{g}_2(K_3,K_4)\circ\mathbf{y}^3)\mathbf{y}^T\right)\mathbf{B}$
4.	r	n	-	$\mathbf{y} = \mathbf{Bu}$	$\triangle\mathbf{B} = \alpha(t)\left(\mathbf{I} - 2\cdot\mathbf{y}\cdot\tanh(\mathbf{y}^T)\right)\mathbf{B}$
5.	r	n	-	$\mathbf{y} = \mathbf{Bu}$	$\triangle\mathbf{B} = \alpha(t)\left(\mathbf{I} - \mathbf{K}_4\cdot\tanh(\mathbf{y})\mathbf{y}^T - \beta\cdot\mathbf{yy}^T\right)\mathbf{B}$
6.	r	n	-	$\mathbf{y} = \mathbf{f(Bu)}$	$\triangle\mathbf{B} = \alpha(t)\left(\left(\mathbf{B}^T\right)^{-1} + \mathbf{g(y)u}^T\right)$
7.	-	n	nl	$y_i = u_i + f_i(u_1,\cdots,u_{i-1},\mathbf{b}_i)$	$\triangle\mathbf{b}_i = -\alpha\cdot\frac{\partial F}{\partial\mathbf{b}_i}$
8.	r+	n	-	$\mathbf{y} = (\mathbf{I}+\mathbf{Q})^{-1}\mathbf{u}$	$\triangle q_{ij} = \alpha\cdot f(y_i)g(y_j),\ i\neq j$
9.	r	n	-	$\mathbf{y} = (\mathbf{I}+\mathbf{Q})^{-1}\mathbf{u}$	$\triangle\mathbf{Q} = -\beta\cdot(\mathbf{I}+\mathbf{Q})\left(\mathbf{\Gamma} - \mathbf{f(y)g(y)}^T\right)$
10.	r	n	ns	$\mathbf{y} = (\mathbf{I}+\mathbf{Q})^{-1}\mathbf{u}$	$\triangle\mathbf{Q} = \alpha\cdot(\mathbf{I}+\mathbf{Q}^T)^{-1}\left(\mathrm{Diag}(\mathbf{yy}^T)^{-1}\mathbf{yy}^T - \mathbf{I}\right)$
11.	fc	y	-	$\mathbf{y} = \mathbf{Wx}$	$\triangle\mathbf{W} = \alpha\cdot\mathbf{g(y)}\left(\mathbf{x}^T - \mathbf{y}^T\mathbf{W}\right)$
12.	fc	y	-	$\mathbf{y} = \mathbf{Wx}$	$\triangle\mathbf{W} = \alpha\cdot\mathbf{g(y)}\left(\mathbf{x}^T - \mathbf{g(y)}^T\mathbf{W}\right)$
13.	fc	y	-	$\mathbf{y} = \mathbf{Wx}$	$\triangle\mathbf{W} = \alpha\cdot\mathbf{g(y)x}^T + \beta\left((\mathbf{I}-\mathbf{WW}^T)\right)\mathbf{W}$
14.	r	y	-	$y_i = \mathbf{w}_i^T\mathbf{x}$	$\triangle\mathbf{w}_i = \pm\alpha\cdot\left(\mathbf{x}y_i^3 + f(t)(1-\|\mathbf{w}_i\|^2) - \gamma\prod_{j=1}^{i-1}y_j^2\mathbf{x}y_i\right)$
15.	r	y	-	$y_i = \mathbf{w}_i^T\mathbf{x}$	$\triangle\tilde{\mathbf{w}}_i = \pm\alpha\cdot g(y_i)\left(\mathbf{I} - \sum_{j=1}^{i-1}\mathbf{w}_j\mathbf{w}_j^T\right)\mathbf{x}$ [5]
16.	r	y	-	$y_i = \mathbf{w}_i^T\mathbf{x}$	$\triangle c_i = -\mu\cdot\left(c_i - y_i^4 + 3\right)$ [6] $\triangle\tilde{\mathbf{w}}_i = \pm\alpha\cdot g(c_i,y_i)\cdot\left(\mathbf{I} - \sum_{j=1}^{i-1}\mathbf{w}_j\mathbf{w}_j^T\right)\mathbf{x}$ [7]

Figure 8: Overview of Activation Functions and Learning Rules of BSS Networks

9. Network of Amari, Cichocki and Yang [1]

10. Network of Matsuoka, Ohya and Kawamoto [41]

11. Network (1) of Oja [46, 45, 33, 34]

12. Network (2) of Oja [46, 45, 33, 34]

13. Network of Wang, Karhunen, Oja, Vigario and Joutsensalo [54, 33, 34]

14. Network of Hyvärinen [29]

15. Network(1) of Freisleben, Hagen, Borschbach [20, 21, 22]

16. Network (2) of Freisleben, Hagen, Borschbach [23]

---

[5] instead of a decay term, an explicit normalization is carried out.
[6] c_i is an additional variable for learning the fourth order cumulant.
[7] instead of a decay term, an explicit normalization is carried out.

Prerequisites for the mixing matrix, the input vector, and the source signals:

A: denotes the prerequisites for the mixing matrix:

- **-** : no prerequisites,
- **fc** : full column rank,
- **r** : regular matrix and
- **r+** : regular matrix with diagonal elements 1

B: denotes the prerequisites for the input vector:

- **y** : the input vector must be standardized, i.e. prewhitened,
- **n** : the input vector is the original mixture vector without preprocessing.

C: denotes additional features of the input vector or the source signals:

- **nl** : the components of the input vector may be nonlinearly correlated and
- **ns** : the distributions of the source signals are *nonstationary*.

Activation and Learning Rules:

1. The functions **g** and **h** are in general nonlinear functions. For example, selecting $g_i(y_i) = y_i$ and $h_i(y_i) = \tanh(y_i)$ allows to separate Sub-Gaussians. Switching these two functions allows to separate Super-Gaussians.

2.+3. The functions $\mathbf{f}_i, \mathbf{g}_i$, $i = 1, 2$ are special functions of the third order cumulants $K_3(\mathbf{y})$ und fourth order cumulants $K_4(\mathbf{y})$, $\circ$ denotes the Hadamard product.

5. The matrix $\mathbf{K}_4$ is the diagonal matrix whose elements are the signs of the fourth order cumulants: $\mathbf{K}_4 = \mathrm{Diag}\left(\mathrm{sgn}(K_4(y_1)), \cdots, \mathrm{sgn}(K_4(y_n))\right)$.

6. The activation function f is nonlinear, e.g. the logistic function $f(a) = (1 + e^{-a})^{-1}$ or $f(a) = \tanh(a)$. The function $g(y_i)$ results from partial differentiation $\frac{\partial}{\partial y_i} \frac{\partial y_i}{\partial a_i}$.

7. The learning rule is a gradient descent of the cost function F, which has its minimum when the components are statistically independent.

8. The network is only for $n = 2$ asymptotically stable. The functions g and h are two different nonlinear functions.

9. The matrix $\mathbf{\Gamma} = \mathrm{Diag}(\gamma_1, \cdots, \gamma_n)$ and f is a special nonlinear function.

11.- 13. The function **g** is a nonlinear odd function, which is selected according to the application. For example, $g(y) = y^3$ is suitable for separating Super-Gaussians and $g(y) = \tanh(y)$ for separating Sub-Gaussians.

14. The sign is selected according to the application: $+$ if Super-Gaussians are to be separated, and $-$ for Sub-Gaussians. The function $f(t)$ is a penalty function, which is growing with t.

15. The function g is used for separating Super-Gaussians ($g(y_i) = y_i^3$) or Sub-Gaussians ($g(y_i) = \tanh(y_i)$).

16. In addition to the learning rule for the weight vector, there is a learning rule for the fourth-order cumulants c_i. The function $g(c_i, y_i)$ is as follows:

$$g(c_i, y_i) = \begin{cases} \tanh(y_i) & : \quad c_i < 0 \\ y_i^3 & : \quad c_i \geq 0 \end{cases} \tag{12}$$

In order to illustrate the use of a network in a source separation task, we present an example where acoustic signals are to be separated. In this example, four independent sound signals, sampled at a rate of 8 kHz were used (see Fig. 9, column (a)). The four signals s_i represent excerpts of four historical speeches given by the following speakers: (1) W. Churchill, (2) J. F. Kennedy, (3) M. Luther-King, and (4) N. Armstrong. They are stationary and have super-Gaussian distributions, e.g. the fourth order cumulants $K_4(s_i) = E((s_i)^4) - 3 \cdot E((s_i)^2)$ are: $K_4(s_1) = 6.155$; $K_4(s_2) = 8.291$; $K_4(s_3) = 9.164$; $K_4(s_4) = 2.293$.

The 4-dimensional input vectors of the neural network, which represent 35392 samples of the linear mixtures, were constructed by the randomly generated regular mixing matrix M shown in Fig. 10. Since M has been chosen to be orthogonal, a prewhithening step is not necessary. The mixtures are shown in column (b) of Fig. 9.

In the training mode, the set of 35392 input vectors was presented 4 times to the network 15 in Fig. 8. The weights of the connections were initially set to random values taken from the interval [-1, 1]. An initial learning rate $\alpha(0) = 0.95$ has been used and successively decreased after each simulation step according to $\alpha(1000) = 0.004 \cdot \alpha(0)$.

After the weight vectors have converged to the columns of the separating matrix, the network is capable of separating the original source signals from the mixtures used as the input, as shown in column (c) of Fig. 9. The acoustic quality of the separated signals is very good. It is impossible for humans to distinguish them from the original signals.

5 Conclusions

In this paper, we have presented a survey of learning rules that are are able to perform a variety of important data analysis tasks, including *Principal Component Analysis* (PCA), *Minor Component Analysis* (MCA) and *Independent Component Analysis* (ICA). These learning rules are variants of unsupervised Hebbian learning rules and operate on two-layer network architectures consisting of an input layer with n units and an output layer with m units, sometimes in conjunction with additional trainable lateral connections between the units in the output layer.

The individual properties of PCA, MCA and ICA/BSS networks proposed in the literature were discussed, and it was demonstrated that the PCA networks can be used for image compression, and the MCA networks can be applied to particular feature extraction tasks, and the BSS networks to acoustic signal separation problems.

There are several areas for future research, such as (a) investigating whether other data analysis tasks (e.g. *exploratory projection pursuit* [25], *sammons mapping* [40] or *nonlinear discrimant analysis* [40]) can be realized by unsupervised learning networks, (b) analyzing whether less mixtures than sources and nonlinear mixtures can be separated by an ICA network, and (c) finding mathematical means to analyze the convergence behaviour and stability of nonlinear Hebbian networks.

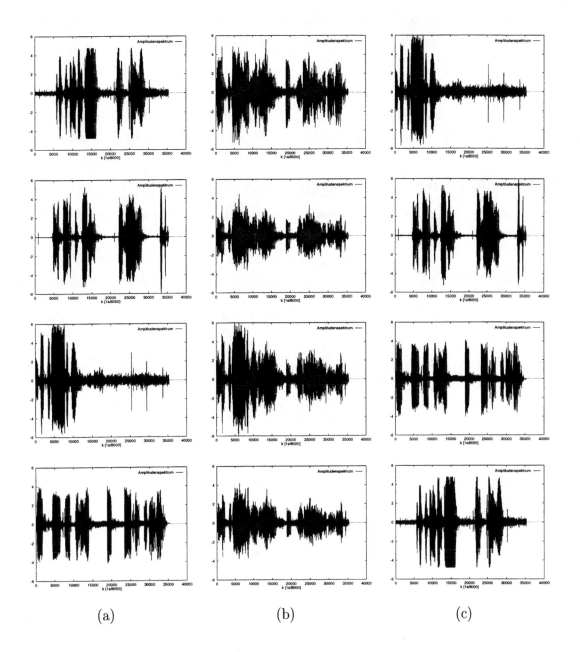

| (a) | (b) | (c) |

Figure 9: Source Signals (a), Linear Mixtures (b), and Separated Signals (c)

$$\begin{pmatrix} -0.246559 & -0.311630 & -0.436252 & 0.898941 \\ -0.199433 & -0.377810 & -0.385189 & -0.416525 \\ 0.426000 & -0.377810 & 0.813212 & 0.412838 \\ -0.228142 & -0.309001 & 0.463215 & 0.354782 \end{pmatrix}$$

Figure 10: Mixing Matrix **M**

References

[1] Amari, S. and A. Cichocki and H. H.Yang. Recurrent Neural Networks For Blind Sep-

aration of Sources. In *Proceedings of the 1995 International Symposium on Nonlinear Theory and Applications*, volume I, pages 37–42, 1995.

[2] Bell, A. J. and T. J. Sejnowski. An Information Maximation Approach to Blind Separation and Blind Deconvolution. *Neural Computation*, 7(1):1129–1159, 1995.

[3] Barlow, H. Sensory Mechanism, the Reduction of Redundancy and Intelligence. In *National Physical Laboratory Symposium N.10: The Mechanisms of thought Processes*, London, 1959.

[4] Barlow, H. Unsupervised Learning. *Neural Computation*, 1(1):295–311, 1989.

[5] Borschbach, M. Separation unabhängiger Signale mit neuronalen Netzen. Diplomarbeit, Universität-Gesamthochschule Siegen, FB Elektrotechnik und Informatik, September 1996.

[6] Brause, R. W. The Minimum Entropy Network-A New Building Block for Clustering Transformations. In *Proceedings of the 1992 International Conference on Artificial Neural Networks (ICANN'92)*, volume 2, pages 1095–1098, Brighton, UK, 1992. Elsevier.

[7] Brause, R. W. A Symmetrical Lateral Inhibition Network for PCA and Feature Decorrelation. In *Proceedings of the 1993 International Conference on Artificial Neural Networks (ICANN'93)*, pages 486–489, Amsterdam, 1993. Springer-Verlag.

[8] Cardoso, J.- F. and B. Laheld. Equivariant Adaptive Source Separation. *IEEE Transactions on Signal Processing*, October 1994.

[9] Comon, P. Independent Component Analysis, A New Concept? *Signal Processing*, 36:287–314, 1994.

[10] Comon, P. and C. Jutten and J. Herault. Blind Separation Of Sources, Part II: Problems Statement. *Signal Processing*, 24:11–20, 1991.

[11] Deco, G. and W. Brauer. Nonlinear Higher-Order Statistical Decorrelation by Volume-Conserving Neural Architectures. *Neural Networks*, 8:525–535, 1995.

[12] Deco, G. and D. Obradovic. *An Information-Theoretic Approach to Neural Computing*. Springer-Verlag, 1996.

[13] Diamantaras, K.I. and S.Y. Kung. *Principal Component Neural Networks – Theory and Applications*. John Wiley & Sons, 1996.

[14] P. Földiák. Adaptive Network for Optimal Linear Feature Extraction. In *International Joint Conference on Neural Networks*, Volume 1, pp. 401–405, IEEE Press, 1989.

[15] Freisleben, B. Feature Extraction Through Entropy Minimization: A Network Model. In *Proceedings of the 1993 World Congress on Neural Networks*, volume IV, pages 194–197, Portland, Oregon, 1993.

[16] Freisleben, B. Parallel Learning Algorithms for Principal Component Extraction. In *Proceedings of the 3rd International Conference on Artificial Neural Networks*, pages 267–271, Brighton, UK, 1993.

[17] Freisleben, B. PCA In A Network With Full Lateral Connections. In *Proceedings of the 1993 International Conference on Artificial Neural Networks (ICANN'93)*, page 618. Springer-Verlag, 1993.

[18] Freisleben, B. and C. Hagen. Analysis of a Network for Minor Component Extraction. In *Proceedings of the 1996 World Congress on Neural Networks (WCNN'96)*, pages 376–379, San Diego, USA, Lawrence Erlbaum Publishers, 1996.

[19] Freisleben, B. and C. Hagen. A Hierarchical Learning Rule for Independent Component Analysis. In *Proceedings of the 1996 International Conference on Artificial Neural Networks (ICANN'96)*, volume 1112 of *Lecture Notes in Computer Science*, pages 525–530, Bochum, Germany, 1996. Springer-Verlag.

[20] Freisleben, B. and C. Hagen and M. Borschbach. A Neural Network for Blind Acoustic Signal Separation. In *Proceedings of the 1996th International Conference on Robotics Vision and Parallel Processing for Industrial Automation (ROVPIA-96)*, volume 1, pages 67–72, Ipoh, Malaysia, 1996.

[21] Freisleben, B. and C. Hagen and M. Borschbach. Blind Separation of Acoustic Signals Using a Neural Network. In *Proceedings of the 15th International Conference on Applied Informatics (IASTED)*, pages 96–99, Innsbruck, Austria, 1997. Iasted-Acta Press.

[22] Freisleben, B. and C. Hagen and M. Borschbach. Blind Source Separation via Unsupervised Learning. In *Proceedings of the 3rd International Conference on Artificial Neural Networks and Genetic Algorithms*, Norwich, UK, 1997. Springer, Wien. (to appear).

[23] Freisleben, B. and C. Hagen and M. Borschbach. A Neural Network for the Blind Separation of Non-Gaussian Sources. submitted for publication, 1997.

[24] Freisleben, B. and M. Mengel. Image Compression with Self-Organizing Networks. In *Proceedings of the 1993 International Workshop on Artificial Neural Networks*, volume 686 of *Lecture Notes in Computer Science*, pages 664–669, Sitges, Spanien, 1993. Springer-Verlag.

[25] Girolami, M. and C. Fyfe. Negentropy and Kurtosis as Projection Pursuit Indices Provide Generalised ICA Algorithms. Technical Report, University of Paisley, Scotland, 1997.

[26] Haykin, S. *Neural Networks: A Comprehensive Foundation*. Macmillan College Publishing, 1994.

[27] Hebb, D.E. *The Organization of Behaviour*. John Wiley, New York, 1949.

[28] Hertz, J.A. and A. Krogh and R. Palmer. *Introduction to the Theory of Neural Computation*. Addison–Wesley, Reading, Massachusetts, 1991.

[29] Hyvärinen, A. Purely Local Neural Principal Component and Independent Component Learning. In *Proceedings of the 1996 International Conference on Artificial Neural Networks (ICANN'96)*, volume 1112 of *Lecture Notes in Computer Science*, pages 139–144. Springer, 1996.

[30] Jain, A.K. Image Data Compression. *Proceedings of the IEEE*, 69(3):349–389, 1981.

[31] Jolliffe, I.T. *Principal Component Analysis*. Springer-Verlag, New York, 1986.

[32] Jutten, C. and J. Herault. Blind Separation of Sources, Part I: An Adaptive Algorithm Based on Neuromimetic Architecture. *Signal Processing*, 24:1–10, 1991.

[33] Karhunen, J. and E. Oja and L. Wang and R. Vigario and J. Joutsensalo. A Class of Neural Networks for Independent Component Analysis. Report A28, Helsinki University of Technology, Faculty of Information Technology, Laboratory of Computer and Information Science, Finland, 1995.

[34] Karhunen, J. and L. Wang and R. Vigario. Nonlinear PCA-Type Approaches for Source Separation and Independent Component Analysis. In *Proceedings of the 1995 IEEE International Conference on Neural Networks (ICNN'95)*, pages 995–1000, Perth, Australia, 1995.

[35] Kung, S. Y. and C.I. Diamantaras. A Neural Network Learning Algorithm for Adaptive Principal Component Extraction (APEX). In *Proceedings of the 1990 International Conference on Acoustics, Speech, and Signal Processing*, volume 2, pages 861–864, 1990.

[36] Leen, T. K. Dynamics of Learning in Linear Feature-Discovery Networks. *Network: Computation in Neural Systems*, 2:85–105, 1991.

[37] Linsker, R. Self-Organization in a Perceptual Network. *IEEE Computer*, 21:105–113, 1988.

[38] Liu, L.-K. and P. A Ligomenides. Unsupervised Orthogonalization Neural Network for Image Compression. *SPIE, Intelligent Robots and Computer Vision*, 1826(11):215–225, 1992.

[39] Luo, F.-L. and R. Unbehauen. Unsupervised Learning of the Minor Subspace. In *Proceedings of the 1996 International Conference on Artificial Neural Networks*, volume 1112 of *Lecture Notes in Computer Science*, pages 489–494. Springer, 1996.

[40] Mao, J. and A. K. Jain. Artificial Neural Networks for Feature Extraction and Multivariate Data Projection. *IEEE Transactions on Neural Networks*, 6(2):296–317, March 1995.

[41] Matsuoka, K. and M. Ohya and M. Kawamoto. A Neural Net for Blind Separation of Nonstationary Signals. *Neural Networks*, 8:411–419, 1995.

[42] Mengel, M. Bildkompression mit Neuronalen Netzen. Diplomarbeit, Technische Hochschule Darmstadt, FB Informatik, August 1992.

[43] Oja, E. A Simplified Neuron Model As a Principal Component Analyzer. *Journal of Mathematical Biology*, 15:267–273, 1982.

[44] Oja, E. Principal Components, Minor Components and Linear Neural Networks. *Neural Networks*, 50:927–935, 1992.

[45] Oja, E. Beyond PCA: Statistical Expansions by Nonlinear Neural Networks. In *Proceedings of the 1994 International Conference on Artificial Neural Networks (ICANN'94)*, volume 2, pages 1049–1054, Sorrento, Italy, 1994. Springer-Verlag.

[46] Oja, E. The Nonlinear PCA Learning Rule and Signal Separation- Mathematical Analysis. Report A 26, Helsinki University of Technology, Faculty of Information Techology, Labaratory of Computer and Information Science, Finland, 1995.

[47] Oja, E. and H. Ogawa and J. Wangviwattana. PCA in Fully Parallel Networks. In *Proceedings of the 1992 International Conference on Artificial Neural Networks (ICANN'92)*, pages 199–202, Amsterdam, 1992. Elsevier.

[48] Peper, F. and H. Noda. A Symmetric Linear Neural Network That Learns Principal Components and Their Variances. *IEEE Transactions on Neural Networks*, 7(4):1042–1047, July 1996.

[49] Plumbley, M. D. Efficient Information Transfer and Anti-Hebbian Neural Networks. *Neural Networks*, 6:823–833, 1993.

[50] Rosen, R. *Anticipatory Systems*. Pergamon Press, 1985.

[51] Rubner, J. and P. Tavan. A Self-Organizing Network for Principal-Component Analysis. *Europhysics Letters*, 10:693–698, 1989.

[52] Sanger, T.D. Optimal Unsupervised Learning in a Single-Layer Linear Feedforward Neural Network. *Neural Networks*, 2:459–473, 1989.

[53] Sorouchyari, E. Blind Separation of Sources, Part III: Stability Analysis. *Signal Processing*, 24:21–29, 1991.

[54] Wang, L. and J. Karhunen and E. Oja. A Bigradient Optimization Approach for Robust PCA, MCA, and Source Separation. In *Proceedings of the 1995 IEEE International Conference on Neural Networks (ICNN'95)*, IEEE International Conference on Neural Networks, pages 1684–1689, Perth, Australia, 1995.

[55] Xu, L. and E. Oja and C. Y. Suen. Modified Hebbian Learning for Curve and Surface Fitting. *Neural Networks*, 5:441–457, 1992.

[56] Yang, H. H. and S. Amari. Two Gradient Descent Algorithms for Blind Signal Separation. In *Proceedings of the 1996 International Conference on Artificial Neural Networks (ICANN'96)*, volume 1112 of *Lecture Notes in Computer Science*, pages 287–292, Bochum, 1996. Springer.

[57] Yaroslavsky, L.P. *Digital Picture Processing*. Springer-Verlag, Berlin, 1985.

[58] Yullie, A.L., D.M. Kammen and D.S. Cohen. Quadrature and the Development of Orientation Selective Cortical Cells by Hebb Rules. *Biological Cybernetics*, 61:183–194, 1989.

[59] Zhang, L. and L. Xu. An Adaptive Nonlinear Decorrelation Learning Algorithm For Blind Separation Of Sources. In *Proceedings of the 1996 World Congress on Neural Networks (WCNN'96)*, pages 646–649, San Diego, USA, 1996.

Single and Multiple Compartment Models in Neural Networks

J. Hoekstra

Delft University of Technology, Dept. Electrical Engineering
P.O. Box 5031, 2600 GA Delft, The Netherlands
e-mail: J.Hoekstra@et.tudelft.nl

Abstract

In this paper some of the most popular neural networks are classified according to their descent of biological models. It is shown that the activation function of the nodes in these networks stems from either the additive activation model or the shunting activation model. In both models the neuron is seen as a single entity (compartment). The models can be integrated by modeling the voltage regulated membrane ion channels with MOS transistors and splitting the neuron into more compartments. SPICE simulations illustrate this. It is then proposed to extend the class of popular neural networks with nodes that have extensive artificial dendrites. These networks are simulated by multi-compartment modeling.

Keywords: (Artificial) Neural Networks, Neural Computing, Compartmental Modeling, Dendritic Computation.

1. Introduction

Since the 1990s, neural networks have become one of the tools in the field of information processing that provide the best results – both regarding quality of the outcome and the ease of implementation – in situations where a model-based or parametric approach is difficult to formulate. Examples of application areas where results obtained are arguably better, in some sens, than those expected from various conventional techniques are (Gelenbe, Bahren ,1996; Hyde, 1996; Cybenko 1996; Pal, Srimani 1996; Haykin 1996; Jain, Mao 1997): nonlinear time-varying control and prediction, telecommunication systems, (medical) image processing, robotics, signal processing, and pattern recognition.

1.1. Neural Network Definition

A general definition of a neural network, based on (Aleksander and Morton, 1991), is: a neural network is a network of adaptable nodes, which, through a process of learning from task examples, store experimental knowledge and make it available for use. It is a computing artefact modeled after neuronal nets of the living brain. The nodes of the brain are adaptable, they acquire knowledge through changes in the function of the node by being exposed to examples. In most models the neuron is viewed as a simple node having a well-defined output fibre, the axon, which transmits pulses when the neuron is firing. The axon fans out to other nodes and makes contact to the input of those nodes at terminals, called synapses. The neuron input consists of the neuron cell body and the dendrites, branch-like protrusions from the cell body. In general, synapses are either excitatory, in which case they tend to fire the neuron – or inhibitory, in which case they tend to prevent the neuron from firing.

CP437, *Computing Anticipatory Systems: CAYS--First International Conference*
edited by Daniel M. Dubois © 1998 The American Institute of Physics 1-56396-827-4/98/$15.00

1.2. McCulloch-Pitts Model

The starting point for most neural network researchers has been a model of the fundamental cell of the brain: the neuron. For example, in 1943 McCulloch and Pitts, inspired by the "all-or-none" character of nervous activity, developed their model, currently known as the McCulloch-Pitts model (McCulloch, Pitts ,1943). In this model it is assumed that firing of a neuron may be represented by 1 and no firing by 0. The "all-or-none" character is based on theoretical neurophysiologic assumptions, at their time, that 1) at any instant a neuron has some threshold, which excitation must exceed to initiate an impulse, and 2) any neuron may be excited by impulses arriving at a sufficient number of neighboring synapses within the period of latent addition.

The simplest node, now, sums a collection of inputs and passes the result through a nonlinearity. The node is typified by an internal threshold θ and by the kind of nonlinearity, e.g., hard limiters, threshold logic, or sigmoidal nonlinearities. The nonlinearity classifies the sum of inputs as being greater or less than the threshold value. In the McCulloch-Pitts model, the node is a binary threshold device and learning is performed by a stochastic algorithm involving sudden 1-0 or 0-1 changes of states of nodes at random times.

1.3. Hebbian Learning

As a theoretical concept, in 1949, Hebb related the function of the synapse to learning and memory (Hebb, 1949). He postulated that: when an axon of cell A is near enough to excite a cell B and repeatedly or persistently takes part in firing it, some growth process takes place in one or both cells such that A's efficiency, as one of the cells firing B, is increased.

In most artificial neural networks the adaptive element is the weight $w_i(t)$. Modification of the interconnections implies modification of the associated weights and occurs whenever the neural network learns in response to new inputs or changes in the environment. The widely accepted mathematical expression that approximates the Hebbian rule is:

$$w_i(t+1) = w_i(t) + cx_i(t)y(t) \tag{1}$$

where $x_i(t)$ is one of the inputs to the node whose output is $y(t)$ and c is a positive constant determining the rate of learning. The output is computed by passing the weighted sum of inputs through a nonlinearity.

Figure 1 shows a diagram of a simple node using a hard-limiting or sigmoid nonlinearity as is adopted in most artificial neural networks, for example in the multi-layer perceptron. A well

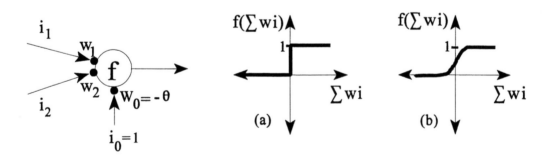

Figure 1 Simple node using (a) hard-limiting or (b) sigmoid nonlinearity

known exception is formed by the Radial Basis Functions (RBF) classifiers, in which the nodes compute radially symmetric functions (usually Gaussians). In most applications, however, RBF classifiers give results comparable to those obtained with multi-layer perceptrons, and will not be considered further in this paper.

Modern research on the synaptic function shows that when a pulse train, propagating along the axon, reaches a synapse – in most neurons – a complex chemical transfer of neurotransmitters carry the signal to the target neuron. Schematically, the process of transmission in this case can be described in the following steps: 1) presynaptic sites (in the axon) release neurotransmitter, 2) the neurotransmitter diffuses across a small synaptic gap between axon and target dendrite, 3) at the postsynaptic receptor site (in the target dendrite) the neurotransmitter opens channels in the membrane, 4) through the open channels a current of, mainly, Na^+ and K^+ ions flows across the membrane inducing a potential pulse in the dendrite. Recently, synaptic receivers with exactly the postulated Hebbian properties have been well established. The current flowing through the so-called NMDA-receptors, a mixture of Na^+ and K^+ with a small fraction of Ca^{2+} ions, depends on the presynaptic input and – in a unique manner – on the postsynaptic potential; it is therefore ideally suited to implement Hebb's rule.

Since neural networks come forth from neurobiological models, we have to answer the question: How much knowledge of the biochemical view of a single neuron or of the topology of groups of neurons do we have to incorporate into our models? In general, first the neuron model is chosen and then the different topologies are examined. Also, the neuron model will, only increase in complexity if the topological issues of the previous neuron models are reasonably understood.

1.4. Organization

In this paper some of the most popular neural networks are classified according to their descent of biological models. It is proposed to extend this group with neural networks having extensive dendrites. The paper is organized as follows: first, a short introduction to signal propagation in and between biological neurons is given. Two different models for the activation of a neuron are discussed: the additive activation model, and the shunting activation model. Second, in section 3, it is shown that the basic nodes of the Perceptron, the Backpropagation networks, the Hopfield network and the ART networks, in essence, stem from the single lump model (or single compartment model) of the neuron, under different approximations. Considering that much of the possibilities and limits of these neural networks are well-known today, the forth section discusses the natural extension of the single lumped models, namely: a synthesis of the two activation models, and the several-lump or multi-compartment model of the neuron; trends in neural networks research that are gaining increasing interest recently.

2. Signal Propagation In and Between Neurons

In the McCulloch-Pitts model each node is a finite-state machine, and operates in discrete moments. At each moment a node has only two possible states: 0 or 1. The moments are assumed synchronous among all nodes, and nodes change state as a consequence of the received inputs. The state of the node at $t+1$ doesn't depend on its state at time t. In real neurons, however, many properties depend on the previous state. For example, one of the notable properties of neurons is that once having fired, there is an interval during which they can't be fired again (called the refractory period). It is, therefore, useful to describe the time evolution of the state of neuron i in terms of two variables, $u_i(t)$ and $T_{ij}(t)$.

In general, $u_i(t)$ will be associated with the postsynaptic potential, and $T_{ij}(t)$ with the synaptic coupling strength (or weight of the connection) from node j to node i. In the world of psychology the variables $u_i(t)$ and $T_{ij}(t)$ are known as: short-term memory (STM) trace and long-term memory (LTM) trace, respectively. In most artificial neural networks equations for $T_{ij}(t)$ follow the Hebbian rule or a variant in which the node output is replaced by a wished value or an error signal. Although equations and learning strategies for $T_{ij}(t)$ are important subjects within the neural network research, they are not part of this paper.

Different equations for $u_i(t)$ are discussed when considering the propagation of signals in and between neurons. Two systems of first order differential equations are discussed. The first system models the neuron membrane and shows communication within the neuron by passive spread of local electrical potentials along dendrites or axons; the second system models the generation of action potentials, which are responsible for most of the communication between neurons (Hall, 1992; Carpenter 1996).

2.1. Membrane Model

A membrane separates the interior of the neuron from the extracellular fluid. Both intracellular and extracellular fluids contain many ions, the most important ones – for this discussion – are Cl^-, Na^+, K^+, and Ca^{2+}, and can produce electrical currents. The membranes are composed of a lipid bilayer with embedded proteins. The phospholipids (Greek: *lipos*, fat) that make up most of the membrane lipids have a nonpolar tail consisting of two carbohydrate chains, but a polar phosphate ester head. The nonpolar tail is hydrophobic, the polar head is hydrophilic. As a consequence of this the membrane lipids form a bilayer. The bilayer is essentially impermeable to polar molecules and ions, except for water.

The detailed operation of the membrane is very complex and not well understood; however, according to the most widely accepted model, called the fluid mosaic model, some small uncharged molecules, such as oxygen, carbon dioxide, and water diffuse freely through the lipid layer, while ions can only pass through "channels" provided by specific proteins embedded in the membrane. In the simplest electrical model, the overall behavior of the passive membrane is characterized by a leaky resistance R, and a capacitance C (see Figure 2).

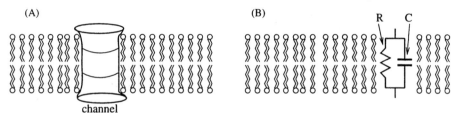

Figure 2 The membrane model (A) and its electrical equivalent (B)

2.2. Additive Activation Model

Quiescent neurons have a potential difference across their surface membrane; the inside of the neuron is about 60 mV negative to the extra cellular fluids outside. This resting potential arises because of the unequal distribution of ions on both sides of the membrane. The difference is maintained by the selective permeability of channels in the membrane and by biological ion pumps that transport Na^+ and K^+ ions against their concentration gradient. The existence of this potential difference and the existence of regulated ion channels are the basis of the electrical signaling.

When in a quiescent neuron a potential pulse $u(t)$ is induced by an external stimulus, it will decay according to the passive electrical properties. This passive membrane decay in a neuron with a resting potential u_0 can be modeled by

$$C \frac{dU}{dt} = -\frac{U}{R}, \quad \text{with} \quad U = u(t) - u_o \tag{2}$$

This leads to:

$$C \frac{du}{dt} = -\frac{u}{R} + \frac{u_0}{R} \tag{3}$$

with solution

$$u(t) = u(0) \, e^{-t/(RC)} + u_0 \, (1 - e^{-t/(RC)}) \tag{4}$$

An induced pulse decays according to the RC-time and the membrane will return to its resting potential. The simplest way the inputs to the neuron can affect the membrane potential is additively, as a variable current entering the membrane. The passive membrane can then be modeled as:

$$C \frac{du}{dt} = -\frac{u}{R} + \sum_j T_j V_j + I \tag{5}$$

where we have set the resting potential $u_0 = 0$ to simplify the equation. The term including the summation represents the inputs from other neurons, the input I represents a direct external input. Equation (5) is called the additive activation equation or additive STM equation. It is shown in section 3.1 that this equation forms the biological origin for the Hopfield, Perceptron, and Backpropagation networks.

2.3. Action Potential

Information may be carried from one part of the neuron to another by passive conduction when the distances are short enough to permit it. Over longer distances information is exchanged by means of action potentials.

Stimulation of nerves show a characteristic transient change in the voltage across the membrane, called action potential. These action potentials show several properties that demonstrate that they are not passive, but actively generated as they pass along the membrane, amplified in such a way as to overcome the losses that they experience through membrane leakage. The action potential from a single electrical stimulus has the same amplitude at different points along the membrane and in fact doesn't decrease at all as a function of the distance; so long as the strength of the stimulus is above a certain threshold value (below which no action potential is seen at all); neither the amplitude nor the speed of the action potential is influenced by the nature of the original stimulus. The action potential thus implements the "all-or-nothing" character of nervous activity.

The action potential is produced by a characteristic and almost invariant sequence of changes in ionic permeabilities, regulated by opening and closing of ion specific channels. Schematically, the process of action potential generation can be described in the following steps: 1) the voltage across the membrane rises due to a transient increase in sodium permeability, this process is called depolarization, 2) after a short delay, channels open through which potassium ions flow from the inside of the membrane to the outside, thereby lowering

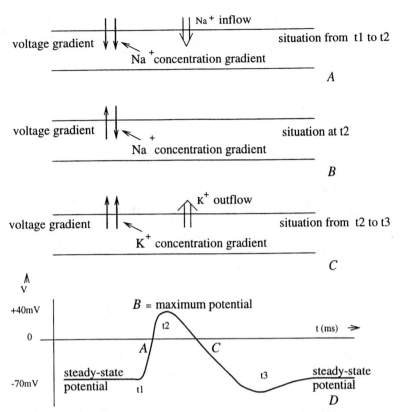

Figure 3 Gradients and potential over the membrane during the generation of an action potential.

the potential towards the resting potential (repolarizing). In some cases this last process actually undershoots the resting potential (the membrane is called to be hyperpolarized), because there is a period at the end of the cycle in which the potassium flow is still elevated, whereas the sodium flow is back to stationary. Figure 3 illustrates this scheme.

2.4. Shunting Activation Model

The active membrane can be described electrically by the following resistances parallel to the membrane capacitance C_m: a constant passive (leaky) resistance R_m, a variable resistance R_{Na} in series with a constant battery V^+ that models the potassium current through a channel, and a variable resistance R_K in series with another constant battery V^- that models the sodium current. Figure 4 which shows the circuit equivalent. The equation representing this model is :

$$C_m \frac{du}{dt} = \frac{(V^+ - u)}{R_{Na}} + \frac{(V^- - u)}{R_K} \qquad (6)$$

which is a special case of the celebrated Hodgkin-Huxley membrane equation.
We obtain the so-called shunting activation equation if we take (Kosko, 1992) $V^+ = 0$, $V^- = B$, $1/R_{Na} = A$, and $1/R_K = \sum_j T_{ij} V_j + I$:

$$C_m \frac{du}{dt} = -Au + (B - u)[\sum_j T_{ij} V_j + I] \qquad (7)$$

In the shunting activation equation the relation between the inputs and the cell potential is not additive as in the additive activation equation, Eq. (5), but multiplicative. As will be described

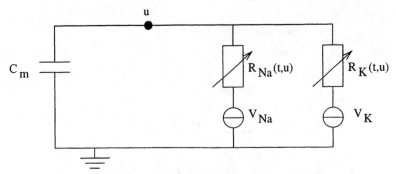

Figure 4 Circuit model of the active membrane

in section 3.2 the shunting activation equation leads to the various ART neural network models.

2.5. Ion Channels

Ion channels are gaps provided by specific proteins across the membrane that allow certain ions to pass from the outside of the membrane to the inside or visa versa. There are numerous channels in the membrane and their density is not uniform. For example, on the dendritic tree the number of channels per area is found to increase towards the cell body, which has an effect on the resting potential, and R_m, of the membrane. There are many different ion channels. Some of those allow a number of different ions to pass, some are always open, but most of them are only receptive to a single ion type.

Based on the mechanisms for opening and closing, the channels can roughly be divided into two groups: neurotransmitter-regulated channels and voltage-regulated channels. In neurotransmitter-regulated channels, neurotransmitters regulate the opening and closing of an ion channel by binding to a side (the receptor side) on the channel protein. The neurotransmitter-regulated channel is considered to be of great importance for the synaptic function of the neuron. A firing neuron releases at the end of its axon (presynaptic side) neurotransmitters that cross the synaptic gap. Upon arriving at the dendrite receptor side they open channels to increase the postsynaptic membrane permeability to Na^+ and K^+. A Na^+ and K^+ ion current through the membrane drive the potential in the target neuron.

A special neurotransmitter-regulated channel is the NMDA-channel. In addition to the regulation by neurotransmitters it also has an unusual form of voltage-regulation in which the open channel occluded at normal resting potential by a magnesium ion in the extracellular fluid. Depolarization drives the magnesium ion in the extracellular fluid, allowing other ions to pass. Thus, the neurotransmitter (here glutamate) is most effective in opening the NMDA channel when the cell is depolarized. Section 1.3 showed that this dependency follows the Hebbian learning rule.

Voltage-regulated ion channels are fundamental for signaling in excitable cells. Because of their importance in propagation of the action potential, much is known of the voltage-regulated sodium channels. Voltage-regulated channels have both activated and inactivated states. The opening and the duration is determined by both the membrane potential and the time. Schematically, channel opening can be described in a simple scheme in which the channels have three states: closed, open (activated), and inactivated. For example, channels in the closed state are opened by membrane depolarization; once the channel is activated it can close by the action of inactivation, a transition to the inactivated state, or it can return to the closed state. In general, only the opening of the channel is voltage dependent, closing is then

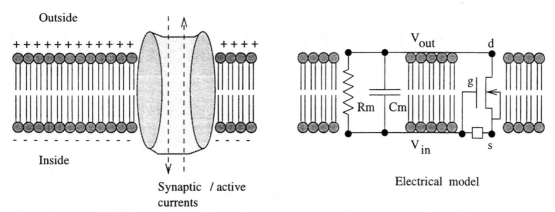

Outside

Inside

Synaptic / active
currents

V_{out}

d

g

Rm Cm

V_{in}

s

Electrical model

Figure 5 Transistor model of a voltage-regulated channel; the white box indicates that some extra circuitry is necessary.

determined by time.

2.6. Simple Transistor Model for Voltage-Regulated Channel

A simple way to model electrically the voltage-regulated channel is by using a MOS-transistor model. For, the transistor operates as a switch driven by the potential of its gate. Also a transistor based implementation may be important if we want to build such networks.

The current through the MOS transistor depends only on the voltage over the gate g. The voltage directs the current from source s to drain d. By placing the battery at the drain and coupling the source of the transistor to the membrane interior the membrane potential directs the current through the channel (transistor), as is illustrated in Figure 5.

3. Single Lump Model

Single lump models consider the neuron as a single entity. The neural network models based on this approach consist of single nodes to which others are connected by weighted connections. Based on the two different activation models two different single lump models can be distinguished. The first category is based on the additive activation model and forms the basis of the nodes present in, among others, the Hopfield networks, the Perceptron, and Backpropagation neural networks. The second category is based on the shunting activation model and forms the basis of the nodes present in, among others, the bidirectional associative memories (BAMs) and adaptive resonance theories (ARTs).

3.1 Hopfield, Perceptron, and Backpropagation Neural Networks

The "all-or-none" character of the nervous activity goes hand in hand with the concept of a threshold. As soon as the potential over the membrane exceeds the threshold, a stereotyped signal, the action potential, is generated at the axon; the neuron is firing. Potential changes below the threshold value are called local potentials, and are described by the passive electrical properties of neurons. The passive electrical properties of a neuron depend on, among others, the capacitance of the membrane and the geometry of the cell. The description of the neuron in terms of its passive electrical properties forms the basis of the artificial neural networks developed in the 1980s. Figure 6 depicts the additive single lump model. The

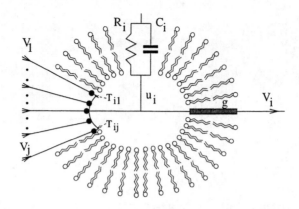

Figure 6 Additive single lump model of the biological neuron. A parallel RC circuit represents the passive electrical behavior of the membrane.

equation of motion for this neuron model, useful for embedding in a network of neurons, is:

$$C_i \frac{du_i}{dt} + \frac{u_i}{R_i} = \sum_j T_{ij} V_j \qquad (8)$$

In the equation u_i is the potential across the membrane, C_i and R_i the capacitance and resistance of the cell membrane, and V_i the output of neuron i. $V_i = g(u_i)$, where g is a 'sigmoid' function, i.e., a function which is almost linear for small values of u_i but which saturates at +1 for large positive values and at -1 for large negative values of u_i. T_{ij} is the weight from neuron j to neuron i. Hopfield (1984) implemented equation (8) in a resistor array, a model which is known as the deterministic Hopfield model. If only the steady state, $du_i/dt=0$, is considered, and R=1 then:

$$V_i = g \left(\sum_j T_{ij} V_j \right) \qquad (9)$$

Equation (9) is recognized as the activation function used in many artificial neural networks, such as Hopfield, Perceptron, and Backpropagation neural networks.

3.2 ART and Kohonen Neural Networks

Adaptive Resonance Theory (ART) architectures are neural networks that carry out stable self-organization of recognition codes for arbitrary sequences of input patterns. The ART architecture consists of two fields of nodes (in the simplest case these two fields are simply two layers) together with control and matching subsystems. An input is initially represented as a pattern of activity across the nodes, or feature detectors, of the first field. The nodes in this field are connected by bi-directional links with nodes of the second field. The pattern of activity across this second field correspondents to the category representation.

The second field implements competitive learning, also called self-organizing feature maps. By internal competitive dynamics the winning node categorizes the input, if the input is sufficiently similar to an already categorized pattern. If the match can't be made a new category is established. In case of competitive learning a single node wins the competition and only weights of the winning node or of nodes within a small region of the winning one adapt

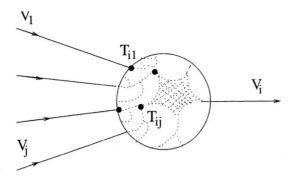

Figure 7 Extensive dendrite in artificial neuron.

their weights. Among the networks that incorporate competitive learning, ART and the Kohonen network are the most important ones.

In ART 3, a model of the chemical synapse is incorporated (Carpenter, Grossberg, 1994). The computational requirements of the ART search process can be fulfilled by formal properties of neurotransmitters, if these properties are appropriately embedded in the total architecture.

Of interest for this paper is the equation governing the node activation:

$$\varepsilon\frac{dx_j}{dt} = -x_j + (A - x_j)[excitatory\ inputs] - (B + x_j)[inhibitory\ inputs]$$

$$= -x_j + (A - x_j)[\Sigma_i v_{ij} + intrafield\ feedback] - (B + x_j)[reset\ signal]$$

(10)

in which x_j is the activity of node j.

Equation (10) is a shunting activation equation, as described in section 2.4. Excitatory inputs drive x_j up towards a maximum depolarization level A; inhibitory inputs drive x_j down towards a minimum hyperpolarization level equal to $-B$, and activity passively decays to a resting level equal to 0 in the absence of inputs. The net effect of the inputs from other nodes is assumed to be excitatory via the term $\bar{\Sigma}_i v_{ij}$. This term represents the amount of neurotransmitter released at the presynaptic side which is assumed to be bound at the postsynaptic side and takes into account both the weight and the value of the presynaptic signal. The control subsystem shuts off the node activation by giving reset signals.

4. Multi-Compartment Model

The single lump model disregards the spatial structure of the neuron's extensive dendritic system. There are a couple of reasons, however, to take a spatial structure into account. First, the time course of the membrane potential at any place in the dendrite depends on the arrival time of an impulse and on the location of the synapse, see Figure 7. Second, the spatial distance of synapses is of importance in case of local learning. The extended structure of dendrites can be included in neuron models by the use of compartmental modeling techniques (Koch, Segev, 1989), i.e., the (artificial) dendrite is subdivided into sufficiently small segments (or compartments) in which the physical properties (e.g., the dendrite diameter,

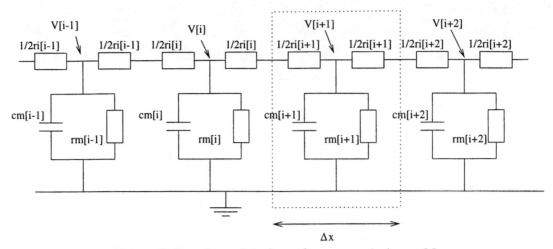

Figure 8 Circuit model of passive transmission cable

specific electrical properties) are spatial uniform and the potential is taken constant. The differences in potential, now, occur between compartments rather than within them. After the introduction of the passive cable model, first a model is discussed in which a synthesis of the two activation models is proposed (Hoekstra, Mantel, 1995), and second an active, single dendrite, multi-compartment artificial neuron is discussed (Hoekstra, van Dongen, 1995).

4.1. Passive Cable Model

In a multi-compartment model, the passive spread of a potential pulse along a tube-like axon or dendrite can be modeled by considering its so-called cable properties. The term cable stems from the mathematical resemblance of the passive nerve fibre with a transmission cable. The cable properties depend both on the membrane capacitance C_m, the membrane resistance R_m and on the longitudinal internal resistance of the membrane R_i. The whole tube can be thought of as consisting of a large number of sufficiently small units, compartments. This tube can electrically represented by the laddernetwork shown in Figure 8.

In case of uniform cable properties, along the whole cable, the circuit model can be described by the so-called cable equation (the membrane thickness is assumed to be neglectible compared with the cylinder diameter):

$$\lambda^2 \frac{\partial^2 V}{\partial x^2} = \tau \frac{\partial V}{\partial t} + V \qquad (12)$$

with

$$\lambda = \sqrt{\frac{\rho_m}{\rho_i} \frac{d_m r}{2}} \qquad (13)$$

and

$$\tau = c_m \rho_m d_m \qquad (14)$$

Here ρ_m and ρ_i are the membrane resistivity and internal resistivity, respectively. r is the cylinder radius, c_m the membrane capacitance per unit area, and d_m the membrane thickness.

This equation can be solved analytically. This method is, however, quite restricted, for the voltages can neither be computed when inputting arbitrary pulses, nor can they be computed when the physical parameters are not constant.

For N compartments the passive cable can be simulated numerically by solving a set of N first order differential equations. Using the Kirchhoff Current Law the equation for compartment i can be obtained:

$$C_{m_i}\frac{\delta V_i}{\delta t} + \frac{V_i}{R_{m_i}} + \frac{V_i - V_{i-1}}{\frac{1}{2}R_{i_i} + \frac{1}{2}R_{i_{i-1}}} + \frac{V_i - V_{i+1}}{\frac{1}{2}R_{i_i} + \frac{1}{2}R_{i_{i+1}}} = 0 \qquad (15)$$

By discretization of the compartmental voltage the set of equations is solved. The obtained system can be solved preferably by an implicit discretion method, such as Backward Euler or Cranc-Nicholson, because these methods are stable for any combination of discretion-intervals Δx and Δt. As an alternative the set of coupled differential equations can be uncoupled using techniques from linear algebra, and subsequently solving independently the uncoupled differential equations. Because in such a method the voltage in a compartment is described as the sum of all neighboring compartments useful approximations can be studied (Hoekstra,1993).

4.2. Combining Additive and Shunting Activation Models: Synaptic Membrane Currents Using SPICE Simulations

The two activation models, the additive activation model and the shunting activation model, describe the two complementary phenomena: passive decay and active pulse generation. If we want to combine both activation models we should be aware of the differences between them, the most important being the fact that in neural networks in which nodes follow the additive activation model, the inputs from other nodes enter as currents while in networks in which nodes follow the shunting activation model, the inputs enter as conductances.

The passive cable is suitable for implementing an additive activation model; the cable adds voltages of input signals (by Ohm's Law the voltages can be translated into currents). As stated is section 2.6 a MOS-transistor can be used to model an active channel. This suggests that a passive cable combined with the transistor circuit is capable of integrating both activation models, as depicted in Figure 9. The typical construction of the shunting part is the voltage source V+ with the parallel resistors S1 and S2. The transistor Mk acts as a switch directed by the channel voltage. Additional circuitry implements a simple notion of the refractory periods.

Figure 10 shows the results of a SPICE (a well-known electronic simulation program) simulation of a cable consisting of five compartments. Rm is infinite. A block-pulse initiates a potential pulse in the first compartment. The figure depicts the potential in the cable at the various compartments. The simulation shows both additive and active behavior. Also, the absolute refractory period, a period after activation during which it is impossible to elicit a response to a second stimulus regardless of its strength, and the relative refractory period during which the membrane potential is resuming its normal (resting) value, can be seen.

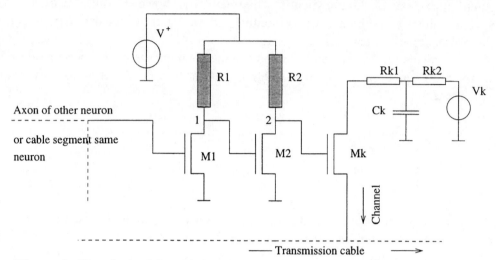

Figure 9 Circuit model used for the simulation of an active membrane, the term transmission cable represents a passive ladder network.

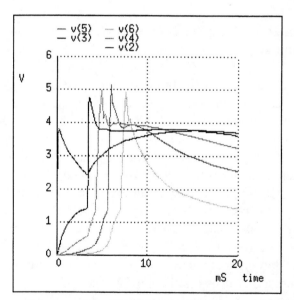

Figure 10 SPICE simulation of active membrane in compartmental neuron

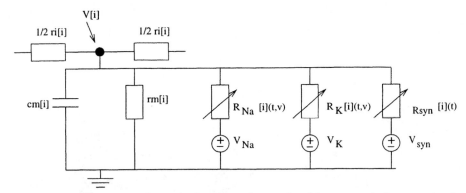

Figure 11 Electrical model of active channel with synaptic input, single compartment

4.3. Active Artificial Dendrites

An alternative for the development of multi-compartmental nodes by general electrical circuits simulation programs is the direct numerical simulation of a set of coupled differential equations. For each compartment there is one equation that relates the time-derivative of the potential to the voltages of the different other compartments. Figure 11 shows the different parts that can be distinguished in such an approach. A number of groups are currently following this line of research. The difficult topics are how to include the nonlinear, time and voltage dependent resistors of the various inputs, and how to design efficient (local) learning rules for these nodes.

A number of different routes for solving the set of differential equations can be chosen, and are also part of the research traject. The underlying philosophy of the compartmental modeling and the different routes for solving is that incoming pulses are faded out while transported passively, so the impact of an incoming pulse can be neglected far away from the point of stimulation. This knowledge cannot easily taken into account when calculating large dendrites, in most cases the voltage of a compartment depends on the voltages and inputs at every other compartment.

In (Hoekstra, van Dongen, 1995) we used the implicit Backward Euler method extended with Picard iterations, as was proposed by Mascagni (1990). Figure 12 shows the voltage distribution in a compound dendrite. The dendrite consists of 30 compartments, from which the first 20 are passive and the last 10 are active, in the first compartment a synaptic input is present. The pulse inserted is strong enough to exceed the threshold potential in the active part. This part reacts by generating a pulse. The passive decay of the initial pulse in compartment 1, just after t=0, as well the passive decay of the internal pulse as a consequence of the active behavior of the last compartments in compartment 1, can be seen clearly. The flat distribution just after the active pulse, represents the absolute refractory period that illustrates the nonlinear character of the resistance function; in this case the actionpotential was generated by step-functions for the resistances of the potassium and sodium channels.

5. Conclusions

In the paper a helicopter-view was presented of neural networks and their relation with biochemical and neurophysiological models of real neurons. The question was raised how

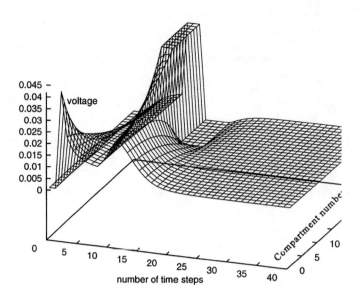

**Figure 12 Compound dendrite with passive
and active channels**

much knowledge of the real neuron do we have to incorporate in our neural network models. The answer is not easy. In the paper it was shown that by taking into account more biological knowledge different neural network concepts can be successfully integrated. For example, the additive activation model, from which the activation function for the Hopfield, Perceptron, and Backpropagation neural networks is derived, can be integrated with the shunting activation model, the biological origin of the ART neural networks, by modeling the voltage regulated membrane channel by a MOS transistor. To make such an integration possible the neuron (or to be more precise: the dendrite) must be split up in compartments.

Many topics were not discussed in this paper. Among others, nothing was said about dendritic spines that are believed to play an important role in synaptic learning, and learning, in general. The last part of the paper shows that by viewing the neuron as a set of compartments a interesting, and exciting field of extensive neural networks can be explored.

6. Acknowledgment

The author would like to thank P. Mantel, M.F. van den Broek, M. Maouli, and A. van Dongen, who were as students involved in projects concerning this research, and Dr. A.J. Klaassen who introduced me in this field of research.

References:

Aleksander I. and Morton H. (1991). An Introduction to Neural Computing. Chapman and Hall, London.

Carpenter R.H.S. (1996). Neurophysiology. Arnold, London.

Carpenter G.A. and Grossberg S. (1994). ART3: Hierarchical Search Using Chemical Transmitters in Self-Organizing Pattern Recognition Architectures. In: Computing with Biological Metaphores, R. Paton (Ed), Chapman-Hall, London.

Cybenko G. (1996). Neural Networks in Computational Science and Engineering. IEEE Computational Science and Engineering, pp. 36-42.

Gelenbe E. and Bahren J. (1996). Scanning the Special Issue on Engineering Applications of Artificial Neural Networks. Proceedings of the IEEE, Vol. 84, pp. 1355-1358.

Hall Z.W. (1992). An Introduction to Molecular Neurobiology, Sinauer Associates, Sunderland.

Haykin S. (1996). Neural Netwroks Expand SP's Horizons. IEEE Signal Processing. Vol. 13, pp. 24-48.

Hebb D.O. (1949). The Organization of Behaviour, Wiley, New York.

Hoekstra J. (1993). Approximation of the Solution of the Dendritic Cable Equation by a Small Series of Coupled Differential Equations. In: New Trends in Neural Computing, J. Mira, J.Cabestany, and A. Prieto (Eds.), Springer Verlag, pp. 43-48.

Hoekstra J. and van Dongen A. (1995). Compartmental Modeling of Telescopic Dendrites for Artificial Neurons. In: Brain Processes, Theories and Models. R. Moreno-Diaz, J. Mira-Mira (Eds), MIT-Press, Cambridge.

Hoekstra J. and Mantel P. (1995). Aspects of Spatiotemporal Learning in Artificial Neural Networks: Modeling Synaptic Membrane Currents Using SPICE Simulations. In: Neural Networks: Artificial Intelligence and Industrial Applications, B. Kappen, S. Gielen (Eds), Springer-Verlag, pp. 47-51.

Hopfield J.J. (1984). Neural Networks and Physical Systems with Graded Response have Collective Properties like those of Two-State Neurons. Proc. National Academy of Science USA, Vol. 81, pp. 3088-3092.

Hyde T. (1996). Neural Computing. Control System, Vol 13, pp. 75-80.

Jain A.K. and Mao J. (1997). Special Issue on Artificial Neural Networks and Statistical Pattern Recognition. IEEE Trans. on Neural Networks, Vol. 8, pp. 1-4.

Koch C. and Segev I. (1989). Methods in Neural Modeling, MIT Press.

Kosko B. Neural Networks and Fuzzy Systems. Prentice-Hall, Englewood Cliffs.

Mascagni M.V. (1990). The Backward Euler Method for Numerical Solution of the Hodgkin-Huxley Equations of Nerve Conduction. SIAM J. Numerical Analysis, pp. 941-962.

McCulloch W.S. and Pitts W. (1943). A Logical Calculus of the Ideas Immanent in Nervous Activity. Bulletin of Mathematical Biophysics, Vol. 5, pp 115-133.

Pal S.K. and Srimani P.K. (1996). Neurocomputing: Motivations, Models, and Hybridization. In: Theme Issue Computer, pp. 24-28.

Asymptotic Behaviour of Neural Networks

M. Loccufier   E. Noldus

University of Gent, Faculty of Applied Sciences,
Technologiepark- Zwijnaarde 9, B-9052 Zwijnaarde-Gent, Belgium,
Tel. : +31 9 264 5587, Fax. +32 9 264 5839,
Email : mia@autoctrl.rug.ac.be

Keywords : Neural Networks, Global Convergence, Liapunov, Trajectory Reversion, Stability Region

Abstract

A new method is presented for analysing the global stability properties of neural networks. An efficient and numerically robust procedure is developed for estimating regions of asymptotic stability. The method combines Liapunov theory, simulation in reverse time and some topological properties of the true stability region. The result is an accurate estimate of the true stability boundary. The main limitation of the method is that a global Liapunov function for the neural network must be found. fortunately large classes of neural networks possess such a Liapunov function.

1   Introduction

Neural networks are nonlinear dynamical systems which exhibit a complex behaviour and often possess many equilibrium points. In case the network is used as a classification network, the equilibrium points constitute the 'prototype vectors' of the different classes. When the network is used as an associative memory, the equilibrium points represent the different memories to be stored, (Atiya A. & Abu-Mostafa, 1993). An essential operating condition for a neural network is that an arbitrary initial condition or input pattern generates a trajectory which converges to the desired equilibrium point. Therefore the system's asymptotic behaviour and the regions of attraction of the equilibrium points are of great interest (Cohen & Grossberg, 1983; Wang & Blum, 1992). We propose a new method which allows to estimate the exact region of attraction of each locally asymptotically stable equilibrium point with any desired accuracy.

In the literature an impressive amount of methods for computing stability regions can be found, each of them trying to find a satisfactory compromise between three more or less conflicting demands. These are : the accuracy of the estimate of the exact stability region, the limitation of the amount of computational effort and the applicability to a broad class of systems. Liapunov type methods, usually based on Lasalle's extension of

CP437, *Computing Anticipatory Systems: CAYS--First International Conference*
edited by Daniel M. Dubois © 1998 The American Institute of Physics 1-56396-827-4/98/$15.00

Liapunov's theory (Lasalle, 1967), result in conservative subsets of the true stability region (Noldus & Loccufier , 1994). Efforts have been made to reduce this conservativeness by constructing sequences of Liapunov functions which produce monotonically improving estimates of the true stability region (Ciliz, 1993). An example of a non-Liapunov method is trajectory reversion. An estimate of the stability region is produced from a set of trajectories obtained by numerical integration of the system's equation in reverse time (Genesio et al., 1985). A number of algorithms based on the same technique have appeared which all share the tendency to become quickly computationally involved when the dimension of the neural network increases (Guttalu & Flahsner, 1988; Psiaki & Luh, 1993). More recent techniques based on a complete topological characterization of the true stability boundary suffer from an iterative trial and error procedure to find initial states for trajectory reversing on the unknown stability boundary (Chiang et al., 1988; Zaborszky et al. , 1988). Moreover a transversality condition has to be satisfied which is difficult or impossible to verify.

In the present paper an alternative trajectory reversing method is presented which overcomes the disadvantages of existing methods due to an unique combination of Liapunov techniques, trajectory reversion and some topological properties of the true stability boundary. The procedure essentially consists of finding a global Liapunov function for the system and generating a number of half inverse trajectories. The Liapunov function is used to compute the starting points for these trajectories. In section 2 a number of basic definitions and a class of neural networks are introduced. The proposed trajectory reversion method is explained is Section 3. In Section 4 we discuss the convergence of the estimated stability boundary to the exact stability boundary as one of the design parameters approaches zero. Section 5 contains a conceptual algorithm of the method, illustrated with a hopfield neural network of dimension 4. We end with some general conclusions in Section 6.

2 Definitions and Basic Assumptions

Consider an autonomous system

$$\dot{x} = f(x) \tag{2.1}$$

with state $x \in R^n$. We shall require that $f(.) \in C^1$ (continuous with continuous partial derivatives w.r.t. the components of x). This condition ensures the existence and the uniqueness of the solution $x(t, x_0)$ for a given initial state x_0 at $t = 0$ (Hartman, 1982). We shall suppose the system has a finite number of equilibrium points $\hat{x}_i$. A nontrivial trajectory of (2.1) is a trajectory different from an equilibrium point. An inverse trajectory is obtained by numerical integration of the state equations in reverse (i.e. decreasing) time. An equilibrium point $\hat{x}_i$ is hyperbolic if the Jacobian matrix $J(\hat{x}_i)$ has no characteristic values on the imaginary axis, where $J(x) \triangleq \frac{\partial f(x)}{\partial x}$. A hyperbolic equilibrium point $\hat{x}_i$ is of index k if $J(\hat{x}_i)$ has exactly k characteristic values in $\{Re\ s > 0\}$. The stable and unstable manifolds of $\hat{x}_i$ will be denoted as $W^s(\hat{x}_i) \triangleq \{x_0 \in R^n;\ x(t, x_0) \to \hat{x}_i \text{ for } t \to +\infty\}$ and $W^u(\hat{x}_i) \triangleq \{x_0 \in R^n;\ x(t, x_0) \to \hat{x}_i \text{ for } t \to -\infty\}$. $W^s(\hat{x}_i)$ and $W^u(\hat{x}_j)$ satisfy the transversality condition if either their tangent spaces span the entire state space in every point of intersection, i.e.

$$T_x[W^s(\hat{x}_i)] + T_x[W^u(\hat{x}_j)] = R^n \quad ; \quad \forall x \in W^s(\hat{x}_i) \bigcap W^u(\hat{x}_j) \tag{2.2}$$

or $W^s(\hat{x}_i)$ and $W^u(\hat{x}_j)$ do not intersect. Let $\hat{x}_s$ be a locally asymptotically stable (l.a.s.) equilibrium point. Let $\Omega(\hat{x}_s) \triangleq W^s(\hat{x}_s)$ be its stability region and $\partial\Omega(\hat{x}_s)$ the stability boundary. $\partial\Omega(\hat{x}_s)$ is invariant. If $\Omega(\hat{x}_s)$ is not dense in R^n then the dimension of $\partial\Omega(\hat{x}_s)$ is $(n-1)$ (Chiang et al., 1988). We shall assume that $\Omega(\hat{x}_s)$ is not dense in R^n. This is always the case if (2.1) has more than one stable equilibrium point. $\partial\overline{\Omega}(\hat{x}_s)$, the boundary of the closure of $\Omega(\hat{x}_s)$, is called the quasi-stability boundary of $\hat{x}_s$ and int$\overline{\Omega}(\hat{x}_s)$ its quasi-stability region (Zaborszky et al., 1988). A global Liapunov function for (2.1) is a real scalar function $V(x) \in C^1$ such that along the solutions of (2.1) :

$$\dot{V}(x) = \left[\frac{\partial V(x)}{\partial x}\right]' f(x) \leq 0 \quad ; \quad \forall x \in R^n \tag{2.3}$$

Regarding the system (2.1) the following assumptions are introduced :

Assumption 1 : All equilibrium points are hyperbolic.

Assumption 2 : There exists a global Liapunov function $V(x)$ such that along any nontrivial solution $x(t)$ the set

$$\{t \in R; \ \dot{V}[x(t)] = 0\}$$

has measure zero in R.

Assumption 3 : All solutions on $\partial\Omega(\hat{x}_s)$ remain bounded for $t > 0$.

Assumption 4 : Let $\hat{x}_i \in \partial\Omega(\hat{x}_s)$ be an equilibrium point of index $k < n$ such that $W^u(\hat{x}_i) \cap \Omega(\hat{x}_s) \neq \emptyset$. Then either :

(i)   Index $\hat{x}_i = 1$
or

(ii)   For any neighbouring equilibrium point $\hat{x}_j$ of $\hat{x}_i$, i.e. any equilibrium point $\hat{x}_j$ that is connected to $\hat{x}_i$ by a complete trajectory on $\partial\Omega(\hat{x}_s)$,

  1. The stable and unstable manifolds of $\hat{x}_i$ and $\hat{x}_j$ satisfy the transversality condition

  2. $W^u(\hat{x}_j) \cap \Omega(\hat{x}_s) \neq \emptyset$

and

(iii) $[W^u(\hat{x}_i) \setminus \{\hat{x}_i\}] \cap \partial\Omega(\hat{x}_s) \neq \emptyset$.

Assumption 1 implies that all equilibrium points are isolated. Assumption 2 ensures that the largest invariant subset of R^n where $\dot{V}(x) = 0$ consists of the set of all equilibria (Chiang, 1991). Hence according to Lasalle (Lasalle, 1967) every solution that remains bounded for $t > 0$ will converge to an equilibrium point as $t \to +\infty$. Assumption 3 can often be verified using a suitable Liapunov function. For example if the system possesses a global Liapunov function that is radially unbounded then all solutions are bounded for $t > 0$ and Assumption 3 holds. Obviously Assumption 3 is also satisfied if it can be shown that $\Omega(\hat{x}_s)$ is bounded. Assumptions 2 and 3 imply that every trajectory on $\partial\Omega(\hat{x}_s)$ converges to an equilibrium point on $\partial\Omega(\hat{x}_s)$ as $t \to +\infty$. Assumption 4 restricts the topological structure of $\partial\Omega(\hat{x}_s)$ and is always satisfied if (2.1) is a second order system. The latter holds as all equilibrium points of second order systems are of index one and of index two. The implication of Assumption 4 for higher order systems will be explained in Sections 3 and 4.

In the literature wide classes of neural networks possessing global Liapunov functions have been described (Cohen & Grossberg, 1983; Grossberg, 1989; Noldus & Loccufier, 1995). We consider a class of competitive neural networks without learning (Cohen & Grossberg 1983), i.e. the weights m_{ij} are fixed during operation. (2.4) can easily be extended to neural networks with adaptive weights m_{ij} (kosko, 1991).

$$\dot{x}_i = a_i(x_i)[b_i(x_i) - \sum_{k=1}^{n} m_{ik}S_k(x_k)] \quad , \quad i = 1 \ldots n \tag{2.4}$$

(2.4) is a one layer neural network, fully connected, with synaptic matrix $M = [m_{ij}]$. The state $x \in R^n$ represents the activation of the neurons. The functions $S_i(.)$ are mostly saturating amplifier characteristics. The corresponding Liapunov function is :

$$V(x) = -\sum_{i=1}^{n} \int_{0}^{x_i} S_{id}(u)b_i(u)du + \frac{1}{2}\sum_{i=1}^{n}\sum_{j=1}^{n} m_{ij}S_i(x_i)S_j(x_j) \tag{2.5}$$

If $M = M'$ and

$$a_i(u) > 0 \qquad ; \qquad \forall u \qquad ; \qquad i = 1 \ldots n$$
$$S_{id} \stackrel{\Delta}{=} \frac{dS_i(u)}{du} > 0 \qquad ; \qquad \forall u \qquad ; \qquad i = 1 \ldots n$$

then $\dot{V}(x)$ can be written as

$$\dot{V}(x) = -\sum_{i=1}^{n} \frac{\dot{x}_i^2}{a_i(x_i)} S_{id}(x_i) \tag{2.6}$$

and

$$\dot{V}(x) = 0 \Leftrightarrow \dot{x} = 0$$

The latter implies that Assumption 2 is satisfied.

Assumption 1 can easily be verified by calculating the eigenvalues of the Jacobian matrix at each equilibrium point. The solutions of (2.4) will be bounded for $t > 0$ if

$$|S_i(u)| \leq h_i < +\infty \quad ; \quad \forall u \quad ; \quad i = 1 \ldots n$$

and

$$\int_{0}^{x_i} S_{id}(u)b_i(u)du \to -\infty \text{ if } |x_i| \to +\infty$$

As Assumption 4 concerns the unknown stability boundary $\partial\Omega(\hat{x}_s)$, we assume it holds without proof. We will need the following lemma which is proved in the Appendix :

Lemma 2.1 *Let $\hat{x}_i$ be a hyperbolic equilibrium point of (2.1) and $V(x)$ be a global Liapunov function which satisfies Assumption 2. Then :*

1. $\forall x_0 \in [W^s(\hat{x}_i) \setminus \{\hat{x}_i\}] : V(x_0) > V(\hat{x}_i)$

2. $\forall x_0 \in [W^u(\hat{x}_i) \setminus \{\hat{x}_i\}] : V(x_0) < V(\hat{x}_i)$

3   A new trajectory reversing method

We consider a system (2.1) of order $n \geq 2$ which satisfies Assumptions 1-4. It holds that

$$\partial\Omega(\hat{x}_s) \subset \bigcup_{\hat{x}_j \in \partial\Omega(\hat{x}_s)} W^s(\hat{x}_j) \tag{3.1}$$

The dimension of $\partial\Omega(\hat{x}_s)$ is $(n-1)$. Since the stable manifold of an index k equilibrium point is uniformly $(n-k)$-dimensional (Zaborszky et al. , 1988), segments of $\partial\Omega(\hat{x}_s)$ will coincide with complete stable manifolds and / or parts of stable manifolds of index one equilibria. The set of index one equilibria on $\partial\Omega(\hat{x}_s)$ can be partitioned in two subsets Υ and Ψ where

$$\Upsilon \triangleq \{\hat{x}_i \in \partial\Omega(\hat{x}_s);\ \text{index } \hat{x}_i = 1,\ W^u(\hat{x}_i) \cap \Omega(\hat{x}_s) \neq \emptyset\} \tag{3.2}$$

$$\Psi \triangleq \{\hat{x}_j \in \partial\Omega(\hat{x}_s);\ \text{index } \hat{x}_j = 1,\ W^u(\hat{x}_j) \cap \Omega(\hat{x}_s) = \emptyset\} \tag{3.3}$$

For every $\hat{x}_i \in \Upsilon$,

$$W^s(\hat{x}_i) \subset \partial\Omega(\hat{x}_s) \tag{3.4}$$

(Zaborszky et al. , 1988), while for every $\hat{x}_j \in \Psi$,

$$[W^u(\hat{x}_j) \setminus \{\hat{x}_j\}] \cap \partial\Omega(\hat{x}_s) \neq \emptyset \tag{3.5}$$

(Chiang et al., 1988). Let $\hat{x}_i \in \Upsilon$ and let $B(\varepsilon)$ be a sufficiently small neighbourhood of $\hat{x}_i$ such that $B(\varepsilon) \rightarrow \{\hat{x}_i\}$ for $\varepsilon \xrightarrow{>} 0$. We analyse the behaviour of the level set

$$S \triangleq \{x;\ V(x) < V(\hat{x}_i)\} \tag{3.6}$$

in $B(\varepsilon)$. Since $W^s(\hat{x}_i)$ is uniformly $(n-1)$-dimensional it divides a well chosen $B(\varepsilon)$ in two open, disjoint and connected subsets $P(\varepsilon)$ and $Q(\varepsilon)$:

$$B(\varepsilon) \setminus W^s(\hat{x}_i) = P(\varepsilon) \cup Q(\varepsilon) \tag{3.7}$$

$W^u(\hat{x}_i)$ consists of $\hat{x}_i$ and exactly two nontrivial trajectories $\{x_i(t),\ -\infty < t < +\infty,\ i = 1,2\}$. Using the fact that $W^u(\hat{x}_i)$ and $W^s(\hat{x}_i)$ intersect transversally (Chiang et al., 1988) we may assume that

$$[\{x_1(t),\ -\infty < t < +\infty\} \cap B(\varepsilon)] \subset P(\varepsilon) \tag{3.8}$$

$$[\{x_2(t),\ -\infty < t < +\infty\} \cap B(\varepsilon)] \subset Q(\varepsilon) \tag{3.9}$$

Since $V(x)$ decreases monotonically along the system's nontrivial trajectories we have that

$$S \cap W^s(\hat{x}_i) = \emptyset \tag{3.10}$$

while by Lemma 2.1,

$$x_i(t) \in S \quad ; \quad -\infty < t < +\infty \ , \ i = 1,2 \tag{3.11}$$

From (3.7) $\rightarrow$ (3.11) it follows that $S \cap B(\varepsilon)$ is the union of two nonempty disjoint subsets :

$$S \cap B(\varepsilon) = [S \cap P(\varepsilon)] \cup [S \cap Q(\varepsilon)]$$

Now we have

647

Lemma 3.1 *If exactly one nontrivial trajectory of $W^u(\hat{x}_i)$ lies in $\Omega(\hat{x}_s)$, say $\{x_1(t), -\infty < t < +\infty\} \subset \Omega(\hat{x}_s)$ then*

$$[S \cap P(\varepsilon)] \subset \Omega(\hat{x}_s) \tag{3.12}$$

while $[S \cap Q(\varepsilon)] \not\subset \Omega(\hat{x}_s)$.
If both nontrivial trajectories of $W^u(\hat{x}_i)$ lie in $\Omega(\hat{x}_s)$ then (3.12) is valid and

$$[S \cap Q(\varepsilon)] \subset \Omega(\hat{x}_s) \tag{3.13}$$

Proof First assume that

$$x_1(t) \to \hat{x}_s \quad \text{for} \quad t \to +\infty \tag{3.14}$$

$$x_2(t) \not\to \hat{x}_s \quad \text{for} \quad t \to +\infty \tag{3.15}$$

Then $P(\varepsilon) \subset \Omega(\hat{x}_s)$. Indeed, (3.8) and (3.14) imply that $P(\varepsilon) \cap \Omega(\hat{x}_s) \neq \emptyset$. If $P(\varepsilon) \not\subset \Omega(\hat{x}_s)$ then $P(\varepsilon) \cap \partial\Omega(\hat{x}_s) \neq \emptyset$ since $P(\varepsilon)$ is connected. But $P(\varepsilon) \cap W^s(\hat{x}_i) = \emptyset$. Hence there exist points $x_0 \in P(\varepsilon)$ which also belong to $\partial\Omega(\hat{x}_s) \backslash W^s(\hat{x}_i)$, arbitrarily close to $\hat{x}_i$. Let $x_0 \to \hat{x}_i$. Then the half trajectory $\{x(t, x_0), t > 0\}$ converges to the trajectory $\{x_1(t), -\infty < t < +\infty\}$. Since $\partial\Omega(\hat{x}_s)$ is closed it follows that $x_1(t) \in \partial\Omega(\hat{x}_s)$, $-\infty < t < +\infty$, which contradicts (3.14). It follows that $P(\varepsilon) \subset \Omega(\hat{x}_s)$, hence (3.12) holds. (3.9), (3.11) and (3.15) imply that $[S \cap Q(\varepsilon)] \not\subset \Omega(\hat{x}_s)$. If $x_1(t) \to \hat{x}_s$ and $x_2(t) \to \hat{x}_s$ for $t \to +\infty$ then it follows in a similar way that (3.12) and (3.13) hold. From Lemma 3.1 we deduce

Theorem 3.1 *If exactly one nontrivial trajectory of $W^u(\hat{x}_i)$ lies in $\Omega(\hat{x}_s)$, say $\{x_1(t), -\infty < t < +\infty\} \subset \Omega(\hat{x}_s)$ then*

$$\Sigma_p \triangleq [\partial S \cap \partial P(\varepsilon)] \backslash \{\hat{x}_i\} \subset \Omega(\hat{x}_s) \tag{3.16}$$

If both nontrivial trajectories of $W^u(\hat{x}_i)$ lie in $\Omega(\hat{x}_s)$ then $\Sigma_p \subset \Omega(\hat{x}_s)$ and

$$\Sigma_q \triangleq [\partial S \cap \partial Q(\varepsilon)] \backslash \{\hat{x}_i\} \subset \Omega(\hat{x}_s) \tag{3.17}$$

Proof We only prove the first part of the theorem. By Lemma 3.1 we have

$$[\partial S \cap \partial P(\varepsilon)] \subset \overline{\Omega}(\hat{x}_s) \tag{3.18}$$

But

$$\partial S \cap \partial P(\varepsilon) \cap \partial\Omega(\hat{x}_s) = \{\hat{x}_i\} \tag{3.19}$$

Indeed $x \in \partial S$ implies that $V(x) = V(\hat{x}_i)$. On the other hand $x \in \partial P(\varepsilon) \cap \partial\Omega(\hat{x}_s)$ implies that $x \in W^s(\hat{x}_i)$ by a similar argument as in the proof of Lemma 3.1. Hence Lemma 2.1 implies that $V(x) > V(\hat{x}_i)$ or $x = \hat{x}_i$. (3.18) and (3.19) show that $[\partial S \cap \partial P(\varepsilon)] \backslash \{\hat{x}_i\} \subset \Omega(\hat{x}_s)$, hence $\Sigma_p \subset \Omega(\hat{x}_s)$.

Theorem 3.1 suggests the following trajectory reversing procedure to estimate $\partial\Omega(\hat{x}_s)$. Each equilibrium point in Υ is considered. Depending on whether exactly one or two nontrivial trajectories of $W^u(\hat{x}_i)$ lie in $\Omega(\hat{x}_s)$ initial points $x_{0_i} \in \Omega(\hat{x}_s)$ for numerical integration in reverse time are chosen on the sets Σ_p or/and Σ_q of dimension $(n-2)$. The numerically computed half trajectories $\{x(t, x_{0_i}), -\infty < t \leq 0, x_{0_i} \in \Sigma_p$ or/and $\Sigma_q\}$ either generate a single $(n-1)$-dimensional set $\Gamma_i \subset \Omega(\hat{x}_s)$ or two such sets $\Gamma_{i_j} \subset \Omega(\hat{x}_s)$, $j = 1, 2$. The inverse trajectories lie in $\Omega(\hat{x}_s)$ as the latter is invariant. If they remain bounded for

$t < 0$ they will converge to an equilibrium point on $\partial\Omega(\hat{x}_s)$ as $t \to -\infty$. The sets of initial points Σ_p resp Σ_q are connected to $\hat{x}_i$ by segments of level set boundaries $\partial S \cap P(\varepsilon)$ resp. $\partial S \cap Q(\varepsilon)$ which also lie in $\Omega(\hat{x}_s)$.
Hence the set $\partial G(\hat{x}_s)$ composed of

- the equilibrium points $\hat{x}_i \in \Upsilon$ and for each $\hat{x}_i$:

- the sets $\overline{\Gamma}_i$ or $\overline{\Gamma}_{i_j}$, $j = 1, 2$;

- the segments of level set boundaries $\partial S \cap P(\varepsilon)$ or/and $\partial S \cap Q(\varepsilon)$

lies completely in $\overline{\Omega}(\hat{x}_s)$.
We illustrate the proposed procedure with neural networks of the form

$$\dot{x}_i = x_i[f_i(x_i) - \sum_{k=1}^{n} m_{ik}x_k] \quad ; \quad i = 1 \ldots n \tag{3.20}$$

with state space $\{x \in R^n; \; x_i \geq 0, \; i = 1 \ldots n\}$. (3.20) constitutes a special case of (2.4). We start with a second order example where

$$\begin{array}{rcl} f_1(x_1) & = & c - dx_1 + ex_1^2 \\ f_2(x_2) & = & -f + gx_2 - hx_2^2 \end{array} \quad ; \quad M = \begin{bmatrix} 0 & -1 \\ -1 & 0 \end{bmatrix} \tag{3.21}$$

For the chosen numerical values of Fig.1, (3.20)-(3.21) possesses 7 equilibrium points. One stable node $\hat{x}_s$, two unstable nodes $\hat{x}_b$, $\hat{x}_c$ and four saddle points $\hat{x}_i$, $\hat{x}_j$, $\hat{x}_k \in \Upsilon$ and $\hat{x}_l \in \Psi$. The exact stability boundary consists of the stable manifolds $W^s(\hat{x}_i)$, $W^s(\hat{x}_j)$, $W^s(\hat{x}_k)$, the part of $W^s(\hat{x}_l)$ that connects $\hat{x}_b$ to $\hat{x}_l$ and the unstable nodes $\hat{x}_b$, $\hat{x}_c$. Assumptions 1 to 4 are satisfied. We apply trajectory reversion in the neighbourhood of all equilibria of Υ. For each $\hat{x} \in \Upsilon$, $W^u(\hat{x})$ has one nontrivial trajectory which belongs to $\Omega(\hat{x}_s)$. Therefore we have two starting points for trajectory reversion per $\hat{x} \in \Upsilon$. Take for example $\hat{x}_i$. Two initial points x_{0_1} and x_{0_2} on Σ_p generate two inverse half trajectories $\{x(t, x_{0_i}), \; -\infty < t \leq 0, \; i = 1, 2\} \in \Omega(\hat{x}_s)$. Both trajectories are bounded and converge to an index two equilibrium point on $\partial\Omega(\hat{x}_s)$ as $t \to -\infty$. $x(t, x_{0_2})$ proceeds from the neighbourhood of $\hat{x}_i$ to $\hat{x}_l$ and converges to $\hat{x}_b$ as $t \to -\infty$. The inverse half trajectories reconstruct approximately segments of $\partial\Omega(\hat{x}_s)$. $\Omega(\hat{x}_s)$ is unbounded as some of the inverse trajectories tend to infinity as $t \to -\infty$. Fig.1 displays $\partial G(\hat{x}_s)$ which consists of Υ, the sets $\overline{\Gamma}_i$, $\overline{\Gamma}_j$, $\overline{\Gamma}_k$ and segments of the level set boundaries $\partial S \cap P(\varepsilon)$ which connect the index one equilibria to their corresponding surfaces $\overline{\Gamma}$. $\partial G(\hat{x}_s) \to \partial\Omega(\hat{x}_s)$ as $\varepsilon \overset{>}{\to} 0$. In this example $\partial\Omega(\hat{x}_s)$ possesses two index one equilibria which are connected by a complete trajectory such that the transversality condition (2.2) is not satisfied. Note that $\partial G(\hat{x}_s)$ is found by the calculation of 6 inverse half trajectories and one Liapunov function. The neighbourhoods $B(\varepsilon)$ of $\hat{x} \in \Upsilon$ where chosen large for the clarity of Fig.1.
As a second example we choose a third order version of (3.20) where

$$f_i(x_i) = -a_i x_i + b_i, \; i = 1 \ldots 3 \quad ; \quad M = \begin{bmatrix} 0 & 1 & 1 \\ 1 & 0 & 1 \\ 1 & 1 & 0 \end{bmatrix} \tag{3.22}$$

For the numerical values of Fig.2, we find 8 equilibrium points : one l.a.s. equilibrium point $\hat{x}_s$, three equilibrium points, $\hat{x}_i$, $\hat{x}_j$, $\hat{x}_k$ of index one, three equilibrium points $\hat{x}_l$,

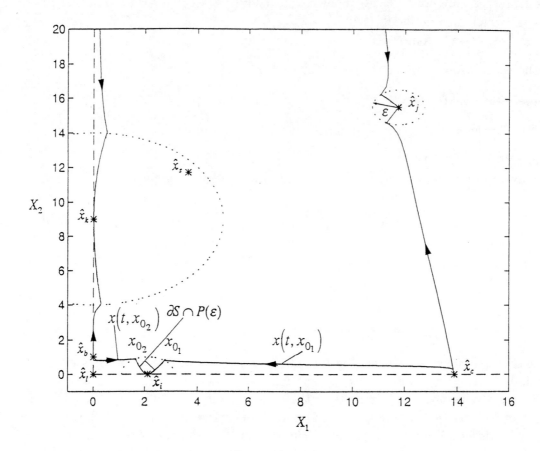

Fig. 1: Trajectory reversion applied to a second order neural network (3.20)-(3.21), $c = 22$, $d = 12$, $e = 0.75$, $f = 1.125$, $g = 1.25$, $h = 0.125$.

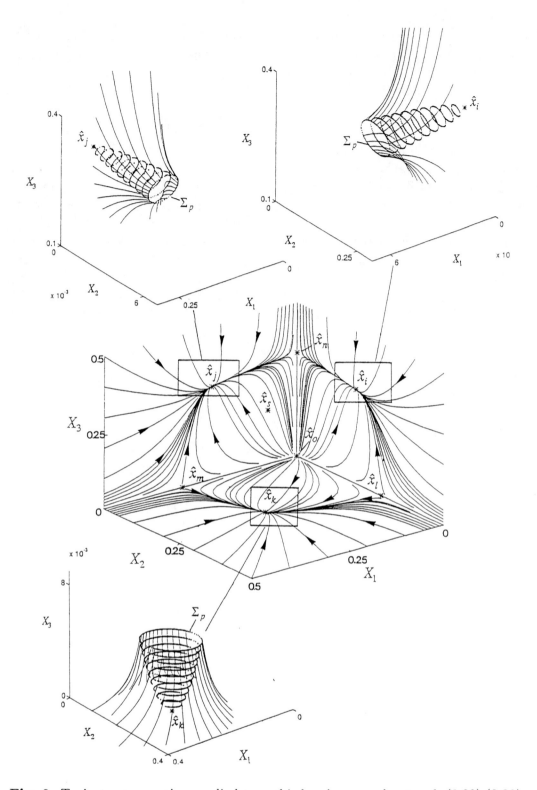

Fig. 2: Trajectory reversion applied to a third order neural network (3.20)-(3.22), $a_1 = 4$, $a_2 = 3.5$, $a_3 = 6.5$, $b_1 = 1.2$, $b_2 = 1$, $b_3 = 2.2$.

$\hat{x}_m$, $\hat{x}_n$ of index two and one equilibrium point $\hat{x}_o$ of index 3. It is easy to see that $\Omega(\hat{x}_s) = \{x \in R^3; \; x_i > 0, \; i = 1 \ldots 3\}$. All solutions on $\partial\Omega(\hat{x}_s)$ remain bounded for $t > 0$. For every equilibrium point $\hat{x}$ on $\partial\Omega(\hat{x}_s)$ it holds that

$$[W^u(\hat{x}) \setminus \{\hat{x}\}] \cap \Omega(\hat{x}_s) \neq \emptyset \quad ; \quad \forall \hat{x} \in \partial\Omega(\hat{x}_s) \tag{3.23}$$

Hence Ψ is an empty set. The simple structure of $\partial\Omega(\hat{x}_s)$ allows us to verify that Assumption 4 is satisfied. The transversality condition is satisfied for every $W^s(\hat{x}_i)$ and $W^u(\hat{x}_j)$ on $\partial\Omega(\hat{x}_s)$. This can easily be illustrated with the index two equilibrium point $\hat{x}_l$ and the index one equilibrium point $\hat{x}_i$. The axis $\{x \in R^3; \; x_1 = 0, \; x_2 > 0, \; x_3 = 0\}$ corresponds to $W^s(\hat{x}_l)$, the plane $\{x \in R^3; \; x_1 = 0, x_2 > 0, x_3 > 0\}$ corresponds to $W^s(\hat{x}_i)$. $\hat{x}_i$ and $\hat{x}_l$ are neighbouring equilibrium points on $\partial\Omega(\hat{x}_s)$ as they are connected by a complete trajectory on $\partial\Omega(\hat{x}_s)$. Part of the intersection $W^s(\hat{x}_i) \cap W^u(\hat{x}_l)$ coincides with this trajectory. This can be seen on Fig.3 where a number of trajectories of $W^u(\hat{x}_l)$ and $W^s(\hat{x}_i)$ are displayed. Fig.3 shows that the tangent spaces $T_x[W^s(\hat{x}_i)]$ and $T_x[W^u(\hat{x}_l)]$ span the entire state space R^3 such that the transveraslity condition holds. It can also be seen that

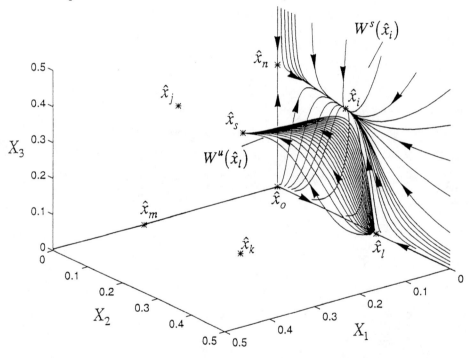

Fig. 3: Intersection of $W^s(\hat{x}_i)$ and $W^u(\hat{x}_l)$ of a third order neural network (3.20)-(3.22).

$$W^s(\hat{x}_l) \subset \partial W^s(\hat{x}_i) \quad ; \quad \text{index } \hat{x}_l > \text{index } \hat{x}_i \tag{3.24}$$

The same reasoning can be repeated for every set of neighbouring equilibrium points on $\partial\Omega(\hat{x}_s)$. (3.24) illustrates the hierarchical structure that is induced on $\partial\Omega(\hat{x}_s)$ by the transversality condition. The transversality condition and (3.23) imply that Assumption 4(ii) is satisfied. Assumption 4(iii) is also satisfied. This example belongs to the practical important class of Morse-Smale systems. Besides the transversality condition it is assumed that the quasi-stability boundary coincides with the exact stability boundary. For these systems the following theorem holds.

Theorem 3.2

$$\partial \Omega(\hat{x}_s) = \bigcup_{\hat{x}_i \in \Upsilon} \overline{W}^s(\hat{x}_i)$$

(Zaborszky et al., 1988)

In this simple example Theorem 3.2 can easily be verified. We apply the trajectory reversion method in the neighbourhood of each index one equilibrium point on $\partial \Omega(\hat{x}_s)$. Fig. 2 shows $\partial G(\hat{x}_s)$ which consists of $\hat{x}_i$, $\hat{x}_j$, $\hat{x}_k$, $\overline{\Gamma}_i$, $\overline{\Gamma}_j$, $\overline{\Gamma}_k$ and the level set boundaries $\partial S \cap P(\varepsilon)$ of each index one equilibrium point. It can be seen that $\partial G(\hat{x}_s) \to \partial \Omega(\hat{x}_s)$ as $\varepsilon \xrightarrow{>} 0$, although $\partial \Omega(\hat{x}_s)$ contains the stable manifolds of index two equilibria.

In the previous examples the trajectory reversion method succeeds in producing an estimate of $\partial \Omega(\hat{x}_s)$ with any desired accuracy. This can intuitively be understood as follows. Let x_0 be an arbitrary point on $\partial \Omega(\hat{x}_s)$ which is not an equilibrium point. With x_0 we associate one or more paths $\gamma \subset \overline{\Omega}(\hat{x}_s)$ which connect x_0 to $\hat{x}_s$. γ is, in the sense of increasing time, constructed as follows :

- the half trajectory $\{x(t, x_0), 0 \le t < +\infty\}$;

- possibly one or more complete trajectories $\{x(t), -\infty < t < +\infty\}$ on $\partial \Omega(\hat{x}_s)$;

- a complete trajectory $\{x(t), -\infty < t < +\infty\}$ lying in $\Omega(\hat{x}_s)$ and connecting an equilibrium point $\hat{x}_i \in \partial \Omega(\hat{x}_s)$ to $\hat{x}_s$.

We need the following lemma which is due to Chiang et al. (1988).

Lemma 3.2 *Let $\hat{x}_i \ne \hat{x}_s$ be a hyperbolic equilibrium point of (2.1). Then*

$$\hat{x}_i \subset \partial \Omega(\hat{x}_s) \Leftrightarrow [W^u(\hat{x}_i) \setminus \{\hat{x}_i\}] \cap \overline{\Omega}(\hat{x}_s) \ne \emptyset$$

Assumption 1-3, Lemma 3.2 and the fact that the number of equilibrium points is finite ensure that for every $x_0 \in \partial \Omega(\hat{x}_s)$ which is not an equilibrium point there always exists at least one associated path γ of the prescribed form. The inverse half trajectories and the level set boundaries reconstruct approximately $\{\gamma \cap \partial \Omega(\hat{x}_s)\} \cup \{x(t, x_0), -\infty < t \le 0\}$. This holds for every possible path γ which connects x_0 to $\hat{x}_s$. The implication of Assumption 4 is the restriction of the possible paths γ. We still need to investigate if Assumptions 1-4 guarantee that the method produces the desired result.

4   Convergence

By convergence we mean that $\partial G(\hat{x}_s)$ converges to $\partial \Omega(\hat{x}_s)$ as $\varepsilon \xrightarrow{>} 0$. $\varepsilon \xrightarrow{>} 0$ implies that the initial states for trajectory reversion approach the index one equilibrium points of Υ along level set boundaries inside $\Omega(\hat{x}_s)$. The following theorem can be proved:

Theorem 4.1 *Let $\hat{x}_s$ be a l.a.s. equilibrium point. Let b be any point on $\partial \Omega(\hat{x}_s)$ that is not an equilibrium point. Let $\partial G(\hat{x}_s)$ be the estimated stability boundary of $\hat{x}_s$ obtained by the method of Section 3 for a system (2.1) that satisfies Assumptions 1-4.*
Then for every $a \in \Omega(\hat{x}_s)$ sufficiently close to b there exists an $\varepsilon > 0$ such that $a \in \partial G(\hat{x}_s)$. Furthermore

$$d \stackrel{\triangle}{=} |b - a| \to 0 \Rightarrow \varepsilon \xrightarrow{>} 0$$

We only give a rough outline of the proof. A rigorous mathematical proof can be found in (Loccufier, 1996). Consider a point $b \in \partial\Omega(\hat{x}_s)$ and a point $a \in \Omega(\hat{x}_s)$ sufficiently close to b as defined in Theorem 4.1. It holds that $x(t,a) \to \hat{x}_s$ as $t \to +\infty$. The proof of Theorem 4.1 consists of two parts.

In the first part it is shown that $\{x(t,a), 0 \le t < +\infty\}$ converges to $\hat{x}_s$ for $t \to +\infty$ along a nontrivial trajectory $\{x(t), -\infty < t < +\infty\} \subset [W^u(\hat{x}_l) \cap \Omega(\hat{x}_s)]$, $\hat{x}_l \in \partial\Omega(\hat{x}_s)$. $x(t,a)$ may reach neighbourhoods of some equilibria $\hat{x}_g$, $\hat{x}_h$, ... before converging to $\hat{x}_s$ along $\{x(t), -\infty < t < +\infty\} \subset [W^u(\hat{x}_l) \cap \Omega(\hat{x}_s)]$. Take for example $b \ne \hat{x}_l$ on $W^s(\hat{x}_l) \cap \partial\Omega(\hat{x}_s)$ on Fig.1. For a sufficiently close to b, $x(t,a)$ will first reach a neighbourhood of $\hat{x}_l$, then a neighbourhood of $\hat{x}_i$ and will finally converge to $\hat{x}_s$ along $W^u(\hat{x}_i) \cap \Omega(\hat{x}_s)$ as $t \to +\infty$. The proof is based on :

- Assumption 2 and 3 which ensure that every trajectory on $\partial\Omega(\hat{x}_s)$ converges to an equilibrium point on $\partial\Omega(\hat{x}_s)$ as $t \to +\infty$.

- Lemma 6.2 of (Loccufier, 1996) which states : if a trajectory $x(t,a)$ reaches a neighbourhood Λ of a hyperbolic equilibrium $\hat{x}_i$ along $W^s(\hat{x}_i)$, it will proceed along $W^u(\hat{x}_i)$ for increasing time t.

As $x(t,a) \in \Omega(\hat{x}_s)$ and the number of equilibrium points is finite, the existence of $\hat{x}_l$ is ensured.

In the second part the index of $\hat{x}_l$ is examined. Two cases can occur.

1. Index $\hat{x}_l = 1$. By Lemma 2.1 $V(b) > V(\hat{x}_l)$. If d is sufficiently small then $V(a) \ge V(\hat{x}_l)$. $V(.)$ decreases monotonically along $\{x(t,a), 0 \le t < +\infty\}$ from $V(a)$ to $V(\hat{x}_s)$. Hence

$$\exists t_0 \ge 0 \; : \; V[x(t_0,a)] = V(\hat{x}_l) \tag{4.1}$$

(4.1) defines t_0 as a continuous function of a. Furthermore

$$a \to b \Rightarrow t_0 \to +\infty \text{ and } x(t_0,a) \to \hat{x}_l \tag{4.2}$$

(4.1) and (4.2) imply that $a \in \Gamma_l$ or Γ_{l_j}, $j = 1$ or $j = 2$ hence $a \in \partial G(\hat{x}_s)$ for a sufficiently small $\varepsilon > 0$.

2. Index $\hat{x}_l > 1$. We need the following lemma's.

Lemma 4.1 *Let $\hat{x}_i$ and $\hat{x}_j$ be equilibria on $\partial\Omega(\hat{x}_s)$ connected by a complete trajectory $\{x(t), -\infty < t < +\infty\}$, such that*

$$\lim_{t \to -\infty} x(t) = \hat{x}_j \quad ; \quad \lim_{t \to +\infty} x(t) = \hat{x}_i$$

If the stable and unstable manifolds of $\hat{x}_i$ and $\hat{x}_j$ satisfy the transversality condition then index $\hat{x}_i <$ index $\hat{x}_j$.

[(Zaborszky et al., 1988), Lemma 4.3]

Lemma 4.2 *Let $\hat{x}_i$ and $\hat{x}_j$ be equilibria on $\partial\Omega(\hat{x}_s)$ whose stable and unstable manifolds satisfy the transversality condition. Then*

$$\hat{x}_j \in \partial W^s(\hat{x}_i) \Leftrightarrow W^s(\hat{x}_j) \subset \partial W^s(\hat{x}_i)$$

654

[(Zaborszky et al., 1988), Theorem 4.6]
Repeatedly applying Assumption 4, Lemma 4.1 and Lemma 4.2 allows us to prove a hierarchical structure of $\partial\Omega(\hat{x}_s)$ where

$$b \in \partial W^s(\hat{x}_n) \tag{4.3}$$

and $\hat{x}_n \in \Upsilon$. (4.3) implies that for a suitable $\varepsilon > 0$, the point a will lie on the closure of Γ_n or Γ_{n_j}, j=1 or j=2, obtained by trajectory reversing from Σ_p/Σ_q in a small neighbourhood of $\hat{x}_n$. It follows that $a \in \partial G(\hat{x}_s)$.

For example take b on $W^s(\hat{x}_m)$ on Fig.2 and the point a sufficiently close to b. $x(t,a)$ converges to $\hat{x}_s$ as $t \to +\infty$ in the neighbourhood of $W^u(\hat{x}_m) \cap \Omega(\hat{x}_s)$ where index $\hat{x}_m = 2$. It holds that $b \in \partial W^s(\hat{x}_j)$ where $\hat{x}_j \in \Upsilon$ such that $a \in \overline{\Gamma}_j$.

Remark : Theorem 4.1 is not valid if b is an equilibrium point on $\partial\Omega(\hat{x}_s)$. If $b \in \Upsilon$ then the theorem is not true for points $a \in W^u(b) \cap \Omega(\hat{x}_s)$. In this case $a \in G(\hat{x}_s)$ instead of $a \in \partial G(\hat{x}_s)$. Theorem 4.1 is not true either if b is an equilibrium point of index n on $\partial\Omega(\hat{x}_s)$. However the analysis above shows that any point $a \in \Omega(\hat{x}_s)$ sufficiently close to $\partial\Omega(\hat{x}_s)$ also lies in $G(\hat{x}_s)$ for $\varepsilon > 0$ and sufficiently small.

5   Algorithm

The algorithm is illustrated by a Hopfield neural network of dimension 4. The dynamics of a Hopfield neural network can be described by

$$\dot{x} = Ax - MS(x) - h \quad ; \quad A = diag[a_i] < 0 \tag{5.1}$$

(5.1) constitutes a special case of (2.4) :

$$a_i(x_i) = 1 \quad ; \quad b_i(x_i) = a_i x_i - h_i \quad ; \quad S(x) = col[S_1(x_1) \dots S_n(x_n)]$$

$S_i(x_i)$ are saturating nonlinearities which implies that all trajectories of (5.1) are bounded. The Liapunov function (2.5) can be written as :

$$V(x) = \frac{1}{2}S'(x)MS(x) - x'AS(x) + \int_0^x S'(u)Adu + h'S(x)$$

In the example we choose $S_i(x_i) = \rho_i \tanh(\alpha_i x_i)$ and :

$$A = \begin{bmatrix} -2 & 0 & 0 & 0 \\ 0 & -5 & 0 & 0 \\ 0 & 0 & -6 & 0 \\ 0 & 0 & 0 & -3.5 \end{bmatrix} \quad , \quad M = -\begin{bmatrix} 1 & 0.25 & 0.5 & 0.1 \\ 0.25 & 3 & 0.25 & 0.5 \\ 0.5 & 0.25 & 3 & 0.1 \\ 0.1 & 0.5 & 0.1 & 2.5 \end{bmatrix}$$

$$h = -\begin{bmatrix} 1 \\ 4 \\ 3 \\ 1.5 \end{bmatrix} \quad , \quad \rho = \begin{bmatrix} 5 \\ 4 \\ 1 \\ 5 \end{bmatrix} \quad , \quad \alpha = \begin{bmatrix} 1 \\ 2 \\ 1 \\ 1.5 \end{bmatrix}$$

The trajectory reversing method of Section 3 leads to the following conceptual algorithm.

Step 1 Find all equilibria of (2.1), i.e. find all solutions of the set of nonlinear equations

$$f(x) = 0$$

In (5.1), $S_i(x_i)$ is approximated by a piecewise continuous linear function. For every combination of 4 pieces a set of linear equations is solved. If the solution lies inside the interval where the linear pieces are valid, then this solution is used as an initial estimate for the equilibrium. The latter follows from solving the set of nonlinear functions with some numerical routine. The numerical example possesses 21 equilibria.

Step 2 Find the corresponding index of all equilibria. The index of $\hat{x}_i$ follows from the eigenvalues of $J(\hat{x}_i) = \frac{\partial f(x)}{\partial x}|_{\hat{x}_i}$.
The 21 equilibria can be ordered as follows : 6 l.a.s. equilibria, 9 equilibria of index 1, 5 equilibria of index 2 and 1 equilibrium of index 3. If the Hopfield network is used as a classification network then for its chosen numerical values it is able to distinguish 6 classes where each l.a.s. equilibrium point represents a prototype vector.

Step 3 For every l.a.s. equilibrium point $\hat{x}_s$ find the set Υ. This can easily be done as follows. For each $\hat{x}_i$ of index one where $V(\hat{x}_i) > V(\hat{x}_s)$ calculate the eigenvector e_1 that corresponds to the unique unstable eigenvalue λ_1 of $J(\hat{x}_i)$. For the initial states

$$x_{0_1} = \hat{x}_i + \varepsilon e_1 \quad ; \quad x_{0_2} = \hat{x}_i - \varepsilon e_1$$

$|\varepsilon| \neq 0$ and sufficiently small, compute the trajectories $\{x(t, x_{0_i}), \ 0 \leq t < +\infty, \ i = 1, 2\}$ by numerical integration of (2.1). $\hat{x}_i \in \Upsilon$ if $x(t, x_{0_1})$ or / and $x(t, x_{0_2})$ converge to $\hat{x}_s$ for $t \to +\infty$.
The index 1 equilibria of (5.1) are classified according to the stability boundary they belong to in Table 1. For every l.a.s. equilibrium point we estimate the exact region

$\hat{x}_g \in \partial\Omega(\hat{x}_a), \ \partial\Omega(\hat{x}_b)$
$\hat{x}_h \in \partial\Omega(\hat{x}_a), \ \partial\Omega(\hat{x}_c)$
$\hat{x}_i \in \partial\Omega(\hat{x}_f), \ \partial\Omega(\hat{x}_b)$
$\hat{x}_j \in \partial\Omega(\hat{x}_b), \ \partial\Omega(\hat{x}_f)$
$\hat{x}_k \in \partial\Omega(\hat{x}_c), \ \partial\Omega(\hat{x}_e)$
$\hat{x}_l \in \partial\Omega(\hat{x}_a), \ \partial\Omega(\hat{x}_d)$
$\hat{x}_m \in \partial\Omega(\hat{x}_d), \ \partial\Omega(\hat{x}_f)$
$\hat{x}_n \in \partial\Omega(\hat{x}_e), \ \partial\Omega(\hat{x}_d)$
$\hat{x}_o \in \partial\Omega(\hat{x}_e), \ \partial\Omega(\hat{x}_f)$

Table 1: Classification of the index one equilibria of a four-dimensional neural network.

of attraction in step 4.

Step 4 For every $\hat{x}_i \in \Upsilon$ compute the set Γ_i (or $\Gamma_{i_j}, \ j = 1, 2$) composed of inverse trajectories starting at points on Σ_p or/and on Σ_q. In (Loccufier & Noldus, 1995) a practical implementation can be found for calculating the initial states based on a

Taylor series expansion of the Liapunov function. The implementation is valid for systems where $\frac{\partial^2 V(x)}{\partial x^2}\big|_{x=\hat{x}_i}$ is nonsingular and can be programmed in a fully automatic computer programme.

Lets start with the computation of $\partial G(\hat{x}_a)$. The index one equilibria $\hat{x}_g$, $\hat{x}_h$ and $\hat{x}_l$ belong to Υ_a. By trajectory reversion the surfaces Γ_g, Γ_h and Γ_l are obtained. The procedure is repeated for every l.a.s. equilibrium point.

To be able to decide whether a point belongs to the exact region of attraction or not, we will use projections of $\partial G(.)$ on the state planes. Fig.4 shows the projection of $\partial G(\hat{x}_a)$ on the $x_2 x_4$-plane. Let $a = (-5, 2, 2, 4)$ be an arbitrarily selected point in state space. Fig.4 indicates that $a \notin G(\hat{x}_a)$. Similar well-selected projections on various two-dimensional phase planes reveal that a does not belong to $G(\hat{x}_b)$, $G(\hat{x}_c)$, $G(\hat{x}_d)$ or $G(\hat{x}_e)$ either. Since the estimated stability regions are close approximations of the true stability regions and since every trajectory converges to an equilibrium point for $t \to +\infty$ it may be concluded that $a \in \Omega(\hat{x}_f)$. Note that a single Liapunov function is used for the computation of $\partial G(.)$ for every l.a.s. equilibrium point of the fourth order neural network.

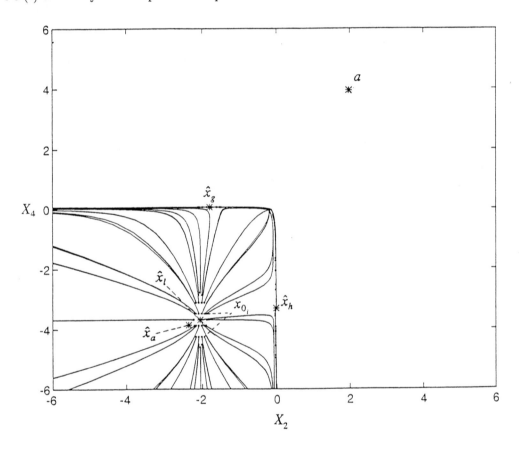

Fig. 4: Projection of a stability region $G(\hat{x}_a)$ on the $x_2 x_4$-plane for a four-dimensional neural network (5.1); $\hat{x}_a = (-2.81, -2.36, -0.32, -3.86)'$, $a \in G(\hat{x}_f)$, $\hat{x}_f = (3.98, 4.00, 1.63, 4.74)'$.

6 Conclusion

A new trajectory reversion method has been presented for the global stability analysis of an important class of neural networks. The method results in an accurate estimate of the exact stability region. The procedure essentially exists of calculating one Liapunov function and a number of inverse trajectories from initial states inside the exact stability region. The main limitation is that a global Liapunov function must be found for the network. This limitation is compensated in two ways. The Liapunov function allows to develop a numerically robust procedure for the calculation of the initial states for trajectory reversion. The initial states lie inside the exact stability region in the neighbourhood of index one equilibria on the stability boundary, but not necessarily very close to it. Secondly the assumptions concerning the topological structure of the stability boundary are weakened compared to other existing methods and are automatically satisfied for second order systems. Methods developed by (Chiang et al. 1988; Zaborszky at al. 1988) can only be applied to systems which satisfy the transversality condition and where the exact stability boundary and the quasi stability boundary coincide. Another important remark is that the number of inverse half trajectories to be calculated is reduced to a minimum. The initial states lie on $(n-2)$-dimensional surfaces, where n is the dimension of the system. In other existing methods based on simulation in reverse time, the initial states lie on $(n-1)$-dimensional surfaces (Genesio et al., 1985; Guttalu & Flashner, 1988; Psiaki & Luh 1993).

Appendix A

Proof of Lemma 2.1
We first prove property 1. For every $x_0 \in W^s(\hat{x}_i)$,

$$V(x_0) \geq V(\hat{x}_i) \qquad \text{(Appendix. A.1)}$$

Suppose that $x_0 \in [W^s(\hat{x}_i) \setminus \{\hat{x}_i\}]$ and

$$V(x_0) = V(\hat{x}_i) \qquad \text{(Appendix. A.2)}$$

Since $x(t, x_0) \to \hat{x}_i$ as $t \to +\infty$, (2.3) and (Appendix. A.2) imply that

$$V[x(t, x_0)] = V(\hat{x}_i) \quad ; \quad 0 \leq t < +\infty$$

which contradicts Assumption 2. Since the stable manifold of $\hat{x}_i$ for the system (2.1) coincides with the unstable manifold of $\hat{x}_i$ if time is reversed, property 1. of $W^s(\hat{x}_i)$ implies property 2. of $W^u(\hat{x}_i)$.

References

Atiya A. & Abu-Mostafa, Y., S. (1993).An Analog Feedback Associative Memory. I.E.E.E. Trans. on Neural Networks. volume 4, pp. 117-126.

Chiang, H., Hirsch, M. & Wu, F. (1988). Stability Regions of Nonlinear Autonomous Dynamical Systems. I.E.E.E. Trans. on Autom. Control, volume 33, pp. 16-27.

Chiang, H. (1991). Analytical Results on Direct Methods for Power System Transient Stability Analysis. In Advances in Control and Dynamic Systems. , volume 43. New York Academic Press. pp. 275-334. Ciliz, M. (1993). Equilibrium Point Basins of Analog Neural Networks. Physics Letters A. volume 173. pp. 25-29.

Cohen, M. A. & Grossberg, S. (1983). Absolute Stability of Global Pattern Formation and Parallel Memory Storage by Competitive Neural Networks. I.E.E.E. Trans. on Systems, Man and Cybernetics. volume 13. pp. 815-826.

Genesio, R., Tartaglia, M. & Vicino, A. (1985). On the Estimation of Asymptotic Stability Regions: State of the Art and New Proposals. I.E.E.E. Trans. on Autom. Control. volume 30. pp. 747-755.

Guttalu, R. S. & Flashner, H. (1988). A Numerical Method for Computing Domains of Attraction for Dynamical Systems. Int. J. for Numerical Methods in Engineering. volume 26. pp. 875-890.

Hartman, P. (1982). Ordinary Differential Equations. Boston : Birkhäuser. Kosko, B. (1991). Structural Stability of Unsupervised Learning in Feedback Neural Networks. I.E.E.E. Trans. on Autom. Control. volume 36. pp. 785-792.

Lasalle, J. P. (1967). An Invariance Principle in the Theory of Stability. In Differential Equations and Dynamical Systems. Edited by J. P. Lasalle and J. K. Hale. Academic Press. New York. pp. 277- 286.

Loccufier, M.& Noldus E. (1995). On the Estimation of Asymptotic Stability Regions for Autonomous Nonlinear Systems. I.M.A. J. of Math. Control and Information. volume 12. pp. 91-109.

Loccufier, M. (1996). Een Nieuwe Trajectoriën-Reversietechniek voor het Stabiliteits Onderzoek van Niet- Lineaire Systemen. Ph. D. Thesis. University of Gent.

Noldus, E. & Loccufier, M. (1994). An Application of Liapunov's Method for the Analysis of Neural Networks. J. of Comp. and Appl. Math., volume 50. pp. 425-432.

Noldus, E. & Loccufier, M. (1995). A New Trajectory Reversing Method for the Estimation of Asymptotic Stability Regions. Int. J. Control. volume 61. pp.917-932.

Psiaki, M. L. & Luh, Y.(1993). Nonlinear System Stability Boundary Approximation by Polytopes in State Space. Int. J. Control. volume 57, pp. 197-224.

Wang, X. & Blum, E., K. (1992). Discrete-time Versus Continuous-time Models of Neural Networks, volume 45. pp. 1-19.

Zaborszky J., Huang G., Zheng, B. & Leung T. (1988). On the Phase Portrait of a Class of Large Nonlinear Dynamic Systems such as the Power System. I.E.E.E. Trans. on Autom. Control. volume 33. pp. 4-15.

Competitive Learning in Decision Trees

Dominique Martinez

Laboratoire d'Analyse et d'Architecture des Systèmes (LAAS-CNRS)

7, Av. Col. Roche

31077 Toulouse - France

Abstract

In this paper, a competitive learning rule is introduced in decision trees as a computationally attractive scheme for adaptive density estimation or lossy compression. It is shown by simulation that the adaptive decision tree performs at least as well as other competitive learning algorithms while being much faster.

keywords: competitive learning, decision tree, density estimation, vector quantization, neural networks.

1 Introduction

The use of decision trees for classification and regression is relatively familiar to statisticians (Breiman, Friedman, Olshen & Stone, 1984). Decision trees are built in a supervised fashion to perform hierarchical partitioning over the input space by means of axis-parallel hyperplanes. Orthogonal hyperplanes offer considerable benefits in terms of computational complexity because encoding requires no multiplications. Decision trees might be helpful for density estimation as well as lossy compression using vector quantization in which conventional design algorithms hit the complexity "barrier" when the input dimensionality is too high. Unfortunately, very little work has been done on combining decision trees with unsupervised learning.

In this paper, competitive learning is introduced in decision trees as a computationally attractive scheme for density estimation or lossy compression. In Section 2, the high resolution theory is used to analyze the performance of decision trees. High resolution theory assumes a large number of partition cells in order to consider the density approximately constant over each cell. This, in turn, allows us to derive a necessary condition for optimality in the case of density estimation or lossy compression.

If density estimation is the objective sought, the proper way of achieving quantization is through maximized quantizer entropy. Thus, the N partition cells possess an equal probability $\frac{1}{N}$ of winning the competition and the density is estimated as $\frac{1}{N V(S_i)}$ in each partition cell S_i of volume $V(S_i)$. If the goal is lossy compression, correct quantization will require minimizing the average distortion because entropy maximization obviates any possibility of further compression by lossless coding. Section 3 describes a competitive learning rule which minimizes the r-th power law distortion in the high resolution case. When $r = 2$, the learning rule minimizes the classical mean squared error distortion, aiming at lossy compression. When $r = 0$, the learning rule yields equiprobable quantization for density estimation purposes.

CP437, *Computing Anticipatory Systems: CAYS--First International Conference*
edited by Daniel M. Dubois © 1998 The American Institute of Physics 1-56396-827-4/98/$15.00

2 High-Resolution Theory in Decision Trees

Let $\mathbf{x} = (x_1, x_2, \cdots x_k)$ be a k-dimensional input vector drawn from a probability density function $p_X(\mathbf{x})$. Assume also that $p_X(\mathbf{x})$ vanishes outside a finite region S so that $\mathbf{x}$ takes on values in S with probability one.

Decision trees perform a hierarchical partitioning of the k-dimensional feature space by means of hyperplanes perpendicular to coordinates axes. An example of hierarchical partitioning is shown in Fig. 1. This results in a binary tree structure in which each internal node t stores two scalar quantities : an index l representing the dimension orthogonal to the hyperplane and a weight w_{tl} representing the location of the hyperplane on this axis. Any input vector $\mathbf{x}$ can be located with respect to this hyperplane through use of a single scalar comparison. If $x_l < w_{tl}$, the feature vector $\mathbf{x}$ is sent to the left child of node i. Similarly, the feature vector is sent to the right child if $x_l \geq w_{tl}$. The leaf nodes are associated with N disjoint partition cells S_i, $1 \leq i \leq N$, whose boundaries are determined by the weights stored above. The geometric boundaries of the cells S_i are defined by the lower and upper boundary point vectors $\mathbf{x}_i^L = (x_{i1}^L \cdots x_{ik}^L)$ and $\mathbf{x}_i^U = (x_{i1}^U \cdots x_{ik}^U)$. The cut value w_{tl} stored at node t may be used for dividing more than two quantization cells. Let $LB(t)$ be the number of cells S_i for which w_{tl} is a lower boundary point, *i.e.* $w_{tl} \equiv x_{il}^L$. Respectively, $UB(t)$ is the number of cells S_i for which w_{tl} is an upper boundary point, *i.e.* $w_{tl} \equiv x_{il}^U$. For the root node of the tree in Fig. 1, w_{11} is an upper boundary point for the cells S_3, S_4 and S_6 and it is a lower boundary point for the cell S_7. Therefore, we have $UB(1) = 3$ and $LB(1) = 1$.

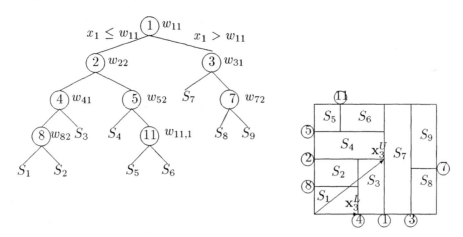

Figure 1: Example of hierarchical partitioning over the input space which could be induced by the binary tree. The lower and upper boundary point vectors of the cell S_3 are given by $\mathbf{x}_3^L = (w_{41}, 0)$ and $\mathbf{x}_3^U = (w_{11}, w_{22})$.

When $\mathbf{x} \in S_i$, it is quantized as $\mathbf{y}_i$ and it results in a quantization error. A frequent goal of competitive learning is the minimization of the average distortion

$$
\begin{aligned}
D_r(k) &= \frac{1}{k} \int_S \|\mathbf{x} - \mathbf{y}\|_\nu^r p_X(\mathbf{x}) d\mathbf{x} \\
&= \frac{1}{k} \sum_{i=1}^N \int_{S_i} \|\mathbf{x} - \mathbf{y}_i\|_\nu^r p_X(\mathbf{x}) d\mathbf{x}
\end{aligned}
\tag{1}
$$

where the distortion measure $||.||_\nu^r$ is the rth power of the l_ν norm

$$||\mathbf{x} - \mathbf{y}_i||_\nu^r = (\sum_{l=1}^{k} |x_l - y_{il}|^\nu)^{\frac{r}{\nu}} \qquad (2)$$

The high resolution theory assumes N large and $p_X(\mathbf{x})$ smooth so that

$$p_X(\mathbf{x}) \approx p_X(\mathbf{y}_i) = \frac{P(S_i)}{V(S_i)} \text{ for } \mathbf{x} \in S_i$$

where $P(S_i) = \int_{S_i} p_X(\mathbf{x})d\mathbf{x}$ and $V(S_i) = \int_{S_i} d\mathbf{x}$ denote the probability and volume of S_i, respectively. Then, Eq. (1) can be approximated as

$$\begin{aligned} D_r(k) &\approx \frac{1}{k}\sum_{i=1}^{N} p_X(\mathbf{y}_i) \int_{S_i} ||\mathbf{x} - \mathbf{y}_i||_\nu^r d\mathbf{x} \\ &= \sum_{i=1}^{N} p_X(\mathbf{y}_i)I(S_i)[V(S_i)]^{1+\frac{r}{k}} \end{aligned} \qquad (3)$$

where $I(S_i)$ denotes the normalized moment of inertia of S_i about the point $\mathbf{y}_i$

$$I(S_i) = \frac{\int_{S_i} ||\mathbf{x} - \mathbf{y}_i||_\nu^r d\mathbf{x}}{k[V(S_i)]^{1+\frac{r}{k}}} \qquad (4)$$

If the distortion is a rth power of the l_2 norm, Zador (1966, 1982) has shown that

$$D_r(k) = A(k,r)N^{-r/k}||p_X(\mathbf{x})||_{k/(k+r)}$$

with the term $A(k,r)$ being independent of $p_X(\mathbf{x})$. Gersho (1979) conjectured that, for large N, most cells are congruent to the polytope with minimum normalized moment of inertia and, thus, considered $A(k,r)$ as the normalized moment of inertia of the optimum polytope. However, the difficulty in utilizing this expression is that $A(k,r)$ is only known with certainty in a few cases.

In the following, we limit our study to single-letter distortion measures for which $\nu = r$ in Eq. (2). Therefore, distortion is the rth power of the l_r norm. This distortion measure is of interest in waveform coding because it allows to place a rth power law on every individual sample. The most commonly used powers are $r = 1$ (mean absolute error) (Kassam, 1978), $r = 2$ (mean squared error) (Gersho & Gray, 1992) and $r = \infty$ (mean maximum absolute error).

Each partition cell S_i of the decision tree is a hyperbox aligned with the coordinate axes and defined by the lower and upper boundary point vectors $\mathbf{x}_i^L = (x_{i1}^L \cdots x_{ik}^L)$ and $\mathbf{x}_i^U = (x_{i1}^U \cdots x_{ik}^U)$. Thus, the normalized moment of inertia of S_i can be expressed as

$$I(S_i) = \frac{1}{k[V(S_i)]^{1+\frac{r}{k}}} \int_{x_{ik}^L}^{x_{ik}^U} \cdots \int_{x_{i1}^L}^{x_{i1}^U} \sum_{l=1}^{k} |x_l - y_{il}|^r dx_1 \cdots dx_k \qquad (5)$$

The high-resolution assumption enables us to take $\mathbf{y}_i$ at the center of its partition cell. Moreover, it is useful to rewrite the absolute value as

$$
\begin{aligned}
\int_{x_{il}^L}^{x_{il}^U} |x_l - y_{il}|^r dx_l &= \int_{x_{il}^L}^{y_{il}} (y_{il} - x_l)^r dx_l + \int_{y_{il}}^{x_{il}^U} (x_l - y_{il})^r dx_l \\
&= \frac{1}{2^r(r+1)}[\delta_{il}]^{r+1}, \text{ with } \delta_{il} = x_{il}^U - x_{il}^L .
\end{aligned}
$$

Then, integrating (5) yields

$$
I(S_i) = \frac{1}{k2^r(r+1)[V(S_i)]^{r/k}} \sum_{l=1}^{k}[\delta_{il}]^r \tag{6}
$$

The inequality property between arithmetic and geometric means gives the following lower bound

$$
I(S_i) \geq \frac{1}{2^r(r+1)} \tag{7}
$$

with equality in case of hypercubical cells with $\delta_{il} = $ constant for all i. With the Gersho's conjecture, the minimum normalized moment of inertia given by Eq. (7) applies to all cells and, using Eq. (3), the distortion can be approximated as

$$
D_r(k) \approx \frac{1}{2^r(r+1)} \sum_{i=1}^{N} p_X(\mathbf{y}_i)[V(S_i)]^{1+\frac{r}{k}} \tag{8}
$$

A necessary condition for minimizing $D_r(k)$ is obtained by using Lagrange's multiplier method, as shown in (Panter & Dite, 1951) for scalar quantization and mean squared error criterion ($r = 2$ and $k = 1$)

$$
P(S_i)[V(S_i)]^{\frac{r}{k}} = P(S_j)[V(S_j)]^{\frac{r}{k}} \text{ for all } i,j \tag{9}
$$

From Eqs. (12), (8) and (9), it follows that every quantization cell S_i produces an identical distortion contribution to the total distortion. This *equidistortion principle* was first reported by Panter and Dite (1951) for scalar quantizers and then extended by Gersho (1979) to vector quantizers with a Voronoi or Dirichlet partition. Eq. (9) shows that the equidistortion principle also applies to decision trees. Also note that $r = 0$ in (9) leads to an equiprobable quantization, $P(S_i) = \frac{1}{N}$ for all i.

3 Unsupervised Learning in Decision Trees

To derive a suitable competitive learning rule, $\mathbb{1}_{S_i}(\mathbf{x})$ is first defined as the code membership function of the partition cell S_i, *i.e.*

$$
\mathbb{1}_{S_i}(\mathbf{x}) = \begin{cases} 1 & \text{if } \mathbf{x} \in S_i \\ 0 & \text{if } \mathbf{x} \notin S_i. \end{cases}
$$

The unsupervised learning rule, called $BAR_r(k)$, is given by

$$\Delta w_{tl} = w_{tl}(n) - w_{tl}(n-1)$$

$$= \eta \left(\sum_{i=1|w_{tl} \equiv x_{il}^L}^{N} [V(S_i)]^{\frac{r}{k}} \frac{1\!\!1_{S_i}}{LB(t)} - \sum_{i=1|w_{tl} \equiv x_{il}^U}^{N} [V(S_i)]^{\frac{r}{k}} \frac{1\!\!1_{S_i}}{UB(t)} \right) \qquad (10)$$

with η the learning rate. The $V(S_i)$s and $1\!\!1_{S_i}$s are defined at the previous time step. Assume that for input $\mathbf{x}$, $1\!\!1_{S_j} = 1$. The learning rule (10) then modifies S_j by increasing the cut values corresponding to the lower boundary point vector $\mathbf{x}_j^L$ and decreasing the cut values corresponding to the upper boundary point vector $\mathbf{x}_j^U$ in proportion to the $\frac{r}{k}$th power of the volume $V(S_j)$. Because this learning rule explicitly modifies the boundaries of the partition cells in a k-dimensional space, it is called a Boundary Adaptation Rule $BAR_r(k)$. Fig. 2 shows an example of adaptation in which $1\!\!1_{S_3} = 1$.

Note that the internal nodes involved in dividing many quantization cells could be more frequently updated than other nodes. Therefore, denominator terms $LB(t)$ and $UB(t)$ in (10) are needed in order to not favor the updates of particular nodes in the tree. It can be shown that $BAR_r(k)$ satisfies the necessary condition Eq. (9) at convergence.

With respect to the scalar case, one gets $k = 1$, $LB(t) = UB(t) = 1$ for all t and the partition cells become intervals : $S_i \equiv [w_i, w_{i-1})$, given an ordered set of boundary points $w_i > w_{i-1}$ for $i = 1 \cdots N$. Eq. (10) reduces to

$$\Delta w_i = \eta(\delta_{i+1}^r 1\!\!1_{S_{i+1}} - \delta_i^r 1\!\!1_{S_i}) \qquad (11)$$

which has been previously introduced for adaptive scalar quantization in (Martinez & VanHulle, 1995). In addition, it is proven that the system of ordinary differential equations associated with Eq. (11) converges to a unique boundary point solution.

4   Experimental Results

$BAR_r(k)$, Eq. (10) with $r \neq 0$, attempts to minimize the r-th power law distortion aiming at lossy compression. If $r = 0$, the learning rule $BAR_0(k)$ attempts to maximize Shannon's entropy aiming at density estimation. At convergence, the partition cells win the competition with an equal probability $\frac{1}{N}$ (*equiprobable quantization*) and the probability density function can be estimated as

$$p_X(\mathbf{x}) \approx p_X(\mathbf{y}_i) = \frac{1}{N \ V(S_i)} \text{ for } \mathbf{x} \in S_i$$

4.1   Density estimation (r=0)

4.1.1   Stationary distributions

In order to demonstrate the performance of decision trees trained with $BAR_0(k)$ in terms of entropy maximization, we will consider both symmetrical and asymmetrical

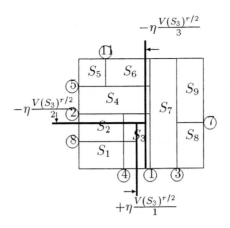

Figure 2: Example of adaptation of the decision tree shown in Fig. 1. It is assumed that for input $\mathbf{x}$, $\mathbb{1}_{S_3} = 1$. Only nodes 1,2 and 4 defining the partition cell S_3 are updated. For lossy compression minimizing the r-th power law distortion ($r \neq 0$), Eq. (10) gives the following modifications: $\Delta w_{11} = -\eta\frac{V(S_3)^{r/2}}{3}$, $\Delta w_{41} = +\eta\frac{V(S_3)^{r/2}}{1}$ and $\Delta w_{22} = -\eta\frac{V(S_3)^{r/2}}{2}$. For density estimation ($r = 0$), the modifications simplify to $\Delta w_{11} = -\frac{\eta}{3}$, $\Delta w_{41} = +\eta$ and $\Delta w_{22} = -\frac{\eta}{2}$.

k-dimensional input probability density functions, for k up to 3. Entropy performance results are listed in Table 1 for Uniform, Gaussian and Exponential densities of unit standard deviation. The values given are averages $\pm$ standard deviations over 10 runs. For each run, the decision tree was first built up in a greedy fashion by considering the splitting of node and axis contributing most to the decrease in average distortion. Then, it was trained on an i.i.d sequence of 20,000 samples and the entropy value was estimated using 40,000 samples. We observe that the relative difference between the entropy value obtained at convergence and the maximum entropy value was observed to be less than $2\%_0$ in every case. In addition, there is no significant difference in entropy performance for symmetrical or asymmetrical densities. These results clearly show the efficiency with which $BAR_0(k)$ converges toward equiprobable quantization.

$BAR_0(k) \rightarrow$ pdf $\downarrow$	k=1, N=8	k=2, N=64	k=3, N=128
Uniform	2.999 $\pm$ 2.3E-4	5.997 $\pm$ 5.7E-4	6.993 $\pm$ 9.1E-4
Gaussian	2.999 $\pm$ 3.0E-4	5.997 $\pm$ 5.9E-4	6.993 $\pm$ 7.6E-4
Exponential	2.999 $\pm$ 2.0E-4	5.989 $\pm$ 1.0E-2	6.993 $\pm$ 8.6E-4

Table 1: The entries listed in the table are entropy values for an input dimension k up to 3. Performance is given for Uniform, Gaussian and Exponential probability density functions (pdf) of unit variance and zero mean. The values are averages over 10 runs with standard deviation. They are estimated by using 40,000 samples drawn from the respective densities. The training set consists of 20,000 samples. The learning rate is linearly decreasing from 0.01 to zero for $BAR_0(1)$ and $BAR_0(2)$ and from 0.03 to zero for $BAR_0(3)$. The number of partition cells is $N = 8$, 64 and 128 for $k = 1,2$ and 3, respectively

Figures 3(A1)-(A3) show the true densities (Uniform, Gaussian and Exponential) and the partitions given by the decision tree ($N = 64$ and $k = 2$) at convergence. The partitions reveal smaller cells in high-density regions and larger cells in low-density regions relative to the current underlying input distribution. Thus, the learning rule provides an adaptive focusing mechanism. Figures 3(B1)-(B3) and 3(C1)-(C3) show the density estimated as a multivariate histogram, constructed with $\hat{p}(\mathbf{x}) = \frac{1}{N} \sum_{i=1}^{N} \frac{\mathbb{1}_{S_i}}{V(S_i)}$, and as a frequency polygon, constructed by linear interpolation, respectively.

4.1.2 Piecewise stationary distributions

Consider the sequence $\{\mathbf{x}(t)\}, t = 1 \cdots 20000$, of independently distributed random variables. The sequence is generated from a two-dimensional (2D) parity problem in which a change occurs when $t = 10000$. Before this change, samples are drawn from unit variance, Gaussian processes with mean vectors $(0,0)$, $(1,1)$. When $t = 10000$, the mean vectors suddenly change to $(0,1)$ and $(1,0)$. We trained the decision tree with $BAR_0(2)$ and a fixed learning rate $\eta = 0.01$ in order to adaptively partition the input space into $N = 64$ equiprobable cells. The starting weight configuration was obtained by growing a decision tree over the entire training sequence, as previously.

The performance of the adaptive neural tree is envisaged in terms of entropy maximization and compared with a number of unsupervised competitive learning algorithms: standard competitive learning (standard CL), the original conscience learning (Conscience 1) (De-Sieno, 1988) and its slightly modified version (Conscience 2) (VanDenBout, 1989), neural gas learning (Martinetz, Berkovich & Shulten, 1993), the original Frequency-Sensitive Competitive Learning (FSCL 1) (Ahalt, Krishnamurthy, Chen & Melton, 1990) and its modified version (FSCL 2) (Galanopoulos & Ahalt, 1996). Since FSCL 2 was known to approximate the density as $p(\mathbf{x})^{(3\beta+1)/(3\beta+3)}$ (Galanopoulos & Ahalt, 1996), a large β value ($\beta = 10$) was chosen for the simulations. Note that FSCL 1 is similar to FSCL 2 when β is set equal to one. For Conscience learning 1 and 2, the conscience factor was 10 and 2, as suggested in (DeSieno, 1988) and (VanDenBout, 1989), respectively. After extensive simulations, a fixed learning rate value $\eta = 0.04$ and $\eta = 0.01$ was chosen for the unsupervised competitive learning algorithms and the decision tree, respectively. At each time step, entropy performance was estimated for the different algorithms with respect to the current input distribution and plotted as a time evolution shown in figure 4(A). Results are averages over 20 runs. We observe that standard CL, neural gas, Conscience 1 and FSCL 1 lead to very poor entropy performance. This is not surprising since standard CL and neural gas, which are stochastic gradient descent algorithms for minimizing the mean squared error distortion, are known to approximate the density as $p(\mathbf{x})^{1/3}$. Thus, they tend to over-estimate the low density regions and under-estimate the high density regions. On the other hand, the decision tree matches the input density more closely and performs at least as well as the best competitive learning algorithms such as Conscience 2 and FSCL 2. Thus, the learning rule provides an adaptive focusing mechanism capable of tracking time-varying distributions. Furthermore, it was found to be at least 20 times faster than any of the other rules. A highly time-consuming part of competitive learning algorithms is due to encoding. Figure 4(B) compares the encoding speed for a set of 20000 k-dimensional input vectors in the different algorithms. The number of partition cells scales as $N = 32\ k$, for k up to 10. The encoding complexity of the competitive

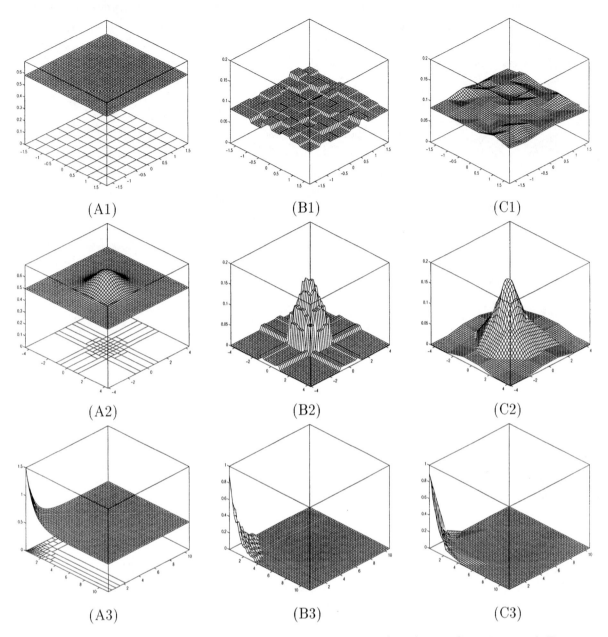

Figure 3: Figures 3(A1)-(A3) show the true densities (Uniform, Gaussian and Exponential) and the partitions obtained from the decision tree with $N = 64$ and $k = 2$ at convergence. Figures 3(B1)-(B3) and 3(C1)-(C3) show the density estimated as a multivariate histogram, constructed with $\hat{p}(\mathbf{x}) = \frac{1}{N} \sum_{i=1}^{N} \frac{\mathbb{1}_{S_i}}{V(S_i)}$, and as a frequency polygon, constructed by linear interpolation, respectively.

neural networks increases exponentially with k since the nearest-neighbor search requires $k\,N$ multiply and accumulate operations. As expected, the slowest algorithm was the neural-gas which requires ranking all the neurons and not only determining the winner. On the other hand, the decision tree is many orders of magnitude faster than the other algorithms because its encoding operation only requires comparisons, whose complexity may be assumed to be negligible.

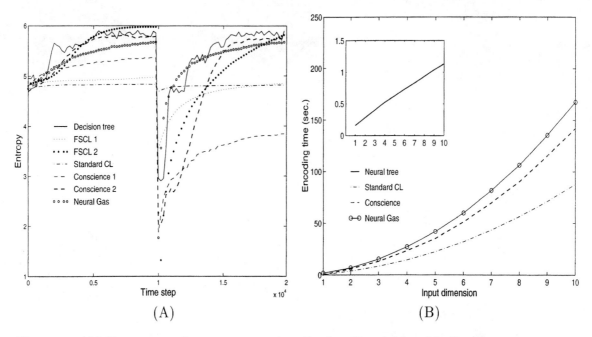

(A)   (B)

Figure 4: (A) Entropy performance over time for $k = 2$ and $N = 64$. Results are averages over 20 runs. (B) Time in sec. taken to encode a set of 20 000 k-dimensional samples generated from a uniform distribution. The number of partition cells is $N = 32\,k$, for k up to 10.

4.2   Vector quantization (r=2)

Vector quantization design using the Generalized Lloyd Algorithm (GLA) involves iterative applications of the nearest neighbor partition and centroid estimation on the training data set so as to monotonically decrease the distortion function towards a local minimum (Linde, Buzo & Gray, 1980). As a result, the quantizer is designed for a given source based on its long term statistical behavior. However, there are many situations in which communication systems may have to carry signals of changing statistics, e.g. speech or images, and the operational data may therefore be quite different from that of the training data. In order to minimize the quantizer-source mismatch due to nonstationnary inputs, the decision tree is continuously adapted with a fixed learning rate value so as to match the observed local statistics of the input sequence.

Table 2 shows the signal-to-noise ratio (SNR) in dB estimated for Uniform, Gaussian and Exponential densities of unit standard deviation. The vector quantizer and the decision tree were optimized by using the GLA and $BAR_2(k)$ for each of these densities, respectively. Because the decision tree is continuously adapted by using $BAR_2(k)$ with a fixed learning rate, the boundaries of the partition cells are modified at each time step.

Therefore, the output $\mathbf{y} = \mathbf{y}_i$ is reconstructed as the midpoint of the active partition cell S_i when $\mathbf{x} \in S_i$.

In the event of no quantizer-source mismatch, the SNR obtained from the GLA is, at the most, 1 dB higher than the SNR obtained from $BAR_2(k)$. However, performance degrades significantly for the GLA when the quantizer is used with sources other than those for which it was optimized. If the training set is drawn from an Exponential density while the test set is generated either from a Uniform or a Gaussian density, the decision tree trained with $BAR_2(k)$ outperforms the GLA by at least 10 dB.

Test → Train ↓	k=1			k=2			k=3		
	U	G	E	U	G	E	U	G	E
U									
GLA	18.01	12.35	3.27	17.92	12.37	3.56	17.87	12.67	3.66
$BAR_2(k)$	18.03	14.00	13.97	17.79	13.89	12.55	17.84	13.89	8.67
G									
GLA	15.00	14.58	6.05	15.76	15.09	6.79	15.90	15.31	7.31
$BAR_2(k)$	18.03	14.00	13.98	16.73	14.05	12.51	16.20	14.40	10.47
E									
GLA	1.75	1.83	15.12	2.19	2.20	16.57	2.40	2.42	16.43
$BAR_2(k)$	15.90	13.51	14.46	13.16	12.60	15.28	11.35	10.40	15.83

Table 2: The entries listed in the table are SNR values in dB for a transmission rate of 3 bits/sample and an input dimension k up to 3. Performance is given for different densities (Uniform (U), Gaussian (G) and Exponential (E)) of unit variance and zero mean. The training and test sets consist of 20,000 and 40,000 samples drawn from the respective densities, respectively. A quantizer-source mismatch occurs when the quantizer is used with densities other than those for which it was optimized. $BAR_2(k)$ is used with a fixed learning rate $\eta = 0.01$ for $BAR_2(1)$ and $\eta = 0.1$ for $BAR_2(2)$ and $BAR_2(3)$.

5 Conclusion

A decision tree performs a hierarchical clustering over the input space by means of axis-parallel hyperplanes and, thus, encoding does not require multiplication. In this paper, a competitive learning rule has been introduced in decision trees as a computationally attractive scheme for adaptive density estimation or lossy compression. Because the proposed learning rule updates the active partition cell after each time step, it can be considered as belonging to neural competitive learning. However, it markedly differs from existing competitive learning algorithms, (e.g Yair, Zeger & Gersho, 1992; Lee & Peterson, 1990; Ahalt, Krishnamurthy, Chen & Melton, 1990; Kohonen, 1995). Instead of finding the centroids of the partition cells, the proposed learning rule does just the opposite by finding the boundary point vectors that separate the partition cells. Since it adapts boundary points directly, the rule is referred to as Boundary Adaptation Rule ($BAR_r(k)$). As demonstrated by simulation, $BAR_r(k)$ provides an adaptive focusing mechanism capable of tracking time-varying distributions and performs at least as well, at high rate, as the best vector quantization algorithms while being much faster. We are currently investigating several lines of research, including the potential for hardware implementation, the possibility for gradually changing the overall tree structure by splitting and merging leaf nodes, and the use of general hyperplanes, albeit at the expense of a higher level of complexity.

References

Ahalt S.C., Krishnamurthy A.K., Chen P. & Melton D.E. (1990). Competitive learning algorithms for vector quantization. *Neural Networks*, **3**, 277–290.

Breiman L., Friedman J.H., Olshen R.A. & Stone C.J. (1984). Classification and regression trees. *The Wadsworth statistics/probability series*, Belmont,CA:Wadsworth.

DeSieno D. (1988). Adding a conscience to competitive learning. *Proc. Int. Conf. on neural networks*, (San Diego), **1**, 117–124.

Galanopoulos A.S. & Ahalt S.C. (1996). Codeword distribution for frequency sensitive competitive learning with one-dimensional input data. *IEEE Trans. Neural Networks*, **7**, 3, 752–756.

Gersho A. & Gray R. M. (1992). Vector quantization and signal compression. *Kluwer Academic Publisher*, Boston/Dordrecht/London.

Gersho A. (1979). Asymptotically optimal block quantization. *IEEE Trans. on Information Theory*, **25**, 4, 373–380.

Kassam S. A. (1978). Quantization based on the mean-absolute-error criterion. *IEEE Trans. on Communication*, **26**, 2, 267–270.

Kohonen T.(1995). Self-Organizing Maps. *Springer-Verlag*.

Lee T-C. & Peterson A. M. (1990). Adaptive vector quantization using a self-development neural network. *IEEE Journal on Selected Arears in Communications*, **8**, 8, 1458–1471.

Linde Y., Buzo A. & Gray R.M. (1980). An algorithm for vector quantizer design. *IEEE Trans. on Communication*, **28**, 1, 84–95.

Martinetz T.M., Berkovich S.G. and Shulten K.J. (1993). Neural-gas' network for vector quantization and its application to time-series prediction. *IEEE Trans. Neural Networks*, **4**, 558–569.

Martinez D. & Van-Hulle M. M. (1995). Generalized Boundary Adaptation Rule for minimizing rth power law distortion in high resolution quantization. *Neural Networks*, **8**, 6, 891–900.

Panter P. F. & Dite W. (1951). Quantization distortion in Pulse-Count modulation with nonuniform spacing of levels. *Proc. IRE*, **39**, 44–48.

Van Den Bout D. E. (1989). TInMANN: The integer markovian artificial neural network. *Proc. Int. Conf. on neural networks*, (San Diego), **2**, 205–211.

Yair E., Zeger K. & Gersho A. (1992). Competitive learning and soft competition for vector quantizer design. *IEEE Trans. on Signal Processing*, **40**, 2, 294–309.

Zador P. L. (1966). Asymptotically quantization error of continuous signals and the quantization dimension. *Unpublished memorandum, Bell Laboratories,*

Zador P. L. (1982). Asymptotically quantization error of continuous signals and the quantization dimension. *IEEE Trans. on Information Theory*, **28**, 2, 139–149.

AUTHOR INDEX

A

Albrecht, R. F., 447
Ali, S. M., 138
Alvarez de Lorenzana, J. M., 116
Araújo, R. A., 364

B

Baginski, B., 375
Bartolini, G., 269
Birk, A., 355
Boxer, P., 157
Brier, S., 182
Broonen, J.-P., 284

C

Ceccarelli, M., 388
Chapurlat, V., 528
Citko, W., 595
Cohen, B., 157
Csányi, V., 259

D

D'Anjou, A., 517
de Almeida, A. T., 364
DeMaris, D. L., 469
Dubois, D. M., 3, 295, 406

E

Elstob, C. M., 138

F

Frachet, J.-P., 495
Freisleben, B., 606

G

Geurts, F., 507
González, A. I., 517
Graña, M., 517
Gross, G. W., 577

H

Hagen, C., 606
Hoekstra, J., 626
Holmberg, S. C., 419
Honkela, T., 563

J

Julià, P., 209

K

Kowalski, J. M., 577

L

Larnac, M., 528
Lenarčič, J., 433
Leydesdorff, L., 309
Loccufier, M., 643
Łuksza, A., 595

M

Magnier, J., 528
Martinez, D., 660
Matsuno, K., 101
Medina Martins, P. R., 49
Mirità, I. I., 195
Moreno Bergareche, A., 202

N

Naranjo, M., 244
Noldus, E., 643
Nunes, U., 364

O

Onana Alima, L., 507

P

Posohina, N. I., 552
Pötter, C., 331
Prem, E., 344

Date Due

			UML 735